Early Miocene Paleobiology in Patagonia
High-Latitude Paleocommunities of the Santa Cruz Formation

Coastal exposures of the Santa Cruz Formation in southern Patagonia have been a fertile ground for recovery of Early Miocene vertebrates for more than 100 years: studies began in the 1840s when specimens were sent for study to Charles Darwin. The formation is noted for yielding remarkably complete specimens, and a richly varied taxonomic assemblage very different from other continents, owing to the long isolation of South America.

This volume presents the most comprehensive compilation yet of important mammalian groups from the Santa Cruz Formation. It includes the most recent fossil finds as well as important new interpretations based on 10 years of fieldwork by the editors. A key focus is placed on the paleoclimate and paleoenvironment during the time of deposition in the Middle Miocene Climatic Optimum (MMCO) between 17 and 15 million years ago – a warm interval dissimilar to the modern climate of Patagonia. Using newly recovered phytoliths, and plant macrofossils, together with invertebrates, amphibians, and reptiles, the authors present the first reconstruction of what climatic conditions were like for this most southerly continental record of the MMCO. They also present important new evidence of the geochronological age, habits, and community structures of fossil bird and mammal species.

Academic researchers and graduate students in paleontology, paleobiology, paleoecology, stratigraphy, climatology, and geochronology will all find this a valuable resource of information about this fascinating geological formation.

SERGIO F. VIZCAÍNO is Professor of Vertebrate Zoology at the Universidad Nacional de La Plata (Argentina) and a researcher of the Consejo Nacional de Investigaciones Científicas y Técnicas working at the Museo de La Plata. His research focuses on the paleobiology of South American fossil vertebrates, mostly mammals, and he has authored approximately 100 research papers and book chapters, and edited one book and several special volumes. Professor Vizcaíno has participated in numerous fieldwork seasons in Argentina and Antarctica. He was the President of the Asociación Paleontológica Argentina (APA), and in 1996 and 2008 he won awards for his publications in *Ameghiniana*, the journal of the APA.

RICHARD F. KAY is Professor of Evolutionary Anthropology and Earth and Ocean Sciences at Duke University, North Carolina, where he has worked since 1973. He has edited six books and authored more than 200 research papers on mammalian and primate paleontology, functional anatomy, adaptations, and phylogenetics. Professor Kay has conducted paleontological field research in seven South American countries since 1982, and is an elected Fellow of the American Association for the Advancement of Science.

M. SUSANA BARGO is a vertebrate paleontologist at the División Paleontología Vertebrados of the Museo de La Plata, Argentina, and a researcher for the Comisión de Investigaciones Científicas de la Provincia de Buenos Aires. Her research focuses on the paleobiology of South American fossil mammals. Dr Bargo has authored about 50 scientific papers and book chapters, and was the Editor in charge of vertebrates for *Ameghiniana*, winning an award in 2008 for a publication in that journal. She has participated in numerous field seasons in Patagonia, Argentina.

Artistic reconstruction by Manuel Sosa.

Early Miocene Paleobiology in Patagonia

High-Latitude Paleocommunities
of the Santa Cruz Formation

EDITED BY

SERGIO F. VIZCAÍNO
Museo de La Plata, Argentina

RICHARD F. KAY
Duke University, North Carolina, USA

M. SUSANA BARGO
Museo de La Plata, Argentina

CAMBRIDGE
UNIVERSITY PRESS

CAMBRIDGE
UNIVERSITY PRESS

University Printing House, Cambridge CB2 8BS, United Kingdom

One Liberty Plaza, 20th Floor, New York, NY 10006, USA

477 Williamstown Road, Port Melbourne, VIC 3207, Australia

4843/24, 2nd Floor, Ansari Road, Daryaganj, Delhi - 110002, India

79 Anson Road, #06-04/06, Singapore 079906

Cambridge University Press is part of the University of Cambridge.

It furthers the University's mission by disseminating knowledge in the pursuit of education, learning and research at the highest international levels of excellence.

www.cambridge.org
Information on this title: www.cambridge.org/9781108445771

First published 2012
First paperback edition 2017

A catalogue record for this publication is available from the British Library

Library of Congress Cataloging in Publication data
Early Miocene paleobiology in Patagonia : high-latitude paleocommunities of the Santa Cruz Formation / edited by Sergio F. Vizcaíno, Richard F. Kay, M. Susana Bargo.
p. cm.
ISBN 978-0-521-19461-7 (Hardback)
1. Paleontology–Miocene. 2. Paleobiology–Argentina–Santa Cruz Formation.
3. Geology, Stratigraphic–Miocene. 4. Santa Cruz Formation (Argentina)
5. Geology–Argentina–Santa Cruz Formation. 6. Geology–Patagonia
(Argentina and Chile) I. Vizcaíno, Sergio F. II. Kay, Richard F.
III. Bargo, M. Susana.
QE739.E27 2012
560´.178709827–dc23

2012001739

ISBN 978-0-521-19461-7 Hardback
ISBN 978-1-108-44577-1 Paperback

Cambridge University Press has no responsibility for the persistence or accuracy of URLs for external or third-party internet websites referred to in this publication, and does not guarantee that any content on such websites is, or will remain, accurate or appropriate.

Contents

Contributors

Maria Alejandra ABELLO
Laboratorio de Sistemática y Biología Evolutiva (LASBE)
Museo de La Plata
Paseo del bosque s/n
B1900FWA La Plata, Argentina

Adriana ALBINO
Departamento de Biología,
Facultad de Ciencias Exactas y Naturales
Universidad Nacional de Mar del Plata
Funes 3250
7600 Mar del Plata, Argentina

Kari L. ALLEN
Dept. Evolutionary Anthropology
Box 90383, Biological Sciences Building
Duke University, Durham
NC 27708–0383, USA

Juan I. ARETA
Laboratorio de Paleontología de Vertebrados
Centro de Investigaciones Científicas CICYTTP-CONICET
Dr. Materi y España s/n
E3105BW Diamante, Entre Rios, Argentina

M. Susana BARGO
División Paleontología Vertebrados
Museo de La Plata
Paseo del bosque s/n
B1900FWA La Plata, Argentina

Thomas M. BOWN
Erathem-Vanir Geological
10350 Dover St., D-32
Westminster
CO 80021, USA

Mariana BREA
Laboratorio de Paleobotánica
Centro de Investigaciones Científicas
CICYTTP-CONICET
Dr. Materi y España s/n
E3105BWA Diamante, Entre Rios, Argentina

Adriana M. CANDELA
División Paleontología Vertebrados
Museo de La Plata
Paseo del bosque s/n
B1900FWA La Plata, Argentina

Guillermo H. CASSINI
Sección Mastozoología
Museo Argentino de Ciencias Naturales
"Bernardino Rivadavia"
Av. Ángel Gallardo 470
1405DJR Buenos Aires, Argentina

Esperanza CERDEÑO
Departamento de Paleontología
IANIGLA, CCT-CONICET-Mendoza
Avda. Ruiz Leal s/n
5500 Mendoza, Argentina

Federico J. DEGRANGE
División Paleontología Vertebrados
Museo de La Plata
Paseo del bosque s/n
B1900FWA La Plata, Argentina

Maria T. DOZO
Laboratorio de Paleontología,
Centro Nacional Patagónico
9120 Puerto Madryn
Chubut, Argentina

Marcos D. ERCOLI
Sección Mastozoología
Museo Argentino de Ciencias Naturales
"Bernardino Rivadavia"
Av. Angel Gallardo 470
C1405DJR Buenos Aires, Argentina

Juan C. FERNICOLA
Sección Paleontología de Vertebrados
Museo Argentino de Ciencias Naturales "Bernardino Rivadavia".
Av. Ángel Gallardo 470
C1405DJR Buenos Aires, Argentina

John G. FLEAGLE
Department of Anatomical Sciences
Stony Brook University
Stony Brook
NY 11794–8081, USA

Analía M. FORASIEPI
Departamento de Paleontología
Museo de Historia Natural de San Rafael
San Rafael
5500 Mendoza, Argentina

Miguel GRIFFIN
División Paleontología Invertebrados
Museo de La Plata
Paseo del bosque s/n
B1900FWA La Plata, Argentina

Matthew T. HEIZLER
New Mexico Geochronology Research Center
New Mexico Bureau of Mines & Mineral Resources
New Mexico Tech, 801 Leroy Place, Socorro
NM 87801–4796, USA

Ari IGLESIAS
División Paleobotánica
Museo de La plata
Paseo del bosque s/n, B1900FWA
La Plata, Argentina

Richard F. KAY
Dept. Evolutionary Anthropology
Box 90383, Biological Sciences Building
Duke University, Durham
NC 27708–0383, USA

E. Christopher KIRK
Department of Anthropology
University of Texas, Austin
TX 78712, USA

Verónica KRAPOVICKAS
Departamento de Ciencias Geológicas
Facultad de Ciencias Exactas y Naturales,
Universidad de Buenos Aires
Intendente Guiraldes 2160 – Pabellon II – 1° Piso
Ciudad Universitaria
C1428EGA Buenos Aires, Argentina

Michael MALINZAK
Dept. Evolutionary Anthropology
Box 90383, Biological Sciences Building
Duke University, Durham
NC 27708–0383, USA

Sergio D. MATHEOS
Centro de Investigaciones Geológicas (CIG)
Calle 1 No. 644, B1900TAC
La Plata, Argentina

Nahuel A. MUÑOZ
División Paleontología Vertebrados
Museo de La plata
Paseo del Bosque s/n, B1900FWA
La Plata, Argentina

Barbara NASH
Department of Geology and Geophysics
University of Utah
115 S, 1460 E, Rm. 383 FASB
Salt Lake City
UT 84112, USA

Jorge I. NORIEGA
Laboratorio de Paleontología de Vertebrados
Centro de Investigaciones Científicas, CICYTTP-CONICET
Dr. Materi y España s/n
Diamante (E3105BWA)
Entre Rios, Argentina

Edgardo ORTIZ-JAUREGUIZAR
Laboratorio de Sistemática y Biología Evolutiva
(LASBE)
Museo de La Plata
Paseo del bosque s/n.
B1900FWA La Plata, Argentina

Ana PARRAS
Facultad de Ciencias Exactas y Naturales
Universidad Nacional de La Pampa
Uruguay 151, 6300 Santa Rosa
La Pampa, Argentina

María E. PÉREZ
Museo Egidio Feruglio
Av. Fontana 140, U9100GYO Trelew
Chubut, Argentina

Michael E. PERKINS
Department of Geology and Geophysics
University of Utah
115 S, 1460 E, Room 383, Frederick A. Sutton Building
Salt Lake City
UT 84112–0102, USA

Jonathan M. G. PERRY
Center for Functional Anatomy and Evolution
The Johns Hopkins University School of Medicine
1830 East Monument Street, Room 301
Baltimore, MD 21205, USA

J. Michael PLAVCAN
Department of Anthropology
University of Arkansas
Fayetteville
AR 72701, USA

Francisco J. PREVOSTI
Sección Mastozoología
Museo Argentino de Ciencias Naturales "Bernardino Rivadavia"
Av. Ángel Gallardo 470
C1405DJR Buenos Aires, Argentina

M. Sol RAIGEMBORN
Centro de Investigaciones Geológicas (CIG)
Calle 1 No. 644
B1900TAC La Plata, Argentina

Luciano L. RASIA
División Paleontología Vertebrados
Museo de La Plata
Paseo del bosque s/n
B1900FWA La Plata, Argentina

Adán A. TAUBER
Facultad de Ciencias Exactas, Físicas y Naturales
Universidad Nacional de Córdoba
Av. Vélez Sarsfield 1611
X5016GCA Córdoba, Argentina

Marcelo F. TEJEDOR
Laboratorio de Paleontología
Centro Nacional Patagónico – CONICET
9120 Puerto Madryn, Chubut, Argentina

Néstor TOLEDO
División Paleontología Vertebrados
Museo de La Plata
Paseo del bosque s/n
B1900FWA La Plata, Argentina

Guillermo F. TURAZZINI
Museo Argentino de Ciencias Naturales "Bernardino Rivadavia"
Av. Angel Gallardo 470
C1405DJR Buenos Aires, Argentina

Amalia L. VILLAFAÑE
Museo Egidio Feruglio
Av. Fontana 140,
U9100GYO Trelew, Chubut, Argentina

Sergio F. VIZCAÍNO
División Paleontología Vertebrados
Museo de La Plata
Paseo del bosque s/n
B1900FWA La Plata, Argentina

Alejandro F. ZUCOL
Laboratorio de Paleobotánica
Centro de Investigaciones Científicas, Diamante
CICYTTP-CONICET
Dr. Materi y España s/n
Diamante (E3105BWA), Entre Ríos, Argentina

Preface

In 1520 Ferdinand Magellan (Fernão Magalhães, 1480–1521), a Portuguese explorer in the service of the Spanish crown, was surveying the coast of a land inhabited by huge natives whom he called *patagones* (big feet). From this name, the land later became known as Patagonia. On September 14 of that same year, the day of the Feast of the Exaltation of the Holy Cross (Exaltación de la Santa Cruz), the fleet reached the mouth of the river that Magellan named Santa Cruz. In Magellan's times, as remains true today, Patagonia east of the Andes was mostly a cold, windy, arid steppe with choiques (Darwin's rheas) and guanacos as its most conspicuous wild animals. Before the arrival of Europeans, this territory was occupied by natives belonging to the *tehuelche* culture, nomadic hunters of choiques and guanacos. Over the course of time, the name Santa Cruz came to be associated with all of the continental territory of the Argentine Republic south of the 46° S parallel.

Santa Cruz has not always been the cold and windy place just described. About 17 million years ago (Early Miocene), the Andes were much lower and the humid winds from the west allowed the eastward extension, to the present Atlantic coast and beyond, of forest and grasslands very similar to those existing today on the piedmont and the lower slopes on both sides of the cordillera. But in contrast to the present, these forests and grasslands were inhabited by a plethora of mammals and birds mostly belonging to long extinct or greatly reduced lineages, and many of the surviving relatives of those earlier occupants are now restricted to the tropical regions of South America. Among the herbivores were bizarre forms such as glyptodonts, ground sloths, and giant tapir-like astrapotheres, and others belonging to extinct orders of mammals resembling small horses, cattle, sheep, and hares. Bear-like marsupials and terror birds reigned among the carnivores. Among the descendants of these early inhabitants are anteaters, armadillos, porcupines, and monkeys. The aim of this volume is to study the biology of the different species of this fauna to interpret their ecological interactions and better understand the environment in which they lived, at a time interval that ranks among the Earth's warmest periods over the past 34 million years, and during which South America was physically separated from other continental land masses.

The remains of this unique fauna are recorded in sedimentary rocks that are abundantly exposed throughout the Province and, not surprisingly, known by geologists as the Santa Cruz Formation. South of the mouth of the Río Santa Cruz, the Atlantic coastal exposures of this formation between the Ríos Coyle and Gallegos provide the best specimens for our purposes.

To achieve our goals we performed continued intensive collecting during the Austral summers of 2003 to the present (2012). The geology and fossils were studied by an assembled team of colleagues from different institutions and with varied fields of expertise, including sedimentology, geochronology, ichnology, and invertebrate and vertebrate paleontology. Paleontologists constituted the largest group, including experts on amphibians, reptiles, birds, marsupials, xenarthrans, ungulates, rodents, and primates. The contributions presented in this volume represent to a large degree original research on subjects that had barely been treated previously, if not completely ignored, for the rocks and the biota of the Santa Cruz Formation. Parts of the research developed in certain chapters constitute the core of doctoral dissertations of several of our former students, who have now become colleagues.

We are grateful to many persons and institutions for making possible the publication of this volume.

To Cambridge University Press for giving us the chance to publish this book, and for the attentive assistance its staff provided over the course of its preparation and publication.

To the authors of the chapters for their contributions, their efforts in meeting our deadlines, and their patience in following our directions.

To the colleagues who reviewed the chapters: Lluís Cabrera (Universitat de Barcelona, España), Victor Ramos (Instituto de Estudios Andinos Don Pablo Groeber UBA – CONICET), Luis Buatois (University of Saskatchewan, Canada), Nicholas Minter (University of Bristol, UK), Claudia Montalvo (Universidad Nacional de La Pampa, Argentina), René Bobe (George Washington University, USA), M. Alejandra Gandolfo (Cornell University, USA), Elisabeth Wheeler (North Carolina State University, USA), Claudia Del Río (Museo Argentino de Ciencias Naturales "B. Rivadavia," Argentina), Sven Nielsen (Christian-Albrechts-University Kiel, Germany), Ana Maria Báez (Universidad de Buenos Aires, Argentina), Thomas LaDuke (East Stroudsburg University, USA), Kenneth E. Campbell (Natural History Museum Los Angeles County, USA), Claudia Tambussi (Museo de La Plata, Argentina), Christine Argot (Muséum national d'Histoire naturelle, France), David Flores (Museo Argentino de Ciencias Naturales "B. Rivadavia", Argentina), Darin Croft

(Case Western Reserve University, USA), Blaire Van Valkenburg (University of California Los Angeles, USA), Richard Fariña (Universidad de la República, Uruguay), H. Gregory McDonald (National Park Service, USA), Gerardo De Iuliis (University of Toronto, Canada), Francois Pujos (Instituto Argentino de Nivología, Glaciología y Ciencias Ambientales, Argentina), Christine Janis (Brown University, USA), M. Guiomar Vucetich (Museo de La Plata, Argentina), Joshua X. Samuels (John Day Fossil Beds National Monument, USA), Jeffrey Meldrum (Idaho State University, USA), Paul Brinkman (North Carolina Museum of Natural Sciences, USA), Eduardo Tonni (Museo de La Plata, Argentina).

To Gerry De Iuliis, Kenneth Campbell, Thomas LaDuke, and Darin Croft for their valuable help with revising the English of some chapters.

To Néstor Toledo and Guillermo Cassini for helpful assistance during the editing of the volume.

To Agustín Abba (Museo de La Plata), and to Ulyses Pardiñas and Daniel Udrizar (Centro Nacional Patagónico, Argentina), who helped to collect information on recent faunas from Patagonia.

To Eric Delson (City University of New York) for advice about geologic nomenclature.

To the Dirección de Patrimonio Cultural de la Subsecretaría de Estado de Cultura de la Provincia de Santa Cruz for support of the fieldwork in Santa Cruz, for permission to conduct expeditions in Santa Cruz Province and for the loan of numerous specimens for study.

To the Museo Regional Provincial Padre Manuel Jesus Molina of Río Gallegos for administrative and logistic support of different aspects of the work.

To all the owners or administrators of the Estancias in the coastal area where we conducted the fieldwork related to this project: Jorge Battini (Estancia La Costa), John Locke Blake (Estancia Killik Aike), Carlos del Río (Estancia Ototel Aike), Leslie and Tamara Hewlett (Estancia Coy Inlet), Mario and Javier Hunicken (Estancia La Angelina), and Robert Lemaire (Estancia Cañadón de Las Vacas).

To the Administración Nacional de Parques Nacionales and the rangers of the Parque Nacional Monte León for permission to conduct and assistance with fieldwork in the Parque Nacional.

To Jorge Battini, and his family (Mónica, Andy, Teddy and Alex) for their hospitality while at Estancia La Costa and the use of their facilities.

To Elena Davidson and Colin Jamieson from the Posada Lemarchand, and the personnel of the Hotel Alonso (Río Gallegos), who helped us in many aspects during our stay with them.

To the many people who collaborated in the fieldwork: Mary Palacios, Sonia Cardozo, Mariella Bruno, Claudia Aguilar, Juan José Moly, Blythe Williams, Carlos Luna, Adán Tauber, Nick Milne, Andy and Teddy Battini, Anne Weil, Francisco Prevosti, Michael Malinzak, Richard Madden, Guillermo Cassini, Néstor Toledo, Gerardo De Iuliis, Lucas Pomi, Sergio Matheos, M. Sol Raigemborn, Josefina Vizcaíno, Michael Griffin, Nahuel Muñoz, Laura Cruz, Siobhán Cooke, and Verónica Krapovickas. We especially thank Jonathan Perry, Juan Carlos Fernicola, and Leonel Acosta for joining us in most of the field seasons and for their permanent collaboration.

To the Museo de La Plata and the División Paleontología Vertebrados, where most of the specimens collected were prepared.

We appreciate the financial support for field expeditions, laboratory preparation of fossils, technical services and study of paleontology collections in the United States, from several agencies, including the Agencia Nacional de Promoción Científica y Tecnológica (ANPCyT – PICT 26219 and 143), Consejo Nacional de Investigaciones Científicas y Técnicas (CONICET – PIP 1054), Universidad Nacional de La Plata (UNLP N/ 474 and 647), National Geographic Society (3181–06) to Sergio F. Vizcaíno, and National Geographic Society and National Science Foundation (0851272 and 0824546) to Richard F. Kay.

Sergio F. Vizcaíno acknowledges the contributions of CONICET, UNLP, and ANPCyT for their continued support for much of his research activity.

Lastly, thanks to our families for their tolerance throughout the duration of this project.

1 Background for a paleoecological study of the Santa Cruz Formation (late Early Miocene) on the Atlantic Coast of Patagonia

Sergio F. Vizcaíno, Richard F. Kay, and M. Susana Bargo

Abstract

For more than 120 years, the coastal exposures of the Santa Cruz Formation have been fertile ground for recovery of vertebrates from the late Early Miocene (~18 to 16 million years ago, Ma). As long ago as the 1840s, Captain Bartholomew Sulivan collected fossils from this region and sent them to Charles Darwin, who passed them to Richard Owen. Carlos Ameghino undertook several explorations of the region starting in the late 1880s. Carlos' specimens were described by his brother Florentino, who believed that many of the species were more ancient than now understood and represented the ancestors of many Holarctic mammalian orders. Ameghino's novel claims prompted William B. Scott to organize fossil collecting expeditions in the Santa Cruz beds led by John B. Hatcher. The fossils were described in a series of exhaustive monographs with the conclusion that the fauna was much younger than Ameghino thought. Several brief expeditions took place during the twentieth century, led by researchers from different institutions. Since 2003, we have undertaken the collection of over 1600 specimens, including large series of relatively complete skeletons. In this edited volume we have gathered together a group of researchers to study the coastal Santa Cruz Formation and its associated flora and fauna to provide a paleobiological reconstruction of the Santacrucian vertebrate community and to place it in its biotic and physical environment.

Resumen

Por más de 120 años, las exposiciones costeras de la Formación Santa Cruz han sido campo fértil para la recuperación de vertebrados del Mioceno Temprano tardío (~18 to 16 Ma). En la década de 1840, el Capitán Bartholomew Sulivan recolectó fósiles que envío a Charles Darwin, quien se los pasó a Richard Owen. Carlos Ameghino llevó a cabo varias exploraciones de la región, comenzando a fines de la década de 1880. Los especímenes colectados por

Early Miocene Paleobiology in Patagonia: High-Latitude Paleocommunities of the Santa Cruz Formation, ed. Sergio F. Vizcaíno, Richard F. Kay and M. Susana Bargo. Published by Cambridge University Press. © Cambridge University Press 2012.

Carlos fueron descriptos por su hermano Florentino, quien pensaba que muchas de las especies eran más antiguas que lo hoy se entiende y representaban los ancestros de muchos órdenes de mamíferos holárticos. Las novedosas propuestas de Ameghino estimularon a William B. Scott a organizar expediciones lideradas por John B. Hatcher para colectar fósiles en los niveles santacrucenses. Los fósiles fueron descriptos en una serie de monografías exhaustivas, concluyendo que las faunas eran mucho más jóvenes que lo que pensaba Ameghino. Varias expediciones breves lideradas por investigadores de diferentes instituciones tuvieron lugar durante el siglo XX. Desde 2003, hemos recolectado más de 1600 especímenes, incluyendo numerosos esqueletos relativamente completos. En este volumen, hemos reunido un grupo de investigadores para estudiar la Formación Santa Cruz de la costa, su flora y fauna asociada, para realizar una reconstrucción paleobiológica de la comunidad de vertebrados santacrucenses y ubicarla en su contexto biótico y ambiental.

1.1 Introduction

Exposures on the Atlantic coast of southern continental Patagonia (Figs. 1.1 and 1.2) have been fertile ground for the recovery of fossil vertebrates of the Santa Cruz Formation for more than 120 years. The fossils come from the late Early Miocene, about 18 to 16 Ma, called the Santacrucian Land Mammal Age (Marshall *et al.*, 1983, Perkins *et al.*, Chapter 2). Through much of the Cenozoic, South American mammals underwent diversification and ecological specialization in isolation from other continents (Simpson, 1980). That isolation was interrupted by two major upheavals, one marked by the arrival of rodents and primates from Africa sometime in the Middle to Late Eocene and the other by the interchange of faunas beginning in the Late Miocene when South America began to establish contact with Central and North America. The Santacrucian marks the peak of known diversification achieved by mammals after the arrival of primates and rodents but before the arrival of North American immigrants (Marshall and Cifelli, 1990).

Between 1887 and 1906, the famous Argentine paleontologist Florentino Ameghino (1853/54 to 1911; Fig. 1.3a) described a plethora of species from a bizarre taxonomic assemblage – very

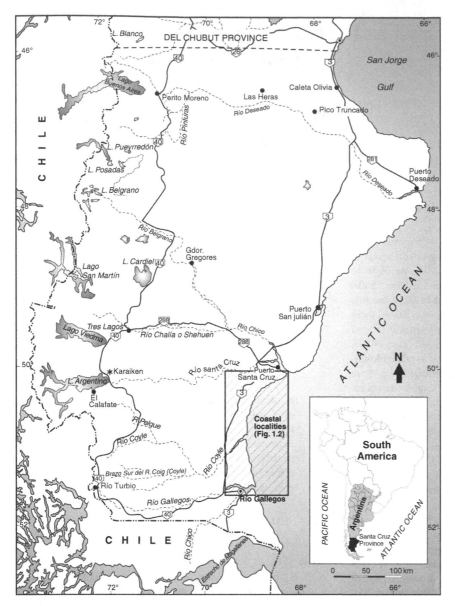

Fig. 1.1. Map of Santa Cruz Province (southern Patagonia, Argentina) and the Santacrucian localities treated in this volume and listed in Appendix 1.1.

different as a consequence of the long isolation of South America from other continents – of large predatory birds, as well as a diverse array of insectivorous, carnivorous, and herbivorous marsupials, glyptodonts, armadillos, sloths, small and large herbivorous hippo-like astrapotheres, rabbit- to cow-sized notoungulates, horse-like and camel-like litopterns, small to medium-sized rodents, and platyrrhine monkeys. Another singular nature of these finds, known even in Ameghino's day, is that the outcrops yield numerous remarkably complete specimens, including skulls with associated skeletons.

Florentino Ameghino believed that many of the species he described were more ancient than we now understand them

to be and that he had documented the ancestors of many mammalian orders in South America, including those of artiodactyls, perissodactyls, and even human beings. Hitherto, the fossil record of many mammalian orders had been restricted to northern continents, and it was supposed by most authorities that most or all orders originated there. Ameghino's novel claims prompted W. B. Scott (1858–1947; see Simpson, 1948) from Princeton University, New Jersey, USA, to organize a series of expeditions (see below) to collect fossils in the Santa Cruz beds where the faunas were best known (see Scott's Preface to the Narrative of the *Princeton University Expeditions to Patagonia* by

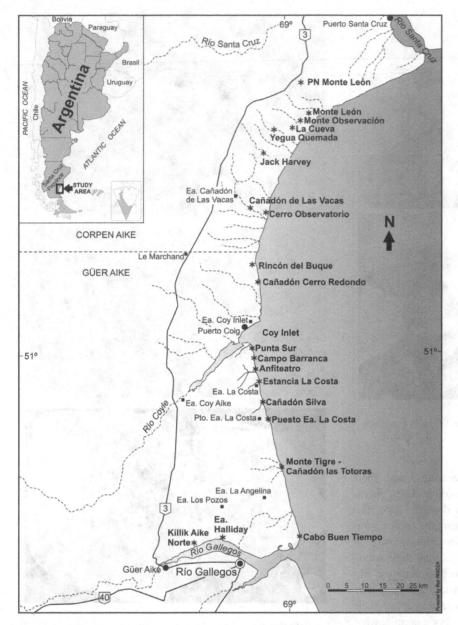

Fig. 1.2. Detailed map of the study area in the coastal Santa Cruz Formation, indicated in Fig. 1.1, and the localities listed in Appendix 1.1. Asterisks indicate localities; filled circles, towns; squares, Estancia (= Ea.).

Hatcher, 1903: vii). The fossils were described in a series of exhaustive monographs (1903–1928) in the *Reports of the Princeton University Expeditions to Patagonia*, edited by Scott between 1903 and 1932 (see Scott, 1903–1905, 1910, 1912, 1928; Sinclair, 1906, 1909; Sinclair and Farr, 1932) with extensive anatomical descriptions and illustrations. The conclusion was that the fauna was much younger than Ameghino had thought, and that most of the similarities that Ameghino identified between the Santa Cruz mammals and their northern counterparts were a consequence of adaptive convergence. In some ways, the size and scope of Scott's edited volumes led later workers to assume that most, if not

all, secrets of this fauna had been revealed. In contrast, we see the relative completeness of the taxonomic coverage as an opportunity, not a curse.

George G. Simpson (1902–1984) claimed that the Santacrucian fauna was particularly important for its wealth of nearly complete skeletons of representatives of a phase in South American mammal history in which the communities consisted of a complex mixture of descendants of ancient lineages of the continent (Marsupialia, Xenarthra, Litopterna, Notoungulata, and Astrapotheria) and new forms from other continents (Rodentia and Primates) (Simpson, 1980). Among these mammals, rather peculiar groups such

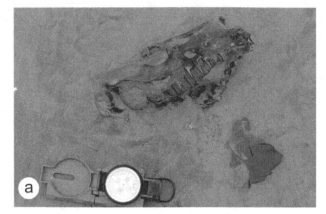

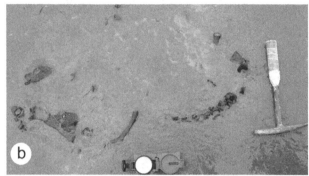

Fig. 1.3. a, Florentino Ameghino (1853/54 to 1911). b, Francisco P. Moreno (1852–1919). c, Carlos Ameghino (1865–1936). d, John Bell Hatcher (1861–1904).

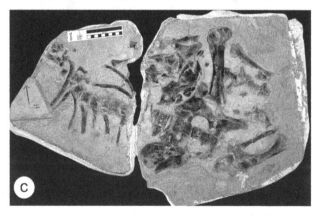

Fig. 1.4. Specimens *in situ* in the field and in the laboratory. a, Skull of the toxodontid *Adinotherium*; b, skeleton of a sloth. c, Specimen (sloth) in preparation at the laboratory of the Museo de La Plata.

as xenarthrans (sloths, armadillos, anteaters, and their relatives) and "archaic" ungulates (notoungulates, astrapotheres, litopterns) are abundant and diverse. Also, the best material of Miocene platyrrhine primates comes from this formation. In many cases the "ancient" forms (such as sloths) have quite different morphologies than their surviving relatives, making it difficult to reconstruct their life habits. In the case of platyrrhine primates, although it is much easier to find living analogs, some aspects of their paleobiology are difficult to reconstruct. For example, what were the adaptations that allowed them to survive in a supposedly highly seasonal environment at a latitude of 52° S, about 20° south of the southern limit of the distribution of the extant members of this group?

Despite the quality of the known Santacrucian fossils, paleobiological reconstructions based on functional morphology, biomechanics, or ecomorphology are in short supply and widely scattered in the specialist literature. There were some important contributions in the 1990s, including the first paleobiological approaches on this fauna, such as studies on the masticatory apparatus of armadillos (Vizcaíno, 1994; Vizcaíno and Fariña, 1994, 1997; Vizcaíno and Bargo, 1998), the locomotory apparatus of Miocene sloths (White, 1993, 1997), or the inference of diets of small

marsupials and notoungulates (Dumont *et al.*, 2000; Tauber, 1996). More recently, Townsend and Croft (2008) inferred the diet of Santacrucian notoungulates based on enamel microwear analysis, and Candela and Picasso (2008) performed a morphofunctional analysis of a Miocene porcupine's limbs. Fleagle *et al.* (1997) provided incidental comments on selected taxa of primates.

Despite continued fieldwork for more than a century, the coastal Santa Cruz Formation, especially south of Río Coyle (also called Coy or Coig Inlet), continues to yield a rich assortment of skulls and articulated skeletons (Fig. 1.4), probably in greater abundance than anywhere else in South America. Because the collected material is often in an excellent state of preservation, the fossils of the Santa Cruz Formation are the best record for interpreting the biological diversity of mammals in the southern part of South America (Patagonia) prior to the Great American Biotic Interchange (GABI), with an approach similar to that already used in another important region with vertebrate fossils at La Venta (Middle Miocene), Colombia, in northern South America (Kay *et al.*, 1997). The spectacular completeness of the fossil remains we recovered from Santacrucian localities of the Atlantic coast allows a detailed examination of the ecological dimensions of pre-interchange mammalian communities. Knowledge of mammalian community structure at this time provides reciprocal illumination on the nature and impact of faunal immigration that occurred later.

In the work described in this edited volume, we bring together a group of researchers to study the ~18–16 Ma coastal Santa Cruz Formation of Patagonia and its flora and fauna. We have the luxury that the systematics of the mammalian taxa is generally agreed upon. Thus, the focus is paleoecological rather than taxonomic. Our main purpose is to reconstruct the paleobiology of the Santacrucian vertebrate community and to place it in its biotic and physical environment.

Some work has been undertaken in the past to reconstruct aspects of the composite paleoecology of the Santa Cruz Formation based on the composition of its mammalian remains. In a landmark paper, Pascual and Ortiz-Jaureguizar (1990) examined faunal change among South American Cenozoic mammals based on the percentages of herbivorous species with different tooth crown heights. For chronologic units they used South American Land Mammal Ages. They included the Oligocene–Miocene Deseadan, Colhuehuapian, and Santacrucian mammal faunas collectively as the "Patagonian Faunistic Cycle," and recognized two subcycles within it – Deseadan and Pansantacrucian. The latter encompasses the Colhuehuapian and the Santacrucian Land Mammal Ages (Fig. 1.5). From low-crowned and rooted to high-crowned, rootless and ever-growing cheek teeth, they recognized four categories: brachydont, mesodont, protohypsodont, and euhypsodont. According to their analysis, by the beginning of the Patagonian Faunistic Cycle (Late Oligocene) many families of mammals had evolved protohypsodont and euhypsodont cheek teeth, a phenomenon they attribute to the increase in the number of grazing species coevolving with the spread of grasslands at mid-Patagonian latitudes. Further changes noted between the Deseadan and Pansantacrucian subcycles included a decrease in brachydont genera from 15% to 6%, and an increase in mesodont taxa from 31% to 48%; protohypsodont forms decline slightly from 31% to 23% and euhypsodont taxa remain

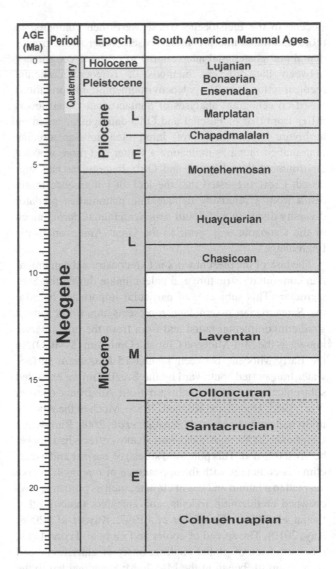

Fig. 1.5. Chronologic chart of the Neogene including the South American Land Mammal Ages (SALMA).

at 23%. These changes are attributed to a shifting balance of grassland and woodland habitats provided by a "park savanna" in Santacrucian times. Such analyses, while broadly useful, lack the kind of stratigraphic and chronologic precision so valuable for an understanding of ecological conditions at a single place and within a narrow range of time. The Santa Cruz Formation offers a unique window for reconstructing the structure of a South American mammalian paleocommunity with precise stratigraphic and geographic control. To date, just one analysis of Santacrucian faunas has been published within this more narrowly restricted scope. Croft (2001) used cenogram analysis (originally, a plot of vertebrate body sizes within a community, a method developed by Valverde, 1964) to interpret paleoenvironmental conditions for some of the best-known South American fossil mammal assemblages from the

Eocene to the Pleistocene. These were then compared to more traditional interpretations (based on herbivore craniodental and postcranial adaptations) to evaluate congruence between the different methods of paleoenvironmental reconstruction. Croft's paleoenvironmental interpretations based on cenogram analyses of Santacrucian faunal levels differ from that of Pascual and Ortiz-Jaureguizar based on herbivore hypsodonty. Croft interpreted the Santacrucian mammalian fauna as indicating a wetter and more wooded environment than Pascual and Ortiz-Jaureguizar had supposed. Croft suggested that the lack of congruence might result from a relatively depauperate mammalian predator diversity distinctive of South American faunas during much of the Cenozoic (e.g. prior to the Great American Biotic Interchange).

The task of the present work is to reconstruct the mammalian community structure and paleoclimate during the Santacrucian. This subject is of particular importance because the Santacrucian assemblage represents the world's most southern continental fauna and flora from the time interval known as the Mid-Miocene Climatic Optimum (MMCO). In the Early Miocene between 17 and 15 Ma, warm surface water transported southward by the Brazil Current extended subtropical conditions southward from Amazonia (Barron *et al.*, 1985; Hodell and Kennett, 1985). Much of the Andean uplift had not yet occurred (Blisniuk *et al.*, 2005; Ramos and Ghiglione, 2008), so Andean rain-shadow effects had not yet been established. This poleward spread of warmer and wetter climates coincides with the appearance of mammalian taxa adapted to a humid and warm climate, such as primates, low-crowned erethizontid rodents, and anteaters (among other faunal elements) (Fleagle *et al.*, 1997; Kay *et al.*, 2008; Kay, 2010). The spread of cooler and more arid conditions, a consequence of global cooling, onset of glaciations, and Andean uplift, began in the Middle Miocene and led to the regional extirpation of many of these faunal elements after about 15.5 Ma (Kay *et al.*, 1998; Tejedor, 2006). Deposits of the Early Miocene Santa Cruz Formation (*sensu lato*) fall within the MMCO interval of warm climate. Nevertheless, the climatic conditions must have been unusual and without modern equivalents, given that seasonality in energy availability at ~50° S latitude must have had an effect upon biotic productivity, as it does today (Kaufman, 1995). A point of emphasis that we asked the contributors to consider is the importance of climatic conditions for the presence of their groups, so as to refine our understanding of mid-Miocene continental climate at southern latitudes.

1.2 Historical background

The nineteenth century saw a rapid rise in the exploration of South America. The industrial revolution fueled the expansionist, colonialist, and imperialist spirit that had

characterized European nations for almost four centuries. Owing to the growing ambition to expand cultural horizons that fostered the prevailing ideals of progress and growth of reason on the one hand and strong demand for raw materials and new consumers on the other, the most powerful states sought new territories to take advantage of their resources and populations. Many exploratory expeditions had their attached naturalists who performed the dual role of collecting information, allowing an expansion of knowledge, while also identifying new natural products for exploitation, thus encouraging commercial growth. The fossils that promised to tell the fascinating remote history of life on Earth were among the many interests that spurred the imagination of naturalists who, after a long journey by sea, reached American shores. Marshall (1976) gave an account of the history of the explorations, early collectors, and major paleontological expeditions to Santa Cruz. In the following compilation that contextualizes the historical background for our research we have avoided duplicating his effort.

Fossil collecting in Santa Cruz began in January 1845, when Captain Bartholomew Sulivan (1810–1890), in command of the HMS *Philomel* and a former shipmate of Charles Darwin (1809–1882) aboard the HMS *Beagle*, discovered fossils on the cliffs of the estuary of the Río Gallegos (Brinkman, 2003). While some of his crew collected fresh water in preparation for a voyage to the Falkland Islands (which had been taken by the British in January 1833, some six months before Darwin collected his first fossils from nearby Bahía Blanca), Sulivan, who was interested in the spectacular bluffs that flank the estuary on its northern bank, landed about 12 miles upriver. Drawing on the training he received from Darwin, Sulivan collected several specimens from fallen blocks and produced a profile detailing his geological observations. He sent his collection to Darwin, who passed the fossils on to Richard Owen (1804–1982) of the Hunterian Museum of the Royal College of Surgeons of London. The great anatomist described what proved to be the first Miocene vertebrates from South America based on these fossils. Among the specimens were those that Owen described as types of the ungulate-like *Nesodon imbricatus* and *Nesodon sulivani*, the latter after its collector. Years later, Florentino Ameghino would regard it as synonymous with the former, thus denying Sulivan his well-deserved tribute (Brinkman, 2003). During the same year Sulivan, as commander of the HMS *Philomel*, participated in the battle of Vuelta de Obligado on the Río Paraná that pitted the small Argentine army against the Anglo-French forces that had blockaded the port of Buenos Aires. Sulivan captured the flag of the Argentine battery, but returned it 38 years later, honoring the courage and bravery of the Argentine soldiers (Vizcaíno, 2008). In subsequent years Sulivan encouraged Darwin to intercede with the British Admiralty to mount a new expedition to collect in the Río Gallegos area. Sulivan himself probably returned to Río Gallegos between 1848 and 1851, and made new collections when he traveled on his own initiative to the Falkland Islands. Sulivan's and Darwin's persistence eventually payed off: in March 1863, the Admiralty sent Sulivan to collect fossils on the banks of the Río Gallegos with his son James Y. F. Sulivan (1838/9–1901) and the naturalist Robert Oliver Cunningham (1841–1918). The specimens were sent to Thomas Henry Huxley (1825–1895). The great naturalist handed them over to Owen's successor as curator of the Royal College of Surgeons, William Henry Flower (1831–1899), who described another ungulate-like mammal, *Homalodotherium cunninghami*.

In 1877, Francisco P. Moreno (1852–1919; Fig. 1.3b) explored Patagonia and collected the first fossil mammals from the Early Miocene Santa Cruz beds along the valley of the Río Santa Cruz; these specimens were described by Florentino Ameghino (Ameghino, 1887, 1889). Until the later nineteenth century, Patagonia was mostly inaccessible owing to poor transportation systems and conflicts with native peoples. The naval transport line, founded in 1880, allowed the establishment of settlements and villages, and was the sole supplier of basic necessities in the region, transporting goods, medicine, fuel, passengers, and news. Taking advantage of this new mode of travel, Florentino's brother, Carlos Ameghino (1865–1936; Fig. 1.3c), then Travelling Naturalist of the Museo de La Plata, first came to Santa Cruz in 1887. Moreno, Director of the Museo de La Plata, had sent Carlos to follow up on his earlier discoveries (1877) along the banks of Río Santa Cruz. Rusconi (1965: 58) noted that on this first trip Carlos collected more than 2000 specimens of Tertiary mammals belonging to more than 120 species, of which only a dozen had been previously known. That same year, Florentino (Ameghino, 1887), then Vice-Director of the Museo de La Plata, published on the fossils collected by Carlos. This journey initiated a remarkable series of geological and paleontological expeditions which lasted for 15 years. Extensive accounts of Carlos' trips are available in Rusconi (1965) and Vizcaíno (2011). In the succeeding decades Carlos discovered many other fossil-bearing localities spread over a wide belt extending from the Gulf of San Jorge to Tierra del Fuego, between the Atlantic coast and the Andes.

By 1890 the Ameghino brothers had become estranged from the Museo de La Plata because of differences with Moreno. On an independently financed trip between June 1, 1890 and July 20, 1891, Carlos, while returning from Puerto Gallegos to the Cordillera, explored the north bank of the Río Gallegos and several areas near its mouth. This second part of the expedition was not accidental, as the brothers knew of Sulivan's finds and had long awaited the opportunity to study the area and collect more fossils. In truth, Carlos was not particularly impressed by Sulivan's collecting area,

despite Carlos' surprising discovery (for those latitudes) of the first Tertiary fossil remains of monkeys. His visit, however, resulted in the discovery of the Atlantic coast localities, which have yielded the abundant and well-preserved Early Miocene continental fossil vertebrates that were crucial to understanding the vertebrate lineages that evolved in South America during its long geographic isolation. These discoveries have inspired and influenced many paleontologists. In a letter to Florentino dated March 10, 1891 (Torcelli, 1935, Vol. 21), Carlos wrote:

> The banks of the Río Gallegos, contrary to my expectations, have not provided much, mainly because of the difficulties experienced in exploring the usually inaccessible cliffs, and I remain amazed by the discovery, many years ago now, of remains of *Homalodontotherium* [sic] and *Nesodontes*, animals very rare in these formations. However, by dint of perseverance, I have been able to recover small pieces with some regularity... (Torcelli, 1935, Vol. 21, page 12, lines 7–14)

> But the most interesting object collected here is a small lower jaw with teeth of an animal that I at first could not refer to any of the known animals, and that kept me puzzled until my last trip, at which point I was able to learn, not without surprise, its true significance: it is nothing less than the first known fossil remains of monkeys of the Republic... (page 12, lines 27–32)

> But if Gallegos has not satisfied all my hopes, I have discovered a site on the sea coast, between Coyle and Gallegos, which, by its wealth and layout, I have no hesitation to declare the most important examined so far in Patagonia, and where I collected truly superb pieces, as you will soon know. (page 13, lines 12–16)

> This site is located on the sea beach, facing the Indian Camp site Corrigenkaik, somewhat closer to the mouth of Coyle. The great feature of this remarkable site is that it is below the level of the sea, in an arrangement quite equal to that of Monte Hermoso, and therefore only recognizable at low tide. (page 13, lines 17–22)

Based on the letters to Florentino Ameghino (Torcelli, 1935; Vizcaíno, 2011), Carlos worked at Corriguen Aike between 1891 and 1893 on at least three and possibly four occasions.

In 1889, Florentino was replaced as the paleontologist of the Museo de La Plata by the French geologist Alcides Mercerat. For a time, Mercerat competed, though not very successfully, with Carlos at collecting fossils in Santa Cruz. Mercerat was appointed Secretary of the Museum in 1891, but left in 1892 over his own disagreements with Moreno. During the last part of his tenure, the Travelling Naturalist was Carlos V. Burmeister who, in 1891, collected Santacrucian fossils from a locality known as Monte Observación (see below). In a brief account, Burmeister (1891) noted that his collection had already been housed in the Museo de

La Plata and would be described in a later publication. In 1892, he collected fossils from a wide region between the Atlantic coast and the Cordillera and the Ríos Santa Cruz and Chalía (Burmeister, 1892; Riccardi, 2008).

The fossils collected in the Santa Cruz beds by Moreno (in 1877) and other Museo de La Plata staff (between 1887 and 1892) were published by Mercerat (1890, 1891a–f), Moreno and Mercerat (1891a, b), and F. Ameghino (1887, 1889). Many of these specimens were reviewed by Lydekker (1893, 1894). The stratigraphic context of these finds was established in detail by Mercerat (1897), who between 1892 and 1895, as researcher at the Museo Nacional de Buenos Aires, produced 10 regional profiles south of the Río Santa Cruz (Mercerat, 1896, 1897). These contributions established the proper relationship between the Santacrucian Land Mammal Age and the marine Patagonian Stage, as they are currently understood.

In letters to Florentino dated October 1892, Carlos mentioned that he had met Mercerat on the ship to Santa Cruz and that he seemed to be a competent geologist. Carlos also pointed out that he was planning to do further work at Monte Observación (see below) and Corriguen Aike to prevent Mercerat finding several specimens that he had left marked in the field during a previous effort. In a letter written at Monte Observación and dated February 1893, Carlos reported that Mercerat was wasting time in Santa Cruz. In a letter in June 1893, Florentino confirmed that Mercerat was back in La Plata; and several months later he asked Carlos about a farm that Mercerat had purchased on the Río Coyle (a farm of 7500 hectares – rather small for Patagonia – named "Pechorok-Aike"; Lenzi, 1980: 355). In March 1894, Carlos confirmed that Mercerat and his family were living at this property, but in May 1895, he reported that Mercerat had gone broke, had been in jail in Río Gallegos for swindling, and had vanished from Santa Cruz. Mercerat's fossil collection was being held by a creditor from Coyle, who was planning to sell it. We do not know what became of this collection.

As mentioned above, Florentino's timely publication of the Santa Cruz collections and, particularly, some of his novel scientific conclusions prompted William B. Scott to mount an expedition to collect fossils in the Santa Cruz beds. This was led by the so-called "King of Collectors," John B. Hatcher (1861–1904; Fig. 1.3d), then Curator of the Department of Vertebrate Paleontology at Princeton University. Three expeditions were launched between 1896 and 1899. The first one extended from March 1896 until July 1897; the second from November 1897 to November 1898; and the third from December 1898 to September 1899. About Corriguen Aike, Hatcher wrote:

> It was in the sandstones of this shelving beach, near Corriguen Aike, that we discovered the rich deposit of fossil bones mentioned above. At this point, as at most

places throughout this beach, erosion has taken place along the bedding planes, so that over considerable areas the surface of the beach represents essentially the same geological horizon. At this particular locality the dark green sandstones in which the bones were imbedded bore evidence of having been deposited over the flood plain of some stream or shallow lake. On walking about over the surface at low tide, there could be seen the skulls and skeletons of those prehistoric beasts protruding from the rock in varying degrees of preservation. At one point the skull and skeleton of *Nesodon* would appear, at another might be seen the limbs or perhaps the teeth of the giant *Astrapotherium* just protruding from the rock, while a little farther on a skull and jaws of the little *Icochilus* grinned curiously, as though delighted with the prospect of being thus awakened from its long and uneventful sleep. On one hand, the muzzle of a skull of one of the larger carnivorous marsupials looked forth, with jaws fully extended and glistening teeth, the characteristic snarl of the living animal still clearly indicated, while at frequent intervals the carapace of a *Glyptodon* raised its highly sculptured shell, like a rounded dome set with miniature rosettes, just above the surface of the sandstones. Throughout eighteen years spent almost constantly in collecting fossil vertebrates, during which time I have visited most of the more important localities of the western hemisphere, I have never seen anything to approach this locality near Corriguen Aike in the wealth of genera, species and individuals. (Hatcher, 1903: page 72, lines 17–22; page 73, lines 1–19)

Princeton University's final expedition started as a collaborative effort with the American Museum of Natural History, New York, represented by paleontologist Barnum Brown (1873–1963). However, by April 1899, Hatcher left Patagonia for fieldwork in Corrientes Province (northeast of Argentina) and Paraguay. Apparently, the first part of the expedition to the west was not a complete success and Hatcher was not interested in continuing to explore the coastal localities where he had conducted so much of his previous work; Brown remained working in the east of Santa Cruz Province until January 1900 (Dingus and Norell, 2010).

Towards the end of the nineteenth century, the famous French paleontologist Albert Gaudry (1827–1908) asked André Tornouër, a young French immigrant to Argentina, to collect fossils from Patagonia for the Muséum national d'Histoire naturelle, Paris. In a letter dated June 1900, Florentino confirmed that Tornouër had made some collections from the Santa Cruz beds.

The frenzy over Santacrucian fossils had apparently subsided by the beginning of the twentieth century. In a letter dated October 1901, Florentino announced that Hatcher had returned home and was not planning to return to Patagonia, that the expedition from the museum in New York had also returned with many Santacrucian fossils, and that the Museo de La Plata had no one working in Patagonia: "*Parece que al fin se han cansado*" (It seems that they finally got tired).

Brief expeditions were led by Handel T. Martin (University of Kansas) just after the turn of the last century (1903–1904), and by Elmer Riggs (1869–1963; Field Museum of Natural History, Chicago) in the 1920s (Marshall, 1975, 1976). Martin arrived in Buenos Aires in late 1903, and traveled from Bahía Blanca to Río Gallegos in January 1904 on the steamer *Chubut* (see Martin, 1904, for a brief narrative of the expedition). His party collected fossils at Killik Aike Norte Estancia (then Felton's Estancia, about 15 km west of Río Gallegos), from the bluffs along the Río Gallegos and in the coastal flats from Cabo Buen Tiempo (Cape Fairweather) and to about 48 km farther north (thus as far as Corriguen Aike). All the specimens collected were shipped to Lawrence, Kansas. Some are still in the Paleontology Section, University of Kansas, and a smaller portion is in the Natural History Museum, London (and elsewhere).

Riggs led one expedition to Patagonia (see Riggs, 1926, 1928 for short reports) from November 1922 to May 1923. As Hatcher and Martin had done, he collected at Felton's Estancia, La Angelina Estancia, and Corriguen Aike, but also 10 miles north of Coy Inlet (= Wreck Flat or Smith's Rock Flat). His efforts produced 282 specimens, cataloged in the Field Museum, belonging to 32 mammal species (according to field determinations) and a few birds. Among these are 177 skulls, as well as a few reasonably complete skeletons (Riggs, 1928).

Most paleontological expeditions to Argentine Patagonia during the past three decades have concentrated on a few easily accessible sites, mostly on the Atlantic coast, while other localities have been neglected. In the early 1980s Rosendo Pascual (who for almost four decades acted as head of the División Paleontología Vertebrados of the Museo de La Plata) led brief forays to several Santacrucian localities. Later in the 1980s and 1990s, Miguel Soria (Museo Argentino de Ciencias Naturales "Bernardino Rivadavia"), John Fleagle (State University of New York, USA), and others made extensive collections north of Río Coyle, at Cerro Observatorio (see below for a clarification of the name of the site), among other places, where fossils are numerous but less complete than those from Corriguen Aike. Collectively, the work of these expeditions helped clarify several aspects of the systematics of these mammals and placed the geochronology of the Santa Cruz Formation on a firm footing.

Few further systematic efforts to collect here were made before the work of Adán A. Tauber (Universidad Nacional de Córdoba, Argentina). Tauber surveyed the geology of the Santa Cruz Formation south of Coy Inlet, reidentified the most productive fossil sites and levels, and made important collections (about 250 specimens, mostly housed at the Museo de Paleontología, Universidad Nacional de Córdoba, Argentina) (Tauber, 1994, 1997). In the past few years, Marcelo Tejedor (Centro Nacional Patagónico, Argentina) and

Laureano González (Universidad Nacional de la Patagonia "San Juan Bosco," Argentina), among others, have conducted expeditions in the western part of Patagonia near the Andes.

Since 2003, we have undertaken the collection of over 1600 specimens, including large series of relatively complete skeletons. All belong to the Museo Regional Provincial Padre M. J. Molina of Río Gallegos (Santa Cruz Province, Argentina). On average, a team of eight people (Fig. 1.6) collected fossils for 15 to 25 days over nine field seasons from several localities situated along an approximately 50-km stretch of the Atlantic coast south of the Río Coyle (Figs. 1.1 and 1.2, and see below for location and description of localities). Virtually all identifiable specimens were collected without bias to size and taxonomic interest. While half of the crew worked recovering medium to large, articulated skeletons, the other half prospected and collected smaller specimens (Fig. 1.7).

1.3 Present environment

While more than 80% of South America lies in the tropical zone (within about 23° of the equator), the southernmost part of the continent reaches about 55° S, not far from the polar circle (~66° S). The uneven distribution of land along this latitudinal gradient and the high elevations of the Andes to the west (which act as a barrier to humid winds from the Pacific Ocean) decisively influence the varied continental landscapes. East of the Andes, tropical and subtropical forests dominate in the north, exceeding all other biomes in the diversity of their flora and fauna. Southward, this dominant biome largely gives way to tropical deciduous forests, woodlands and temperate grasslands, and these in turn are replaced south of the Río Colorado by the steppes of Patagonia that extend to the southernmost tip of continental South America. A geographic unit of just more than one million square kilometers, Patagonia borders the Andes on the west and includes the Tierra del Fuego archipelago at its southern extent. Extra-Andean Patagonia is now cold and arid, and it has been characterized as one of the windiest regions in the world: strong, dry, and westerly winds are among its characteristic climatic features, emphasizing the semi-arid or arid nature of the region. Yearly rainfall is highest in the Patagonian Andes, exceeding 2000 mm. In the central part of Patagonia, annual precipitation ranges from 125 mm in the central-east to 500 mm in the west, concentrated in the coldest months of the year (April to September). The average annual temperature is ~12 °C and annual rainfall is ~300 mm. The decrease in rainfall from west to east influences a gradient of vegetation types: forest, grassy steppe, grassy-shrublands, and steppe (Leon *et al.*, 1998). The flora is dominated by shrubs and herbs (steppe) with morphological and physiological features associated with environmental stress (Barreda and

Fig. 1.6. Field teams who worked during the summer seasons between 2003 and 2011. a, **2003**: from left to right, (back row) Jonathan Perry, Carlos Luna, Sergio Vizcaíno, Juan Moly, Adán Tauber, (front row) Susana Bargo, Mariella Bruno, Sonia Cardozo, Mary Palacios. b, **2004**: (back row) Jorge Battini, C. Luna, J. Perry, Nick Milne, S. Vizcaíno, Richard Kay, S. Bargo, A. Tauber, (front row) Andy, Teddy and Alex Battini. c, **2005**: Leonel Acosta, Juan Fernicola, Anne Weill, S. Vizcaíno, R. Kay, Francisco Prevosti; (in front) S. Bargo. d, **2006**: F. Prevosti, S. Vizcaíno, Michael Malinzak, S. Bargo, R. Kay, J. Fernicola; (front) L. Acosta. e, **2007**: (back row) F. Prevosti, Lucas Pomi, S. Vizcaíno, S. Bargo; (front row) L. Acosta, Néstor Toledo, J. Perry, J. Fernicola. f, **2008**: (back row) Guillermo Cassini, N. Toledo, J. Fernicola; (middle row) S. Vizcaíno, S. Bargo, R. Kay; (front row) Gerry De Iuliis, L. Acosta. g, **2009**: R. Kay, S. Bargo, L. Acosta, J. Perry, S. Vizcaíno. h, **2010**: J. Perry, N. Toledo, Nahuel Muñoz, L. Acosta, J. Fernicola, R. Kay, S. Bargo, S. Vizcaíno. i, **2011**: S. Vizcaíno, N. Muñoz, J. Perry, L. Acosta, S. Bargo, J. Fernicola, Verónica Krapovickas, Laura Cruz, Siobhan Cooke.

Palazzesi, 2007). Only in western Patagonia, close to the Andes, is rainfall sufficient to allow the development of forests with a dense understory of tree ferns, vines, and shrubs.

The aridification of Patagonia is a fairly recent geological occurrence that resulted from the interplay of regional paleogeographic and tectonic events, and global paleoclimatic changes. The isolation of Antarctica from Australia and South America caused a general trend towards cooler conditions and led to the development of major ice sheets in Antarctica. Uplift of the Patagonian Andes produced an important (and still present) orographic rain shadow to the east of this mountain belt during Miocene times, causing a progressive increase in aridity (Blisniuk *et al.*, 2005). All of these changes led to the expansion of a cooler and drier climate throughout the Patagonian landscape. For further information on the vegetational history of Patagonia during the Cenozoic see Barreda and Palazzesi (2007).

Fig. 1.7. Fossil collecting at the fossil levels during low tide at Puesto Estancia La Costa. a–c, Extraction of a glyptodont (carapace and skeleton). d, e, Collecting small specimens.

Our area of study is under the influence of the two southernmost continental volcanic arcs of the Andean Volcanic Belt, the Southern and Austral Volcanic Zones (SVZ and AVZ, respectively), separated by the Patagonic Volcanic Gap. The SVZ extends for 1400 km along the southern margin where the Nazca Plate is subducted beneath South America, from 33.4° to 45.9° S latitude, and contains more than 60 volcanoes considered to have been active in the Holocene. The AVZ extends south of the Patagonian

Volcanic Gap to the Tierra del Fuego archipelago, a distance of well over 1000 km (600 miles). The arc was formed as the Antarctic Plate was subducted under the South American Plate.

Vulcanism in the AVZ is less vigorous than in the SVZ. Even so, Andean volcanic activity had an impact on the Patagonian coast. Although the mountains lie more than 500 km from the Patagonian coast, the prevailing winds (currently from west-to-east, as must also have been the case during the Early Miocene based upon continental position) can easily transport ash across this distance. Beginning in August 1991, the eruption of Mt Hudson, in Chile, deposited ash over a wide area of Santa Cruz for about 4 months, resulting in mass death and destruction of cattle and field crops throughout the province. The ash plume deposited an ash layer, several centimeters thick, on the Atlantic coast near the town of San Julián, approximately 535 km away and about 100 km north of Monte León, one of the better-known Santacrucian sites (Crawford *et al.*, 2008). This modern-day example demonstrates that ash produced by the vulcanism that occurred in the Andes region and Austral basin during the Early Miocene could have reached the area of the present Atlantic coast by wind transport. In addition to transportation by rivers and streams, these ashes generated the materials that constitute most of the sediments of the lower Estancia La Costa Member of the Santa Cruz Formation in our area of study (see Matheos and Raigemborn, Chapter 4, and references therein).

Our area of study extends about 70 km from north to south on the Atlantic coast between Río Coyle and Río Gallegos, at about 52° S latitude, barely 100 km from the Strait of Magellan (Fig. 1.1). Today, the area is dominated by xeric grassy steppes on a gently undulating landscape. It belongs to the Distrito Magallánico of the Provincia Fitogeográfica Patagónica (León *et al.*, 1998), and is characterized by a cold maritime climate. In the city of Río Gallegos, the mean annual temperature ranges from 3 to 13 °C, and the mean annual rainfall is 270 mm (data for the period 1981–1990). The grassy steppe is dominated by coirón (*Festuca gracillima*), forming a fairly closed layer, 30 to 40 cm in height, which constitutes 85–90% of the local biomass. Shrubs of the species mata negra fueguina (*Chiliotrichum diffusum*) and murtilla (*Empetrum rubrum*) are also common.

1.4 The age of the fauna of the coastal Santa Cruz Formation

The Santa Cruz Formation extends through much of southern Patagonia, Argentina, from the eastern area of the South Patagonian Cordillera (Cordillera Patagónica Austral) southward from Lago Buenos Aires to the Río Turbio area, and eastward from southern extra-Andean Patagonia to the Atlantic Ocean, between Golfo de San Jorge and northern Tierra del Fuego (Fig. 1.1; see Matheos and Raigemborn, Chapter 4, and references therein).

Recently, Tonni (2009) claimed that in Ameghino's time a *formación* (Formation) was a unit of time inferred by its fossiliferous content; lithology was not the main, let alone only, criterion used to define it. Ameghino divided his *formaciones* into *pisos* (stages) or *horizontes* (horizons).

There is some confusion in the literature about the use of what is currently considered the Santa Cruz Formation in relation to Ameghino's *formación Santacruzeña* and *piso Santacruceño*. Ameghino (1889: 16) defined the *formación Santacruzeña* as including the marine *piso Sub-Patagónico* and the terrestrial or subaerial *piso Santacruzeño* " … composed of layers of sand, clay, calcareous and more or less compact rocks of volcanic origin, arranged in horizontal beds, with numerous terrestrial fossils, particularly bones of mammals, and a thickness of 100 to 200 meters" (*formado por estratos de arenas, arcillas, calcareos y rocas más o menos compactas de origen volcánico, dispuestas en lechos horizontales, con numerosos fósiles terrestres, particularmente huesos de mamíferos, y con un espesor de 100 hasta 200 metros*). This description was based on Carlos' exploration along the Río Santa Cruz (Ameghino, 1889: 17), but in later publications, Florentino included the exposures on the Atlantic coast (e.g. Ameghino, 1891; 1894; 1900–1902). In summary, it is Ameghino's *piso Santacruceño* that is today considered equivalent to the Santa Cruz Formation (Marshall, 1976).

After 1889, Ameghino published much of his work in French (Spanish versions are available in the Obras Completas posthumously compiled by Torcelli, 1935). Ameghino (1906: 201, footnote 1) claimed that in the French literature he used the ending "*ien*" or "*ienne*" (e.g. "*Santacrucienne*") for his *formaciones* (formations) and "*éen*" (e.g. "*Santacruzéen*") for his *pisos* (étages). However, in his important contribution (Ameghino, 1900–1902) on the ages of the sedimentary formations of Patagonia, Ameghino used the ending *ien* for the *étage* (e.g. Santacruzien), which certainly caused confusion.

On the basis of its fossil vertebrate content, Ameghino (1906: 228–232, 498) recognized two continental *étages* within the *formation santacruzien*: the *notohippidéen* and the *santacruzéen*. The *étage notohippidéen* is typically represented in Karaiken, near Lago Argentino (Marshall and Pascual, 1977; Riccardi and Rolleri, 1980) and contains the *Notohippus* fauna, traditionally considered as part of the Santacrucian. González Ruiz and Scillato-Yané (2009) summarized the background of controversies on the age of this fauna. The "Notohippidian" (*étage notohippidéen*) was recognized by Ameghino (1900–1902: 54, 179–180, 220–224) in reference to his "*couches à* Notohippus." Ameghino (1900–1902) concluded that this fauna indicated an evolutionary stage generally less advanced than the typical Santacrucian (*étage santacruzéen*). Kraglievich (1930) named Ameghino's *Notohippus* beds as "Karaikense" (after the locality Karaiken). According to Simpson (1940) the Karaiken beds might be considered simply as a local facies of the Early Santa Cruz Formation, and the *Notohippus* fauna, as described by Ameghino (1904, 1906), represents an early phase of the Santa Cruz fauna. However, Simpson

(1940: 664) claimed that "the distinction of the Karaiken formation and of a Karaikenian stage from the Santa Cruz and the Santacrucian is thus not clear-cut but may be accepted tentatively." Marshall *et al.* (1983: 31) also regarded the small and fragmentary faunas from Karaiken as representing early Santacrucian local faunas. Most of the Karaiken fauna has not been revised in detail since Ameghino's original work, and the validity and provenance of the species are usually uncertain (Gonzáles Ruiz and Scillato-Yané, 2009).

The *étage santacruzéen* is typically represented in the outcrops of the Santa Cruz Formation in the eastern region of Santa Cruz Province along the Atlantic coast (e.g. Marshall and Pascual, 1977). According to Ameghino (1906), this stage is the younger. As described in Chapter 4 (Matheos and Raigemborn), the outcrops of this formation in this area are arranged parallel to the coast between Parque Nacional Monte León and Río Gallegos as an active sea cliff with numerous slumps, slips, and scourings, but representing most of the thickness of the Santa Cruz Formation. In this place the beds of the Formation show a general dip of 1–3° SE (Tauber, 1994, 1997). The basal limit of the Santa Cruz Formation lies below an oyster bed (*Crassostrea dorbignyi*) (Griffin and Parras, Chapter 5), and is located 23 m above current sea level at Punta Norte of the mouth of Río Coyle (Feruglio, 1938), at sea level near Rincón del Buque and Cerro Redondo, and below sea level south of Río Coyle. The Santa Cruz Formation is covered by marine sediments of the Cape Fairweather Formation and/or the Rodados Patagónicos (Patagonian Shingle Formation).

The assemblage of fossil mammals of the Santa Cruz Formation was used by Pascual *et al.* (1965) to establish the *Edad Mamífero Santacrucense*, known in the English literature as the Santacrucian South American Land Mammal Age (SALMA). In the past two decades a series of contributions has discussed the theoretical framework in which the concept of SALMAs was developed (see Tonni, 2009, and references therein). A general conclusion of these contributions is that, although based on poorly defined stages according to current geosystematics (Cione and Tonni, 1995), SALMAs are formal ages. As a consequence, contributions in this volume will refer to the age of the biota of the Santa Cruz Formation as Santacrucian SALMA or Santacrucian Age.

Another fauna originally described by Ameghino (1906) requires further consideration here. The *étage Astrapothériculéen*, characterized by the presence of *Astrapothericulus* and its associated fauna, outcrops in the northwestern part of Santa Cruz Province, along the banks of the Río Pinturas (Kramarz and Bellosi, 2005). Ameghino (1906: 226, 498) included it in the top of this *Formation Patagonienne*. The deposits corresponding to this *étage* have been designated formally as the Pinturas Formation (Bown *et al.*, 1988;

Bown and Larriestra, 1990; Fleagle, 1990; Fleagle *et al.*, Chapter 3).

Marshall *et al.* (1986) restricted the Santacrucian vertebrate fauna to the interval between 18 and 15 Ma. Later, Bown and Fleagle (1993) and Fleagle *et al.* (1995) reported an age of 16.6–16 Ma for the Santa Cruz Formation at Monte León. Finally, based on new ^{40}Ar/^{39}Ar analyses, Perkins *et al.* (Chapter 2) indicate that the exposures of this formation in the Atlantic coastal area are ~18–16 Ma. According to these dates, the Santa Cruz Formation is late Early Miocene in age (Fig. 1.5). Biostratigraphic studies and radiometric dates of the Pinturas Formation (~18 Ma) suggested that part of this unit is contemporaneous with the lowest part of Santa Cruz Formation at the typical coastal localities, i.e. Early Miocene (Bown and Fleagle, 1993; Fleagle *et al.*, 1995; Kramarz and Bellosi, 2005; Ré *et al.*, 2010).

1.5 Localities

Despite the enormous impact that Santacrucian fossils have had on the paleontology of the last part of the nineteenth and the first half of the twentieth century, it was not until the mid 1970s that an attempt was made to synthesize precise information on Santacrucian localities from both published and unpublished records (Marshall, 1976). Marshall provided a historical account of paleontological expeditions to the Santa Cruz Formation, including three maps and an annotated list of 20 localities then included in the formation. He summarized coordinates on the most recent topographic maps, distances to the most conspicuous reference points noted in the literature, general characteristics of the outcrops, and the fate of the previously made collections. This contribution has been, and will continue to be, the starting point for any researcher interested in Santacrucian fossils. We will not repeat information given in that article. Instead, we update some of its contents with, in some cases, more accurate data generated since its publication, in particular for the localities found in eastern Santa Cruz Province, from Monte León in the north to Río Gallegos in the south. We also summarize all this information on a map (Figs. 1.1 and 1.2) and a list of localities (Appendix 1.1).

According to Tauber (1991, following C. Ameghino in Torcelli, 1935), the Indian camp site known as Corriguen Aike in Ameghino's and Hatcher's times was close to the present Puesto de la Estancia La Costa, next to the creek that provides access to the strand and fossiliferous rocks in the intertidal zone (Fig. 1.8). This canyon is at 51° 11′ S, 69° 05′ W, 25 km south of Punta Sur, the feature that marks the southern limit of the mouth of the Río Coyle. Using a map published in 1968, Marshall (1976) also located Corriguen Aike at approximately 25 km (some 15 miles) south of Coyle Inlet, noting that Hatcher (1903: 72) listed this locality as "12 miles S of Coy Inlet" (some 19 km). Tauber

Fig. 1.8. Puesto Estancia La Costa locality (= Corriguen Aike) during low tide. The canyon provides access to the fossiliferous rocks in the intertidal zone.

(1991) suggested that the information given by Hatcher and Marshall produced some inconsistencies in relation to the location of the Corriguen Aike site discovered by C. Ameghino and, in order to avoid uncertainty over the provenance of his fossils, decided to name his locality as Puesto de la Estancia La Costa (or Puesto Estancia La Costa, PLC in Tauber, 1997 and subsequent publications). Certainly, a distance measured from Punta Sur of "12 miles S of Coy Inlet" would fall in Cañadón Silva, the next canyon north of PLC, identified by Tauber (1997) as a different locality (Cañadón Silva or CS). Actually, what Hatcher described in his Narrative as "12 miles S of Coy Inlet" was the site where he camped inland, not necessarily where he collected on the beach, and he mentioned that his party collected in "an area about one and a half miles in length, with an average breadth of perhaps three hundred yards" (Hatcher, 1903: 73, line 35). Furthermore, Hatcher (1897) had previously provided a sketch of the mouth of a canyon "in bluffs of coast at *Corriken Aike*," claiming that it was "18 miles South of Coy Inlet" (about 29 km). Although the distance does not match with that mentioned in 1903, his sketch almost perfectly represents the downcutting gravel layer resting on the Santa Cruz Formation as it occurs in the canyon of Tauber's Puesto Estancia La Costa, and no other similar canyon is present for the next 13 km southward. We do not know how Hatcher measured distances, nor what his reference point was at Río Coyle. Consequently, old references based on distances should be interpreted with caution.

In a previous contribution (Vizcaíno *et al.*, 2010), we located Puesto Estancia La Costa (which is equivalent to Tauber's PLC) at approximately 51° 12′ S, and 69° 04′ W, about 12 km south of the Estancia La Costa locality. The area of exposure of the PLC locality is approximately 3 km from

north to south and 300 to 400 m east to west, depending on the tide. At this locality, Tauber (1997) identified two main fossiliferous levels (FL): FL 6, composed of tuffaceous silty sands, and FL 7, composed of greenish silty clays. More recently, Tauber *et al.* (2004) identified a new and particularly rich fossiliferous level, FL 5.3, subjacent to FL 6, composed of greenish silty clays with carbonatic concretions, which probably became evident in the last decade – a consequence of the removal of part of FL 6 through erosion (see also Vizcaíno *et al.*, 2010: Fig. 3, and Matheos and Raigemborn, Chapter 4). Owing to strong erosion by the ocean, we cannot be certain that the specimens collected 100 years ago are from the same rocks we are sampling today. As an example, Hatcher (1900: 107, 1903: 73) mentions not only the striking abundance of skeletons observed at the beach during low tides, but also fossil footprints in several places:

> … series of such foot prints were occasionally seen extending for upwards of 100 feet, each track distinctly impressed so that it would have been possible to count the number of steps taking in covering a given distance and to measure the exact stride of the animal (Hatcher 1900: 107, lines 19–23).

After 10 uninterrupted years of fieldwork in this same area, we have only recently discovered vertebrate footprints at Puesto Estancia La Costa and Monte Tigre. Owing to the rate of erosion at PLC, these surely are not the ones Hatcher reported and we have not yet had the chance to study them.

Although we argue that Tauber's Puesto Estancia La Costa locality is to a great extent what Ameghino, Hatcher, and Marshall referred to as Corriguen Aike, for the purposes of this book we mostly follow Tauber (1994, 1997) for the names of the localities between the Ríos Coyle and

Gallegos: Cañadón del Indio (CI), Estancia La Costa (ELC), Cañadón Silva (CS), Puesto Estancia La Costa (PLC), Monte Tigre (MT), Cañadón las Totoras (CT) and Cabo Buen Tiempo (CBT) (Fig. 1.2).

Recently, we identified two new fossil vertebrate localities north of the area of Estancia La Costa locality and south of the Río Coyle (Kay *et al.*, 2008) that we named Campo Barranca (CB) and Anfiteatro (ANF) (Fig. 1.2). Campo Barranca, named for a parcel of land within the Estancia La Costa, is in the intertidal zone ~7 km north of the entrance to the beach at ELC locality. Anfiteatro, named for the semicircular shape of the cliff wall produced after a large slump on the cliff above the fossiliferous beach exposures, is located in the intertidal zone ~3 km north of the entrance to the beach at ELC. We claimed that CB and ANF are noteworthy for being the stratigraphically lowest localities of any precisely recorded in coastal Santa Cruz Province; this remains true only for Campo Barranca.

The names of two sites north of the Río Coyle also require clarification. Important collections were made by C. Ameghino, especially during 1890–91 and 1892–93, in what F. Ameghino (1906) referred to as Monte Observación (Marshall, 1976: 1139). This geographic feature is labeled "Co. (= Cerro) Observatorio" on all recent maps. The site is located within the lands of Estancia Cañadón de Las Vacas, and should not be confused with "Co. Monte Observación," a large hill 7 km southwest of Monte León within the current Parque Nacional Monte León (Fig. 1.2). However, the name Monte Observación was still used by Bown and Fleagle (1993) for the outcrops in Estancia Cañadón de las Vacas. To avoid confusion, in the present volume we use the names as explained above, i.e. Monte Observación is the hill at the Parque Nacional Monte Léon, and Cerro Observatorio the one at Estancia Cañadón de Las Vacas (~30 km south of the park) where Ameghino's Santacrucian collections came from.

1.6 Organization of this volume

For this volume we gathered together a varied group of scientists from different institutions. One particular strength of this volume is that the list of contributors includes several Ph.D. and postdoctoral students. Younger contributors worked primarily on the paleobiology of specific groups of vertebrates, and many of them participated in the fieldwork. We also invited specialists in the systematics of certain groups to collaborate with these young people.

In this introductory Chapter 1 we provide the requisite contextual framework for the remaining chapters. Chapter 2 and Chapter 3 summarize the geochronology of the Santa Cruz and Pinturas Formations based on new radiometric dates and tephrostratigraphic correlations among different fossiliferous localities throughout Santa Cruz Province.

The paleoenvironmental outline on which we have made our paleoecological considerations about the faunas of the lower member (Estancia La Costa Member) of the Santa Cruz Formation in our area of study is established in the next three chapters. Chapter 4 provides a stratigraphic framework for the large collection of fossil vertebrates recovered by us since 2003, describes the sedimentary environments of the Santa Cruz Formation in the area of the Atlantic coast between the Ríos Coyle and Gallegos, and compares them with those of other localities north of Río Coyle in which the lowest levels of the formation and the contact with the underlying marine Monte León Formation are exposed. Chapter 5 briefly describes the invertebrate fauna that appears mostly at the base of the Santa Cruz Formation. This fauna was adapted to the peculiar paleoenvironment that seems to have been prevalent at the onset of the sedimentary history of the formation. Chapter 6 provides additional paleoenvironmental evidence based on trace fossils. The previously little known Early Miocene paleoflora of coastal Southern Patagonia is treated in Chapter 7, based on newly discovered fossil leaves, wood, and siliceous microfossils collected in coastal localities between Ríos Coyle and Gallegos.

The next nine chapters supply an up-to-date summary of vertebrate diversity recorded in the lower member of the Santa Cruz Formation and provide reconstructions of the paleobiology of each taxon. Chapter 8 offers an overview of the sparse information available on heterothermic vertebrates (frogs and lizards) and their paleoenvironmental significance. The other eight chapters summarize and present new data on the postulated habits of birds and mammals based on their morphology, estimated body size, dietary inferences from the masticatory apparatus, and inferences from the appendicular skeleton concerning locomotion and use of substrate. Chapter 9 deals with birds, Chapter 10 with small non-carnivorous marsupials (paucituberculates and microbiotheres), Chapter 11 with carnivorous sparassodont marsupials, Chapter 12 with armadillos and glyptodonts (armoured xenarthrans), Chapter 13 with anteaters and sloths (pilosan xenarthrans), Chapter 14 with native extinct South American ungulates (astrapotheres, notoungulates, and litopterns), Chapter 15 with caviomorph rodents, and Chapter 16 with primates.

In the final Chapter 17 we summarize our findings on the chronological framework and the environmental information provided by the geology, trace fossils, and paleobotany. Then we present a synthesis of the implications of the assemblage of vertebrates for the paleoecology of the lower Santa Cruz Formation. The results are compared with equivalent studies for other latitudes and ages.

1.7 Future directions

The compilation of the following chapters represents our efforts to revive the study of the Santacrucian fauna, which, as noted by Simpson (1980), is particularly important for understanding a phase in the history of South American mammals in which the communities consisted of a complex mixture of descendants of ancient lineages of the continent and new taxa (primates and rodents) from other land masses (probably Africa). It may be said that any fossiliferous deposit provides a unique window into the past, but rarely are we presented with so precious an opportunity for such a profound and comprehensive understanding of the paleobiology and paleoecology of a region as afforded by the Santa Cruz Formation. It is for this reason that we believe that the already recovered fauna, and that still entombed in Santacrucian sediments, merits further and more detailed attention, and we hope, with all humility, that the current work will act as the catalyst for future endeavors.

There is now a clearer framework in which to study the paleoecology of this fauna. As we reconstruct the paleoenvironment of Santacrucian times, it is essential to keep in mind that we are examining an interval of time when environments must have fluctuated. Therefore, it would be an oversimplification to include every taxon referred to the Santa Cruz Formation as being part of a single fauna. We have been more precise and selected a series of localities that are penecontemporaneous and well sampled with good specimens of vertebrates and plants. Under these constraints, at the moment, only the coastal fossil localities from Anfiteatro to Puesto Estancia La Costa represent a sufficiently restricted interval of time to develop a paleosinecological analysis and at the same time contain enough fossil material to undertake paleobiological studies including body size, feeding, and locomotion of the fauna as a whole. These localities also include paleobotanical records that assist this endeavor. Our conclusions are, then, restricted to the lower member (Estancia La Costa Member) of the Santa Cruz Formation in the coastal area, and for the most part, only the localities from Puesto Estancia La Costa, Cañadón Silva, and Estancia La Costa, with some further evidence from Anfiteatro, Campo Barranca, and Punta Sur (a paleobotany site).

Obvious gaps in our knowledge remain, of course. Some have already been noted in the various chapters, and every other researcher will surely identify different questions that need answering. We will end this introductory chapter by highlighting several of the broad topics that we believe are particularly deserving of future attention.

In the near future we will be expanding our approach to a more complete geographic and chronologic range of the Santa Cruz Formation. In doing so, we do not need to redo the paleoautoecological analysis for each taxon, except as warranted by new material, by new taxa, or when novel approaches present themselves. We will, instead, record the different assemblages at different levels and evaluate the ecological changes that occurred during the time of deposition of the formation in different areas.

Several localities to the north and west are very promising for this kind of approach, and many of them have not been revisited or systematically sampled for 65 to 100 years. A major objective of our research is to revisit two particularly promising but under-sampled areas, Barrancas Blancas (a region located on the south bank of the Río Santa Cruz, which was worked in the 1890s by Carlos Ameghino and John Hatcher) and Rincón del Buque (= Wreck Flat; identified in the unpublished field notes of Elmer Riggs' Field Museum expedition of 1923), which have not been systematically collected since the 1940s. We will begin the study of biostratigraphic correlations with other better-documented localities. Finally, we will integrate the new fossils into our paleoecological study of fossil vertebrates of the Santa Cruz Formation.

A critical aspect of our studies is to better establish the biostratigraphic framework in which the successive Santacrucian faunas evolved (Tauber, 1997; but see Kay *et al.*, Chapter 17). Within this context it will be possible to separate the effects of local paleoenvironmental effects on community structure from those introduced by the sampling of different intervals of geologic time. The chronostratigraphic framework provided by Perkins *et al.* (Chapter 2) and Fleagle *et al.* (Chapter 3) is essential to accomplishing this goal and, though it is as current as possible, it will require further refinement and expansion.

An important contribution of our project is a controlled recovery of paleontological heritage, with precise information on geographic and stratigraphic provenance, and the uses of appropriate preparation techniques of the specimens recovered. For historical reasons, the best collections in quality of preservation of specimens of this unique fauna have been those housed in US institutions (Yale Peabody Museum, American Museum of Natural History, Field Museum of Natural History). Now, an equally spectacular (and perhaps even more comprehensive) collection is available in the Museo Regional Provincial Padre Manuel Jesus Molina of Río Gallegos (Santa Cruz Province, Argentina). Articulated specimens with their peculiar morphology, so different from those of the modern floras and faunas of Patagonia, should form a world-class museum display in Rio Gallegos for the benefit both of the local population and of international tourists.

ACKNOWLEDGMENTS
We thank the Dirección de Patrimonio Cultural, the Museo Regional Provincial "Padre M. J. Molina" (Río Gallegos, Santa Cruz Province), and the Battini family for its hospitality during the fieldwork. We thank the reviewers G. De Iuliis and P. Brinkman for their valuable suggestions and comments. This is a contribution to the projects PICT 26219 and 0143, UNLP N647, PIP-CONICET 1054 and National Geographic Society 8131–06 to SFV; and also NSF 0851272 and 0824546 to RFK.

Appendix 1.1

List of the localities in Santa Cruz Province with fauna of Santacrucian age mentioned in this volume. The most relevant are shown in Figs. 1.1 and 1.2. Abbreviations are given only for those that are used in some chapters in this volume. In most cases we provide a single coordinate as a reference; in some cases we provide ranges of extension of the exposures.

Institutional abbreviations

AMNH, American Museum of Natural History, New York, USA.
DU, Duke University, Durham, NC, USA.
MACN, Museo Argentino de Ciencias Naturales "B. Rivadavia," Buenos Aires, Argentina.
MLP, Museo de La Plata, La Plata, Argentina.

Atlantic coastal and nearby localities

We describe the localities from north to south in three groups: mostly inland localities near the coast north to Río Coyle; localities on the Río Coyle and on the Atlantic coast from the mouth of the Río Coyle (Coy inlet) to Cabo Buen Tiempo (Cape Fairweather, north point of the Gallegos Inlet); and localities near to Río Gallegos.

Localities north from Río Coyle
These localities were mostly prospected by Carlos Ameghino, John Bell Hatcher, Barnum Brown, and Elmer Riggs during the late nineteenth and early twentieth centuries.

Monte León (ML). 50° 19′ S, 68° 54′ W (Marshall, 1976). The name of this locality comes from Cerro Monte León, a hill located at Parque Nacional Monte León, about 70 km north of the mouth of Río Coyle. A large part of the Ameghino collection is at MLP and MACN (Marshall, 1976). Some specimens collected by Barnum Brown are at the AMNH. References: Chapters 2, 3, 5, 6, 8, 9, 10, 11, 12, and 16.

Monte Observación (= Cerro Observación). 50° 21′ S, 68° 58′ W. Hill located at Parque Nacional Monte León, about 7 km south of Cerro Monte León. To our knowledge there are no records of fossils coming from this hill, except a report by A. Parras *et al.* (unpublished data) and our own observations. Because of its proximity to Monte León, the possibility exists that some of Ameghino's specimens from Monte León really come from here. According to Marshall (1976), Cerro Observación may be the same as Ameghino's "La Cueva" locality. References: Chapter 5.

La Cueva (LC). 50° 21′ S, 68° 57′ W (Marshall, 1976). According to Marshall (1976) and Bown and Fleagle (1993), the exact location is unknown. The coordinates that Marshall (1976) provided place it within the current Parque Nacional Monte León. As mentioned above, it could be the same as Cerro Observación. Various specimens of the Ameghino collection come from this locality. References: Chapters 8, 9, 10, 11, and 12.

Yegua Quemada (= Cañadón de la Yegua Quemada). 50° 21′ S, 69° 02′ W (Marshall, 1976). A canyon within the Parque Nacional Monte León. Various specimens of the Ameghino collection come from this locality. References: Chapters 5, 9, 10, 11, and 12.

Yack Harvey (= Cañadón Jack; Yak-Harvey). 50° 26′ S, 69° 05′ W (Marshall, 1976). The southermost large canyon within Parque Nacional Monte León. Material at the AMNH labeled as "15 miles South of Monte León" (Cassini *et al.*, Chapter 14) may come from this locality. References: Chapters 11, 12, and 14.

Cañadón de Las Vacas. 50° 33′ S, 69° 13′ W (Marshall, 1976). A canyon within Estancia Cañadón de Las Vacas. References: Chapters 9, 11, and 14.

Monte Observación of Ameghino (MO) (= Cerro Observatorio). 50° 36′ S, 69° 05′ W (Marshall, 1976). Hill located close to the coast, in lands of Estancia Cañadón de Las Vacas. One of the most extensive areas of exposures (about 8 km) of the Santa Cruz Formation, with more than 100 recorded fossil sites (Bown and Fleagle, 1993). This is the only locality in the area that has been extensively prospected in the past 80 years (Bown and Fleagle, 1993). References: Chapters 2, 3, 5, 6, 8, 10, 11, 13, 14, and 16.

Rincón del Buque (RB) (= Media Luna; Wreck Flat of Riggs). 50° 45′ S, 69° 11′ W (Marshall, 1976). Large amphitheater located mostly in lands of Estancia Ototel Aike and partly in Estancia Cañadón de Las Vacas, extending from ~ 50° 39′ S, 69° 06′ W in the north, to 50° 46′ S, 69° 08′ W in the south. Fossils are abundant at different sites along the cliffs that surround the amphitheater (Marshal, 1976; Vizcaíno *et al.*, 2009). Marshall (1976) suggests that the locality "10 miles N of Coy Inlet" may correspond to this specific locality. References: Chapters 4, 5, and 11.

Cañadón Cerro Redondo (CCR). 50° 51′ S, 69° 08′ W. Canyon that reaches the beach, north to the Punta Norte lighthouse, within the Estancia Coy Inlet. Santacrucian fossils have been found in the canyon and in the coastal platform during low tide (Vizcaíno *et al.*, 2009). A locality called "5 miles North of Coyle Inlet (= Cobaredonda)" by Marshall (1976) probably corresponds to Cañadón Cerro Redondo. References: Chapters 4, 5, and 14.

Another locality under the ambiguous name "Santa Cruz" may belong to this section. According to Marshall (1976), it would correspond to "Cañadón Corto," a canyon between Cañadón de la Yegua Quemada and Cañadón Yack Harvey within the current Parque Nacional Monte Léon. To date, we have found no decisive reasons to support this idea or any other. References: Chapters 10, 13, and 15.

Localities between the Río Coyle (Coy Inlet) and Cabo Buen Tiempo

The great majority of the fossils from these localities come from the platform below the sea cliffs exposed only at low tide. In a few cases specimens come from the sea cliffs or the cliffs on the banks of the Río Coyle.

Coy Inlet ("Boca del Coyle" of Ameghino). 50° 57′ S, 69° 12′ W. Mouth of the Río Coyle to the Atlantic Ocean. A large number of sites worked by C. Ameghino, J. B. Hatcher, B. Brown, and E. S. Riggs are near the Coy Inlet, both up and down the coast, or on the banks of the Río Coyle (Marshall, 1976). They usually include references in miles north or south of Coy Inlet and may refer to other localities mentioned above (e.g. Cañadón Cerro Redondo or Rincón del Buque) or below (Cañadón del Indio-Punta Sur, Campo Barranca, Anfiteatro, Estancia La Costa, Puesto Estancia La Costa, Monte Tigre-Cañadón Totoras), or be simply labeled as "Río Coyle." References: Chapters 11, 12, 13, and 14.

Cañadón del Indio (CI)–Punta Sur (PS). 50° 58′ S, 69° 10′ W. Punta Sur is the southern point of the mouth of the Río Coyle, next to a broad canyon known as Cañadón del Indio. Tauber (1997) recognized it as a locality when reporting fossils from the cliffs in the area. Recently, abundant plant remains and a few bones have been collected from the platform in the intertidal zone (Fernicola *et al.*, 2009; Brea *et al.*, 2010). References: Chapters 2, 3, 4, 6, 7, and 12.

Campo Barranca (CB). Locality that extends about 4 km from north to south (51° 00′ S, 69° 09′ W to 51° 02′ S, 69° 09′ W; Kay *et al.*, 2008). More than 150 specimens have been recovered from this locality by the MLP–DU expeditions since its discovery in 2006. References: Chapters 4, 6, 9, 10, 11, 12, 13, 14, and 15.

Anfiteatro (ANF). 51° 03′ S, 69° 08′ W (Kay *et al.*, 2008). See text for a description. As for CB, several specimens (~130) have been collected here since 2006. It may include the localities listed as "5 miles [or 8 km] south of Coy Inlet" in old collections and literature. References: Chapters 4, 6, 9, 11, 14, and 15.

Estancia La Costa (ELC). 51° 05′ S, 69° 08′ W (Tauber, 1996, 1997). It may include localities listed as "La Costa" or "7 or 8 miles south of Coy Inlet" in the old collections and literature. References: Chapters 2, 3, 4, 6, 8, 11, 12, 13, 14, and 16.

Cañadón Silva (CS). 51° 09′ S, 69° 05′ W (Tauber, 1996, 1997). Cañadón Silva may include localities listed as "10 miles [or 15 km] south of Coy Inlet" in the old collections and literature. References: Chapters 2, 3, 4, 6, 11, 12, 13, and 14.

Puesto Estancia La Costa (PLC) (= Corriguen Aike). 51° 11′ S, 69° 05′ W (Tauber, 1996, 1997). The issue of the correspondence of these names and this locality is discussed in detail in the text. It may include localities listed as "15 miles (or 16 km) south of Coy Inlet" in the old collections and literature. References: Chapters 4, 6, 9, 10, 11, 12, 13, 14, 15, and 16.

Monte Tigre–Cañadón Totoras (MT–CT 51° 20′ S, 69° 01′ W. (Tauber, 1996, 1997). This corresponds to the "Estancia La Angelina Locality" of Marshall (1976). According to this author, Estancia La Angelina is not cited specifically in the old literature although J. B. Hatcher, L. H. Martin, and E. S. Riggs probably collected in the area. References: Chapters 2, 3, 6, 11, and 12.

Cabo Buen Tiempo (CBT) (= Cape Fairweather). 51° 34′ S, 68° 57′ W (Marshall 1976). Also known as Rudd's Estancia (Marshall, 1976), actually Estancia Cabo Buen Tiempo. Exposures are mostly on the tidal platform, now mostly covered by silt, although some specimens come from the base of the cliff. Some fossils were collected by Tauber at a place he called Cañadon Palo correspond to this locality (A. A. Tauber, personal communication). References: Chapters 2, 3, and 14.

Localities near the Río Gallegos

Estancia Halliday (= Estancia Los Pozos). 51° 35′ S 69° 14′ W (Marshall, 1976). Exposures on the north bank of the Río Gallegos. References: Chapters 9 and 14.

Killik Aike Norte (KAN) (= Estancia Felton). 51° 34′ S, 69° 25′ W (Marshall, 1976). Exposures in a gulch that enters the Río Gallegos from the north; the exposures are continuous with those of Estancia Halliday on the north bank of the Río Gallegos (Marshall, 1976). It may include the localities labeled "Río Gallegos" or "Killik Aike, Río Gallegos" in the old collections and literature. References: Chapters 2, 3, 10, 11, 13, and 14.

Güer Aike. This may refer to a continuation to the west of the exposures of Killik Aike Norte on the north bank of the Río Gallegos. References: Chapters 2, 13 and 14.

Arroyo Aike. There are no specific references in the literature on this locality. There is no Arroyo (stream) called

Aike in the Santa Cruz Province. Aike was the Indian word for place and is present in the names of numerous geographic sites all over the province. References: Chapter 13.

Central and western localities

Barrancas del Río Santa Cruz (= Río Santa Cruz). Extensive and discontinuous exposures of Santa Cruz beds along the middle part of the Río Santa Cruz, between the Lago Argentino and the town of Santa Cruz (Marshall, 1976). We have rediscovered the four localities where Carlos Ameghino collected: "Primeras barrancas blancas", Segundas barrancas blancas," "Yaten Huageno", and "Río Bote" (Fernicola *et al.*, 2010). References: Chapters 11, 12, and 14.

Shehuén (= Río Sehuén or Río Chalía). Extensive badlands along the Río Sehuén (Marshall, 1976), especially in Estancia Viven Aike and Los Sauces, between Estancia La Rosita (48° 31′ S, 69° 43′ W) on the west and Estancia La Julia (48° 4′ S, 67° 4′ W) on the east (Bown and Fleagle, 1993). References: Chapters 2, 3, 10, 11, and 12.

Río Chico. Rio Shehuén flows into Rio Chico to the west. Carlos Ameghino made some collections in the cliffs of Rio Chico. References: Chapter 12.

Karaiken. 50° 03′ S, 71° 47′ W (Marshall, 1976). This is the type locality of "Notohippidense" of Ameghino and is discussed in the text. References: Chapters 2, 3, and 11.

Lago Argentino. This probably includes different localities in proximity to Lago Argentino, such as Karaiken, Río Bandurrias (it may correspond to Arroyo de las Bandurrias), Estancia Quien Sabe and Estancia 25 de Mayo. References: Chapters 5, 9, and 12.

Lago Pueyrredón and Lago Posadas. Approximately 47° 30′ S, 71° 53′ W. Impressive steep exposures on top of a plateau. Few and small collections have been made. References: Chapters 2, 3, 5, 9, 11, 13, and 14.

Lago Cardiel. Exposures adjacent to the eastern shore of the Lago Cardiel (approximately 48° 55′ S, 71° 04′ W) mapped as Santa Cruz Formation by Ramos (1982). Bown and Fleagle (1993) found no correlations with the Pinturas Formation. References: Chapters 2, 3, and 10.

La Cañada (= Estancia La Cañada). At approximately 48° 1′ S, 70° 10′ W, with rocks strongly resembling those of the Santa Cruz Formation (Bown and Fleagle, 1993), it would correlate with the lower levels of the Santa Cruz Formation (Kramarz, 2009). References: Chapters 2, 3, and 14.

Gobernador Gregores. Southwest of the town of Gobernador Gregores (48° 45′ S, 70° 15′ W), tentatively assigned to the Pinturas Formation (Bown and Fleagle, 1993). References: Chapters 2, 3, and 10.

Río Pinturas. At least the top of the Pinturas Formation, and the base of the coastal Santa Cruz Formation are approximately the same age; for information about the fossiliferous localities from the valley of the Río Pinturas see Chapters 2 and 3.

REFERENCES

Ameghino, F. (1887). Enumeración sistemática de las especies de mamíferos fósiles coleccionados por Carlos Ameghino en los terrenos eocenos de la Patagonia. *Museo de La Plata, Boletín*, **1**, 1–26.
Ameghino, F. (1889). Contribución al conocimiento de los mamíferos fósiles de la República Argentina. *Academia Nacional de Ciencias, Actas* **6**, 1–1028.
Ameghino, F. (1891). Nuevos restos de mamíferos fósiles descubiertos por Carlos Ameghino en el Eoceno inferior de la Patagonia austral. Especies nuevas, adiciones y correcciones. *Revista Argentina de Historia Natural*, **1**, 289–328.
Ameghino, F. (1894). Enumération synoptique des espéces de Mammifères fossiles des formations Eocènes de Patagonie. *Boletín de la Academia Nacional de Ciencias de Córdoba*, **13**, 259–455.
Ameghino, F. (1900–1902). L'age des formations sédimentaires de Patagonie. *Anales de la Sociedad Científica Argentina*, **50**, 109–130, 145–165, 209–229 (1900); **51**, 20–39, 65–91 (1901); **52**, 189–197, 244–250 (1901); **54**, 161–180, 220–249, 283–342 (1902).
Ameghino, F. (1904). Nuevas especies de mamíferos cretáceos y terciarios de la Republica Argentina. *Anales de la Sociedad Científica Argentina*, **LVI**, 193–208; **LVII**, 162–175, 327–341; **LVIII**, 35–41, 56–71, 182–192, 225–291.
Ameghino, F. (1906). Les formations sédimentaires du crétacé superieur et du tertaire de Patagonie avec un parallele entre leurs faunes mammalogiques et celles de l'ancien continent. *Anales del Museo Nacional de Buenos Aires*, **15**, 1–568.
Barreda, V. and Palazzesi, L. (2007). Patagonian vegetation turnovers during the Paleogene–Early Neogene: origin of arid-adapted floras. *The Botanical Review*, **73**, 31–50.
Barron, J. A., Keller, G. and Dunn, D. A. (1985). A multiple microfossil biochronology for the Miocene. In *The Miocene Ocean: Paleoceanography and Biogeography*, ed. J. P. Kennett, Geological Society of America Memoir 163, pp. 21–36.
Blisniuk, P. M., Stern, L. A., Chamberlain, C. P., Idleman, B. and Zeitler, P. K. (2005). Climatic and ecologic changes during Miocene surface uplift in the Southern Patagonian Andes. *Earth and Planetary Science Letters*, **230**, 125–142.
Bown, T. M. and Fleagle, J. F. (1993). Systematics, biostratigraphy, and dental evolution of the Palaeothentidae, Later Oligocene to Early-Middle Miocene (Deseadan–Santacrucian) Caenolestoid marsupials of South America. *Journal of Paleontology*, **67**, 1–76.
Bown, T. M. and Larriestra, C. N. (1990). Sedimentary paleoenvironments of fossil platyrrhine localities, Miocene Pinturas Formation, Santa Cruz Province, Argentina. *Journal of Human Evolution*, **19**, 87–119.
Bown, T. M., Larriestra, C. N. and Powers, D. W. (1988). Análisis paleoambiental de la Formación Pinturas (Mioceno

Inferior), Provincia de Santa Cruz. *Segunda Reunión Argentina de Sedimentología*, **1**, 31–35.

Brea, M., Zucol, A. F. and Bargo, M. S. (2010). Estudios paleobotánicos en la Formación Santa Cruz (Mioceno), Patagonia, Argentina. *X Congreso Argentino de Paleontología y Bioestratigrafía y VII Congreso Latinoamericano de Paleontología, Actas*, 140.

Brinkman, P. (2003). Bartholomew James Sulivan's discovery of fossil vertebrates in the Tertiary beds of Patagonia. *Archives of Natural History*, **30**, 56–74.

Burmeister, C. V. (1891). Breves datos sobre una excursión a Patagonia. *Revista del Museo de La Plata*, **2**, 381–394.

Burmeister, C. V. (1892). Nuevos datos sobre el Territorio Patagónico de Santa Cruz. *Revista del Museo La Plata*, **4**, 227–256, 337–352.

Candela, A. M. and Picasso, M. B. J. (2008). Functional anatomy of the limbs of Erethizontidae (Rodentia, Caviomorpha): indicators of locomotor behaviour in Miocene porcupines. *Journal of Morphology*, **269**, 552–593.

Cione, A. L. and Tonni, E. P. (1995). Chronostratigraphy and "land mammal-ages": the Uquian problem. *Journal of Paleontology*, **69**, 135–159.

Crawford, R. S., Casadío, S., Feldmann, R. M. *et al.* (2008). Mass mortality of fossil decapods within the Monte León Formation (Early Miocene), Southern Argentina: victims of andean volcanism. *Annals of Carnegie Museum*, **77**, 259–287.

Croft, D. (2001). Changing environments in South America as indicated by mammalian body size distributions (cenograms). *Diversity and Distributions*, **7**, 271–287.

Dingus, L. and Norell, A. (2010). *Barnum Brown. The Man who Discovered* Tyrannosaurus rex. University of California Press.

Dumont, E. R., Strait, S. G. and Friscia, A. R. (2000). Abderitid marsupials from the Miocene of Patagonia: an assessment of form, function, and evolution. *Journal of Paleontology*, **74**, 1161–1172.

Fernicola, J. C., Vizcaíno, S. F. and Bargo, M. S. (2009). Primer registro de *Vetelia puncta* Ameghino (Xenarthra, Cingulata) en la Formación Santa Cruz (Mioceno temprano) de la costa atlántica de la provincia de Santa Cruz, Argentina. *Ameghiniana*, **46**, 77R.

Fernicola, J. C., Vizcaíno, S. F. and Bargo, M. S. (2010). Localidades fosilíferas descubiertas por Carlos Ameghino en 1887 en la margen derecha del río Santa Cruz, provincia de Santa Cruz, Argentina. *X Congreso Argentino de Paleontología y Bioestratigrafía y VII Congreso Latinoamericano de Paleontología, Actas*, 164.

Feruglio, E. (1938). El Cretácico Superior del Lago San Martín (Patagonia) y de las regiones adyacentes. *Physis*, **12**, 293–342.

Fleagle, J. G. (1990). New fossil platyrrhines from the Pinturas Formation, southern Argentina. *Journal of Human Evolution*, **19**, 61–85.

Fleagle, J. G., Bown, T. M., Swisher III, C. C. and Buckley, G. A. (1995). Age of the Pinturas and Santa Cruz formations. *VI Congreso Argentino de Paleontologia y Bioestratigrafia, Actas*, 129–135.

Fleagle, J. G., Kay, R. F. and Anthony, M. R. L. (1997). Fossil New World monkeys. In *Mammalian Evolution in the Neotropics*, ed. R. F. Kay, R. H. Madden, R. L. Cifelli and J. J. Flynn. Washington DC: Smithsonian Institution Press, pp. 473–495.

González Ruiz, L. R. and Scillato-Yané, G. J. (2009). A new Stegotheriini (Mammalia, Xenarthra, Dasypodidae) from the "Notohippidian" (early Miocene) of Patagonia, Argentina. *Neues Jahrbuch für Geologie und Palaontologie / Abhandlungen*, **252**, 81–90.

Hatcher, J. B. (1897). On the geology of southern Patagonia. *American Journal of Science (1880–1910)*, **4**, 327–354.

Hatcher, J. B. (1900). Sedimentary rocks of southern Patagonia. *American Journal of Science (1880–1910)*, **9**, 85–108.

Hatcher, J. B. (1903). Narrative of the expedition. In *Reports of the Princeton University Expeditions to Patagonia, 1896–1899*. Vol. 1, *Narrative and Geography*, ed. W. B. Scott. Princeton: Princeton University Press. pp. 1–296.

Hodell, D. A. and Kennett, J. P. (1985). Miocene paleoceanography of the South Atlantic Ocean. In *The Miocene Ocean: Paleoceanography and Biogeography*, ed. J. P. Kennett, Geological Society of America Memoir 163, pp. 317–337.

Kaufman, D. M. (1995). Diversity of New World mammals – universality of the latitudinal gradients of species and bauplans. *Journal of Mammalogy*, **76**, 322–334.

Kay, R. F. (2010). A new primate from the Early Miocene of Gran Barranca, Chubut Province, Argentina: paleoecological implications. In *The Paleontology of Gran Barranca: Evolution and Environmental Change through the Middle Cenozoic of Patagonia*, ed. R. H. Madden, A. A, Carlini, G. M. Vucetich and R. F. Kay. Cambridge: Cambridge University Press, pp. 220–239.

Kay, R. F., Madden, R. H., Cifelli, R. L. and Flynn, J. J. (1997). *Vertebrate Paleontology in the Neotropics. The Miocene Fauna of La Venta, Colombia*. Washington DC: Smithsonian Institution Press.

Kay, R. F., Johnson, D. J. and Meldrum, D. J. (1998). A new pitheciin primate from the middle Miocene of Argentina. *American Journal of Primatology*, **45**, 317–336.

Kay, R. F., Fleagle, J. G., Mitchell, T. R. T. *et al.* (2008). The anatomy of *Dolichocebus gaimanensis*, a stem platyrrhine monkey from Argentina. *Journal of Human Evolution*, **54**, 323–382.

Kraglievich, L. (1930). La formación Friasiana de Río Frías, Rio Fénix, Laguna Blanca, etc., y su fauna de mamíferos. *Physis*, **10**, 127–161.

Kramarz, A. G. (2009). Adiciones al conocimiento de *Astrapothericulus* (Mammalia, Astrapotheria): anatomía cráneo-dentaria, diversidad y distribución. *Revista Brasileira de Paleontologia*, **12**, 55–66.

Kramarz, A. G. and Bellosi, E. S. (2005). Hystricognath rodents from the Pinturas Formation, Early–Middle Miocene of Patagonia, biostratigraphic and paleoenvironmental implications. *Journal of South American Earth Sciences*, **18**, 199–212.

Lenzi, J. H. (1980). *Historia de Santa Cruz*, ed. A. R. Segovia. Río Gallegos: Santa Cruz.

León, R. J. C., Bran, D., Collantes, M., Paruelo, J. M. and Soriano, A. (1998). Grandes unidades de vegetación de la Patagonia extra andina. *Ecología Austral*, **8**, 75–308.

Lydekker, R. (1893). Contribuciones al conocimiento de los vertebrados fósiles de la Argentina. I. *Museo La Plata, Anales* (Paleontología) **1**(2), 118 pp., 43 plates.

Lydekker, R. (1894). Contribuciones al conocimiento de los vertebrados fósiles de la Argentina. II. *Museo La Plata, Anales* (Paleontología) **1**(3), 128 pp., 64 plates.

Marshall, L. G. (1975). The Handel T. Martin paleontological expedition to Patagonia in 1903. *Ameghiniana*, **12**, 109–111.

Marshall, L. G. (1976). Fossil localities for Santacrucian (Early Miocene) mammals, Santa Cruz Province, Southern Patagonia, Argentina. *Journal of Paleontology*, **50**, 1129–1142.

Marshall, L. G. and Cifelli, R. L. (1990). Analysis of changing diversity patterns in Cenozoic land mammal age fauna, South America. *Palaeovertebrata*, **19**, 169–210.

Marshall, L. G. and Pascual, R. (1977). Nuevos marsupials Caenolestidae del "Piso Notohippidense" (SW de Santa Cruz, Patagonia) de Ameghino. Sus aportaciones a la cronología y evolución de las comunidades de mamíferos sudamericanos. *Publicaciones del Museo Municipal de Ciencias Naturales de Mar del Plata "Lorenzo Scaglia,"* **2**, 91–122.

Marshall, L. G., Hoffstetter, R. and Pascual, R. (1983). Mammals and stratigraphy: geochronology of the continental mammal-bearing Tertiary of South America. *Paleovertebrata, Mém. Extraordinaire*, 1–93.

Marshall, L. G., Drake, R. E., Curtis, G. H. *et al.* (1986). Geochronology of type Santacrucian (middle Tertiary) Land Mammal Age, Patagonia, Argentina. *Journal of Geology* **94**, 449–457.

Martin, H. T. (1904). A collecting trip to Patagonia, South America. *Transactions of the Kansas Academy of Science*, **19**, 101–104.

Mercerat, A. (1890). Notas sobre la Paleontología de la República Argentina, I, II, III. *Museo La Plata, Revista* **1**, 241–255, 381–442, 447–470.

Mercerat, A. (1891a). Datos sobre restos de Mamíferos fósiles pertenecientes a los Bruta. *Museo La Plata, Revista* **2**, 5–46.

Mercerat, A. (1891b). Caracteres diagnósticos de algunas especies del género *Theosodon* conservadas en el Museo de La Plata. *Museo La Plata, Revista* **2**, 47–49.

Mercerat, A. (1891c). Caracteres diagnósticos de algunas especies de Creodonta. *Museo La Plata, Revista* **2**, 51–56.

Mercerat, A. (1891d). Fórmula dentaria del Gen. *Listriotherium. Museo La Plata, Revista* **2**, 72.

Mercerat, A. (1891e). Sobre la presencia de restos de Monos en el Eoceno de Patagonia. *Museo La Plata, Revista*, **2**, 73–74.

Mercerat, A. (1891f). Apuntes sobre el género *Typotherium. Museo La Plata, Revista*, **2**, 74–80.

Mercerat, A. (1896). Essai de classification des Terrains sédimentaires du versant oriental de la Patagonia Australe. *Museo Nacional Buenos Aires, Anales*, **5**, 105–130.

Mercerat, A. (1897). Coupes géologiques de la Patagonia Australe. *Museo Nacional Buenos Aires, Anales*, **5**, 309–319.

Moreno, F. P. and Mercerat, A. (1891a). Catálogo de los pájaros fósiles de la República Argentina conservados en el Museo de La Plata. *Museo La Plata, Anales* (Paleontología Argentina) **1**, 7–71.

Moreno, F. P. and Mercerat, A. (1891b). Nota sobre algunas especies de un género aberrante de los Dasypoda (Eoceno de Patagonia). *Museo La Plata, Revista* **2**, 57–63.

Pascual, R. and Ortiz-Jaureguizar, E. (1990). Evolving climates and mammal faunas in Cenozoic South America. *Journal of Human Evolution*, **19**, 23–60.

Pascual, R., Ortega Hinojosa, E. J., Gondar, D. and Tonni, E. P. (1965). Las Edades del Cenozoico mamalífero de la Argentina, con especial atención a aquellas del territorio bonaerense. *Anales de la Comisión de Investigaciones Científicas de la Provincia de Buenos Aires*, **6**, 165–193.

Ramos, V. A. (1982). Geología de la región del lago Cardiel, provincia de Santa Cruz. *Asociación Geológica Argentina, Revista*, **37**(1), 23–49.

Ramos, V. A. and Ghiglione, M. C. (2008). Tectonic evolution of the patagonian Andes. In *The Late Cenozoic of Patagonia and Tierra Del Fuego*, ed. J. Rabassa. *Developments in Quaternary Sciences*, **11**, 57–72.

Ré, G. H., Bellosi, E. S., Heizler, M. *et al.* (2010). A geochronology for the Sarmiento Formation at Gran Barranca. In *The Paleontology of Gran Barranca: Evolution and Environmental Change through the Middle Cenozoic of Patagonia*, ed. R. H. Madden, A. A. Carlini, G. M. Vucetich and R. F. Kay. Cambridge: Cambridge University Press, pp. 46–58.

Riccardi, A. C. (2008). El Museo de La Plata en el avance del conocimiento geológico a fines del Siglo XIX. *Serie Correlación Geológica*, **24**, 109–126.

Riccardi, A. C. and Rolleri, E. O. (1980). Cordillera Patagónica Austral. *Segundo Simposio de Geología Regional Argentina*, **2**, 1173–1306.

Riggs, E. S. (1926). Fossil hunting in Patagonia. *Natural History*, **26**, 536–544.

Riggs, E. S. (1928). Work accomplished by the Field Museum Paleontological Expeditions to South America. *Science, New Series*, **67**, 585–587.

Rusconi, C. (1965). Carlos Ameghino. Rasgos de su vida y su obra. *Revista del Museo de Historia Natural de Mendoza*, **17**, 1–162.

Scott, W. B. (1903–1905). Mammalia of the Santa Cruz beds. Part I. Edentata. Part II. Insectivora. Part III. Glires. In *Reports of the Princeton University Expeditions to Patagonia 1896–1899*. Vol. 5, *Paleontology* 2, ed. W. B. Scott. Princeton: Princeton University Press.

Scott, W. B. (1910). Mammalia of the Santa Cruz beds. Part I. Litopterna. In *Reports of the Princeton University Expeditions to Patagonia, 1896–1899*. Vol. 7, Paleontology 4, ed. W. B. Scott. Princeton: Princeton University Press.

Scott, W. B. (1912). Mammalia of the Santa Cruz beds. Part II. Toxodonta. In *Reports of the Princeton University Expeditions to Patagonia, 1896–1899*. Vol. 3, *Paleontology* 6, ed. W. B. Scott. Princeton: Princeton University Press.

Scott, W. B. (1928). Mammalia of the Santa Cruz beds. Part V. Primates. In *Reports of the Princeton University*

Expeditions to Patagonia, 1896–1899. Vol. 3, *Paleontology* 6, ed. W. B. Scott. Princeton: Princeton University Press.

Simpson, G. G. (1940). Review of the mammal-bearing Tertiary of South America. *Proceedings of the American Philosophical Society,* **83**, 649–709.

Simpson, G. G. (1948). Biographical memoir of William Berryman Scott 1858–1947. *Biographical Memoirs, National Academy of Sciences of the United States of America,* **25**, 173–203.

Simpson, G. G. (1980). *Splendid Isolation: The Curious History of South American Mammals.* New Haven: Yale University Press.

Sinclair, W. J. (1906). Mammalia of the Santa Cruz beds. Part III. Marsupialia of the Santa Cruz beds. In *Reports of the Princeton University Expeditions to Patagonia, 1896–1899.* Vol. 4, Paleontology 1, ed. W. B. Scott. Princeton: Princeton University Press.

Sinclair, W. J. (1909). Mammalia of the Santa Cruz beds. Part I. Typotheria. In *Reports of the Princeton University Expeditions to Patagonia, 1896–1899.* Vol. 6, Paleontology 3, ed. W. B. Scott. Princeton: Princeton University Press.

Sinclair, W. J. and Farr, M. S. (1932). Mammalia of the Santa Cruz beds. Part II. Aves of the Santa Cruz beds. In *Reports of the Princeton University Expeditions to Patagonia, 1896–1899.* Vol. 7, Paleontology 4, ed. W. B. Scott. Princeton: Princeton University Press.

Tauber, A. A. (1991). *Homunculus patagonicus* Ameghino, 1891 (Primates, Ceboidea), Mioceno temprano, de la costa Atlántica austral, provincia de Santa Cruz, Argentina. *Academia Nacional de Ciencias (Córdoba),* **82**, 1–32.

Tauber, A. A. (1994). Estratigrafía y vertebrados fósiles de la Formación Santa Cruz (Mioceno inferior) en la costa atlántica entre las rías del Coyle y Río Gallegos, Provincia de Santa Cruz, República Argentina. Unpublished thesis, Facultad de Ciencias Exactas, Físicas y Naturales, Universidad Nacional de Córdoba, República Argentina.

Tauber, A. A. (1996). Los representantes del género *Protypotherium* (Mammalia, Notoungulata, Interatheridae) del Mioceno Temprano del sudoeste de la Provincia de Santa Cruz, República Argentina. *Academia Nacional de Ciencias (Córdoba),* **95**, 1–29.

Tauber, A. A. (1997). Bioestratigrafía de la Formación Santa Cruz (Mioceno inferior) en el extremo sudeste de la Patagonia. *Ameghiniana,* **34**, 413–426.

Tauber, A. A., Vizcaíno, S. F., Kay, R. F., Bargo, M. S. and Luna, C. (2004). Aspectos biostratigráficos y paleoecológicos de la Formación Santa Cruz (Mioceno temprano-medio) de Patagonia, Argentina. *Ameghiniana,* **41**, 64.

Tejedor, M. F., Tauber, A. A., Rosenberger, A. L., Swisher, C. C. III and Palacios, M. E. (2006). New primate genus from the Miocene of Argentina. *Proceedings of the National Academy of Sciences USA,* **103**, 5437–5441.

Tonni, E. P. (2009). Los mamíferos del Cuaternario de la región pampeana de Buenos Aires, Argentina. In *Quaternario do Rio Grande do Sul. Integrando Conhecimentos,* ed. A. M.

Ribeiro, S. Girardi Bauermann and C. Saldanha Scherer. *Monografías da Sociedades Brasileira de Paleontologia,* pp. 207–216.

Torcelli, A. J. (1935). Correspondencia entre Don Florentino Ameghino y Don Carlos Ameghino. In *Obras Completas y Correspondencia Científica de Florentino Ameghino.* Segunda década (1881 a 1890), XX, 117–181; Tercera Década (1891 a 1900), XXI, 7–115; Cuarta Década (1901 a 1911), XXII, 7–24.

Townsend, K. E. B. and Croft, D. A. (2008). Diets of notoungulates from the Santa Cruz Formation, Argentina: new evidence from enamel microwear. *Journal of Vertebrate Paleontology,* **28**, 217–230.

Valverde, J. A. (1964). Remarques sur la structure et l'évolution des communautés de vertébrés terrestres. I. Structure d'une communuaté. II. Rapports entre prédateurs et proies. *La Terre et la Vie,* **111**, 121–154.

Vizcaíno, S. F. (1994). Mecánica masticatoria de *Stegotherium tessellatum* Ameghino (Mammalia, Xenarthra) del Mioceno temprano de Santa Cruz (Argentina). Algunos aspectos paleoecológicos relacionados. *Ameghiniana,* **31**, 283–290.

Vizcaíno, S. F. (2008). Historias de barcos y fósiles. *Museo, Revista de la Fundación Museo de La Plata,* **3**, 29–37.

Vizcaíno, S. F. (2011). Cartas para Florentino desde la Patagonia. Crónica de la correspondencia édita entre los hermanos Ameghino (1887–1902). *Publicación Especial Asociación Paleontológica Argentina,* **12**, 51–67.

Vizcaíno, S. F. and Bargo, M. S. (1998). The masticatory apparatus of *Eutatus* (Mammalia, Cingulata) and some allied genera. Evolution and paleobiology. *Paleobiology,* **24**, 371–383.

Vizcaíno, S. F. and Fariña, R. A. (1994). Caracterización trófica de los armadillos (Mammalia, Xenarthra, Dasypodidae) de Edad Santacrucense (Mioceno temprano) de Patagonia (Argentina). *Acta Geologica Leopoldensia,* **39**, 191–200.

Vizcaíno, S. F. and Fariña, R. A. (1997). Diet and locomotion of the armadillo *Peltephilus*: a new view. *Lethaia,* **30**, 79–86.

Vizcaíno, S. F., Bargo, M. S., Kay, R. F. *et al.* (2009). Localidades portadoras de mamíferos en los niveles inferiores de la Formación Santa Cruz (Mioceno temprano, Edad Santacrucense) al norte del Río Coyle. *Ameghiniana,* **46**, 55R.

Vizcaíno, S. F., Bargo, M. S., Kay, R. F. *et al.* (2010). A baseline paleoecological study for the Santa Cruz Formation (late-Early Miocene) at the atlantic coast of Patagonia, Argentina. *Palaeogeography, Palaeoclimatology, Palaeoecology,* **292**, 507–519.

White, J. L. (1993). Indicators of locomotor habits in Xenarthrans: evidence for locomotor heterogeneity among fossil sloths. *Journal of Vertebrate Paleontology,* **13**, 230–242.

White, J. L. (1997). Locomotor adaptations in Miocene Xenarthrans. In *Vertebrate Paleontology in the Neotropics. The Miocene Fauna of La Venta, Colombia,* ed. R. F. Kay, R. H. Madden, R. L. Cifelli and J. J. Flynn. Washington DC: Smithsonian Institution Press, pp. 246–264.

2 Tephrochronology of the Miocene Santa Cruz and Pinturas Formations, Argentina

Michael E. Perkins, John G. Fleagle, Matthew T. Heizler, Barbara Nash, Thomas M. Bown, Adán A. Tauber, and Maria T. Dozo

Abstract

The Santa Cruz and Pinturas Formations (SCF and PF) are two partially coeval formations in the southern part of Santa Cruz Province, Argentina, that were deposited during the Early to Middle Miocene. The SCF underlies the coastal plain between 47.0° and 51.6° S and extends from the Atlantic Coast into the Andean foothills. The PF has a more restricted distribution centered on eastern tributaries of the Rio Pinturas along the northern perimeter of the SCF. Both formations have abundant tephra and tuffaceous sediments with likely sources in volcanoes associated with emplacement of the late Cenozoic South Patagonian batholith. This study re-evaluates the age of the SCF and the relationship of the SCF to the PF, adding some radiometric dates to those previously published and using the methods of tephrochronology. Tephra samples were collected from 26 localities in the SCF and PF. Glass shards were analyzed by electron microscopy. Ten tephra samples were analyzed by the $^{40}Ar/^{39}Ar$ method: nine from the SCF and one from the PF. Results of these analyses, in conjunction with previous studies, indicate that there are at least 38 individual tephra layers in the SCF, while there are likely many more tephra than the six analyzed from the PF. Of the 38 tephra layers in the SCF, 16 are shared by two or more sections, with one key tephra, the Toba Blanca, present in eight and possibly nine localities from 51.6° S northward to 47.0° S, over a distance of ~525 km. Integrating results of the tephra correlations and radiometric ages indicates that the SCF spans the interval ~18 Ma to 16 Ma in the Atlantic coastal plain and ~19 to 14 Ma in the Andean foothills, with a chronologic overlap between the PF and lower part of the SCF. With this tephrochronology in place, studies of space-time variations such as rates of sediment accumulation, composition of mammalian faunas, facies changes, and other aspects of the SCF and PF can be fruitfully pursued.

Resumen

Las Formación Santa Cruz (SCF) y la Formación Pinturas (PF) en la parte austral de la provincia de Santa Cruz, Argentina, son dos formaciones parcialmente coetáneas depositadas durante el Mioceno Temprano a Medio. La SCF subyace la planicie costera entre 47.0° y 51.6° S y se extiende desde la costa atlántica hasta el piedemonte andino. La PF tiene una distribución más restringida centrada en los tributarios orientales del Río Pinturas a lo largo del perímetro norte de la SCF. Ambas formaciones tienen abundantes tefras y sedimentos tobáceos con fuentes probables en los volcanes asociados al el emplazamiento del batolito del Cenozoico tardío del sur de Patagonia. Este estudio reevalúa la edad de la SCF y la relación de ésta con la PF, agregando dataciones radimétricas a las ya publicadas y utilizando los métodos de la tefrocronología. Se colectaron muestras de tefras de 26 localidades en ambas formaciones. Se analizaron fragmentos de vidrio utilizando microscopio electrónico. Se analizaron diez muestras de tefras (nueve de SCF y una de PF) por el método de $^{40}Ar/^{39}Ar$. En conjunto con estudios previos los resultados de estos análisis indican que hay al menos 38 capas individuales de tefras en la SCF y posiblemente muchas más tefras en la PF que las 6 analizadas. Dieciseis de las 38 capas de tefras de la SCF están compartidas en dos o más secciones con una tefra clave, la Toba Blanca, presente en 8, y posiblemente 9, localidades desde 51.6° S al norte a 47.0° S al sur, por una distancia de ~525 km. La integración de los resultados de la correlación de tefras y las edades radimétricas indica que la SCF abarca el intervalo ~18 a 16 Ma en las planicies de la costa Atlántica y ~19 a 14 Ma en el piedemonte andino, con una superposición cronológica entre la PF y la parte inferior de la SCF. Con el establecimiento de esta tefrocronología, los estudios de variación espacio-temporal, como tasas de acumulación de sedimento, composición de las faunas de mamíferos, cambios de facies y otros aspectos de las SCF y PF prometen ser fructíferos.

2.1 Introduction

The Santa Cruz and Pinturas Formations (SCF and PF) are two partially coeval formations in the southern part of the Santa Cruz Province, Argentina (Fig. 2.1). The SCF is a widespread unit formed by coalescing and aggrading alluvial deposits of sandstone and mudstone with some estuarine intercalations at its base. The SCF is exposed locally from the Atlantic coast westward into the Andean foothills and northwestward to the headwaters of the Río Pinturas.

Early Miocene Paleobiology in Patagonia: High-Latitude Paleocommunities of the Santa Cruz Formation, ed. Sergio F. Vizcaíno, Richard F. Kay and M. Susana Bargo. Published by Cambridge University Press. © Cambridge University Press 2012.

Excellent cliff exposures of the SCF extend eastward along the lower reaches of Río Gallegos then northward 140 km along the Atlantic coast from Cabo Buen Tiempo to Monte León. Inland there are locally good exposures of the SCF as far west as the Andean foothills. The PF is a more localized unit and is only exposed extensively in the upper reaches of the Río Pinturas. Here the PF, unlike the SCF, is generally a non-fluvial complex of tuffaceous and pedogenically overprinted deposits draped over a rolling paleotopography

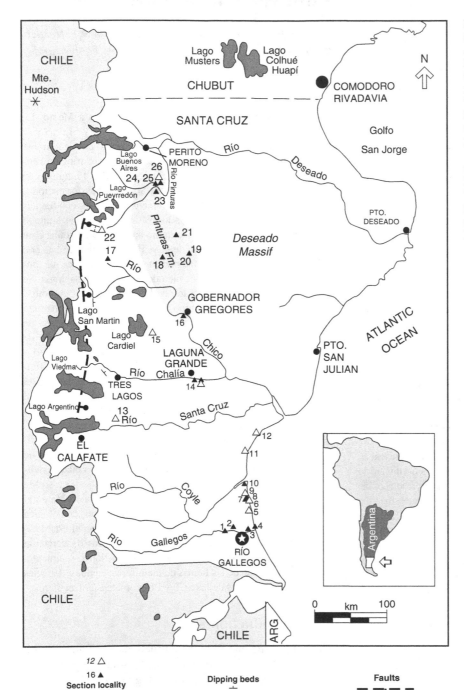

Fig. 2.1. Tephra sample localities in the Southern Patagonia (Santa Cruz Province, Argentina). Locality numbers: 1, Güer Aike Bridge; 2, Killik Aike Norte; 3, Estancia Cabo Buen Tiempo; 4, Cabo Buen Tiempo; 5, Monte Tigre; 6, Puesto Estancia La Costa; 7, Drill Pad; 8, Cañadón Silva; 9, Estancia La Costa; 10, Cañadón del Indio; 11, Cerro Observatorio; 12, Monte León; 13, Karaiken; 14, Río Chalía; 15, Lago Cardiel; 16, Gobernador Gregores; 17, Estancia La Olguita; 18, Estancia La Pluma; 19, Estancia La Cañada; 20, Estancia La Cañada W; 21, Estancia La Península; 22, Lago Posadas; 23, Estancia El Carmen; 24, Cerro de Los Monos; 25, Loma de la Lluvia; 26, Portezuelo Sumich Norte.

developed on the underlying volcanic and sedimentary rocks of the Jurassic Bahía Laura Formation (Fm.).

For more than a century the SCF and PF have yielded thousands of fossil mammals from dozens of localities. However, with the exception of studies carried out over the past 20 years, most of these fossils have little stratigraphic context other than the name of the locality from which they were collected. In addition, the stratigraphic and temporal relationships among different localities are largely unknown and the subject of conflicting views. To clarify the stratigraphic and chronologic context of SCF mammals, and the relationship of PF mammals to SCF mammals, we carried out expeditions to southern Patagonia in 2000, 2004, and 2007 to measure sections and collect the often numerous glass-shard-rich tephra at 26 key localities enumerated in Fig. 2.1, including many localities known to yield fossil mammals (for more detailed locality information, see Vizcaíno *et al.*, Chapter 1; Fleagle *et al.*, Chapter 3). The strata at these localities are mostly mapped as SCF, but six localities are in strata mapped as PF, and these include localities in the type-section PF in the Pinturas Valley and PF-like strata to the south of the Pinturas Valley.

Using electron probe microanalyses of tephra glass shards, we have identified 38 unique tephra in the sampled localities. Of these unique tephra, 15 are recognized at two or more localities. Assuming, as is likely, that the biotite-bearing tephra in the Lago Posadas section (LP) is the Toba Blanca (TB) tephra, then 14 of these 15 "correlative" tephra link sections at 19 of the 26 localities into a space-stratigraphic network (Fig. 2.2). The fifteenth tephra, the Cerro Monos (CM) tephra, links only two nearby sections and there are, as yet, no links between these two sections and the 19 linked sections. The remaining seven localities are placed approximately within the linked network using one of more of the following characteristics: (1) isotopic ages; (2) elevation; (3) a thin oyster zone near the base of the SCF; or (4) general lithologic characteristics. Using the correlative tephra or the other proxy reference horizons, it is now possible to place Santacrucian fossil horizons at any of the localities in their approximate position in this space-stratigraphic network (Fleagle *et al.*, Chapter 3). Argon (^{40}Ar/^{39}Ar) ages or ages linked to the global magnetic polarity timescale provide the primary chronologic calibration for this network. Ages of other tephra are estimated by interpolation/extrapolation (secondary ages) relative to the primary ages. Other horizons, including the tops and bottoms of sections, are also estimated by interpolation or extrapolation relative to a combination of primary and secondary ages. Using the tephrochronology developed in this study, important aspects of both the SCF and PF that have, to varying degrees, eluded geologists and paleontologists in the past are brought into focus. These include: (1) the space-time variation sediment accumulation rates; (2) the space-time variation of mammalian faunas; and (3) the pattern of facies changes within the SCF.

2.2 Methods

The findings of this study are based on: (1) field sample collection, section measurement, and photography; (2) laboratory sample preparation, electron probe analyses, and ^{40}Ar/^{39}Ar analyses of plagioclase, biotite, and hornblende; and (3) examination and measurements based on field photography and Google Earth™ imagery.

2.2.1 Field methods

Fieldwork was carried out in southern Argentina, during expeditions in 2000, 2004, and 2007. In total, 35 days were spent collecting samples of tephra and measuring stratigraphic sections in the SCF and PF.

Sample collection The tephra in the SCF and PF are typically reworked and strongly overprinted by pedogenesis, so care was taken to collect the least contaminated samples with the least alteration of glass shards to clay. Also, for tephra that were candidates for ^{40}Ar/^{39}Ar analyses, samples were taken from the coarsest and least contaminated horizons in the tephra.

Section measurements Sections were measured at all stratigraphic profiles. Measurements were made either directly with a Jacob's Staff or indirectly with a Global Positioning System (GPS) receiver. Only a Jacob's Staff was used in the year 2000, as at that time selected availability was still applied to the GPS signals, so location accuracy was only good to about ±100 m and elevation accuracy was even lower. In 2004 and 2007, section measurements were often done with a GPS receiver but supplemented with Jacob's Staff measurements. These measurements were later checked with the imagery and elevation model of Google Earth™. Since about 2009 the quality of the Google Earth™ imagery of the region has improved substantially with resolution in many areas as good as a few meters. The Google Earth™ elevations are best in areas with low gradient slopes and can degrade substantially on steeper slopes.

2.2.2 Laboratory methods

Sample preparation and electron probe analyses About 50 g of oven-dried sample was crushed and treated with a 10% solution of HNO_3 to remove carbonates, followed by treatment with a 3% solution of HF to remove clay. Most tephra are quite clay-rich, and the HF treatment was often repeated three to four times before the glass shards were free of adhering clay. The final processed sample generally weighed only ~40% of the starting sample.

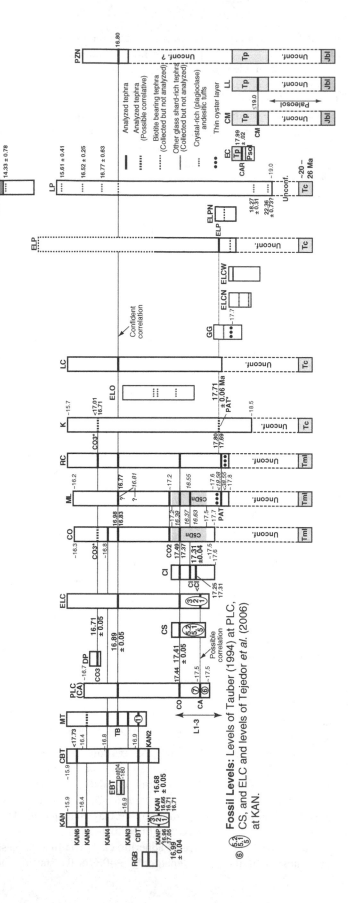

Fig. 2.2. Tephra correlation diagram. The diagram shows tephra identified at two or more localities as well as tephra at single localities that have an isotopic age determination. Tephra in stratigraphic columns with boldly highlighted boundaries link together in a space-stratigraphic–time network. Other localities can be tied approximately into this network using variously: isotopic ages, elevation, an oyster layer, or lithologic characteristics. Uncircled numbers are ages. Ages preceded by ~ are based on sedimentation rates. Ages preceded by ≤ or < are maximum $^{40}Ar/^{39}Ar$ ages. Tephra marked by bold lines are analyzed glass-shard-rich tephra. Tephra marked by asterisks are plagioclase-rich andesitic tephra (Lago Posadas [LP] and Estancia La Olguita [ELO] only). Unshaded stratigraphic columns are included in the Santa Cruz Fm. Other units as follows: Jbl = Bahía Laura Fm.; Tml = Monte León Fm.; Tc = Centinela Fm.; Tp = Pinturas Fm.; Psol = paleosol; shaded intervals in the Santa Cruz Fm. at Cerro Observatorio and Monte León mark Chron C5Dn. See Table 2.3 for full names of localities and Table 2.4 for full names of tephra. L1–3 is the generic designation of a group of closely spaced tephra at the base of the coastal Santa Cruz Fm.

The glass shards, typically 12 to 20 shards per sample, were analyzed on the Cameca SX-50 electron microprobe at the University of Utah using the analytical conditions of Nash (1992). All samples were analyzed for Si, Ti, Al, Fe, Mn, Mg, Ca, Ba, Na, K, Cl, and F and most were analyzed for O as well. The elements Ti, Fe, Mg, Ca, and Cl are most useful for the correlation of SCF and PF tephra. These elements show the largest variation relative to analytical precision (Table 2.1).

Sample comparisons/correlations Sample comparisons and correlations were done using a statistical distance function D as discussed by Perkins et al. (1995, 1998). The function D^2 has a chi-square distribution and for a five-element comparison (Fe, Ti, Mg, Ca, and Cl) D will, ideally, be less than 3.33 for 95% of the analyses from a given tephra with a uniform glass shard composition. But in practice, samples from the same tephra analyzed during separate analytical runs will show more variability and may have D values as high as ~4 to 5 for correlative tephra. The particular expression used for comparing tephra with two or more analyses is given in Table 2.1 along with the estimated analytical precision for the analyzed elements.

As shown by the probe analyses, many of the samples contain two or more compositional modes with some samples clearly showing compositional patterns indicative of a reworked mixture of two or more distinct tephra. So in this study correlations are based on matching the composition of primary modes in the tephra (either a single mode or less commonly two modes with approximately the same number of analyzed shards). The dominant modes of most of the different tephra are well separated in compositional space (mostly $D > 6.0$) and correlation can generally be made with confidence with the criteria $D < \sim4$. Besides the compositional restriction of $D < \sim4$, a second requirement is stratigraphic compatibility of tephra pairs: if two tephra in one section match two in another section, the stratigraphic order must match as well. Numerous examples of pair matches from one section to another are shown in the correlation chart (Fig. 2.2).

Argon (^{40}Ar/^{39}Ar) analyses Primary age control is provided by ^{40}Ar/^{39}Ar analyses of minerals separated from the tephra. Four studies have reported ^{40}Ar/^{39}Ar analyses of tephra from the SCF and PF (Fleagle et al., 1995; Blisniuk et al., 2005; Tejedor et al., 2006; and this study). Analyzed minerals include plagioclase, hornblende, and biotite (Table 2.2). All dated samples have plagioclase ages, but only two have biotite ages and only one has a hornblende age. For consistency we use plagioclase ages to chronologically calibrate the tephrostratigraphy. Only fusion analyses of small groups of grains were carried out by previous workers, but we did both fusion and step

heating analyses for this study. Analyses were carried out in M. T. Heizler's laboratory, with analytical details available on request.

The ^{40}Ar/^{39}Ar ages in this study and those of Fleagle et al. (1995) and Tejedor et al. (2006) used Fish Canyon Tuff sanidine (FCs) as a neutron flux monitor and calculated ages based on an age 27.84 Ma for this monitor. These ages, listed in Table 2.2, are followed by revised ages, recalculated relative to an age of 28.201 Ma for the FCs. The Blisniuk et al. (2005) study used the GA 1550 biotite as neutron flux monitor and calculated ages based on an age of 98.79 Ma for this monitor. This is equivalent to an age of 28.02 Ma for the FCs monitor. These ages are also listed in Table 2.2, followed by ages recalculated relative to an age of 99.43 Ma for GA 1550 biotite (equivalent to 28.201 Ma for FCs). The ages relative to FCs = 28.201 Ma are the only ages used in this study. Recent work indicates this latter age may be a more appropriate age for the FCs monitor (Kuiper et al., 2008, Rivera et al., 2011, Smith et al., 2011) as it brings the ^{40}Ar/^{39}Ar ages into synchronicity with the astronomically tuned Neogene magnetic polarity timescale ATNTS2004 of Lourens et al. (2004).

Photography and Google Earth™ imagery To supplement the fieldwork on the geologic sections made in the field, individual photos and photo panoramas were taken of most sections. The field measurements and descriptions were augmented with analysis of these photos. Google Earth™ imagery in most areas was also detailed enough to add to the post-fieldwork measurements and description of the sections.

2.3 Results

2.3.1 Tephra types and distribution patterns

Tephra in the SCF and PF are mostly medium to light-colored glass-shard-rich silicic tephra of various shades including gray to yellowish gray, brown, and sometimes pink. Locally layers of gray to dark gray lithic-rich andesitic tephra occur in some SCF localities, and light colored crystal-rich (plagioclase, ± hornblende, ± pyroxene, ± magnetite) tephra are abundant in the SCF at the Estancia La Olguita and Lago Posadas localities in the Andean foothills (Fig. 2.1). The frequency of the SCF silicic tephra appears to be highest in the localities south of the Río Chalía. In the more complete exposures of the SCF along the Atlantic coast, up to 19 separate tephra units are identified with many fewer silicic tephra found at SCF localities north and west of the coastal localities. This suggests a generally southern source for the silicic tephra. In contrast, a reverse pattern is suggested for the andesitic and crystal-rich tephra. Few or none are found in most SCF localities, but there are about a dozen such tephra at Estancia La

Table 2.1. Electron microprobe analyses of glass shards from selected tephra

Tephra	s	n	SiO$_2$	TiO$_2$*	Al$_2$O$_3$	Fe$_2$O$_3$*	MnO	MgO*	CaO*	BaO	K$_2$O	Na$_2$O	Cl*	F	H$_2$O	Sum	D
σ			0.66	0.018	0.14	0.043	0.008	0.007	0.018	0.010	0.17	0.54	0.010	0.03	- -	- -	
TB	14	15	74.16	0.078	11.99	0.73	0.067	0.063	0.443	0.043	4.42	3.26	0.176	0.095	4.43	99.87	- -
KAN	2	14	74.14	0.077	12.03	0.80	0.072	0.073	0.626	0.072	4.75	2.93	0.131	0.031	4.32	100.00	6.8
KAN3	2	15	73.62	0.122	12.16	1.15	0.040	0.190	1.300	0.053	3.98	2.74	0.139	0.055	4.38	99.87	25.0
CBT	4	19	73.49	0.215	11.68	1.25	0.031	0.171	0.966	0.041	4.12	3.01	0.160	0.142	4.70	99.87	12.6
KAN4	15	10	72.09	0.290	12.49	1.56	0.056	0.276	1.323	0.032	3.39	3.53	0.176	0.108	4.78	100.01	15.1
KAN5	2	15	72.36	0.220	12.58	1.25	0.045	0.167	0.824	0.043	4.78	3.21	0.290	0.138	4.78	100.56	18.9
KAN6	6	19	72.56	0.273	12.85	1.34	0.048	0.188	0.737	0.044	4.63	3.57	0.262	0.176	4.43	100.96	4.6
CI	2	10	72.08	0.262	12.31	1.32	0.056	0.190	0.844	0.042	4.06	3.25	0.217	0.105	5.13	99.76	4.2
CO	7	18	72.67	0.295	12.41	1.48	0.053	0.239	1.012	0.031	3.79	3.16	0.206	0.111	4.65	100.02	7.0
CO2	2	13	72.46	0.067	12.78	1.43	0.058	0.041	0.697	0.074	3.99	3.24	0.147	0.105	4.81	99.82	22.7
CO3	4	15	73.15	0.077	12.49	1.64	0.058	0.054	0.941	0.067	3.67	3.44	0.178	0.106	4.40	100.18	8.7
ELP	2	7	74.55	0.129	12.02	1.74	0.040	0.094	0.982	0.053	3.06	2.92	0.118	0.120	3.99	99.74	4.6
KAN2	5	17	71.95	0.276	13.69	2.07	0.086	0.263	1.198	0.039	3.77	4.05	0.183	0.108	3.87	101.47	15.5
CA	2	6	72.28	0.177	11.86	2.20	0.064	0.132	1.103	0.028	2.91	3.09	0.089	0.120	5.17	99.15	11.0
PAT	4	17	71.76	0.207	12.45	2.51	0.085	0.139	1.206	0.036	3.33	3.28	0.138	0.078	4.86	100.02	5.0
PAT*	2	13	71.85	0.271	12.32	2.60	0.079	0.188	1.191	0.029	3.05	3.02	0.117	0.096	4.96	99.65	5.2
CM	2	16	72.49	0.402	11.24	2.60	0.065	0.309	1.779	0.001	2.28	2.29	0.146	0.122	5.21	98.85	22.6

Notes: σ – mean precision of an analysis of 20 shards; Tephra – abbreviation of a tephra (see Table 2.4 for key to abbreviations); s – number of analyzed samples; n – average number of analyzed shards per sample; * – element used in comparison of tephra; H$_2$O – estimated by oxide balancing; tephra IDs highlighted with bold font are identified at two or more localities; tephra groups with similar composition are demarked by underlying and overlying dotted lines; D is the statistical distance between one analysis and the overlying analysis; tephra with D below about 4 are regarded as statistically identical; D is the square root of the chi-square test: $D^2 = \Sigma \{(C_{ik} - C_{i-1,k})^2/[(20/n_i + 20/n_{i-1})\sigma_k^2]\}$ where $i = i$th sample, n_i = number of analyses in the ith sample, $k = k$th element, and C_{ik} = concentration of kth element in the ith sample.

Table 2.2. *Summary of $^{40}Ar/^{39}Ar$ analyses*

Sample	Lab.[a]	Run no.	Section[b]	Fm.	Tephra[c]	Mineral[d]	Analysis[e]	Fusions[f]	Age[g] (Ma)	Error (1σ) (Ma)	Revised age[h] (Ma)	Error (1σ) (Ma)	Age[i] (Ma)	Error (Ma)	Publication[j]
LP-I≥504.5 m	LUNGL	- - -	Southwest of LP	SCF	- - -	plag	Fusion	3	14.26	0.78	14.35	0.78	- - -	- - -	D
LP 331.5 m	LUNGL	- - -	LP	SCF	- - -	plag	Fusion	2	15.51	0.41	15.61	0.41	- - -	- - -	D
pat04-179	NMGRL	55450-01	CBT	SCF	KAN6	plag	SH	- - -	<17.50	0.12	<17.73	0.12	- - -	- - -	A
pat04-192	NMGRL	55453-01	K	SCF	CO3*	plag	SH	- - -	16.8	0.07	17.01	0.07	- - -	- - -	A
pat04-192	NMGRL	55453	K	"	CO3*	plag	Fusion	8 of 8	16.5	0.05	16.71	0.05	16.71	0.05	A
251.7m	LUNGL	- - -	LP	SCF	- - -	plag	Fusion	3	16.45	0.25	16.56	0.25	- - -	- - -	D
LP 181.2 m	LUNGL	- - -	LP	SCF	- - -	plag	Fusion	4	16.71	0.63	16.82	0.63	- - -	- - -	D
pat00-8	NMGRL	52260-01	CO	SCF	TB	biotite	SH	- - -	<17.25	0.51	<17.47	0.52	- - -	- - -	A
pat00-8	NMGRL	52257-05	CO	"	TB	plag	SH	- - -	16.61	0.27	16.83	0.27	- - -	- - -	A
pat00-8	NMGRL	52259	CO	"	TB	plag	Fusion	10 of 12	16.76	0.07	16.98	0.07	- - -	- - -	B
91CS-MLBT	BGC	5501	ML	"	TB	plag	Fusion	4	16.56	0.11	16.77	0.11	- - -	- - -	B
91CS-P16	BGC	5487	PZN	"	TB	plag	Fusion	4	16.58	0.10	16.80	0.10	16.86	0.05	B
91CS-P16	BGC	5486	PZN	"	TB	hbl	Fusion	4	16.43	0.16	16.64	0.16	- - -	- - -	B
91CS-MLBT	BGC	5502	ML	"	TB	biotite	Fusion	4	16.28	0.23	16.49	0.23	- - -	- - -	B
91CS-ML24	BGC	5509	ML	"	TB?	plag	Fusion	3	16.59	0.59	16.81	0.60	- - -	- - -	B
- - -	BGC	- - -	KAN	SCF	KAN	plag	SH	- - -	16.45	0.14	16.66	0.14	- - -	- - -	C
- - -	BGC	- - -	KAN	"	KAN	plag	SH	- - -	16.50	0.20	16.71	0.20	- - -	- - -	C
- - -	BGC	- - -	KAN	"	KAN	plag	SH	- - -	16.50	0.30	16.71	0.30	16.68	0.05	C
pat04-183	NMGRL	55451-01	KAN	SCF	KANP	plag	SH	- - -	16.84	0.07	17.05	0.07	- - -	- - -	A
pat04-183	NMGRL	55451	KAN	"	KANP	plag	Fusion	12 of 12	16.74	0.05	16.96	0.05	16.99	0.04	A
pat00-1	NMGRL	52255-02	CO	SCF	CO	plag	SH	- - -	17.27	0.25	17.49	0.26	- - -	- - -	A
pat04-142	NMGRL	55448-01	PLC	"	CO	plag	SH	- - -	17.15	0.09	17.37	0.09	- - -	- - -	A
91CS-MO206	BGC	5512	CO	"	CO	plag	Fusion	4	16.18	0.61	16.39	0.62	- - -	- - -	B
pat00-1	NMGRL	52255	CO	"	CO	plag	Fusion	2 of 8	17.20	0.18	17.42	0.20	- - -	- - -	A
pat04-142	NMGRL	55448	CA	"	CO	plag	Fusion	9 of 9	17.2	0.06	17.42	0.06	17.41	0.05	A
pat04-150	NMGRL	55449-01	CI	SCF	<CI	plag	SH	- - -	17.03	0.12	17.25	0.12	- - -	- - -	A
pat04-150	NMGRL	55449	CI	"	<CI	plag	Fusion	5 of 5	17.09	0.04	17.31	0.04	17.31	0.04	A
91CS-MO64	BGC	5510	CO	SCF	L1-3	plag	Fusion	4	16.16	0.27	16.37	0.27	- - -	- - -	B
91CS-MO206	BGC	5512	CO	"	L1-3 (CO)	plag	Fusion	4	16.18	0.61	16.39	0.62	- - -	- - -	B
91CS-ML18	BGC	5506	ML	"	L1-3	plag	Fusion	4	16.34	0.35	16.55	0.35	- - -	- - -	B
90JF-MO2	BGC	2269	CO	"	L1-3	plag	Fusion	4	16.42	0.23	16.63	0.23	16.43	0.20	B
pat04-188	NMGRL	55452-01	K	SCF	PAT*	plag	SH	- - -	17.57	0.15	17.80	0.15	- - -	- - -	A
pat04-188	NMGRL	55452	K	"	PAT*	plag	Fusion	9 of 11	17.46	0.06	17.69	0.06	17.71	0.06	A

Table 2.2. (cont.)

Sample	Lab[a]	Run no.	Section[b]	Fm.	Tephra[c]	Mineral[d]	Analysis[e]	Fusions[f]	Age[g] (Ma)	Error (1σ) (Ma)	Revised age[h] (Ma)	Error (1σ) (Ma)	Age[i] (Ma)	Error (Ma)	Publication[j]
pat00–12	NMGRL	52258	ML	"	PAT	plag	Fusion	2 of 6	<35.42	0.31	<35.88	0.31	---	---	A
91CS-ML3	BGC	5503	ML	"	PAT	plag	Fusion	5	19.33	0.18	≤19.58	0.18	≤19.58	0.18	B
pat00–57	NMGRL	52256	CM	PF	CM	plag	Fusion	3 of 16	<18.74	0.30	<18.98	0.30	<18.98	0.30	A
91CS-PT7	BGC	5492	EC	PF	CAR	plag	Fusion	7	17.76	0.02	17.99	0.02	17.99	0.02	B
91CS-PT7	BGC	5496	EC	"	CAR	biotite	Fusion	4	17.67	0.16	17.90	0.16	---	---	B
LP 58.8 m	LUNGL	---	LP	SCF	---	plag	Fusion	3	18.15	0.31	18.27	0.31	---	---	D
LP 10.0 m	LUNGL	---	LP	SCF	---	plag	Fusion	3	22.36	0.73	22.51	0.73	---	---	D

The ^{40}Ar/^{39}Ar age of 91CS-ML24 suggests this sample may be the TB tephra. However, the exact position of this sample in the Monte Léon section is unknown and its glass shard composition was never analyzed, hence the uncertainty of identification.

a BGC – Berkeley Geochronology Center; LUNGL – Lehigh University Noble Gas Lab; NMGRL – New Mexico Geochronology Laboratory.

b Section code; see Table 2.4 for full name.

c Tephra listed in stratigraphic order; see Table 2.4 for full name; * symbol indicates sample at approximate level of listed tephra while < symbol indicates sample somewhat beneath listed tephra.

d plag – plagioclase, hbl – hornblende.

e SH – step heating.

f Number of fusions with three or more grains per fusion; number of fusions included in age determination is also shown.

g Age reported by laboratory with age relative to Fish Canyon Tuff sanidine = 27.84 Ma or GA 1550 biotite = 98.79 Ma (LUNGL only).

h Revised ages used in this study are relative to Fish Canyon Tuff sanidine = 28.201 Ma or GA 1550 biotite = 99.17 Ma (LUNGL only); underlined plagioclase ages are used in this study and if more than one such age is available for a given horizon then the weighted mean age is used.

i Weighted mean age for given horizon.

j A – this study; B – Fleagle et al. (1995); C – Tejedor et al. (2006); D – Blisniuk et al. (2005).

Olguita (P. M. Blisniuk, personal communication, 2007) and many more in the section at Lago Posadas (Blisniuk et al., 2005). This likely reflects a nearby source for these tephra near these two localities that the sources involved only small volumes of tephra.

The thickest SCF tephra are the silicic tephra. They range up to ~4 m in thickness, but most are 0.5–1.5 m thick with even thinner beds at some localities. In contrast, the andesitic and crystal-rich tephra are mostly a few centimeters to a few tens of centimeters in thickness (Blisniuk et al., 2005). Primary silicic ash-fall deposits are uncommon, with most tephra layers displaying varying degrees of fluvial reworking, bioturbation, and pedogenic overprinting. Many silicic tephra layers contain introduced sand-sized quartz, feldspar, and lithic grains as a result of fluvial reworking and bioturbation. In addition, these tephra have a substantial amount of clay, which commonly makes up more than 50% of the deposit. This clay is likely the result of pedogenic alteration of glass shards, the dominant component of the less contaminated tephra layers. Typical exposures of SCF tephra along the Atlantic coast are shown in Fig. 2.3.

In the PF, discrete tephra layers are present in basal beds but higher in the formation consist mostly of an amalgam of glass-shard-rich sediment of eolian and fluvial origin, often thoroughly overprinted by pedogenic processes. The color of these tuffaceous sediments is similar to the SCF tephra, i.e. shades of gray to yellowish gray, to brown and sometimes pink. The degree of argillization of the tuffaceous PF sediments commonly exceeds that of SCF silicic tephra. We did no detailed sampling of these tuffaceous sediments, but do note that we generally observed that glass shards are sparse to absent in these intensely argillized PF tuffaceous sediments.

2.3.2 Tephrostratigraphy

The locations of the 26 sample localities are shown in Fig. 2.1, and their geographic coordinates, names, and associated abbreviations are listed in Table 2.3. Sixteen compositionally correlative tephra samples link together two or more localities. The analyses of these tephra are listed in Table 2.1. In combination, these links tie 16 of the 26 sections together in relative stratigraphic order. These key sections are highlighted with bold boundaries in Fig. 2.2, as are the ages for section bases and tops when they can be estimated with some confidence. The relative stratigraphic positions and abbreviations for 18 named tephra are also shown. Tephra names and associated abbreviations are given in Table 2.4, as are the ages of the named tephra, based either on direct isotopic determination, correlation with a dated tephra, or interpolation or extrapolation based on sedimentation rates.

2.3.3 Age calibration of the tephrostratigraphy

Since the early 1960s, ^{40}Ar/^{39}Ar, K–Ar, and fission track ages have been reported from the SCF and PF (Tables 2.2,

(a)

(b)

Fig. 2.3. Coastal Santa Cruz Formation deposits at Monte León and Cerro Observatorio. a, View eastward towards the Atlantic Ocean showing the basal 50 m of the Santa Cruz Fm. at Monte León. Tephra CO (Cerro Observatorio), CO$_2$ (Cerro Observatorio 2), and PAT (Patagonia) are identified in other localities. The Santa Cruz Fm. rests unconformably on the Early Miocene, marine Monte León Fm. b, View northward showing the upper ~100 m of the Santa Cruz Fm. at Cerro Observatorio. Important features include: (1) the TB tephra (Toba Blanca; 16.89 ± 0.05 Ma) and the CO$_3$ tephra (Co. Observatorio 3; ~16.7 Ma); (2) SS – a layer of dark andesitic sandstone, and (3) YB (yellow band) a zone of yellowish thin bedded sandstone.

2.5, and 2.6). In general, each of these three methods of isotopic age determination indicates that the SCF is late Early Miocene to Middle Miocene in age and that the PF is late Early Miocene in age. Of the three methods the most recently developed method, the ^{40}Ar/^{39}Ar method, yields the most precise and potentially most accurate ages. A total of 40 ^{40}Ar/^{39}Ar ages are now reported from the SCF and PF (Table 2.2) from 18 horizons. Winnowing out ages that are likely too old or too young leaves 13 horizons useful in establishing ages for the SCF (12 horizons) and the PF (1 horizon). For consistency only plagioclase ages are used for this age calibration of these 13 horizons. The ^{40}Ar/^{39}Ar

Table 2.3. *Tephra sample localities*

Loc.	Locality name	Abbrev.	South latitude	West longitude	Loc.	Locality name	Abbrev.	South latitude	West longitude
1	Güer Aike (= Río Gallegos Bridge)	RGB	51.62547	69.61700	14	Río Chalía E-2	RC	49.64690	70.01300
2	Killik Aike Norte	KAN	51.57420	69.44580	14	Río Chalía W	RC	49.62500	70.13633
3	Estancia Cabo Buen Tiempo	EBT	51.56450	69.05603	15	Lago Cardiel	LC	48.98445	70.93265
4	Cabo Buen Tiempo	CBT	51.53390	68.95082	16	Gobernado Gregores	GG	48.78170	70.32028
5	Monte Tigre	MT	51.34372	69.03832	17	Estancia La Olguita	ELO	47.97045	71.82677
6	Puesto Estancia La Costa	PLC	51.19655	69.09168	18	Estancia La Pluma	ELP	47.99735	70.66640
7	Drill Pad	DP	51.18266	69.16155	19	Estancia La Cañada	ELCN	47.97551	70.16332
8	Cañadón Silva	CS	51.15675	69.11214	20	Estancia La Cañada W	ELCW	47.97835	70.18889
9	Estancia La Costa	ELC	51.08509	69.13430	21	Estancia La Peninsula	ELPN	47.72646	70.38501
10	Cañadón del Indio	CI	50.98415	69.18342	22	Lago Posadas	LP	47.58990	71.82403
11	Cerro Observatorio	CO	50.56478	69.15127	23	Estancia El Carmen	EC	47.21240	70.57888
12	Monte León	ML	50.31875	68.89494	24	Cerro de Los Monos	CM	47.04413	70.73009
13	Karaiken	K	50.06280	71.76550	25	Loma de la Lluvia	LL	47.02973	70.71881
14	Río Chalia E-1	RC	49.61600	69.99800	26	Portezuelo Sumich Norte	PZN	46.95840	70.66760

Geographic coordinates referenced to WGS 84.

Table 2.4. *Tephra names and ages*

Abbreviation (sorted by age of tephra)[a]	Tephra name	Age (Ma)[b]	Abbreviation (sorted alphabetically)	Tephra name	Age (Ma)
KAN6	Killik Aike Norte 6	~16.4	CA	Corriguen Aike	~17.5
KAN5	Killik Aike Norte 5	~16.4	CAR	Estancia El Carmen	17.99
CO3/CO3*	Cerro Observatorio 3	~16.7	CBT	Cabo Buen Tiempo	~16.9
KAN4	Killik Aike Norte 4	~16.8	CI	Cañadón del Indio	~17.3
TB	Toba Blanca	16.86	<CI	Below CI	17.31
KAN3	Killik Aike Norte 3	~16.9	CM	Cerro de Los Monos	≤19.0
CBT	Cabo Buen Tiempo	~16.9	ELP	Estancia La Pluma	~17.7
KAN2	Killik Aike Norte 2	~16.9	KAN	Killik Aike Norte	~17.0
KAN	Killik Aike Norte	~17.0	KAN2	Killik Aike Norte 2	~16.9
CO2	Cerro Observatorio 2	~17.2	KAN3	Killik Aike Norte 3	~16.9
CO	Cerro Observatorio	~17.3	CM	Cerro de Los Monos	≤19.0
CI	Cañadón del Indio	~17.3	CO	Cerro Observatorio	~17.4
<CI	Below CI	17.31	CO2	Cerro Observatorio 2	~17.3
CA	Corriguen Aike	~17.5	CO3/CO3*	Cerro Observatorio 3	16.7
PAT/PAT*	Patagonia	~17.7	KAN4	Killik Aike Norte 4	~16.8
ELP	Estancia La Pluma	~17.7	KAN5	Killik Aike Norte 5	~16.4
CAR	Estancia Carmen	17.99	KAN6	Killik Aike Norte 6	~16.4
CM	Cerro de Los Monos	≤19.0	PAT/PAT*	Patagonia	~17.7

[a] Abbreviations used in Fig. 2.2; *after a tephra abbreviation indicates other tephra at about the same stratigraphic level as the named tephra.
[b] ~: Symbol for interpolated/extrapolated age estimates.

ages are listed in Table 2.2 and those used in age calibration are shown in Fig. 2.2 at the location of the dated sample. In addition, a short normal polarity event in the SCF was identified by Fleagle *et al.* (1995) near the base of the Cerro Observatorio locality, formerly called Monte Observacíon (see Chapter 1) and likely by Marshall *et al.* (1976) in a similar position in the Monte León locality. As discussed below this normal event is, in the context of the $^{40}Ar/^{39}Ar$ ages, Chron C5Dn (17.235–17.533 Ma with reference to ATNTS2004).

Santa Cruz Formation As noted above, the $^{40}Ar/^{39}Ar$ ages for SCF tephra include both plateau and fusion ages of biotite (2), hornblende (1), and plagioclase (32), with plagioclase ages available for all the dated tephra. As noted above we use the plagioclase to establish ages for SCF tephra. These ages are plotted against relative stratigraphic position in Fig. 2.4. As shown in this figure most of the ages fall in a narrow band extending from about 18.0 to 16.7 Ma. The tephra in this acceptance band are used to establish the age of the SCF. Below, we start with a discussion of the key SCF tephra, the regionally distributed Toba Blanca (TB), and then evaluate the validity of ages for other tephra in sections in context of the age of the Toba Blanca.

The Toba Blanca The TB (Figs. 2.1, 2.2) was first recognized by Tauber (1994) as a key marker in the cliff exposures of the SCF along the Atlantic coast south of the Río Coyle. In our investigation we show that the TB is of regional extent and present in at least eight localities. These extend from the Atlantic coast northward ~500 km to the Portezuelo Norte (PZN) locality. The TB tephra is white to pink to pale brown in color, contains biotite, and has a distinctive glass shard composition (Table 2.1). In addition, a biotite bearing tephra at the expected elevation of the TB was identified at Lago Posadas (LP) 10–20 m below the LP 181.2 m tuff dated at 16.77 Ma (Fig. 2.5). As shown in Table 2.2 there are eight $^{40}Ar/^{39}Ar$ ages for the TB, ranging from 16.49 to 17.24 Ma from three localities (CO, ML, and PZN). They include four concordant plagioclase ages in age range 16.77 to 16.98 Ma. These four ages have a weighted mean age of 16.89 ± 0.05 Ma. Given the narrow ranges of ages determined for these widely separated samples we feel that any potential contamination by older plagioclase is minimal and the mean age determined for the TB is close to its actual age.

Sediment accumulation rates Sediment accumulation rates provide one test of the validity of other ages within the acceptance band. We find that sections with the TB and

Michael E. Perkins et al.

Table 2.5. *Potassium–argon ages*

Sample	Laboratory[a]	Section	Laboratory number	Tephra unit	Mineral	Age (Ma)	Error (1σ) (Ma)	Publication
KA 1252	BGC	Killik Aike N	KA 1252	- - -	ash	21.7	0.3	Evernden *et al.* (1964)
83–5	BGC	Monte León	4619	- - -	glass	16.5	0.4	Marshall *et al.* (1986)
83–6	BGC	Monte León	4623	- - -	plag	17.4	0.26	Marshall *et al.* (1986)
83–6	BGC	Monte León	4627	- - -	glass	18.5	1.3	Marshall *et al.* (1986)
83–6	BGC	Monte León	4627R	- - -	glass	18.2	1.0	Marshall *et al.* (1986)
R-7	BGC	Monte León	2944	- - -	glass	18.4	0.9	Marshall *et al.* (1986)
83–7	BGC	Monte León	4593	Toba Blanca	glass	17.3	0.3	Marshall *et al.* (1986)
83–7	BGC	Monte León	4869	Toba Blanca	biotite	19.9	1.2	Marshall *et al.* (1986)
83–4	BGC	Karaiken	4618	- - -	basalt	2.66	0.14	Marshall *et al.* (1986)
83–2	BGC	Karaiken	4626	- - -	plag	31.6	3.2	Marshall *et al.* (1986)
83–1A	BGC	Karaiken	4558	- - -	plag	16.8	0.7	Marshall *et al.* (1986)
83–1A	BGC	Karaiken	4611	- - -	plag	16.9	1.5	Marshall *et al.* (1986)
83–1B	BGC	Karaiken	4587	- - -	plag	16.5	1.1	Marshall *et al.* (1986)
83–3	BGC	Karaiken	4594	- - -	plag	31.1	3.2	Marshall *et al.* (1986)
	BGC	Monte Leon						

[a] Berkeley Geochronology Center.

Table 2.6. *Fission track ages*

Sample	Lab[a]	Section[b]	Lab number	Tephra unit	Mineral	Method[c]	Age (Ma)	Error (Ma)	Publication
83–3	USGS	K	DF-1220	Unnamed	zircon	FT	15.7	1.8 (2σ)	Marshall *et al.* (1986)
- - -	USGS	LL	- - -	Unnamed	zircon?	FT?	13.3	3.5 (2σ?)	Bown and Larriestra (1990)
- - -	USGS	CAR?	- - -	Estancia El Carmen?	zircon?	FT?	16.6	1.5 (2σ?)	Bown *et al.* (1988)

[a] USGS – United States Geological Survey, Denver, CO. Analysis by C. Naeser, J. Obradovich, and K. Tabbutt.
[b] K – Karaiken; LL – Loma de Luvia; CAR – Estancia El Carmen.
[c] FT – Fission track; FT? – unknown; fission track or K–Ar?

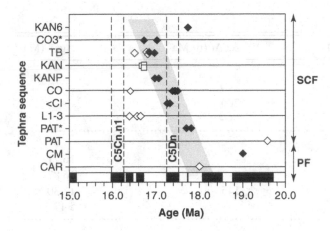

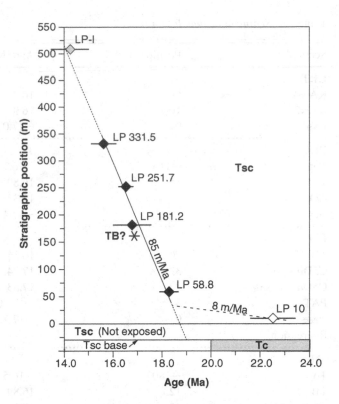

Fig. 2.4. Relative stratigraphic position versus age. Symbols mark individual ^{40}Ar/^{39}Ar ages. Open diamonds are analyses of Fleagle *et al.* (1995), open squares are analyses of Tejedor *et al.* (2006), and filled diamonds are analyses from this study. Tephra in the SCF are in relative stratigraphic order. The two tephra in the PF are both near the local base of the formation, but their relative stratigraphic positions are unknown. The ATNTS2004 magnetic polarity events are shown at the bottom of this plot.

Fig. 2.5. Revised interpretation of sediment accumulation curve at Lago Posadas. Diamond symbols show stratigraphic positions and ^{40}Ar/^{39}Ar ages of tephra dated by Blisniuk *et al.* (2005) with 1σ error bars. The date for the tephra at the base of the section (open diamond) is likely too old. The solid line through the four tephra from 18.3 to 15.6 Ma (black filled diamonds) is the linear weighted best-fit sediment accumulation through these tephra. The gray filled diamond near the top of the section is for an ~14.3 Ma tephra in an area 10 km southeast of the Lago Posadas locality. The upper finely dashed line extends the sediment accumulation line up to this tephra while the lower finely dashed line extends it down to the base of the SCF. Tsc = Santa Cruz Fm. and Tc = Centinela Fm.

other dated tephra layers within the band show coherent patterns of age vs. stratigraphic position, as would be expected if ages were accurate (Fig. 2.4). In addition, with the exception of the Lago Posadas locality, for localities where three or more dated tephra are present average sediment accumulation rates are similar (130 to 220 m Ma^{-1}), and where two sequential dated tephra are separated by 50 m or more these accumulation rates are all in the range of 90 to 220 m Ma^{-1} (Table 2.7). Given the likely imprecision in thickness measurements in combination with the analytical errors of the ages, the actual range of accumulation rates between two sequential tephra probably falls into a more restricted range. For the Lago Posadas locality, the independent dating by Blisniuk *et al.* (2005) shows a lower average sediment accumulation rate of ~70 m Ma^{-1} (Fig. 2.5). These observations support the validity of the ages within the acceptance band.

Anomalous ages The ^{40}Ar/^{39}Ar ages of several tephra are at significant odds with the general patterns of age vs. stratigraphic position. These are the Killik Aike Norte #6 tephra (KAN6), some of the L1–3 group of tephra, the Patagonia tephra (PAT), and the Cerro Observatorio tephra (CO). In addition, the Cañadón del Indio (CI) and the Killik Aike Norte (KAN) tephra display some minor deviations with respect to underlying or overlying tephra. These anomalous tephra ages are discussed below.

The KAN6 tephra. The age for the KAN6 tephra obtained in this study is clearly too old with respect to tephra in the acceptance band (Fig. 2.4). Older contaminant plagioclase in the sample is the likely explanation for this anomaly.

The L1–3 tephra. The four ages that Fleagle *et al.* (1995) report for L1–3 tephra range from 16.4 to 16.6 Ma and are clearly too young. The reason for these anomalous results is unknown.

The PAT tephra. The PAT tephra is a 3-m-thick tephra interbedded with typical gray mudstone, sandstone, and paleosols found in the basal SCF along the coast. It lies about 3 m above the brown, open marine siltstone and sandstone of the Monte León Fm. The contact at the top of the Monte León Fm. is sharp and clearly visible both in the field (Fig. 2.3) and on Google Earth™ imagery. The age reported by Fleagle *et al.* (1995) for PAT is 19.58 Ma, indicating that it is some 3 Ma older than the overlying L1–3 tephra (~16.5 Ma). Hence, they proposed that a package of sediment enclosing PAT was at the top of the Monte León Fm. Our results indicate otherwise. The PAT tephra is in close stratigraphic proximity to two other tephra

Table 2.7. *Sediment accumulation rates*

Section[a]	Position (m)[b]	Age (Ma)	$\Delta z/\Delta t$ (m Ma^{-1})[c]	Age* (Ma)[d]
CBT				
KAN4[e]	50	16.7	- - -	- - -
KAN2	10	16.9	200	- - -
Base	0	~17.0	- - -	- - -
PLC				
Top	~115	~16.7	- - -	16.71
TB	85	16.89	- - -	16.87
CO	5	17.3	145	17.30
Base	<0	~17.4	- - -	>17.33
CO				
TB	90	16.86	- - -	16.83
C5Dn – top	33	17.24	153	17.37
C5Dn – base	22	17.53	17	17.48
PAT	3	~17.70	59	17.66
Base	0	~17.7	- - -	17.69
Regression	- - -	- - -	~100	- - -
ML				
Top	~180	~16.5	- - -	- - -
TB	123	16.89	- - -	16.85
C5Dn – top	42	17.24	213	17.38
C5Dn – base	23	17.53	66	17.51
PAT*	6	17.70	100	17.62
Base	0	~17.8	- - -	17.51
Regression	- - -	- - -	136	17.66
RC-E				
Top	380	~16.0	- - -	- - -
TB	260	16.86	- - -	- - -
PAT	150	~17.70	~130	- - -
Base	135	~17.8	- - -	- - -
RC-W				
CO3	270	16.71	- - -	- - -
PAT	140	~17.70	~130	- - -
Base	120	~17.8	- - -	- - -
K				
Top	1080	~16.0	- - -	- - -
CO3	940	16.71	- - -	- - -
PAT*	785	17.69	160	- - -
Base	650	~18.5	- - -	- - -
LP				
Top	508	- - -	- - -	~14.3
LP 331.5	331	15.51	- - -	15.65
LP 251.7	252	16.45	88	16.40
LP 181.2	181	16.82	218	17.05
TB?	~165	- - -	- - -	~17.2
LP 58.8	59	18.15	91	18.19
LP 10.0	10	22.51?	11?	18.54

Table 2.7. (*cont.*)

Section[a]	Position (m)[b]	Age (Ma)	$\Delta z / \Delta t$ (m Ma^{-1})[c]	Age* (Ma)[d]
Base	−30	- - -	- - -	18.92
Regression	- - -	~19.0	110	~19.0

[a] Section abbreviation. See Table 2.3 for section name. RC-E = Río Chalia East; RC-W = Río Chalia West.

[b] Stratigraphic position.

[c] Sedimentation rate.

[d] Ages based on linear regression of age vs. position for section with three or more tephra with ^{40}Ar/^{39}Ar ages or a tie to the magnetic polarity timescale. At Lago Posadas the ages of the uppermost and lowermost LP samples of Blisniuk *et al.* (2005) are excluded (see text).

[e] Tephra abbreviation; see Table 2.4 for full names.

labeled PAT* (i.e. tephra at about the level of PAT). One of these PAT* tephra is at the Karaiken locality and was dated for this study. Its weighted mean age is 17.71 ± 0.06 Ma. Thus, PAT is also ~17.7 Ma old.

The CO tephra. The Cerro Observatorio tephra lies within Chron C5Dn at both the Cerro Observatorio and Monte León localities (Fig. 2.6). The weighted mean age of two samples of CO is 17.41 Ma. This age appears to be slightly too old for two reasons. First, the CO tephra plots somewhat to the right of the sediment accumulation trends across Chron C5Dn at both the Cerro Observatorio and Monte León. This suggests its ^{40}Ar/^{39}Ar age is ~0.1 Ma too old. In addition, CO lies about 10 m above the <CI tephra (a separate tephra that is just below the CI tephra). The age for the <CI tephra age is 17.31 ± 0.04 Ma. Accepting the ATNTS2004 age for the top of C5Dn (17.235 Ma), the CO tephra is probably about 17.2–17.3 Ma.

The KAN tephra. The Killik Aike Norte tephra outcrops along the north bank of Río Gallegos in the vicinity of Estancia Killik Aike Norte. This tephra was dated by Tejedor *et al.* (2006) at 16.68 ± 0.11 Ma, ~0.22 Ma younger than the overlying TB tephra (16.89 ± 0.05 Ma). This may reflect some small interlaboratory differences. However, we do note that KAN is about 25 m below the TB. Thus, it would be 0.1 to 0.3 Ma older than TB on the basis of sediment accumulation rates. Relative to our age for TB we estimate the age of KAN to be about 17.0 to 17.2 Ma. This is in concordance with an age of 17.06 ± 0.07 Ma we determined for a pumice clast within the KAN tephra (KANP in Table 2.2). But it is important to point out that the composition of the pumice glass does not match that of KAN shards. So this clast could pre-date the eruption of the KAN tephra.

Sediment accumulation rates, secondary ages, and age range of the SCF By using tephra ages within the acceptance band in conjunction with measurements of stratigraphic thickness, the following can be estimated: (1) sediment accumulation rates; (2) secondary ages for undated tephra by interpolation/extrapolation; and (3) the

ages of tops and bottoms of section which provide estimates of the age range of the SCF.

Sedimentation rates are calculated for sections with two or more tephra with ^{40}Ar/^{39}Ar ages. These rates are listed in Table 2.7 and also shown graphically for three well-dated sections (CO, ML, and LP; see Figs. 2.5, 2.6). Secondary ages for tephra layers identified in two or more sections are listed in Table 2.4 and shown in Fig. 2.2. The age ranges of individual sections are also shown in Fig. 2.2.

As seen in Table 2.7, sedimentation rates for the SCF are relatively high. They are commonly in the range of ~100 to ~200 m Ma^{-1} with somewhat lower rates of ~60 to 80 m Ma^{-1} in the basal part of the CO and ML sections. The overall age range for the SCF is ~19–14 Ma, though only the LP locality extends across this full age range.

In the coastal area, three sections (CO, ML, and RC) extend to or nearly to the base of the SCF. These sections indicate an age of ~17.7 to 17.8 Ma for the base for these sections. Exposures in other coastal sections (PLC, ELC, and CI) extend almost to the base with lowest exposed outcrops having estimated ages of ~17.5 to 17.6 Ma.

In the inland sections, older basal ages are estimated using extrapolations of sedimentation rates. At Karaiken (K) the projected age for the base of the SCF is ~18.5 Ma and at Lago Posadas it is ~19 Ma.

At the Gobernador Gregores (GG) locality, we found a thin oyster layer similar to those seen at the base of coastal sections and at Río Chalía (RC). This layer is ~17.7 Ma old. The base at GG is not exposed but it may be ≥18 Ma at this section. While the GG section has been included in the PF by Bown and Fleagle (1993), it has a mixture of SCF-like and PF-like characteristics as emphasized by the presence of the oyster layer. Other inland sections (LC, ELP, and ELPN), while lacking the thin oyster layer, have bedding characteristics more like those of the "classic" coastal SCF than those of the type PF.

Finally, the age of tops can be estimated for most localities (Fig. 2.2). They are generally in the age range of 16.0 to 16.5

(a)
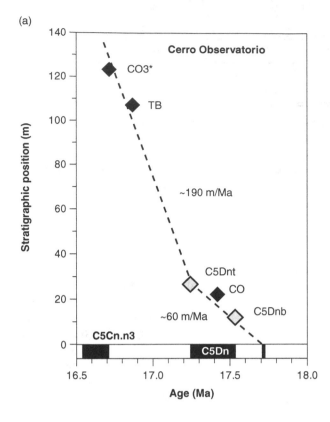

(b)
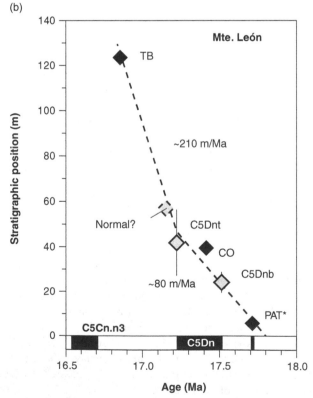

Ma. The exception, mentioned earlier, is in the vicinity of the LP locality. Here, about 10 km southeast of the Lago Posadas locality, a 14.35 ± 0.78 Ma tephra outcrops just 3 m below the base of an unconformably overlying 12.1 Ma basalt (Blisniuk *et al.*, 2005).

Comments on the Lago Posadas section Our interpretation of the age of the basal part of the LP section (Fig. 2.5) differs from that of Blisniuk *et al.* (2005). These workers accepted the age of 22.51 ± 0.73 Ma for their sample LP10.0m, near the base of the section, as valid, but we question the validity of this age. It is likely too old for several reasons. First, at Lago Posadas the SCF rests on the marine Centinela Fm., the inland equivalent of the coastal Monte León Fm. Recently, Parras *et al.* (2009) reported that $^{87}Sr/^{86}Sr$ ages determined for oyster shells in the Centinela Fm. range from 21.24 to 26.38 Ma. They further noted that a sample from the Centinela Fm. yielded a $^{40}Ar/^{39}Ar$ age of 20.48 ± 0.27 Ma. On the face of it these results rule out an age of ~22.5 for sample LP10.0m. Second, the age reported for LP10.0m requires an extremely low sediment accumulation rate of no more than ~12 m Ma^{-1} and more likely less than 8 m Ma^{-1} in the basal part of the section. This requires roughly a 10-fold increase in the sediment accumulation rate somewhere between LP 10.0 and LP 58.8. There is no apparent change in the character of the strata between these two samples. The sequence is composed of siltstone, thin- to thick-bedded sandstones and a few thin tuffs. Only above LP 58.8 do the very thick channel sandstones that characterize the upper part of the section make their appearance (Blisniuk *et al.*, 2005). Finally, our field notes for the thin LP 10.0 tuff (2–10 mm) indicate "it appears to be somewhat sandy and unsuitable for an age." Based on the downward extension of the sediment accumulation curve, we estimate that the LP 10.0 tuff is ~18.3 Ma and the base of the Lago Posadas section is ~19 Ma.

Pinturas Formation Compared with the relative lithologic monotony of the SCF, the PF is a mixture of varied lithologies including basal strata of tuffaceous mudstone overprinted by very mature paleosols, overlying eolian sandstone, and capping massive, poorly bedded tuffaceous mudstone (Bown and Larriestra, 1990). All these units are deposited over a rolling upland paleotopography developed on the volcanic rocks of the Jurassic Bahía Laura Group.

Fig. 2.6. Stratigraphic position vs. age at Cerro Observatorio and Monte León localities. Four tephra and a zone of magnetically normal strata show the coherence of the tephra data in these two sections. The position of the normal interval relative to the tephra ages indicates it is likely Chron C5Dn.

A deep, reddish-brown, very mature paleosol is developed on this paleotopography.

Distinct layers of tephra are generally absent in PF but one has been identified at the Estancia El Carmen locality and dated by Fleagle *et al.* (1995) at ~18 Ma based on concordant biotite and plagioclase ages (Table 2.2). Also, a series of up to five gray silicic tephra are found in mudstones overlying the paleosol developed on the Bahía Laura Group at Cerro de Los Monos and Loma de la Lluvia. The lowest of this group of tephra, the Cerro de Los Monos (CM) tephra yielded a minimum ^{40}Ar/^{39}Ar fusion age of 19.89 ± 0.30 Ma (weighted mean of three fusions, Table 2.2). However, most fusions for the CM tephra yielded older ages ranging from about 20.5 Ma to 33.5 Ma, with the two oldest fusions yielding ages of ~51 and 65 Ma. Clearly older plagioclase is incorporated into this tephra, and the fusion of the three multi-grain samples at ~19.9 Ma age may include older plagioclase as well.

Reconnaissance sampling of basal PF tuffaceous mudstone showed that some contain unaltered shards. Glass shards from one of these samples were analyzed and the analysis showed a single mode. This suggests the possibility in future of obtaining a useful tephra stratigraphy from the tuffaceous sediments of the PF.

Finally, at the Portezuelo Sumich Norte (PZN) locality we analyzed glass shards from a white biotitic tuff. This tuff, likely within one of the most northerly exposures of the SCF, lies near the base of a sequence of gray sandstones and mudstones. These SCF strata rest (unconformably) on the PF. The result of this analysis shows the tuff to be the 16.89 Ma Toba Blanca. A sample from this outcrop of the Toba Blanca was dated directly by Fleagle *et al.* (1995) and yielded a plagioclase age of 16.80 ± 0.10 Ma, in agreement with the weighted mean age of 16.89 ± 0.05 Ma for the Toba Blanca.

2.4 Discussion and conclusions

Our reconnaissance study of tephra in the Santa Cruz and Pinturas Formations in southern Patagonia establishes a high-precision tephrochronology for these important Miocene fossil-rich formations. In particular, on the basis of glass-shard composition of silicic tephra, we have identified 15 tephra in these two formations that link two or more localities. Fourteen of these tephra, all in the Santa Cruz Formation, link 19 of the 26 localities into a space-stratigraphic network that extends from the mouth of the Río Gallegos on the Atlantic coast (51.6° S) north to the Río Pinturas headwaters (47.0° S) and westward across the coastal plain into the Andean foothills. Argon (^{40}Ar/^{39}Ar) ages for these and other Santa Cruz tephra provide the first regional age calibration of this formation. In the Pinturas Formation one tephra, the CM tephra, links two nearby localities, the Cerro de Los Monos and Loma de la Lluvia

localites. While information on the Pinturas Formation tephra is still sparse, analyses of glass shards from a limited number of Pinturas tephra and glass-shard-rich mudstones suggest the feasibility of establishing a useful tephrochronology for this formation as well.

Our tephrostratigraphy, in combination with the measurement of sections and ^{40}Ar/^{39}Ar ages, provides the first definitive information on the age range and sediment accumulation rates in the SCF. East of the Andean foothills, the more complete sections indicate that the initial influx of sediment in this region began at ~17.8 Ma and continued uninterrupted until at least 16 Ma, the age of the youngest preserved strata. In the Andean foothills at the Karaiken and Lago Posadas localities, the initial influx occurred somewhat earlier, at ~18.5 Ma (K) and ~19.0 Ma (LP). The youngest known SCF strata are near the Lago Posadas with a ~14.3 Ma andesitic tephra preserved under a 12.1 Ma basalt. With the exception of Lago Posadas, average rates of sediment accumulation are generally in the range of ~100 to 200 m Ma^{-1}. These relatively high rates reflect the influx of sediment from both coeval volcanism and uplift in Andes. At Lago Posadas, accumulation rates are somewhat lower at ~70 m Ma^{-1}. The general paucity of tuffaceous sediments at Lago Posadas, with the exception of thin andesitic tuffs, may account for this lower rate. There is also the suggestion, at the coastal localities of Cerro Observatorio and Monte León, that accumulation rates in the basal Santa Cruz are lower, in the range of 60–80 m Ma^{-1}.

Much less is known about the Pinturas Formation. The age of 17.99 ± 0.02 Ma near the base of the formation at the Estancia El Carmen may be close to the age of the oldest Pinturas strata. The 16.89 Ma Toba Blanca unconformably overlies the upper Pinturas beds at Portezuelo Sumich Norte. Clearly the PF is coeval with the lower portion of the SCF, but more work is needed to establish both the relative and chronologic ages of the complex array of facies in the PF.

ACKNOWLEDGMENTS
The fieldwork upon which this chapter was based involved the efforts of many people, including Bobby Taylor, Hormiga Diaz, Mario Cozzuol, Elise Schloeder, Julio César Rúa, Amber Bayker, Siobhan Cooke, Frederico Degrange, Laureano Gonzalez-Ruiz, Keith Jones, Sebastian Pintos, Alfred Rosenberger, Sergio Vincon, and many generous estancia owners. Paleontological fieldwork in the Santa Cruz and Pinturas Formations over several decades relied on the efforts of others too numerous to name, but many are listed in Bown and Fleagle (1993). This research was supported by funds from the US National Science Foundation, the National Geographic Society, and the LSB Leakey Foundation. The authors thank the editors and anonymous reviewers for helpful suggestions.

REFERENCES

Blisniuk, P. M., Stern, L. A., Chamberlain, C. P., Idleman, B. and Zeitler, P. K. (2005). Climatic and ecologic changes during Miocene surface uplift in the Southern Patagonia Andes. *Earth and Planetary Science Letters*, **230**, 125–142.

Bown, T. M. and Fleagle, J. G. (1993). Systematics, biostratigraphy, and dental evolution of the Palaeothentidae, later Oligocene to early–middle Miocene (Deseandan–Santacrucian) Caenolestoid marsupials of South America. *Journal of Paleontology, Memoir* **29**, 1–76.

Bown, T. M. and Larriestra, C. N. (1990). Sedimentary paleoenvironments of fossil platyrrine localities, Miocene Pinturas Formation, Santa Cruz Province, Argentina. *Journal of Human Evolution*, **19**, 87–119.

Evernden, J. F., Savage, D. E., Curtis, G. H. and James, G. T. (1964). Potassium-argon dates and the Cenozoic mammalian chronology of North America. *American Journal of Science*, **262**, 145–198.

Fleagle, J. G., Bown, T. M., Swisher, C. C. III and Buckley, G. (1995). Age of the Pinturas and Santa Cruz Formations. *VI Congreso Argentino de Paleontología y Bioestratigrafía, Actas*, 129–135.

Kuiper, K. F., Deino, A., Hilgen, F. J. *et al.* (2008). Synchronizing rock clocks of Earth history. *Science*, **320**, 500–504.

Lourens, L., Hilgen, F., Shackleton, N. J., Laskar, J. and Wilson, J. (2004). Appendix 2 – Orbital tuning calibrations and conversions for the Neogene Period. In *A Geologic Time Scale 2004*, ed. F. M. Gradstein, J. G. Ogg and A. G. Smith. Cambridge: Cambridge University Press.

Marshall, L. G., Pascual, R., Curtis, G. H. and Drake, R. E. (1976). South American geochronology: Radiometric time scale for middle to late Tertiary mammal-bearing horizons in Patagonia. *Science*, **195**, 1325–1328.

Marshall, L. G., Drake, R. E., Curtis, G. H. *et al.* (1986). Geochronology of the type Santacrucian (middle Tertiary) Land Mammal Age, Patagonia, Argentina. *The Journal of Geology*, **94**, 449–457.

Nash, W. P. (1992). Analysis of oxygen with the electron microprobe: application to hydrous glass and minerals. *American Mineralogist*, **77**, 453–457.

Parras, A., Griffin, M., Feldmann, R. *et al.* (2009). Correlation of marine beds based on Sr- and Ar-date determinations and faunal affinities across the Paleogene/Neogene boundary in southern Patagonia, Argentina. *Journal of South American Earth Sciences*, **26**, 204–216.

Perkins, M. E., Nash, W. P., Brown, F. H. and Fleck, R. J. (1995). Fallout tuffs of Trapper Creek, Idaho: a record of Miocene explosive volcanism in the Snake River Plain volcanic province. *Geological Society of America Bulletin*, **107**, 1484–1506.

Perkins, M. E., Brown, F. H., Nash, W. P., Williams, S. K. and McIntosh, W. (1998). Sequence, age, and source of silicic fallout tuffs in middle to late Miocene basins of the northern Basin and Range province. *Geological Society of America Bulletin*, **110**, 344–360.

Rivera, T., Storey, M., Zeeden, C., Kuiper, K. and Hilgen, F. (2011). Support for the astronomically calibrated ^{40}Ar/^{39}Ar Age of Fish Canyon Sanidine: evidence from the Quaternary. *Geophysical Research Abstracts*, **13**, EGU2011–10378.

Smith, M. E., Chamberlain, K. R., Singer, B. S. and Carroll, A. R. (2011). Eocene clocks agree: coeval ^{40}Ar/^{39}Ar, U–Pb, and astronomical ages from the Green River Formation. *Geology*, **38**, 527–530.

Tauber, A. A. (1994). Estratigrafía y vertebrados fósiles de la Formación Santa Cruz (Mioceno Inferior) en la costa Atlántica entre las rías del Coyle y Gallegos, Provincia de Santa Cruz, Republica Argentina. Unpublished doctoral thesis, Universidad Nacional de Córdoba, Argentina.

Tejedor, M. F., Tauber, A. A., Rosenberger, A. L., Swisher, C. C. III and Palacios, M. E. (2006). New primate genus from the Miocene of Argentina. *Proceedings of the National Academy of Sciences of USA*, **103**, 5437–5441.

3 Absolute and relative ages of fossil localities in the Santa Cruz and Pinturas Formations

John G. Fleagle, Michael E. Perkins, Matthew T. Heizler, Barbara Nash, Thomas M. Bown, Adán A. Tauber, Maria T. Dozo, and Marcelo F. Tejedor

Abstract

Santa Cruz Province, Argentina, has some of the richest fossil mammal localities in the world. However, the absolute and relative ages of its fossil localities have long been a source of confusion and debate. In particular, there has been longstanding disagreement about the relative ages of the fossils from the western part of the province in deposits of the Pinturas Formation compared with those from the numerous localities of the Santa Cruz Formation along the Atlantic coast. Drawing on recent studies of the tuffaceous sediments in many classic fossil localities, and studies of fossil representatives of marsupials, rodents, and primates, we provide a synthesis of the temporal relationship among fossil localities throughout the province. There is broad agreement between the results of the tephrochronology and mammalian paleontology. Both tephra correlations and paleontological comparisons indicate that the lower units of the Pinturas Formation are older than the sections of the Santa Cruz Formation preserved at Monte León and Cerro Observatorio, supporting Ameghino's suggestion that part of the Pinturas Formation represents a distinct faunal zone. However, the upper unit of the Pinturas Formation seems to correspond in age with the lower part of the sections at Monte León and Cerro Observatorio.

Resumen

En la provincia de Santa Cruz, Argentina se encuentran algunas de las localidades fosilíferas más ricas del mundo. Sin embargo, las edades absolutas y relativas de estas localidades han sido fuente de confusión y debate por mucho tiempo. En particular, ha habido desacuerdos de larga data acerca de las edades relativas de los fósiles de la Formación Santa Cruz en la parte occidental en comparación con las de las numerosas localidades de la Formación Santa Cruz a lo largo de la costa atlántica. En base a estudios recientes sobre los sedimentos tobáceos varias localidades clásicas y estudios de fósiles de diferentes

grupos, incluyendo marsupiales, roedores y primates, aportamos una síntesis de la relación temporal entre las localidades fosilíferas de la provincia. Hay una amplia concordancia entre los resultados de la tefrocronología y la paleontología de mamíferos. Tanto las correlaciones de las tefras como las comparaciones paleontológicas indican que las unidades inferiores de la Formación Pinturas son más antiguas que las secciones de la Formación Santa Cruz preservadas en Monte León y Cerro Observatorio, apoyando la sugerencia de Ameghino de que parte de la Formación Pinturas representa una zona faunística diferente. Sin embargo, el límite superior de la Formación Pinturas parece corresponderse en edad con la parte más baja de las secciones en Monte León y Cerro Observatorio.

3.1 Introduction

Sedimentary deposits attributed to the Santacrucian Land Mammal Age are widespread throughout much of southernmost Argentina from Chubut Province in the north to Tierra del Fuego in the south, and from the Andean Precordillera in the west to the Atlantic coast in the east (e.g. Feruglio, 1938; Matheos and Raigemborn, Chapter 4). Through paleontological research for over a century, these sediments have yielded a remarkable abundance and diversity of fossil mammals from numerous individual localities. In this paper we briefly review previous efforts to determine the absolute age of deposits assigned to the Pinturas and Santa Cruz formations, summarize the results of recent studies of tephrochronology of numerous stratigraphic sections in these two formations (Perkins *et al.*, Chapter 2), and correlate the results of tephrochronology with paleontological studies on the mammals from the Pinturas and Santa Cruz Formations in order to assess the stratigraphic relationships and relative ages of individual fossil localities within these formations.

The earliest discovery of fossils from what would be called the Santa Cruz Formation is attributed to Commander Bartholomew James Sulivan who recovered fossils from the banks of the Río Gallegos in 1845 (Brinkman, 2003; Vizcaíno *et al.*, Chapter 1) and reported this to Charles Darwin. In later editions of *The Voyage of the Beagle* (e.g.

Early Miocene Paleobiology in Patagonia: High-Latitude Paleocommunities of the Santa Cruz Formation, ed. Sergio F. Vizcaíno, Richard F. Kay and M. Susana Bargo. Published by Cambridge University Press. © Cambridge University Press 2012.

1862), Darwin reported that "Captain Sulivan, R.N., has found numerous bones, embedded in regular strata on the banks of the Río Gallegos." The noted anatomist Sir Richard Owen described these fossils in 1846. In addition, Darwin (1846) observed sedimentary deposits of what is known to be the Santa Cruz Formation along the course of the Río Santa Cruz.

However, the naming of the Santa Cruz Formation (Ameghino, 1889), the description of its relationship to other fossiliferous deposits in Patagonia, and the documentation of the extraordinary fossils to be found in these deposits came primarily from the decades of fieldwork by Carlos Ameghino and the voluminous research and publications of his brother Florentino beginning in 1887 and continuing for the next 25 years. For a variety of reasons, Florentino Ameghino (e.g. 1906; reviewed in Simpson, 1940, 1984) believed that the deposits of the Santa Cruz Formation, as well as those of two other deposits from the Andean Precordillera that he called the *Notohippidense* beds near Lago Argentino and the *Astrapothericulense* beds in the valley of the Río Pinturas, were contemporary with Eocene deposits in other parts of the world. However, by the beginning of the twentieth century, invertebrate faunas from the marine Patagonian beds that underlie many exposures of the coastal Santa Cruz Formation were being correlated with those near the Oligocene–Miocene boundary in other parts of the world, indicating that, contrary to the view of Ameghino, the Santa Cruz Formation was no older than Early Miocene in age (e.g. Simpson, 1940, 1984; Patterson and Pascual, 1972; Marshall *et al.*, 1983).

The first absolute date for the Santa Cruz Formation was a K–Ar age of 21.7 ± 3 Ma based on a tuffaceous matrix glass concentrate taken from a glyptodont carapace recovered from deposits at the well-known site of Felton's Estancia (now Killik Aike Norte) (Evernden *et al.*, 1964; Marshall *et al.*, 1986a). The age corroborated the proposed correlation of the Santa Cruz Formation with strata from other continents attributed to the Early Miocene based on the content of the marine fauna. In 1976, Marshall published a review of the history of paleontological research and a gazetteer of major localities of the Santa Cruz Formation. Beginning in 1977, Marshall and many colleagues published a series of ground-breaking papers in which they undertook to provide a radiometric timescale for the South American Land Mammal Ages (SALMA) based on K–Ar dating of samples from many parts of Argentina, including deposits of the Santa Cruz Formation (e.g. Marshall *et al.*, 1977, 1981, 1983, 1986a, b). In 1986 Marshall and colleagues published a paper devoted to the age of the Santa Cruz Formation based on K–Ar ages on samples from four localities: Felton's Estancia (= Killik Aike Norte), Monte León, Rincón del Buque, and Karaiken (= Estancia La

Laurita). In that paper they concluded that the strata of the Santa Cruz Formation itself range in age from "about 17.6 Ma to perhaps 16.0 Ma... [and that] the Santacrucian Land Mammal Age ranges from about 18.0 Ma to about 15 Ma" (Marshall *et al.*, 1986a: 449).

Because of its very different lithology, Bown and Larriestra (1990) gave the name Pinturas Formation to the fossiliferous deposits in the valley of the Río Pinturas that previous authors (Ameghino, 1906; Castellanos, 1937; Feruglio, 1949; de Barrio *et al.*, 1984) had called the "*Astrapothericulense*" beds or the "*Astrapothericulense*" level of the Santa Cruz Formation. Studies of the mammalian fauna by numerous workers (Pascual and Odreman-Rivas, 1971; de Barrio *et al.*, 1984; Fleagle *et al.*, 1987; Fleagle, 1990; Bown and Fleagle, 1993; Kramarz and Bellosi, 2005) have led them to concur with the view of Ameghino that at least some of the Pinturas Formation mammals are different from those of the "typical" Santa Cruz Formation along the Atlantic Coast and seem likely to be older. Based on samples collected and analyzed by Carl Swisher III, Bown and Fleagle (1993) and Fleagle *et al.* (1995) reported Ar–Ar radiometric ages for two sites in the Pinturas Formation. The Pinturas Formation is divided into three sedimentary sequences separated by unconformities, and the above authors reported average ages of 17.76 Ma on plagioclase samples and 17.67 Ma on biotite samples from a vitreous tuff in the lower part of the lower sequence at the site of Estancia El Carmen. They reported an age of 16.58 Ma from plagioclase samples, and 16.43 Ma from hornblende samples from rocks at the top of the middle sequence, the latter two dates coming from rocks above most of the fossiliferous deposits in the Pinturas Formation.

These same authors also reported ages for several levels in the extensive stratigraphic sections of the Santa Cruz Formation at Monte León and Cerro Observatorio (formerly Monte Observación) as well as magnetic polarity data for the stratigraphic section at Cerro Observatorio based on samples collected and analyzed by Dr. Greg Buckley. The dated samples for the sections of the Santa Cruz Formation at Monte León and at Cerro Observatorio yielded ages between roughly 17.0 and 16.0 Ma and were stratigraphically correlated with the stratigraphic interval at Cerro Observatorio showing normal polarity. Fleagle *et al.* (1995) concluded that most of the section at these two sites correlated with Chron 5 of the Global Magnetic Polarity Timescale. Both the radiometric ages and the morphological patterns among fossil marsupials recovered from the Pinturas and Santa Cruz Formations supported the earlier views that most of the Pinturas fossils were older than most of the sediments and fossils of the Santa Cruz Formation exposed along the Atlantic coast. However, the ages for individual horizons within the Santa Cruz Formation at Cerro Observatorio and Monte León did not yield an orderly pattern (i.e. were not concordant with superposition), and there was considerable overlap in the confidence intervals for the ages.

Based on his study of the stratigraphy and paleontology in the late 1980s and early 1990s, Adán Tauber (1991, 1994, 1996, 1997a, b) proposed lithologic and biostratigraphic divisions in the coastal exposures of the Santa Cruz Formation in Santa Cruz Province, between the Río Coyle in the north and the Río Gallegos in the south. He divided the Santa Cruz Formation in this region into two members (lower Estancia La Costa Member and upper Estancia Angelina Member), based on lithology, the regional southeastward dip of the beds, and the relative position of tuffaceous horizons. On the basis of the faunal changes in 22 fossiliferous levels, he reconstructed the paleoecology of the Santa Cruz Formation during the time covered by this section and proposed evidence of a climate change between the two members (Tauber, 1997b).

In 2006, Tejedor *et al.* (2006) described a new primate from the site of Killik Aike Norte (Felton's Estancia of previous authors). They also provided a stratigraphic profile of the site and reported new plagioclase Ar–Ar dates from a vitric tuff at the site from which the fossil monkey was recovered. They reported ages of 16.5, 16.5, and 16.45 Ma. They further suggested that the ratio of calcium to potassium in the glass component of the Killik Aike Norte tuff (KAN) compared best with a sample called CO2 from Cerro Observatorio (see below for further details), but they considered this correlation to be imprecise, so the KAN tuff might also correlate with other tuffs at Cerro Observatorio or Monte León. They noted that this age was significantly younger than the age of 17.7 Ma reported by Marshall and colleagues (1986a) from the same site, but accepted the dates reported from the more northern sites by Bown and Fleagle (1993), and Fleagle *et al.* (1995, 2004).

Kramarz *et al.* (2010) described fossils from the Upper Fossil Zone (UFZ) at Gran Barranca in Chubut Province, Argentina, that were similar to Pinturas fossils. They informally named a new South American Land Mammal Age (SALMA), the Pinturan, as a stage intermediate between the Colhuehuapian and Santacrucian Land Mammal Ages, to include the UFZ fossils and those from the Astrapothericulan Zone of the Pinturas Formation. They suggested a date of 19.7–18.7 Ma for the UFZ and 18.75–16.5 Ma for the Pinturas fossils. This was based on U/Pb ages of 19.1 and 18.7 Ma reported by Dunn *et al.* (2009; also Ré *et al.*, 2010) for the UFZ at the Gran Barranca in which the Pinturan fossils seemed to lie near the younger limits of that time period. They noted uncertainty about whether there was temporal overlap between the Astrapothericulan section of the Pinturas Formation (Lower and Middle sequences), and the lowest part of the Santacrucian levels at Monte León and Cerro Observatorio.

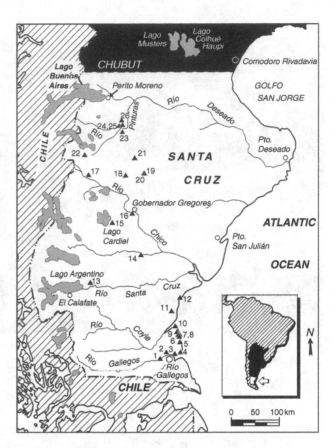

Fig. 3.1. Map of Santa Cruz Province showing the localities discussed in the text.

In an effort to help clarify the stratigraphic and temporal relationships of fossil vertebrates from localities in the Santa Cruz Formation and their temporal relationship to vertebrates from localities in the Pinturas Formation, we visited southern Patagonia in 2000, 2004, and 2007 and collected glass-shard-rich tephra at 26 stratigraphic profiles in several localities (Fig. 3.1; see also Perkins *et al.*, Chapter 2, Fig. 2.1), most of which have previously yielded fossil mammals (e.g. Marshall, 1976; Tauber, 1997a, b; Bown and Fleagle, 1993). As discussed below, the strata in some of these stratigraphic profiles have been attributed to the Santa Cruz Formation, while others are in sediments attributed to the Pinturas Formation including the type Pinturas Formation in the Pinturas Valley, and Pinturas Formation-like strata to the south of the Pinturas Valley (e.g. de Barrio *et al.*, 1984; Bown and Fleagle, 1993).

3.2   Tephra sample sites

Santa Cruz Formation

Marshall (1976) surveyed the history of fossil collecting in the Santa Cruz Formation. He summarized the paleontological and geological studies of B. Sulivan, Carlos Ameghino,

(a)

Estancia Killik Aike Norte

andesitic sandstone   KAN 4

andesitic tuff   KAN =~17.0

(b)

Cabo Buen Tiempo

Fig. 3.2. Fossil localities near the
mouth of the Río Gallegos. a,
A geological exposure along the north
bank of the Río Gallegos at Estancia
Killik Aike Norte (= Felton's
Estancia) showing the position of
several tephra layers. b, A geological
exposure at Cabo Buen Tiempo
showing the position of a tephra layer.
c, A geological exposure at Monte
Tigre with the position of the Toba
Blanca identified.

(c)

Monte Tigre

Toba Blanca

J. B. Hatcher, H. T. Martin, H. E. Riggs, and many others. In
the 1980s and 1990s, research groups from Stony Brook
University and the Museo Argentino de Ciencias Naturales
in Buenos Aires collected fossils and studied the geology of
Cerro Observatorio and Monte León, as well as doing extensive reconnaissance in the Pinturas Formation. Also in the late
1980s through 2008, A. A. Tauber collected extensively at
localities between Río Coyle and Río Gallegos. Between 2002
and 2011 a research group from Museo de la Plata and Duke
University made extensive geological and paleontological
collections, especially at the localities between Río Coyle

and Río Gallegos (Vizcaíno *et al.*, Chapter 1). The sites
sampled for tephrochronology studies reported by Perkins
et al. in Chapter 2 are as follows, with coordinates indicating
where the tephra samples were collected.

1. Río Gallegos Bridge (S 51.62547°; W 69.617°). This locality is where National Highway 3 crosses the Río Gallegos
   at Estancia Güer Aike. There are vertebrate fossils reported
   from Río Gallegos, but none from this precise place.
2. Estancia Killik Aike Norte (formerly Felton's Estancia)
   (S 51.57420°; W 69.44580°) on the north bank of the Río
   Gallegos. This locality (Fig. 3.2a) has been the site of

paleontological work for over a century by many expeditions and has been identified, described, and discussed by numerous authors (e.g. Hatcher, 1903; Martin, 1904; Riggs, 1926; Marshall, 1976; Marshall *et al.*, 1986a; Tejedor *et al.*, 2006). It is the type locality of *Killikaike blakei* (Tejedor *et al.*, 2006). Tauber *et al.* (2004) and Tejedor *et al.* (2006) published a stratigraphic section of the site.

3 and 4. Cabo Buen Tiempo (= Cape Fairweather) (S 51.53390°, W 68.95082°) and Estancia Cabo Buen Tiempo (S 51.56450°, W 69.05603°) are located on the northeast bank of Río Gallegos where it enters the Atlantic Ocean (Fig. 3.2b). Like Estancia Killik Aike Norte, localities from this area have been described and identified by numerous researchers for over a century (Hatcher, 1903; Martin, 1904; Marshall, 1976; Tauber, 1997a, b; Vizcaíno *et al.*, Chapter 1).

5. Monte Tigre (S 51.34372°, W 69.03832°), north of Cabo Buen Tiempo, within the Estancia Angelina (Tauber, 1997a, b). This site (Fig. 3.2c) was not mentioned by name in the early literature but was probably visited by paleontologists as they moved along the coast (e.g. Marshall, 1976).

6. Puesto Estancia La Costa (= Corriguen Aike, Corriken Aike, Corriguen Kaik) (S 51.19655°, W 69.09168°). This site (Fig. 3.3a) is probably the most famous fossil locality in the Santa Cruz Formation and one of the richest fossil mammal sites in the world. It is well known for the remarkable number of relatively complete specimens recovered from the intertidal levels. It has been identified, described, and discussed by numerous researchers (Hatcher, 1903; Ameghino, 1906; Marshall, 1976; Simpson, 1984; Tauber, 1991, 1997a, b; Kay *et al.*, 2008; Perry *et al.*, 2010; Vizcaíno *et al.*, 2010; Vizcaíno *et al.*, Chapter 1). The site has yielded numerous primate and other mammalian fossils, including many specimens of *Homunculus patagonicus* Ameghino 1887 (Ameghino, 1893; Tauber, 1991; Tejedor and Rosenberger, 2008; Perry *et al.*, 2010; Kay *et al.*, Chapter 16).

7. Drill Pad (S 51.18266°, W 69.16155°). This is the site of an oil pump above the costal cliffs with a tuffaceous exposure next to it. There are no fossils from this site.

8. Cañadón Silva (S 51.15675°, W 69.11214°). This locality (Fig. 3.3b) was identified and described by Tauber (1997a, b) and has yielded numerous fossils described in this volume.

9. Estancia La Costa (S 51.08509°, W 69.13430°). This site (Fig. 3.3c) is approximately 13 km south of Coy Inlet or the mouth of Río Coyle. This site has yielded many fossil vertebrates and has been identified and discussed by numerous workers for over a century (e.g. Hatcher, 1903; Riggs, 1926; Marshall, 1976; Tauber 1997a, b; Kay *et al.*, 2008; Vizcaíno *et al.*, 2010; Vizcaíno *et al.*, Chapter 1).

10. Cañadón del Indio (= Coy Inlet, Boca del Coyle) (S 50.98415°, W 69.15127°). This site (Fig. 3.3d) is on the south bank of the Río Coyle. Paleontologists for over a century have collected from numerous sites near the mouth of the Río Coyle (Hatcher, 1903; Ameghino, 1906; see discussion in Marshall, 1976; Tauber, 1997a, b; Vizcaíno *et al.*, Chapter 1).

11. Cerro Observatorio (= Monte Observación) (S 50.56478°, W 69.15727°). Monte Observación is the name that has been used historically in the literature to refer to this locality (e.g. Marshall, 1976; Bown and Fleagle, 1993; Fleagle *et al.*, 1995) but the name of the topographic feature closest to the sea is Cerro Observatorio; Mt. Observación is ascribed to a different feature further north (Vizcaíno *et al.*, Chapter 1). This is a large and very rich series of localities (Figs. 3.4 and 3.5; also Bown and Fleagle, 1993: Fig. 6a) in a thick profile that has been collected by paleontologists for more than a century (Ameghino, 1906; Hatcher, 1903; see discussion in Marshall, 1976; Bown and Fleagle, 1993; Anderson *et al.*, 1995). Detailed geological sections and numerous individual localities have been identified throughout the profile at Cerro Observatorio and Monte León to the north (see below) (Fig. 3.5; Bown and Fleagle, 1993; Fleagle *et al.*, 1995). Radiometric dates come from both profiles and a paleomagnetic record from Cerro Observatorio (Bown and Fleagle, 1993; Fleagle *et al.*, 1995; Anderson *et al.*, 1995). There are thousands of fossils from this "locality" *sensu lato*, including many partial mandibles and isolated teeth attributed to *Homunculus patagonicus* (Fleagle *et al.*, 1988; Kay *et al.*, Chapter 16).

12. Monte León (S 50.31875°, W 68.89494°). This is another large locality (Fig. 3.6a; also Bown and Fleagle, 1993: Fig. 8) (Ameghino, 1906; Tournouër, 1903; Feruglio, 1938; Simpson, 1940; Marshall, 1976; Marshall *et al.*, 1986a; Bown and Fleagle, 1993; Fleagle *et al.*, 1995). Marshall *et al.* (1986a) published a stratigraphic section with radiometric ages and a paleomagnetic profile (see also Fleagle *et al.*, 1995).

13. Karaiken (= Estancia La Laurita, Carr-aiken, the "*Notohippidense*" horizon of Ameghino) (S 50.06280°, W 71.76550°). This is a small site (Fig. 3.6b) about 20 km northeast of the outlet of Río Santa Cruz from Lago Argentino. Its fauna is thought to be slightly older than those from the coastal sites (e.g. Ameghino, 1906; Hatcher, 1903; Simpson, 1940; Feruglio, 1944; Marshall, 1976; Marshall and Pascual, 1977; Marshall *et al.*, 1986a). Marshall *et al.* (1986a) published a stratigraphic section for the rocks at this site and a series of radiometric ages.

14. Río Chalía (= Sehuen, Río Sehuen, Chalía) (S 49.61600°, W 69.99800°, and S 49.64690°, W 70.1300°). This site name refers to an extensive set

(a) Puesto Estancia La Costa (= Corriguen Aike) — Toba Blanca

(b) Cañadón Silva — MO

(c) Estancia La Costa — Toba Blanca, andesitic sandstone channels, fossil level 3, fossil level 2

(d) Cañadón del Indio — MO, CI

Fig. 3.3. Fossil localities south of the Río Coyle. a, A paleontologist removing fossils from deposits in the intertidal zone at Puesto Estancia La Costa (= Corriguen Aike). A geological section is visible in the background with the Toba Blanca identified. b, A geological exposure at Cañadón Silva. c, A geological exposure at Estancia La Costa with fossil levels identified in the foreground and the Toba Blanca identified in the cliff behind. d, A geological exposure at Cañadón del Indio at the mouth of the Río Coyle with two tephra layers identified.

of localities south of Río Chalía (or Sehuen) just west of where it joins the Río Chico. The localities (Fig. 3.6c) are on the north face of the large meseta that lies between the Río Santa Cruz to the south and the Río Chalía to the north. It is probably where Carlos Ameghino collected in 1890–1892. The outcrops appear to be those described by Hatcher (1903) as Sehuen, as their position seems to accord with his narrative of how he reached them by wagon; however,

they are farther east than the description given in his text as "30 miles east of Cordillera along the Río Sehuen" (Marshall, 1976).

15. Lago Cardiel (S 48.98445°, W 70.93265°). East of Lago Cardiel is a large exposure of rocks (Fig. 3.6d) that were mapped by Ramos (1982) as Santa Cruz Formation.

16. Gobernador Gregores (S 48.78170°, W 70.32028°). A series of exposures (Fig. 3.7a) south of the town of Gobernador Gregores, adjacent to Highway 27. Fossil

(a)

Fig. 3.4. Part of the geological section at Cerro Observatorio (= Monte Observación). a, Upper part of the geological section at Cerro Observatorio (= Monte Observación) with tephra layers labeled. b, Lower part of the section at Cerro Observatorio (= Monte Observación) with tephra layers labeled.

(b)

marsupials from this site were discussed by Bown and Fleagle (1993).

17. Estancia La Olguita (S 47.97045°, W 71.82677°). This is a series of exposures north of Highway 37 in the Andean Precordillera just east of Parque Nacional F.P. Moreno. It has yielded numerous fossil mammals (e.g. González *et al.*, 2005, González and Scillato-Yané, 2007; González, 2010).

18. Estancia La Pluma (S 47.99735°, W 70.66640°). This is one of several fossiliferous deposits identified by de Barrio *et al.* (1984) around the edges of the vast, largely empty area east of Bajo Caracoles in the central part of

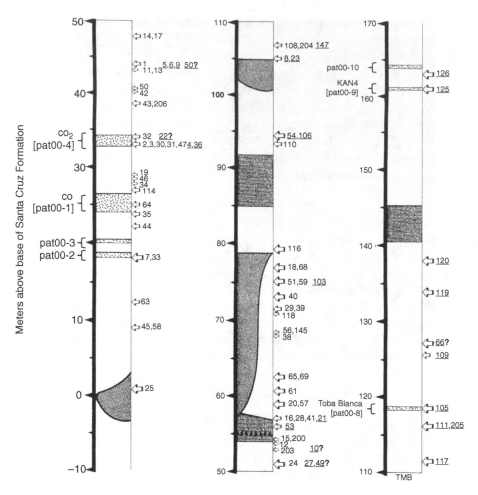

Fig. 3.5. A drawing of the stratigraphic section at the Cerro Observatorio (= Monte Observación) locality with individual fossil levels and tephra layers identified (see also Bown and Fleagle, 1993).

Santa Cruz Province. A vertical section of rock stands adjacent to the estancia house. In addition to the work by de Barrio and colleagues, brief paleontological and geological expeditions have been made there since the late 1980s, but have recovered only a few fossils.

19 and 20. Estancia La Cañada (S 47.97551°, W 70.16332°). This is another site first described by de Barrio and colleagues (1984). The fossil site (Fig. 3.7b) is a small deflation crater exposing tuffaceous deposits. Paleontological Expeditions from Stony Brook University, the Museo Argentino de Ciencias Naturales, and the Centro Nacional Patagónico visited localities 14–16 and 18–20 in the 1980s and 1990s and made small collections of fossil mammals (Bown and Fleagle, 1993; Rae *et al.*, 1996). A few kilometers from the fossil site is a more extensive vertical section (Fig. 3.7c) from which tephra samples were taken (S 47.97835°, W 70.18889°).

21. Estancia La Península (S 47.72646°, W 70.38501°). This is a series of exposures identified by de Barrio and colleagues (1984) on the western border of the

Macizo del Deseado in central Santa Cruz Province. This site has yielded an unusual glyptodont fossil (González Ruiz *et al.*, 2011).

22. Lago Posadas (= Lago Pueyrredón) (S 47.58990°, W 71.82403°). On top of a large mountain south of Lago Posadas is a broad exposure of rocks of the Santa Cruz Formation (Fig. 3.7d) (Ameghino, 1906; Hatcher, 1903; Marshall, 1976; Blisniuk *et al.*, 2006). Blisniuk *et al.* (2006) described a stratigraphic section of over 500 m and reported radiometric dates ranging from ~14 to ~22 Ma.

Pinturas Formation

Fossiliferous localities from the valley of Río Pinturas contain a mammalian fauna that is generally considered to be slightly older than localities in the coastal Santa Cruz Formation (Ameghino, 1906; Castellanos, 1937; Simpson, 1940; Marshall, 1976; de Barrio *et al.*, 1984; Kramarz and Bellosi, 2005). Since 1985, expeditions from Stony Brook University, the Museo Argentino de Ciencias Naturales, the Centro Nacional Patagónico, and the Universidad Nacional de la

Fig. 3.6. a, Geological exposure at the locality of Monte León with several tephra layers identified as well as the boundary between the Santa Cruz Formation and the underlying Monte León Formation. b, The geological exposure at Karaiken (= Estancia La Laurita). c, A geological exposure at the Río Chalía locality showing the position of a tephra layer. d, A geological exposure at the locality of Lago Cardiel with two tephra layers identified.

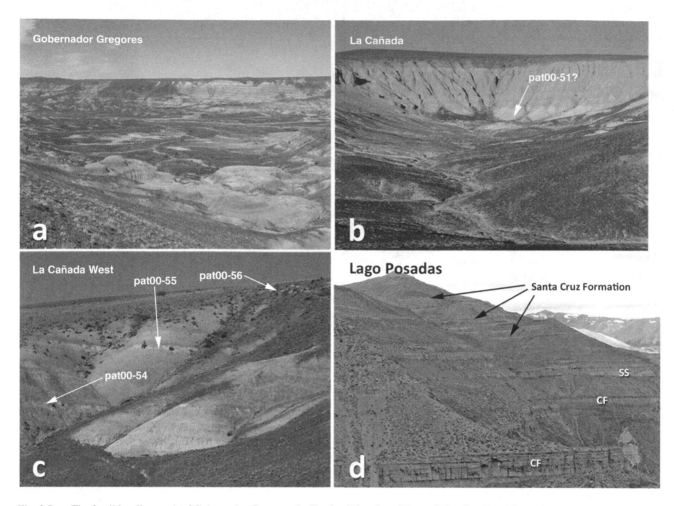

Fig. 3.7. a, The fossil locality south of Gobernador Gregores. b, The fossil locality of Estancia La Cañada with a tephra layer identified. c, The geological exposure west of the fossil locality of La Cañada from which tephra samples were collected. d, The exposures of the Santa Cruz Formation south of Lago Posadas, lying above an extensive exposure of Mesozoic rocks and the Centinela Formation (CF).

Patagonia "San Juan Bosco" (Esquel) have conducted extensive paleontological and geological research in this area and recovered a large and diverse mammalian fauna from numerous widespread and isolated separate localities, including at least four primate species, *Soriacebus ameghinorum* Fleagle *et al.*, 1987, *Soriacebus adriani* Fleagle, 1990, *Carlocebus carmenensi* Fleagle, 1990, and *Carlocebus intermedius* Fleagle, 1990, and also many distinctive marsupial and rodent taxa (e.g. Bown and Larriestra, 1990; Fleagle *et al.*, 1987, 1995; Fleagle, 1990; Bown and Fleagle, 1993; Kramarz and Bellosi, 2005). In 1990, Bown and Larriestra described these deposits as the Pinturas Formation because of a distinctive lithology from that characterizing the Santa Cruz Formation on the coast. Fleagle and colleagues (1995; also Bown and Fleagle, 1993) reported radiometric ages for two parts of the Pinturas Formation. We examined tephra samples from four sites in the Pinturas Formation:

23. Estancia El Carmen (S 47.21240°; W 70.57888°). This is a visually prominent deflation crater adjacent to the

valley of the Río Pinturas, northeast of the town of Bajo Caracoles and very near the archeological site Cueva de Las Manos. The site (Fig. 3.8a) was noted by de Barrio *et al.* (1984) and has yielded numerous fossil mammals including primates and marsupials (Fleagle, 1990; Bown and Fleagle, 1993). It is the type locality for *Carlocebus carmenensis*. Radiometric dates have been published from this site by several researchers (Bown and Larriesta, 1990; Fleagle *et al.*, 1995).

24. Cerro de los Monos (S 47.04413°, W 70.73009°). This is a small conical hill (Fig. 3.8b) adjacent to National Highway 40, south of the town of Perito Moreno and just south of Portezuelo Sumich. Numerous fossil mammals have come from this site (Anapol and Fleagle, 1988; Fleagle, 1990; Bown and Fleagle, 1993).

25. Loma de la Lluvia (S 47.02973°, W 70.71881°). This is a low white hill (Fig. 3.8c) very close to National

(a)

(b)

(c)

Fig. 3.8. Three fossil localities in the valley of the Río Pinturas and its tributaries. a, The fossiliferous deposits at the locality of Estancia El Carmen with the dated tephra layer and the fossiliferous horizons labeled. b, The locality of Cerro de los Monos. c, The locality of Loma de la Lluvia.

Highway 40, south of the town of Perito Moreno and just south of Portezuelo Sumich. Numerous fossil mammals have come from this site (Fleagle, 1990; Bown and Fleagle, 1993).

26. Portezuelo Sumich Norte (S 46.95840°, W 70.66760°). This is a rich fossil locality lying low in a prominent and extensive set of exposures of the Pinturas Formation (Fig. 3.9) draped over the underlying Bahia Laura Formation, adjacent to Highway 40 in Arroyo Feo, south of the town of Perito Moreno and just north of Portezuelo Sumich. This is the type locality for several primate species, including *Soriacebus ameghinorum* and *Carlocebus intermedius* (Fleagle *et al.*, 1987). The tephra samples were taken on the side of a hill well above the fossil site.

3.3 Tephrochronology

As described in more detail by Perkins *et al.* (Chapter 2), during fieldwork conducted in 2000, 2004, and 2007 we sampled tephra from the localities listed above for tephrochronological correlations. We identified 38 unique tephra in measured stratigraphic profiles, including 16 that linked two or more sections. These 16 "correlative" tephra link the sections at 19 of the 26 stratigraphic profiles into a space-stratigraphic network (Fig. 3.10; also Perkins *et al.*, Chapter 2: Fig. 2.2). The remaining seven stratigraphic profiles have been placed within this network using various other data, including isotopic ages, elevation, an oyster layer, or lithologic characteristics. The chronologic calibration for this network is based on $^{40}Ar/^{39}Ar$ ages or ages linked to the magnetic polarity timescale. Ages of other tephra are interpolation/extrapolation ages relative to these primary ages. Other horizons, including the tops and bottoms of sections, are estimated by interpolation or extrapolation relative to primary ages.

One volcanic deposit deserves special attention, the Toba Blanca (Tauber, 1994). This is the most widespread of the tephra identified. It was identified at eight localities extending from Monte Tigre in the south, northward 490 km to Portezuelo Sumich Norte in the northwest. In addition, biotite-bearing tephra at Estancia La Península and Lago Posadas may well be the Toba Blanca, but this has yet to be confirmed by probe analyses.

3.4 Stratigraphic context of fossils and the relationship of the Pinturas and Santa Cruz Formations

The relationship between the fossil mammals of the coastal Santa Cruz Formation and those from the Pinturas Formation in the Andean Precordillera has been the subject of discussion for over a century. In the original description of fossil mammals from the area of the Rio Pinturas, Ameghino (1906) argued that they represent an earlier stage, the "*Astrapothericulus* Stage," older than the fossil mammals from the coastal localities. In contrast, Marshall (1976), following Wood and Patterson (1959), opined: "Examination of the scanty, and fragmentary Pinturas fauna reveals, however, that there is really no good evidence for considering it to be anything but a Santacrucian local fauna." With the expanded number of mammalian fossils that have been recovered during the past three decades, more recent workers have tended to support the initial view of Ameghino. De Barrio *et al.* (1984) described numerous fossils, including xenarthrans and astrapotheres, primarily from the Arroyo Feo region, and agreed with Ameghino (1906), but emphasized the need for more fossils from the region and studies of other groups.

In a detailed study of palaeothentid marsupials, Bown and Fleagle (1993) argued that the fossils from the Pinturas Formation were more primitive than those of the coastal Santa Cruz Formation. Using a temporal scale based on the relative development of paleosols and measurement on the palaeothentid teeth, they concluded that the upper part of the Pinturas Formation was temporally equivalent to the lower part of the Santa Cruz Formation, with the most prominent break occurring in the lower part of the Santa Cruz Formation at Cerro Observatorio at a large ash fall (CO, Figs. 3.4 and 3.5). Anderson *et al.* (1995) made similar observations on the basis of morphometric analyses of the rodent *Spaniomys* from the two formations. Dumont *et al.* (2000) examined the abderitid marsupials in the Pinturas Formation,

Fig. 3.9. A geological exposure in Portezuelo Sumich Norte of sediments of the Pinturas Formation draped over the underlying Jurassic Bahia Laura Formation with the upper, middle, and lower units of the Pinturas Formation labeled, as well as the Toba Blanca and the fossiliferous horizon.

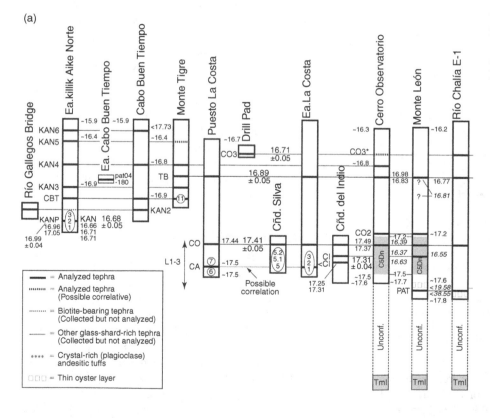

Fig. 3.10. A modified version of Fig. 2.2 from Perkins *et al.* (Chapter 2) showing the tephra correlations and age determinations for the geological profiles discussed in the text.

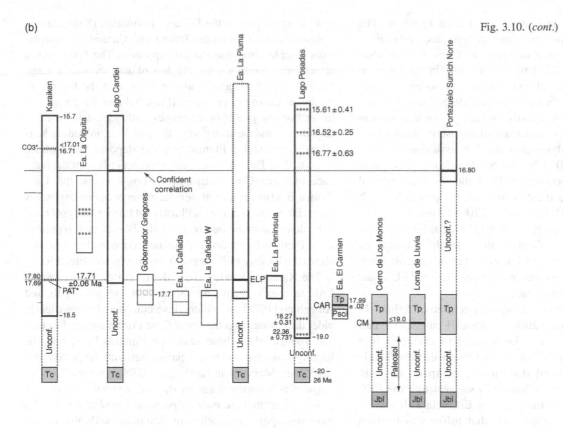

Fig. 3.10. (*cont.*)

and for two taxa found a break between those of the Pinturas Formation and those from the Santa Cruz Formation.

In a recent review of caviomorph rodents from the Pinturas Formation, Kramarz and Bellosi (2005) demonstrated that the rodents from the lower and middle sequences of the Pinturas Formation differ from those of the upper sequence. However, the rodents from the upper sequence are similar to rodents from Karaiken (Notohippidian Zone) and the lower levels at Cerro Observatorio and Monte León below the large tuff (CO, Figs. 3.4 and 3.5; see also Anderson *et al.*, 1995). The localities in the upper sequence include the widely separated localities of La Cañada, Gobernador Gregores, and Lago Cardiel as well as the locality of Los Toldos Sur in the Pinturas Valley. Similarly, Kramarz and Bond (2005) found that litopterns from the lower and middle sequences of the Pinturas Formation were more similar to Deseadan and Colhuehuapian litopterns than to those of the Santa Cruz Formation, and observed that macraucheniids are known only from the upper sequence.

The primates, as currently understood, seem to show a complete separation at the generic level between the taxa in the Pinturas Formation and those of the Santa Cruz Formation, although detailed comparisons have not been made of the many specimens attributed to *Carlocebus carmenensis*, or of the many isolated teeth from the two formations (but see Tejedor, 2002). In addition, the primates show a separation within lower and middle sequences of the Pinturas Formation.

One species, *Soriacebus ameghinorum*, is restricted to the two lowest localities (Estancia el Carmen and Portezuelo Sumich Norte) and a separate species, *S. adrianae*, is found in the localities higher in the stratigraphy (e.g. Portezuelo Sumich Sur, Cerro de los Monos, and Loma de las Ranas (= Estancia Ana Maria)). There are no identified primates from any of the sites attributed to the upper sequence of the Pinturas Formation. There is no record of *Soriacebus* from the Santa Cruz Formation, but there is a closely related taxon from Colhuehuapian levels at the Gran Barranca (Kay, 2010).

3.5 Integrating tephrochronology and fossil mammals

Having developed a series of tephrostratigraphic correlations and a tephrochronology linking many geological sections across Santa Cruz Province that contain fossiliferous horizons, we can now evaluate how the biostratigraphic estimates of the relative ages of different sites and fossil-bearing horizons compare with the results of the tephrostratigraphy and tephrochronology.

3.5.1 Coastal localities of the Santa Cruz Formation

There are numerous tephra samples from the extensive and very fossiliferous exposures of the Santa Cruz Formation along the Atlantic coastline of Santa Cruz Province. The tephra analyses support previous stratigraphic correlations of the numerous tuffaceous layers at Cerro Observatorio and Monte León that have been used in analyses of evolutionary patterns of marsupials and rodents (Bown and Fleagle, 1993; Anderson *et al.*, 1995; Fleagle *et al.*, 1995). However, tephra analyses suggest some modifications of stratigraphic correlations of exposures south of the Río Coyle. Our analyses indicate that fossiliferous levels (FL) 1, 2, 3 identified by Tauber (1997a) are the same sequence as those identified as FL 5 and also those labeled as FL 6 and 7 (Fig. 3.10; Perkins *et al.*, Chapter 2: Fig. 2.2; also Fleagle *et al.*, 2004; Kay *et al.*, 2008). Thus, the lowest seven levels recognized by Tauber (1997a) are just a single sequence repeated at (from north to south) Estancia La Costa, Cañadón Silva, and Puesto Estancia La Costa, and there is no temporal distinction between FL 3 and FL 7, just a repeat of the same sequence.

In their description of the type specimen of *Killikaike blakei* Tejedor *et al.* (2006) cautiously suggested that the tephra from Killik Aike seemed to correlate best with the CO2 tephra from Cerro Observatorio (Fleagle *et al.*, 1995) but noted that further comparisons were needed. Perkins *et al.* (Chapter 2) suggest that the tuffaceous level yielding the fossil at Estancia Killik Aike does not clearly correlate with that tuffaceous horizon or any of the tephra from Cerro Observatorio, and the radiometric date for the Estancia Killik Aike tephra suggests a much younger age, similar to that for the Toba Blanca (Fig. 3.10; Perkins *et al.*, Chapter 2: Fig. 2.2).

3.5.2 Correlations between coastal localities and inland localities

The long geological section at the Río Chalía (Sehuen) locality (approximately 150 km to the northwest of Monte León) contains numerous tephra levels that can be correlated with the levels at Cerro Observatorio and Monte León, supporting the suggestion by Bown and Fleagle (1993: 12) that "certain stratigraphic marker units may be directly correlatable with their counterparts at Monte Observación (= Cerro Observatorio) and Monte León." However, a stratigraphic section of the Río Chalía site showing the positions of the numbered localities remains to be published.

A series of exposures near the eastern side of Lago Cardiel were mapped as Santa Cruz Formation by Ramos (1982). Bown and Fleagle (1993) suggested that they resemble deposits of the Pinturas Formation in some lithologic features but differed from the main deposits around the Río Pinturas. Among the few mammals from the site are rodents that are comparable to those from the Santa Cruz Formation

and the upper part of the Pinturas Formation (Kramarz and Bellosi, 2005). Two tephra from Lago Cardiel are correlatives of tephra from the coastal exposures. The Toba Blanca is definitely present near the top of the section at Lago Cardiel (Fig. 3.6d) and is also present near the top of the section at Cerro Observatorio (Figs. 3.4 and 3.5); a second lower horizon probably correlates with tephra from other localities in the western part of the province, including Karaiken, Estancia La Pluma, and the deposits west of La Cañada. The Toba Blanca correlation accords with the presence of Santa Cruz rodents. It is possible that the Toba Blanca is also present at several other western exposures (Lago Posadas, Estancia La Pluma) but further analyses need to be done. As noted below, the Toba Blanca is also present in the Pinturas Formation high in the section stratigraphically above the locality of Portezuelo Sumich Norte (Fig. 3.9).

The Karaiken locality, near the origin of the Río Santa Cruz as it exits from Lago Argentino, yielded fossils that Ameghino (1906) and others have considered to be slightly older than those from the Santa Cruz Formation on the coast and younger than those from the Pinturas beds, and the locality was assigned to a separate horizon, the Notohippidean Zone. Marshall and colleagues (1986a) reported a wide range of radiometric dates for the site. Kramarz and Bellosi (2005) found that the rodent *Spaniomys modestus* offered a biostratigraphic correlation of Karaiken with the lowest levels at Cerro Observatorio and Monte León as well as with the upper levels of the Pinturas Formation. According to Marshall and Pascual (1977; see also Feruglio, 1944), the fossils come from the lower part of the section. We found that tephra from the lower part of the Karaiken section correlate with tephra from the lowest levels at Monte León and that the uppermost part of the section contains tephra that correlate with the CO3 tephra from the uppermost part of the section at Cerro Observatorio. The correlation of the lowest levels of the Karaiken deposits with the very lowest levels at Monte León is consistent with primitive nature of the Karaiken fossils relative to those from the coast.

The deposits south of Lago Posadas (= Lago Pueyrredón) (Fig. 3.7d) identified by Hatcher (1903) as Santacrucian in age were considered by him to be younger than the deposits from the coastal exposures. However, it is unclear whether Hatcher's assessment was based on fossils or topography. Blisniuk *et al.* (2005, 2006) reported a series of radiometric dates from this section ranging from 22.5 near the bottom to 14.4 Ma at the top. This suggests that the Lago Posadas section could encompass all of the time range of the Pinturas and Santa Cruz Formations as well as much younger times. We collected tephra and fossils from this locality in 2007; however, further analyses of fossils and tephra from this site are needed to confirm the relationships of these exposures. We did not record any tephra correlations for the section at Estancia La Olguita.

3.5.3 The Pinturas Formation and the Santa Cruz Formation

As discussed above, fossils from the Pinturas Formation, especially those from the basin of the Río Pinturas and its tributaries such as Arroyo Feo, have long been regarded by many paleontologists (e.g. Ameghino, 1906; de Barrio *et al.*, 1984; Bown and Fleagle, 1993; Kramarz and Bellosi, 2005) as more primitive, and presumably older than fossils from the coastal exposure of the Santa Cruz Formation. More specifically, Kramarz and Bellosi (2005) have argued that the fossils from the lower and middle sequences of the Pinturas Formation are older and more primitive than fossils from Monte León and Cerro Observatorio, but that fossils from the upper sequence are more similar to those from the lowest part of the coastal Santa Cruz exposures, below the large tuffaceous unit 8 at Cerro Observatorio (= CO, Figs. 3.4 and 3.5). Our limited results from the Pinturas Formation are consistent with this view. The tephra from Estancia El Carmen (Fig. 3.8a) which is attributed to the lower sequence of the Pinturas Formation and has distinctive fossils is older (17.99 Ma) than any of the tephra from the coastal Santa Cruz exposures and does not correlate with any of the younger tephra from those sections. Portezuelo Sumich Norte, the other Pinturas locality with primitive fossils, similar to those from Estancia El Carmen, lies well below the Toba Blanca in a sequence of very mature paleosols (Fig. 3.9; Bown and Larriestra, 1990: Fig. 6). Thus, it is likely much older than 16.9 Ma. Two fossil localities from the middle sequence of the Pinturas Formation, Cerro de los Monos and Loma de la Lluvia (Fig. 3.8b, c), lie above a tephra level that is likely younger than 19.2 Ma, but does not correlate with any of the younger tephra from the Santa Cruz Formation. Thus, the tephra data are consistent with the view that the fossils from the lower and middle sequences of the Pinturas Formation are older than fossils from the coastal localities. For several other Pinturas sites, including Portezuelo Sumich Sur, Ana Maria (= Loma de las Ranas), and Los Toldos Sur, there are no suitable tephra to analyze.

From the vast area of central Santa Cruz Province south of the Río Pinturas and north of the Río Chalía, de Barrio *et al.* (1984) and Bown and Fleagle (1990) identified several fossiliferous exposures that were attributed to either the Santa Cruz or Pinturas Formations, including La Cañada, Estancia La Pluma, Estancia La Península, and Gobernador Gregores. Bown and Fleagle (1993) and Rae *et al.* (1996) attributed the La Cañada locality (Fig. 3.7b) to the upper-to-middle sequence of the Pinturas Formation. Kramarz and Bellosi found that the rodents from La Cañada were typical of the Santa Cruz Formation and argued that the marsupials were also more comparable to Santa Cruz species. Tephra from the paleontological site of La Cañada have not yet been analyzed. However, one tephra from a nearby exposure, La Cañada West, is the same as that found in an exposure at Estancia La Pluma, which dates to approximately 18 Ma, among the oldest rocks in either formation. This seems to contradict the paleontological evidence. The locality south of Gobernador Gregores (Fig. 3.7a) was tentatively assigned a correlation with the lower sequence of the Pinturas Formation by Bown and Fleagle (1993), but Kramarz and Bellosi (2005) argued that the rodents support a correlation with the upper sequence of the Pinturas Formation or lowermost Santa Cruz Formation. The relatively old age (17.8 Ma) suggested for the Gobernador Gregores site appears to conflict with the evidence from the rodents. Further analyses of the tephra from that site may help resolve this contradiction.

Overall, the tephrostratigraphic and tephrachronology data from the type area of the Pinturas Formation (including Estancia El Carmen) is largely consistent with the faunal evidence in suggesting that the lower sequences of the Pinturas Formation are older than the exposures of the Santa Cruz Formation along the Atlantic Coast (Fig. 3.10). The upper sequence of the Pinturas Formation seems to be equivalent to the lowest levels of the coastal Santa Cruz Formation. For some of the sites, especially those outside of the drainage of the Río Pinturas, there is contradictory evidence from tephra and fossils, or reliable tephra data are not available.

3.6 Conclusions

In general, the results of tephrochronology and paleontology offer compatible views of the temporal relationships among the different localities and support previous suggestions based on stratigraphy or analysis of fossils. However, there are several instances in which these data disagree with previous interpretations or seem to suggest contradictory relationships. In the exposures of the Santa Cruz Formation, the tephra data indicate that several of the lowest biostratigraphic levels identified by Tauber (1994, 1997a) are not a temporal sequence, as originally supposed, but represent the same two or three levels spanning no more than 5 meters of thickness repeated in adjacent localities. Also, suggested correlations of the fossil-bearing tephra at Estancia Killik Aike Norte with the lower levels at Cerro Observatorio (= Monte Observación) are not supported by tephra analyses, and it seems likely that the tephra at Estancia Killik Aike Norte are younger than Tejedor *et al.* (2006) suggested.

Compared with the numerous correlated tephra along the Atlantic coast, there are fewer distinct tephra linking coastal localities with fossiliferous localities farther inland. Only the widespread Toba Blanca offers a common benchmark. However, the Río Chalía localities can be clearly linked with the coastal localities, as can the deposits at Lago

Cardiel. In both cases the tephra correlations accord with correlations indicated by rodents. There are virtually no common tephra linking the localities in central Santa Cruz Province with either the coastal localities to the east or the Pinturas Formation to the northwest. And there are no tephra currently analyzed for resolving the temporal position of fossil and stratigraphic sections at Estancia La Olguita or Lago Posadas. Although there are no tephra that correlate individual localities within the Pinturas Formations, radiometric dates and faunal evidence agree in placing the lower units of the Pinturas Formation below most or all of the Santa Cruz Formation at Monte León and Monte Observación (= Cerro Observatorio). The presence of the Toba Blanca high in the stratigraphic section well above the fossiliferous beds at Portezuelo Sumich Norte is consistent with paleontological analyses indicating that only the highest part of the Pinturas Formation correlates with the lower part of the Santa Cruz Formation. Hopefully, further analyses, both geological and paleontological, will enable further resolution of these issues.

ACKNOWLEDGMENTS
The fieldwork upon which this chapter is based was made possible through the efforts of many people, including Bobby Taylor, Hormiga Diaz, Mario Cozzuol, Elise Schloeder, Julio César Rúa, Amber Bayker, Siobhan Cooke, Federico Degrange, Laureano González-Ruiz, Keith Jones, Sebastian Pintos, Alfred Rosenberger, Sergio Vincon, and many generous estancia owners. Paleontological fieldwork in the Santa Cruz and Pinturas Formations over several decades relied on the efforts of others too numerous to name, but many are listed in Bown and Fleagle (1993). Ian Wallace, Andrea Baden, Rich Kay, M. S. Bargo, and S. F. Vizcaíno greatly improved the manuscript, and Luci Betti-Nash and Andrea Baden helped with many of the figures. This research was supported by funds from the US National Science Foundation, the National Geographic Society, and the LSB Leakey Foundation. Many of the locality photographs are courtesy of R. F. Kay, S. F. Vizcaíno, and M. S. Bargo.

REFERENCES
Ameghino, F. (1889). Contribución al conocimiento de los mamíferos fósiles de la República Argentina. *Actas Academia Nacional de Ciencias en Córdoba*, **6**, 1–1027.
Ameghino, F. (1893). Sobre la presencia de vertebrados de aspecto Mesozoico en la Formación Santacruceña de Patagonia austral. *Revista del Jardín Zoológico de Buenos Aires*, **1**, 75–84.
Ameghino, F. (1906). Les formations sédimentaires du crétacé supérieur et du tertaire de Patagonie avec un parallélle entre leurs faunes mammalogiques et celles de l'ancien continent. *Anales del Museo Nacional de Buenos Aires*, **15**, 1–568.
Anapol, F. and Fleagle, J. G. (1988). Fossil platyrrhine forelimb bones from the early Miocene of Argentina. *American Journal of Physical Anthropology*, **76**, 417–428.
Anderson, D. K., Damuth, J. and Bown, T. M. (1995). Rapid morphological change in Miocene marsupials and rodents associated with a volcanic catastrophe in Argentina. *Journal of Vertebrate Paleontology*, **15**, 640–649.
Blisniuk, P. M., Stern, L. A., Chamberlain, C. P., Idelman, B. and Zeitler, P. K. (2005). Climatic and ecologic changes during Miocene surface uplift in the southern Patagonian Andes. *Earth and Planetary Science Letters*, **230**, 125–142.
Blisniuk, P. M., Stern, L. A., Chamberlain, C. P. *et al.* (2006). Links between mountain uplift, climate, and surface processes in the southern Patagonian Andes. In *The Andes: Frontiers in Earth Science* Part III. Berlin, Heidelberg: Springer, pp. 429–440.
Bown, T. M. and Fleagle, J. G. (1993). Systematics, biostratigraphy, and dental evolution of the Palaeothentidae, later Oligocene to early–middle Miocene (Deseadan–Santacrucian) caenolestoid marsupials of South America. *Journal of Paleontology* **67**, 1–76.
Bown, T. M. and Larriestra, C. N. (1990). Sedimentary paleoenvironments of fossil platyrrhine localities, Miocene Pinturas Formation, Santa Cruz Province, Argentina. *Journal of Human Evolution*, **19**, 87–119.
Brinkman, P. (2003). Bartholomew James Sulivan's discovery of fossil vertebrates in the Tertiary beds of Patagonia. *Archives of Natural History*, **30**, 56–74.
Castellanos, A. (1937). Ameghino y la antigüedad del hombre sudamericano. *Asociación Cultural de Conferencias de Rosario, Ciclo de Caracter General*, **2**, 47–192.
Darwin, C. (1846). *Geological Observations on South America: Being the Third Part of the Geology of the Voyage of the Beagle, Under the Command of Capt. Fitzroy, R.N. During the Years 1832 to 1836*. London: Smith, Elder and Co.
Darwin, C. (1862). *The Voyage of the Beagle*. New York: Doubleday and Co. (First published in 1839).
de Barrio, R. E., Scillato-Yané, G. J. and Bond, M. (1984). La Formación Santa Cruz en el borde occidental del macizo del Deseado (Provincia de Santa Cruz) y su contenido paleontológico. *Actas IX Congreso Geológico Argentino*, **4**, 539–556.
Dumont, E. R., Strait, S. G. and Friscia, A. R. (2000). Abderitid marsupials from the Miocene of Patagonia: an assessment of form, function and evolution. *Journal of Paleontology*, **74**, 1161–1172.
Dunn, R. E., Kohn, M. J., Madden, R. H., Stromberg, C. E. and Carlini A. A. (2009). High precision U/Pb geochronology of Eocene-Miocene South American land mammal ages at Gran Barranca, Argentina. *Eos Transactions, American Geophysical Union*, **90**, Fall Meeting Suppl., Abs. GP236–0791.
Evernden, J. F., Savage, D. E., Curtis, G. H. and James, G. T. (1964). Potassium–argon dates and the Cenozoic

mammalian chronology of North America. *American Journal of Science*, **262**, 145–198.

Feruglio, E. (1938). Relaciones estratigráficas entre el Patagoniano y el Santacruciano en la Patagonia austral. *Revista del Museo de La Plata, Sección Geológica*, **1**, 129–159.

Feruglio, E. (1944). Estudios geológicos y glaciológicos en la región del Lago Argentino (Patagonia). *Boletín de la Academia Nacional de Ciencias, Córdoba*, **37**, 3–255.

Feruglio, E. (1949). *Descripción Geológica de la Patagonia*, 3 vols. Buenos Aires: Dirección General de Yacimientos Petrolíferos Fiscales.

Fleagle, J. G. (1990). New fossil platyrrhine from the Pinturas Formation, southern Argentina. *Journal of Human Evolution*, **19**, 61–68.

Fleagle, J. G., Powers, D. W., Conroy, G. C. and Watters, J. P. (1987). New fossil platyrrhine from Santa Cruz province, Argentina. *Folia Primatologica*, **48**, 65–77.

Fleagle, J. G., Buckley, G. A. and Schloeder, M. E. (1988). New Primate Fossils from Monte Observación (Lower Miocene) Santa Cruz Province, Argentina. *Journal of Vertebrate Paleontology*, **8**, 14A

Fleagle, J. G., Bown, T. M., Swisher, C. C. III and Buckley G. A. (1995). Age of the Pinturas and Santa Cruz Formations. *Actas del VI Congreso Argentino de Paleontología y Bioestratigrafía*, 129–135.

Fleagle, J. G., Perkins, M. E., Bown, T. M. and Tauber, A. A. (2004). Tephrostratigraphy and fossil mammals of the Santa Cruz Formation, Argentina. *Journal of Vertebrate Paleontology*, **24**, 58A.

González, L. R. (2010). Los Cingulata (Mammalia, Xenarthra) del Mioceno temprano y medio de Patagonia (edades Santacrucense y "Friasense"). Revisión sistemática y consideraciones bioestratigráficas. Unpublished Ph.D. thesis, Universidad Nacional de La Plata, Argentina.

González, L. R. and Scillato-Yané, G. J. (2007). El genero *Vetelia* Ameghino (Xenarthra, Dasypodidae). Distribución cronológica y geográfica durante el Mioceno de Patagonia, Argentina. *Ameghiniana*, **44**, 21R.

González, L. R., Scillato-Yané, G. J. and Tejedor, M. F. (2005). Diversidad de Cingulata (Mammalia, Xenarthra) en dos nuevas localidades santacrusenses (Mioceno) del oeste de la Provincia de Santa Cruz. *Ameghiniana*, **42**, 31R.

González Ruiz, L. R., Zurita, A., Fleagle, J. G. *et al.* (2011). The southernmost record of a Neuryurini Hoffstetter, 1958 (Mammalia, Xenarthra, Glyptodontidae). *Paläontologische Zeitschrift*, **85**, 155–161.

Hatcher, J. B. (1903). Narrative of the expedition. In *Reports of the Princeton University Expeditions to Patagonia, 1896–1899*. Vol. 1, *Narrative and Geography*, ed. W. B. Scott. Princeton: Princeton University Press, pp. 1–314.

Kay, R. F. (2010). A new primate from the early Miocene of Gran Barranca, Chubut Province, Argentina: paleoecological implications. In *The Paleontology of Gran Barranca: Evolution and Environmental Change through the Middle Cenozoic of Patagonia*, ed. R. H. Madden, A. A. Carlini, M. G. Guiomar Vucetich, and R. F. Kay. Cambridge: Cambridge University Press, pp. 220–239.

Kay, R. F., Vizcaíno, S. F., Bargo, M. S. *et al.* (2008). Two new fossil vertebrate localities in the Santa Cruz Formation (late Early–early Middle Miocene, Argentina), ∼51 degrees South latitude. *Journal of South American Earth Sciences*, **25**, 187–195.

Kramarz, A. G. and Bellosi, E. S. (2005). Hystricognath rodents from the Pinturas Formation, Early–Middle Miocene of Patagonia, biostratigraphic and paleoenvironmental implications. *Journal of South American Earth Sciences*, **18**, 199–212.

Kramarz, A. G. and Bond, M. (2005). Los Astrapotheriidae (Mammalia) de la Formacion Cerro Bandera, Mioceno temprano de Patagonia septentrional. *Ameghiniana*, **42**, 72R–73R.

Kramarz, A. G., Vucetich, M. G., Carlini, A. A. *et al.* (2010). A new mammal fauna at the top of the Gran Barranca sequence and its biochronological significance. In *The Paleontology of Gran Barranca: Evolution and Environmental Change through the Middle Cenozoic of Patagonia*, ed. R. H. Madden, A. A. Carlini, M. G. Vucetich and R. F. Kay. Cambridge: Cambridge University Press, pp. 264–277.

Marshall, L. G. (1976). Fossil localities for Santacrucian (early Miocene) mammals, Santa Cruz Province, southern Patagonia, Argentina. *Journal of Paleontology*, **50**, 1129–1142.

Marshall, L. G. and Pascual, R. (1977). Nuevos marsupials Caenolestidae del "Piso Notohipidense" (SW de Santa Cruz, Patagonia) de Ameghino. Sus aportaciones a la cronologia y evolución de las comunidades de mamíferos sudamericanos. *Publicaciones del Museo Municipal de Ciencias Naturales "Lorenzo Scaglia"*, **2**, 91–122.

Marshall, L. G., Pascual, R., Curtis, G. H. and Drake, R. E. (1977). South American geochronology: radiometric time scale for middle to late Tertiary mammal-bearing horizons in Patagonia. *Science*, **195**, 1325–1328.

Marshall, L. G., Butler, R. F., Drake, R. E. and Curtis, G. H. (1981). Calibration of the beginning of the age of mammals in Patagonia. *Science*, **212**, 43–45.

Marshall, L. G., Hoffstetter, R. and Pascual, R. (1983). Geochronology of the continental mammal-bearing Tertiary of South America. *Palaeovertebrata, Mémoire Extraordinaire*, 1–93.

Marshall, L. G., Drake, R. E., Curtis, G. H. *et al.* (1986a). Geochronology of the type Santacrucian (Middle Tertiary) Land Mammal Age, Patagonia, Argentina. *Journal of Geology*, **94**, 449–457.

Marshall, L. G., Cifelli, R. L., Drake, R. E. and Curtis, G. H. (1986b). Vertebrate paleontology, geology, and geochronology of the Tapera de López and Scarritt Pocket, Chubut Province, Argentina. *Journal of Paleontology*, **60**, 920–951.

Martin, H. T. (1904). A collecting trip to Patagonia, South America. *Transactions of the Kansas Academy of Science*, **19**, 101–104.

Pascual, R. and Odreman Rivas, E. O. (1971). Evolución de las comunidades de vertebrados del Terciario argentino. Los aspectos zoogeográficos y paleoclimáticos relacionados. *Ameghiniana*, **8**, 372–412.

Patterson, B. and Pascual, R. (1972). The fossil mammal fauna of South America. In *Evolution, Mammals, and Southern Continents*, eds. A. Keast, F. C. Erk and B. Glass. Albany: State University of New York Press, pp. 247–309.

Perry, J. M. G., Kay, R. F., Vizcaíno, S. F., and Bargo, M. S. (2010). Tooth root size, chewing muscle leverage, and the biology of *Homunculus patagonicus* (Primates) from the late early Miocene of Patagonia. *Ameghiniana* **47**, 355–371.

Ramos, V. A. (1982). Geología de la región del Lago Cardiel, Provincia de Santa Cruz. *Revista de la Asociación Geológica Argentina*, **37**, 23–49.

Rae, T. C., Bown, T. M. and Fleagle, J. G. (1996). New palaeothentid marsupials (Caenolestoidea) from the early Miocene of Patagonian Argentina. *American Museum Novitates*, **3164**, 1–10.

Ré, G. H., Geuna, S. E. and Vilas, J. F. (2010). Paleomagnetism and magnetostratigraphy of Sarmiento Formation (Eocene-Miocene) at Gran Barranca, Chubut, Argentina. In *The Paleontology of Gran Barranca: Evolution and Environmental Change through the Middle Cenozoic of Patagonia*, eds. R. H. Madden, A. A. Carlini, M. G. Vucetich and R. F. Kay. Cambridge: Cambridge University Press, pp. 32–45.

Riggs, E. S. (1926). Fossil hunting in Patagonia. *Natural History*, **26**, 536–544.

Simpson, G. G. (1940). Review of the mammal-bearing Tertiary of South America. *Proceedings of the American Philosophical Society*, **83**, 649–709.

Simpson, G. G. (1984). *Discoverers of the Lost World: An Account of Some of Those Who Brought Back to Life South American Mammals Long Buried in the Abyss of Time*. New Haven: Yale University Press.

Tauber, A. A. (1991). *Homunculus patagonicus* Ameghino, 1891 (Primates, Ceboidea), Mioceno temprano, de la costa Atlántica austral, Prov. de Santa Cruz, República Argentina. *Miscellanea – Academia Nacional de Ciencias (Córdoba)*, **82**, 1–32.

Tauber, A. A. (1994). Estratigrafía y vertebrados fósiles de la Formación Santa Cruz (Mioceno Inferior) en la costa atlántica entre las rías del Coyle y Río Gallegos, provincia de Santa Cruz, República Argentina. Unpublished Ph.D. thesis, Universidad Nacional de Córdoba, Argentina.

Tauber, A. A. (1996). Los representantes del genero *Protypotherium* (Mammalia, Notoungulata, Interatheriidae) del Mioceno temprano del sudeste de la Provincia de Santa Cruz, República Argentina. *Miscellanea – Academia Nacional de Ciencias (Córdoba)*, **95**, 1–29.

Tauber, A. A. (1997a). Bioestratigrafía de la Formación Santa Cruz (Mioceno Inferior) en el extremo sudeste de la Patagonia. *Ameghiniana*, **34**, 413–426.

Tauber, A. A. (1997b). Paleoecología de la Formación Santa Cruz (Mioceno inferior) en el extremo sudeste de la Patagonia. *Ameghiniana*, **34**, 517–529.

Tauber, A. A., Kay, R. F. and Luna, C. (2004). Killik Aike Norte, una localidad clásica de la Formación Santa Cruz (Mioceno temprano-medio), Patagonia, Argentina. *Ameghiniana*, **41**, 63–64R.

Tejedor, M. F. (2002). Primate canines from the early Miocene Pinturas Formation, southern Argentina. *Journal of Human Evolution*, **43**, 127–141.

Tejedor, M. F. and Rosenberger, A. L. (2008). A Neotype for *Homunculus patagonicus* Ameghino, 1891, and a new interpretation of the taxon. *PaleoAnthropology*, 68–82.

Tejedor, M. F., Tauber, A. A., Rosenberger, A. L., Swisher, C. C. III and Palacios, M. E. (2006). New primate genus from the Miocene of Argentina. *Proceedings of the National Academy of Sciences USA*, **103**, 5437–5441.

Tournouër, A. (1903). Note sur la géologie et al paléontologie de la Patagonie. *Bulletin de la Société Géologique de France*, **4**, Série 3, 463–473.

Vizcaíno, S. F., Bargo, M. S., Kay, R. F. *et al.* (2010). A baseline paleoecological study for the Santa Cruz Formation (late-early miocene) at the Atlantic Coast of Patagonia, Argentina. *Palaeogeography, Palaeoclimatology, Palaeoecology*, **292**, 507–519.

Wood, A. E. and Patterson, B. (1959). The rodents of the Deseadan Oligocene of Patagonia and the beginnings of the South American rodent evolution. *Bulletin of the Museum of Comparative Zoology*, **120**, 281–428.

4 Sedimentology and paleoenvironment of the Santa Cruz Formation

Sergio D. Matheos and M. Sol Raigemborn

Abstract

The Santa Cruz Formation in the coastal area of Santa Cruz Province, Austral Patagonia, Argentina, is a sedimentary sequence consisting of mainly fine and tuffaceous sediments in the lower section (Estancia La Costa Member) and coarser and siliciclastic sediments towards the top (Estancia La Angelina Member). This formation rests upon and is transitional with marine sediments of the Monte León Formation. The lower member bears a rich mammalian association of Santacrucian Age (late Early Miocene). Compositional and paleoenvironmental analysis of the lower member allows the differentiation of three sections: lower, middle, and upper. The lower section crops out in the northern part of our study area and comprises a basal part with bioclastic sandstones and bioturbated heterolithic facies of continental-marine transitional environment. This is gradually replaced by fine primary and reworked tuffs and massive silty sandstones from volcanic source, with immature paleosols deposited in a relatively low-energy fluvial system with vegetated floodplains with a high sedimentation rate. The middle section shows an evident increase in the coarse facies and paleosols, and a decrease in pyroclastic materials, which suggests a higher energy for the system compared with the lower section, with a lower sedimentation rate. The upper section is compositionally even coarser and is exclusively siliciclastic in origin, and evinces even more energetic conditions of the fluvial system but with fluctuating flow regimes. During the deposition of the Estancia La Costa Member, the climate changed from warm with somewhat cooler and/or drier intervals (lower and middle sections) to cool and dry conditions towards the top.

Resumen

La Formación Santa Cruz en la zona costera de la provincia de Santa Cruz, Patagonia Austral, Argentina, es una unidad sedimentaria compuesta principalmente por material volcaniclástico fino en su sección inferior (Miembro Estancia La Costa) con mayor participación de material silicoclástico hacia la sección superior (Miembro Estancia La Angelina). Dicha formación cubre transicionalmente a

Early Miocene Paleobiology in Patagonia: High-Latitude Paleocommunities of the Santa Cruz Formation, ed. Sergio F. Vizcaíno, Richard F. Kay and M. Susana Bargo. Published by Cambridge University Press. © Cambridge University Press 2012.

los sedimentos marinos de la Formación Monte León. Su miembro inferior posee una riquísima fauna de mamíferos de Edad Santacrucense (late Early Miocene). Sobre la base del análisis composicional y paleoambiental desarrollado exclusivamente en el miembro inferior, se diferenciaron tres secciones: inferior, media y superior. La sección inferior, aflorante en la parte norte del área de estudio, posee en su base sedimentos bioclásticos y heterolíticos bioturbados depositados en un ambiente transicional. Éstos son gradualmente reemplazados por tobas primarias y retrabajadas y areniscas limosas masivas con procedencia de arco volcánico, con frecuentes niveles de paleosuelos inmaduros que representa un sistema fluvial de energía relativamente baja con llanuras de inundación vegetadas, acumulado con una tasa de sedimentación relativamente alta. La sección media muestra un incremento en las facies de mayor granulometría, una disminución en la participación de material piroclástico y una mayor participación de niveles edafizados, indicando un ambiente con energía relativamente mayor que la sección inferior, acumulado bajo una tasa de sedimentación relativamente baja. Por último, la sección superior resulta exclusivamente de origen silicoclástico y de facies gruesas, lo cual evidencia condiciones aún de mayor energía del sistema fluvial con un régimen de flujo fluctuante. Durante la depositación del Miembro Estancia La Costa el clima osciló desde cálido, con algunos intervalos fríos y/o secos (sección media e inferior) a condiciones más frías y secas hacia la parte superior.

4.1 Introduction

Outcrops of the continental Santa Cruz Formation in the coastal area of Santa Cruz Province (late Early Miocene), Austral Patagonia, Argentina, present a record of fluvial deposits which are mainly composed of volcaniclastic material, and are very rich in vertebrate fossils (Tauber, 1994, 1997; Kay *et al.*, 2008, among others). The Santa Cruz Formation is the most widespread and most richly fossiliferous of all Patagonian non-marine Tertiary formations of southern Argentina (Tauber, 1994; Tejedor *et al.*, 2006; Vizcaíno *et al.*, 2006, 2010; Tambussi, 2011). For this reason, the unit has been intensively studied since the 1890s from a paleontological and biostratigraphical point of view (see references in Vizcaíno *et al.*, Chapter 1), but the sedimentological aspects of the formation have been only briefly described (Bown and Fleagle, 1993; Tauber, 1994; Fleagle *et al.*, 1995).

The deposition of the Santa Cruz Formation in the Austral geologic basin can be understood in the context of regional tectonic, geographical, and climatic changes (Blisniuk *et al.*, 2005, 2006; Barreda and Palazzesi, 2007; Tassone *et al.*, 2008; Ramos and Ghiglione, 2008; Madden *et al.*, 2010; Quattrochio *et al.*, 2011) during part of the time interval (17–15 Ma) that experienced a global climatic event known as the Mid-Miocene Climatic Optimum (Zachos *et al.*, 2001).

This sedimentological study of the Santa Cruz Formation has the following objectives: (1) to characterize compositionally the Estancia La Costa Member and to establish the source area of its sediments; (2) to present a description of the facies and facies association within the formation; (3) to reconstruct the paleoenvironment of deposition; (4) to infer the paleoclimatic conditions that prevailed in Austral Patagonia during the late Early Miocene; and (5) to consider the regional effects of volcanism on the sedimentary environment.

4.2   Geological setting and stratigraphy

The Santa Cruz Formation is part of the fill of the cratonic area of the foreland Austral or Magellan Basin in southernmost South America (Russo *et al.*, 1980; Olivero and Malumián, 2002, 2008; Menichetti *et al.*, 2008). The basin follows a preferential NNW–SSE axis and is limited by the mountain range of the Andes towards the west, and by the Alto de Río Chico to the east (Biddle *et al.*, 1986; Corbella, 2002; Peroni *et al.*, 2002). The latter is the southern extension of the Deseado Massif that served as the basement of the Cretaceous sedimentation, separating the Austral–Magellan Basin from the Malvinas Basin (Rossello *et al.*, 2008). Our study area comprises the southeastern part of this basin.

The geological history of the Austral Basin begins in the Triassic with the initiation of rifting between South America and Africa (Biddle *et al.*, 1986). The Bahía Laura Group consists of volcaniclastic rocks deposited in grabens and half-grabens during the Jurassic (Uliana *et al.*, 1985). During the Late Jurassic to Early Cretaceous, the margin of the basin was closed by the uplift of the western arc (Arbe, 1987), and the deposition of the Springhill Formation occurred (Corbella, 2002). Subsequently, from the Early Cretaceous into the Miocene, the basin experienced repeated transgressions and regressions of the sea. Towards the east the marine Palermo Aike Formation (Campanian–Maastrichtian) was deposited, while to the north marine and continental deposition alternated: continental San Julián Formation (Late Eocene–Early Oligocene), marine Monte León Formation (Late Oligocene–Early Miocene),

continental Santa Cruz Formation (late Early Miocene), and finally, the Pliocene marine terraces and continental Rodados Patagónicos (Late Pliocene–Quaternary) (Uliana *et al.*, 1985; Corbella, 2002; Peroni *et al.*, 2002). The continental deposits of the Santa Cruz Formation occurred during an important phase of deformation and uplift of the cordillera in the west. Its sedimentation clearly indicates that the regression of the marine deposits of the underlying Late Oligocene to Early Miocene Centinela Formation (= 25 de Mayo Formation; Cuitiño and Scasso, 2010) in the west was forced by the cordillera uplift (Ramos and Ghiglione, 2008). The deposition of the Santa Cruz Formation occurred in a foreland basin where sediment supply exceeded the accommodation space. As a result, a prograding sequence expanded to the Atlantic coast (Nullo and Combina, 2002).

4.2.1   Distribution of Santa Cruz Formation and study area

The Santa Cruz Formation is developed over much of the south of Patagonian Argentina both in surface exposures and in drill logs, from the eastern area of the South Patagonian Cordillera (Cordillera Patagónica Austral) (Bande *et al.*, 2008; Cuitiño and Scasso, 2010) southward from Lago Buenos Aires to the Río Turbio area, and eastward from extra-Andean Patagonia to the Atlantic Ocean between Golfo San Jorge and northern Tierra del Fuego (Marshall, 1976; Bown and Fleagle, 1993; Tauber, 1997; Malumián, 1999). These continental deposits are thickest along the northern foothills of the Patagonian Cordillera, where they reach thicknesses up to 1500 m, decreasing to 225 m along the Atlantic coast (Riggi, 1979; Riccardi and Rolleri, 1980; Tauber, 1997).

The outcrops of the Santa Cruz Formation in the study area are arrayed parallel to the Atlantic coast of Santa Cruz Province between Parque Nacional Monte León and Río Gallegos (see Vizcaíno *et al.*, Chapter 1, Figs. 1.1 and 1.2). The following localities were studied: just to the north of the Río Coyle, Rincón del Buque (= Media Luna; S 50° 39′ 41″, W 69° 06′ 52″) and Cañadón del Cerro Redondo (S 50° 51′ 25″, W 69° 08′ 03″) (Fig. 4.1a); south of the Río Coyle from north to south, Punta Sur–Cañadón del Indio (S 50° 58′ 59″, W 69° 09′ 56″), Campo Barranca (S 51° 00′ 36″, W 69° 09′ 15″), Anfiteatro (S 51° 03′ 12″, W 69° 08′ 20″, Estancia La Costa (S 51° 04′ 42.6″, W 69° 08′ 02.2″), Cañadón Silva (S 51° 11′ 03″, W 69° 05′ 55″), Puesto Estancia La Costa (S 51° 11′ 47″; W 69° 05′ 34″) (Fig. 4.1b–e) and Monte Tigre–Cañadón las Totoras (S 51° 20′ 32″, W 69° 02′ 08″).

This study area has an active sea cliff with numerous slumps, slips, and scouring, but which subsumes most of the thickness of the Santa Cruz Formation with excellent surface exposures. In this area the unit shows dips of 1–3° southeast (Tauber, 1994, 1997). Lithologically, the exposures of the coastal Santa Cruz Formation (225 m thick on average) are a succession of superimposed mudstones, fine

Fig. 4.1. a, Photograph showing the lower and middle section of the Estancia La Costa Member at Punta Sur of Rincón del Buque locality. White arrows indicate the first conspicuous tuff level of the unit, close to the last oyster bank that outcrops (black arrow). b, View of the lower section of the Estancia La Costa Member at Punta Sur–Cañadón del Indio locality. Arrows indicate the first three more conspicuous tuff levels of the unit. The person at the lower right is working at the beds that contain the fossil logs (detail in panel c). The dashed line marks the contact between Santa Cruz Formation and the Rodados Patagónicos. d, Outcrops of the lower and middle section of the Estancia La Costa Member at Campo Barranca. Arrows indicate the same tuffaceous levels shown in panel b. e, General view of the middle and upper section of the Estancia La Costa Member and the lower levels of the Estancia La Angelina Member at Puesto Estancia La Costa locality. The contact between the two members is indicated by the dashed line.

to medium sandstones of volcaniclastic origin, and tuffs with light colors, containing immature paleosols laid down on a coastal plain incised by sand bodies, with some pebbles, representing river channels (Bown and Fleagle, 1993; Tauber, 1994, 1997). According to Feruglio (1938), the uppermost level with *Crassostred orbignyi* represents the limit between the Santa Cruz Formation and the

underlying "*Patagoniano*" (currently Monte León Formation; see Chapter 5). The same author mentions that this oyster level is located 23 m above current sea level at Punta Norte at the mouth of the Río Coyle and buried south of Río Coyle. In contrast, Griffin and Parras (Chapter 5) place the basal limit of the Santa Cruz Formation below the uppermost bed of *Crassostrea orbignyi*. The oyster bed is closer

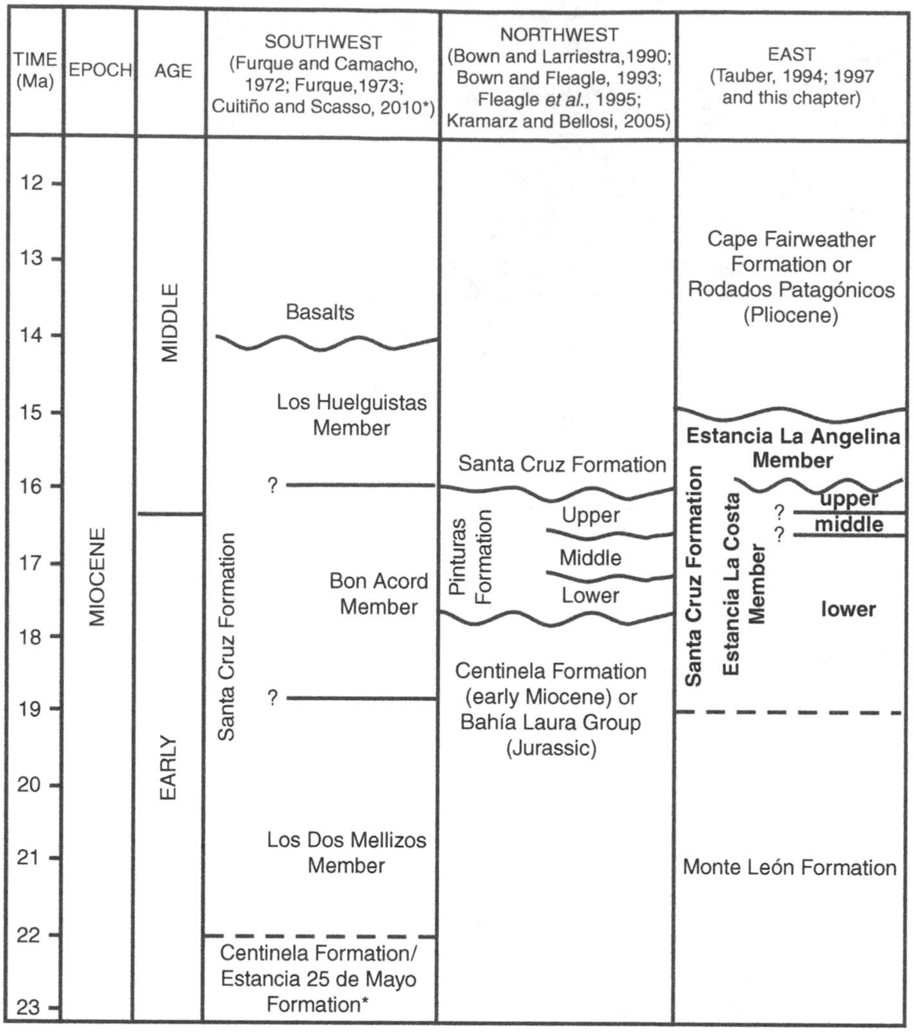

Fig. 4.2. Proposed lithostratigraphic correlation of units exposed in the southwest, northwest, and east part of the Santa Cruz Province. The different ages of the stratigraphic units are according to the authors mentioned in the chart, to Blisniuk *et al.* (2006) and to Parras *et al.* (2008).

to sea level near Rincón del Buque and Cerro Redondo, and buried south of Río Coyle (Fig. 4.1a). The contact between the units in this area is considered transitional and conformable (Tauber, 1994; Matheos *et al.*, 2008). The upper limit of the Santa Cruz Formation is marked by the marine sediments of the Cape Fairweather Formation and/or by the Rodados Patagónicos (Figs. 4.1b and 4.2).

4.2.2 Lithostratigraphy

The Santa Cruz Formation was formally named near Lago Argentino, located approximately 300 km west of the coastal area, by Furque and Camacho (1972) and Furque (1973), and recently by Cuitiño and Scasso (2010). Furque and Camacho (1972) and Furque (1973) distinguished three members (Fig. 4.2). They are, from oldest to youngest, the Los Dos Mellizos Member (250 m thickness of gray, yellowish, and greenish-gray clays), the Bon Acord Member

(150 m thickness of conglomeratic sandstones and blue tuffs), and the Los Huelguistas Member (95 m thickness of sandstones and conglomerates).

Tauber (1994, 1997) recognized two members in the coastal outcrops of the Santa Cruz Formation south of Río Coyle: a lower very fossiliferous unit, the Estancia La Costa Member, with a predominance of pyroclastic deposits with claystones and mudstones, and an upper fossil-poor unit, the Estancia La Angelina Member, chiefly composed of claystones, mudstones, and sandstones (Figs. 4.2, 4.3, and 4.4). The two members are separated by a discontinuous surface evident on the sea cliffs from the north of Estancia La Costa southward to Cabo Buen Tiempo. Tauber (1994) and Matheos *et al.* (2008) state that in the coastal area, the base of the Santa Cruz Formation is transitional with the marine Monte León Formation (Griffin and Parras, Chapter 5); meanwhile the top of the unit is unconformably overlaid by marine deposits of the Cape Fairweather Formation or the Rodados Patagónicos (Fig. 4.2).

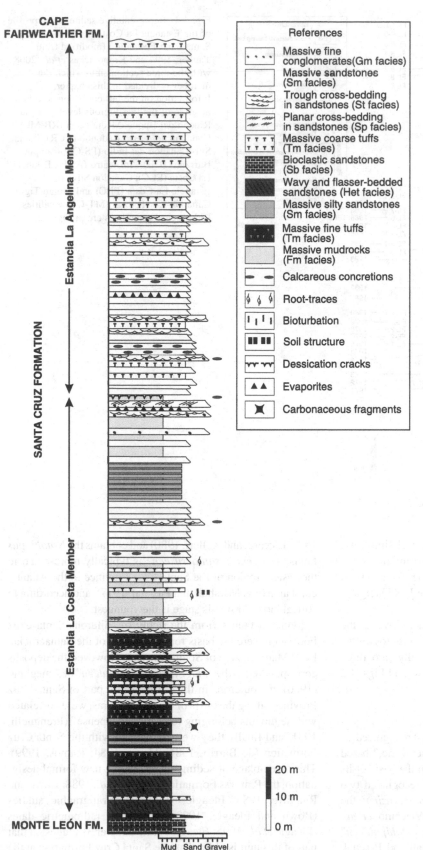

Fig. 4.3. Representative sedimentary profile of coastal Santa Cruz Formation, modified from Tauber (1994).

References

Massive fine conglomerates (Gm facies)

Massive sandstones (Sm facies)

Trough cross-bedding in sandstones (St facies)

Planar cross-bedding in sandstones (Sp facies)

Massive coarse tuffs (Tm facies)

Bioclastic sandstones (Sb facies)

Wavy and flasser-bedded sandstones (Het facies)

Massive silty sandstones (Sm facies)

Massive fine tuffs (Tm facies)

Massive mudrocks (Fm facies)

Calcareous concretions

Root-traces

Bioturbation

Soil structure

Dessication cracks

Evaporites

Carbonaceous fragments

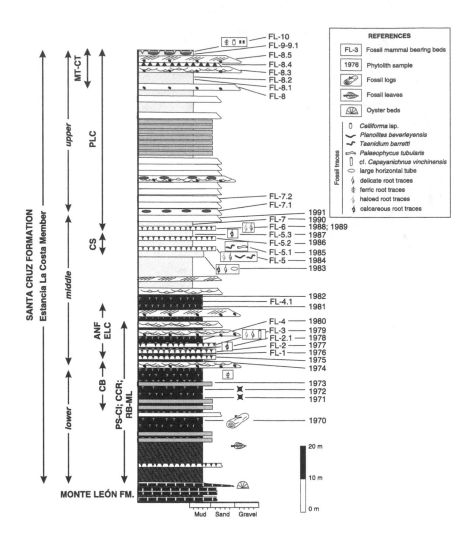

Fig. 4.4. Representative sedimentary profile of the Estancia La Costa Member of the Santa Cruz Formation (modified from Tauber, 1994 and Krapovickas *et al.*, 2008) with the three sections into which the member is divided in this chapter. Information on the left corresponds to the position of the fossiliferous levels (FL) at Rincón del Buque–Media Luna (RB-ML), Cañadón del Cerro Redondo (CCR), Punta Sur–Cañadón del Indio (PS-CI), Campo Barranca (CB), Anfiteatro (ANF), Estancia La Costa (ELC), Cañadón Silva (CS), Puesto Estancia La Costa (PLC) and Monte Tigre–Cañadón Las Totoras (MT-CT) localities. Lithological references are in Fig. 4.3.

Tauber (1994), on the basis of the lithological similarity, suggested that the Estancia La Costa Member might correspond to the Bon Acord Member, and the Estancia La Angelina Member could be equivalent to the Los Huelguistas Member of Furque (1973) (Fig. 4.2).

On the basis of the sedimentological characteristics of the Estancia La Costa Member and the location of its fossiliferous levels, we divided the member informally into three sections: lower, middle, and upper (see below, and Figs. 4.2 and 4.4).

4.2.3 Biostratigraphy–biochronology

The continental Santa Cruz Formation was first recognized by Ameghino (1902) who named it "Piso Santacruzeño," based on its fossil vertebrate content. Later, also on the basis of its fossil vertebrate content, Ameghino (1906) recognized two continental *étages* within the *formation santacruzien*: the *notohippidéen* and the *santacruzéen* (see Vizcaíno *et al.*, Chapter 1, and references therein). The "*étage notohippidéen*" is typically represented in Karaiken (Marshall and Pascual,

1977; Riccardi and Rolleri, 1980) and contains the *Notohippus* fauna. The "*étage santacruzéen*" is typically represented in the eastern region of the Santa Cruz Province in the Atlantic coast area (e.g. Marshall and Pascual, 1977), and according to Ameghino (1906), this stage is the youngest.

Mammal faunas from the richly fossiliferous Santa Cruz Formation were the basis for recognition of the Santacrucian Land Mammal Age of South America. However, the deposits corresponding to the "*étage astrapothericuléen*" of Ameghino (1906) that outcrops in the northwestern part of Santa Cruz Province, along the cliffs of the Río Pinturas, were correlated with sediments belonging to "Colhuehuapense" (Frenguelli, 1931) and finally they were associated with the Santa Cruz Formation (de Barrio, 1984; Marín, 1984; Ramos, 1999). This assemblage of sediments received a new formal designation, the Pinturas Formation (Bown *et al.*, 1988; Bown and Ratcliffe, 1988; Fleagle, 1990). Biostratigraphic studies (Bown and Fleagle, 1993) and recent radiometric dates (Fleagle *et al.*, 1995; Fleagle *et al.*, Chapter 3) suggested that part of this unit is older than the Santa Cruz Formation at the typical coastal localities (Kramarz and Bellosi, 2005). The last authors mentioned that the lower and middle sequences of the Pinturas Formation are older than the base of the Santa Cruz Formation at Monte León and, consequently, older than the Santacrucian age of Simpson (1940). Soria (2001) suggested that all the Miocene fauna from the Pinturas Formation must be regarded as Early Santacrucian in age, whereas those from the Santa Cruz Formation are late Santacrucian. A new date for the Pinturense (Upper Fossil Zone of the Colhue-Huapi Member of the Sarmiento Formation), which crops out at Gran Barranca, south of Chubut province, is 18.7 to 19.7 Ma (Ré *et al.*, 2010).

Tauber (1997) described two biozones for the coastal Santa Cruz Formation, the *Protypotherium attenuatum* biozone (the younger) and the *Protypotherium australe* biozone (the older).

4.2.4 Geochronology and chronostratigraphy

The age of the Santa Cruz Formation has been a subject of debate for more than a century. Marshall *et al.* (1986) concluded that the Santa Cruz Formation ranges from 17.6 to 16.1 Ma (K–Ar dating from Monte León and Rincón del Buque localities). Later, Fleagle *et al.* (1995) dated levels from Monte Observación (i.e. Cerro Observatorio at Cañadón de Las Vacas) and Monte León localities, with Ar–Ar, and obtained ages of 16.5 Ma for the Santa Cruz Formation and 19.33 Ma for the top of the Monte León Formation (Fig. 4.2). At the same time, Fleagle *et al.* (1995) dated samples of the Pinturas Formation using Ar–Ar dating, and obtained ages of 17.76 Ma for the lower sequence of the unit, and 16.5 Ma for the top and middle sequence of the unit (Fig. 4.2). These data and biochronologic evidence (see above) indicate that the Pinturas Formation lies below the levels of the Santa Cruz Formation at Cerro Observatorio and that the Monte León Formation is older than the Pinturas Formation. Fleagle *et al.* (1995) indicated that the Pinturas Formation is intermediate in age between the Monte León and Santa Cruz formations. They proposed that the age of the Santa Cruz Formation at Monte Observación (i.e. Cerro Observatorio) and Monte León could be correlated with Chron 5Cn. The long section of reversed magnetic polarity above the dated levels would correlate with Chron C5Br which ranges from 15.2 to 16.0 Ma. In this volume, Perkins *et al.* (Chapter 2) provide integrated results of the tephra correlations and radiometric ages indicating that the Santa Cruz Formation spans the interval ~18 to 16 Ma in the Atlantic coastal plain and ~19 to 14 Ma in the Andean foothills, with a chronologic overlap between the Pinturas Formation and the lower part of the Santa Cruz Formation.

The radiometric dates mentioned above permit the assignment of the bulk of the coastal Santacrucian faunas to the late Early Miocene. At the same time, the beds below the Santa Cruz Formation, the Monte León Formation, are consistent with an Early Miocene age of the mollusc assemblage from that unit (del Río, 2004). In addition to this, new ages obtained by Parras *et al.* (2008) from the Centinela and San Julián Formations, suggest that the age of the Monte León Formation along the coast should lie between 24.12 and 19.33 Ma (Fig. 4.2).

However, more recently Ar–Ar ages obtained by Blisniuk *et al.* (2005) for the Santa Cruz Formation, at the south of Lago Posadas, yielded 22.36 Ma for the base of the unit, and 14.24 Ma for the upper section (Fig. 4.2). Ramos and Ghiglione (2008) interpreted these data to mean that the continental foreland basin began at about 23 Ma along the foothills, and that progradation had reached the Atlantic coast by about 19 Ma.

4.3 Methodology and techniques

4.3.1 Composition

In order to define the composition of the Estancia La Costa Member of the Santa Cruz Formation, to establish the source area and to evaluate its relationship with the paleoclimatic context, mudstones, sandstones, and tuff samples of different localities were collected for diffractometric analyses and petrographic studies.

A total of 24 samples, including siliciclastic, reworked, and primary pyroclastic material from Campo Barranca, Anfiteatro and Estancia La Costa localities, was analyzed using X-ray diffraction (XRD). Soft grinding with a rubber mortar was used to disaggregate the more indurated samples, followed by repeated washes in distilled water until deflocculating occurred. The <4 μm fraction was separated by gravity settling in suspension, and oriented mounts were prepared on glass slides. Clay mineralogy was determined from diffraction patterns obtained by using samples that were air-dried, ethylene glycol solvated, and heated to 550 °C for 2 hours (Brindley and Brown, 1980). Diffractograms were run on an X PAN analytical model X'Pert PRO diffractometer (Centro de Investigaciones Geológicas, Universidad Nacional de La Plata, Argentina), using Cu/Ni radiation and generation settings of 40 kV and 40 mA. Routine air-dried mounts were run between 2 and 32° 2θ at scan speed of 2° 2θ per min. Ethylene glycol solvated and heated samples were run from 2 to 27° 2θ and 3 to 15° 2θ, respectively, at a scan speed of 2° 2θ per min. Semiquantitative estimates of the clay mineral relative concentrations were based on the peak area method (Biscaye, 1965) on glycolated samples (17 Å for smectite, 10 Å for illite, and 7 Å for chlorite and kaolinite). The relative percentages of each clay mineral were determined with the application of empiric factors (Moore and Reynolds, 1989). Semiquantification was considered sufficient to define clay mineral composition because the presence/absence of minerals or dominant/subordinate relationships clearly allowed significant groups to be established.

Rocks for petrographic studies included a total of 42 samples with 14 thin sections of very fine to coarse-grained

Table 4.1. *Clay mineral composition from the Estancia La Costa Member in the size fraction <4 μm*

Locality	Sample	Relative percentage					Clay mineral assemblage	Facies	FA
		Sm	K	I	Chl	I/S			
Campo Barranca	CB-2	100	0	0	0	0	S1	Tm	FA-4
	CB-1	70	0	0	30	0	S3	Gm	FA-1
	CB-O	100	0	0	0	0	S1	Tm	FA-4
	CB-T-2	100	0	0	0	0	S1	Tm	
	CB-T-3–4	100	0	0	0	0	S1	Tm	
	CB-22	70	0	20	10	0	S3	St	FA-1
	CB-21	80	0	15	5	0	S3	Fm	FA-3
	CB-19	65	0	25	10	0	S3	Fm	
Anfiteatro	ANF-3	100	0	0	0	0	S1	Tm	FA-4
	ANF-2	100	0	0	0	0	S1	Tm	
	ANF-1	100	0	0	0	0	S1	Tm	
	MR-5	95	0	0	0	5	S2	Tm	
	MR-6	95	0	0	0	5	S2	Tm	
	MR-7	100	0	0	0	0	S1	Tm	
	MR-8	95	0	0	0	5	S2	Tm	
	MR-9	100	0	0	0	0	S1	Tm	
Estancia La Costa	MR-1	90	0	5	0	5	S2	Tm	FA-4
	MR-2	90	0	5	0	5	S2	Tm	
	MR-3	100	0	0	0	0	S1	Tm	
	MR-4	100	0	0	0	0	S1	Tm	

Notes: **Sm**: smectite; **K**: kaolinite; **I**: illite; **Chl**: chlorite; **I/S**: illite/smectite mixed-layer. **S1**: smectitic assemblage; **S2**: smectite/illite and mixed-layer clays assemblage; **S3**: smectite–illite/chlorite assemblage. **Gm**: massive clast or matrix-supported fine conglomerates; **Tm**: massive coarse and fine vitric tuffs; **St**: trough cross-bedded lithic sandstones; **Fm**: massive very fine sandstones, tuff or reworked tuff, and mudrocks. **FA-1**: fluvial channel facies association; **FA-3**: distal floodplain facies association; **FA-4**: pyroclastic-fall deposits facies association.

sandstone and 28 thin sections of reworked and primary tuffs. These samples were analyzed with a Nikon Eclipse E-200 petrographic microscope at the Centro de Investigationes Geológicas (La Plata, Argentina. Criteria used to distinguish lithic types, matrix types, and other components are those used by Dickinson (1970). A qualitative approach was made to characterize the main constituent of the rocks, matrix, and cement. Both roundness and sphericity of the clasts were estimated by visual comparison with the chart of Pettijohn *et al.* (1987).

4.3.2 Facies analysis and paleoenvironmental interpretation

With the aim of understanding the different sedimentary paleoenvironments and sedimentary conditions where the Estancia La Costa Member was deposited, facies and facies associations following Miall (1996) and Bridge (2003) with modification for pyroclastic deposits (Smith, 1987) were observed. The shapes of the sandy bodies were classified following Friend *et al.* (1979) as tabular bodies

(width/thickness ratio > 15), and lenticular or lens-shaped bodies (ratio < 15).

4.4 Results

4.4.1 Composition

Clay mineralogy Clay minerals identified in the early Estancia La Costa Member of the Santa Cruz Formation include smectite (Sm), illite (I), chlorite (Chl), and mixed-layer clays containing illite/smectite (I/S). The results obtained from the XRD analysis in the <4 μm fraction are listed in Table 4.1. The non-clay minerals identified in this fraction, in decreasing order of abundance, are quartz and feldspars (plagioclases > K-feldspar), and small amounts of amorphous silica (opal), zeolites (clinoptilolite), carbonates, and gypsum. In all the stratigraphic levels studied, the smectite is the most frequent clay mineral, and is present in almost all the analyzed samples, while illite, chlorite, and mixed-layer clays are less frequent.

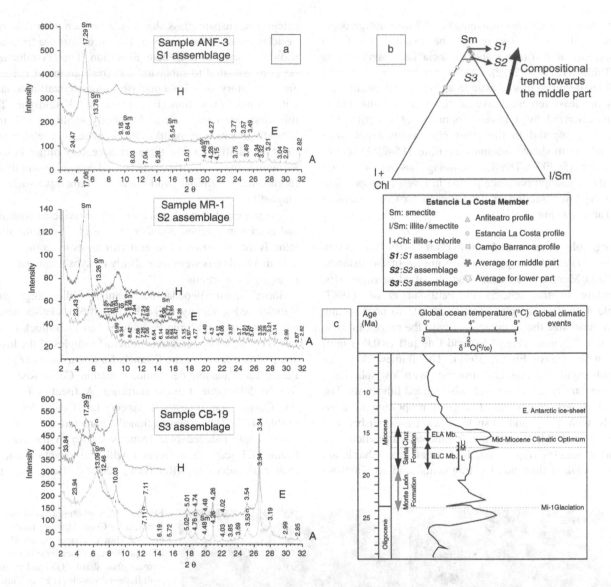

Fig. 4.5. a, Representative X-ray diffraction patterns (A: air-dried; E: ethylene-glycol solvated; H: heated to 550 °C) of each clay-mineral assemblage recognized in the <4 μm fraction in the lower and middle section of the Estancia La Costa Member. Sm: smectite; C: chlorite; I: illite; IS: illite/smectite mixed layer. b, Triangular sketch showing the distribution of clay minerals and the clay-mineral proportion in the samples analyzed. c, Neogene global ocean temperature curve after Zachos *et al.* (2001) showing the possibly geochronological position (see Fig. 4.2) of the Monte León and Santa Cruz Formations, the two members of the former unit and the three sections of the Estancia La Costa Member, in relation to global climatic events. ELC Mb: Estancia La Costa Member; ELA Mb: Estancia La Angelina Member; L: lower section; M: middle section; U: upper section.

On the basis of the presence, type, and relative amount of the above-mentioned clay minerals in the studied samples, three clay mineral assemblages have been defined (Table 4.1). The S1 assemblage is composed exclusively of Sm; the S2 assemblage is composed of Sm>I, I/S; and the S3 assemblage is composed of Sm>I, Chl. Stratigraphic distribution and proportion of clay minerals and the X-ray diffraction patterns of the <4 μm fraction representative of each assemblage are shown in Fig. 4.5a.

1. Smectitic assemblage (S1): this assemblage is characterized by a total dominance of smectite (100%) and is best represented in the lower and middle section of Estancia La Costa Member at Campo Barranca, Anfiteatro, and Estancia La Costa localities (Table 4.1, Fig. 4.5a–c).
2. Smectite/illite and mixed clays assemblage (S2) the high proportion of smectite (90–95%) associated with illite (5%) and mixed-layer clays of illite/smectite (5%)

characterizes this assemblage. The S2 assemblage dominates the lower section of the Estancia La Costa Member at Anfiteatro and Estancia La Costa sections (Table 4.1, Fig. 4.5a–c).

3. Smectite/illite/chlorite assemblage (S3): this assemblage is the least representative of the studied units and is characterized by a lower proportion of smectite (65–80%) compared to the other clay assemblages, with small to moderate amounts of illite (15–25%) and/or chlorite (5–30%). The S3 assemblage has been identified only at Campo Barranca profile in levels corresponding to the lower section of the Estancia La Costa Member (Table 4.1, Fig. 4.5a–c).

Petrography of reworked and primary tuffs Thin-section analysis revealed that pyroclastic samples of the Estancia La Costa Member are similar and classified as vitric tuffs, according to the scheme of Pettijohn *et al.* (1987) (Table 4.2). The samples show moderate to poor sorting and a mean size that is coincident with the range of coarse tuff (0.063–2 mm) (Fig. 4.6a) and fine tuff (<0.063 mm) (Fig. 4.6b) (*sensu* Fisher, 1961). The framework grains are sub-angular to angular vitroclasts with low sphericity, sub-angular crystaloclasts and sub-rounded lithoclasts. The vitroclasts are composed of a great proportion of glass shards with platy and cuspate types (*sensu* Fisher and Schmincke, 1984) and pumice fragments with flattened or rounded vesicles (Fig. 4.6a). In general, glass shards are more abundant than pumice fragments, and within these

categories, cuspate glass shards and pumice with flattened vesicles are the most common. The vitroclasts are fresh and show few signs of diagenetic alteration. The crystaloclasts occur as euhedral to subhedral and fresh grains of twinned and oscillatory zoned plagioclases, monocrystalline quartz with straight extinction (Fig. 4.6b), and K-feldspar. The lithoclasts are composed of frequently unaltered volcanic lithic fragments with pilotaxitic, trachytic, and felsitic textures. In addition to this, there are accessory minerals such as biotite, pyroxenes (hyperstene and augite), amphiboles (hornblende), zircon, and opaque minerals such as magnetite.

The principal cements are clay and Fe-oxide as coatings, and calcite in patches. Another feature of diagenetic alteration is the distortion of several vitroclasts by compaction. The analyzed tuffs were altered only by synsedimentary and pedogenic conditions.

Some micropaleoedaphic features like aggregates (blocky peds), variety of voids (Fig. 4.6b), clay-coatings, and concentric and typic Fe-nodules (*sensu* Bullock *et al.*, 1985) can be recognized in certain tuff samples of the lower and middle section of the Estancia La Costa Member at the Punta Sur–Cañadón del Indio, Cañadón Cerro Redondo, Rincón del Buque, Campo Barranca, Anfiteatro, Estancia La Costa, and Puesto Estancia La Costa localities (Table 4.2). The aforementioned characteristics and the macroscopic paleoedaphic features shown in the outcrop (Table 4.3) can be associated with the development of incipient paleosols (Retallack, 2001). However, the

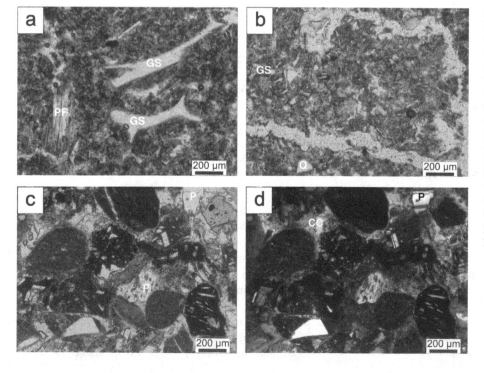

Fig. 4.6. Microphotographs of the Estancia La Costa Member rocks. a, General view of coarse vitric tuff (Campo Barranca locality) showing cuspate glass shards (GS) and pumice with flattened vesicles (PF). b, View of a fine vitric tuff with micropaleoedaphic features (voids), a quartz fragment (Q) and fine glass shards (GS) (Campo Barranca locality). c and d, General view of a lithic sandstone (Puesto Estancia La Costa locality) showing different volcanic lithic fragments, plagioclase fragments (P) and sparry calcite cement (CC). Panels a, b, and c are in plane-polarized light, d in cross-polarized light.

Table 4.2. *Classification of sandy and tuffaceous samples of the Estancia La Costa Member*

Locality	Sample	Classification	Facies	Facies association
Punta Sur–Cañadón del Indio	CÑ-1	medium vitric reworked tuff	Tm	FA-4: pyroclastic-fall deposits
	CÑ-2	medium vitric tuff	Tm	
	CÑ-3	medium vitric tuff with paleoedaphic features	Tm	
	CI-1	fine vitric reworked tuff with paleoedaphic features	Tm	
	CI-2	fine vitric tuff with paleoedaphic features	Tm	
Cañadón Cerro Redondo	CCR-1	fine vitric tuff with paleoedaphic features	Tm	FA-4: pyroclastic-fall deposits
	CCR-3	very fine vitric tuff	Fm	FA-3: distal floodplain
	CCR-10	medium vitric tuff	Tm	FA-4: pyroclastic-fall deposits
Punta Norte of Rincón del Buque	PRB-ML-1	medium vitric tuff	Tm	FA-4: pyroclastic-fall deposits
	PRB-10	medium vitric tuff with paleoedaphic features	Tm	
	PRB-11	very fine vitric tuff	Fm	FA-3: distal floodplain
	PRB-12	fine to medium litharenite	St	FA-1: fluvial channel
	PRB-13	medium vitric tuff	Tm	FA-4: pyroclastic-fall deposits
	PRB-14	medium vitric tuff	Tm	
Campo Barranca	MR-9	medium vitric tuff	Tm	FA-4: pyroclastic-fall deposits
	CB-2	fine vitric reworked tuff	Tm	
	CB-3–4	medium vitric tuff	Tm	
	CB-CM	coarse litharenite	St	FA-1: fluvial channel
	CB-24	medium litharenite	St	
	CB-23	fine vitric tuff with paleoedaphic features	Tm	FA-4: pyroclastic-fall deposits
	CB-22	medium litharenite	St	FA-1: fluvial channel
	CB-20	medium litharenite	St	
	CB-19-b	very fine litharenite with paleoedaphic features	Fm	FA-3: distal floodplain
Anfiteatro	MR-8	medium vitric tuff	Tm	FA-4: pyroclastic-fall deposits
	MR-7	medium vitric tuff	Tm	
	MR-6	medium vitric tuff	Tm	
	MR-5	medium vitric tuff	Tm	
	ANF-4	medium vitric tuff	Tm	
	ANF-3	medium vitric tuff	Tm	
	ANF-2	fine vitric tuff with paleoedaphic features	Tm	
Estancia La Costa	MR-2	fine vitric reworked tuff	Tm	FA-4: pyroclastic-fall deposits
	MR-1	medium vitric tuff with paleoedaphic features	Tm	
Puesto Estancia La Costa	PPLC-4	medium vitric tuff with paleoedaphic features	Tm	FA-4: pyroclastic-fall deposits
	PPLC-2	fine litharenite	Sm	FA-2: proximal floodplain (overbank and sheet-flood)
	PPLC-OB	medium litharenite	St	FA-1: fluvial channel

Note: Tm: Massive coarse and fine vitric tuffs; **Sm:** massive lithic tuffs; **St:** trough cross-bedded lithic sandstones; **Fm:** massive very fine sandstones, tuff or reworked tuff, and mudrocks; **FA-1:** fluvial channel facies association; **FA-2:** proximal floodplain facies association; **FA-3:** distal floodplain facies association; **FA-4:** pyroclastic-fall deposits facies association.

Table 4.3. *Lithofacies identified in the Estancia La Costa Member*

Facies code	Lithology and composition	Sedimentary structures	Thickness and boundaries	Fossil content	Paleoedaphic features/bioturbation/others	Interpretation
Gm	Clast or matrix-supported fine grained conglomerates with intraclasts	Massive	0.2–0.3 m. Erosive, concave upward lower boundary	Frogs	–	Lag deposits, small cores of longitudinal bars. Moderate to high flow regime
Tm	Coarse to fine vitric tuffs and reworked vitric tuffs	Massive to poor lamination	0.2–1.5 m. Sharp boundaries	Vertebrates, logs, and phytoliths	Paleoedaphic features, bioturbation, and carbonaceous remains	Ash-fall events. Pedogenic processes
Sm	Medium lithic sandstones and silty sandstones	Massive	<2.5 m. Sharp or slightly concave upward lower boundary and sharp or transitional upper boundary	Phytoliths	Paleoedaphic features (calcareous concretions)	Stream-flow during high discharge conditions with subaerial exposure and pedogenic modification
St	Fine to very coarse lithic sandstones	Trough cross-bedding, fining-upward trend	<3.5 m. Sharp or slightly concave upward lower boundary and sharp or transitional upper boundary	Vertebrates	Intraclasts in the bases	Subaqueous sandy 3-D dunes. Lower flow regime
Sp	Fine to medium lithic sandstones	Planar cross-bedding, fining-upward trend	0.1–3 m. Slightly concave upward lower boundary or sharp boundaries	–	Paleoedaphic features (iron-rich concretions)	Deposits of sandy 2-D bars. Transitional to lower flow regime. Subaerial exposure and pedogenic modification
Het	Fine sandstones interbedded with mudstones	Horizontal and cross-lamination (wavy and flaser bedding)	0.1–1 m. Transitional boundaries	Marine fossil traces. Leaves	Marine bioturbation. Carbonaceous remains. Soft deformation	Alternation of bedload transport and suspension fallout during slack-water periods

Facies	Lithology	Sedimentary structures	Geometry/Boundaries	Fossils	Other features	Interpretation
Sb	Biogenic beds with sandy fine to medium matrix	Massive	<0.5 m. Concave upward lower boundary and sharp upper boundary	Oysters	—	Bioclastic accumulation with oysters in situ and in place reworked and redeposited
Fl	Very fine sandstones and mudrocks	Horizontal lamination	<1 m. Sharp boundaries	—	Bioturbation, paleoedaphic features (calcareous concretions and root traces). Carbonaceous remains	Settling from suspension with subaerial exposure, pedogenetic processes and bioturbation
Fm	Very fine sandstones, tuff or reworked tuff, and mudrocks	Massive	<1 m. Sharp or transitional boundaries	Vertebrates and phytoliths	Bioturbation, paleoedaphic features (calcareous concretions, root traces, blocky peds) and dessication cracks	Settling from suspension with subaerial exposure, pedogenetic processes, and bioturbation. Aeolian deposition of pyroclastic origin

Note: Gm: massive conglomerates; Tm: massive tuffs and reworked tuffs; Sm: massive sandstones and silty sandstones; St: trough cross-stratified sandstones; Sp: planar cross-stratified sandstones; Het: heterolithic facies; Sb: bioclastic sandstones; Fl: laminated very fine sandstones and mudrocks; Fm: massive very fine sandstones, tuff or reworked tuff, and mudrocks.

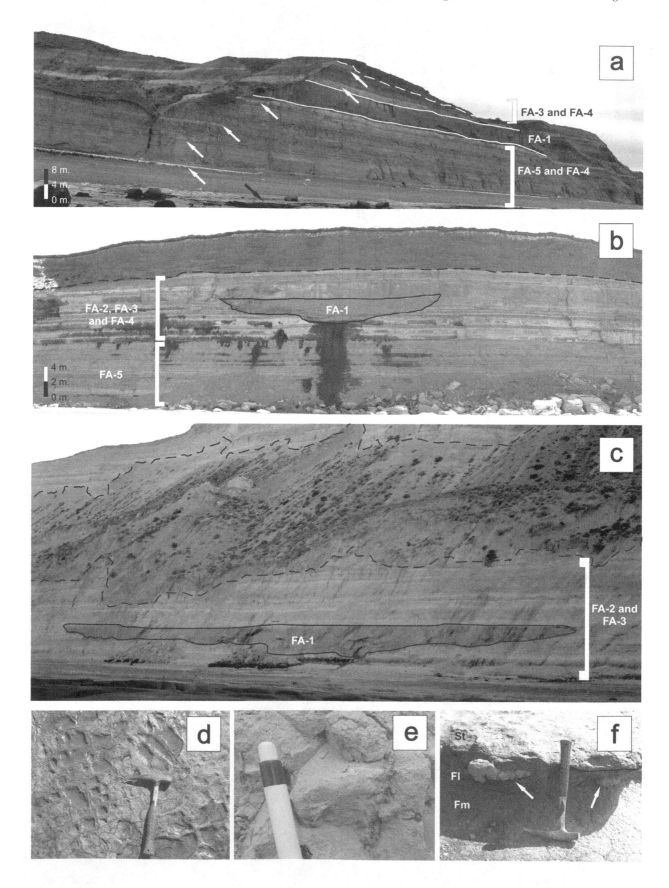

presence of blocky peds suggests a better developed and more slowly forming paleosol (Retallack, 2001). Tauber (1994) classified the first paleosols as probably mollisol and immature, and the second ones as more mature.

Sandstone petrography Sandy samples with scant matrix were studied from the Rincón del Buque, Campo Barranca, and Puesto Estancia La Costa localities (Table 4.2). In general, the samples are moderate to well sorted, and the framework grains are rounded to sub-angular with low sphericity (Fig. 4.6c, d). The texture of the sandstones is clast-supported, and contacts between the grains are tangential or planar (Fig. 4.6c, d). The samples can be classified as litharenites (*sensu* Folk *et al.*, 1970) corresponding to undissected to transitional arc provenance and lithic recycled orogen (*sensu* diagrams of Dickinson *et al.*, 1983).

The framework grains are composed, in decreasing order of abundance, of rock fragments, feldspars, and quartz. The lithic fragments are volcanic in origin and have pilotaxitic, trachytic, and felsitic textures with fresh aspect (Fig. 4.6c, d). However, partially altered volcanic lithics are also present. Sedimentary fragments are identified in smaller proportions. The dominant feldspar is fresh plagioclase crystals with oscillatory zonation and twinning (Fig. 4.6d), but in a few cases the crystals show signs of dissolution (Fig. 4.6c, d) and replacement by clay minerals or carbonates. K-feldspar is very rare or non-existent. The quartz is generally monocrystalline with straight extinction and inclusion-free. Polycrystalline quartz grains or monocrystalline grains with undulose extinction are detected less frequently. Some biotite, glauconite, hornblende, hyperstene, and opaque minerals are also present.

The matrix is sparse and is of the orthomatrix and pseudomatrix type. The principal cements are clay and Fe-oxide coatings, but in several samples sparry calcite patches (Fig. 4.6d) and siliceous cement are observed. These diagenetic products suggest that the burial conditions were very shallow and coincident with synsedimentary or eodiagenetic conditions (Harwood, 1991; Morad *et al.*, 2000; Worden and Burley, 2003).

Matheos *et al.* (2008) described similar petrographic and diagenetic characteristics for the sandstones and tuffs of the levels that form the transition between Monte León and Santa Cruz Formations in the area of the Parque Nacional Monte Léon.

4.4.2 Facies analysis

According to the lithology, sedimentary structures, thickness, and geometry of the sedimentary bodies, nine principal lithofacies are identifiable in the Estancia La Costa Member: (1) massive clast or matrix-supported fine conglomerates (Gm); (2) massive coarse and fine vitric tuffs (Tm); (3) massive lithic sandstones (Sm); (4) trough cross-bedded lithic sandstones (St); (5) planar cross-bedded lithic sandstones (Sp); (6) heterolithic sandstone–mudstone beds deformed in some places (Het); (7) bioclastic sandstones (Sb); (8) laminated mudrocks (Fl); and (9) massive very fine sandstones, tuff or reworked tuff, and mudrocks (Fm). The characteristics of facies and a dynamic interpretation of each one are summarized in Table 4.3.

4.4.3 Facies associations

The deposits of the Estancia La Costa Member have been grouped into five facies associations (FA).

Facies Association 1 (FA-1) is composed of lens- or tabular-shaped gravel-sandstones with cross-bedding, fining-upward bodies and slightly erosive-based with intraclasts (facies Gm, St, Sp), under- and overlain by fine-grained facies (Sm, Fl, Fm, Tm) (Fig. 4.7a, b and c). FA-1 is mainly greenish-gray and brown to yellowish-green. The matrix of the coarser facies of this FA is composed of smectite and chlorite (S3 clay mineral assemblage). The sandstones (St facies) are rich in lithic clasts (litharenite), smectite, illite, and chlorite (S3 clay mineral assemblage). In some sandy levels incipient paleosols and bioturbation are recognized. FA-1 is especially evident towards the middle and upper section of the Estancia La Costa Member (Fig. 4.7b, c). This facies association can be interpreted as the infill of fluvial channels. The absence of lateral accretion surfaces implies that the channels were laterally stable. On the other hand, the absence of internal erosional surfaces suggests deposition in a single channel (Miall, 1996). The presence of fine intraclasts on the base of the bodies is the result of the reworking of components of underlying fine

Caption for Fig. 4.7. Selected photographs of facies and facies associations of Estancia La Costa Member. a, Tabular bodies of FA-2, 3, and 4, composed of facies Sm, Fm, Fl, and Tm, and a tabular sandstone body of FA-1 composed of Gt, St, and Sp facies. Black arrow marks the oyster bank (Sb facies), white arrows point to the Tm facies, and the dotted line indicates the upper contact of the unit with the Quaternary deposits. This picture is representative of the lowest section of the unit (Punta Norte of Rincón del Buque locality). b, Outcrops of the lower and middle section of the unit at Punta Sur-Cañadón del Indio locality. The lower section is characterized by tabular bodies of heterolithic sandstones (Het facies) interbedded with massive or laminated pyroclastic and siliciclastic mudrocks (Fm, Fl facies) (FA-5). To the top of the profile, tabular bodies of Tm, Sm, and Fm facies (FA-2, 3, and 4) enclose a lenticular coarse sand body (Gt, St facies) of FA-1. The dashed line indicates the upper contact with the Quaternary deposits. c, Photograph of the middle and upper section of the member at Puesto Estancia La Costa locality. Here the composition is essentially siliciclastic and the FA-1 is covered by FA-2 and 3. The dashed line delimits a slumping zone. d, Dessication cracks in fine massive sediments (Fm facies) towards the top of the Estancia La Costa Member (Monte Tigre–Cañadón Las Totoras locality). e, Detail of dark yellowish orange rizoliths in an edaphized reworked tuff (Tm facies) at Rincón del Buque locality. f, Calcareous concretion in a paleoedaphized Fl facies at Campo Barranca locality.

sediments (Fm, Fl, Tm facies). These channel deposits are present as single bodies enclosed in floodplain facies (Fig. 4.7b and c); less frequently, channels are multiple units that can crop out at the same stratigraphic horizon, suggesting the presence of multiple active channels (Makaske, 2001). The presence of pedogenic features and bioturbation in the upper levels of the filling indicates subaerial exposure, suggesting that the water flow was ephemeral, not permanent. In particular, the development of iron-rich concretions in the Sp facies indicates mobilization of iron, which suggests frequent fluctuation in groundwater levels (Retallack, 2001).

Facies Association 2 (FA-2) is composed of tabular or lens-shaped sandstones with planar or slightly erosive bases and sharp or convex-upward tops. The facies also includes gray massive beds of lithic sandy or silty sandy composition (facies Sm) where paleoedaphic features and phytoliths remains are evident. FA-2 is commonly found in association with FA-3 and FA-4 and each is typically truncated by channel bodies (Fig. 4.7b, c). This facies is common in the middle and upper sections of the Estancia La Costa Member. This association is interpreted as having been deposited during periods of overbank flooding or sheet flooding and deposited in a proximal floodplain as crevasse splay deposits (lobe to sheet geometry), crevasse channel deposits (lenticular geometry) and sheet-flood deposits (sheet geometry). Following deposition, these bodies were leveled by subaerial exposure and pedogenesis.

Facies Association 3 (FA-3) consists of tabular bodies of massive or laminated gray and bluish-green very fine lithic sandstones and mudrocks (facies Fm and Fl), in part of volcaniclastic origin (vitric tuffs and reworked tuffs), bioturbated, pedogenically modified and with desiccation cracks (Fig. 4.7a–d). The clay-mineral composition of FA-3 is characterized by the presence of smectite, illite, and chlorite (S3 clay mineral assemblage). Vertebrates and phytoliths are frequently found in this facies. FA-3 is the dominant association towards the middle and top sections of the Estancia La Costa Member. FA-3 is interpreted as material settling from suspension in the distal setting of the floodplain. The presence of immature paleosols and desiccation cracks indicates fluctuating wet to dry surface conditions (Ghazi and Mountney, 2009) or desiccation of superficial sediments between flooding events (Collinson, 1996). The gray-green color could be associated with periods of water-logged, vegetated floodplain setting, indicative of an oxygen-poor environment and a ferrous state of the iron in the paleosol levels (Besly and Fielding, 1989). However, the presence of calcareous concretions (Fig. 4.7f) can be interpreted as an excess of alkaline solutes and precipitation due to fluctuations in groundwater table in a well-drained environment (Retallack, 2001). In addition, fine volcaniclastic levels may be aeolian deposits (loess) (Tauber, 1994, 1997).

Facies Association 4 (FA-4) consists of tabular massive bodies composed by Tm facies, non-graded, with planar basal surfaces (Fig. 4.7a, b, e). The finest tuffs are grayish-white or greenish-white in color and on some occasions show carbonaceous remains; FA-4 also contains coarse white tuffs. Compositionally the Tm facies are vitric and rich in smectite (S1 clay mineral assemblage) or rich in smectite with small proportions of illite and chlorite (S2 clay mineral assemblage). Several beds are weakly pedogenized (Fig. 4.7e) and show bioturbation. Fossil vertebrates and the remains of carbonized logs and phytoliths (Brea et al., 2010 and Chapter 7) are present in this facies association. FA-4 levels have great lateral continuity and constitute guide beds where the principal fossiliferous levels are found. In fine Tm facies at Killik Aike Norte, Monte León, and Cerro Observatorio localities, Genise and Bown (1994) and A. Tauber (personal communication, 2010) found burrows of solitary bees and nests of scarabeid dung-beetles (see also Krapovickas, Chapter 6). FA-4 is present at the lower and middle section of the Estancia La Costa Member but is most common in the lower section (Fig. 4.7a, b). This facies association could be formed by intermittent deposition of tabular ash-fall beds in low-gradient, subaerial and subaqueous areas in a poorly drained floodplain (Cas and Wright, 1987). After deposition, pyroclastic-fall deposits were slightly modified by pedogenesis and bioturbation. The color of the finest tuff could suggest that these paleosols were developed owing to slight variation in drainage characteristics across the floodplain. The presence of calcareous steinferns of trees (Fleagle et al., 1995), carbonized fossil logs, phytoliths, and carbonaceous remains attests to the availability of standing water on the floodplain (Roberts, 2007) and indicates that this floodplain was vegetated. However, the presence of burrows of bees and scarab beetles suggests that at least some more open areas were developed (Genise and Bown, 1994; Tauber, 1997).

Facies Association 5 (FA-5) includes lenticular bodies composed of grayish-green bioclastic sandstones with oysters (Sb facies) (see Chapter 5) and greenish-yellow tabular heterolithic deposits consisting of alternating horizontally or cross-laminated fine sandstones and mudstones, displaying wavy and flaser bedding (Het facies) in place with soft deformation, and fining-upward trend. Marine trace fossils are very abundant in the lowest levels of this heterolithic facies at Cerro Redondo and Rincón del Buque localities (V. Krapovickas, personal communication, 2011). Fossil leaves and carbonaceous remains are encountered in the uppermost levels of this heterolithic facies. FA-5 is exclusive of the lower section of the Estancia La Costa Member where it is interbedded and overlain by FA-4 (Fig. 4.7a, b). Facies Association 5 is attributed to sedimentation in a fluvial–marine transitional environment near the coastline because of the presence of heterolithic facies with fining-upward arrangement, the disappearance of marine trace fossils and oysters upward, and the appearance of fossil leaves and carbonaceous remains towards the top of the succession. It is probable that the deposition of FA-5 occurred in an estuarine environment, as was proposed by Cuitiño and Scasso (2010).

4.5 Implications for paleoenvironmental conditions of the Estancia La Costa Member

Sedimentological and compositional data of the Estancia La Costa Member provide a clearer understanding of the paleoenvironmental and paleoclimatic conditions and the nature of the source material that persisted during deposition in southern coastal Santa Cruz Province.

4.5.1 Depositional environments
Depositional processes, facies, and facies associations can be used to interpret the paleoenvironmental setting for the succession and to reconstruct the depositional history of the unit (e.g. Paredes *et al.*, 2007; Roberts, 2007).

The depositional paleoenvironment of the Estancia La Costa Member in the study area was previously interpreted by Tauber (1994, 1997) as a low-energy fluvial system developed under warm and humid climatic conditions. On the other hand, he interpreted the upper member of the Santa Cruz Formation (Estancia La Angelina Member) as a higher-energy fluvial system with sheet-like deposits and meandering channels formed under a more arid climate. Outcrops of the Santa Cruz Formation northwest of the study area (Lago Posadas) were interpreted by Bande *et al.* (2008) as representing a meandering or anastomosing fluvial system. At Lago Argentino, to the west of the study area, Cuitiño and Scasso (2010) interpret the lower section of Santa Cruz Formation as a meandering fluvial system. In the San Jorge Basin, north of the study area, Bellosi (1998) suggested that the Santa Cruz Formation was deposited on a vegetated coastal and alluvial floodplain with sheet-like channels changing to meandering channels towards the top of the unit.

We interpret deposition of the lower part of the Estancia La Costa Member as being dominated by a fluvial–marine transitional environment that grades upward to a fluvial environment characterized by sheet-flooding, overbank-flooding ash-fall deposits, and single laterally stable channels. The preponderance of sheet-flood and overbank deposits, overlying within-channel deposits, indicates a broad floodplain across which the channel was built. The common primary and reworked tuff deposits in the member reflect the influence of ash-fall events. The vertical accretion of the floodplains was by deposition from crevasse-

splays and crevasse-channels, sheet-floods, and settling from suspension. These deposits eventually became desiccated, allowing the formation of subaerial floodplain soils (incipient paleosols) exhibiting bioturbation. The development of paleosols was influenced by fluctuations in the groundwater table. In this sense, the presence of grayish colors and the occurrence of calcareous and ferruginous concretions and desiccation cracks suggest slight variation in drainage characteristics across the floodplain with fluctuating wet to dry surface conditions. The presence of phytoliths, carbonaceous remains, fossil leaves, logs, and steinferns of trees confirms the existence of large vegetation indicating the constant availability of water on the floodplains; also see Brea *et al.*, Chapter 7.

However, some paleoenvironmental changes in the Estancia La Costa Member are noted. Informally, we consider the member to contain three sections (Fig. 4.4). A lower section of this member, outcropping in Rincón del Buque (= Media Luna), Cañadón del Cerro Redondo, Punta Sur–Cañadón del Indio, and Campo Barranca, had a greater input of pyroclastic material and a relatively lower energy for the fluvial system, based on the greater proportion of tuff beds interbedded with mudrock or silty sandstone beds deposited by ash-fall events with sheet-flooding and overbank-flooding. In this part of the unit the paleosols are less evident than in the middle and upper sections. The pyroclastic composition of the facies suggests a higher sedimentation rate and a relatively continuous volcanic source. Fossil leaves and logs are exclusively present in this interval, and phytoliths remains are common. This situation can be linked with paleoenvironments with both arboreal and shrub vegetation (interspersed woodlands and grasslands).

Throughout the middle section of the member (upper outcrops of Rincón del Buque and outcrops of Anfiteatro, Estancia La Costa, Cañadón Silva, and Puesto Estancia La Costa) there is an evident increase in the participation of coarse facies (Tm, Gm, St, Sp), which suggests a relatively higher energy for the system. However, toward the upper zone of the middle section we again see a gradual decrease in the proportion of sandstones, an increased participation of pyroclastic material, and an increase in mudrocks. We can interpret this as the resumption of a paleoenvironment with lower energy and with more siliciclastic source. More developed paleosols and bioturbated beds are frequent in the middle and upper parts of the middle section. This situation could reflect a lower sedimentation rate compared with the lower section of the member, and probably several interruptions in the volcanic source, which favored increased landscape maturity and pedogenesis. The principal fossiliferous levels and the phytolith remains are concentrated in the basal and upper parts of the middle section (Tauber's fossil levels FL 1 to FL 4.1 in the former, and FL

5 to FL 7 in the latter). The existence of these fossils can be linked with densely vegetated paleoenvironments.

The base of the upper section of the Estancia La Costa Member (outcropping at Puesto Estancia La Costa and Monte Tigre–Cañadón Las Totoras) shows more energetic conditions with a greater participation of sandstones which are finer toward the middle zone of the upper section. The top of the upper section is composed of both sandstones and mudrocks. The absence of pyroclastics and the presence of phytolith remains suggests respectively that those environments only received siliclastic material and that they were more openly vegetated (grasslands). Paleosols and fossiliferous levels are found in the basal (FL 7.1 and FL 7.2) and upper zones in the upper section (FL 8 to FL 10). The most remarkable feature is that the paleosols have a clear color and are marked by the presence of calcareous concretions and desiccation cracks consistent with a fluctuating flow regime.

4.5.2 The record of paleoclimatic change

Clay minerals in sedimentary sequences provide important information on pre- and post-burial conditions. Pre-burial controls include source area lithology, paleoclimate (chemical and physical weathering), depositional environment, and topography, among others (e.g. Chamley, 1989; Inglés and Ramos-Guerrero, 1995). Although this information can be altered by diagenesis that changes the original clay mineral composition (e.g. Egger *et al.*, 2002), the study of the change of clay minerals in sequences that did not undergo intense diagenesis becomes an important tool in understanding the environmental conditions of deposition (Net *et al.*, 2002; Raucsik and Varga, 2008; Do Campo *et al.*, 2010). The abundance of smectite with limited amounts of illite and chlorite throughout the studied sections in the Santa Cruz Formation, together with the low abundance of mixed-layer clays and the compositional-diagenetic aspects of sandstones and tuffs, indicates that these deposits were not affected by deep-burial diagenesis. This suggests that diagenetic effects (post-burial conditions) on the composition of clay mineral assemblages were minimal, and that the assemblages reflect the conditions at deposition.

Given the existence of a volcanic source and the recurrence of tuffaceous beds, the high proportion of smectite, together with opal and clinoptilolite registered in the <4 μm fraction, suggests the partial devitrification and alteration of volcanic ash under relatively warm and wet climatic conditions (de Ros *et al.*, 1997; Dingle and Lavelle, 2000). Smectite is typically formed by chemical weathering in weakly drained soils under warm and seasonal climates with alternating wet and dry conditions (Chamley, 1989; Robert and Kennett, 1994; Thiry, 2000). On the other hand, illite, chlorite, and mixed-layer illite/smectite can be developed in immature soils that have undergone little chemical weathering. In general, illite and chlorite are indicators of mechanical erosion that affected the parent rocks either as a result of cool and dry climatic conditions or as a consequence of a pronounced relief (Robert and Kennett, 1994; Thiry, 2000).

In this context, the three clay-mineral assemblages recognized (S1, S2, and S3) can be related mainly to slightly different weathering histories of a similar source material. The presence of smectite with illite, chlorite, and illite/smectite mixed layers (S2 and S3 assemblages) in the lower section of the Estancia La Costa Member (Campo Barranca, Anfiteatro and Estancia La Costa localities) probably indicates periods with cool or cool and dry climatic conditions (Chamley, 1989; Robert and Kennett, 1994; Thiry, 2000). Nevertheless, the presence of smectite as sole clay mineral towards the middle section of Estancia La Costa Member (S1 assemblage) could indicate a shift to warmer and seasonal climatic conditions (Fig. 4.5c). In addition, the fact that the upper levels of the Estancia La Costa Member are rich in smectite and illite was interpreted by Matheos *et al.* (2010) as indicating a more persistent cool and dry climate.

Paleofloristic evolution of the Late Oligocene–Early Miocene of southern Patagonia demonstrated that forest elements are gradually replaced by other shrubby and herbaceous forms, signaling the beginning of the expansion of xerophytic environments determined by cooling and drying trends (Barreda and Palazzesi, 2007). This is concordant with the existence of both closed and open environments for the Santa Cruz Formation (Vizcaíno *et al.*, 2010; Tambussi, 2011) and would be comparable to temperate forests and bushland with less than 1000 mm of annual rainfall (Vizcaíno *et al.*, 2010).

Isotopic studies from pedogenic carbonate contained in outcrops of the Santa Cruz Formation at Lago Posadas (northwest of Santa Cruz Province) demonstrated that surface uplift of the southern Andes at 16.5 Ma led to a climatic deterioration with a strong aridification in the eastern foreland and presumably, strongly increased precipitation rates on the windward western side of the mountains (Blisniuk *et al.*, 2005; 2006). These conditions were apparently accompanied by a transition from balanced subtropical woodlands and grasslands to predominantly grasslands (Blisniuk *et al.*, 2005), corresponding with the lower and middle section of the Estancia La Costa Member, and with the upper section of this member and the Estancia La Angelina Member, respectively. In this scenario, the change trend towards the youngest zone of the section may reflect that the initial phase of this uplift caused an increased supply of moisture to the eastern foreland prior to the rain-shadow effect becoming much stronger at 15–14 Ma.

The late Early Miocene clay-mineral composition of the coastal Estancia La Costa Member indicates some cooling and/or drying alternating with warmer, more humid periods registered globally during the Early Miocene (Zachos *et al.*, 2001) (Fig. 4.5c). A shift to warmer and seasonal climatic conditions showing at the middle section of the member probably will be coincident with part of the global Middle Miocene warmth known as the Mid-Miocene Climatic Optimum (Zachos *et al.*, 2001) which took place between 17 and 15 Ma (Fig. 4.5c). Tauber (1997) interpreted this stratigraphic interval, where fossiliferous levels from 1 to 4 occur, as indicating a wooded environment (woodlands and grasslands) with more humid and warmer climatic conditions than the upper interval (fossiliferous levels from 8 to 11). The return to cooler and drier conditions towards the upper levels of the member (*c.* 16 Ma) is consistent with the interpretation of Tauber (1997) and Vizcaíno *et al.* (2006). These authors proposed an environment with open vegetation developed in relatively dry conditions with marked seasonality, similar to that recorded today in the Chaqueña Biogeographic Province, for the upper levels of the Estancia La Costa Member and part of the Estancia La Angelina Member (*c.* 16 to 14 Ma). Moreover, Brea *et al.* (2010 and Chapter 7) demonstrated that the prevailing climate during the lower part of Estancia La Costa Member was temperate to warm, and subhumid to dry with marked seasonality. In addition, ichnoassemblages of the Estancia La Costa Member defined by Krapovickas (Chapter 6) show a general trend from the bottom to the top of the unit from more humid to drier conditions.

The decreasing temperatures implied by the fossil and isotopic record and clay-mineral composition may be related to gradual global cooling at ~14.5 Ma (Fig. 4.5c), which was associated with rapid Antarctic ice-sheet growth, and major biogeographic changes on all continents (Zachos *et al.*, 2001). At the same time, the change in clay-mineral composition registered at the analyzed section of the Estancia La Costa Member could be correlated with changes in erosion rates identified by Blisniuk *et al.* (2005). The presence of illite together with smectite towards the upper section of the member (Matheos *et al.*, 2010) could be a consequence of an increase of precipitation and erosion rates on the humid windward western side of the mountains. Blisniuk *et al.* (2005) indicated that the increasing surface elevation would have blocked more moisture from reaching the leeward eastern side of the mountains.

4.5.3 Source areas and volcanic influence

The compositional analysis of sandstones is a useful tool to characterize the source area of sediments (Dickinson and Suczek, 1979; Dickinson *et al.*, 1983). Furthermore, the compositional study of volcaniclastic and pyroclastic rocks helps to clarify the clast-forming processes and establish the character, composition, and setting of the volcanic source (McPhie *et al.*, 1993). It is important to bear in mind,

however, that diagenesis and weathering are the major factors in altering volcanic glass and minerals, and destroying depositional texture (Tucker, 1996).

The great proportion of lithic fragments of volcanic origin, together with the preponderance of plagioclases, and the presence of inclusion-free quartz with straight extinction in the sandstone samples, indicates that the source-area for these deposits was dominated by volcanic rocks (Folk, 1964; Dickinson and Suczek, 1979), as confirmed by the abundance of plagioclase crystals with oscillatory zonation (e.g. Spalletti *et al.*, 1993). In this context, the presence of volcanic fragments with pilotaxitic and trachytic textures is linked with an intermediate volcanic source-rock, and volcanic fragments with felsitic texture with a more acidic volcanic source-rock (Dickinson, 1970; Scasso and Limarino, 1997). Both the fresh aspect and bigger size of the volcanic fragments compared with the other components of the sandstones suggest that the lithics and the volcanism are contemporaneous (Critelli and Ingersoll, 1995). Similarly, the small contribution of quartz and feldspar crystals in the lithic fragments confirms active volcanism (Scasso and Limarino, 1997). At the same time, the sedimentary lithics and the polycrystalline quartz fragments are derived from deformed sedimentary units of the thrust belt (Critelli and Le Pera, 1994). Sources from magmatic arc and recycled orogen frequently coexist, especially when the evolution of the orogenic belt included active volcanism (Critelli and Ingersoll, 1994). The dominance of vitric fragments with glass shards (cuspade and platy shape) and pumice fragments (flattened and rounded vesicles) in the tuffaceous samples is linked with Plinian-like explosive eruptions of felsic to intermediate magmas (McPhie *et al.*, 1993; Mazzoni, 1996; Umazano *et al.*, 2009).

Based on the inferred composition, the tectonic setting of the source-area (volcanic arc), and the characteristic of the volcanism that could have generated the tuffaceous material, the main source-area for the sediments of the Estancia La Costa Member must be related to the Miocene volcanic cycle described by Ardolino *et al.* (1999). This volcanism was explosive and fragmentary (acidic to intermediate in composition) and took place in the Patagonian Andes region and Austral Basin during the Early Miocene (Ardolino *et al.*, 1999). However, no Miocene volcanic unit crops out in this area, and the only evidence of this coeval Miocene volcanism is given by the pyroclastic detritus in the clastic sedimentary beds of the Santa Cruz Formation. At the same time, the emplacement of subvolcanic bodies (18–16 Ma) and the contemporary intrusion of granitic and dioritic rocks in the Andes region (Argentina and Chile) were interpreted by Ramos (1999) and Nullo and Otamendi (2002) as the magmatic phase related with the uplifting of the Austral Patagonian Andes. In addition, Hatcher (1897,

1903) described several craters that cross the lower beds of Santa Cruz Formation; he considered the possible existence and activity of these craters before and during the sedimentation of the unit, and proposed that this volcanism could be the source of the youngest tuffaceous material of the unit.

In this geological context, Ramos (1999) and Ramos and Ghiglione (2008) interpreted the succession of Santa Cruz Formation as synorogenic deposits that mark the uplifting of the Austral Patagonian Andes. Bellosi (1998) mentioned that the paleocurrent orientation in the Santa Cruz Formation north of our study area is predominantly eastward, and this is in concordance with the uplifting of the Austral Patagonian Andes and foothill areas.

Considering the presence of subduction processes at the western margin of Patagonia during the Early Miocene (Ramos, 1999; Ardolino *et al.*, 1999; Nullo and Otamendi, 2002; Ramos and Ghiglione, 2008; among others), contemporaneous with the deposition of the Santa Cruz Formation, the tuffaceous material of this unit is likely to represent the distal eruptive phase of a simultaneously occurring explosive volcanism generated in an arc magmatic setting. This is related to the uplift of the Austral Patagonian Andes, located about 300 km westward of the study area. The ash was probably transported towards the coast by the westerly winds.

4.6 Conclusions

The coastal Santa Cruz Formation has two members: the Estancia La Costa Member (fine siliciclastic and pyroclastic deposits, rich in fossils), and the Estancia La Angelina Member (fossil-poor claystones, mudstones, and sandstones). On the basis of sedimentological features we divided the first member informally into a lower, middle, and upper section.

The basal part of the lower section of the Estancia La Costa Member is dominated by bioclastic and heterolithic facies deposited in a continental–marine transitional environment, whereas the upper part of this lower section of the Estancia La Costa Member is dominated by pyroclastic material deposited in a relatively low-energy fluvial system with a high sedimentation rate. The middle section of this member shows an evident increase in coarse facies which suggests a higher energy for the system compared with the lower section, but a lower sedimentation rate. The upper section of the member evinces even more energetic conditions of the fluvial system, but with a fluctuating flow regime and places where the vegetation was more open.

The climate and the source of volcaniclastic material played important roles in the evolution of this system. Based on the clay-mineral assemblages and paleoedaphic features, we conclude that during the deposition of the Estancia La Costa Member the climate was seasonal and warm, trending to cooling and/or drying stages toward the top. This trend is related to the Middle Miocene climatic deterioration. The volcanic deposits in the Estancia La Costa Member reflect distal pyroclastic events, at about 300 km westward of the study area and related to the uplift of the Austral Patagonian Andes.

ACKNOWLEDGMENTS

The authors thank Sergio Vizcaíno, Richard Kay and María Susana Bargo for their invitation to participate in this book. We are grateful to two anonymous reviewers for their constructive comments. Our special thanks to S. Vizcaíno and his group of collaborators for logistical support and to S. Vizcaíno, S. Bargo, A. Tauber. J. C. Fernicola, N. Muñoz, V. Krapovickas, and J. Vizcaíno for their help with the fieldwork. This is a contribution to the UNLP N549 and PIP-CONICET No. 114–200801–00066 projects to SDM, and NSF grants 0851272 and 0824546 to Richard F. Kay.

REFERENCES

Ameghino, F. (1902). Notices préliminaires sur des mammifères nouveaux des terrains crétacés de Patagonie. *Boletín de la Academia Nacional de Ciencias*, **17**, 5–70.

Ameghino, F. (1906). Les formations sédimentaires du Crétacé Supérieur et du Tertiaire de Patagonie avec un parallelé entre leurs faunes mammalogiques et celles de l´ancien continent. *Anales del Museo Nacional de Buenos Aires*, **8**, 1–568.

Arbe, H. A. (1987). El Cretácico de la Cuenca Austral. *Boletín de Informaciones Petroleras*, **9**, 91–110.

Ardolino, A., Franchi, M., Remesal, M. and Salani, F. (1999). El vulcanismo en la Patagonia Extraandina. In Geología Argentina, ed. R. Caminos. *Anales 29 de la Subsecretaría de Minería de la Nación, Servicio Geológico Minero Argentino e Instituto de Geología y Recursos Minerales*, **18**, 579–612.

Bande, A. E., Scasso, R. and Porfiri, G. (2008). Análisis paleoambiental de las formaciones Patagonia y Santa Cruz, en la zona del Lago Posadas, provincia de Santa Cruz. *XII Reunión Argentina de Sedimentología, Actas*, p. 37.

Barreda, V. and Palazzesi, L. (2007). Patagonian vegetation turnovers during the Paleogene–Early Neogene: origin of arid-adapted floras. *The Botanical Review*, **73**, 31–50.

Bellosi, E. (1998). Depósitos progradantes de la Formación Santa Cruz, Mioceno de la Cuenca del Golfo San Jorge. *VII Reunión Argentina de Sedimentología, Actas*, pp. 110–111.

Besly, B. M. and Fielding, C. R. (1989). Palaeosols in Westphalian coal-bearing and red-bed sequences, central and northern England. *Palaeogeography, Palaeoclimatology and Palaeoecology*, **70**, 303–330.

Biddle, K. T., Uliana, M. A., Mitchum, R. M. Jr., Fitzgerald, M. G. and Wright, R. C. (1986). The stratigraphic and structural evolution of the central and eastern Magallanes Basin, southern South America. In *Foreland Basins*, ed. P. A. Allen and P. Homewood, *Special Publication of the International Association of Sedimentologists* **8**, 41–61.

Biscaye, P. (1965). Mineralogy and sedimentation of recent deep-sea clay in the Atlantic Ocean and adjacent seas and oceans. *Geological Society of American Bulletin*, **76**, 803–832.

Blisniuk, P. M., Stern, L. A., Chamberlain, C. P., Idleman, B. and Zeitler, K. P. (2005). Climatic and ecologic changes during Miocene surface uplift in the southern Patagonian Andes. *Earth and Planetary Science Letters*, **230**, 125–142.

Blisniuk, P. M., Stern, L. A., Chamberlain, C. P. *et al.* (2006). Links between mountain uplift, climate, and surface processes in the southern Patagonian Andes. In *The Andes: Active Subduction Orogeny*, ed. O. Oncken *et al.* Frontiers in Earth Science Series 1. Berlin: Springer, 425–436.

Bown, T. M. and Fleagle, J. G. (1993). Systematics, biostratigraphy, and dental evolution of the Palaeothentidae, later Oligocene to early–middle Miocene (Deseadan–Santacrucian) caenolestoid marsupials of South America. *Journal of Paleontology*, **67**, 1–76.

Bown, T. M. and Larriestra, C. (1990). Sedimentary paleoenvironments of fossil platyrrhine localities, Miocene Pinturas Formation, Santa Cruz Province, Argentina. *Journal of Human Evolution*, **19**, 87–119.

Bown, T. M. and Ratcliffe, B. C. (1988). The origin of *Chubutolithes ihering*, ichnofossils from the Eocene and Oligocene of Chubut Province, Argentina. *Journal of Paleontology*, **62**, 163–167.

Bown, T. M., Larriestra, C. N. and Powers, D. W. (1988). Análisis paleoambiental de la Formación Pinturas (Mioceno Inferior), Provincia de Santa Cruz. *Segunda Reunión Argentina de Sedimentología*, **1**, 31–35.

Brea, M., Zucol, A. F. and Bargo, M. S. (2010). Estudios paleobotánicos en la Formación Santa Cruz (Mioceno), Patagonia, Argentina. *X Congreso Argentino de Paleontología y Bioestratigrafía y VII Congreso Latinoamericano de Paleontología, Actas*, 66–67.

Bridge, J. (2003). *Rivers and Floodplains*, 1st Edn. Blackwell Science.

Brindley, G. and Brown, G. (1980). Crystal structures of clay minerals and their X-ray identification. *Mineralogical Society Monograph* **5**, 1–495.

Bullock, P., Fedoroff, N., Jongerius, A., Stoops, G. and Tursina, T. (1985). *Handbook for Soil Thin Section Description*. Albrighton: Waine Research Publications.

Cas, R. A. F. and Wright, J. V. (1987). *Volcanic Successions: Modern and Ancient*. London: Unwin Hyman.

Chamley, H. (1989). *Clay Sedimentology*. Berlin: Springer-Verlag.

Collinson, J. D. (1996). Alluvial sediments. In *Sedimentary Environments: Processes, Facies and Stratigraphy*, ed. H. G. Reading, 3rd Edn. Oxford: Blackwell, pp. 37–82.

Corbella, H. (2002). El campo volcánico-tectónico de Pali Aike. In *Geología y Recursos Naturales de Santa Cruz*, ed. M. J. Haller. *Relatorio XV Congreso Geológico Argentino*, **I**(18), 285–301.

Critelli, S. and Ingersoll, R. (1994). Sandstone petrology and provenance of the Siwalik Group (Northwestern Pakistan and Western–Southeastern Nepal). *Journal of Sedimentary Research*, **A64**, 815–823.

Critelli, S. and Ingersoll, R. (1995). Interpretation of neovolcanic versus palaeovolcanic sand grains: an example from Miocene deep-marine sandstone of the Topanga Group (Southern California). *Sedimentology*, **42**, 783–804.

Critelli, S. and Le Pera, E. (1994). Detrital modes and provenance of Miocene sandstones and modern sands of the Southern Apennines Thrust-Top basins (Italy). *Journal of Sedimentary Research*, **A64**, 824–835.

Cuitiño, J. I. and Scasso, R. A. (2010). Sedimentología y paleoambientes del Patagoniano y su transición a la Formación Santa Cruz al sur del Lago Argentino, Patagonia Austral. *Revista de la Asociación Geológica Argentina*, **66**, 406–417.

de Barrio, R. (1984). *Descripción geológica de la Hoja 53c, Laguna Olín, provincia de Santa Cruz*. Servicio Geológico Nacional, pp. 54.

de Ros, L., Morad, S. and Al-Aasm, I. (1997). Diagénesis of silicoclastic and volcaniclastic sediments in the Cretaceous and Miocene sequences of the NW African margin (DSDP Leg 47A, Site 397). *Sedimentary Geology*, **112**, 137–156.

del Río, C. (2004). Tertiary marine molluscan assemblages of eastern Patagonia (Argentina): a biostratigraphic analysis. *Journal of Paleontology*, **78**, 1097–1122.

Dickinson, W. (1970). Interpreting detrital modes of graywacke and arkose. *Journal of Sedimentary Petrology*, **40**, 695–707.

Dickinson, W. and Suczek, C. (1979). Plate tectonics and sandstone compositions. *American Association of Petroleum Geologists Bulletin*, **63**, 2164–2182.

Dickinson, W., Breard, L., Brakenridge, G. *et al.* (1983). Provenance of North American Phanerozoic sandstones in relation to tectonic setting. *Geological Society of America Bulletin*, **94**, 222–235.

Dingle, R. and Lavelle, M. (2000). Antarctic Peninsula Late Cretaceous–Early Cenozoic palaeoenvironments and Gondwana palaeogeographies. *Journal of African Earth Sciences*, **31**, 91–105.

Do Campo, M., del Papa, C., Nieto, F., Hongn, F. and Petrinovic, I. (2010). Integrated analysis for constraining palaeoclimatic and volcanic influences on clay-mineral assemblages in orogenic basins (Palaeogene Andean foreland, Northwestern Argentina). *Sedimentary Geology*, **228**, 98–112.

Egger, H., Homayoun, M. and Schnabel, W. (2002). Tectonic and climatic control of Paleogene sedimentation in the Rhenodanubian Flysch basin (Eastern Alps, Austria). *Sedimentary Geology*, **152**, 247–262.

Feruglio, E. (1938). El Cretácico Superior del Lago San Martín (Patagonia) y de las regiones adyacentes. *Physis*, **12**, 293–342.

Fisher, R. V. (1961). Proposed classification of volcaniclastic sediments and rocks. *Geological Society of America Bulletin*, **72**, 1409–1414.

Fisher, R. V. and Schmincke, H. (1984). *Pyroclastic Rocks*. Berlin: Springer-Verlag.

Fleagle, J. G. (1990). New fossil platyrrhines from the Pinturas Formation, Southern Argentina. *Journal of Human Evolution*, **19**, 61–85.

Fleagle, J. G., Bown, T. M., Swisher, C. C. III and Buckley, G. A. (1995). Age of the Pinturas and Santa Cruz formations. *VI Congreso Argentino de Paleontologia y Bioestratigrafia, Actas*, 129–135.

Folk, R. L. (1964). *Petrology of Sedimentary Rocks*. Austin: Hemphill.

Folk, R. L., Andrews, P. B. and Lewis, D. W. (1970). Detrital sedimentary rock classification and nomenclature for use in New Zealand. *New Zealand Journal of Geology and Geophysics*, **13**, 937–968.

Frenguelli, J. (1931). Nomenclatura estratigráfica patagónica. *Anales de la Sociedad Científica*, **3**, 115.

Friend, P. F., Slater, M. J. and Williams, R. C. (1979). Vertical and lateral building of river sandstones bodies, Ebro basin, Spain. *Journal of the Geological Society of London*, **136**, 39–46.

Furque, G. (1973). Descripción geológica de la Hoja 58b Lago Argentino. *Boletín del Servicio Nacional Minero y Geológico*, **140**, 1–51.

Furque, G. and Camacho, H. H. (1972). El Cretácico Superior y Terciario de la región austral del Lago Argentino (provincia de Santa Cruz). *Cuartas Jornadas Geológicas Argentinas*, **3**, 61–75.

Genise, J. F. and Bown, T. M. (1994). New Miocene scarabeid and hymenopterous nests and Early Miocene (Santacrucian) paleoenvironments, Patagonian Argentina. *Ichnos*, **3**, 107–117.

Ghazi, S. and Mountney, N. P. (2009). Facies and architectural element analysis of a meandering fluvial succession: The Permian Sandstone Warchha, Salt Range, Pakistan. *Sedimentary Geology*, **221**, 99–126.

Harwood, G. (1991). Microscopic techniques: II, principles of sedimentary petrography. In *Techniques in Sedimentology*, ed. M. Tucker. Oxford: Blackwell, pp. 108–173.

Hatcher, J. B. (1897). On the geology of southern Patagonia. *American Journal of Science*, **4**, 321–354.

Hatcher, J. B. (1903). Narrative of the expedition. In *Reports of the Princeton University Expeditions to Patagonia, 1896–1899*. Vol. 1, *Narrative and Geography*, ed. W. B. Scott. Princeton: Princeton University Press.

Inglés, M. and Ramos-Guerrero, E. (1995). Sedimentological control on the clay mineral distribution in the marine and non-marine Paleogene deposits of Mallorca (Western Mediterranean). *Sedimentary Geology*, **94**, 229–243.

Kay, R. F., Vizcaíno, S. F., Bargo, M. S. *et al.* (2008). Two new fossil vertebrate localities in the Santa Cruz Formation (late early – early middle Miocene, Argentina), 51° South latitude. *Journal of South American Earth Sciences*, **25**, 187–195.

Kramarz, A. G. and Bellosi, E. S. (2005). Hystricognath rodents from the Pinturas Formation, Early–Middle Miocene of Patagonia, biostratigraphic and paleoenvironmental implications. *Journal of South American Earth Sciences*, **18**, 199–212.

Krapovickas, J. M., Tauber, A. A. and Rodriguez, P. E. (2008). Nuevo registro de *Protypotherium australe* Ameghino, 1887: implicancias bioestratigráficas en la Formación Santa Cruz. *XVII Congreso Geológico Argentino*, **III**, 1020–1021.

Madden, R. H., Kay, R. F., Vucetich, M. G. and Carlini, A. A. (2010). Gran Barranca: a 23 million-year record of middle Cenozoic faunal evolution in Patagonia. In *The Paleontology of Gran Barranca: Evolution and Environmental Change through the Middle Cenozoic of Patagonia*, ed. R. H. Madden, A. A. Carlini, G. M. Vucetich and R. F. Kay. London: Cambridge University Press, pp. 423–439.

Makaske, B. (2001). Anastomosing rivers: a review of their classification, origin and sedimentary products. *Earth Science Review*, **53**, 149–196.

Malumián, N. (1999). La sedimentación y el volcanismo terciarios en la Patagonia Extraandina, 1. La sedimentación en la Patagonia Extraandina. In *Geología Argentina*, ed. J. Caminos. *Anales del Instituto de Geología y Recursos Minerales*, **29**, 557–578.

Marín, G. (1984). Descripción geológica de la Hoja 55c, Gobernador Gregores, Provincia de Santa Cruz. *Servicio Geológico Nacional* (Inédito), pp. 67.

Marshall, L. G. (1976). Fossil localities for Santacrucian (Early Miocene) mammals from Santa Cruz Province, southern Patagonia, Argentina. *Journal of Paleontology*, **50**, 1129–1142.

Marshall, L. G. and Pascual, R. (1977). Nuevos marsupiales Caenolestidae del "Piso Notohippidense" (SW de Santa Cruz, Patagonia) de Ameghino. Sus aportaciones a la cronología y evolución de las comunidades de mamíferos sudamericanos. *Publicaciones del Museo Municipal de Ciencias Naturales "Lorenzo Scaglia"*, **2**, 91–122.

Marshall, L. G., Cifelli, R. L., Drake, R. E. and Curtis G. H. (1986). Geochronology of type Santacrucian (middle Tertiary) Land Mammal Age, Patagonia, Argentina. *Journal of Geology*, **94**, 449–457.

Matheos, S. D., Raigemborn, M. S., Vizcaíno, S. F., Bargo, M. S. and Vizcaíno J. (2008). Sedimentología de la transición marino-continental del Mioceno temprano en el Parque Nacional Monte León, Santa Cruz, Argentina. *XVII Congreso Geológico Argentino*, **II**, 876–877.

Matheos, S. D., Raigemborn, M. S., Gomez Peral, L. and Tauber, A. A. (2010). Paleoclimatic interpretation from clay minerals in the early–middle Miocene of Southeast Patagonia, Argentina. *18th International Sedimentological Congress, Actas*, 247.

Mazzoni, M. M. (1996). *Vulcanismo y rocas volcaniclásticas. Curso de postgrado*. Puerto Alegre, Brasil. Unpublished document.

McPhie, J., Doyle, M. and Allen, R. (1993). *Volcanic Textures: A Guide to the Interpretation of Textures in Volcanic Rocks*. University of Tasmania: Centre for the Ore Deposit and Exploration Studies.

Menichetti, M., Lodolo, E. and Tassone, A. (2008). Structural geology of the Fuegian Andes and Magallanes fold-and-thrust belt – Tierra del Fuego Island. *Geologica Acta*, **6**, 19–42.

Miall, A. (1996). *The Geology of Fluvial Deposits: Sedimentary Facies, Basin Analysis and Petroleum Geology*. Berlin and Heidelberg: Springer-Verlag.

Moore, D. and Reynolds, R. Jr. (1989). *X-ray Diffraction and the Identification and Analysis of Clay Minerals*. New York: Oxford University Press.

Morad, S., Ketzer, J. and De Ros, L. (2000). Spatial and temporal distribution of diagenetic alterations in siliciclastic rocks: implications for mass transfer in sedimentary basins. *Sedimentology*, **47**, 95–120.

Net, L., Alonso, S. and Limarino, C. (2002). Source rock and environmental control on clay mineral associations, Lowe section of Paganzo Group (Carboniferous), Northwest Argentina. *Sedimentary Geology*, **152**, 183–199.

Nullo, F. E. and Combina, A. M. (2002). Sedimentitas terciarias continentales. In *Geología y Recursos Naturales de Santa Cruz*, ed. M. J. Haller. *Relatorio XV Congreso Geológico Argentino*, **I**, 245–258.

Nullo, F. E. and Otamendi, J. (2002). El batolito Patagónico. In *Geología y Recursos Naturales de Santa Cruz*, ed. M. J. Haller. *Relatorio XV Congreso Geológico Argentino*, **I**, 365–387.

Olivero, E. B. and Malumián, N. (2002). Upper Cretaceous–Cenozoic clastic wedges from the Austral-Malvinas foreland basins, Tierra del Fuego, Argentina: eustatic and tectonic controls. *Third European Meeting on the Palaeontology and Stratigraphy of Latin America*, Toulouse, France, Addendum (6–9).

Olivero, E. B. and Malumián, N. (2008). Mesozoic–Cenozoic stratigraphy of the Fuegian Andes, Argentina. *Geologica Acta*, **6**, 5–18.

Paredes, J. M., Foix, N., Colombo Piñol, F. *et al.* (2007). Volcanic and climatic control on fluvial style in a high-energy system: the Lower Cretaceous Matasiete Formation, Golfo San Jorge basin, Argentina. *Sedimentary Geology*, **202**, 96–123.

Parras, A., Griffin, M., Feldmann, R. *et al.* (2008). Correlation of marine beds based on Sr- and Ar-date determinations and faunal affinities across the Paleogene/Neogene boundary in southern Patagonia, Argentina. *Journal of South American Earth Sciences*, **26**, 204–216.

Peroni, G., Cagnolatti, M. and Pedrazzini, M. (2002). Cuenca Austral: marco geológico y reserve histórica de la actividad petrolera. In *Rocas reservorio de las cuencas productivas de la Argentina*, eds. M. Schiuma, G. Hinterwimmer and G. Vergani. V Congreso de Exploración y Desarrollo de Hidrocarburos, pp. 11–26.

Pettijohn, F., Potter, P. and Siever, R. (1987). *Sand and Sandstones*, 2nd Edn. New York, Berlin: Springer-Verlag.

Quattrochio, M. E., Volkheimer, W., Borromei, A. M. and Martínez, M. A. (2011). Changes of the palynobiotas in the Mesozoic and Cenozoic of Patagonia: a review. *Biological Journal of the Linnean Society*, **103**, 380–396.

Ramos, V. A. (1999). Plate tectonic setting of the Andean Cordillera. *Episodes*, **22**, 183–190.

Ramos, V. A. and Ghiglione, M. C. (2008). Tectonic evolution of the Patagonian Andes. In *The Late Cenozoic of Patagonia and Tierra Del Fuego*, ed. J. Rabassa. Developments in Quaternary Sciences, **11**, 57–72.

Raucsik, B. and Varga, A. (2008). Climato-environmental controls on clay mineralogy of the Hettangian–Bajocian successions of the Mecsek Mountains, Hungary: An evidence for extreme continental weathering during the early Toarcian

oceanic anoxic event. *Palaeogeography, Palaeoclimatology, Palaeoecology*, **265**, 1–13.

Ré, G. H., Bellosi, E. S., Heizler, M. *et al.* (2010). A geochronology for the Sarmiento Formation at Gran Barranca. In *The Paleontology of Gran Barranca: Evolution and Environmental Change through the Middle Cenozoic of Patagonia*, ed. R. H. Madden, A. A. Carlini, M. G. Vucetich and R. F. Kay. Cambridge: Cambridge University Press, pp. 46–59.

Retallack, G. (2001). *Soils of the Past. An Introduction to Paleopedology*, 2nd Edn. London: Blackwell Science Ltd.

Riccardi, A. C. and Rolleri, E. O. (1980). Cordillera Patagónica Austral. *Segundo Simposio de Geología Regional Argentina*, **2**, 1173–1306.

Riggi, J. C. (1979). Nomenclatura, categoría litoestratigráfica y correlación de la Formación Patagonia en la costa atlántica. *Revista Asociación Geológica Argentina*, **34**, 243–248.

Robert, C. and Kennett, J. (1994). Antartic subtropical humid episode at the Paleocene-Eocene boundary: clay-mineral evidence. *Geology*, **22**, 211–214.

Roberts, E. (2007). Facies architecture and depositional environments of the Upper Cretaceous Kaiparowits Formation, southern Utah. *Sedimentary Geology*, **197**, 207–233.

Rossello, E. A., Haring, C. E., Cardinali, G. *et al.* (2008). Hydrocarbons and petroleum geology of Tierra del Fuego, Argentina. *Geologica Acta*, **6**, 69–83.

Russo, A., Flores, M. and Di Benedetto, H. (1980). Patagonia austral extraandina. In *Geología Regional Argentina, Segundo Simposio Geología Regional Argentina, Academia Nacional de Ciencias* **2**, Córdoba: Academia Nacional de Ciencias, pp. 1431–1462.

Scasso, R. and Limarino, C. (1997). *Petrología y diagénesis de rocas clásticas*. Asociación Argentina de Sedimentología, Publicación Especial 1, Buenos Aires.

Simpson, G. G. (1940). Review of the mammal-bearing Tertiary of South America. *Proceedings of the American Philosophical Society*, **83**, 649–709.

Smith, G. A. (1987). Sedimentology of volcanism-induced aggradation in fluvial basins: examples from the Pacific Northwest, USA. In *Recent Developments in Fluvial Sedimentology*, ed. F. G. Ethridge, R. M. Flores and M. G. Harvey. Society of Economic Paleontologists and Mineralogists Special Publication. American Association of Petroleum Geologists, **39**, 217–228.

Soria, M. F. (2001). Los Proterotheriidae (Litopterna, Mammalia), sistemática, origen y filogenia. *Monografías del Museo Argentino de Ciencias Naturales*, **1**, 1–167.

Spalletti, L., Merodio, J. and Matheos, S. (1993). Geoquímica y significado tectónico-deposicional de las pelitas y margas cretácico-terciarias del noreste de la Patagonia argentina. *Revista Geológica de Chile*, **20**, 3–13.

Tambussi, C. (2011). Palaeoenvironmental and faunal inferences based on the avian fossil record of Patagonia and Pampa: what works and what does not. *Biological Journal of the Linnean Society*, **103**, 458–474.

Tassone, A., Lodolo, E., Menichetti, M. et al.(2008). Seismostratigraphic and structural setting of the Malvinas Basin and its southern margin (Tierra del Fuego Atlantic offshore). *Geologica Acta*, **6**, 55–67.

Tauber, A. A. (1994). Estratigrafía y vertebrados fósiles de la Formación Santa Cruz (Mioceno Inferior) en la costa atlántica entre las rías del Coyle y de Río Gallegos, Provincia de Santa Cruz, República Argentina. Unpublished Ph.D. thesis, Universidad Nacional de Córdoba, Argentina.

Tauber, A. A. (1997). Bioestratigrafía de la Formación Santa Cruz (Mioceno inferior) en el extremo sudeste de la Patagonia. *Ameghiniana*, **34**, 413–426.

Tejedor, M. F., Tauber, A. A., Rosenberger, A. L., Swisher, C. C. and Palacios, M. E. (2006). New primate genus from the Miocene of Argentina. *Proceedings of the National Academy of Sciences USA*, **103**, 5437–5441.

Thiry, M. (2000). Paleoclimatic interpretation of clay minerals in marine deposits: an outlook from the continental orogin. *Earth Science Reviews*, **49**, 201–221.

Tucker, M. (1996). *Sedimentary Petrology: An Introduction to the Origin of Sedimentary Rocks*, 2nd Edn. Blackwell Sciences, London.

Tunbridge, L. P. (1984). Facies model for a sandy ephemeral streams and clay playa complex: the Middle Devonian Trentishoe Formation of north Devon, UK. *Sedimentology*, **31**, 697–715.

Uliana, M., Biddle, K., Phelps, D. and Gust, D. (1985). Significado del vulcanismo y extensión mesojurásicos en el extremo meridional de Sudamérica. *Revista de la Asociación Geológica Argentina*, **40**, 231–253.

Umazano, A. M., Bellosi, E. S., Visconti, G., Jalfin, G. A. and Melchor, R. N. (2009). Sedimentary record of a Late Cretaceous volcanic arc in central Patagonia: petrography, geochemistry and provenance of fluvial volcaniclastic deposits of the Bajo Barreal Formation, San Jorge Basin, Argentina. *Cretaceous Research*, **30**, 749–766.

Vizcaíno, S. F., Bargo, M. S., Kay, R. F. and Milne, N. (2006). The armadillos (Mammalia, Xenarthra, Dasypodidae) of the Santa Cruz Formation (Early–Middle Miocene): an approach to their paleobiology. *Palaeogeography, Palaeoclimatology, Palaeoecology*, **237**, 255–269.

Vizcaíno, S. F., Bargo, M. S., Kay, R. F. *et al.* (2010). A baseline paleoecological study for the Santa Cruz Formation (late–Early Miocene) at the Atlantic coast of Patagonia, Argentina. *Palaeogeography, Palaeoclimatology, Palaeoecology*, **292**, 507–519.

Worden, R. and Burley, S. (2003). Sandstone diagenesis: the evolution of sand to stone. In *Sandstone Diagenesis: Recent and Ancient*, eds. S. Burley and R. Worden. International Association of Sedimentologists, **4**, 2–44.

Zachos, J., Pagani, M., Sloan, L., Thomas, E. and Billups, K. (2001). Trends, rhythms, and aberrations in global climate 65 Ma to present. *Science*, **292**, 686–693.

5 Oysters from the base of the Santa Cruz Formation (late Early Miocene) of Patagonia

Miguel Griffin and Ana Parras

Abstract

The Santa Cruz Formation (late Early Miocene of Patagonia, Argentina) has been traditionally considered a continental unit and its contact with the underlying Early Miocene Monte León Formation as transitional or unconformable according to different authors. Invertebrates – mainly molluscs – present in the purportedly transitional beds were used as an argument to refer them to the marine Monte León Formation. This was generally accepted, but the lithological features shared with the continental Santa Cruz Formation indicate that they should be included within this unit instead. Analysis of the fauna revealed that the only *in situ* species in these beds is *Crassostrea orbignyi* (Ihering, 1897). This oyster belongs in a group adapted to inhabiting marginal marine settings with hard or soft bottom, and tolerating a wide range of salinity and subaereal exposure. The other accompanying members of the fauna are reworked marine molluscs referable to taxa included in the underlying Monte León Formation.

Resumen

La Formación Santa Cruz (Mioceno Temprano tardío de Patagonia, Argentina) ha sido tradicionalmente considerada como de origen exclusivamente continental y su contacto con la infrayacente Formación Monte León (Mioceno Temprano) como de naturaleza transicional o discordante, según diferentes autores. La presencia de invertebrados, principalmente moluscos, en los niveles supuestamente transicionales se usó como argumento para incluirlos en la Formación Monte León, de origen marino. Esto fue ampliamente aceptado; sin embargo, las características litológicas comunes con la Formación Santa Cruz indican que deberían ser incluidos en esta unidad. Un análisis de la fauna reveló que la única especie *in situ* en estos niveles es una ostra, *Crassostrea orbignyi* (Ihering, 1897), perteneciente a un grupo adaptado a vivir en ambientes marinos marginales, tanto de fondos duros como blandos y tolerantes a un amplio rango de salinidad y exposición subaérea. Los otros integrantes de la fauna son moluscos marinos retrabajados, todos pertenecientes a taxones incluidos en la infrayacente Formación Monte León.

Early Miocene Paleobiology in Patagonia: High-Latitude Paleocommunities of the Santa Cruz Formation, ed. Sergio F. Vizcaíno, Richard F. Kay and M. Susana Bargo. Published by Cambridge University Press. © Cambridge University Press 2012.

5.1 Introduction

The Santa Cruz Formation (late Early Miocene of Patagonia, Argentina) is known for its rich fauna of fossil vertebrates – mainly mammals and birds. A complete account of the vertebrate fauna it carries and a discussion of its biostratigraphic, paleoecologic, evolutionary, and palaeobiogeographic implications are given elsewhere in this volume.

In contrast to the vertebrate fauna, the invertebrate fauna contained in the Santa Cruz Formation is far more restricted from both the stratigraphic and the taxonomic perspectives. It is restricted to the lower 40 m of the unit. Two taphonomically different assemblages can be distinguished: (1) *in situ* concentrations of an oyster species unique to these beds, and (2) material – mainly molluscs – redeposited from the underlying Early Miocene Monte León Formation and generally showing evidence of diverse taphonomic signatures.

The earliest records of these invertebrates were by Ihering (1897), who described a number of species that were collected by Carlos Ameghino from the "*Étage Superpatagonienne*" of the "*Formatio Santacruzensis*." Part of the material was sent by Florentino Ameghino to Maurice Cossmann, who described a few gastropods based on generally poorly preserved material (Cossmann, 1899).

5.2 The Superpatagonian

During the Late Cretaceous, Paleogene, and Neogene, the surface of Patagonia was covered repeatedly by Atlantic transgressions that deposited marine rocks referred nowadays to a number of stratigraphic units. These transgressions are intercalated with non-marine rocks and intervals of non-deposition and erosion (Feruglio, 1949; Malumián, 1999).

Of these transgressions, two informally known as "*Patagoniano*" occurred between the Late Oligocene and the Early Miocene. The rocks deposited during these transgressions are currently included in the Late Oligocene–Early Miocene? San Julián Formation and in the Early Miocene Monte León Formation along the Atlantic coast of Santa Cruz Province, and in the Late Oligocene–Early Miocene Centinela Formation (= 25 de Mayo Formation; Cuitiño and Scasso, 2010) along the Andean foothills in this province (Austral Basin) (Parras *et al.*, 2008). The Patagonian transgressions were also identified in the Early Miocene Chenque and Gaiman Formations, in the San Jorge Basin, north of the Deseado Massiff (Bellosi, 1995; Malumián, 1999). In southern Chile, these rocks are known as the Guadal Formation, and are exposed south of Lago General Carrera/Lago Buenos Aires (Frassinetti and Covacevich, 1999).

The marine invertebrate fossils included in the Santa Cruz Formation ("*Superpatagonian*") may be better understood when the history of the successive explorations of the underlying marine rocks referred to the "*Patagoniano*" is properly accounted for.

Alcide d'Orbigny (1842) published the earliest record of marine rocks in Patagonia. He explored this part of South America during the years 1826–1833, but d'Orbigny himself never went ashore further south than the coast of northern Patagonia. He did, however, name all the deposits that constitute the cliffs along the coast between Carmen de Patagones (Río Negro) and the Strait of Magellan as "*Tertiaire Patagonien*." Although he was unable to study the cliffs further south, he did have available the large oysters from San Julián. D'Orbigny believed they were identical to those he had collected in Río Negro and further north in Entre Ríos.

The second explorer to survey the geological features of the cliffs along the coast of Patagonia was Charles Darwin. He used d'Orbigny's terminology, but noted that there were differences between the faunas from the southern and northern areas in which Darwin's "*Patagonian Tertiary Formation*" was exposed (Darwin, 1846). After these initial surveys the area remained largely unexplored for many years from a geological and paleontological point of view. Ramón Vidal Gormaz collected a few fossils from the cliffs just two kilometers west from the present city of Puerto Santa Cruz; these were described by Philippi (1887). The only geological information attached to these fossils was that they were collected from the Tertiary marine beds exposed at Santa Cruz. As the nineteenth century drew to a close, this region was surveyed again and its geology described in more detail. Three early explorers were responsible for assembling the data that for many years were used to support diverse stratigraphic, paleogeographic, and paleobiogeographic interpretations of the rocks and fossil biota in the area.

Carlos Ameghino – brother of Florentino Ameghino, the famed Argentine paleontologist – surveyed the area and collected vertebrate and invertebrate fossils that were subsequently studied either by his brother (vertebrates) or other renowned paleontologists or malacologists of the time (invertebrates). Although C. Ameghino did not publish his results extensively, his collections and his keen and detailed geological observations were used by Florentino to formulate the earliest biostratigraphic subdivisions of the "*Tertiaire Patagonien*" or the "*Great Patagonian*

Formation." Based on the fossil content (mainly molluscs) of the marine beds at the mouth of the Río Santa Cruz and in the environs of Puerto San Julián, Ameghino (1898) split the marine beds he had named "*Formación Patagónica*" (Ameghino 1896) into two chronostratigraphic units he called "*Juliense*" and "*Leonense.*" Overlying these, Ameghino (1900–1902) recognized the continental "*Formación Santacruceña*," at the base of which he could distinguish a marine stage he called "*Superpatagónico.*" The fauna of marine invertebrates that was the basis of Ameghino's subdivisions was collected by Carlos and later described by Cossmann (1899) and Ihering (1897, 1899, 1902, 1907, 1914).

During the same years, the area was visited by John Bell Hatcher, geologist in charge of the Princeton University Expedition to southern Patagonia. His observations were published in the reports and geology of this expedition (Hatcher, 1903) and in a series of papers in the *American Journal of Science* (Hatcher, 1897, 1900), and his collection of Tertiary marine fossils was described by Ortmann (1902). Hatcher's conclusions on the geology of the area differed from those of Florentino Ameghino, as he did not recognize the existence of the "*Superpatagónico.*" Hatcher stated that the fauna described by Ortmann (1902) included essentially the same taxa as those present in the underlying "*Formación Patagónica.*" This initial disagreement pervaded all further work on these units and has never been satisfactorily resolved. Poor knowledge of the sedimentology and stratigraphy at the localities involved, inaccurate location of some of these localities, and outdated taxonomic placement of some of the fossil material on which the biostratigraphy was based – among others – were reasons for continued controversy.

In 1897 Beniamino Bicego was sent by Ihering to Patagonia with the task of assembling further collections at the localities initially visited by C. Ameghino. His large collection of molluscs was described by Ihering, but Bicego himself did not provide further geological data beyond that mentioned by Ameghino. Thus, geological interpretations by Ihering were largely based on F. Ameghino's work.

Wichmann (1922), Windhausen (1931), and Feruglio (1949) published detailed accounts on the geological features of Patagonia, including the marine Cenozoic rocks. Their sections and geological observations are generally easy to corroborate in the field. Again, interpretations largely reflect the ideas of Ameghino and Ihering.

Henceforth, the use of "*Patagonian Formation*" for the Cenozoic marine rocks in southern South America was widespread, although the exact chronostratigraphic and lithostratigraphic meaning of this formational name remained somewhat unstable.

At the mouth of the Río Santa Cruz, rocks included in the "*Patagonian Formation*" were formally named Monte León Formation by Bertels (1970), to include mainly

Fig. 5.1. Exposures of the lower beds of the Santa Cruz Formation at Cerro Observación in the Monte León National Park, Santa Cruz Province. Note the intercalated lenticular *in situ* oyster beds (*c*. 1 m thick) of *Crassostrea orbignyi* (Ihering, 1897). Photo by Mariana Martínez, National Park Service Argentina.

yellowish-gray siltstones and fine sandstones with high pyroclastic content and intercalated shell beds. The rocks record a relatively high sea stand reflecting the greatest extent of the Patagonian Cenozoic Sea, thus allowing input of corrosive Antarctic water onto the present Argentine continental shelf (Malumián, 2002).

It is generally agreed that these rocks were deposited in a shallow marine environment under littoral and neritic conditions (Panza *et al.*, 1995; Barreda and Palamarczuk, 2000). Bertels (1980) suggested outer shelf conditions for the lower part of this formation. Panza *et al.* (1995) stated that towards the top of the unit conditions became progressively shallower, ending in a marshy environment just before the onset of the continental environment in which the overlying Santa Cruz Formation was deposited. Parras and Griffin (2009) stated that the different fossil concentrations contained in the Monte León Formation originated in different settings, ranging from inner shelf to nearshore subtidal to intertidal environments.

Accounts of the history of the subdivisions of the Patagonian Formation (i.e. the Monte León Formation as understood here) can be found in Camacho (1974, 1979, 1995), del Río (2004), and Parras and Griffin (2009). The latter authors concluded that the use of the term "*Superpatagonian*" is superfluous.

5.3 Oysters of the Santa Cruz Formation

The most conspicuous invertebrate elements at the base of the Santa Cruz Formation are oyster beds, which appear to be widespread throughout Patagonia. These oyster beds are always present wherever the base of the Santa Cruz Formation is exposed. The oysters appear either as sedimentologic concentrations or else as *in situ* oyster beds, which show variable degrees of continuity and thickness. Initially described from Yegua Quemada, a locality now lying within the Monte León National Park, these oysters are known also from Cerro Monte León, Cerro Observación (Fig. 5.1), Monte Observación (of C. Ameghino; now known as Cerro Observatorio), Rincón del Buque, and Cerro Redondo, all localities along the Atlantic coast of Santa Cruz between Monte León and the mouth of the Río Coyle, where the uppermost oyster beds are visible only at sea level (Fig. 5.2). Beyond this area, they have also been observed by us in Estancia Quien Sabe and Estancia 25 de Mayo (south of Lago Argentino), and at Lago Pueyrredón. Localities are given in Vizcaíno *et al.* (Chapter 1, Appendix 1.1, and Figs. 1.1 and 1.2).

The oyster beds are usually patchy, especially the *in situ* beds with oysters in life position (Fig. 5.3). Sedimentologic concentrations form more uniform beds, in which the shells are variably broken or reworked. This oyster belongs in a taxonomic group adapted to inhabiting marginal marine settings, with hard or soft bottom, and tolerating a wide range of salinity and subaerial exposure. The resulting different ecophenotypes show great variation and are very irregularly distributed, but are always easily recognizable owing to the peculiarities shown by the shell of this oyster. Shell characters allow taxonomic assignment to *Crassostrea* Sacco, 1897

Fig. 5.2. Exposures of the lower beds of the Santa Cruz Formation at Cerro Redondo, Santa Cruz Province, with sedimentological concentrations at the base and *in situ* oyster beds of *Crassostrea orbignyi* (Ihering, 1897) overlying it.

Fig. 5.3. Detail of a shell bed at Cerro Redondo, showing nesting specimens of *Crassostrea orbignyi* (Ihering, 1897) in life position.

[type species: *Ostrea* (*C.*) *virginica* (Gmelin) (1791)], within the subfamily Crassostreinae Scarlato and Starobogatov, 1979, as understood by Malchus (1990).

A detailed account of this genus was published by Lawrence (1995). Following a careful analysis of the taxa usually referred to the Crassostreinae, Lawrence concluded that none of the characters used to separate different nominal genera is valid, and that all belong to *Crassostrea*.

Among the shell characters used by Lawrence (1995) in his diagnosis of the genus *Crassostrea* are the presence of a left-valve umbonal cavity, left-valve cupping, dorsoventral elongation, and posterior and/or ventral displacement of the adductor muscle scar. He also stated that these characters could appear singly or in combination in a given species, and that their expression was not constant throughout the genus. Other shell characters he considered

Fig. 5.4. *Crassostrea orbignyi* (Ihering, 1897), from Yegua Quemada, Santa Cruz Province, Argentina – the type locality of the species. All illustrated specimens are housed in the División Paleozoología Invertebrados, Museo de La Plata, Argentina. a, b, MLP-27217, right valve; a, interior; b, exterior. c, d, MLP 27215, left valve; c, exterior; d, interior. e, MLP 27219, right valve, interior. f, g, MLP 27216, left valve, f, interior; g, exterior. Scale bar = 3 cm.

are the non-orbicular shape of the adductor muscle scar and the prominent valve chambers, together with the non-vesicular chalky deposits, when these are present. Based on such characters, the oyster of the Santa Cruz Formation, i.e. *Ostrea orbignyi* Ihering, can be placed in *Crassostrea*, along with other large species of oysters from Patagonia. Therefore, it should be correctly referred to as *Crassostrea orbignyi* (Ihering, 1897) (Fig. 5.4).

This taxon has a very confusing nomenclatural history. Ihering published the first illustration and description of a specimen connected with the name *Ostrea orbignyi* in 1897 (Ihering, 1897: 222, pl. 9, fig. 52), although there seems to have been some typographical error, as the description is under the name *Ostrea patagonica* d'Orbigny, 1842. However, the name "*orbignyi*" is written immediately below as well as in the figure captions. Ihering himself later (Ihering, 1907: 16) acknowledged that he had introduced the name.

Since that time, several different taxa have been incorrectly allocated to *Ostrea orbignyi*. As restricted here, the species should include only the specimens from the marginal marine intercalation at the base of the Santa Cruz Formation, which overlies the Monte León Formation along the Atlantic coast, and the Centinela Formation along the foothills of the Andes. The shells of most specimens of this species show well-developed chomata on the margins near the umbo. Some specimens show also delicate radial striae on the outer surface of the right valve. This may lead to mistaken inclusion of these specimens in *Striostrea* Vyalov, 1936. Close inspection, however, shows that these striae are nothing but the traces of chomata revealed upon partial decortication of shell surface.

After Ihering's 1897 paper, this species was confused with *Ostrea philippii* Ortmann, 1897, a quite different taxon that appears in the Monte León Formation, and is synonymous with *Ostrea hatcheri* Ortmann, 1897. This is the case for Ihering's illustration of *Ostrea orbignyi* of 1907 (as *Ostrea d'Orbignyi*; Ihering, 1907, pl. 2, fig. 9), which is actually a right valve of *Ostrea hatcheri*. He did not state where it came from, but it seems likely that it was collected in western Santa Cruz Province (Sierra de los Baguales).

Just south of Trelew, in Chubut Province, a very similar oyster can be found at the top of the Puerto Madryn Formation, a Late Miocene unit exposed in northern Patagonia. The identity of this material with *Crassostrea orbignyi* still needs to be confirmed.

5.4 Other molluscs in the Santa Cruz Formation

Ihering (1907) reported 141 species of molluscs in the "*Superpatagonian.*" However, almost all of them come from the shell beds at the top of the Monte León Formation, while a few of them are specimens reworked into the overlying Santa Cruz Formation. All the shell beds in the Monte León Formation are incredibly rich, and the material is in general exquisitely preserved, many gastropods showing the nacreous shell layer, details of micro-ornamentation, and even perfectly well-preserved protoconchs.

Del Río (2004) recognized two different mollusc assemblages (RSP and PA) in the area south of the mouth of the Río Santa Cruz. Her Assemblage PA is represented at Cerro Monte León, Cerro Observación, and Las Cuevas, but the presence of *Crassostrea orbignyi* was not recorded by her at any of these three localities (del Río, 2004: Appendix 2). However, this oyster is by far the most common taxon in the exposures at Cerro Monte León and Cerro Observación. It occurs together with reworked specimens of *Ostrea hatcheri* and of other groups of molluscs (mainly micromolluscs). *Crassostrea orbignyi* is not present at Las Cuevas, while *Ostrea hatcheri* occurs there among the rich mollusc fauna. This fauna, together with the oyster, is mostly parautochthonous or autochthonous and shows little evidence of reworking. Thus, Assemblage PA of del Río (2004) contains two suites of molluscs, i.e. one including the species contained in the shell-beds at Las Cuevas, and a second one with the autochthonous *Crassostrea orbignyi* and the specimens of reworked molluscs. These are taxonomically identical to species appearing in the underlying shell beds exposed nearby (at Las Cuevas and other localities along the coastal cliff).

As at Cerro Observación and Cerro Monte León, at other localities above the uppermost shell beds of the Monte León Formation the lithology changes and exposures are far less common. The cliffs retreat inland (from a few dozen meters to a few kilometers), where the Santa Cruz Formation reaches up to the flat tops of the Patagonian tablelands. The marginal marine intercalations are exposed at the base of this second cliff. It is in these beds that – besides the sedimentological and *in situ* beds of *Crassostrea orbignyi* – remains of other marine invertebrates occur. These are generally abraded and broken, but are clearly part of the fauna from the top beds of the Monte León Formation, which have been reworked into the base of the Santa Cruz Formation. While most of the specimens still retain their aragonitic shells to different degrees, these show signs of decortication, fragmentation, abrasion, and corrosion, which render identification difficult. Among the species represented by reworked specimens in these beds at the base of Cerro Monte León and Cerro Observación are *Nuculana*? sp., *Yoldia*? sp., *Pseudoportlandia glabra* (Sowerby, 1846), *Glycymeris cuevensis* (Ihering, 1897), *Ostrea hatcheri* Ortmann, 1897, *Mactra*? sp., *Pleuromeris cruzensis* (Ihering, 1907), *Eosolen crucis* (Ihering, 1907), *Caryocorbula hatcheri* (Ortmann, 1900), *Valdesia dalli* (Ihering, 1897), *Crepidula gregaria* Sowerby, 1846, *Trochita* sp., *Adelomelon (Pachycymbiola) ameghinoi* (Ihering, 1896), and Terebratulidae indet.

5.5 Conclusions

Any biostratigraphic interpretation based on the fossil content of the purportedly transitional beds between the Monte León Formation and the Santa Cruz Formation, or at the base of the latter, should be carefully assessed. The fossils it contains – except *Crassostrea orbignyi* – show signs of reworking, and all identifiable taxa are also found in the underlying Monte León Formation. Thus the composite faunal list of the "*Superpatagonian*" beds cannot be used to draw paleoecological and/or chronostratigraphic interpretation on the Santa Cruz Formation. However, they may be highly significant from a taphonomic point of view.

Careful sampling of the bottom beds of the Santa Cruz Formation and identification of the taphonomic signature of the reworked molluscs it contains (in addition to the *in situ Crassostrea orbignyi* beds), together with detailed sedimentological data, may allow conclusions to be drawn on the environmental conditions prevailing at the onset of sedimentation of the Santa Cruz Formation, at the time that continentalization of Patagonia began during the late Early Miocene.

Crassostrea orbignyi is the only autochthonous marine invertebrate fossil contained in the rocks constituting the base of the Santa Cruz Formation at several localities along the coast (Yegua Quemada, Cerro Monte León, Cerro Observación, Monte Observación (= Co. Observatorio), Rincón del Buque, and Cerro Redondo) and along the foothills of the Andes (localities in the Lago Argentino and Lago Pueyrredón areas). The rocks at the base of the Santa Cruz Formation exposed at these localities are not necessarily of the same age. Therefore, the presence of *Crassostrea orbignyi* is not time-related, but facies-related instead, suggesting similar environmental conditions at the onset of deposition of the Santa Cruz Formation.

REFERENCES

Ameghino, F. (1896). Notas sobre cuestiones de geología y paleontología argentina. *Boletín del Instituto Geográfico Argentino*, **17**, 87–119.

Ameghino, F. (1898). Sinopsis Geológica-Paleontológica. *Segundo Censo de la República Argentina*, **1**, 111–225.

Ameghino, F. (1900–1902). L'age des formations sédimentaires de Patagonie. *Anales de la Sociedad Científica Argentina*, **50**, 109–130, 145–165, 209–229 (1900); **51**, 20–39, 65–91 (1901); **52**, 189–197, 244–250 (1901); **54**,161–180, 220–249, 283–342 (1902).

Barreda, V. and Palamarczuk, S. (2000). Palinomorfos continentales y marinos de la Formación Monte León en su área tipo, provincia de Santa Cruz, Argentina. *Ameghiniana*, **37**, 3–12.

Bellosi, E. (1995). Paleografía y cambios ambientales de la Patagonia Central durante el Terciario Medio. *Boletín de Informaciones Petroleras*, **44**, 50–83.

Bertels, A. (1970). Sobre el "Piso Patagoniano" y la representación de la época del Oligoceno en Patagonia austral, República Argentina. *Revista de la Asociación Geológica Argentina*, **25**, 495–501.

Bertels, A. (1980). Estratigrafía y foraminíferos (Protozoa) bentónicos de la Formación Monte León (Oligoceno) en su área tipo, provincia de Santa Cruz, República Argentina. *2° Congreso Argentino de Paleontología y Bioestratigrafía y 1° Congreso Latinoamericano de Paleontología, Actas*, **2**, 213–273.

Camacho, H. H. (1974). Bioestratigrafía de las formaciones marinas del Eoceno y Oligoceno de la Patagonia. *Anales de la Academia de Ciencias Exactas, Físicas y Naturales*, **26**, 39–57.

Camacho, H. H. (1979). Significados y usos de "Patagoniano", "Patagoniense", "Formación Patagonica", "Formación Patagonia" y otros términos de la estratigrafía del Terciario marino argentino. *Revista de la Asociación Geológica Argentina*, **34**, 235–242.

Camacho, H. H. (1995). La Formación Patagonica (F. Ameghino, 1894): Su actual significación estratigráfica y paleontológica. *Anales de la Academia Chilena de Ciencias*, **5**, 117–151.

Cossmann, M. (1899). Description des quelques coquilles de la formation Santacruzienne en Patagonie. *Journal de Conchiliologie*, **47**, 223–241.

Cuitiño, J. I. and Scasso, R. A. (2010). Sedimentología y paleoambientes del Patagoniano y su transición a la Formación Santa Cruz al sur del Lago Argentino, Patagonia Austral. *Revista de la Asociación Geológica Argentina*, **66**, 406–417.

Darwin, C. (1846). *Geological Observations on South America: Being the Third Part of the Geology of the Voyage of the Beagle, Under the Command of Capt. Fitzroy, R.N. During the Years 1832 to 1836*. London: Smith, Elder and Co.

del Río, C. J. (2004). Tertiary marine molluscan assemblages of eastern Patagonia (Argentina): a biostratigraphic analysis. *Journal of Paleontology*, **78**, 1097–1122.

d'Orbigny, A. (1842). *Voyage dans l'Amérique Méridionale (le Brésil, la République Orientale de l'Uruguay, la République Argentine, la Patagonie, la République du Chili, la République du Bolivia, la République du Perou), exécuté pendant les années 1826, 1827, 1828, 1829, 1830, 1831, 1832 et 1833*, Vol. 3, parts 3 and 4. Paris: P. Bertrand; Strasbourg: V. Levrault.

Feruglio, E. (1949). Descripción Geológica de La Patagonia. Ministerio de Industria y Comercio de La Nación, *Dirección General de Yacimientos Petrolíferos Fiscales*, **2**, 1–349.

Frassinetti, D. and Covacevich, V. (1999). Invertebrados fósiles marinos de la Formación Guadal (Oligoceno Superior – Mioceno Inferior) en Pampa Castillo, Región de Aisén, Chile. *Boletín del Servicio Nacional de Geología y Minería (Chile)*, **51**, 1–96.

Hatcher, J. B. (1897). On the geology of southern Patagonia. *American Journal of Science*, **4**, 327–354.

Hatcher, J. B. (1900). Sedimentary rocks of southern Patagonia. *American Journal of Science*, **9**, 85–107.

Hatcher, J. B. (1903). Geography of southern Patagonia. In *Reports of the Princeton University Expeditions to Patagonia 1896–1899*. Vol. 1, *Narrative and Geography*, ed. W. B. Scott. Princeton: Princeton University Press, pp. 1–296.

Ihering, H. von (1897). Os Molluscos dos terrenos terciarios da Patagonia. *Revista do Museo Paulista*, **2**, 217–382.

Ihering, H. von (1899). Die Conchylien der patagonischen Formation. *Neues Jahrbuch für Mineralogie, Geologie und Paläontologie*, **1899**, 1–46.

Ihering, H. von (1902). Historia de las ostras argentinas. *Anales del Museo Nacional de Buenos Aires*, **7**, 109–123.

Ihering, H. von (1907). Les mollusques fossiles du Tertiaire et du Crétacé superieur de l'Argentine. *Anales del Museo Nacional de Buenos Aires, serie* **3**, 1–611.

Ihering, H. von (1914). Catálogo de molluscos Cretáceos e Terciarios da Argentina da collecçao do auctor [with appendix in German]. *Notas Preliminares da Revista do Museu Paulista*, **1**, 1–148.

Lawrence, D. R. (1995). Diagnosis of the genus *Crassostrea* (Bivalvia, Ostreidae). *Malacologia*, **36**, 185–202.

Malchus, N. (1990). Revision der Kreide-Austern (Bivalvia: Pteriomorphia) Ägyptens (Biostratigraphie, Systematik). *Berliner geowissenschaftliche Abhandlungen*, A **125**, 1–321.

Malumián, N. (1999). La sedimentación y el volcanismo terciarios en la Patagonia extraandina. 1. La sedimentación en la Patagonia extraandina. In *Geología Argentina*, ed. R. Caminos, Anales del Servicio Geológico Minero Argentino y del Instituto de Geología y Recursos Minerales, **29**, 557–612.

Malumián, N. (2002). El Terciario marino. Sus relaciones con el eustatismo. In *Geología y Recursos Naturales de Santa Cruz*, ed. M. J. Haller. Relatorio 15° Congreso Geológico Argentino, **I-15**, 237–244.

Ortmann, A. E. (1902). Tertiary invertebrates. In *Reports of the Princeton University Expeditions to Patagonia, 1896–1899*. Vol. 4, *Paleontology*, ed. W. B. Scott. Princeton: Princeton University Press, pp. 45–332.

Panza, J. L., Irigoyen, M. V. and Genini, A. (1995). Hoja Geológica 4969-IV, Puerto San Julián, provincia de Santa Cruz, República Argentina. *Boletín de la Secretaría de Minería de la Nación, Dirección Nacional del Servicio Geológico*, **211**, 1–77.

Parras, A. and Griffin, M. (2009). Darwin's Great Patagonian Formation at the mouth of the Santa Cruz River: a reappraisal. In *Darwin in Argentina*, ed. B. Aguirre-Urreta, M. Griffin and V. A. Ramos, *Revista de la Asociación Geológica Argentina*, **64**, 70–82.

Parras, A., Griffin, M., Feldmann, R. *et al.* (2008). Correlation of marine beds based on Sr- and Ar-date determinations and faunal affinities across the Paleogene/Neogene boundary in southern Patagonia, Argentina. *Journal of South American Earth Sciences*, **26**, 204–216.

Philippi, R. A. (1887). *Los fósiles terciarios i cuartarios de Chile*. Leipzig: F. A. Brockhaus; Santiago: Museo Nacional de Historia Natural.

Wichmann, R. (1922). Observaciones geológicas en el Gran Bajo de San Julián y sus alrededores (Territorio de Santa Cruz). *Boletín Dirección General de Minas*, **30**B, 1–34.

Windhausen, A. (1931). *Geología Argentina. Segunda Parte: Geología Histórica y Regional del territorio Argentino*. Buenos Aires: Editorial Peuser.

6 Ichnology of distal overbank deposits of the Santa Cruz Formation (late Early Miocene): paleohydrologic and paleoclimatic significance

Verónica Krapovickas

Abstract

This chapter deals with the trace fossils, also called ichno-fossils, preserved in the lower Estancia La Costa Member of the Santa Cruz Formation (late Early Miocene) along the Atlantic coast, between the Río Gallegos and Río Coyle, Santa Cruz Province, Argentina. The succession is mostly composed of fluvial deposits. Extensive overbank areas record trace fossils in a variety of deposits such as floodplain water bodies and paleosols that developed under variable climatic conditions. From bottom to top, it records a general trend of paleosol development from humid to drier climatic conditions, with waterlogged areas developed in the middle portion of the succession. Floodplain water bodies record *Taenidium barretti* and *Palaeophycus tubularis* which correspond to a "pre-desiccation suite" of the *Scoyenia* ichnofacies that developed in soft substrates. Also, root traces are preserved when the time between depositional events is long enough to allow colonization by plants but not so much as to obliterate animal traces. Integrated ichnology and sedimentology suggests that paleosols that record abundant cf. - *Capayanichnus vinchinensis*, fine and haloed root traces, and the less common occurrence of *Taenidium barretti* and *Planolites beverleyensis* were episodically waterlogged and are considered moderately drained. Other paleosols that record abundant ferric root traces suggest that they were moderately well-drained and developed under more humid climatic conditions. Similar moderately well-drained paleosols record abundant calcareous rhizoconcretions and a dwelling burrow attributed to a mammal. A third type of paleosol contains cells of solitary digging bees (*Celliforma* isp.) and ferric root traces, and is interpreted as being moderately well-drained and developed under drier climatic conditions.

Resumen

Este capítulo trata sobre trazas fósiles, también llamadas icnófosiles, preservadas en el Miembro inferior (Miembro Estancia La Costa) de la Formación Santa Cruz (Mioceno Temprano tardío) en la costa atlántica, entre el Río Gallegos y el Río Coyle, provincia de Santa Cruz, Argentina. La sucesión está mayormente compuesta por depósitos fluviales. Sus extensas planicies de inundación presentan trazas fósiles preservadas en una variedad de depósitos, como cuerpos de agua y paleosuelos desarrollados bajo condiciones climáticas variables. Estratigráficamente, de abajo hacia arriba, se registra una tendencia general al desarrollo de paleosuelos bajo condiciones climáticas húmedas a más secas, con áreas inundadas hacia la porción media de la sucesión. Los cuerpos de agua desarrollados en planicies de inundación registran *Taenidium barretti* y *Palaeophycus tubularis* correspondientes al conjunto de pre-desecación de la icnofacies de *Scoyenia* desarrollada en sustratos blandos. Adicionalmente, se preservan trazas de raíces cuando el tiempo transcurrido entre eventos depositacionales fue suficientemente largo para permitir la colonización por plantas, pero no demasiado como para obliterar las trazas de animales. Estudios icnológicos y sedimentológicos integrados sugieren que los palesuelos con abundantes cf. *Capayanichnus vinchinensis*, trazas de raíces delicadas y con halo y con pocos *Taenidium barretti* and *Planolites beverleyensis* fueron temporalmente anegados y son considerados moderadamente drenados. El registro de trazas de raíces férricas en palesuelos, sugiere que estos fueron moderadamente bien drenados y desarrollados bajo condiciones climáticas húmedas. Otros paleosuelos moderadamente bien drenados, registran abundantes rizoconcreciones y una excavación de habitación atribuida a mamíferos. Un tercer tipo de paleosuelo contiene celdas de abejas cavadoras solitarias (*Celliforma* isp.) y trazas de raíces férricas. Éste ha sido interpretado como moderadamente bien drenado y desarrollados bajo condiciones climáticas más secas.

6.1 Introduction

The Santa Cruz Formation (late Early Miocene) is a continental succession regionally exposed from the Andean Precordillera to the Atlantic coast of southern Patagonia (Matheos and Raigemborn, Chapter 4). Along the Atlantic coast of the Santa Cruz Province those deposits contain a rich assemblage of fossil mammals (Vizcaíno *et al.*, Chapter 1).

Most studies have focused on systematic, biostratigraphic, paleobiologic, and paleoecologic aspects of its mammalian fauna, which constitute the Santacrucian South American Land Mammal Age (SALMA) (e.g. Tauber, 1997a, b, 1999; Croft, 2001; Kay *et al.*, 2008; Vizcaíno *et al.*, 2006; 2010). In contrast, the ichnological content of the Santa Cruz Formation is poorly known, and there is a lack of detailed studies focused on the trace fossils preserved from the most important fossiliferous levels. The only documented trace fossils are from Monte Observación (= Cerro Observatorio; see Vizcaíno *et al.*, Chapter 1), approximately 60 km north of the study area where a few bee nest specimens (*Celliforma rosellii* Genise and Bown, 1994), some dung beetle balls, and root traces were reported (Genise and Bown, 1994). In addition, Hatcher (1900, 1903) mentioned the presence of fossil footprints in levels of the Santa Cruz Formation at Corriguen Aike (= Puesto Estancia La Costa; see Vizcaíno *et al.*, Chapter 1). Unfortunately, neither a description nor a figure was included. Vizcaíno *et al.* (Chapter 1) report newly discovered footprints at Puerto Estancia La Costa and Monte Tigre–Cañadón Las Totoras that are currently under study. The study of trace fossils is a very useful source of paleoenvironmental and paleoecological data, integrating aspects of both sedimentology and paleobiology. Thus, trace fossil analysis of the Santa Cruz Formation represents a potential framework to complement paleoecologic and paleoenvironmental interpretation of the Santacrucian fauna. Here, I analyze the ichnofauna preserved in the most important fossiliferous levels of the Santa Cruz Formation on the Atlantic coast, between the Río Coyle and Río Gallegos.

6.2   Geologic framework and location of the study area

The Santa Cruz Formation at the Atlantic coast is up to 225 m in preserved thickness. It is mainly composed of sandy river channels and overbank deposits upon which paleosols are developed (Bown and Fleagle, 1993; Tauber, 1994, 1997a). North of the Río Coyle, at Monte León and Monte Observación (= Cerro Observatorio), Bown and Fleagle (1993) described the lithology of the Santa Cruz Formation. To the south, Tauber (1997a) studied the stratigraphy between the Río Coyle and the Río Gallegos and defined two members. The lower Estancia La Costa (ELC) Member is dominated by pyroclastic deposits, claystones, and mudstones, encompassing the fossiliferous levels where most of the fossil vertebrates are preserved. The upper Estancia La Angelina Member is composed mainly of claystones, mudstones, and sandstones, containing less-abundant fossil vertebrate remains. Tauber (1994, 1997a, b) interpreted that the ELC Member was deposited by a low-energy fluvial system of integrated channels with low width/depth ratios, lenticular shapes, and lateral stability. This is consistent with Matheos and Raigemborn's

interpretation (Chapter 4) of a fluvial environment characterized by sheet-flooding, overbank-flooding, ash-fall deposits, and single laterally stable channels.

Extensive overbank areas of the ELC Member record diverse trace fossils mostly preserved in floodplain water bodies and in paleosols in distal overbank deposits. For this study, I focus on the localities that involve most of the fossiliferous levels of the ELC Member (see Matheos and Raigemborn, Chapter 4, Fig. 4.3). All are exposed in the present-day intertidal zone and include, from north to south (see Vizcaíno *et al.*, Chapter 1, Figs. 1.1 and 1.2), Punta Sur (Cañadón del Indio), Campo Barranca, Anfiteatro, Estancia La Costa, Cañadón Silva, Puesto Estancia La Costa (= Corriguen Aike), and Monte Tigre–Cañadón las Totoras (in Estancia La Angelina).

6.3   Materials

The specimens studied are mostly preserved in the field. The most relevant material was collected and placed in the Museo Regional Provincial Padre M. J. Molina (MPM-PIC), Santa Cruz Province, Argentina.

MPM-PIC 3632–3634, *Celliforma* isp. Horizon and locality: Santa Cruz Formation, Estancia La Costa Member; Cañadón las Totoras–Monte Tigre, Santa Cruz Province, Argentina.

MPM-PIC 3635–3637, cf. *Capayanichnus vinchinensis*. Horizon and locality: Santa Cruz Formation, Estancia La Costa Member, Fossiliferous Level 3; Estancia La Costa, Santa Cruz Province, Argentina.

MPM-PIC 3638, delicate root trace. Horizon and locality: Santa Cruz Formation, Estancia La Costa Member, Fossiliferous Level 3; Estancia La Costa, Santa Cruz Province, Argentina.

MPM-PIC 3639, ferric root trace with horizontally inclined secondary root traces. Horizon and locality: Santa Cruz Formation, Estancia La Costa Member; Cañadón del Indio, Santa Cruz Province, Argentina.

MPM-PIC 3640, calcareous rhizoconcretion. Horizon and locality: Santa Cruz Formation, Estancia La Costa Member, Fossiliferous Level 5.3; Puesto Estancia La Costa, Santa Cruz Province, Argentina.

6.4   Ichnologic composition

Ichnogenera are placed in alphabetic order, and are followed by those described in open nomenclature. The ethological categories proposed by Seilacher (1964) and Genise and Bown (1994) are followed.

Celliforma **isp.** Material of *Celliforma* isp. consists of strongly indurated internal casts of more or less cylindrical cells (Fig. 6.1a). The casts are oval-shaped with one end rounded and the other flattened. The cell dimensions of the three recorded specimens are 3–5 mm wide and 8–10 mm long. The specimens can be confidently assigned to the

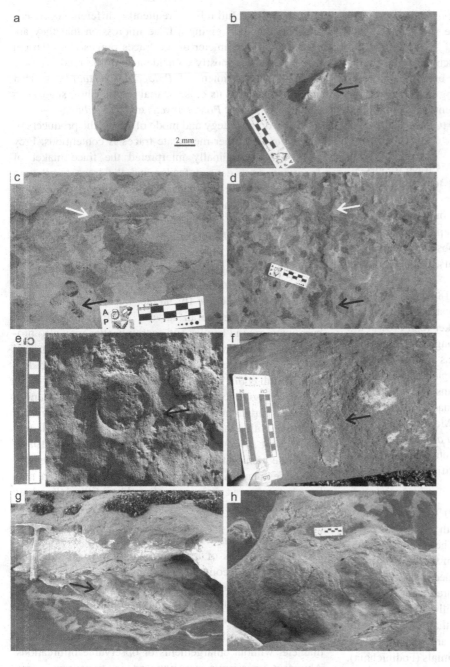

Fig. 6.1. Invertebrate and vertebrate trace fossils. a, *Celliforma* isp. (MPM-PIC 3632). b, *Planolites beverleyensis*. c, *Taenidium barretti* (black arrow) and *Palaeophycus tubularis* (white arrow). d, Intensely bioturbated *Taenidium barretti* (black arrow) and *Palaeophycus tubularis* (white arrow) bed. e, f, cf. *Capayanichnus vinchinensis* (arrow), in transverse (e, MPM-PIC 3635) and longitudinal (f, MPM-PIC 3636) cross-section. g, h, Large horizontal tube (arrow) in lateral (g) and planar (h) view. In g and h, note the horizontally disposed Y-shaped burrow.

ichnogenus *Celliforma* Brown, 1934 as defined by Brown (1934) and amended by Retallack (1984): "vasiform, globular or subcylindrical chambers or internal moulds of chambers; distal or inner end rounded; proximal or outer end either truncated irregularly or capped by a flat or conical closure, bearing spiral or concentric grooves on its inner surface; walls polished and smooth so that internal mould

is easily separated from rock matrix." Based on the morphology of the cells, Genise and Bown (1994) proposed a classification of *Celliforma* into various ichnospecies. In the present case, however, it is preferable to classify the structures at the ichnogeneric level, as a more detailed analysis would be needed for precise taxonomic classification. The material does not clearly fit with the established ichnospecies: it resembles *Celliforma spirifer* Brown, 1934 in the ovoid shape of the cells, but the spiral cap of that species is not observed.

The cells represent nesting structures constructed from modified substrate material for breeding purposes (Calichnia). A great diversity of digging bees may be the trace makers of *Celliforma*; at the moment it is not possible to establish a more specific trace maker (Genise and Bown, 1994; Genise, 2000).

Palaeophycus tubularis Halls, 1847 This trace fossil consists of simple sub-horizontal, straight to slightly curved burrows, with smooth walls and a thin lining (Fig. 6.1c, d). The burrow infill is similar to the surrounding rock and it is preserved as full relief or as a convex epirelief. The traces are locally abundant and are usually observable owing to the contrast with the overlying sediments. Diameters range from 0.4 cm to 1.4 cm.

The ichnogenus *Palaeophycus* Hall, 1847 was originally interpreted as a passively filled dwelling structure (Domichnia) produced by a worm-like trace maker (e.g. Frey *et al.*, 1984; Pemberton and Frey, 1982), although horizontal sub-superficial burrows on Recent floodplains are also produced by semi-aquatic insects including orthopterans (Tridactylidae), hemipterans (Salidae), coleopterans (Heteroceridae), and terrestrial coleopterans (Cicindelinae) (Ratcliffe and Fagerstrom, 1980; Bachmann and Mazzucconi, 1995; Mazzucconi *et al.*, 1995; Krapovickas *et al.*, 2008).

Palaeophycus are abundant at the Santa Cruz Formation and constitute the majority of the identifiable traces of the background bioturbation.

Planolites beverleyensis (Billings, 1862) This trace consists of straight to slightly curved burrows with unlined walls and sub-horizontal disposition (Fig. 6.1b). The infilling is without structure and different in color from the host sediment but is of the same grain size. The specimens preserved are uncommon; one specimen measured had diameters ranging from 0.6 to 0.8 cm. The infilling of *Planolites* Nicholson, 1873 represents sediment that was processed by the trace maker; consequently, they are generally interpreted as burrows of deposit-feeding animals (Fodinichnia), including oligochaete annelids, terrestrial insects, and millipedes (Pemberton and Frey, 1982; Hembree and Hasiotis, 2007; Minter *et al.*, 2007). The material of Santa Cruz is similar in size to *Palaeophycus* and *Taenidium* Heer, 1877, so they were probably produced by the same animals.

Taenidium barretti (Bradshaw, 1981) These traces consist of straight to sinuous simple burrows characterized by a thinly segmented meniscate backfill and having no lining (Fig. 6.1c, d). The menisci are of alternating grain size, subtly arcuate, and 1 to 2 mm thick. They are less distinctive where the burrow-fill is of uniform composition. Traces are preserved as full relief. Frequently, different specimens cross one another, giving a false impression that they are branching. The diameter of the traces ranges from 0.6 cm to 1.3 cm and is mostly coincident with the recorded size range of the specimens of *Palaeophycus tubularis*. That fact, together with its close spatial relationship, suggests a common origin for *Palaeophycus* and *Taenidium*.

The feeding strategy and mode of life of the producers of *Taenidium* and other meniscate traces is contentious. Frey *et al.* (1984) originally interpreted the trace maker of *Taenidium* as a deposit feeder and the trace fossil as a feeding structure (Fodinichnia). Accordingly, oligochaete annelids have been identified as one of the possible *Taenidium* trace makers (Bown and Kraus, 1983; D'Alessandro and Bromley, 1987; Schlirf *et al.*, 2001). However, arthropods tend to backfill their burrows mechanically, rather than deposit feeding. Recent and fossil animals that produce meniscate traces include terrestrial myriapods and insects like tiger beetles, crane-fly larvae, burrower bugs, cicada nymphs, and coleopteran larvae. Adults and larvae of burrowing ground beetles and scarab beetles may also produce meniscate traces (Frey *et al.*, 1984; O'Geen and Busacca, 2001; Gregory *et al.*, 2004; Morrissey and Braddy, 2004; Smith *et al.*, 2008; Counts and Hasiotis, 2009).

The configuration, texture, and composition of menisci indicate a particular combination of sediment ingestion and/or sorting within the digestive apparatus of the animal, and/or sediment sorting by its appendages (Frey *et al.*, 1984). When the meniscate backfill comprises thinly and densely packed segments, as in the Santa Cruz specimens, there may have been a more continuous external backfilling (Keighley and Pickerill, 1994).

cf. *Capayanichnus vinchinensis* Melchor *et al.*, 2010 This trace consists of a predominantly vertical simple to L-shaped burrow, commonly 3 cm wide and up to 18 cm in length. Burrows are unlined or thinly lined. Externally, one often sees an irregular halo, perhaps produced by the animal when the burrow was excavated and the external material compacted (Fig. 6.1e, f). The burrow has an almost constant diameter without enlargements or observable bifurcations. Abundant specimens are exposed in transverse cross-section and more rarely in longitudinal cross-section. They are preserved in very fine sandstone and reworked tuff deposits at Estancia La Costa. The infill is coincident with the host rock, suggesting a passive infilling of the open burrow. Because it is preserved in longitudinal and transverse cross-section, the distinctive surface texture of *Capayanichnus vinchinensis* is not observed in the Santa Cruz material (Melchor *et al.*, 2010). It has been interpreted as a dwelling structure (Domichnia) having been produced by freshwater crabs (Melchor *et al.*, 2010) and I accept that interpretation for the Santa Cruz material.

Large horizontal tube This large structure consists of one forked burrow of horizontal disposition up to 5 to 7 cm in height and 9 to 12 cm wide. It consists of two branches that converge in a U-shape union, with a third branch connected to the base of the U, forming a horizontally disposed Y-shaped burrow (Fig. 6.1g, h). The burrow is elliptical in cross-section. The infill of the burrow is a very fine sandstone and reworked tuff from an overlying level, distinct from the claystone of the host rock, and suggesting it was passively filled at the time that the overlying fine tuff was deposited. The dimensions and the overall architectural morphology of the burrow suggest that it was a mammalian dwelling burrow (Domichnia). Burrow dimensions have a tight correspondence with the size of the trace maker. The diameter of the burrow should be as small as possible because minimizing the dimensions of the burrow reduces the energetic cost of excavation (Anderson, 1982; White, 2005). In this case, the size of the burrow is more coincident with a tetrapod than with an invertebrate. In addition, other mammalian dwelling burrows show a main horizontal component with numerous interconnected tunnels (see Krapovickas *et al.*, in press). Among recent examples, mammalian burrows of similar form are used for food storage, as latrines, and for breeding. They are typically permanent residences for their producer (Kinlaw, 1999).

Root traces A variety of root traces (rhizoliths *sensu* Klappa, 1980) are preserved in the ELC Member. They include mainly (1) delicate root traces, (2) ferric root traces, (3) haloed root traces, and (4) calcareous rhizoconcretions.

Delicate or filamentous root traces are 0.5–3 mm in diameter. They taper downwards and lack infill, probably owing to selective erosion of the infilling material (Fig. 6.2a, b).

Ferric root traces have a red to yellow-brown color within the root infill and on its margins. The red and yellow-brown colors result from a mix of iron oxides (hematite and goethite). A redder color indicates more hematite (e.g. Schwertmann and Taylor, 1977; Scheinost and Schwertmann, 1999; Kraus and Hasiotis, 2006). Ferric root traces vary from delicate or filamentous to those with greater size and diverse branching pattern. Some consist of vertical primary roots that taper downwards (5–2 mm wide) and have secondary roots (1–2 mm wide) that branch downwards with an inclination of 45° (Fig. 6.2d). Others have an almost constant diameter subtly tapering downwards from 7 to 4 mm and horizontally inclined secondary root traces (Fig. 6.2c).

Haloed root traces represent delicate rhizoliths with drab haloes extending out into the paleosol matrix (Fig. 6.2e).

Calcareous rhizoconcretions are mainly vertically disposed tubular structures that comprise concentric layers of carbonate deposited on the ancient root margins

(rhizosphere), with a central hole filled by the host rock, where the ancient root was placed (Fig. 6.2f).

6.5 Sedimentary facies and distribution of trace fossils

6.5.1 Fine tuff and subordinate claystone and limestone

A facies consisting of fine tuffs and subordinate claystone and limestone is observed, with its associated trace fossils, in the lower part of the ELC Member along the coastal cliffs at Cañadón del Indio (Matheos and Raigemborn, Chapter 4, Fig. 4.4). It consists of light grayish-green to light green fine tuffs of massive structure, with locally interbedded massive and laminated claystones and limestones (Fig. 6.3a, b). It presents some structural elements typical of fossil soils such as a variety of voids (Matheos and Raigemborn, Chapter 4), blocky peds (soil aggregates), irregular ferric mottling, and abundant ferric root traces of the filamentous, long downward branching, and horizontally branching types (Fig. 6.2c, d; Table 6.1). Coarser white tuffs have fewer ferric root traces and ferric mottling. This facies formed by intermittent deposition of ash-fall in subaerial and subaqueous areas of a floodplain (Cas and Wright, 1987), with subsequent modification by biogenic activity and pedogenetic processes.

6.5.2 Very fine sandstone, claystone, and reworked tuff

Very fine sandstone, claystone, and reworked tuff facies and its associated trace fossils are mostly found through the middle part of the ELC Member in the present-day intertidal zone at Anfiteatro, Estancia La Costa, Cañadón Silva, and Puesto Estancia La Costa localities. They consist of light brown to light greenish-gray, massive, very fine sandstone, locally mottled light bluish-green massive claystone, and reworked tuff (Fig. 6.3c). The sandstones correspond to the fossiliferous levels 3 (Estancia La Costa), 5 (Cañadón Silva), and 6 (Puesto Estancia La Costa) defined by Tauber (1997a) (Matheos and Raigemborn, Chapter 4, Fig. 4.4). They have paleoedaphic features (Matheos and Raigemborn, Chapter 4), root traces, and desiccation cracks (Fig. 6.3c, e), and they contain abundant filamentous and haloed root traces and fewer calcareous rhizoconcretions. Invertebrate trace fossils are uncommon and consist of isolated dwelling burrows and meniscate specimens (*Palaeophycus tubularis* and *Taenidium barretti*, respectively) and locally abundant crab dwelling burrows (cf. *Capayanichnus vinchinensis*) (Table 6.1). The claystone beds correspond to the fossiliferous levels 2.1 (Estancia La Costa), 5.1 (Cañadón Silva), and 5.3 (Puesto Estancia La Costa) defined by Tauber (1997a) and Krapovickas *et al.* (2008) (Matheos and Raigemborn, Chapter 4, Fig. 4.4). They are commonly less

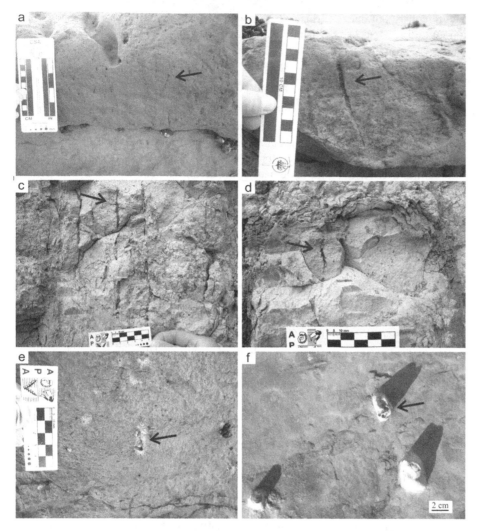

Fig. 6.2. Root traces. a, Fine root
traces. b, Detail of a fine root trace
(MPM-PIC 3638). c, d, Ferric root
traces with secondary roots of
horizontal disposition (c, MPM-PIC
3639) and branching down (d). e,
Haloed root traces. f, Calcareous root
traces.

bioturbated than the sandstones and have abundant calcareous
concretions and rhizoconcretions and locally developed haloed
root traces (Figs. 6.2e, f, 6.3d). Locally, there is a record of a
mammalian dwelling burrow (a large horizontal tube); invertebrate bioturbation is widespread (Table 6.1). The latter is
mostly represented by background bioturbation, and when
identifiable, by dwelling (*Palaeophycus tubularis*) and meniscate (*Taenidium barretti*) structures. These levels correspond
to distal floodplain deposits settling from suspension with
subaerial exposure and paleosol development, and aeolian
deposition of pyroclastic material (loess) (Tauber, 1994;
Matheos and Raigemborn, Chapter 4).

6.5.3 Claystone and limestone
Claystone and limestone facies and its associated trace
fossils are recorded at the top of the upper section of the
ELC Member at the coastal cliff of Monte Tigre–Cañadón
las Totoras locality (Matheos and Raigemborn, Chapter 4,

Fig. 4.4). It is mainly composed of a light greenish-gray and
pink claystone, ranging from horizontally laminated to massive, with slickensides and abundant ferric root traces (Fig.
6.3f, Table 6.1). Light brown to light greenish-gray massive
limestones are also observed with subspherical concretions,
slickensides, yellowish-brown mottling, delicate ferric root
traces, and cells of solitary digging bees (*Celliforma*) (Fig.
6.3g, h; Table 6.1). These levels represent distal floodplain
environments where overbank-flooding, settling from suspension, subsequent subaerial exposure, and development
of paleosols occurred (Tauber, 1994; Matheos and Raigemborn, Chapter 4).

6.6   Environmental implications of
trace-fossil assemblages

The overbank deposits of the Estancia La Costa Member
between Río Coyle and Río Gallegos record a series of

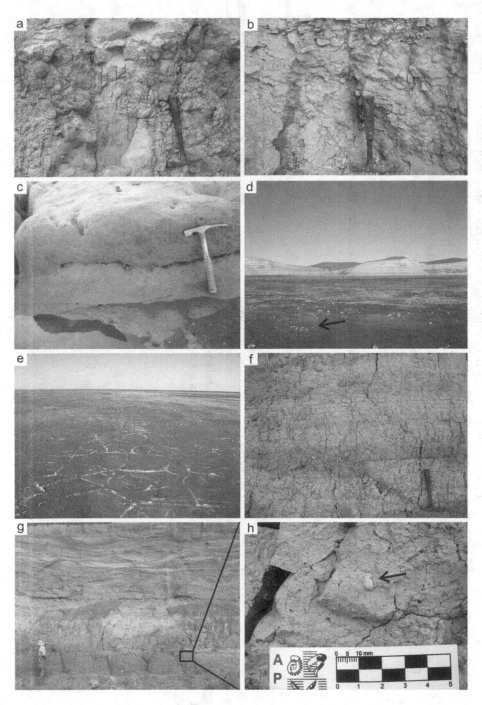

Fig. 6.3. a, Light grayish-green fine tuff paleosol, at Cañadón del Indio. b, Light green fine tuffs with locally interbeded claystones and limestones paleosol, at Cañadón del Indio. c, Light brown to light greenish-gray very fine sandstone and reworked tuff paleosol (above) and light bluish-green claystone paleosol, at Cañadón Silva. d, Light bluish-green claystone paleosol with abundant calcareous concretions and rhizoconcretions, at Puesto Estancia La Costa. e, Desiccation cracks of light bluish-green claystone paleosol, at Puesto Estancia La Costa. f, Light greenish-gray and pink claytone paleosols, at Monte Tigre–Cañadón las Totoras. g, h, Light brown to light greenish-gray limestone paleosol (black square), at Monte Tigre–Cañadón las Totoras. h, Detail of the paleosol with a *Celliforma* specimen (arrow) preserved *in situ* (MPM-PIC 3633).

trace-fossil assemblages in a variety of environments where floodplain water bodies and paleosols developed under variable climatic conditions.

6.6.1 Floodplain water bodies

Taenidium–Palaeophycus **ichnoassemblage** Ephemeral floodplain water bodies of the middle portion of the ELC Member record an assemblage predominantly composed of horizontal to gently inclined simple and meniscate trace fossils (*Palaeophycus tubularis*, *Taenidium barretti*). The trace fossil assemblage corresponds to a "pre-dessication suite" characterized by traces without ornamentation developed in soft substrates (i.e. Buatois and Mángano, 2004). Associated physical structures, in the form of

Table 6.1. *Trace fossils recorded at the ELC Member of the Santa Cruz Formation*

	Animal traces						Root traces				IA	L	I
	Ci	Pt	Pb	Tb	Cv	Ht	Dr	Fr	Hr	Cr			
Cañadón del Indio								C			Ferric root traces	White coarse tuffs	Moderately well-drained soils, humid climate
								A			Ferric root traces	Fine tuffs with interbedded mudrocks	Moderately well-drained soils, humid climate
Anfiteatro		A		C						S	*Taenidium – Palaeophycus*	Massive claystone	Floodplain water bodies, soft substrate
Estancia La Costa					A		S		S		*Capayanichnus*	Very fine sandstone, reworked tuff	Moderately drained soils, high water table
										S	Large horizontal tube	Massive claystone	Moderately well-drained soils
Cañadón Silva			S	S			C				*Capayanichnus*	Very fine sandstone, reworked tuff	Moderately drained soils, high water table
		A		C			S		S		*Taenidium – Palaeophycus*	Massive claystone	Floodplain water bodies, soft substrate
						S				S	Large horizontal tube	Massive claystone	Moderately well-drained soils
Puesto Estancia La Costa							S		S		*Capayanichnus*	Very fine sandstone, reworked tuff	Moderately drained soils, high water table
										A	Large horizontal tube	Massive claystone	Moderately well-drained soils
Cañadón las Totoras–Monte Tigre								C			*Celliforma*	Laminated claystone	Moderately well-drained soils, dry climate
	S							C			*Celliforma*	Massive limestone	Moderately well-drained soils, dry climate

IA, ichnoassemblages; **L**, lithology; **I**, interpretation; **Ci**, *Celliforma* isp.; **Pt**, *Palaeophycus tubularis*; **Pb**, *Planolites beverleyensis*; **Tb**, *Taenidium barretti*; **Cv**, cf. *Capayanichnus vinchinensis*; **Ht**, large horizontal tube; **Dr**, delicate root traces; **Fr**, ferric root traces; **Hr**, haloed root traces; **Cr**, calcareous root traces; A, abundant; C, common; S, scarce.

desiccation cracks, are indicative of subaerial exposure and loss of water content in the beds (Fig. 6.3e, Fig. 6.4).

The ichnofauna present in these desiccated floodplain water bodies is an example of the *Scoyenia* ichnofacies. This ichnofacies represents slow-diversity ichnocoenoses, mainly composed of meniscate trace fossils (Frey *et al.*, 1984). In addition, it characterizes low-energy deposits periodically aerially exposed and inundated, situating this ichnofacies between aquatic and terrestrial environments (Frey *et al.*, 1984). In fluvial systems, the *Scoyenia* ichnofacies is commonly seen in abandoned river channels and in

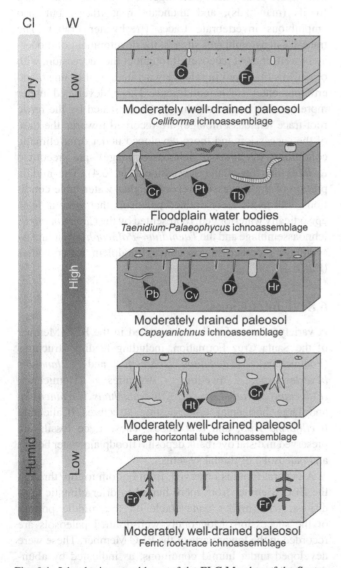

Fig. 6.4. Ichnologic assemblages of the ELC Member of the Santa Cruz Formation. Ichnoassemblages are organized stratigraphically from bottom to top. W, water table; Cl, climate. C, *Celliforma* isp.; Pt, *Palaeophycus tubularis*; Pb, *Planolites beverleyensis*; Tb, *Taenidium barretti*; Cv, cf. *Capayanichnus vinchinensis*; Ht, large horizontal tube; Dr, delicate root traces; Hr, haloed root traces; Fr, ferric root traces, Cr, calcareous root traces.

overbank deposits, and encompasses a wide variety of environments, such as ponds, levees, sand bars, and crevasse splays (Buatois and Mángano, 1995, 2002, 2004; Frey and Pemberton, 1984; Frey *et al.*, 1984).

Desiccated floodplain water bodies also record fine root traces and some calcareous rhizoconcretions together with trace fossils of the *Scoyenia* ichnofacies, denoting its later colonization by plants in the early stages of paleosol development. That may have occurred when the time between depositional events was long enough to allow colonization by plants but not so long as to obliterate animal traces (Maples and Archer, 1989).

6.6.2 Moderately drained paleosols

Capayanichnus **ichnoassemblage** The distal floodplain paleosols recorded in the middle portion of the ELC Member have abundant cf. *Capayanichnus vinchinensis*, root traces, and a few examples of *Taenidium barretti* and *Planolites beverleyensis*. The specimens of cf. *Capayanichnus vinchinensis* represent dwelling structures of crustaceans that excavated through the soil to a position below the water table (Table 6.1). Crab burrows are common where there is ample water at the surface or where the water table is relatively near to the surface (Hasiotis *et al.*, 2007; Melchor *et al.*, 2010). Fine root traces are very common, and haloed root traces are also present. The root haloes represent mineral-depleted areas that formed around the roots as a result of fluctuating soil-moisture and the decay of roots (Retallack, 2001; Kraus and Hasiotis, 2006). Desiccation cracks are present at the top of the bed, denoting subaerial exposure and variable loss of water content of the bed. Invertebrate feeding traces are scarce (*Taenidium barretti* and *Planolites beverleyensis*), probably because of intense bioturbation and/or obliteration by plant colonization.

6.6.3 Moderately well-drained paleosols

Ferric root-trace ichnoassemblage Soils forming on volcanic ash and containing abundant and diverse ferric root traces occur in the lower part of the ELC Member (Fig. 6.4). Recent volcanic soils are highly fertile and rich in organic matter, and have high porosity, high phosphate fixation capacity, and low bulk density (Tan, 1984). They are found in a variety of humid and subhumid climates but are rare in semi-arid and arid climates (Tan, 1984; Mack *et al.*, 1993). Root traces are diverse and abundant, with different growing patterns and depth of penetration suggesting a well-developed plant cover in a steady distal overbank. The presence of ferric root traces implies that the B horizon of the soil is moderately well-drained and oxidized (Kraus and Hasiotis, 2006). Iron is reduced and mobilized when the soil is water-saturated, and oxidized and immobilized when the soil dries, precipitating in places open to oxygen circulation (Duchaufour, 1987; Kraus and Aslan, 1993; PiPujol and Buurman, 1994).

Large horizontal tube ichnoassemblage Paleosols with many calcareous rhizoconcretions and other calcareous concretions are recorded in the middle section of the ELC Member (Fig. 6.4). Today, soils rich in calcium carbonate are common in arid and semi-arid regions where rainfall is insufficient to leach it from the soil, but also in humid coastal areas where carbonate is available owing to grains of broken seashells (Retallack, 2001). Calcareous rhizoconcretions indicate episodic saturation of the ancient soil (Retallack, 2001; Kaus and Hasiotis, 2006). That is because repeated cycles of wetting and drying of the soil precipitate numerous fine layers of carbonate on the root margins; then the root dies and the remaining hole is filled by other material (Esteban and Klappa, 1983). Other trace fossils are almost absent in these paleosols. The presence of an open dwelling burrow (large horizontal tube) produced by a medium-sized mammal suggests that the paleosol was relatively dry seasonally, allowing its construction and use.

Celliforma **ichnoassemblage** The paleosols occurring at the top of the ELC Member have scarce trace fossils, and the ichnologic assemblage recorded is characterized by delicate ferric root traces mainly of short depth of penetration, and by cells of solitary digging bees (*Celliforma*) (Fig. 6.4).

As previously mentioned, ferric root traces and mottling occur in well-drained soils of humid climates (Retallack, 2001). However, burrowing bees generally nidify in more arid conditions or in soils with only moderate plant growth, good drainage, and limited exposure to the sun (Ratcliffe and Fagerstron, 1980; Genise and Bown, 1994). Given all these factors, seasonal shifts between subhumid and semi-arid climatic conditions may be inferred, producing moderate plant growth.

6.7 Vertical variation of ichnofossils and its paleoclimatic and paleohydrologic significance

Most of the trace fossils recorded on the ELC Member of the Santa Cruz Formation are preserved in distal overbank deposits (floodplain water bodies and paleosols) of a fluvial system dominated by sheet-flooding, overbank-flooding, and ash-fall deposits, rather than being found in within-channel deposits of single laterally stable channels. The pedogenic processes that occurred in overbank deposits of the Santa Cruz Formation were mainly influenced by the distance from a sediment source, the position and fluctuation of the groundwater profile, the composition of biotic

communities, and the climatic setting with regard to temperature and precipitation. From the trace-fossil assemblages preserved, it is possible to link trace-making biota and paleosols to the ancient soil-water balance and climatic conditions (Fig. 6.4). Floodplain water bodies record invertebrate trace fossils that are an example of the *Scoyenia* ichnofacies (*Taenidium–Palaeophycus* ichnoassemblage), most likely displaying semi-aquatic insect and millipede trace fossils emplaced in soft substrates that subsequently dried out. Ancient soils record trace fossils with variable relationship to the height of the water table: fully terrestrial invertebrate traces (soil-digging bees), vertebrate trace fossils (mammals), and abundant root traces, but also amphibious invertebrate traces (freshwater crabs). The trace-fossil assemblages suggest the development of moderately well-drained paleosols throughout the succession, with organisms that lived in relatively aerated, well-drained, and environmental stable conditions. Those developed under more humid climatic conditions (as indicated by the ferric root-trace ichnoassemblage) are recorded towards the base of the succession and those developed under drier climatic conditions (*Celliforma* ichnoassemblage) are recorded towards the top of the succession (Fig. 6.4). The middle portion of the succession records higher water table conditions with moderately drained paleosols that were at least episodically waterlogged, represented by the *Capayanichnus* ichnoassemblage and the *Taenidium–Palaeophycus* ichnoassemblage recorded in associated floodplain water bodies (Fig. 6.4).

6.8 Conclusions

A variety of trace fossils are recorded in the ELC Member of the Santa Cruz Formation, including feeding structures (Fodinichnia) (*Taenidium barretti* and *Planolites beverleyensis*); dwelling structures (Domichnia) (*Palaeophycus tubularis*, cf. *Capayanichnus vinchinensis*), and large horizontal tube; nesting structures (Calichnia) (*Celliforma* isp.); and root traces. Those trace fossils are preserved in distal overbank deposits (floodplain water bodies and paleosols) of fluvial systems.

A general trend is observed, from bottom to top, through the ELC Member from more humid to drier climatic conditions, with higher water table on the middle portion of the succession. Moderately well-drained paleosols are recorded in the lower parts of the ELC Member. These were developed under humid conditions, as indicated by abundant ferric root traces. Also, moderately well-drained soils are recorded in the middle portion of the ELC Member, denoted by the large horizontal tube ichnoassemblage. In addition, there are records of moderately drained soils related to a higher water table as evidenced by the *Capayanichnus* ichnoassemblage. Finally, floodplain water bodies evidently occurred in the middle portion of the ELC Member record, as indicated by trace fossils corresponding to a "pre-desiccation suite" of the *Scoyenia* ichnofacies, developed in soft substrates that subsequently were colonized by plants at the initial stages of soil development. The upper parts of the ELC Member record cells of solitary digging bees (*Celliforma*) and ferric root traces, and were interpreted as developed under drier conditions.

ACKNOWLEDGMENTS
I would like to thank S. F. Vizcaíno, M. S. Bargo, and R. F. Kay for their kind invitation to contribute to this book. I am grateful to the reviewers L. Buatois and N. Minter for their comments and suggestions which greatly improved this manuscript, to L. Palazzesi for his comments and discussion, and to J. Genise and L. Sarzetti for their comments on some specimens. I am also grateful to S. F. Vizcaíno, M. S. Bargo, J. C. Fernicola, L. Acosta, L. Cruz, S. Matheos, and S. Raigemborn for their help during fieldwork. Funding for this research was provided by UBACyT-X133, PICT 0143 to Sergio F. Vizcaíno and NSF 0851272, 0824546 to Richard F. Kay. This is contribution R-45 of the Instituto de Estudios Andinos Don Pablo Groeber.

REFERENCES
Anderson, D. C. (1982). Below ground herbivory: the adaptive geometry of geomyid burrows. *American Naturalist*, **119**, 18–28.

Bachmann, A. O. and Mazzucconi, S. A. (1995). Insecta Heteroptera (= Hemiptera s. str.). In *Ecosistemas de Aguas Continentales*, ed. E. C. Lopretto and G. Tell. La Plata: Ediciones Sur, pp. 1291–1325.

Bown, T. M. and Fleagle, J. G. (1993). Systematics, biostratigraphy, and dental evolution of the Palaeothentidae, later Oligocene to early–middle Miocene (Deseadan–Santacrucian) caenolestoid marsupials of South America. *Journal of Paleontology*, **67**, 1–76.

Bown, T. M. and Kraus, M. J. (1983). Ichnofossils of the alluvial Willwood Formation (lower Eocene), Bighorn Basin, northwest Wyoming, USA. *Palaeogeography, Palaeoclimatology, Palaeoecology*, **43**, 95–128.

Brown, R. W. (1934). *Celliforma spirifer*, the fossil larval chambers of mining bees. *Journal of the Washington Academy of Sciences*, **24**, 532–539.

Buatois, L. A. and Mángano, M. G. (1995). The paleoenvironmental and paleoecological significance of the lacustrine *Mermia* ichnofacies: an archetypical subaqueous nonmarine trace fossil assemblage. *Ichnos*, **4**, 151–161.

Buatois, L. A. and Mángano, M. G. (2002). Trace fossils from Carboniferous floodplain deposits in western Argentina: implications for ichnofacies models of continental environments. *Palaeogeography, Palaeoclimatology, Palaeoecology*, **183**, 71–86.

Buatois, L. A. and Mángano, M. G. (2004). Animal–substrate interactions in freshwater environments: applications of ichnology in facies and sequence stratigraphic analysis of fluvio-lacustrine successions. In *The Application of Ichnology to Palaeoenvironmental and Stratigraphic Analysis*, ed. D. McIlroy. Geological Society Special Publications, London, **228**, 311–335.

Cas, R. A. F. and Wright, J. V. (1987). *Volcanic Successions: Modern and Ancient*. London: Unwin Hyman.

Counts, J. W. and Hasiotis, S. T. (2009). Neoichnological experiments with masked chafer beetles (Coleoptera: Scarabaeidae): implications for backfilled continental trace fossils. *PALAIOS*, **24**, 74–91.

Croft, D. A. (2001). Cenozoic environmental change in South America as indicated by mammalian body size distributions (cenograms). *Diversity and Distributions*, **7**, 271–287.

D'Alessandro, A. and Brombley, R. G. (1987). Meniscate trace fossils and the *Muensteria–Taenidium* problem. *Palaeontology*, **30**, 743–763.

Duchaufour, P. (1987) *Manual de Edafología*. Masson, S.A.: Barcelona.

Esteban, M. and Klappa, C. F. (1983). Subaerial exposure environment. In *Carbonate Depositional Environments*, ed. P. A. Scholle, D. G. Bebout and C. H. Moore. Memoirs, American Association of Petroleum Geologists, **33**, 1–54.

Frey, R. W. and Pemberton, S. G. (1984). Trace fossils facies models. In *Facies Models*, ed. R. G. Walker. St. John's: Geosciences Canada, reprint series, pp. 189–207.

Frey, R. W., Pemberton, S. G. and Fagerstrom, J. A. (1984). Morphological, ethological and environmental significance of the ichnogenera *Scoyenia* and *Ancorichnus*. *Journal of Paleontology*, **58**, 511–528.

Genise, J. F. (2000). The ichnofamily Celliformidae for *Celliforma* and allied ichnogenera. *Ichnos*, **7**, 267–284.

Genise, J. F. and Bown, T. M. (1994). New Miocene scarabeid and hymenopterous nests and Early Miocene (Santacrucian) paleoenvironments, Patagonian Argentina. *Ichnos*, **3**, 107–117.

Gregory, M. R., Martin, A. J. and Campbell, K. A. (2004). Compound trace fossils formed by plant and animal interactions: Quaternary of northern New Zealand and Sapelo Island, Georgia (USA). *Fossils and Strata*, **51**, 88–105.

Hasiotis, S. T., Kraus, M. J. and Demko, T. M. (2007). Climate controls on continental trace fossils. In *Trace Fossils: Concepts, Problems, Prospects*, ed. W. Miller III. Amsterdam: Elsevier, pp. 172–195.

Hatcher, J. B. (1900). Sedimentary rocks of southern Patagonia. *American Journal of Science (1880–1910)*, Ser. 4, **9**, 85–108.

Hatcher, J. B. (1903). Narrative and geography. In *Reports of the Princeton University Expeditions to Patagonia, 1896–1899*. Vol. 1, *Narrative and Geography*, ed. W. B. Scott. Princeton: Princeton University Press, pp. 1–296.

Hembree, D. I. and Hasiotis, S. T. (2007). Paleosols and ichnofossils of the White River Formation of Colorado: insight into soil ecosystems of the North American midcontinent during the Eocene-Oligocene transition. *PALAIOS*, **22**, 123–142.

Kay, R. F., Vizcaíno, S. F., Bargo, M. S. *et al.* (2008). Two new fossil vertebrate localities in the Santa Cruz Formation (late Early Miocene, Argentina), ~51 degrees South latitude. *Journal of South American Earth Sciences*, **25**, 187–195.

Keighley, D. G. and Pickerill, R. K. (1994). The ichnogenus *Beaconites* and its distinction from *Ancorichnus* and *Taenidium*. *Palaeontology*, **37**, 305–337.

Kinlaw, A. (1999). A review of burrowing by semi-fossorial vertebrates in arid environment. *Journal of Arid Environments*, **41**, 127–145.

Klappa, C. F. (1980). Rhizoliths in terrestrial carbonates: classification, recognition, genesis, and significance. *Sedimentology*, **26**, 613–629.

Krapovickas, J. M., Tauber, A. A. and Rodriguez, P. E. (2008). Nuevo registro de *Protypotherium* australe Ameghino, 1887: implicancias bioestratigráficas en la Formación Santa Cruz. *Actas XVII Congreso Geológico Argentino*, **3**, 1020–1021.

Krapovickas, V., Mángano, M. G., Mancuso, A., Marsicano, C. A. and Volkheimer, W. (2008). Icnofaunas triásicas en abanicos aluviales distales: evidencias de la Formación Cerro Puntudo, Cuenca Cuyana, Argentina. *Ameghiniana*, **45**, 463–472.

Krapovickas, V., Mancuso, A. C., Marsicano, C. A., Domnanovich, N. S. and Schultz, C. (in press). Large tetrapod burrows from the Middle Triassic of Argentina: a behavioural adaptation to seasonal semi-arid climates? *Lethaia*.

Kraus, M. J. and Aslan, A. (1993) Eocene hydromorphic paleosols: significance for interpreting ancient floodplain processes. *Journal of Sedimentary Petrology*, **63**, 453–463.

Kraus, M. J. and Hasiotis, S. T. (2006). Significance of different modes of rhizolith preservation to interpreting paleoenvironmental and paleohydrologic settings: Examples from paleogene paleosols, Bighorn Basin, Wyoming, U.S.A. *Journal of Sedimentary Research*, **76**, 633–646.

Mack, G. H., James, W. C. and Monger, H. C. (1993). Classification of paleosols. *Geological Society of America Bulletin*, **105**, 129–136.

Maples, C. G. and Archer, A. W. (1989). The potential of Paleozoic nonmarine trace fossils for paleoecological interpretations. *Palaeogeography, Palaeoclimatology, Palaeoecology*, **73**, 185–195.

Mazzucconi, S. A., Bachmann, A. O. and Trémouilles, E. R. (1995). Insecta Saltatoria (= Orthoptera s. str.). In *Ecosistemas de aguas continentales*, ed. E. C. Lopretto and G. Tell. La Plata: Ediciones Sur, pp. 1113–1121.

Melchor, R. N., Genise, J. F., Farina, J. L. *et al.* (2010). Large striated burrows from fluvial deposits of the Neogene Vinchina Formation, La Rioja, Argentina: a crab origin suggested by neoichnology and sedimentology. *Palaeogeography, Palaeoclimatology, Palaeoecology*, **291**, 400–418.

Minter, N. J., Krainer, K., Lucas, S. G., Braddy, S. J. and Hunt, A. P. (2007). Palaeoecology of an Early Permian playa lake trace fossil assemblage from Castle Peak, Texas, USA. *Palaeogeography, Palaeoclimatology, Palaeoecology*, **246**, 390–423.

Morrissey, L. B. and Braddy, S. J. (2004). Terrestrial trace fossils from the Lower Old Red Sandstone, southwest Wales. *Geological Journal*, **39**, 315–336.

O'Geen, A. T. and Busacca, A. J. (2001). Faunal burrows as indicator of paleo-vegetation in eastern Washington, USA. *Palaeogeography, Palaeoclimatology, Palaeoecology*, **169**, 23–37.

Pemberton, S. G. and Frey, R. W. (1982). Trace fossil nomenclature and the *Planolites Palaeophycus* dilemma. *Journal of Paleontology*, **56**, 843–881.

PiPujol, M. D. and Buurman, P. (1994). The distinction between ground-water gley and surface-water gley phenomena in Tertiary paleosols of the Ebro Basin, NE Spain. *Palaeogeography, Palaeoclimatology, Palaeoecology*, **110**, 103–113.

Ratcliffe, B. C. and Fagerstrom, J. A. (1980). Invertebrate lebensspuren of Holocene floodplains: their morphology, origin and palaeoecological significance. *Journal of Paleontology*, **54**, 614–630.

Retallack, G. J. (1984). Trace fossils of burrowing beetles and bees in an Oligocene paleosol, Badlands National Park, South Dakota. *Journal of Paleontology*, **58**, 571–582.

Retallack, G. J. (2001). *Soils of the Past: An Introduction to Paleopedology*, 2nd Edn. Oxford: Blackwell.

Scheinost, A. C. and Schwertmann, U. (1999). Color identification of iron oxides and hydroxysulfates: use and limitations. *Soil Science Society of America, Journal*, **63**, 1463–1471.

Schlirf, M., Uchman, A. and Kümmel, M. (2001). Upper Triassic (Keuper) non-marine trace fossils from the Haßberge area (Franconia, south-eastern Germany). *Paläontologische Zeitschrift*, **75**, 71–96.

Schwertmann, U. and Taylor, R. M. (1977). Iron oxides. In *Minerals in Soil Environments*, ed. J. B. Dixon and S. B Weed. Madison, Wisconsin: American Society of Agronomy and Soil Science, pp. 145–180.

Seilacher, A. (1964). Sedimentological classification and nomenclature of trace fossils. *Sedimentology*, **3**, 253–256.

Smith, J. J., Hasiotis, S. T., Kraus, M. J. and Woody, D. T. (2008). *Naktodemasis bowni*: new ichnogenus and ichnospecies for adhesive meniscate burrows (AMB), and paleoenvironmental implications, Paleogene Willwood Formation, Bighorn Basin, Wyoming. *Journal of Paleontology*, **82**, 267–278.

Tan, K. H. (1984). *Andosols*. New York: Van Nostrand Reinhold Company Inc.

Tauber, A. A. (1994). Estratigrafía y vertebrados fósiles de la Formación Santa Cruz (Mioceno Inferior) en la costa atlántica entre las rías del Coyle y Río Gallegos, Provincia de Santa Cruz, República Argentina. Unpublished Ph.D. thesis, Facultad de Ciencia Exactas, Físicas y Naturales, Universidad Nacional de Córdoba, Argentina.

Tauber, A. A. (1997a). Bioestratigrafía de la formación Santa Cruz (Mioceno inferior) en el extremo sudeste de la Patagonia. *Ameghiniana*, **34**, 413–426.

Tauber, A. A. (1997b). Paleoecología de la Formación Santa Cruz (Mioceno inferior) en el extremo sudeste de la Patagonia. *Ameghiniana*, **34**, 517–529.

Tauber, A. A. (1999). Los vertebrados de la Formación Santa Cruz (Mioceno inferior-medio) en el extremo sureste de la Patagonia y su significado paleoecológico. *Revista Española de Paleontología*, **14**, 173–182.

Vizcaíno, S. F., Bargo, M. S., Kay, R. F. and Milne, N. (2006). The armadillos (Mammalia, Xenarthra) of the Santa Cruz Formation (early–middle Miocene). An approach to their paleobiology. *Palaeogeography, Palaeoclimatology, Palaeoecology*, **237**, 255–269.

Vizcaíno, S. F., Bargo, M. S., Kay, R. F. *et al.* (2010). A baseline paleoecological study for the Santa Cruz Formation (late Early Miocene) at the Atlantic coast of Patagonia, Argentina. *Palaeogeography, Palaeoclimatology, Palaeoecology*, **292**, 507–519.

White, C. R. (2005). The allometry of burrow geometry. *Journal of Zoology*, **265**, 395–403.

7 Fossil plant studies from late Early Miocene of the Santa Cruz Formation: paleoecology and paleoclimatology at the passive margin of Patagonia, Argentina

Mariana Brea, Alejandro F. Zucol, and Ari Iglesias

Abstract

This chapter analyses plant fossils including phytoliths, carbonized wood, and leaf compressions, along with microcharcoal, chrysophycean stomatocysts, sponge spicules, and diatoms, recovered from the Estancia La Costa Member, Santa Cruz Formation (late Early Miocene), in the Atlantic margin of Patagonia, Argentina. The floristic composition and paleoclimatic inferences based on this fossil plants assemblage from the late Early Miocene are presented. The fossil flora is characterized by the presence of herbaceous components including chloridoid, panicoid, danthonioid, pooid, and festucoid grasses. The arboreal elements include members of the Araucariaceae, Lauraceae, Arecaceae, Nothofagaceae, Myrtaceae, Cunoniaceae, Fabaceae (Faboideae), and ?Proteaceae. The vegetation was a mixture of open temperate semi-arid forests and humid warm-temperate forests. The integrated analysis of multiple sets of proxy data suggests that southeastern Patagonia, during late Early Miocene, was characterized by a temperate to warm-temperate and semi-arid to humid climate, where seasonal low levels of precipitation served as a limiting factor for plant growth.

Resumen

Este capítulo analiza el registro paleobotánico preservado como fitolitos, leños fósiles y compresiones foliares, junto con microcarbones, estomatocistos de crisostomatáceas, espículas de esponjas y diatomeas, hallados en el Miembro Estancia La Costa de la Formación Santa Cruz (Mioceno Temprano tardío), en el margen atlántico de Patagonia, Argentina. Se discute la composición florística y las interpretaciones paleoclimáticas en base a la asociación de plantas fósiles restringidas al Mioceno Temprano tardío. La flora estaba caracterizada por la presencia de componentes herbáceos tales como gramíneas cloridoideas, panicoideas, dantoneas, pooideas y festucoideas. Los elementos

arbóreos estaban integrados por Araucariaceae, Lauraceae, Arecaceae, Nothofagaceae, Myrtaceae, Cunoniaceae, Fabaceae (Faboideae) y Proteaceae? La vegetación representa un ambiente heterogéneo integrado por bosques abiertos templados semiáridos y bosques templado-cálidos y húmedos. El análisis integral de datos múltiples sugiere que en la región sudeste de Patagonia, durante el Mioceno Temprano tardío, el clima era templado a templado-cálido y semiárido a húmedo, con bajas precipitaciones estacionales que actuaron como uno de los factores limitantes para el crecimiento vegetal.

7.1 Introduction

Late Early Miocene continental outcrops from the Santa Cruz Formation are found along the Atlantic coast of southernmost Patagonia, Argentina, between the Río Coyle and Río Gallegos (c. S 51° 00′, W 69° 10′; see Vizcaíno *et al.*, Chapter 1, Figs. 1.1 and 1.2). The Santa Cruz Formation, known by an exceptionally rich mammal fossil fauna, is composed of two members: the lower Estancia La Costa Member, with a predominance of pyroclastic deposits, and the upper Estancia La Angelina Member, mainly composed of claystones, mudstones, and sandstones (Tauber, 1997, 1999; Matheos and Raigemborn, Chapter 4).

The Santa Cruz Formation has been dated by $^{40}Ar/^{39}Ar$ between 16.56 Ma and 16.18 Ma, at Monte Observación (= Co. Observatorio) and Monte León localities (Fleagle *et al.*, 1995). Integrating results of the tephra correlations and radiometric ages indicates that the Santa Cruz Formation spans the interval ∼18 to 16 Ma in the Atlantic coastal plain, indicating an Early Miocene age (see Perkins *et al.*, Chapter 2).

Fossil wood and silica microfossil records from the Miocene in Patagonia are scarce (Schönfeld, 1954; Hünicken, 1995; Schöning and Bandel, 2004; Zucol *et al.*, 2007). However, palynological information documents a flora characterized by the occurrence of xerophytic and halophytic shrubby-herbaceous plants belonging to the extant families Asteraceae, Chenopodiaceae, Ephedraceae,

Early Miocene Paleobiology in Patagonia: High-Latitude Paleocommunities of the Santa Cruz Formation, ed. Sergio F. Vizcaíno, Richard F. Kay and M. Susana Bargo. Published by Cambridge University Press. © Cambridge University Press 2012.

Convolvulaceae, Fabaceae, and Poaceae. Podocarpaceae, Araucariaceae, and Nothofagaceae dominated forest communities, with a few tropical elements including Arecaceae, Symplocaceae, and Lauraceae (Romero, 1970; Barreda, 1993, 1996; Barreda and Palazzesi, 2007; Palazzessi and Barreda, 2007). During the Middle to Late Miocene, palynological records show an increase in diversity, with increasingly abundant xerophytic taxa (including members of the Asteraceae, Chenopodiaceae, Anacardiaceae, and Convolvulaceae) and sclerophyll trees (Casuarinaceae, Fabaceae, and Proteaceae), but forest and riparian forest communities as well (Barreda and Palazzesi, 2007; Palazzesi and Barreda, 2007; Barreda *et al.*, 2007). The development of xerophytic plant communities is probably linked with a water deficit in open forest areas or along marginal coastlines (Barreda and Palazzesi, 2007, 2010, and references therein).

The possible causes for the expansion of xerophytic elements with a pronounced ecological change in the Late Miocene include both global and local factors. On the one hand there was the uplift of Patagonian Andes in the west, which presumably established a pronounced rain shadow by the Middle Miocene (Blisniuk *et al.*, 2005); on the other hand the global cooling since ~14.5 Ma, which was associated with rapid growth of the Antarctic ice sheet and with major biogeographic changes on all southern continents (Zachos *et al.*, 2001).

Analysis of microscopic silica from continental sediments can be valuable for clarifying paleoenvironmental conditions. Phytolith studies in particular can contribute to understanding factors related to ecosystem diversity, climate, and paleoecology (Kondo *et al.*, 1988).

Fossil dicotyledonous woods can be used to establish paleoecological and paleoclimate conditions (Carlquist, 1977; Wiemann *et al.*, 1998, 1999, 2001; Wheeler and Bass, 1991, 1993; Poole and van Bergen, 2006). The approach proposed by Wiemann *et al.* (1998, 1999) has demonstrated that temperature-related climate variables, particularly mean annual temperature (MAT), can be estimated from anatomical features of wood. Also, calculations of Vulnerability Index (V) and Mesomorphy Ratio (M) show strong reliability because they are strongly correlated with rainfall and temperature (Carlquist, 1977). The Vulnerability Index is used as an indicator of hydric conductivity. A low value indicates greater capacity for withstanding water stress or freezing. High values of M denote mesomorphic wood structure (Carlquist, 1977).

In this chapter we present the first study of the Estancia La Costa Member of the Santa Cruz Formation based on the paleobotanical record from the Atlantic coast of Patagonia, analyzing phytoliths, carbonized wood, and leaf compressions, as well as records of microcharcoals, chrysophycean stomatocysts, sponge spicules, and diatom remains. These data are used to reconstruct the plant communities and paleoclimatic conditions. The records fall within the temporal interval called the Mid-Miocene Climatic Optimum (MMCO) (Zachos *et al.*, 2001; Blisniuk *et al.*, 2005), a period of time when the world was much warmer than today and plant communities experienced marked changes in floristic composition.

7.2 Institutional abbreviations

BAPb ex CIRGEO PALEOB: Colección de Paleobotánica del Museo Argentino de Ciencias Naturales "Bernardino Rivadavia," Buenos Aires, Argentina.

TK: Laboratory of Phylogenetic Botany, Faculty of Science, Chiba University, Japan.

MPM-PB: Colección Paleobotánica del Museo Regional Provincial Padre Manuel Jesús Molina, Río Gallegos, Argentina.

CIDPalbo: Colección Laboratorio de Paleobotánica, Centro de Investigaciones Científicas y Transferencia de Tecnología a la Producción, Diamante, Argentina.

MNHN: Collection des végétaux fossiles du Muséum National d'histoire naturelle de Paris.

DJ: British Antarctic Survey Paleontological Collections, Cambridge, UK.

LPPB: Colección Paleobotánica, Museo de La Plata, La Plata, Argentina.

S: Swedish Museum of Natural History, Stockholm, Sweden.

A-PF: Colección Paleoxilológica Torres, Instituto Antártico Chileno, Santiago, Chile.

P: British Antarctic Survey, Cambridge, UK.

7.3 Material and methods

Four localities were examined for silica microremains at Punta Sur (PS, sample 1970), Campo Barranca (CB, samples 1971–1973), Anfiteatro (ANF, samples 1974–1975), and Estancia La Costa (ELC, samples 1976–1982), all from the lower levels of the Estancia La Costa Member (see Table 7.1). This represents the lower unit of the Santa Cruz Formation, late Early Miocene, which crops out along the Atlantic coast of southernmost continental Patagonia in Argentina (see Vizcaíno *et al.*, Chapter 1, Figs. 1.1 and 1.2, and Matheos and Raigemborn, Chapter 4, Figs. 4.3 and 4.4). A total of 13 sedimentary samples contained silica microremains.

The silica microremains were extracted using standard wet oxidation and heavy-liquid flotation techniques following the protocol described in Zucol *et al.* (2010a, 2010b). Silica microremains were mounted for viewing on microscope slides in both liquid (oil immersion) and solid (Canada balsam) media, and examined using a Nikon Eclipse E200 light microscope. Photomicrographs were taken with a Nikon Coolpix 990 digital camera.

Table 7.1. *Microremains abundance, and isolated phytolith morphotypes in each sample*

Locality	ELC							ANF		CB			PS	Botanical affinities
Sample	1982	1981	1980	1979	1978	1977	1976	1975	1974	1973	1972	1971	1970	
Microcharcoal	NO	R	R	NO	F	S	F	S	NO	F	NO	VF	S	
Stomatocyst	NO	NO	NO	S	S	S	NO	NO	S	NO	NO	NO	NO	
Sponge spicule	F	R	R	S	NO	S	R	NO	NO	S	NO	S	R	
Diatom	NO	NO	R	S	NO	S	S	S	S	NO	NO	NO	R	
Articulated phytolith	R	R	VF	NO	S	NO	NO	NO	S	NO	NO	NO	S	
Isolated phytolith	VF	VF	F	VF	VF	VF	VF	VF	VF	VF	NO	VF	VF	
Isolated phytolith types														
Point-shaped	NO	X	NO	X	X	X	NO	NO	NO	X	NO	X	X	(1) [a,b]
Saddle	NO	NO	NO	NO	NO	X	NO	NO	NO	NO	NO	NO	X	(2) [a,b]
Cylindrical vascular element	NO	NO	NO	X	NO	NO	NO	NO	NO	NO	NO	NO	X	(1) [a,b]
Dumbbell and Cross	NO	NO	NO	X	NO	X	NO	NO	X	X	NO	X	X	(3) [a,b]
Bilobate stipa-type	NO	NO	NO	X	NO	NO	NO	NO	NO	NO	NO	NO	NO	(4) [c]
Fan-shaped	NO	NO	NO	NO	NO	X	X	NO	NO	NO	NO	X	X	(1) [a,b]
Globular	NO	NO	X	X	NO	X	NO	X	NO	NO	NO	NO	X	(5) [a,d,e]
Lobular	NO	NO	NO	X	NO	NO	NO	NO	NO	NO	NO	NO	X	(6)[a]
Elongate	X	X	X	X	X	X	X	X	X	X	NO	X	X	(1)[a,b]
Truncated cone	NO	NO	NO	X	X	X	NO	X	X	X	NO	X	X	(7)[a]
Conical	NO	NO	NO	X	NO	NO	NO	NO	NO	X	NO	NO	NO	(8)[a,d]
Polyhedral bulliforms	NO	X	NO	X	X	X	NO	X	NO	NO	NO	X	X	(1)[a,b]
Festucoid boat	NO	NO	NO	NO	X	X	X	NO	NO	NO	NO	NO	X	(9)[a]
Oblong/Circular/Square	NO	NO	NO	X	NO	X	NO	NO	NO	NO	NO	NO	X	(10)[a,b]
With modified surface	NO	X	NO	X	NO	NO	NO	X	NO	X	NO	NO	X	--
Other	X	NO	NO	X	X	X	X	X	X	X	NO	X	X	--

VF: very frequent; F: frequent; R: rare; S: scarce; NO: not observed; X: presence; PS: Punta Sur locality; CB: Campo Barranca locality; ANF: Anfiteatro locality; ELC: Estancia La Costa locality.
Botanical affinities of isolated phytolith types: 1. Non-diagnostic types; 2. Chloridoideae phytoliths; 3. Panicoideae phytoliths; 4. Stipeae phytoliths; 5. Echinate globular: Arecaceae phytoliths (I and IV), Smooth globular: Nothofagaceae phytolith (I and V), Cunoniaceae, Lauraceae and Myrtaceae phytolith (I); 6. Dicot phytoliths; 7. Danthonoidea phytoliths; 8. Cyperaceae phytoliths; 9. Festuceae (Pooideae) phytoliths; 10. Pooideae phytoliths.
[a] Kondo *et al.* (1994), [b] Twiss (1992), [c] Fredlund and Tieszen (1994), [d] Bertoldi de Pomar (1971), [e] Whang and Hill (1995).

The various phytolith morphotypes present in each sample were recorded, and counts were made of the number of phytoliths belonging to each. Phytolith morphotypes were classified according to existing classifications (Zucol *et al.* 2010c and references therein) considering their botanical affinities (see Table 7.1) and described using the morphotype descriptors proposed by the IPCNWG (2005). Chrysophycean stomatocysts were described following the terminology used in Rull and Vegas-Vilarrúbia (2000) and Coradeghini and Vigna (2001).

To quantify the relative abundances of analyzed microremains, a scale was used that references the comparative abundance of each morphotype, where elements with relative frequency values greater than 30% in the sample are categorized as very frequent (VF); 20–29.9% as frequent (F); 10–19.9% as rare (R); and less than 9.9% as scarce (S) (Zucol *et al.*, 2010b).

Processed sediment samples and microscope slides have been deposited in the Laboratorio de Paleobotánica of the CICYTTP, Diamante (Argentina), with the label heading CIDPalbo.

Leaf compressions were collected from the lowermost levels of the Estancia La Costa Member that crop out in the intertidal shore platform at Rincón del Buque, north of the Río Coyle (see Vizcaíno *et al.*, Chapter 1, Fig. 1.2 and Matheos and Raigemborn, Chapter 4, Fig. 4.4). The leaves are preserved in poorly consolidated medium to fine sands and muddy sediments right above an oyster bed. Whole

Table 7.2. *Fossil wood species from Punta Sur locality (PS) used in the multivariate anatomical analyses*

Taxa	Wood anatomical character											
	(1) mult	(2) spir	(3) <100 μm	(4) sept	(5) >10 ser	(6) heter4+	(7) stor	(8) abs	(9) marg	(10) RP	(11) para	(12) homo
Laurinoxylon atlanticum (Lauraceae)	0	0	1	0	0	0	0	0.5	0	0	1	0
Nothofagoxylon triseriatum (Nothofagaceae)	0.5	0	1	0	0	0	0	1	0	0.5	0	0
Myrceugenia chubutense (Myrtaceae)	0	0	1	0	0	0	0	0	0	0.5	0	0
Eucryphiaceoxylon eucryphioides (Cunoniaceae)	0.5	0	1	0	0	0	0	0	0	0.5	0	0
Doroteoxylon vicenti-perezii (Proteaceae?)	0	0	1	0	0.5	0	0	0	0	1	1	0
aff. *Xilotype 3* Pujana, 2008 (Fabaceae)	0	1	1	0	0	0	0	0.5	0	0.5	1	1

Abbreviations: (1) Vessels with multiple perforations (mult). (2) Spiral thickenings present in the vessels (spir). (3) Vessel mean tangential diameter less than 100 μm (<100 μm). (4) Fibers septate (sept). (5) Rays commonly more than 10 cells wide (>10 ser). (6) Rays heterocellular with four or more rows of upright cells (het4+). (7) Rays storied (stor). (8) Axial parenchyma absent or rare (abs). (9) Marginal parenchyma present (marg). (10) Wood ring-porous (RP). (11) Paratracheal parenchyma (para). (12) Homocellular rays (homo). Note: 0 = absent, 1 = present, 0.5 = absent/present.

leaves are preserved with petiole and apex but no good detail in vein architecture. Because of the fragmentary character of the matrix, large blocks of rock were taken from the bed and carried to the laboratory wrapped in film to conserve humidity and retard drying. The best leaf samples were mechanically cleaned using needles under a stereoscope, and photographed at the Museo de La Plata.

The wood materials consist of 20 carbonized specimens from the Punta Sur locality, which were recovered at 20 m below the tuff levels in the Estancia La Costa Member (see Vizcaíno *et al.*, Chapter 1, Fig. 1.2 and Matheos and Raigemborn, Chapter 4, Fig. 4.4). Only 13 specimens have well-preserved anatomical features. Thin-sections were made using petrographic techniques for the three characteristic xylem sections: transverse (TS), tangential longitudinal (TLS), and radial longitudinal (RLS). The terminology used for the wood descriptions follows the IAWA Lists of Features Suitable for Hardwood Identification (IAWA Committee, 1989, 2004), with some terms also taken from Carlquist (2001). The number of vessels per square millimeter was counted individually following Wheeler (1986). The fossils were identified using the wood atlases and descriptions of Wagemann (1948), Tortorelli (1956), Rancusi *et al.* (1987),

and Pujana (2008, 2009a, 2009b), as well as the InsideWood database (2004 onwards, accessed on March 2011).

One of the main functions of wood is water conduction, so climatic differences that affect water availability may be reflected in wood structure. Studies of ecological adaptations reflected in the anatomical characteristics of wood, from various vegetation communities and geographical areas, have demonstrated that wood anatomy can be used to estimate environmental parameters (Baas, 1976; Van der Graaff and Baas, 1974; Wheeler and Baas, 1991, 1993; Baas and Wheeler, 1996).

The relation between wood anatomy and climatic factors was proposed by pioneer plant anatomists (Bailey and Tupper, 1918; Frost 1930a, b, 1931) and then confirmed by various authors (see Baas, 1986; Wheeler and Baas, 1991; Lindorf, 1994; Baas *et al.*, 2004 and references therein). Furthermore, Wiemann *et al.* (1998, 1999) proposed equations for estimating some paleoclimate factors using simple and multiple regressions. The climate variables calculated by wood features (Table 7.2) in the present work were: mean annual temperature (MAT), mean annual range in temperature (MART), cold month mean temperature (CMMT), mean annual precipitation (MAP), and length of dry season (DRY). Two tested equations (Eq. (12) and (15) of Wiemann

Table 7.3. *Summary of the quantitative paleoclimate data from Punta Sur locality (PS)*

Wood physiognomy (WP)	Punta Sur locality
Temperature	
MAT (°C ± 1.7 °C)	
(Eq. 12)	19.31
$\quad$ 24.78 + 36.57(stor) − 15.61(marg) − 16.41(abs)	
(Eq. 15)	9.35
$\quad$ 17.07 + 25.23(stor) − 23.17(abs) + 13.79(sept)	
MART (°C ± 2.4 °C)	4.26
$\quad$ 4.16 + 0.319(spir) + 0.135(<100 µm) − 0.373(> 10 ser) − 0.154(het4+)	
CMMT (°C ± 2.6 °C)	9.80
$\quad$ 9.91 − 0.355(spir) − 0.098(< 100 µm) + 0.845(> 10 ser) + 0.368(het4+) + 0.528(stor) − 0.210(abs)	
DRY (months ± 1.4 months)	6.80
$\quad$ 6.81 − 0.186(mult) − 0.122(sept)	
Precipitation	
MAP (mm, ± 940 mm)	
−6.06 +6.332(sept) + 7.901(abs)	869

MAT, mean annual temperature; **MART**, mean annual range in temperature; **CMMT**, cold month mean temperature; **DRY**, length of dry season; **MAP**, mean annual precipitation. Equations (12) and (15) are equations in Wiemann *et al.* (1999) and Poole *et al.*(2005). For codification of wood anatomical characters see Table 7.2.

et al., 1999 and Poole *et al.*, 2005; see Table 7.3) were considered best for estimating MAT based on how well they worked as MAT predictors in present-day sites (Wiemann *et al.*, 1999). The standard errors for MAT, MART, CMMMT, MAP, and DRY are 1.7 °C, 2.4 °C, 2.6 °C, 940 mm, and 1.4 months, respectively. These wood physiognomy methods are appropriate for using fossil wood assemblages to determine terrestrial paleotemperatures, but they do not possess the same level of accuracy in determining precipitation levels. Nevertheless, the results of all of the analyses proposed by Wiemann *et al.* (1998), including precipitation, are included here. The sample size recommended by Wiemann *et al.* (1998) is 20–25 woods of different types from each locality. Unfortunately, with these Miocene fossil woods this prerequisite could not be satisfied. However, an interpretation based upon diverse bioproxies, such as that proposed in this chapter, would reduce the potential error.

The Vulnerability Index (V = mean vessel diameter divided by mean vessel frequency) and Mesomorphy Ratio (M = Vulnerability Index multiplied by mean vessel member length) were calculated using the equations of Carlquist (1977). Vulnerability Index is indicative of the sensitivity to the risks of embolisms. Low Vs indicate woods with numerous, narrow vessels and high safety in conduction, whilst high Vs are typical of efficient water conductors with wide, infrequent vessels. Mesomorphy is used as a measure of the water availability of the species, with high values being typical for species with a mesic

ecology. Mesic habits have a moderate or a well-balanced supply of moisture (Baas, 1986).

A diverse range of wood anatomical characteristics were also analyzed in terms of their ability to serve as paleoecological indicators, including vessel diameter, number of vessels per square milimeter, degree of vessel grouping, vessel arrangement, length of vessel elements, perforation plate type, porosity, presence of vasicentric and vascular tracheids, axial parenchyma, and growth rings (Carlquist and Hoekman, 1985; Wheeler and Baas, 1991, 1993; Lindorf, 1994; Woodcock, 1994).

The terms tropical, subtropical, and temperate are used to indicate broad geographic/macroclimatic zones. "Tropical" implies non-seasonal, equable warm and wet climates, while "temperate" implies markedly seasonal climates, with seasonality in temperature and/or water availability following Wheeler and Baas (1993). For temperature inferences based on grass phytoliths, the terms megathermal (>24 °C MAT), meso-megathermal (20–24 °C), mesothermal (14–20 °C), and microthermal (<14 °C) are used following Greenwood (1994) and Burkart (1975).

The quantitative values provided in the wood anatomical description are averages of 25 measurements. The average is cited first, followed by the minimum and maximum values, which are given in brackets.

The photomicrographs of the fossil wood sections were taken with a Nikon Coolpix S4 digital camera. The wood fossil materials, including thin-section slides and leaf compressions, were deposited at the Museo Regional Provincial Padre Manuel Jesús Molina, Río Gallegos, Santa Cruz Province, Argentina, labeled as MPM-PB 4381–4393 (wood fossils) and MPM-PB 4928–4931 (leaf compressions). Thin-section slides bear the specimen numbers followed by a lower-case letter indicating the particular section: a, transverse section; b, radial longitudinal section; and c, tangential longitudinal section.

7.4 Results

7.4.1 Microfossil records

For the analysis of microremains from the Estancia La Costa Member, 13 sedimentary samples from four localities were selected (PS, CB, ANF, and ELC) (Matheos and Raigemborn, Chapter 4: Fig. 4.4). This lower part of the Member also contained carbonized wood and plant remains that permit comparative analysis. All but one of the 13 samples (1972 – a clay from CB locality) contained microremains that possess heterogeneous assemblages classified as: microcharcoals, chrysophycean stomatocysts, sponge spicules, diatoms, and phytoliths (which occur in both articulated and isolated forms). A low abundance of microremains was observed in the other samples from the CB locality (1971 and 1973), as well as in the upper samples from the ELC locality (1980–1982). The samples with the highest phytolith abundance and variability, on the other hand, were from PS (1970), ANF (1975), and ELC (1977 and 1979) localities.

The microcharcoal remains (Table 7.1) are considered for the purpose of contextualizing the presence of the carbonized wood macroremains, because in some samples only these small carbonized fragments were present. In some cases these had indeterminate forms, whereas in others individual cells and tissue fragments could be recognized. Microcharcoal is abundant in the lower samples of the analyzed sequence (especially in PS and CB samples) and in the basal samples of the ELC section (1976 and 1978). Correspondingly, similar high abundances and variability of phytolith assemblages were observed in the PS and CB samples while the ELC basal samples showed less phytolith variability.

Chrysophycean stomatocysts (Table 7.1) were mainly present in the ELC basal samples (1976–1979), with a variety of types (Fig. 7.1o). These stomatocysts had either smooth or scrobiculate surfaces, simple or ornamented operculi, and spherical to ovoid shapes.

Sponges (Table 7.1) were represented mainly by whole or fragmented macroscleres (Fig. 7.2g), with either smooth or, rarely, lightly echinate surfaces. These were present in great abundance, mainly as whole elements, in the PS sample (1970), and in the lowest ELC samples (1977 and 1979). In the CB samples (1971 and 1973) as well as in the upper

ELC samples (1976 and 1980–1982) the sponge remains were scarce, occurring either whole or fragmented.

Diatom remains (Table 7.1) were present in the form of either whole or fragmented frustules (Fig. 7.2f). These were most abundant in the PS sample, and were found in very low proportions in samples from the ANF (1974 and 1975) and ELC (1976, 1977, 1979, and 1980) localities. The frustules that remained whole were small or medium types, while large diatoms were observed in fragmented forms.

Articulated phytoliths (Fig. 7.2e) formed by multiple cells were abundant in the PS sample and in the upper samples from the ELC locality, but in some samples (as in ANF 1974) phytoliths consisting of sections of articulated silicified epidermal tissue were found. The great majority of these articulated phytoliths are graminoid in origin, and are made up of epidermal and subepidermal elements such as long and short cells, hair and prickle cell elements, and buliform cells (fan-shaped and polyhedral types).

Isolated phytoliths were abundant and variable, some well preserved, some heavily eroded (especially in the 1975 sample and to a lesser degree 1979) (Fig. 7.1n). The greatest variability of isolated phytoliths is seen at the PS and CB localities.

The most abundant phytolith morphotypes are non-diagnostic elongated forms, including small elongated prismatic (Fig. 7.1j) and elongated prismatic with diverse contours (Fig. 7.2a). On the other hand, point-shaped phytoliths (Fig. 7.1k) along with their variants are present in samples from the PS, CB, and ELC localities; while fan-shaped (Fig. 7.2d) and polyhedral bulliforms (Fig. 7.2b) phytoliths are mainly present in samples from PS and lower samples from ELC.

Among the diagnostic phytolith morphotypes, saddles (Fig. 7.1c) are present in the samples 1970 and 1977; dumbbells (Fig. 7.1a), crosses (Fig. 7.1b), and polylobate dumbbells (Fig. 7.1e) are present in samples 1977 and 1979 from ELC; while sample 1979 (not shown) is the only one that contains stipa-type bilobate phytoliths. Truncated cones (Fig. 7.1l) are distributed throughout most of the sequence (1970–1979) with greatest abundance in samples 1973–1977. Other phytolith types that are more abundant in the lower section from ELC (samples 1976–1979) and scarce in sample 1970 are the festucoid boats (Fig. 7.1f), circular to oblong (Fig. 7.1h), square (Fig. 7.1d), and small conical (Fig. 7.1g) phytoliths.

Cylindrical vascular element (Fig. 7.2c) and lobular (Fig. 7.1i) phytoliths were found in samples 1970 and 1979; while globular phytoliths (Fig. 7.1m) are abundant in samples from PS and ANF, and in the lower ELC samples. These phytoliths are spherical to slightly elliptical in contour and differentiated by their surface texture, which can be either smooth (Fig. 7.1m, upper row) or echinate (Fig. 7.1m, lower row). The globular echinate phytoliths are more abundant in the majority of the cases, whereas smooth surfaced types are relatively abundant in samples 1970, 1977, and 1979.

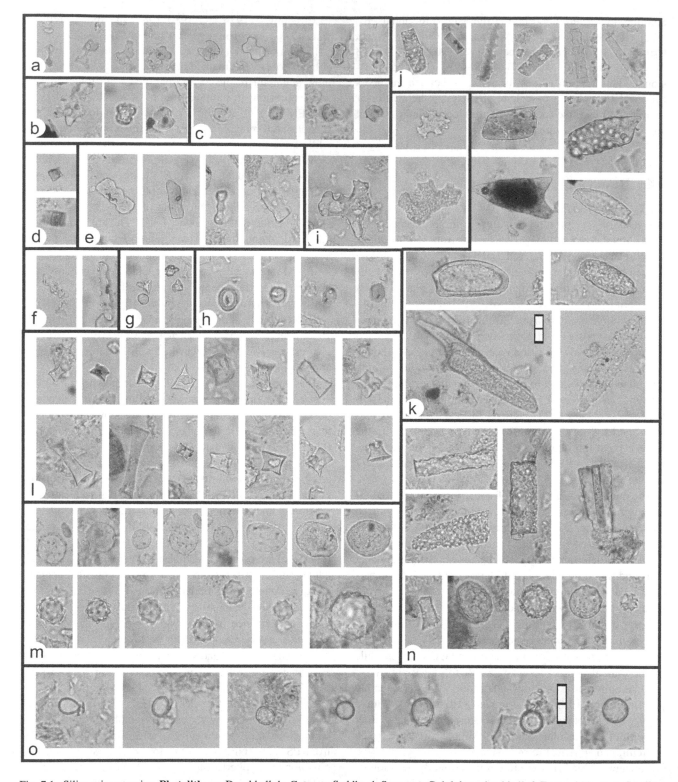

Fig. 7.1. Silica microremains. **Phytoliths**: a, Dumbbell. b, Cross. c, Saddle. d, Square. e, Polylobate dumbbell. f, Festucoid boat. g, Small conical. h, Rondel and oblong. i, Lobate. j, Small elongate. k, Point-shaped. l, Truncated cones. m, Smooth and echinate globular. n, Microremains with modified surface. **Chrysophycean stomatocysts**: o, Different stomatocyst types. Scale bar in k = 20 μm (applies to a–n); scale bar in o = 20 μm.

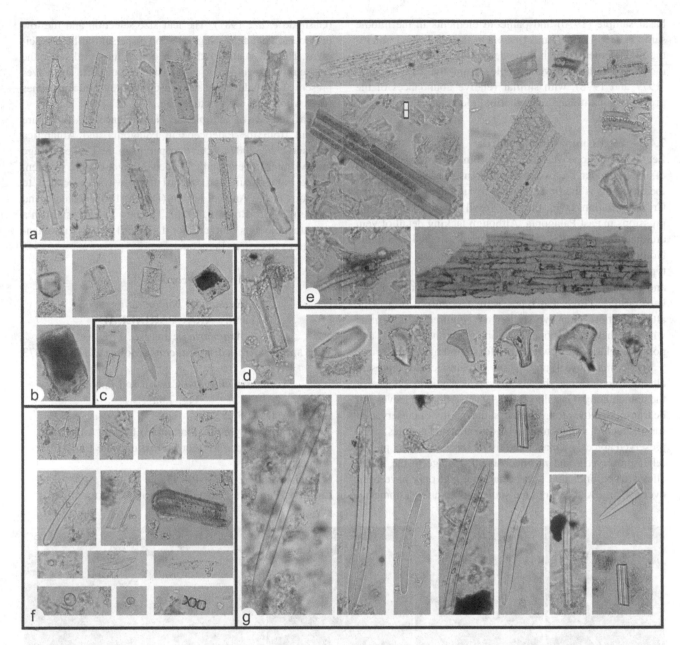

Fig. 7.2. Silica microremains. **Isolated phytoliths**: a, Elongate. b, Polyhedral bulliforms. c, Cylindrical vascular elements. d, Fan-shaped. **Articulated phytoliths**: e, Different phytoliths originated in epidermal and bulliform cells. **Diatoms**: f, Fragmented and entire diatoms. **Sponge spicules**: g, Sponge macroscleres. Scale bar = 20 μm.

Several of the phytolith morphotypes, such as panicoid (dumbbells, crosses, and polylobate dumbbells), chloridoid (saddles), danthonioid (truncated cones), stipoid (stipa-type bilobate), pooid, and festucoid (circular to oblong, square, small conical and festucoid boat) elements, indicate the presence of graminoid vegetation (Table 7.1). Other types are associated with dicotyledonous vegetation elements (lobular), while still others have arecoid affinity, such as globular echinate phytoliths. The globular smooth types are linked by some authors with the Nothofagaceae (Kondo *et al.*, 1994; Whang

and Hill, 1995), some with size and surface variations also are assigned to Cunoniaceae, Lauraceae, and Myrtaceae (Kondo *et al.*, 1994). The distribution of this particular smooth globular type jointly with the globular echinate phytoliths in the middle–lower section of the Estancia La Costa Member sequence indicates the presence of arecoid/nothofagoid components.

Danthonioid grass elements characterize the ANF locality. Phytoliths representing the megathermal grass groups are abundant in the samples from ELC sequence, with a predominance of chloridoid grass elements found in the basal sample (1970) and panicoid elements in the middle samples (1974, 1977, and 1979). On the other hand, microthermal elements (such stipoid, pooid, and festucoid) are most abundant in lower samples from the ELC locality (1977 and 1978), with similar relative abundances of the previously mentioned megathermal elements.

7.4.2   Leaf compression records

Leaf compressions were collected from the lowermost levels of the Estancia La Costa Member at Rincón del Buque locality. This is the first record of fossil leaves for the Santa Cruz Formation, being the only fossil record for leaf compressions in east Patagonia. Although lacking detail of vein architecture, all leaves can be unequivocally assigned to the Nothofagaceae family based on leaf shape and size, tooth type and venation, primary and secondary venation pattern, and leaf surface plication (Romero, 1980; Romero and Dibbern, 1985). Three leaf types were recognized (Fig. 7.3), although more intensive studies are required to determine whether they correspond to more than one species, and possible relationships with other fossil and modern species.

While microfossil and wood remains could represent allochthonous floras, the leaf record is a good evidence of the autochthony (Gastaldo *et al.*, 1987). Although leaves can be transported over long distances, the quality of preservation for this record, with complete blades, attached petioles, well-preserved apices, no macerated tissue (coarse texture, plicate relief preserved), and no mechanical damage to the blade, allows us to infer that transportation could only have been over a short distance from the source, up to a few kilometers

(Gastaldo *et al.*, 1987). The leaf size selection and the low presence of leaf debris suggest wind dispersal for these specimens. The low quality of the vein architecture is related to the coarse rock matrix, not to long-distance transport. The leaves preserved in muddy lenses are best preserved, and sometimes show whole-leaf architecture (Fig. 7.3).

Leaf records of indubitable Nothofagaceae affinity are present from the Paleocene in Patagonia and Antarctica (Iglesias *et al.*, 2007). After the opening of the Drake Passage in the Upper Eocene and Oligocene, they are more frequent in the Patagonian fossil record but always associated to outcrops in the Andean region (Dusén, 1907; Romero and Dibbern, 1985; Romero, 1986). The new fossils establish a parautochthonous character for the record of the Nothofagaceae trees in eastern Patagonia.

7.4.3   The fossil wood record

Systematic paleontology
Family **ARAUCARIACEAE**
Araucaria de Jussieu, 1789
Type species: ***Araucaria araucana*** (Molina) K. Koch, 1873

Araucaria marensii Cantrill and Poole, 2005
Fig. 7.4a–l
Holotype: DJ 1057.53.
Type locality: Seymour (Marambio) Island (Antarctica).
Type formation: La Meseta Formation (late Early Eocene to latest Eocene).
Material studied: MPM-PB 4388.
Locality: Punta Sur, Santa Cruz Province, Argentina.

Fig. 7.3. Three leaf types of Nothofagaceae. a and b, Leaf type 1 (a = MPM-PB 4928, b = MPM-PB 4929); note rounded base, eight pairs of secondary veins, and percurrent tertiary veins with tertiary angles decreasing exmedially from the axis of symmetry. c, Leaf type 2 (MPM-PB 4930); note small size, plicate blade surface, and uniform angles in tertiary veins. d and e, Leaf type 3, part and counterpart (MPM-PB 4931); note asymmetric cuneate base, and nine or more secondary veins. Scale bars = 1 cm.

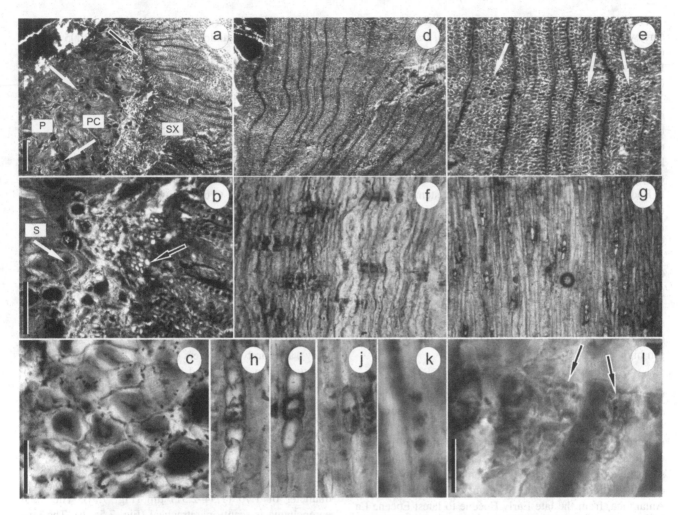

Fig. 7.4. *Araucaria marensii*. a, TS in general view, showing heterogeneous pith (P), sclereids in clusters (white arrows), parenchyma cells (PC), primary xylem (black arrow) and secondary xylem (SX). b, TS, details of sclereid cell (S, white arrow), tracheids in radial files of the primary xylem (black arrow). c, TS, details of ticked walled sclereid from pith. d, TS in general view, showing growth rings slightly demarcated and the gradual transition from earlywood to latewood. e, TS, showing axial parenchyma (white arrows). f, RLS in general view, illustrating the distribution of pitting and homogeneous rays. g, TLS in general view, showing the uniseriate rays distribution. h, TLS showing uniseriate ray composed of five cells. i, TLS showing uniseriate ray composed of three cells. j, TLS showing uniseriate ray composed of two cells. k, TLS showing the tracheid radial walls with uniseriate bordered pitting. l, RLS, cross-field regions with three to seven crowded araucarioid pits (black arrows). Scale bars: a, d = 300 μm; b, e–g = 200 μm; c, h–j = 50 μm; k, l = 30 μm.

Formation: Santa Cruz Formation, Estancia La Costa Member.

Age: late Early Miocene.

Description: this specimen is a single piece of gymnosperm twig wood. Transverse section: the preserved twig is eccentric (54.23 mm × 36.61 mm in diameter) and 8 cm in length. Pith, primary xylem, and secondary xylem were preserved in this specimen. The heterogeneous pith is subcircular with a diameter of 4.89 mm × 4.27 mm, composed of parenchyma and sclerenchyma tissues. Parenchyma consists of thin-walled individual cells ranging from 59–108 μm in diameter (Fig. 7.4a). Sclereid clusters range from 274 to 470 μm in diameter with thick walls and are hexagonal in cross-section (Fig. 7.4a–c). In transverse section, the primary xylem is dispersed in wedge-shaped bundles at the periphery of the pith, and is composed of endarch protoxylem and centrifugal metaxylem, with tracheids in radial files showing a gradual transition from primary xylem to first-formed secondary xylem (Fig. 7.4a, b). The cells of the primary tracheids are thin-walled and range from 15 to 28 μm in diameter (Fig. 7.4b). Secondary xylem is centrifugal and radially arranged (Fig. 7.4a, d, e). Growth rings are characterized by distinct ring boundaries; each ring has a gradual transition from earlywood to latewood, and the latewood has a few layers of thick-walled, radially narrower tracheids.

Tracheids are quadrate to rectangular in cross-section. In earlywood, the mean tangential diameter of tracheids is 29 (23–38) μm and the mean radial diameter is 30 (23–35) μm. In latewood, the mean tangential diameter of tracheids is 22 (15–28) μm and the mean radial diameter is 19 (13–30) μm. Tracheids arranged in regular radial rows are separated by rays (Fig. 7.4d, e). The rays are separated from each other by 1–11 rows of tracheids; with an average of 5 rows. The mean number of rays per linear millimeter is 7 (4–9 per linear mm) (Fig. 7.4d, e). Axial parenchyma rarely observed (Fig. 7.4e).

Radial longitudinal section: the radial system is homogeneous, comprising procumbent parenchyma cells (Fig. 7.4f). Radial walls have uniseriate-bordered pits (Fig. 7.4k), occasionally biseriate. The bordered pits are circular and have a separate arrangement. Where biseriate, pits are alternate. Cross-field regions with 3–7, usually crowded araucarioid pits (Fig. 7.4l).

Tangential longitudinal section: rays are uniseriate, short and squat, ranging from one to seven cells in height (Fig. 7.4h–j). Rays are 21 (15–30) μm wide by 79 (30–135) μm high.

Comments The material studied here is closely related to *Araucaria marensii* by the presence of heterogeneous pith with sclerenchyma and parenchyma, wood pycnoxylic with uniseriate or alternate biseriate pits, cross-field pitting araucarioid with four to six pits per cross-field, and short and squat rays (one to ten cells tall). Cantrill and Poole (2005) described this species at Seymour (Marambio) Island, Antarctica, from the late Early Eocene to latest Eocene La Meseta Formation. The holotype (DJ 1057.53) is a preserved trunk 28.5 cm long by 15.2 cm in diameter with its outer surface covered by a layer of bark. The specimen described here (MPM-PB 4388) possessed well-preserved features of the primary xylem.

Family LAURACEAE

Laurinoxylon Felix, 1883 *emend* Dupéron, Dupéron-Laudoueneix, Sakala and De Franceschi 2008
Type species: *Laurinoxylon diluviale* (Unger) Felix, 1883 *emend* Dupéron, Dupéron-Laudoueneix, Sakala and De Franceschi, 2008.
Holotype: MNHN 8652 to 8655.
Type locality: Jáchymov Mine, Bohemia.
Type formation: unknown? (Oligocene).
Synonyms
Ulminium diluviale Unger, 1842.
Betulinium diluviale (Unger) Felix, 1882.
Perseoxylon diluviale (Unger) Felix, 1887.
Betuloxylon diluviale (Unger) Lakowitz, 1890.
Laurinium diluviale (Unger) Edwards, 1931.

Laurinoxylon atlanticum (Romero)
Dupéron-Laudoueneix and Dupéron, 2005
Fig. 7.5a–l
Holotype: LPPB 7812 a-m; pm LPPB 774–797.
Type locality: Bahía Solano, Chubut, Argentina.
Type formation: unknown (Eocene).
Synonyms
Ulminium atlanticum Romero, 1970

Material studied: MPM-PB 4384–4386, and MPM-PB 4393.
Locality: Punta Sur, Santa Cruz Province, Argentina.
Formation: Santa Cruz Formation, Estancia La Costa Member.
Age: late Early Miocene.
Affinity: Angiospermae, Laurales, Lauraceae, more closely with *Persea* Mill, and less closely with *Beilschmiedia* Ness, *Nectandra* Rottb., and *Ocotea* Aubl.
Description: indistinct growth rings (Fig. 7.5a). Diffuse porous. Vessels frequently solitary (47%), rarely in radial multiples of two to three elements (6%), some in radial multiples of four to five elements (4% and 2% respectively), and rarely in clusters (Fig. 7.5a, b, f–h). The vessels are circular in outline and are thick-walled (Fig. 7.5f–h). The mean vessel tangential diameter is 62 (25–115) μm and mean vessel radial diameter is 57 (20–110) μm. Mean vessel element length is 155 (60–290) μm. Many vessels have tyloses (Fig. 7.5a, b, d, f, g). The mean vessel density is 21 (17–18)/mm^2. Perforation plates are simple (Fig. 7.5i), with oblique or straight end walls (Fig. 7.5c–e). Intervessel pits are bordered, alternate, and polygonal with a mean diameter of 1.70 (1.34–2.31) μm (Fig. 7.5j). The axial parenchyma is scanty paratracheal (Fig. 7.5a, b). The rays are multiseriate, two to three cells wide (73% and 7% respectively), or uniseriate (20%) (Fig. 7.5c). The mean number of rays per linear millimeter is 16 (12–22 per linear mm). Rays are heterocellular, body composed of procumbent cells, with one row of upright cells (Fig. 7.5d, e). Mean ray width is 22 (13–30) μm for multiseriate rays and 19 (15–25) μm for uniseriate rays. Mean ray height is 265 (85–510) μm for multiseriate rays and 125 (70–200) μm for uniserite rays. Vessel-ray pits, poorly preserved, with reduced borders, similar to intervessel pits in size and shape, with a mean diameter of 2.37 (3.17–1.84) μm (Fig. 7.5k). Some rays appear to contain idioblastic secretory cells ("oil cells"), most common in the ray margins (Fig. 7.5e, l). Fibers are non-septate, quadrangular to rectangular in outline, and have a mean diameter of 26 (20–36) μm. They are thick-walled, with a mean wall thickness of 8 (6–10) μm. Pits were not observed.

Comments The combination of solitary vessels, multiple radial vessels, simple perforation plates, alternate intervessel pits, vessel-ray pits similar to intervessel pits in size and

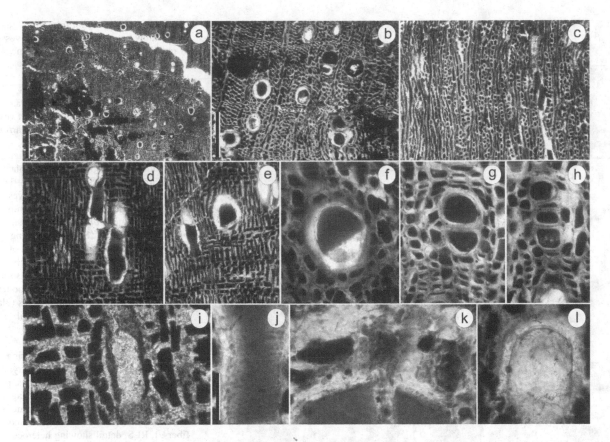

Fig. 7.5. *Laurinoxylon atlanticum*. a, TS, general view. b, TS, showing solitary and radial multiple vessels. c, TLS, detail of multiseriate and uniseriate rays. d, RLS, detail of heterocellular rays and vessel elements with oblique end walls and simple perforation plate. e, RLS, details of vessel element, heterocellular rays and idioblasts in ray margins. f, TS, detail showing circular outline and thick-walled vessels. g, TS, detail of radial multiple of two elements. h, TS, detail of radial multiple of four elements. i, TLS, showing a simple perforation plate. j, TLS, showing intervessel pits pattern. k, RLS, vessel-ray parenchyma pits with reduced borders. l, RLS, detail showing idioblastic secretory cells ("oil cells"). Scale bars: a = 300 μm; b–e = 200 μm; f–i, l = 50 μm; j, k = 30 μm.

shape, and idioblastic secretory cells (probable "oil cells") are diagnostic of *Laurinoxylon*, which is related to the Lauraceae family (Dupéron-Laudoueneix and Dupéron, 2005; Dupéron *et al.*, 2008).

Family NOTHOFAGACEAE
Nothofagoxylon Gothan, 1908 *emend* Poole, Hunt and Cantrill, 2001
Type species: *Nothofagoxylon scalariforme* Gothan, 1908 *emend* Poole, 2002.
Lectotype: Specimen No. 14 (S004067).
Syntypes: Specimen No. 13 (S004067) and No. 18 (S004068).
Type locality: Marambio (Seymour) Island, Antarctica.
Other materials and synonyms: see Poole (2002) and Pujana (2009b).

Nothofagoxylon triseriatum
Torres and Lemoigne, 1988 *emend* Poole, 2002
Fig. 7.6a–e
Holotype: A-PF-51.

Paratype: A-PF-53.
Type locality: Rey Jorge Island, Antarctica.
Type formation: Arctowski Cove Formation (Middle Eocene).
Other materials: MPM-PB 1966 and 1967 (Pujana, 2009b).
Synonyms
Nothofagoxylon paleoglauca Torres and Lemoigne (1988)
Material studied: MPM-PB 4392.
Locality: Punta Sur, Santa Cruz Province, Argentina.
Formation: Santa Cruz Formation, Estancia La Costa Member.
Age: late Early Miocene.
Affinity: Angiospermae, Fagales, Nothofagaceae, *Nothofagus* Blume.
Description: Growth rings distinct, marked by a reduction of vessel radial diameter of the last latewood fibers (Fig. 7.6a). Semi-ring porous to diffuse. Vessels roughly circular to slightly oval in transverse section, solitary, in radial files up to six vessels and in clusters (Fig. 7.6a).

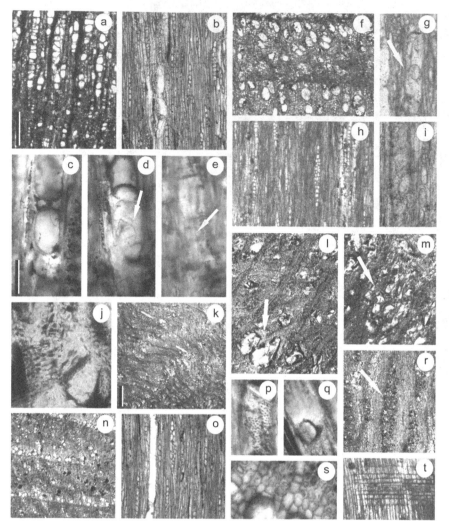

Fig. 7.6. *Nothofagoxylon triseriatum* (a–e): a, TS, showing growth ring, solitary vessels, in radial files up to six vessels and in clusters. b, TLS, general view shows uniseriate and biseriate rays. c, RLS, detail of a simple perforation plate. d, RLS, showing a scalariform perforation plate (white arrow). e, TLS, showing short vessel elements and intervessel pits (white arrow). *Myrceugenia chubutense* (f–i): f, TS, showing growth ring and vessel elements. g, TLS, showing a simple perforation plate (black arrow). h, TLS, showing uniseriate and triseriate rays. i, TLS, showing vessels elements with intervessel pits and simple perforation plate. *Doroteoxylon vicenti-perezii* (j–m, r): j, TLS, detail showing intervessel pits. k, TS, general view, of growth rings and large solitary vessels (black arrows). l, TS, detail of large solitary vessels (black arrows) in the earlywood. m, TS, detail of small vessels in clusters (black arrow) in the latewood. r, TLS, detail showing multiseriate rays. *Eucryphiaceoxylon eucryphioides* (n–q, s, t): n, TS, general view showing growth ring and vessel mainly solitary. o, TLS, general view showing ray types. p, TLS, detail showing intervessel pitting. q, RLS, detail of simple perforation plate. s, TS, detail showing vessel elements and fibers. t, RLS, detail showing heterocellular rays. Scale bars: k, n = 300 μm; a, b, f, h, l, m, o, r, t = 200 μm; c–e, g, i, j, p, q, s = 50 μm.

The mean tangential vessel diameter is 41 (25–55) μm and mean radial vessel diameter is 42 (25–70) μm. Mean vessel element length is 99 (50–205) μm. Mean vessel density is 58 (47–72)/mm^2. Tyloses are present in many vessels. Perforation plates are simple and horizontal (Fig. 7.6c), rarely scalariform with few bars (Fig. 7.6d), and oblique perforation plates are present in the latewood vessels. Intervessel pits are circular to polygonal with an alternate to opposite arrangement with a mean diameter of 16 (11–19) μm (Fig. 7.6e). Axial parenchyma is mainly apotracheal, rare, and diffuse (Fig. 7.6a). Rays are triseriate, biseriate, and uniseriate (Fig. 7.6b), heterocellular, and composed of procumbent body cells with one or two rows of square to upright marginal cells. Mean ray width is 24 (15–35) μm and mean ray height is 243 (145–400) μm. The mean number of rays per linear millimeter is 11 (6–17 per linear

mm). Fibers form a ground mass and are quadrangular or roughly circular in cross-section. They are non-septate and thick-walled, with a mean diameter of 17 (10–25) μm. Fiber pitting was not observed.

Comments The presence of predominantly triseriate and unfused vertical rays, simple occasionally scalariform perforation plates, diffuse and opposite axial parenchyma alternate and scalariform intervessel pits justify assignment to *Nothofagoxylon triseriatum* from an Antarctic Eocene flora. *Nothofagoxylon palaeoglauca*, which has been found in Rey Jorge Island, Antarctica, in the Arctowski Cove Formation (Middle Eocene) (Torres and Lemoigne, 1988) has been synonymized with *N. triseriatum* based upon the fact that the only major distinguishing features between these two species are the presence of septate fibers and helical thickenings in the vessels, as well as differences in the predominance of biseriate rays. This synonymization is supported by the hierarchical cluster analysis performed by Poole (2002).

Family **MYRTACEAE**
Myrceugenia O. Berg, 1855
Type species: *Myrceugenia myrtoides* O. Berg, 1857.
Holotype: BAPb ex CIRGEO PALEOB. No. 139, 140, 141-PB No. 408.
Type locality: Puesto Álvarez, Chubut Province, Argentina.
Type formation: Salamanca Formation, Bustamante Member (Paleocene).

Myrceugenia chubutense Ragonese, 1980
Fig. 7.6f–i
Material studied: MPM-PB 4387.
Locality: Punta Sur, Santa Cruz Province, Argentina.
Formation: Santa Cruz Formation, Estancia La Costa Member.
Age: late Early Miocene.
Affinity: Angiospermae, Myrtales, Myrtaceae, *Myrceugenia* O. Breg.
Description: this specimen is based on one stem or trunk piece of wood of unknown diameter. Growth rings are demarcated by reduction of the radial diameter of the last two or four rows of latewood fibers (Fig. 7.6f). Diffuse porous with a tendency to semi-ring-porosity. Vessels are mainly solitary (70%), rarely in radial pairs (18%) and in tangential pairs (12%). Vessels are radially elliptic to circular in cross-section (Fig. 7.6f). Vessels are thin-walled. Mean tangential vessel diameter is 53 (30–80) μm and mean radial vessel diameter is 84 (50–120) μm. Mean vessel element length is 239 (160–330) μm with oblique or straight end walls (Fig. 7.6i). Some vessels have tyloses. Vessels are numerous, with a mean vessel density of 62 (57–67)/mm². Perforation plates are exclusively simple (Fig. 7.6g). Intervessel pits are circular to horizontally elliptic and have an alternate arrangement. Ray-vessel pits appear to be small and circular. Axial parenchyma is diffuse and diffuse-in-aggregate. Rays are heterocellular composed of procumbent body cells with one to two upright marginal cells. Rays are short, with a mean height of 301 (110–480) μm, and are 12 (4–21) cells high. Mean ray width is 31 (13–53) μm. The mean number of rays per linear millimeter is 9 (6–11 per linear mm). Rays are mainly biseriate with uniseriate portions (57%), with some uniseriate (22%) and triseriate (23%) rays also present (Fig. 7.6h). The cells in the biseriate portions are smaller than those in the uniseriate portions. Fibers are non-septate and circular in outline, and fiber pitting was not observed. Mean fiber diameter is 20 (10–38) μm and fiber walls have a mean thickness of 4 (4–11) μm.

Comments Five fossil species of Myrtaceae are known from southern Argentina, Chile, and Antarctica: *Myrceugenellites maytenoides* (Nishida *et al.*, 1988); *Myrceugenellites antarticum* (Poole *et al.*, 2001; Pujana, 2009a); *Myrceugenellites oligocenicum* (Pujana, 2009a); *Myrceugenelloxylon pseudoapiculatum* (Nishida, 1984a, b) and *Myrceugenia chubutense* (Ragonese, 1980). The anatomy of this new fossil wood allows the material to be assigned to the Patagonian species *Myrceugenia chubutense*. This species is recognized by having almost exclusively solitary vessels that also occur in tangential and radial pairs, exclusively simple perforation plates, numerous and narrow vessels, mainly biseriate rays, and diffuse and diffuse-in-aggregate axial parenchyma.

Family **CUNONIACEAE**
Eucryphiaceoxylon Poole, Mennega and Cantril, 2003
Type species: *Eucryphiaceoxylon eucryphioides* (Poole *et al.*, 2001) *comb. nov.* Poole, Mennega and Cantrill, 2003.
Basionym: *Weinmannioxylon eucryphiodes* Poole, Hunt and Cantrill, 2001.
Holotype: P.3023.11.
Type locality: Fildes, Peninsula, King George (25 de Mayo) Island, Antarctic Peninsula.
Type formation: Middle unit of the Fildes Formation (Middle Eocene).
Additional material: DJ.1057.62, D.J.1057.57, DJ.1057.63A, DJ.1055.32 (group 1); DJ.1054.1, DJ.1056.56, DJ.1056.14, DJ.1056.22 (group 2); DJ.1057.38 (group 3).

Eucryphiaceoxylon eucryphioides
(Poole *et al.*, 2001) *comb. nov.*
Poole, Mennega and Cantrill, 2003
Fig. 7.6n–q, s, t
Material studied: MPM-PB 4391.
Locality: Punta Sur, Santa Cruz Province, Argentina.
Formation: Santa Cruz Formation, Estancia La Costa Member.
Age: late Early Miocene.
Affinity: Angiospermae, Rosales, Cunoniaceae, *Eucryphia* Cav.
Description: growth rings are distinct and demarcated by a narrow zone of radially flattened fibers and a decrease in vessel diameter and abundance (Fig. 7.6n). Diffuse porous or occasionally semi-ring-porous. Vessels are roughly circular to radially elliptical in cross-section. Vessels are mainly solitary or paired, but also occur in radial group of three elements (Fig. 7.6n). Mean tangential vessel diameter is 49 (25–70) μm and mean radial vessel diameter is 44 (20–65) μm. Mean vessel element length is 230 (130–385) μm with oblique or straight end walls. Some vessels have tyloses. Mean vessel density is 45 (35–55)/mm². Perforation plates are simple (Fig. 7.6q) or rarely scalariform with five to

nine fine bars, which are sometimes branched. Intervessel pits are circular and opposite (Fig. 7.6p) to transitional and scalariform. Axial parenchyma is diffuse and diffuse-aggregated (Fig. 7.6n). Rays are heterocellular composed of procumbent body cells with one to three rows of square and upright marginal cells (Fig. 7.6t). Mean ray height is 269 (80–385) µm and mean ray width is 36 (15–60) µm. The mean number of rays per linear millimeter is 7 (6–9 per linear mm). Rays are mainly uniseriate and biseriate, or may also be triseriate (Fig. 7.6o). Fibers are non-septate and circular in outline, and fiber pitting was not observed. Fibers are thick- walled and have a mean diameter of 13 (10–15) µm (Fig. 7.6s).

Comments The combination of characters – predominantly uniseriate rays; scalariform and simple perforation plates; circular, opposite, transitional, and scalariform intervessel pitting; and transitional to scalariform vessel ray pitting – is characteristic of *Eucryphia* Cav. of the Cunoniaceae (or Eucryphiaceae). The King George (25 de Mayo) Island specimens previously assigned to *Weinmannioxylon eucryphiodes* by Poole *et al.* (2001) were transferred to *Eucryphiaceoxylon eucryphioides* by Poole *et al.* (2003). The reason was that these specimens differ from other members of *Weinmannioxylon* described from South America and Antarctica by their ray characteristics in having entirely scalariform perforation plates (Poole *et al.*, 2000). The specimen studied here shows the greatest anatomical similarity with Group 2, especially with DJ. 1054.1 described by Poole *et al.* (2003). These divisions are predominantly on the basis of ray width and difference in relative abundance of scalariform vs. simple perforation plates.

Family **FABACEAE**
Subfamily **FABOIDEAE**
Xilotype 3 Pujana, 2008

Type species: *Xilotype 3* Pujana, 2008
Holotype: MPM-PB 2112.
Paratypes: MPM-PB 2110, 2115, 2117, 2120, and 2122.
Other specimen: MPM-PB 2102.
Type locality: Cerro Calafate, Santa Cruz Province, Argentina.
Type formation: upper section of the Río Leona Formation (Oligocene).

aff. *Xilotype 3* Pujana, 2008
Fig. 7.7a–l

Material studied: MP-MPB 4390.
Locality: Punta Sur, Santa Cruz Province, Argentina.
Formation: Santa Cruz Formation, Estancia La Costa Member.
Age: late Early Miocene.
Affinity: Angiospermae, Fabales, Fabaceae, Faboidea, *Sophora* section *Edwardsia*.

Description: growth rings distinct. Diffuse porous to semi-ring-porous (Fig. 7.7a, b). Vessels are mainly solitary (50%), in radial and tangential series of two to three elements (13%), or in clusters (17%) (Fig. 7.7a–c). Tangential bands of vessels, and diagonal and irregular dendritic patterning of vessels are also present (Fig. 7.7a–c). Vessels are circular to oval in cross-section. Mean tangential vessel diameter is 58 (30–90) µm and mean radial vessel diameter is 57 (30–90) µm. Mean vessel element length is 214 (160–285) µm. Mean vessel density is 137 (114–181)/mm^2. Perforation plates are mainly simple (Fig. 7.7d, g, l) or rarely reticulate (Fig. 7.7f). Intervessel pits are bordered round to irregular in outline with alternate arrangement with a mean diameter of 6 (3–9) µm (Fig. 7.7i, j). Vascular tracheids are present around the vessels, irregular in shape. The mean tracheid length is 112 (89–141) µm. The vascular tracheids have bordered pits with a mean diameter of 7 (4–10) µm (Fig. 7.7l). Axial parenchyma is vasicentric scarce and apotracheal diffuse (Fig. 7.7a–c). Vessel-ray parenchyma pits are similar in size and shape to intervascular pits (Fig. 7.7g). Helicoidal thickenings present throughout the length of the vessel elements (Fig. 7.7k). Rays are homocellular composed of procumbent cells (Fig. 7.7d). The rays are uniseriate (Fig. 7.7e), biseriate (Fig. 7.7e), and multiseriate, four to six cells wide (Fig. 7.7e). The mean height of uniseriate rays is 133 (55–165) µm and 5 (3–7) cells high, and the mean width of uniseriate rays is 22 (20–25) µm. The mean height of biseriate rays is 205 (160–260) µm and 8 (6–11) cells high, and the mean width of biseriate rays is 39 (37–40) µm. The mean height of multiseriate rays is 474 (195–725) µm and 18 (9–28) cells high, and the mean width of multiseriate rays is 71 (55–90) µm and 5 (4–6) cells wide. Mean number of rays per linear millimeter is 8 (6–11 per linear mm). Fibers are non-septate, and hexagonal in cross-section (Fig. 7.7h). They are thick- to very thick-walled, with a mean thickness of 2 (2.5–5) µm and a mean diameter of 13 (9–20) µm.

Comments The combination of vessels in tangential bands to irregular dendritic patterns, narrow vessels, simple perforation plates, presence of vascular tracheids, uni-biseriate rays and multiseriate rays, and vasicentric axial parenchyma, also occur in Pujana's *Xilotype 3* Fabaceae-Faboideae specimen (Pujana, 2008 and R. Pujana, personal communication, 2010). This is represented by material collected from the Cerro Calafate locality, Santa Cruz Province, Argentina, from the Oligocene Río Leona Formation. However, *Xilotype 3* differs from the specimen described here by the presence of diffuse axial apotracheal parenchyma, occasionally reticulate perforation plates, helical thickenings in vessel elements and bordered, alternated, and the round to irregular outline of the intervessel pits.

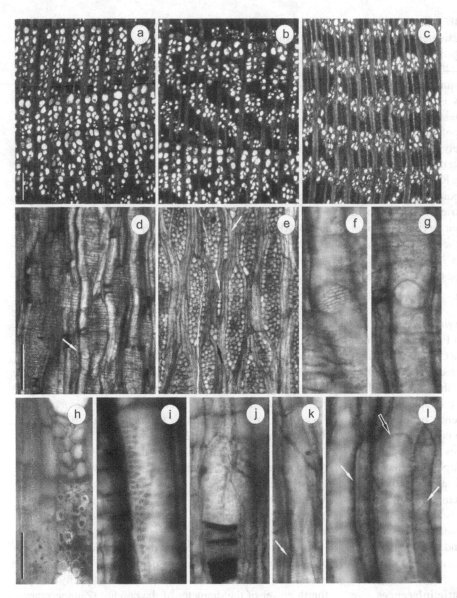

Fig. 7.7. a–l, aff. *Xilotype 3*. a, TS, showing growth ring boundary and vessels in irregular dendritic pattern. b, TS, showing vessels in diagonal patterning. c, TS, showing tangential bands patterning of vessels. d, RLS, showing homocellular rays and simple perforation plate (white arrow). e, TLS, detail of uniseriate (white arrows), biseriate (black arrow) and multiseriate rays. f, RLS, showing detail of reticulate perforation plate. g, RLS, detail of simple perforation plate and vessel-ray parenchyma pitting. h, TS, detail of axial parenchyma cells, ray cells, and fibers. i, j, RLS, showing bordered, round to irregular in outline, intervessel pits with alternate arrangement. k, RLS, showing vessel elements with helicoidal thickenings (white arrow). l, RLS, showing vascular tracheids (white arrows) and simple perforation plate (black arrow). Scale bars: a–c = 300 µm; d, e = 200 µm; f–l = 50 µm.

Family **PROTEACEAE?**

Doroteoxylon Nishida, Nishida and Ohsawa, 1989
Type species: *Doroteoxylon vicenti-perezii* Nishida, Nishida and Ohsawa, 1989
Holotype: TK No. 797848.
Type locality: Cerro Dorotea, Última Esperanza, Chile.
Type formation: Mina Chilena Formation (Late Oligocene–Early Miocene).
Other materials: MP-MPB 1955 (Pujana, 2009a).

Doroteoxylon vicenti-perezii Nishida,
Nishida and Ohsawa, 1989
Fig. 7.6j–m, r

Material studied: MPM-PB 4381–4384.
Locality: Punta Sur, Santa Cruz Province, Argentina.
Formation: Santa Cruz Formation, Estancia La Costa Member.
Age: late Early Miocene.
Affinity: Angiospermae, Fabales, Fabaceae, Caesalpinoideae.
Description: this description is based upon three pieces of stem or trunk wood of unknown diameter. The material is not well preserved. Growth rings are demarcated by reduction of the radial diameter of the last two to four rows of latewood fibers, reduction in vessel size, and the presence of vessel clusters in the latewood (Fig. 7.6m). Ring porous. Vessels occur in two distinct sizes. Larger vessels

are solitary or in tangential pairs in the earlywood (Fig. 7.6k, l), and have a mean tangential diameter of 140 (50–245) µm. Narrower vessels are arranged in clusters, and have a mean tangential diameter of 56 (40–100) µm. Mean vessel element length is 225 (59–323) µm with oblique or straight end walls. Mean vessel density is 12 (9–16)/mm^2. Vessel elements are short. Narrow vessels have helical thickenings, and large vessels commonly have abundant tyloses. Perforation plates are exclusively simple. Intervessel pits are hexagonal and have an alternate arrangement with a mean diameter of 10 (9–13) µm. Ray-vessel pits are similar to the intervessel pits. Axial parenchyma is vasicentric. Rays are heterocellular composed of procumbent cells forming the body with two to four upright marginal cells. Rays are moderately long, and their mean height is 771 (392–1078) µm. Mean ray width is 131 (88–196) µm. Rays are multiseriate, 5 to 13 cells wide (Fig. 7.6r). Fibers are non-septate, and circular in outline.

Comments The holotype of *Doroteoxylon vicenti-perezii* comes from the Cerro Dorotea, Chilean Patagonia (Nishida *et al.*, 1989), from sediments of the Mina Chilena Formation (Late Oligocene–Early Miocene). Terada *et al.* (2006) described additional specimens from the same locality. These authors consider that the most probable affinity is to the Proteaceae, because of the presence of multiseriate rays and the tendency to form tangential bands of earlywood vessels. Pujana (2009a) describes an exceptionally well-preserved new specimen from Argentina that gives more anatomical details. The storied and pitted fibers, extent of vessel density, and vessel element length values suggest to Pujana (2009a) that this fossil species shares greater affinity with members of the subfamily Caesalpinoideae, in particular with *Gleditsia* L. and *Robinia* L., both genera native to the Northern Hemisphere.

7.4.4 Physiognomic and paleoclimatic inferences obtained from the dicotyledonous wood structure

The modern climate at the Atlantic coast of Santa Cruz Province between the Río Coyle and the Río Gallegos is characterized as a cold semi-desert area, with windy and cold winters, and dry and warm summers. Climatic values for Río Gallegos locality are: mean annual temperature (MAT) 7.6 °C, mean annual range in temperature (MART) 12.5 °C, cold month mean temperature (CMMT) 0.9 °C, annual precipitation (MAP) 242 mm, and length of the dry season (DRY) *c.* 3 months during winter and spring. The modern dominant vegetation is of medium height (20–80 cm), and low density (1 plant every 6 m^2), in an arid shrubby-herbaceous steppe. An intensive W–E gradient in the floristic composition is a consequence of the rain-shadow effects of the modern Cordillera interacting with the

Antarctic anticyclone weather system. Grasses replace the low bushes as precipitation decreases (León *et al.*, 1998; Coronato *et al.*, 2008). However, in the Mid-Miocene Climatic Optimum, between 17 and 15 Ma (Zachos *et al.*, 2001), climatic conditions were very different from those of today. The added effect of the Andean rain shadow on changing climate during the Miocene (Blisniuk *et al.*, 2005), discussed by Kay *et al.* in Chapter 17, is seen in the changing taxonomical composition of the paleoflora studied here.

Twelve anatomical characteristics were used to estimate climate variables (Tables 7.2, 7.3). The MAT estimates for PS locality were 19.31 ± 1.7 °C and 9.35 ± 1.7 °C from equations (12) and (15) respectively (Table 7.3). CMMT and MART values were 9.8 ± 2.6 °C and 4.26 ± 2.4 °C respectively (Table 7.3). However, CMMT and MART estimates are considered less precise than those for other climate variables. The DRY value was 6.8 ± 1.4 months, and MAP was estimated at 869 ± 940 mm. The MAP value obtained from the multivariable anatomical analyses must be considered as potentially less accurate than the MAT estimate (Wiemann *et al.*, 1998). These data will be discussed further below.

The Vulnerability Index (V) and Mesomorphy Ratio (M) values range from 0.43 to 8.17 and from 70 to 1837 respectively. *Nothofagoxylon triseriatum*, *Eucryphiaceoxylon eucryphioides*, *Myrceugenia chubutense* and aff. *Xilotype 3* have V values less than 2.5, and M values less than 1500 (Table 7.4). These values suggest that these woods possess well-developed xeromorphic features. However, *Laurinoxylon atlanticum* and *Doroteoxylon vicenti-perezii* have mesomorphic features (Table 7.4), indicating that these taxa would have lived in more humid areas and/or areas with greater soil humidity.

Vessel diameter is one of the most significant indicators of hydraulic efficiency in angiosperm wood, because of the relationship that exists between conductive capacity and the fourth power of the diameter of the conduit (Zimmermann, 1983). All the fossil specimens from the PS locality have mean tangential vessel diameters less than 100 µm (Table 7.4). In extant woods, values such as these are common in cool temperate or high montane tropical species (Wheeler and Baas, 1991). Based upon these modern ecological patterns, this would indicate that these Miocene fossil woods developed in cool and/or xeric climate (Lindorf, 1994).

High vessel densities (40/mm^2 or more) are typical of cool temperate to arctic species, tropical high montane species, and xerophytes. High vessel density is seen in 67% of the Miocene woods studied here (Table 7.4). In addition, many of these woods from the PS locality have simple perforation plates, which are considered an adaption to more efficient water transport (Baas, 1976; Carlquist, 1977; Wheeler and Baas, 1993; Lindorf, 1994). *Xilotype 3* (Pujana, 2008) shows tangential bands of vessels and

Table 7.4. *Anatomical wood data used for determinate paleoclimate signals, and to calculate Vulnerability Index and Mesomorphy Ratio*

Taxa	Growth rings	Vessel tangential diameter (μm)	Vessels per mm²	Vessel element length (μm)	Perforation plates	Porosity	Axial parenchyma	Fiber type	Vulnerability Index (V)	Mesomorphy Ratio (M)
Laurinoxylon atlanticum	P	62 (25–115)	21 (17–28)	155 (60–290)	SI	D	Vs	n-sp	2.95	458
Nothofagoxylon triseriatum	P	41 (25–55)	58 (47–72)	99 (50–205)	SI, SC	D	D	n-sp	0.71	70
Myrceugenia chubutense	P	53 (30–80)	62 (57–67)	239 (160–330)	SI	D to S	D to D-DA	n-sp	0.85	287
Eucryphiaceoxylon eucryphioides	P	49 (25–70)	45 (35–55)	230 (130–385)	SI, SC	S to D	D to D-DA	n-sp	1.09	251
Doroteoxylon vicenti-perezii	P	98 (42–172)	12 (9–16)	225 (59–323)	SI	R	Vs	n-sp	8.17	1837
aff. *Xilotype 3* Pujana, 2008	P	58 (30–90)	135 (114–181)	214 (160–285)	SI	S	Vs	n-sp	0.43	92

Growth rings: P, present. **Perforation plates:** SI, simple perforation plate, SC, scalariform plate. **Porosity:** D, diffuse-porous wood; R, ring-porous wood. **Axial parenchyma:** D, diffuse; D-DA, diffuse-in-aggregates; Vs, vasicentric scarce. **Fiber type:** n-sp, non-septate. **Vulnerability Index** (V) = mean vessel diameter/number of vessels per mm². **Mesomorphy Ratio (M)** = (mean vessel diameter/number of vessels per mm²) × vessel element length.

diagonal and irregular dendritic vessel patterns, a feature largely restricted to woods from temperate to subtropical regions (Wheeler and Baas, 1991, 1993). The occurrence of vessels in clusters is more common in subtropical to temperate floras than in tropical floras. More than 50% of the woods from the fossil assemblages studied here have grouped vessels, whether in radial multiples or clusters.

The presence of helical thickenings in narrow vessels in *Doroteoxylon vicenti-perezii* is associated with regions that experience water stress created by drought or freezing. It might diminish the danger of cavitation, aid in refilling of vessels, or increase vessel wall strength (Carlquist, 2001; Wheeler *et al.*, 2007).

The relation between porosity type and the evergreen habit in angiosperm trees has also been used for climatic interpretation (Studhalter, 1955; Boura and De Franceschi, 2007; Wheeler *et al.*, 2007). The presence of semi-ring- to ring-porosity, as seen in *Eucryphiaceoxylon eucryphiodes*, *Myrceugenia chubutense*, *Doroteoxylon vicenti-perezii*, and aff. *Xilotype 3*, could be related to existence of semi-arid and arid climatic conditions (Woodcock, 1994; Moglia and Giménez, 1998). Furthermore, these two characteristics (semi-ring- porosity and ring- porosity) are usually, but not always, associated with temperate climates and/or deciduousness (Chowdhury, 1964; Wheeler and Baas, 1993; Boura and De Franceschi, 2007). Deciduousness could be correlated to seasonality in precipitation, temperature, or day length (Wheeler and Manchester, 2002; Poole and van Bregen, 2006). The presence of ring-porosity in fossil specimens can be used to infer seasonality of the paleoclimate (Boura and De Franceschi, 2007). Ring-porous wood appears to be more narrowly adapted to strongly seasonal environments where growing conditions are suitable for only part of the year (Baas *et al.*, 2004). All Miocene woods studied here show distinct growth rings (Table 7.4), which indicates growth in a seasonal environment (Wheeler and Baas, 1991).

Seasonality in temperature or in rainfall is also correlated with significant reductions in vessel element length (Baas *et al.*, 1983). The vessel element lengths measured in these Miocene woods are short (99–239 μm; see Table 7.4), suggesting xeric conditions, at least seasonally (Baas *et al.*, 1983; Lindorf, 1994).

Vasicentric or vascular tracheids are present in aff. *Xilotype 3*. This feature provides additional safety in water transport and is especially common in taxa subject to seasonally arid conditions (Wheeler and Baas, 1991).

Diffuse, diffuse-in-aggregates, or vasicentric scarce axial parenchyma are more common in temperate floras than in the tropical ones. On the other hand, septate fibers and abundant parenchyma are more frequently found in tropical floras and not in temperate ones. Extrapolating this pattern to the PS flora would indicate a seasonal temperate climate,

an inference that is also supported by the presence of growth rings in all of the fossil specimens.

7.5 Discussion

Microremain assemblages show the presence of a variety of components. In the majority of cases these are dominated by phytoliths, but some samples also include microcharcoal, sponge spicules, diatoms, and chrysophycean stomatocysts. The phytolith assemblages are characterized by the presence of articulated grass phytoliths, which are particularly abundant in the PS samples and upper samples from the ELC locality, as well as isolated phytoliths with the presence of graminoid (pooid, festucoid, stipoid, danthonioid, chloridoid, and panicoid), arecoid, nothofagoid, and general dicotyledonous elements.

Microremain assemblages representing this particular type of floristic composition have also been reported in some present-day soils from New Zealand's North Island (Kondo *et al.*, 1994). This comparison is based not only on the coexistence of megathermic and microthermic grasses, but also on the pronounced abundance of arecoid (palm), nothofagoid, and dicotyledonous elements, floristic assemblages absent in present-day South America.

The coexistence of microthermal and megathermal elements is better represented in the PS locality than at ELC. Although the percentage of microthermic grass components (pooid, festucoid, and stipoid) is lower, in most cases, than the percentage of megathermal elements, the occurrence of panicoid and chloridoid elements in association with the arecoid palm phytoliths suggests warm but somewhat arid climatic conditions. The profusion of whole freshwater sponge spicules and diatoms also suggests the presence of water courses in the local environment of the PS and ELC localities. The PS and CB samples contain the greatest abundance of microcharcoal. The elements that would indicate the presence of freshwater streams – sponge spicules, diatoms, and stomatocysts – are absent at the CB locality. At the CB and ANF localities, danthonioid grass phytoliths dominate the assemblages. The 1975 sample (Table 7.1) yielded a high abundance of silica microremains with superficial damage and only a few aquatic elements, which when considered in conjunction with the sample's granulometric characteristics (see Matheos and Raigemborn, Chapter 4) seems to indicate an aeolian source for the sediments, which may have occurred in a semi-arid temperate climate.

In the basal and middle samples from the ELC locality, high relative abundances of pooid, stipoid, and festucoid linked phytoliths are observed, with an increase in the 1977–1979 samples of the sponge spicules and diatoms, together with the presence of chrysopycean stomatocysts. This type of assemblage suggests humid cold-temperate climatic conditions, with seasonality in water availability.

In this locality's middle levels, an increase of panicoid elements is seen along with a decrease in pooid/festucoid phytoliths, indicating increase in temperature. Also, a pronounced presence of palm and nothofagoid elements is observed in the samples 1977 and 1979 from the ELC locality (Matheos and Raigemborn, Chapter 4, Fig. 4.4). The upper section of the ELC locality has relatively few phytoliths, but those that are present tend to be large, well preserved, and articulated.

The megafossil record found in the PS locality includes several pieces of wood and leaf compressions. Although the fossil woods were transported, the complete leaf preservation containing tips, margins, and petioles, and the articulated grass phytoliths suggest that the fossil plants were living close to the final deposition area, or only transported short distances from the growth site. This is the first macrofossil documentation of a floristically diverse forest with components of Araucariaceae, Lauraceae, Nothofagaceae, Cunoniaceae, Myrtaceae, Fabaceae, and ?Proteaceae, so far from its modern Andean distribution, demonstrating that forests of the *Nothofagidites* Province covered southern Patagonia, as hitherto noted in palynological records (Barreda and Palazzesi, 2007; Barreda *et al.*, 2007; Iglesias *et al.*, 2011).

Fossil wood assemblages similar to those described here have also been reported from Late Cretaceous and early Cenozoic context in Antarctica (Poole and Cantrill, 2006 and references therein; Poole *et al.*, 2005). Growth rings and angiosperm anatomical analyses from the Paleocene/Eocene interval in Antarctica have demonstrated the existence of seasonally warm humid rainforests with estimated MAT of 7 to 15 °C, but these data also indicate that conditions deteriorated through the Late Eocene, when cold seasonal climates were emplaced (Francis and Poole, 2002). Paleogene floras from the Antarctic Peninsula are very similar to the extant cool temperate rainforests of the Valdivia region in Chile, where characteristic elements such as Nothofagaceae, Myrtaceae, Cunoniaceae, Lauraceae, Monimiaceae, Cupressaceae, and Podocarpaceae have been described (Poole *et al.*, 2001, 2003).

The Early Miocene macroflora from Bacchus Marsh in Victoria, Australia, is poorly known in general, but leaves of Nothofagus, Myrtaceae, Araucariaceae, and Podocarpaceae dominate fossil assemblages. The same vegetation is still found in the Victoria region today, suggesting that similar climatic conditions prevailed in the Miocene. Similarly the Miocene flora from Latrobe Valley, Australia, contains elements including Araucariaceae, Podocarpaceae, Nothofagaceae, Lauraceae, Cunoniaceae, and Proteaceae (Greenwood, 1994). Also, Pole (1989) found Early Miocene plant macrofossil assemblages from the Manuherikia Group, Central Otago, New Zealand with Nothofagaceae, Myrtaceae, Casuarinaceae, Fabaceae, and Arecaceae.

Closer to the Andean region in Patagonia, similar vegetation assemblages to the paleoflora present at the PS are recorded from the Oligocene Río Leona Formation (Pujana, 2007, 2008, 2009a, 2009b; Barreda *et al.*, 2009), with fossil woods being assigned to the families Nothofagaceae, Cunoniaceae, Leguminosae, Proteaceae, Anacardiaceae, Myrtaceae, Araucariaceae, and Podocarpaceae. Palynological assemblages indicate an abundance of freshwater algae, rushes, and sedges. Rainforests composed of Nothofagaceae, Podocarpaceae, Araucariaceae, Myrtaceae, Fabaceae, Proteaceae, and Casuarinaceae have also been reconstructed, with other elements including Gunneraceae, Onagraceae, Poaceae, and Asteraceae. Presence of pteridophytes and bryophytes suggest coastal streams or lowland regions with swampy areas (Barreda *et al.*, 2009).

Seasonal *Nothofagus* forests composed of taxa with varying ecological requirements and phytogeographic origins occur in the Middle and Late Miocene on the Pacific coast of central Chile (Hinojosa and Villagrán, 1997, 2005). The Climate Leaf Analysis Multivariate Program (CLAMP) analyses from their studies produced MATs between 15.9 (±2.4) °C and 21.4 (±2.1) °C for the Navidad-Goretones and Navidad-Boca Pupuya fossil localities, respectively. Mean annual precipitation (MAP) was estimate at 93.1 cm and the estimates of Miocene mean dry season precipitation (MPD) for two members of the Navidad Formation were 17.2 (±15.3) and 20.0 (±15.3) cm.

On the Arauco Peninsula in central Chile, a diverse Miocene flora contains Anacardiaceae, Boraginaceae, Euphorbiaceae, Fagaceae, Lauraceae, Fabaceae, Monimiaceae, Myristicaceae, Myrtaceae, and Proteaceae. These Arauco paleoforests are thought to have possessed a rich flora with mostly dicotyledonous plants, with apparent regression of gymnosperms occurring during the Miocene. Moreover, the occurrence of predominantly tropical angiosperms in fossil assemblages suggests a stronger influence of Amazonian paleofloras in the southern Andean coastal range during the Miocene (Schöning and Bandel, 2004). In the present study, Miocene paleoforests located in high latitudes at the Atlantic coast of Patagonia are inferred to be growing in a climate with estimated MATs of 19.31± 1.7 °C and 9.35± 1.7 °C, with a low thermal amplitude (<10 °C). This pattern is typically associated with coastal marine climates, such as present-day vegetation communities in central Chile, southeastern Australia, and New Zealand's North Island (Hueck, 1978; San Martín *et al.*, 1984; Donoso Zegers, 1993; Hill, 1994; Kondo *et al.*, 1994; Armesto *et al.*, 1995).

In particular, the Mediterranean area of central Chile is characterized by the presence of mesophite vegetation, with precipitation of 200–1000 mm/year and average temperatures between 15 and 20 °C. The heterogeneous floristic composition of this area is characterized by the presence of the most australly distributed palm, *Jubaea chilensis* (Molina) Baill, which coexists at the same latitude with xerophytic forests composed of Rosaceae, Lauraceae, Cunioniaceae, Winteraceae, Myrtaceae, and Nothofagaceae (Hueck, 1978; San Martín *et al.*, 1984; Donoso Zegers, 1993; Armesto *et al.*, 1995). *Jubaea chilensis* reaches a southern limit of around 37° S. Thus, the presence of palms at the PS locality indicate that during the late Early Miocene this region was much warmer than today.

The V and M Index values suggest that the climate at the PS locality was characterized by a season with low water availability, and according to the results obtained in this study this is consistent with other palynological inferences (Barreda and Palazzesi, 2010), which suggest that throughout the Early and early Middle Miocene, angiosperms with megathermal affinities persisted. Many of these have been reported from both southern and northern Patagonia. During the Early Miocene, xerophytic taxa were dominant, as Casuarianaceae, Ephedraceae, Chenopodiaceae, Anacardiaceae, and Fabaceae. Members of the Poaceae and Asteraceae were also present. The development of these floras was probably linked with a restricted water supply in open forests or coastal regions (Palazzesi and Barreda, 2007, and references therein).

7.6 Conclusions

Based upon fossil floral assemblages, including fossil wood and leaf compressions, phytoliths, microcharcoal, chrysophycean stomatocysts, sponge spicules, and diatoms, we have obtained the first detailed evidence of the floristic composition and paleoclimatic conditions of the continental sequences of the Santa Cruz Formation that crop out along the Atlantic coast of Patagonia. These bioproxies have enabled us to reconstruct paleovegetation patterns and climate variables at several localities by means of comparative analysis using nearest living relative (NLR) and physiognomy-based methodologies. This latter approach estimates paleoclimatic parameters by assuming that the relationship between wood anatomical characteristics and climatic factors were the same in the past as they are in the present day (Poole and van Bergen, 2006).

The Punta Sur locality yielded the most complete information for the analyzed bioproxies. The plant terrestrial community at PS had a herbaceous component dominated by chloridoid and panicoid grasses, but also contained less common danthonioid, pooid, and festucoid grasses, and an arboreal component that included members of the Araucariaceae, Arecaceae, Nothofagaceae, Myrtaceae, Cunoniaceae, and Fabaceae (Faboideae). A second plant community at PS is characterized by marked fluvial influences, with characteristic elements including members of the Lauraceae and Proteaceae?, and indication of fluvial

processes based upon the presence of microremains including sponge spicules and diatoms. The forests at PS had evergreen and deciduous elements with growth rings in their woods. For this particular late Early Miocene locality, a mean annual temperature (MAT) is variously estimated to have been 19.31 ± 1.7 °C and 9.35 ± 1.7 °C. The V and M Index values, and many of the anatomical characters observed in the fossil woods, are likely to have contributed to hydraulic safety, reflecting dry temperate and temperate-warm climatic conditions. The results based on wood physiognomy methods are provisional, and more data and fieldwork are needed to confirm the inferences proposed here.

Floras similar in composition and anatomical characteristics with the coexistence of micro- and megathermal floristic elements identified at PS are found today in several regions of the Southern Hemisphere including central Chile, southeastern Australian, and New Zealand, three regions with pronounced marine influence on their climates, especially the moderation of seasonal extremes. Floral microremains at Anfiteatro suggest multiple contributing sources for these deposits, with an abundance of allochthonous elements that in some cases show considerable superficial damage.

At the Estancia La Costa locality, the basal and mid-level samples (1976–1979) have similar microremain assemblages to those from the Punta Sur locality, but with some variation in composition, such as a greater abundance of temperate grasses in the basal samples, which give way in mid-level samples to more tropical panicoid grasses. This may reflect changes from colder, humid conditions to warmer humid conditions during a short depositional lapse of this sedimentary sequence. Also, the presence of chrysophycean stomatocysts (Bertoldi de Pomar, 1973) suggests that moisture conditions were not constant, with the presence of periods of low water availability, possibly with seasonal occurrence.

The new records of leaf compressions at Rincón del Buque locality establish a parautochthonous character for the fossil record described here, being the first record of fossil leaves for the Santa Cruz Formation and for the eastern side of Patagonia.

This is the first record of diverse forest with components of Araucariaceae, Lauraceae, Nothofagaceae, Cunoniaceae, Myrtaceae, Fabaceae, and ?Proteaceae in the Atlantic margin, so far from the Andean modern distribution, corroborating a previous hypothesis (Barreda and Palazzesi, 2007; Barreda *et al.*, 2007) about forests covering Patagonia in the Oligocene–Miocene (today a steppe) from the Pacific up to the Atlantic coast.

The record of palms in Punta Sur locality provides concrete evidence of warmer conditions during this period of time. This record is 15° of latitude further south than the southernmost record of a palm in South America today.

ACKNOWLEDGMENTS

The authors thank Sergio F. Vizcaíno, M. Susana Bargo, and Richard F. Kay for their invitation to contribute to this volume and for their discussion and critical review of the final manuscript. We are also grateful to Roberto Pujana (Museo Argentino de Ciencias Naturales "B. Rivadavia," Buenos Aires, Argentina) and to Juan Enrique Bostelmann (Museo Nacional de Historia Natural, Montevideo, Uruguay) for their valuable comments and for providing bibliographical materials. Three anonymous reviewers provided constructive suggestions that improved the clarity of the original manuscript. This work was supported by CONICET, Agencia Nacional de Promoción Científica y Tecnológica (ANPCyT) PICT 2007–13864, PICT 0143, and NSF 0851272 and 0824546.

REFERENCES

Armesto, J. J., Villagrán, C. and Arroyo, M. K. (1995). *Ecología de los bosques nativos de Chile*. Santiago de Chile: Editorial Universitaria.

Baas, P. (1976). Some functional and adaptative aspects of vessel member morphology. In *Wood Structure in Biological and Technological Research*, ed. P. Baas, A. J. Bolton and D. M. Catling. Leiden Botanical Series, Leiden (Holland): Leiden University Press, pp. 157–181.

Baas, P. (1986). Ecological patterns of xylem anatomy. In *On the Economy of Plant Form and Function*, ed. T. J. Givnish. Cambridge: Cambridge University Press, pp. 327–353.

Baas, P. and Wheeler, E. A. (1996). Parallelism and reversibility in xylem evolution: a review. *IAWA Journal*, 17, 351–364.

Baas, P., Werker, E. and Fahn, A. (1983). Some ecological trends in vessel characters. *IAWA Bulletin*, 4, 141–159.

Baas, P., Ewers, F. W., Davis, S. D. and Wheeler, E. A. (2004). Evolution of xylem physiology. In *The Evolution of Plant Physiology*, ed. A. R. Hemsley and I. Poole. Linnean Society Symposium Series 21. London: Elsevier, pp. 273–295.

Bailey, I. W. and Tupper, W. W. (1918). Size variation in tracheary cells. I. A comparison between the secondary xylems of vascular cryptogams, gymnosperms and angiosperms. *Proceedings of the American Academy of Arts and Sciences*, 54, 149–204.

Barreda, V. (1993). Late Oligocene?–Miocene pollen of the families Compositae, Malvaceae and Polygonaceae from the Chenque Formation, Golfo San Jorge basin, southeastern Argentina. *Palynology*, 17, 169–186.

Barreda, V. (1996). Bioestratigrafía de polen y esporas de la Formación Chenque, Oligoceno tardío?–Mioceno de las provincias de Chubut y Santa Cruz, Patagonia, Argentina. *Ameghiniana*, 33, 35–56.

Barreda, V. and Palazzesi, L. (2007). Patagonian vegetation turnovers during the Paleogene–Early Neogene: origin of arid-adapted floras. *The Botanical Review*, 73, 31–50.

Barreda, V. and Palazzesi, L. (2010). Vegetation during the Eocene–Miocene interval in central Patagonia: a context of mammal evolution. In *The Paleontology of Gran Barranca: Evolution and Environmental Change through the Middle Cenozoic of Patagonia*, ed. R. Madden, A. Carlini, M. Vucetich and R. F. Kay. Cambridge: Cambridge University Press, pp. 375–382.

Barreda, V., Anzótegui, M. L., Prieto, A. R. *et al.* (2007). Diversificación y cambios de las angiospermas durante el Neógeno de Argentina. In *Publicación Especial 11, Ameghiniana 50° Aniversario*, ed. S. Archangelsky, T. Sánchez and E. P. Tonni. Asociación Paleontológica Argentina, pp. 173–191.

Barreda, V., Palazzesi, L. and Marenssi, S. (2009). Palynological record of the Paleogene Río Leona Formation (southernmost South America): stratigraphical and paleoenvironmental implications. *Review of Palaeobotany and Palynology*, 154, 22–33.

Bertoldi de Pomar, H. (1971). Ensayo de clasificación morfológica de los silicofitolitos. *Ameghiniana*, 8, 317–328.

Bertoldi de Pomar, H. (1973). Crisostomatáceas en sedimentos de fondo de la laguna Guadalupe. *Asociación de Ciencias Naturales del Litoral*, 4, 73–86.

Blisniuk, P. M., Stern, L. A., Chamberlain, C. P., Idleman B. and Zeitler, P. K. (2005). Climatic and ecologic changes during Miocene surface uplift in the Southern Patagonian Andes. *Earth and Planetary Sciences Letters*, 230, 125–142.

Boura, A. and De Franceschi, D. (2007). Is porous wood structure exclusive of deciduous trees? *Comptes Rendus Palevol*, 6, 385–391.

Burkart, A. (1975). Evolution of grasses and grasslands in South America. *Taxon*, 24, 53–66.

Cantrill, D. and Poole, I. (2005). A new Eocene *Araucaria* from Seymour Island, Antarctica: evidence from growth form and bark morphology. *Alcheringa*, 29, 341–350.

Carlquist, S. (1977). Ecological factors in wood evolution: a floristic approach. *American Journal of Botany*, 64, 887–896.

Carlquist, S. (2001). *Comparative Wood Anatomy. Systematic, Ecological, and Evolutionary Aspects of Dicotyledon Wood*. Springer Series in Wood Science. Berlin/Heidelberg: Springer.

Carlquist, S. and Hoekman, D. A. (1985). Ecological wood anatomy of the woody southern Californian flora. *IAWA Bulletin*, 6, 319–347.

Chowdhury, K. A. (1964). Growth rings in tropical trees and taxonomy. *Journal of the Indian Botanical Society*, 43, 334–342.

Coradeghini, A. and Vigna, M. S. (2001). Flora de quistes crisofíceos fósiles en sedimentos recientes de Mallín Book, Río Negro (Argentina). *Revista Española de Micropaleontología*, 33, 163–181.

Coronato, A. J., Coronato, F., Mazzoni, E. and Vázquez, M. (2008). The physical geography of Patagonia and Tierra del Fuego. In *The Late Cenozoic of Patagonia and Tierra del Fuego*, ed. J. Rabassa. Developments in Quaternary Science Series 11. Amsterdam: Elsevier, pp. 13–56.

Donoso Zegers, C. (1993). *Bosques templados de Chile y Argentina. Variación, Estructura y Dinámica*. Santiago de Chile: Editorial Universitaria.

Dupéron, J., Dupéron-Laudoueneix, M., Sakala, J. and De Franceschi, D. (2008). *Ulminium diluviale* Unger: historique de la découverte et nouvelle étude. *Annales de Paléontologie*, **94**, 1–12.

Dupéron-Laudoueneix, M. and Dupéron, J. (2005). Bois fossiles de Lauraceae: nouvelle découverte au Cameroun, inventaire et discussion. *Annales de Paléontologie*, **91**, 127–151.

Dusén, P. (1907). Über die tertiärie Flora der Magellansländer I. Wissenschaftlidie Ergebnisse Schwedische Expedition Magellansländ 1895–97. Estocolmo, Bd I, 87–108.

Fleagle, J. G., Bown, T. M., Swisher, C. C. and Buckley, G. A. (1995). Age of the Pinturas and Santa Cruz Formations. *VI Congreso Argentino de Paleontología y Bioestratigrafía. Actas*, 129–135.

Francis, J. and Poole, I. (2002). Cretaceous and early Tertiary climates of Antarctica: evidence from fossil wood. *Palaeogeography, Palaeoclimatology, Palaeoecology*, **182**, 47–64.

Fredlund, G. G. and Tieszen, L. T. (1994). Modern phytoliths assemblages from North American Great Plains. *Journal of Biogeography*, **21**, 321–335.

Frost, F. H. (1930a). Specialization in secondary xylem of Dicotyledons. I. Origin of vessel. *Botanical Gazette*, **89**, 67–94.

Frost, F. H. (1930b). Specialization in secondary xylem of Dicotyledons. II. Evolution of end wall of vessel segment. *Botanical Gazette*, **90**, 198–212.

Frost, F. H. (1931). Specialization in secondary xylem of Dicotyledons. III. Specialization of lateral wall of vessel segment. *Botanical Gazette*, **91**, 88–96.

Gastaldo, R. A., Douglass, D. P. and McCarroll, S. M. (1987). Origin, characteristic, and provenance of plant macrodetritus in a Holocene crevasse splay, mobile delta, Alabama. *Palaios*, **2**, 229–240.

Greenwood, D. R. (1994). Palaeobotany evidence for Tertiary climates. In *History of the Australian Vegetation. Cretaceous to Recent*, ed. R. S. Hill. Cambridge: Cambridge University Press, pp. 44–59.

Hill, R. S. (ed.). (1994). *History of Australian Vegetation. Cretaceous to Recent*. Cambridge: Cambridge University Press.

Hinojosa, L. F. and Villagrán, C. (1997). Historia de los bosques del sur de Sudamérica, I: antecedentes paleobotánicos, geológicos y climáticos del Terciario del cono sur de América. *Revista Chilena de Historia Natural*, **70**, 225–239.

Hinojosa, L. F. and Villagrán, C. (2005). Did South American mixed paleofloras evolve under thermal equability or in the absence of an effective Andean barrier during the Cenozoic? *Palaeogeography, Palaeoclimatology, Palaeoecology*, **217**, 1–23.

Hueck, K. (1978). *Los Bosques de Sudamérica. Ecología, Composición e Importancia Económica*. Berlin: Sociedad Alemana de Cooperación Técnica GTZ.

Hünicken, M. A. (1995). Floras Cretácicas y Terciarias. In *Revisión y actualización de la obra paleoboánica de Kurtz en la República Argentina*, eds. P. N. Stipanicic and M. A. Hünicken. *Actas de la Academia Nacional de Ciencias (Córdoba)*, **11**, 199–219.

IAWA Committee (1989). IAWA list of microscopic feature for hardwood identification. *IAWA Journal*, **10**, 219–332.

IAWA Committee (2004). List of microscopic features for softwood identification. *IAWA Journal*, **25**, 1–70.

ICPNWG (2005). International Code for Phytolith Nomenclature 1.0. *Annals of Botany*, **96**, 253–260. doi:10.1093/aob/mci172.

Iglesias, A., Wilf, P., Johnson, K. R. *et al.* (2007). A Paleocene lowland macroflora from Patagonia reveals significantly greater richness than North American analogs. *Geology*, **35**, 947–950.

Iglesias, A., Artabe, A. and Morel, E. (2011). The evolution of Patagonian climate and vegetation from the Mesozoic to the present. In *Special Issue, Palaeogeography and Palaeoclimatology of Patagonia: Implications for Biodiversity. Biological Journal of the Linnean Society of London*, **103**, 409–422.

InsideWood. (2004 onwards). Available at http://insidewood.lib.ncsu.edu/search [last accessed February 2012].

Kondo, K., Childs, C. and Atkinson, I. (1994). *Opal Phytolith of New Zealand*. Lincoln, Canterbury, New Zealand: Manaki Whenua Press.

Kondo, R., Sase, T. and Kato, Y. (1988). Opal phytolith analysis of Andosols with regard to interpretation of paleovegetation. In *Proceedings of the 9th International Soil Classification Workshop, Japan*, ed. D. I. Kinloch, pp. 520–534.

León, R., Bran, D., Collantes, M., Paruelo, J. and Soriano, A. (1998). Grandes unidades de vegetación de la Patagonia Extra Andina. *Ecología Austral*, **8**, 125–144.

Lindorf, H. (1994). Eco-anatomical wood features of species from a very dry tropical forest. *IAWA Journal*, **15**, 361–376.

Moglia, G. and Giménez, A. M. (1998). Rasgos anatómicos característicos del hidrosistema de las principales especies arbóreas de la región Chaqueña argentina. *Investigaciones Agrarias: Sistemas de Recursos Forestales*, **7**, 53–71.

Nishida, M. (1984a). The anatomy and affinities of the petrified plants from the Tertiary of Chile. III. Petrified woods from Mocha island, Central Chile. In *Contributions to the Botany in the Andes I*, ed. M. Nishida. Tokyo: Academia Scientific Book Inc., pp. 96–110.

Nishida, M. (1984b). The anatomy and affinities of the petrified plants from the Tertiary of Chile. IV. Dicotyledonous woods from Quiriquina Island, near Concepción. In *Contributions to the Botany in the Andes I*, ed. M. Nishida. Tokyo: Academia Scientific Book Inc., pp. 111–121.

Nishida, M., Nishida H. and Nasa, T. (1988). Anatomy and affinities of the petrified plants from the Tertiary of Chile (V). *Botanical Magazine*, **101**, 293–309.

Nishida, M., Nishida, H. and Ohsawa, T. (1989). Comparison of the petrified woods from the Cretaceous and Tertiary of Antarctica and Patagonia. *NIPR Symposium on Polar Biology, Proceedings* **2**, 198–212.

Palazzesi, L. and Barreda, V. (2007). Major vegetation trends in the Tertiary of Patagonia (Argentina): a qualitative paleoclimatic approach based on palynological evidence. *Flora*, **202**, 328–337.

Pole, M. (1989). Early Miocene floras from central Otago, New Zealand. *Journal of the Royal Society of New Zealand*, **19**, 121–125.

Poole, I. (2002). Systematics of Cretaceous and Tertiary *Nothofagoxylon*: implications for southern hemisphere biogeography and evolution of the Nothofagaceae. *Australian Systematic Botany*, **15**, 247–276.

Poole, I. and Cantrill, D. J. (2006). Cretaceous and Cenozoic vegetation of Antarctica integrating the fossil wood record. In *Cretaceous–Tertiary High-Latitude Palaeoenvironments, James Ross Basin, Antarctica*, ed. J. E. Francis, D. Pirrie and J. A. Crame. Geological Society, London. Special Publications, **258**, 63–81.

Poole, I. and van Bergen, P. F. (2006). Physiognomic and chemical characters in wood as paleoclimate proxies. *Plant Ecology*, **182**, 175–195.

Poole, I., Cantrill, D. J., Hayes, P. and Francis, J. E. (2000). The fossil record of Cunoniaceae: new evidence from Late Cretaceous wood of Antarctica? *Review of Palaeobotany and Palynology*, **111**, 127–144.

Poole, I., Hunt, R. J. and Cantrill, D. J. (2001). A fossil wood flora from King George Island: ecological implications for an Antarctic Eocene vegetation. *Annals of Botany*, **88**, 33–54.

Poole, I., Mennega, A. M. W. and Cantrill, D. J. (2003). Valdivian ecosystems in the Late Cretaceous and Early Tertiary of Antarctica: further evidence from myrtaceous and eucryphiaceous fossil wood. *Review of Palaeobotany and Palynology*, **124**, 9–27.

Poole, I., Cantrill, D. J. and Utescher, T. (2005). A multi-proxy approach to determine Antarctic terrestrial palaeoclimate during the Late Cretaceous and Early Tertiary. *Palaeogeography, Palaeoclimatology, Palaeoecology*, **222**, 95–121.

Pujana, R. R. (2007). New fossil woods of Proteaceae from Oligocene of southern Patagonia. *Australian Systematic Botany*, **20**, 119–125.

Pujana, R. R. (2008). Estudio paleoxilológico del Paleógeno de Patagonia austral (Formaciones Río Leona, Río Guillermo y Río Turbio) y Antártida (Formación La Meseta). Unpublished Ph.D. thesis, Universidad de Buenos Aires, Buenos Aires.

Pujana, R. R. (2009a). Fossil woods from the Oligocene of southwestern Patagonia (Río Leona Formation). Atherospermataceae, Myrtaceae, Leguminosae and Anacardiaceae. *Ameghiniana*, **46**, 523–535.

Pujana, R. R. (2009b). Fossil woods from the Oligocene of southwestern Patagonia (Río Leona Formation). Rosaceae and Nothofagaceae. *Ameghiniana*, **46**, 621–636.

Ragonese, A. M. (1980). Leños fósiles de dicotiledoneas del Paleoceno de Patagonia, Argentina. I. *Myrceugenia chubutense* n. sp. (Myrtaceae). *Ameghiniana*, **17**, 297–311.

Rancusi, M. H., Nishida, M. and Nishida, H. (1987). Xylotomy of important Chilean woods. In *Contributions to the Botany in the Andes II*, ed. M. Nishida. Tokyo: Academia Scientific Book Inc., pp. 68–158.

Romero, E. J. (1970). *Ulminium atlanticum* n. sp. Tronco petrificado de Lauraceae del Eoceno de Bahía Solano, Chubut, Argentina. *Ameghiniana*, **7**, 205–224.

Romero, E. J. (1980). Arquitectura foliar de las especies sudamericanas de *Nothofagus* Bl. *Boletín de la Sociedad Argentina de Botánica*, **19**, 289–308.

Romero, E. J. (1986). Fossil evidence regarding the evolution of *Nothofagus* Blume. *Annals of Missouri Botanical Garden*, **73**, 276–283.

Romero, E. J. and Dibbern, M. (1985). A review of the species described as *Fagus* and *Nothofagus* by Dusén. *Palaeontographica, B*, **197**, 123–137.

Rull, V. and Vegas-Vilarrúbia, T. (2000). Chrysophycean stomacysts in a Caribbean mangrove. *Hydrobiologia*, **428**, 145–150.

San Martín, J., Figueroa, H. and Ramírez, C. (1984). Fitosociología de los bosques de ruil (*Nothofagus alessandri* Espinosa) en Chile central. *Revista Chilena de Historia Natural*, **57**, 171–200.

Schönfeld, E. (1954). Ueber eine fossile liane aus Patagonien. *Paleontographica, B* **97**, 23–25.

Schöning, M. and Bandel, K. (2004). A diverse assemblage of fossil hardwood from the Upper Tertiary (Miocene?) of the Arauco Peninsula, Chile. *Journal of South American Earth Sciences*, **17**, 59–71.

Studhalter, R. A. (1955). Tree growth. I. Some historical chapters. *The Botanical Review*, **21**, 1–72.

Tauber, A. A. (1997). Bioestratigrafía de la Formación Santa Cruz (Mioceno inferior) en el extremo sudeste de la Patagonia. *Ameghiniana*, **34**, 413–426.

Tauber, A. A. (1999). Los vertebrados de la Formación Santa Cruz (Mioceno inferior-medio) en el extremo sureste de la Patagonia y su significado paleoecológico. *Revista Española de Paleontología*, **14**, 173–182.

Terada, K., Nishida, H., Asakawa, T. O. and Rancusi, M. (2006). Fossil wood assemblage from Cerro Dorotea, Última Esperanza, Magallanes (XII) region, Chile. In *Post-Cretaceous floristic changes in Southern Patagonia, Chile*, ed. H. Nishida. Chuo University Faculty of Science and Engineering, pp. 67–90.

Torres, T. and Lemoigne, Y. (1988). Maderas fósiles terciarias de la Formación Caleta Arctowski, Isla Rey Jorge, Antártica. *Serie Científica INACH*, **37**, 69–107.

Tortorelli, L. A. (1956). *Maderas y Bosques Argentinos*. Buenos Aires: Editorial Acme.

Twiss, P. C. (1992). Predicted world distribution of C_3 and C_4 grass phytoliths. In *Phytoliths Systematics: Emerging Issues*, ed. G. Rapp and S. Mulholland. New York: Plenum Press, pp. 113–128.

Van der Graaff, N. A. and Baas, P. (1974). Wood anatomical variation in relation to latitude and altitude. *Blumea*, **22**, 101–121.

Wagemann, W. (1948). Maderas Chilenas. Contribución a su anatomía e identificación. *Lilloa*, **16**, 263–375.

Whang, S. S. and Hill, R. S. (1995). Phytolith analysis in leaves of extant and fossil populations of *Nothofagus* subgenus *Lophonzonia*. *Australian Systematic Botany*, **8**, 1055–1065.

Wheeler, E. A. (1986). Vessels per square millimetre or vessel groups per square millimetre? *IAWA Bulletin*, **7**, 73–74.

Wheeler, E. A. and Baas, P. (1991). A survey of the fossil record for dicotyledonous wood and its significance for evolutionary and ecological wood anatomy. *IAWA Bulletin*, **12**, 275–332.

Wheeler, E. A. and Baas, P. (1993). The potentials and limitations of dicotyledonous wood anatomy for climatic reconstructions. *Paleobiology*, **19**, 487–498.

Wheeler, E. A. and Manchester, S. R. (2002). Woods of the Middle Eocene Nut Beds Flora, Clarno Formation, Oregon, USA. *IAWA Journal Supplement*, **3**, 1–188.

Wheeler, E. A., Baas, P. and Rodgers, S. (2007). Variation in dicot wood anatomy: a global analysis based on the Insidewood Database. *IAWA Journal* **28**, 229–258.

Wiemann, M. C., Wheeler, E. A., Manchester, S. R. and Portier, K. M. (1998). Dicotyledonous wood anatomical characters as predictors of climate. *Palaeogeography, Palaeoclimatology, Palaeoecology*, **139**, 83–100.

Wiemann, M. C., Manchester, S. R. and Wheeler, E. A. (1999). Paleotemperature estimation from dicotyledonous wood anatomical characters. *Palaios*, **14**, 459–474.

Wiemann, M. C., Dilcher, D. L. and Manchester, S. R. (2001). Estimation of mean annual temperature from leaf and wood physiognomy. *Forest Sciences*, **47**, 141–149.

Woodcock, D. W. (1994). Occurrence of woods with a gradation in vessel diameter across the ring. *IAWA Journal*, **15**, 377–385.

Zachos, J., Pagani, M., Sloan, L., Thomas, E. and Billups, K. (2001). Trends, rhythms, and aberrations in global climate 65 Ma to present. *Science*, **292**, 686–693.

Zimmermann, M. H. (1983). *Xylem Structure and the Ascent of Sap*. Springer, Berlin.

Zucol, A. F., Brea, M., Madden, R. H. *et al.* (2007). Preliminary phytolith analysis of Sarmiento Formation in the Gran Barranca (Central Patagonia, Argentina). In *Plants, People and Places: Recent Studies in Phytolithic Analysis*, ed. M. Madella and D. Zurro. Oxford: Oxbow Books, pp. 197–203.

Zucol, A. F., Passeggi, E., Brea, M. *et al.* (2010a). Phytolith analysis for the Potrok Aike Lake Drilling Project: sample treatment protocols for the PASADO Microfossil Manual. In *1a Reunión Internodos del Proyecto Interdisciplinario Patagonia Austral y 1er Workshop Argentino del Proyecto Potrok Aike Maar Lake Sediment Archive Drilling Project*, ed. H. Corbella and N. I. Maidana. Buenos Aires: Proyecto Editorial PIPA, pp. 81–84.

Zucol, A. F., Colobig, M. M., Patterer, N. I. *et al.* (2010b). Phytolith analysis for the Potrok Aike Lake Drilling Project: general methodologies for analysis. In *1a Reunión Internodos del Proyecto Interdisciplinario Patagonia Austral y 1er Workshop Argentino del Proyecto Potrok Aike Maar Lake Sediment Archive Drilling Project*, ed. H. Corbella and N. I. Maidana. Buenos Aires: Proyecto Editorial PIPA, pp. 85–88.

Zucol, A. F., Brea, M. and Bellosi, E. (2010c). Phytolith studies in Gran Barranca (central Patagonia, Argentina): the middle–late Eocene. In *The Paleontology of Gran Barranca: Evolution and Environmental Change through the Middle Cenozoic of Patagonia*, ed. R. Madden, A. Carlini, M. Vucetich and R. F. Kay. Cambridge: Cambridge University Press, pp. 317–340.

8 Amphibians and squamate reptiles from the Santa Cruz Formation (late Early Miocene), Santa Cruz Province, Argentina: paleoenvironmental and paleobiological considerations

Juan C. Fernicola and Adriana Albino

Abstract

The herpetological diversity recorded in the Santa Cruz Formation (late Early Miocene) is low when compared with that of birds and mammals. It includes the calyptocephalellid anuran *Calyptocephalella*, an indeterminate "leptodactylid," indeterminate pleurodont iguanians (including those previously assigned to the extinct genus "*Erichosaurus*"), the tupinambine teiid *Tupinambis*, and indeterminate "colubrids." The presence of *Calyptocephalella* in the Estancia La Costa locality represents its southernmost record and might indicate the occurrence of permanent lowland lakes, ponds, and quiet streams, possibly developed in a forested area. The presence of *Tupinambis* and "colubrids" at around 50° S represents the southernmost record in their respective evolutionary histories, suggesting warmer and probably more humid conditions in the late Early Miocene than those prevailing in southern Patagonia at present. Based upon the diets of extant *Calyptocephalella* and "colubrids" we consider the Santacrucian Miocene representatives to be small carnivorous vertebrates. Santacrucian pleurodont iguanians should be included in the insectivorous and/or herbivorous groups, whereas *Tupinambis* would have been a generalist omnivorous reptile.

Resumen

La diversidad herpetológica registrada en la Formación Santa Cruz (Mioceno Temprano tardío) es pobre comparada con la de aves y mamíferos Entre los anuros se registran el caliptocefalélido *Calyptocephalella* y un "Leptodactylidae" indeterminado, mientras que entre los reptiles se encuentran iguanios pleurodontes indeterminados (incluyendo aquellos previamente asignados al género

extinto *Erichosaurus*), el teido *Tupinambis* y colúbridos indeterminados. La presencia de *Calyptocephalella* en la localidad Estancia La Costa representa su registro más austral e indicaría la existencia de lagos de tierras bajas, lagunas y arroyos, posiblemente desarrollados en un área boscosa. La presencia de *Tupinambis* y "colúbridos" alrededor de los 50° S constituye el registro más austral en sus respectivas historias evolutivas y sugiere condiciones más cálidas y probablemente más húmedas en el sur de la Patagonia durante el Mioceno Temprano tardío que en la actualidad. Los hábitos alimentarios que actualmente poseen *Calyptocephalella* y los "colúbridos" permiten considerar a sus representantes miocénicos como parte del elenco de pequeños vertebrados carnívoros del Santacrucense. Por otra parte, los iguanios pleurodontes integrarían los grupos de pequeños vertebrados insectívoros y/o herbívoros del Mioceno, mientras que *Tupinambis* habría sido un reptil omnívoro generalista.

8.1 Introduction

The continental deposits of the Santa Cruz Formation that crop out along the Atlantic coast of southernmost Patagonia (Argentina) between the Ríos Santa Cruz and Gallegos are known for their extraordinary fossil mammal record. According to the genera listed in Bondesio (1986), Albino (1996a), and Degrange *et al.* (Chapter 9), mammals represent about 84%, birds 13%, squamate reptiles 2%, and anuran amphibians 1% of the vertebrates collected there.

Florentino Ameghino (1893) described the first fossil remains of squamates from the Santa Cruz Formation based on specimens collected by his brother Carlos at the Monte León locality. He named two new genera and species, "*Diasemosaurus occidentalis*" Ameghino, 1893 and "*Dibolosodon typicus*" Ameghino, 1893, whose systematic relationships remained unclear (Ameghino, 1898, 1899; Romer, 1967; Estes, 1983; Gasparini and Báez, 1975; Báez and

Early Miocene Paleobiology in Patagonia: High-Latitude Paleocommunities of the Santa Cruz Formation, ed. Sergio F. Vizcaíno, Richard F. Kay and M. Susana Bargo. Published by Cambridge University Press. © Cambridge University Press 2012.

a

c

b

d

e

g

i

f

h

j

k

m

o

l

n

p

q

s

t

u

r

v

w

x

Fig 8.1.

Gasparini, 1977, 1979; Donadío, 1984; Gasparini *et al.*, 1986). At present, these taxa are considered invalid and referable to the living teiid *Tupinambis* (Brizuela and Albino, 2008a; Brizuela, 2010). Ameghino (1899) also described three species of an extinct iguanian genus from the Santa Cruz Formation discovered in the La Cueva locality by Carlos Ameghino, "*Erichosaurus diminutus*" Ameghino, 1899, "*Erichosaurus debilis*" Ameghino, 1899 and "*Erichosaurus bombimaxilla*" Ameghino, 1899. Although these lizards were frequently mentioned in the literature (Gasparini and Báez, 1975; Báez and Gasparini, 1977, 1979; Gasparini *et al.*, 1986; Estes, 1983), the type material has not been revised until now. The presence of snakes was first reported by Albino (1996b) based on specimens collected at the Ameghino's Monte Observación locality (now Cerro Observatorio; see Vizcaíno *et al.*, Chapter 1) and assigned to "Colubridae."

The occurrence of anuran remains at Estancia La Costa locality was first reported by Tauber (1994, 1999), who referred them to *Caudiverbera* sp. (= *Calyptocephalella*; for nomenclatural changes see Russell and Bauer, 1988; Myers and Stothers, 2006) and "Leptodactilidae" sp.

This chapter provides an updated overview of the herpetological diversity of the Santa Cruz Formation, based on a reassessment of the specimens collected since the nineteenth century and new fossils recovered along the Atlantic coast of Santa Cruz Province (see Vizcaíno *et al.*, Chapter 1, and Matheos and Raigemborn, Chapter 4). In particular, we evaluate the validity of the lizard genus "*Erichosaurus*." Photographs of these fossils are published for the first time. We also comment on the paleoenvironmental and paleobiological implications of the material studied herein.

8.2 Institutional abbreviations

CORD-PZ: Museo de Paleontología de la Universidad Nacional de Córdoba, Colección Paleozoología, Argentina.

MACN-A: Museo Argentino de Ciencias Naturales "Bernardino Rivadavia," Colección Nacional Ameghino, Ciudad Autónoma de Buenos Aires, Argentina.

MACN-Pv: Museo Argentino de Ciencias Naturales "Bernardino Rivadavia," Colección Nacional de Paleovertebrados, Ciudad Autónoma de Buenos Aires, Argentina.

MPM-PV: Museo Regional Provincial Padre Manuel Jesús Molina, Río Gallegos, Santa Cruz Province, Argentina.

8.3 Systematic paleontology

AMPHIBIA Gray, 1825
ANURA Fischer von Waldheim, 1813
NEOBATRACHIA Reig, 1958
AUSTRALOBATRACHIA Frost *et al.*, 2006
CALYPTOCEPHALELLIDAE Reig, 1960
Calyptocephalella Strand, 1928
Calyptocephalella sp.
(Fig. 8.1a–d)

Referred material Complete left frontoparietal, incomplete right frontoparietal and complete right maxilla (MPM-PV 3712); incomplete right frontoparietal (MPM-PV 3507); incomplete right maxilla (CORD-PZ 1232) (the latter two not figured).

Locality and horizon Estancia La Costa, Santa Cruz Province, Argentina. MPM-PV 3712 and MPM-PV 3507 from Fossiliferous Level 3 and CORD-PZ 1232 from Fossiliferous Level 4, Estancia La Costa Member, Santa Cruz Formation.

Description The description is based on the specimen MPM-PV 3712. The maxilla (Fig. 8.1a, b) is heavily ossified and completely ornamented with low conical tubercles, except for the external surface of the *pars dentalis* which bears lower ridges. Teeth are present from the anterior end of the bone to the anterior two-thirds of the articular facet for the quadratojugal. The *pars facialis* forms the lateral margin of the orbit. On the inner side of the maxilla, between the *pars facialis* and the ascending maxillary palatine process, there is a groove for articulation with the nasal. Anteriorly, the *pars dentalis* invests for a short length the *pars dentalis* of the premaxilla. Posteriorly, the maxilla articulates with the quadratojugal for approximately the posterior third of the length of the maxilla. At the level of the anterior border of the subtemporal fenestra (*sensu* Trueb, 1974), the *pars palatina* bears a short pterygoid process that articulates with the anterior ramus of the pterygoid, and anteriorly, at the level of the anterior margin part of the orbit, bears an ascending maxillary palatine process for articulation with the palatine.

The frontoparietals (Fig. 8.1c, d) are wide and heavily ossified. The dorsal surface of these bones has a

Caption for Fig. 8.1. a–d, *Calyptocephalella* sp. MPM-PV 3712, (a) maxilla in external view, (b) maxilla in internal view, (c) frontoparietal in dorsal view, (d) frontoparietal in ventral view; e–n, Pleurodonta indet.: e, f, MACN-A 2272, fragmentary right maxilla previously described as "*Erichosaurus diminutus*," in labial (e) and lingual (f) views; g, h, MACN-A 5807, fragmentary right dentary and articulated suprangular previously described as "*Erichosaurus debilis*," in labial (g) and lingual (h) views; i–n, MACN-A 2283a, b, and c, fragments of toothed bones presumably corresponding to a sole individual previously described as "*Erichosaurus bombimaxilla*," in labial (i, k, m) and lingual (j, l, n) views; o–s, *Tupinambis* sp. (see Brizuela and Albino, 2010): o, p, MACN-A 621, fragmentary dentary in labial (o) and lingual (p) views; q, r, MACN-A 5806-b fragmentary maxilla in labial (q) and lingual (r) views; MACN-A 5806-a, fragmentary dentary in occlusal view (s); t–x, MACN-Pv SC 3317, isolated precloacal vertebra of "Colubridae" in anterior (t), posterior (u), ventral (v), dorsal (w), and lateral (x) views. a–d: scale bar = 10 mm; e–x: scale bar = 1 mm.

tuberculated ornamentation. The anterior margin of each frontoparietal has a smooth tongue-like process for articulation with the nasal; posterolaterally, another smooth tongue-like process articulates with the squamosal. The contact between the frontoparietal and the posteromedial margin of the nasal is concave. Ventrally, this bone exhibits a projecting flange, the *lamina perpendicularis*, which overlapped the dorsolateral surface of the braincase and the most dorsal part of the anterior wall of the otic capsule. Laterally, the frontoparietal extends beyond the *lamina perpendicularis* as a bony shelf. The posterior surface of the frontoparietal has lateral and medial foramina 10 mm apart for arterial vessels.

Comparisons The specimens described here are attributed to the extant neobatrachian *Calyptocephalella* because of unique derived features not shared by other South American extinct or extant anurans that have ornamentation. *Calyptocephalella* differs from the living genera *Lepidobatrachus* and *Ceratophrys* in the broad participation of the maxilla in the lateral margin of the orbit (instead of short or no participation; see Perí, 1993 and Wild, 1997). It differs from the living genera *Chacophrys* and *Ceratophrys* in having a broad contact between frontoparietal and squamosal (instead of short contact; see Perí, 1993); from *Wawelia gerholdi* Casamiquela, 1963, from the Miocene of Río Negro Province, Argentina (Báez and Perí, 1990), in having tuberculated instead of vermiform ornamentation (Báez and Perí, 1990); and from *Baurubatrachus pricei* Báez y Perí, 1989, from the Upper Cretaceous of Minas Gerais in Brazil (Báez and Perí, 1989), in having a long occipital channel for the occipital artery (instead of a short channel, see Báez and Perí, 1989).

Comments Tauber (1999) assigned a fragment of an ornamented maxilla (CORD-PZ 1232) to *Caudiverbera* sp., a junior synonym of *Calyptocephalella* sp. Fernicola and Vizcaíno (2006) referred the frontoparietal fragment MPM-PV 3507 to the genus *Ceratophrys*, based on the absence of posterolateral contact between the frontoparietal and squamosal. However, a more detailed evaluation of the specimen shows that this portion of the frontoparietal is slightly eroded and therefore the contact between frontoparietal and squamosal cannot be determined. Moreover, the presence of two foramina in the posterior margin of the frontoparietal MPM-PV 3507, as described for the new specimen MPM-PV 3712 and identified in *Calyptocephalella* by Reinbach (1939), allows us to refer specimen MPM-PV 3507 to this genus.

NEOBATRACHIA Reig, 1958
"LEPTODACTYLIDAE"

Referred material Left premaxilla, left maxilla, left frontoparietal and presacral vertebrae (CORD-PZ 1230).

Locality and horizon Estancia La Costa, Santa Cruz Province, Argentina. Fossiliferous Level 1, Estancia La Costa Member, Santa Cruz Formation.

Comments Tauber (1999) assigned this specimen to "Leptodactylidae." Because this taxon is paraphyletic (Frost *et al.*, 2006) a more detailed analysis of this fossil specimen will be needed.

REPTILIA Laurenti, 1768
SQUAMATA Oppel, 1811
IGUANIA Cope, 1864
PLEURODONTA Cope, 1864
Indeterminate genus and species
(Fig. 8.1e–n)

Referred material MACN-A 2272 (Fig. 8.1e, f), fragmentary right maxilla, originally described as "*Erichosaurus diminutus*" holotype; MACN-A 5807 (Fig. 8.1g, h), fragmentary right dentary and articulated surangular, originally described as "*Erichosaurus debilis*" holotype; MACN-A 2283a, b and c (Fig. 8.1i–n), three fragmentary toothed bones originally described as "*Erichosaurus bombimaxila*" holotype.

Locality and horizon La Cueva, Corpen Aike Department, Santa Cruz Province; Santa Cruz Formation.

Comments Ameghino (1899: 706, lines 37–40) erected the extinct iguanian genus "*Erichosaurus*" on the basis of the following characters: "*dental de borde inferior derecho, con dentadura pleurodonte: dientes largos, de base elíptica, comprimida de adelante hacia atrás, corona comprimida lateralmente, con un cono central más grande y dos laterales pequeños*" [dentary bone with straight lower edge, pleurodont long teeth with elliptical base compressed from front to back, laterally compressed crown with three cones, the center one larger than the laterals]. Ameghino (1899) recognized three species of the genus, "*E. diminutus*," "*E. bombimaxila*," and "*E. debilis*," which Estes (1983) considered *nomina dubia*. Estes (1983) was unable to locate the material corresponding to "*E. diminutus*," the type species of the genus, but all of the specimens studied by Ameghino (1899), identified by the species names, specimen numbers, year of collection, and collector (C. Ameghino), have now been located in the MACN-A collection.

The original diagnosis of "*E. diminutus*" (Ameghino, 1899: 707, lines 31–34) is reproduced here: "*es de tamaño muy pequeño, el dental con una línea de numerosas perforaciones emisarias y cara externa ligeramente convexa. Hacia la mitad de su largo tiene sólo 1,5 mm de altura, y en un espacio de 5 mm tiene 13 dientes*" [very small dentary with many open perforations along a line and slightly convex external face. It has a height of only 1.5 mm at the middle of its length, and 13 teeth along 5 mm length]. Ameghino (1899) interpreted the specimen as a dentary, but it is actually a fragmentary right maxilla (MACN-A 2272, Fig. 8.1e, f). The perforations mentioned by Ameghino (1899) are labial foramina. The dental series is almost complete, with 13 tooth positions, although only the fourth and the sixth to thirteenth tooth positions are preserved.

The diagnostic characters of "*E. debilis*" given by Ameghino (1899: 707, lines 34–36) are compared with features present in "*E. diminutus*": "*es de tamaño mayor, de cara externa deprimida y de una altura casi uniforme, tiene 2–3 mm de alto y en un espacio longitudinal de 8,5 mm tiene implantados 15 dientes*" [it is bigger, with a depressed external face and a uniform height of 2–3 mm, and it has 15 teeth along its 8.5 mm length]. The material corresponding to "*E. debilis*" includes two dentary fragments and an articulated suprangular (MACN-A 5807, Fig. 8.1g, h), which exactly correspond to the description of Ameghino (1899). The two fragments are now joined with glue. There are 15 complete teeth but probably there were two additional tooth positions on the missing anterior part. There is another box cataloged as "*E. debilis*" under the number MACN-A 5808, but it is empty. The box MACN-A 5807 also includes an edentate fragment which is not identifiable as belonging to a lizard but probably was the specimen kept in box MACN-A 5808.

The diagnostic features of "*E. bombimaxila*" mentioned by Ameghino (1899: 707, lines 36–38) are: "*de talla aproximada a la anterior, de la cual se distingue por el dental, cuya altura aumenta hacia atrás más rápidamente y tiene la cara externa muy convexa*" [size similar to the previous remains, but can be distinguished by the structure of the dentary, which increases in height more abruptly from front to back and has a very convex external face]. The material includes three specimens (MACN-A 2283a, b, and c, Fig. 8.1i–n), of which at least two are dentary fragments. They might correspond to a single mostly broken dentary.

Examination of the remains described as "*Erichosaurus*" by Ameghino (1899) demonstrates that the tooth crowns are compressed and tricuspid, with a higher central cusp. These characters, emphasized in the original diagnosis of Ameghino (1899), are not appropriate for a definition of a particular genus, because tricuspid teeth with pleurodont implantation characterize the Iguania Pleurodonta clade as a whole. Currently, the specimens studied by Ameghino (1899) are being exhaustively compared with extant iguanian genera (A. Albino and S. Brizuela, unpublished data). The homogeneous morphology of the specimens indicates that they correspond to a single species.

In addition to the materials from the Ameghino's collection, there are several undescribed remains of pleurodont iguanians from coastal localities of the Santa Cruz Formation north to the Río Coyle including maxillae and dentaries with teeth. The localities are Monte León (specimens MACN-Pv 3326, 3327, 3328) and Cerro Observatorio (specimens MACN-Pv SC3310, SC3311, SC3312, SC3313, SC3314, SC3320, SC3321, SC3322, SC3323, SC3324, SC3325).

SCLEROGLOSSA Estes, de Queiroz y Gauthier, 1988
SCINCOMORPHA Camp, 1923
TEIIOIDEA Estes, de Queiroz y Gauthier, 1988
TEIIDAE Gray, 1827
TUPINAMBINAE Presch, 1974
Tupinambis Daudin, 1802
Tupinambis sp.
(Fig. 8.1o–s)

Referred material MACN-A 621 (Fig. 8.1o, p), fragmentary right dentary, originally described as "*Dibolosodon typicus*" holotype; MACN-A 5806-a (Fig. 8.1s), fragmentary left dentary; and MACN-A 5806-b (Fig. 8.1q, r), fragmentary maxilla both originally described as "*Diasemosaurus occidentalis*" holotype.

Locality and horizon Monte León, Corpen Aike Department, Santa Cruz Province; Santa Cruz Formation.

Comments Ameghino (1893) erected two lizard taxa based on three fragmentary remains from Monte León (Corpen Aike Department, Santa Cruz Province). He named two of the specimens as "*Diasemosaurus occidentalis*" and the third as "*Dibolosodon typicus*," but he never described the specimens or figured them. "*Diasemosaurus*" is based on a fragmentary left dentary (MACN-A 5806-a, Fig. 8.1s) and a fragmentary maxilla (MACN-A 5806-b, Fig. 8.1q, r). "*Dibolosodon*" is based on a fragmentary right dentary (MACN-A 621, Fig. 8.1o, p). Some authors referred to these genera but without a revision of the holotypes (Romer, 1967; Estes, 1983; Gasparini and Báez, 1975; Báez and Gasparini, 1977, 1979; Donadío, 1984). Redescriptions and re-evaluation of the taxonomic affinities of both "*Diasemosaurus*" and "*Dibolosodon*" have been recently made by Brizuela and Albino (2008a). In accord with this re-evaluation, all three specimens belong to an indeterminate species of the tupinambine teiid *Tupinambis*. The specimen MACN-A 621 has bicuspid teeth whereas MACN-A 5806-a and -b have large, molariform posterior teeth. These differences made the taxonomic assignment confusing (Brizuela and Albino, 2008a); nevertheless, according to examination of ontogenetic series of osteological specimens of *Tupinambis merianae* (Brizuela and Albino, 2010), the molariform posterior teeth present in MACN-A 5806-a and -b are from larger and presumably older individuals whereas bicuspid teeth of the dentary MACN-A 621 are characteristic of smaller and younger individuals (Brizuela and Albino, 2008a).

OPHIDIA Brongniart, 1800
ALETHINOPHIDIA Nopcsa, 1923
CAENOPHIDIA Hoffstetter, 1939
COLUBROIDEA Oppel, 1811
"COLUBRIDAE" Oppel, 1811
Indeterminate genus and species
(Fig 8.1t–x)

Referred material MACN-Pv SC3315, SC3316, SC3317 (Fig. 8.1t–x), SC3318, SC3319, incomplete trunk vertebrae.

Locality and horizon Cerro Observatorio, Corpen Aike Department, Santa Cruz Province; Santa Cruz Formation.

Comments The specimens MACN-Pv SC3316, SC3317 (Fig. 8.1t–x), SC3318, and SC3319 consist of mid-trunk vertebrae, whereas MACN-Pv SC3315 is a vertebra from the anterior region or between the anterior and mid-region. As in "Colubridae," the vertebrae are elongate, having a greater length than height, with a long and narrow centrum, cotyle and condyle small and rounded, paracotylar foramen present, and subcentral ridges well defined (Albino, 1996b).

8.4 Paleoenvironmental and paleobiological considerations

Calyptocephalella is an endemic South American neobatrachian genus restricted to the northern temperate region of Chile from Coquimbo (at approximately 30° S) to Puerto Montt (40° S) (Cei, 1962). At present it is represented by a single living species, *Calyptocephalella gayi*. This taxon is an aquatic to semi-aquatic frog that dwells in lowland lakes, ponds, and slow-moving streams (Cei, 1962) up to 500 m above sea level (Veloso *et al.*, 2008), in general associated with the *Nothophagus* forest (Muzzopappa and Báez, 2009). Several fossil remains assigned or considered related to this genus have been recovered from sediments that range from the uppermost Cretaceous in Argentina (Los Alamitos Formation in Río Negro Province; Báez, 1987) to the Pleistocene in Chile (Laguna de Tagua Tagua archeological locality, in O'Higgins Province; Casamiquela, 1976; Jiménez-Huidobro *et al.*, 2009). The record of *Calyptocephalella* in Estancia La Costa locality at Santa Cruz Province constitutes the southernmost record for this taxon and indicates the presence of a permanent body of water, possibly developed in a forested area.

Calyptocephalella gayi is a voracious carnivore, whose adult individuals eat aquatic insect larvae, fish, frogs, and even small birds and mammals (Oliver Schneider, 1930; Lira Lira, 1948; Cei, 1962; Duellman, 2003). Their total body length varies from 80 to 320 mm (Oliver Schneider, 1930; Cei, 1962) and adult females are sometimes greater than 1 kg (Acuña Ortiz and Vélez-R, 2010). According to Cei (1962), the head represents one-third of the total body length. Therefore, the fossil maxilla described herein, whose length is 82 mm, would correspond to an individual whose total length was at least 250 mm, falling in the range of the highest values of body size and mass of the living species. By analogy with *Calyptocephallela gayi*, the Miocene *Calyptocephalella* was evidently in the small carnivore guild.

Among reptiles, only squamates (Squamata) are recorded in the Santa Cruz Formation: pleurodont Iguania and Teiidae, and a "Colubridae" snake. Pleurodont iguanians have a modern distribution as far south as Santa Cruz Province, whereas the modern distributional limit of teiids and "colubrids" is much further north than where they occur in the late Early Miocene. *Tupinambis* is the largest genus of tupinambine teiid, with the broadest range of distribution. They are found mostly in tropical and subtropical environments throughout much of South America from northern Amazonia to northern Patagonia, east of the Andes. The current southernmost limit of the distribution of the genus, and of tupinambine teiids in general, is Río Negro (Cei, 1986), with the exception of isolated populations of *T. rufescens* that live further south, in the Bajo del Gualicho area of the Río Negro Province (Cei and Scolaro, 1982), approximately at a latitude of 40° S.

The largest species of *Tupinambis* live in the tropics where high temperatures facilitate the evolution of large body sizes among air-breathing animals whose body temperatures are dependent on environmental temperatures (poikilothermy) (Makarieva *et al.*, 2005a, b). The sizes of the specimens found in the Miocene of Patagonia are a little smaller than that of the present-day species, which reach a body length of around 600 mm. Thus, the Miocene specimens are of similar size to modern *Tupinambis* from both tropical and temperate regions. Based on the present-day distribution of *Tupinambis* species in South America, these lizards exploit areas with mean annual temperatures above 14 °C, more than 200 mm/year of rainfall, and a hydric index above –40 (Brizuela and Albino, 2008b; Brizuela, 2010). Also, the optimal feeding and digestive temperature for one of the species distributed in most temperate regions (*T. merianae*) has been estimated to be above 22 °C (Giambelluca and Casciaro, 1999). The Monte León locality in the Santa Cruz Province is the southernmost site where fossil specimens of *Tupinambis* have been found (Brizuela and Albino, 2008a). This late Early Miocene locality is approximately at 50° S, far south of the present limit of distribution of the genus. Therefore, based on the modern distribution of *Tupinambis* species and its climatic requeriments, the teiid remains reported in this paper indicate that the environmental conditions during the deposition of the Santa Cruz Formation in southernmost Patagonia would have been warmer, and probably more humid, than those at present. The same paleoclimatic conclusions are suggested by the record of late Early Miocene "colubrids" in Cerro Observatorio (Albino, 1996b), where representatives of this group are not found today. Present-day "colubrids" do not exceed a latitude of around 44° S (Cei, 1986); therefore, these Miocene specimens constitute the southernmost records for "colubrid" snakes, suggesting climatic conditions warmer than today to make their survival possible (Albino, 1996b).

Regarding the paleobiology of reptiles, lizards of the genus *Tupinambis* are generalist consumers with a diet of diverse vertebrates (fish, amphibians, reptiles, birds, and small mammals) and invertebrates (millipedes, arachnids, insects, and molluscs), as well as seeds, fruits, eggs, carrion, and mushrooms (Toledo *et al.*, 2004). "Colubrids" are strictly carnivorous vertebrates, and pleurodont iguanians are insectivorous and/or herbivorous.

According to Pascual *et al.* (1996) and Ortiz-Jaureguizar and Cladera (2006), South America underwent an extensive development of woodlands throughout the Miocene, coexisting with open areas, constituting a park-savanna equilibrium under a humid and warm climate. These authors concluded that mammal history in South America would have been severely affected during the Miocene by geodynamic events, sea level changes, and related climatic fluctuations, which would have caused the extinction of some higher-latitude groups (e.g. primates) and an increase of diversity of the large cursorial forms because of expansion of the plains. These changes would have significantly influenced the evolutionary history and biogeography of South American anuran and reptilian taxa, as they did those of mammals.

The earliest record of *Tupinambis* indicates a probable origin of the genus in a tropical or subtropical environment, and that it spread throughout South America in the middle Tertiary (Brizuela and Albino, 2004; Brizuela, 2010). In southern South America, the progressive aridity and decreasing temperatures of Patagonia that developed especially during the Middle and Late Miocene (Pascual *et al.*, 1996) might have pushed the distribution of *Tupinambis* to its present more northerly position. The present conditions at the latitude where late Early Miocene remains of *Tupinambis* were found are too cold and dry for the genus. The southernmost population of *T. rufescens* found at the Gran Bajo del Gualicho depression (Río Negro Province; Cei and Scolaro, 1982) is probably a relic of the more widespread Miocene distribution. "Colubrids" also would have suffered a restriction of distribution due to paleoclimatical and paleoenvironmental changes in southern Argentina.

Disappearance of forests and of permanent standing water on the eastern side of the Patagonian Andes would have caused the extinction of *Calyptocephalella* from this region.

8.5 Conclusions

Although scant, the herpetological record in the Santa Cruz Formation that crops out along the Atlantic coast of southernmost Patagonia is relevant because it includes taxa with a more northerly extra-Patagonian distribution. These taxa include the anuran *Calyptocephalella*, the teiid lizard *Tupinambis*, and the "colubrid" snakes; on the other hand, pleurodont iguanians remain today, but are not restricted to this region. This distribution of herpetofauna during the Early to Middle Miocene implies paleoclimatic and paleoenvironmental conditions different than those present today. The presence of *Tupinambis* and "colubrids" in Monte León and Cerro Observatorio, respectively, suggests a warmer and perhaps more humid climate, whereas *Calyptocephalella* indicates the presence of permanent standing water in a forested environment for the lower levels of the Santa Cruz Formation at Estancia La Costa locality. Paleoclimatical and paleoenviromental changes occurring in southern Patagonia during the Neogene would have produced the retreat of *Tupinambis* and "colubrids" to more northern regions. Disappearance of forests on the eastern side of the Patagonian Andes would have brought about the regional extirpation of *Calyptocephalella*.

Finally, the feeding information about extant *Calyptocephalella* and "colubrids" allows us to consider their Miocene representatives as part of the Santacrucian small carnivorous vertebrate guild. Pleurodont iguanians could be included in the Miocene insectivorous and/or herbivorous group, whereas *Tupinambis* would have been a generalist, omnivorous reptile.

ACKNOWLEDGMENTS
We thank the editors for their kind invitation to contribute to this book. A. Kramarz (Museo Argentino de Ciencias Naturales "Bernardino Rivadavia," Buenos Aires) facilitated access to the fossil material at the MACN collections. This is a contribution to the grants PICT 26219 and 0143, UNLP N647 and PIP-CONICET 1054 to Sergio F. Vizcaíno, NSF 0851272 and 0824546 to Richard F. Kay, and UNLu CDD-CD 281–09 to Juan C. Fernicola.

REFERENCES
Acuña Ortiz, P. L. and Vélez-R., C. M. (2010). Descripción de manejos y variables ambientales encontradas en desoves ocurridos en cautiverio de la rana grande chilena *Calyptocephalella gayi*, en la zona central de chile. *XVII Congreso de la Asociación Latinoamericana de Parques Zoológicos y Acuarios, Resúmenes* 57.

Albino, A. M. (1996a). The South American fossil Squamata (Reptilia: Lepidosauria). In *Contributions of Southern South America to Vertebrate Paleontology*, ed. G. Arratia. Münchner Geowiss Abhandlungen (A) **30**, 9–72.

Albino, A. M. (1996b). Snakes from the Miocene of Patagonia (Argentina). Part II. The Colubroidea. *Neues Jahrbuch für Geologie und Paläontologie, Abhandlungen*, **200**, 353–360.

Ameghino, F. (1893). Sobre la presencia de vertebrados de aspecto Mesozoico en la Formación Santacruceña de Patagonia austral. *Revista del Jardín Zoológico de Buenos Aires*, **1**, 75–84.

Ameghino, F. (1898). Sinopsis geológico-paleontológica. In *Segundo Censo de la República Argentina*, **1**, 111–255.

Ameghino, F. (1899). Sinopsis geológico-paleontológica. Segundo Censo Nacional de la República Argentina. Suplemento. Adiciones y correcciones. *Obras completas* **12**, 706.

Báez, A. M. (1987). Anurans. In *The Late Cretaceous fauna of Los Alamitos, Patagonia, Argentina*, ed. J. F. Bonaparte. Revista del Museo Argentino de Ciencias Naturales Bernardino Rivadavia, Paleontología, **3**, 121–130.

Báez, A. M. and Gasparini, Z. (1977). Orígenes y Evolución de los Anfibios y Reptiles del Cenozoico de América del Sur. *Acta Geológica Lilloana*, **14**, 149–232.

Báez, A. M. and Gasparini, Z. (1979). The South America herpetofauna: an evaluation of the fossil record. In *The South American Herpetofauna: Its Origin, Evolution, and Dispersal*, ed. W. E. Duellman. Monograph of the Museum of Natural History, The University of Kansas, **7**, 29–54.

Báez, A. M. and Perí, S. (1989). *Baurubatrachus pricei*, nov. gen. et sp., un Anuro del Cretácico Superior de Minas Gerais, Brasil. *Annais Academia Brasilera de Ciências*, **61**, 447–458.

Báez, A. M. and Perí, S. (1990). Revisión de *Wawelia gerholdi*, un anuro del Mioceno de Patagonia. *Ameghiniana*, **27**, 379–386.

Bondesio, P. (1986). Lista sistemática de los vertebrados terrestres del Cenozoico de Argentina. *Actas IV Congreso Argentino de Paleontología y Bioestratigrafía*, **2**, 187–190.

Brizuela, S. (2010). Los lagartos continentales fósiles de la Argentina (excepto Iguania). Unpublished Ph.D. thesis, Facultad de Ciencias Naturales y Museo, Universidad Nacional de La Plata.

Brizuela, S. and Albino, A. M. (2004). The earliest *Tupinambis* teiid from South America and its palaeoenvironmental significance. *Journal of Herpetology*, **38**, 113–119.

Brizuela, S. and Albino, A. M. (2008a). Re-evaluation of the type material of "*Diasemosaurus occidentalis*" Ameghino and "*Dibolosodon typicus*" Ameghino (Squamata: Teiidae) from the Miocene of Argentina. *Journal of Vertebrate Paleontology*, **28**, 253–257.

Brizuela, S. and Albino, A. M. (2008b). Tupinambine teiids from the middle Miocene of north-western Patagonia (Argentina). *Amphibia-Reptilia*, **29**, 425–431.

Brizuela, S. and Albino, A. M. (2010). Variaciones dentarias en *Tupinambis merianae* (Squamata: Teiidae). *Cuadernos de Herpetología*, **24**, 5–16.

Casamiquela, R. M. (1976). Los vertebrados fósiles de Tagua-Tagua. *Actas Primer Congreso Geológico Chileno*, **1**, C87–C102.

Cei, J. M. (1962). *Batracios de Chile*, 1st Edn. Santiago de Chile: Ediciones de la Universidad de Chile.

Cei, J. M. (1986). Reptiles del Centro, Centro-oeste y Sur de la Argentina. Herpetofauna de las Zonas Áridas y Semiáridas. *Monograph of the Museo Regionale di Scienze Naturali*, **4**, 527.

Cei, J. M. and Scolaro, J. A. (1982). A population of *Tupinambis* from northern Patagonia, south of the Río Negro River. *Herpetological Review*, **13**, 26.

Donadío, O. E. (1984). Teidos del Mioceno temprano (Squamata, Sauria, Teiidae) de la Provincia de Santa Cruz, Argentina. *Circular Informativa de la Asociación Paleontológica Argentina*, **12**, 2–3.

Duellman, W. E. (2003). Helmeted water toad, *Caudiverbera caudiverbera*. In *Grzimek's Animal Life Encyclopedia, Amphibians*, Vol. 6, ed. M. Hutchins, W. E. Duellman and N. Schlager. Detroit: Gale, p. 170.

Estes, R. (1983). Sauria terrestria, Amphisbaenia. In *Handbuch der Paläoherpetologie* Part 10A, ed. P. Wellnhofer. Stuttgart: Gustav Fischer Verlag Press, pp. 1–249.

Fernicola, J. C. and Vizcaíno, S. F. (2006). Sobre la posible presencia de *Ceratophrys* (Anura: Ceratophryinae) en la Formación Santa Cruz. *Ameghiniana*, **43**, 38R.

Frost, D. R., Grant, T., Faivovich, J. *et al.* (2006). The amphibian tree of life. *Bulletin of the American Museum of Natural History*, **297**, 1–370.

Gasparini, Z. and Báez, A. M. (1975). Aportes al conocimiento de la herpetofauna Terciaria de la Argentina. *Actas Primer Congreso Argentino de Paleontología y Bioestratigrafía*, **2**, 377–415.

Gasparini, Z., de la Fuente, M. S. and Donadío, O. E. (1986). Los Reptiles Cenozoicos de la Argentina: implicancias paleoambientales y evolución biogeográfica. *Actas IV Congreso Argentino de Paleontología y Bioestratigrafía*, **2**, 119–130.

Giambelluca, L. A. and Casciaro, M. (1999). Manejo de Lagarto Overo y Lagarto Colorado en ECAS. *Revista Argentina de Producción Animal*, **19**, 471–480.

Jiménez-Huidobro, P., Gustein, C. S., Sallaberry, M. and Rubilar-Rogers, D. (2009). Anuros del Pleistoceno de Chile. *Ameghiniana*, **46**, 32R–33R.

Lira Lira, E. (1948). Límite de saciedad y sensación de repleción en *Calyptocephalus gayi*. *Biologica*, **5**, 29–40.

Makarieva, A. M., Gorshkov, V. G. and Li, B. L. (2005a). Temperature associated upper limits to body size in terrestrial poikilotherms. *Oikos*, **111**, 425–436.

Makarieva, A. M., Gorshkov, V. G. and Li, B. L. (2005b). Gigantism, temperature and metabolic rate in terrestrial poikilotherms. *Proceedings of the Royal Society London*, B **272**, 2325–2328.

Muzzopappa, P. and Báez A. M. (2009). Systematic status of the mid-Tertiary neobatrachian frog *Calyptocephalella canqueli* from Patagonia (Argentina), with comments on the evolution of the genus. *Ameghiniana*, **46**, 113–125.

Myers, C. W. and Stothers, R. B. (2006). The myth of Hylas revisited: the frog name *Hyla* and other commentary on

Specimen medicum (1768) of J. N. Laurenti, the "father of herpetology". *Archives of Natural History*, **33**, 241–266.

Oliver Schneider, C. (1930). Observaciones sobre batracios chilenos. *Revista Chilena de Historia Natural*, **34**, 220–223.

Ortiz-Jaureguizar, E. and Cladera, G. A. (2006). Paleoenvironmental evolution of southern South America during the Cenozoic. *Journal of Arid Environments*, **66**, 498–532.

Pascual, R., Ortiz-Jaureguizar, E. and Prado, J. L. (1996). Land mammals: paradigm for Cenozoic South American geobiotic evolution. In *Contributions of Southern South America to Vertebrate Paleontology*, ed. G. Arratia. München Geowissenschftliche Abhandlungen, **30**, 265–319.

Perí, S. I. (1993). Relaciones evolutivas de las especies de la Subfamilia Ceratophryinae (Anura: Leptodactylidae). Unpublished Ph.D. thesis, Facultad de Ciencias Naturales y Museo, Universidad Nacional de La Plata.

Reinbach, W. (1939). Untersuchungen über die Entwicklung des Kopfskeletts von *Calyptocephalus gayi*. *Jenaische Zeitschrift Für Naturwissencschaft*, **72**, 211–362.

Romer, A. S. (1967). *Vertebrate Paleontology*, 1st Edn. Chicago: University of Chicago Press.

Russell, A. P. and Bauer, A. M. (1988). An early description of a member of the genus *Phelsuma* (Reptilia: Gekkonidae),

with comments on names erroneously applied to *Uroplatus fimbriatus*. *Amphibia-Reptilia*, **9**, 107–116.

Tauber, A. A. (1994). Estratigrafía y vertebrados fósiles de la Formación Santa Cruz (Mioceno Inferior) en la costa atlántica de Santa Cruz, República Argentina. Unpublished Ph.D. thesis, Universidad Nacional de Córdoba.

Tauber, A. A. (1999). Los vertebrados de la Formación Santa Cruz (Mioceno Inferior) en el extremo sureste de la Patagonia y su significado paleoecológico. *Revista Española de Paleontología*, **14**, 173–182.

Toledo, L. F., Peralta de Almeida Prado, C. and Vieira Andrade, D. (2004) *Tupinambis merianae* (Tegu Lizard). Fungivory. *Herpetological Review*, **35**, 174.

Trueb, L. (1974). Systematic relationships of neotropical horned frogs, genus *Hemiphractus* (Anura: Hylidae). *Occasional Papers of the Museum of Natural History, University of Kansas*, **29**, 1–60.

Veloso, A., Formas, R. and Gerson, H. (2008). *Calyptocephalella gayi*. In *IUCN Red List of Threatened Species*. Version 2010.3. http://www.iucnredlist.org.

Wild, E. R. (1997). Description of the adult skeleton and developmental osteology of the hyperossified horned frog, *Ceratophrys cornuta* (Anura: Leptodactylidae). *Journal of Morphology*, **232**, 169–206.

9 Diversity and paleobiology of the Santacrucian birds

Federico J. Degrange, Jorge I. Noriega, and Juan I. Areta

Abstract

This chapter presents the state of knowledge of the avian diversity recorded in Santacrucian beds (late Early Miocene) with an updated systematic summary of all taxa. Phorusrhacids outnumber seriemas, rheas, and basal falconiforms in diversity and abundance. More fragmentary occurrences are reported of pelecaniforms, anseriforms, gruiforms, and ciconiiforms. Body masses of fossil forms are inferred from the dimensions of their hindlimb bones (i.e. femur, tibiotarsus) based on logarithmic equations previously modeled from living analogs. In some cases, body sizes of the extinct species are also inferred from the relative sizes of other bones of similar extant species. Inferences about diet and foraging strategies are based on the size and shape of the limb elements and structural details of the cranial elements, by analogy with extant birds. The predator niche is represented by falconids, four species of phorusrhacids and a seriemid. Phorusrhacids and seriemids probably lived in open areas because of their cursorial capabilities. However, birds such as the waterfowls, limpkins, spoonbills, and darters indicate the presence of temporarily flooded savannas or permanent water bodies in forested areas. Habitat preferences of extant seriemas, rheas, tinamous, and the falconid *Herpetotheres* are consistent with Chacoan-like conditions, and they are useful to infer by extrapolation similar Santacrucian paleoenvironments for their extinct analogs. Santacrucian environments were probably characterized by seasonality in temperature and rainfall and the presence of alternating areas of herbaceous vegetation with shrubby or wooded areas.

Resumen

Este capítulo contiene una síntesis actualizada del conocimiento sobre el registro paleontológico de las aves del Santacrucense (Mioceno Temprano tardío). Los fororracos superan a las chuñas, ñandúes y falconiformes en diversidad y abundancia. Especímenes fragmentarios de pelecaniformes, anseriformes, gruiformes y ciconiiformes son también reportados. Las estimaciones de las masas corporales de taxones fósiles aquí presentadas, a partir de las dimensiones de los huesos de sus miembros posteriores (fémur, tibiotarso), están basadas en ecuaciones logarítmicas previamente modeladas en análogos vivientes. En algunos casos, el tamaño corporal de las especies extinguidas es inferido a partir del tamaño relativo de otros huesos en las especies actuales análogas. Las interpretaciones sobre dietas y estrategias de alimentación están basadas en el tamaño y la forma de los elementos de los miembros y los detalles estructurales del cráneo y mandíbula, por analogía con las aves actuales. El nicho de los depredadores se encuentra representado por los falcónidos, cuatro especies de fororracos y una chuña, los cuales habrían habitado ambientes abiertos debido a sus capacidades locomotoras cursoriales. Sin embargo, aves como patos, caraúes, espátulas y anhingas indican la presencia de sabanas temporalmente inundadas o de cuerpos de agua permanentes en áreas más forestadas. Las preferencias de habitat de las actuales chuñas, ñandúes, perdices y *Herpetotheres* son consistentes con condiciones de tipo chaqueño y son útiles para inferir similares paleoambientes santacrucenses al extrapolarlas a sus análogos fósiles. Los escenarios santacrucences habrían estado caracterizados por la estacionalidad en las temperaturas y las lluvias, así como por la presencia de áreas de vegetación herbácea alternando con zonas arbustivas o incluso boscosas.

9.1 Introduction

The late Early Miocene vertebrate fauna of the Santa Cruz Formation along the Atlantic coast of southern Patagonia (see Vizcaíno *et al.*, Chapter 1, Figs. 1.1 and 1.2) comprises one of the richest fossil assemblages known in the continent (Hatcher, 1903; Tauber, 1997a, b; Vizcaíno *et al.*, 2006; Vizcaíno *et al.*, 2010). The collection of avian fossils began in the nineteenth century with Carlos Ameghino. Some of his specimens were originally deposited at the Museo de La Plata and Museo Argentino de Ciencias Naturales "B. Rivadavia" (Argentina), but some were later sold to the British Museum of Natural History (England), as is evident through the analysis of Ameghino's catalog at the latter institution. Florentino Ameghino described several new taxa based on these specimens. In the 1890s, John B. Hatcher made extensive collections for Princeton University, and this was followed by Elmer S. Riggs' expeditions in the 1920s for the Field Museum of Natural History in Chicago. These collections included numerous well-preserved bird remains. Some of these older collections have been revised recently (Alvarenga and

Early Miocene Paleobiology in Patagonia: High-Latitude Paleocommunities of the Santa Cruz Formation, ed. Sergio F. Vizcaíno, Richard F. Kay and M. Susana Bargo. Published by Cambridge University Press. © Cambridge University Press 2012.

Höfling, 2003; Agnolín, 2006a, b, 2007, 2009; Noriega *et al.*, 2008, 2011; Cenizo and Agnolín, 2010; Tambussi, 2011; C. Tambussi and F. Degrange, unpublished data, 2011).

Scattered bird remains were also reported from localities farther northwest in Santa Cruz Province, in sediments of the Pinturas Formation (Chiappe, 1991; Noriega and Chiappe, 1993; Bertelli and Chiappe, 2005). The Pinturas Formation (Early Miocene) is regarded as slightly older than the Santa Cruz Formation (Ameghino, 1906; Frenguelli, 1931; Barrio *et al.*, 1984; Fleagle *et al.*, 1995). The distinctiveness of the Pinturas fauna from the Santacrucian fauna has been questioned, with some authors referring the former to the Santa Cruz Formation (Pascual and Odreman Rivas, 1971; Marshall, 1976), but recent radiometric dates for a Pinturan fauna at Gran Barranca show that the Pinturas Formation is definitely older than the coastal Santa Cruz rocks (Ré *et al.*, 2010).

9.2 The fossil record

The record of Santacrucian birds comprises at least 18 species included in 15 genera and nine families. Most holotypes are very fragmentary and eroded. Brief descriptions and poor illustrations make the reassessment of the known taxa challenging (Olson, 1981, 1985; Tonni, 1980; Tambussi and Noriega, 1996). What follows is a commentary about the material. The main localities where all these remains were recovered are shown in Vizcaíno *et al.* (Chapter 1, Figs. 1.1 and 1.2).

Rheiformes Rheids are represented by *Opisthodactylus patagonicus* Ameghino, 1891 (Fig. 9.1a), a taxon initially placed in its own family, Opisthodactylidae (Ameghino, 1895; Brodkorb, 1963), but transferred to Rheidae because of its strong resemblance to living rheas (Patterson and Kraglievich, 1960; Tonni and Tambussi, 1986; Tambussi, 1995).

Tinamiformes At least two species coming from the localities of Monte Observación (= Cerro Observatorio), Monte León, and Cañadón de Las Vacas are recognized (Bertelli and Chiappe, 2005). They represent two distinct morphotypes of Nothurinae, but they cannot be assigned to any known taxon or described as new taxa because of their fragmentary state.

"Gruiformes" A distal end of tarsometatarsus (Fig. 9.1b) described as *Anisolornis excavatus* Ameghino, 1891 was originally assigned to the "Pelecyornidae" (= Phorusrhacidae Psilopterinae). Later it was tentatively assigned to the Phasianidae (Galliformes) by Ameghino (1895). Alternatively, it has been considered a seriema or a tinamou (Brodkorb, 1964), a gruiform of the family Aramidae (Cracraft, 1973), or closely related to the Psophiidae (Olson, 1985; see Tambussi and Noriega, 1996).

Anseriformes Two species of waterfowl (Anatidae) have been described: *Eutelornis patagonicus* Ameghino, 1891 (based on a distal end of a humerus and a proximal end of a tibiotarsus, Fig. 9.1c) and *Eoneornis australis* Ameghino, 1891 (represented by a distal end of radius, Fig. 9.1d) (Ameghino, 1895). Recently, Worthy (2008) considered both taxa as *incertae sedis*, but Cenizo and Agnolín (2010) stated that *Eoneornis australis* is similar to basal screamers (Anhimidae) and that *Eutelornis patagonicus* also belongs to a basal anseriform clade. *Ankonetta larriestrai* Cenizo and Agnolín (2010) was recently described as a basal anatid on the basis of an incomplete tarsometatarsus.

The anseriform *Brontornis burmeisteri* Moreno and Mercerat, 1891 is represented by a mandibular symphysis, a quadrate, fragmentary remains of a femur, tibiotarsus, tarsometatarsus (Fig. 9.2a–c), and several phalanges. This taxon was considered a phorusrhacid (Brodkorb, 1967; Mourer-Chauviré, 1981; Alvarenga and Höfling, 2003; Alvarenga *et al.*, 2011), or related to the Anseriformes (Moreno and Mercerat, 1891; Agnolín, 2007). The hindlimb of *Brontornis* Moreno and Mercerat, 1891 resembles that of *Cygnus* (Anatidae), and Dolgopol de Sáez (1927) even created a separate order for the genera *Brontornis* and *Rostrornis* Moreno and Mercerat, 1891 (synonym of the first *sensu* Alvarenga and Höfling, 2003) based on features of the tarsometatarsus and phalanx.

Pelecaniformes *Liptornis hesternus* Ameghino, 1895 was originally described, with doubt, as a pelecaniform based on a cervical vertebra (Fig. 9.1e). Our study of a cast and photographs of the type specimen allowed us to confirm that *L. hesternus* is an anhingid. Cenizo and Agnolín (2010) reported the presence of an indeterminate species of the extinct genus *Macranhinga* Noriega, 1992 at Río Bandurrias from the Santa Cruz Formation in the homonymous Argentinian province. Santacrucian anhingids are also known from sediments of the Cura Mallín Formation (Early to Middle Miocene) in Chile (Alvarenga, 1995). These records are the oldest for the family in South America, earlier than the important radiation that took place in the Middle to Late Miocene at northern latitudes in Argentina, Uruguay, Brazil, Colombia, and Peru (Rasmussen and Kay, 1992; Noriega, 1995, 2002; Campbell, 1996; Rinderknecht and Noriega, 2002; Areta *et al.*, 2007; Noriega and Agnolín, 2008).

Ciconiiformes *Protibis cnemialis* Ameghino, 1891, based on a distal end of tibiotarsus (Fig. 9.1f), has been considered a plataleid (Threskiornithidae). The lack of cranial or complete appendicular remains referred to this purported spoonbill species makes its systematic assignment highly speculative.

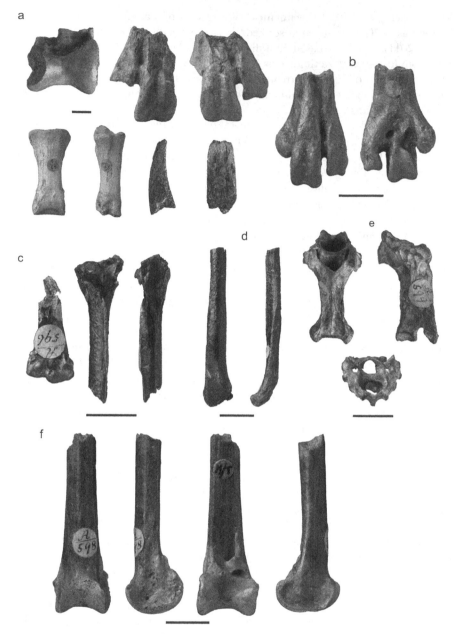

Fig. 9.1. Original specimens described by F. Ameghino in erecting several of his Santacrucian bird taxa. a, *Opisthodactylus patagonicus*, BMNH-A586, BMNH-A587, and BMNH-A588, including tibiotarsus, right and left tarsometatarsus, phalanges, and beak. b, *Anisolornis excavatus*, BMNH-A594, distal portion of left tarsometatarsus. c, *Eutelornis patagonicus*, BMNH-A596, distal end of a right humerus and proximal end of a right tibiotarsus. d, *Eoneornis australis*, BMNH-A595, distal end of radius. e, *Liptornis hesternus*, BMNH-A599, cervical vertebra. f, *Protibis cnemialis*, BMNH-A598, distal end of right tibiotarsus. Scale bar = 1 cm.

Falconiformes The birds of prey are known by three species of the family Falconidae: *Badiostes patagonicus* Ameghino, 1895, *Thegornis musculosus* Ameghino, 1895, and *T. debilis* Ameghino, 1895. The first was described on the basis of a very damaged and fragmentary specimen (Fig. 9.3a) from the La Cueva locality and it was referred with doubt to the family Strigidae. However, Brodkorb (1964) assigned *B. patagonicus* to the Falconidae, a view also shared by Olson (1985). The type material belonging to this taxon is currently under review (J. Noriega and H. Alvarenga, unpublished data, 2011). The species of *Thegornis* were both erected on fragments of distal tarsometatarsi (*T. debilis*, Fig. 9.3b, and *T. musculosus*, Fig. 9.3c) and referred to the Falconidae. The holotype of *T. musculosus* was collected at the locality Yegua Quemada, whereas *T. debilis* came from the Corriguen Aike (= Puesto Estancia La Costa) locality. The recent find of a very well-preserved, complete specimen (Fig. 9.3d–i) referable to *T. musculosus* at Puesto Estancia La Costa reconfirms Ameghino's placement of this

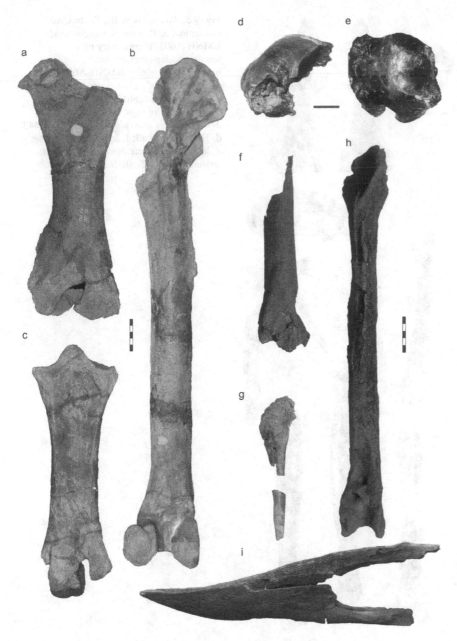

Fig. 9.2. a–c, Hindlimb bones of *Brontornis burmeisteri*, MLP 88–91. a, Femur; b, tibiotarsus. c, tarsometatarsus. Scale bar = 5 cm. d, e, *Cariama santacrucensis*, MPM-PV 3511, basicranium. d, lateral view; e, dorsal view. Scale bar = 1 cm. f–i, *Phorusrhacos longissimus*, MPM-PV 4241. f, left femur; g, left fibula; h, left tibiotarsus; i, fragmentary jaw in lateral view. Scale bar = 5 cm.

taxon within the Falconidae (Noriega *et al.*, 2008; Noriega *et al.*, 2011). A cladistic analysis places *T. musculosus* as the sister-group of the Laughing Falcon, *Herpetotheres cachinnans*, within the basal clade of the Herpetotherinae (Noriega *et al.*, 2011).

Cariamiformes The seriemas (Cariamidae) are represented only by *Cariama santacrucensis* Noriega *et al.*, 2009, which is based on an incomplete cranium (Fig. 9.2d, e) and

two unassociated fragments of tibiotarsi, all from Puesto Estancia La Costa (= Corriguen Aike).

The "terror birds" (Phorusrhacidae) are the world's largest known terrestrial carnivorous birds. Their biological and morphological design represents an extreme of terrestrial vertebrate evolution. They are a very interesting group not only for this reason, but also for the presence of an enormous and rigid skull, with a high and narrow beak, which is a peculiar, unique morphology among birds (Degrange *et al.*, 2010a).

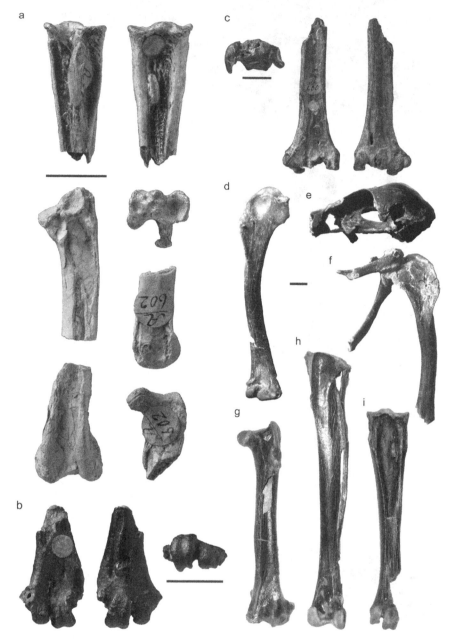

Fig. 9.3. Falconids of the Santa Cruz Formation. a, *Badiostes patagonicus*, BMNH-A602, fragmentary ulna, femur, and tarsometatarsus; b, *Thegornis debilis*, BMNH-A601, distal portion of right tarsometatarsus; c, *Thegornis musculosus*, BMNH-A600, distal portion of right tarsometatarsus. d–i, *Thegornis musculosus*, MPM-PV-3443. d, Humeri; e, skull; f, left pectoral girdle; g, left femur; h, left tibiotarsus; i, right tarsometatarsus. Scale bar = 1 cm.

The fossil record of phorusrhacids extends from the Middle Paleocene (Alvarenga, 1985) to the Late Pleistocene (Alvarenga *et al.*, 2010). Their greatest diversity occurred during the late Early Miocene (Santacrucian).

Among the several systematic proposals for the subdivision of the Phorusrhacidae (Dolgopol de Sáez, 1927; Patterson and Kraglievich, 1960; Brodkorb, 1967; Alvarenga and Höfling, 2003; Agnolín, 2009), we recognize four of the five subfamilies proposed by Alvarenga and Höfling (2003): Mesembriornithinae, Phorusrhacinae, Patagornithinae, and Psilopterinae, of which only the latter three are represented in the Santacrucian.

1. Phorusrhacinae: *Phorusrhacos longissimus* Ameghino, 1887 (Fig. 9.2f–i) is known for the Santacrucian and represented by fragments of skulls, jaws, femora, tibiotarsi, tarsometatarsi, and phalanges. Most of these

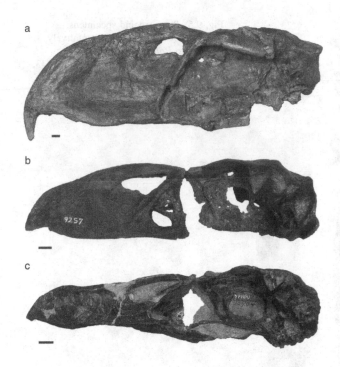

Fig. 9.4. Skulls of Santacrucian phorusrhacids. a, *Patagornis marshi*, BMNH-A516; b, *Psilopterus lemoinei*, AMNH 9257; c, *Psilopterus bachmanni*, YPM-VPPU15904. Scale bar = 1 cm.

materials were widely, although ambiguously, described by F. Ameghino (1887, 1889, 1895).

2. Patagornithinae: *Patagornis marshi* Moreno and Mercerat, 1891 is known by very abundant and well-preserved fossil remains (Figs. 9.4a, 9.5a). A very extensive description of *P. marshi* can be found in Andrews (1899).

3. Psilopterinae: two species of psilopterines come from Santacrucian sediments: *Psilopterus lemoinei* (Moreno and Mercerat, 1891) (Fig. 9.4b) and *P. bachmanni* (Moreno and Mercerat, 1891) (Figs. 9.4c, 9.5b), both widely, but ambiguously, described by Sinclair and Farr (1932). *Psilopterus lemoinei* is one of the best-known phorusrhacids, being represented by several specimens, including skulls.

9.3 Materials and methods

9.3.1 Institutional abbreviations

AMNH, American Museum of Natural History, New York, USA.

FMNH, Field Museum of Natural History, Chicago, USA.

MLP, Museo de La Plata, La Plata, Argentina.

MPM-PV, Museo Regional Provincial Padre Manuel Jesús Molina, Río Gallegos, Santa Cruz Province, Argentina.

YPM-VPPU, Yale Peabody Museum, Princeton University Collection, New Haven, USA.

The material studied is listed in Appendix 9.1.

9.3.2 Body mass estimation

Campbell and Marcus (1992) developed a method to estimate body masses of extinct birds based on regression equations derived from the minimum shaft circumferences of the femur (LFC) and tibiotarsus (LTC) of extant birds of known body mass representing 89 families and spanning several ecologically (and morphologically) distinct subgroups (Campbell and Marcus, 1992: 397); data available in Campbell and Marcus (1992: 405; Table 1). The femur and tibiotarsus dimensions are better estimators of body masses than the tarsometatarsus, evidently because the former have a more direct role in body support whereas the tarsometatarsus tends to be more associated with feeding and locomotor styles than with body support (Campbell and Marcus, 1992).

In most cases, for body mass estimates we employed Campbell and Marcus data for "long-legged" birds because phorusrhacids, seriemas, and rheas had relatively long legs. We used Campbell and Marcus "predatory" flying birds to estimate the body mass of *Thegornis musculosus*.

The logarithmic estimating equation is as follows:

$$\log_{10}(\text{mean mass for subgroup})/\log_{10}(\text{LFC or LTC for subgroup}) = \log_{10}(\text{mass for fossil})/\log_{10}(\text{LFC or LTC of fossil})$$

(Eq. 9.1)

Several Santacrucian bird taxa are not represented by femora or tibiotarsi. In such cases, we used measurements taken from other bones and body masses of living analogs to predict the body weight of fossil forms, assuming constant proportions (i.e. isometry). Geometric similarity between a fossil species and its living analog implies that the length of homologous parts will be proportional to linear dimensions, and the surface will increase to the square and the volume to the cube.

The isometric weight estimation is thus calculated as:

$$\text{mass (fossil)}/\text{mass (living analog)} = (L\,\text{fossil}/L\,\text{living analog})^3$$

(Eq. 9.2)

where *L* is a linear dimension.

Accuracy of estimations using this method depends directly on the presence of allometric differences or similarities along the evolutionary lineages involved in each case.

9.3.3 Bite force estimation

Published *in vivo* bite force data for birds is largely restricted to passerines (van der Meij and Bout,

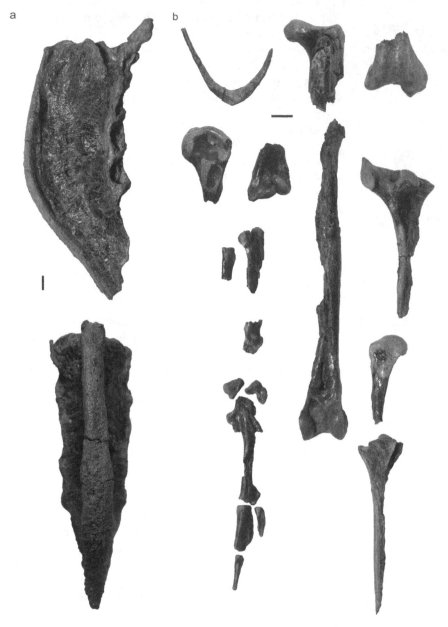

a

b

Fig. 9.5. Phorusrhacid specimens. a, MPM-PV 4242, sternum tentatively referred to *Patagornis marshi*; b, *Psilopterus bachmanni* MPM-PV4243, furcula, left forelimb, and left hindlimb. Scale bar = 1 cm.

2004, 2006). These studies show that bite force is related to skull morphology and geometry, as well as to the capacity of contraction of the jaw muscles (van der Meij and Bout, 2004; Herrel *et al.*, 2005a, b). It has been demonstrated that in Galápagos finches, beak size, and especially head width, are strongly correlated with bite force, and head size closely correlates with jaw muscle dimensions (Herrel *et al.*, 2001, 2002). However, passerines are a poor model for the estimation of bite force in phorusrhacids because of differences in

phylogeny, skull size, and morphology. Instead, we use the linear model presented by Degrange *et al.* (2010a) to estimate bite force of the Santacrucian phorusrhacids, *Brontornis*, and *Thegornis*. This model considers the relationship between bite force and body mass as follows:

$$\log_{10} \text{BF} = 1.3988 + 0.4541 \log_{10} \text{BM} \qquad \text{(Eq. 9.3)}$$

where BF is the bite force (in newtons) and BM is the body mass (in kilograms).

Table 9.1. *Allometric and isometric body mass and bite force estimations for selected Santacrucian birds*

| Species | Body mass (kg) | | | Bite force (N) |
| | Allometric calculation | | Isometric calculation | |
	Femur	Tibiotarsus		
Opisthodactylus patagonicus	–	–	56.95	–
Anisolornis excavatus	–	–	1.43	–
Brontornis burmeisteri	319	350	–	343
Liptornis hesternus	–	–	0.97	–
Protibis cnemialis	–	1.76	–	–
Thegornis musculosus	1.9	2.3	–	33.6
Badiostes patagonicus	–	–	0.217	–
Phorusrhacos longissimus	93	153	–	196
Patagornis marshi	26	34	–	110
Psilopterus lemoinei	8	13	–	64
Psilopterus bachmanni	4.5	–	–	49.5
Cariama santacrucensis	–	–	1.5	–

9.4 Results

9.4.1 Body mass estimations

Table 9.1 lists the body mass estimates of the taxa analyzed. The body mass predictions using Campbell and Marcus (1992) model revealed that phorusrhacids show a wide variation in body size. The Santacrucian species range from 4.5 kg in *Psilopterus bachmanni* (Psilopterinae) to 93 kg in *Phorusrhacos longissimus* (Phorusrhacinae), with intermediate weights of 26 kg exhibited by *Patagornis marshi* (Patagornithinae).

The estimations obtained by means of ratios between linear dimensions and body masses in extant analog forms are also presented in Table 9.1. We used the lengths of the tenth vertebral body (from *facies articularis cranialis* through *facies articularis caudalis*) as a scaling variable for *Liptornis hesternus* based on the extant *Anhinga anhinga* (21.5 mm; Alvarenga and Guilherme, 2003) and the average mass of *A. anhinga* (1.2 kg; Owre, 1967). In a similar way, the values of the greatest distal tarsometatarsal widths through trochleae and mean body weights in living *Rhea americana* and *Aramus guarauna* were used to estimate the body mass of *Opisthodactylus patagonicus* and *Anisolornis excavatus*. Likewise, the greatest width of the proximal end of the tarsometatarsus (5.75 mm) and the average body mass (0.125 kg) in the extant American Kestrel (*Falco sparverius*), together with the tarsometatarsal proximal end width (10 mm) of *Badiostes patagonicus*, allowed us to estimate the mass of the latter. Finally, the cranial width through the postorbital apophyses of *Cariama santacrucensis* (c. 50 mm), which is nearly identical to the same measurement in *Cariama cristata*, and the average weight of the latter (1.5 kg; Jutglar, 1992: 239) were employed to estimate the body mass of the former. The material

belonging to fossil waterfowl is too fragmentary and eroded to be useful in scaling. The same is true for *Thegornis debilis*.

9.4.2 Bite force estimations

Bite force estimations are summarized in Table 9.1. These forces were estimated based solely on the mass predictions using the femur.

While the results presented here show that bite force is correlated with body mass in birds ($R = 0.77912$, $R^2 = 0.607$, $P < 0.0001$), the extrapolations based on body mass for the phorusrhacids produced relatively low values. The lower values are found among the psilopterines and the higher values occur among phorusrhacines. The Santacrucian psilopterines show values between 49 (*P. bachmanni*) and 64 N (*P. lemoinei*). *Patagornis marshi* had a bite force of 110 N and *Phorusrhacos longissimus* had a bite force of 196 N. The bite of *Brontornis* (343 N) is difficult to interpret in the absence of more cranial data.

Thegornis musculosus (33.6 N) had a stronger bite than extant accipitrids, but weaker than extant falconids (Sustaita and Hertel, 2010).

9.5 Discussion

9.5.1 Paleobiology of the Santacrucian birds

Rheiformes *Opisthodactylus patagonicus* had proportions similar to modern rheas, although it was almost twice as heavy. It is probable that they were mainly grazers, roaming open grasslands, scrub forest, or chaparral in search of different vegetation items, as well as feeding on insects and small animals in the same fashion as living rheas (Folch, 1992).

Federico J. Degrange et al.

Fig. 9.6. Body shapes and sizes of some
representative Santacrucian birds.
Scale bar = 3 m.

Tinamiformes The Santacrucian tinamous appear to be related to the extant nothurine forms of open habitats (Bertelli and Chiappe, 2005). Tinamous are exclusively ground foragers, feeding on seeds, buds, roots, and insects. They prefer to run rapidly when threatened, but they can fly for up to 100 m when alarmed (Feduccia, 1996; Cabot, 1992; Davies, 2002).

"Gruiformes" *Anisolornis excavatus* represents a most interesting record from paleobiological and paleoenvironmental perspectives. This taxon has been considered to be a trumpeter (Psophiidae) or a limpkin (Aramidae) by different authors. Trumpeters, a relictual Amazonian family of medium-sized terrestrial birds, comprise five living species in the genus *Psophia* (Oppenheimer and Silveira, 2009). They have fairly long necks and legs, short, stout bills, and a hump-backed appearance. They are inhabitants of dense areas of tropical forest, feeding on fallen fruits and insects (Sherman, 1996). On the other hand, the living limpkin (*Aramus guarauna*) is the sole member of the family Aramidae, and it is thought to represent a primitive crane-like form (Olson, 1985; Feduccia, 1996). It is a medium to large wading bird about 70 cm in height, intermediate in size between true cranes and rails. Limpkins generally live in wetlands, preferring marshy wooded areas (Bryan, 1996). They are resident in open freshwater marshes, swamp forests bordering slow-moving rivers, lake or pond shores, and mangroves. They feed mainly on large freshwater snails that they obtain by walking in shallow water, rarely in deeper water.

Notwithstanding the controversy about the family allocation of *A. excavatus*, it is clear that the holotypical distal tarsometatarsus corresponds to a basal gruiform that was a medium-sized, long-legged bird with terrestrial habits, but perhaps with facultative wading abilities as well.

Anseriformes *Brontornis burmeisteri* is a bulky-bodied bird that surpassed 2 m in height (Fig. 9.6). It is characterized by its short, wide, and deep mandibular symphysis, and its short but wide tarsometatarsus which barely reaches half the length of the tibiotarsus (Alvarenga and Höfling, 2003). This feature and its heavy body were interpreted as characterizing a slow bird of scavenging habits (Tonni, 1977; Tambussi, 1997). However, based on the morphology of its jaw and its relationship with the anseriforms, Agnolín (2007) proposed that *B. burmeisteri* was herbivorous. A similar diet was also proposed for the giant Holarctic *Diatryma gigantea* Cope (Watson, 1976; Andors, 1988, 1992). Unfortunately, cranial remains of *Brontornis* are scarce and very fragmentary, making all assumptions of dietary habits speculative.

Santacrucian waterfowl are either of highly dubious identity or known only through very fragmentary material that does not provide any paleobiological information.

Pelecaniformes The size and predicted body mass of *Liptornis hesternus* are similar to those of extant *Anhinga anhinga*. Modern anhingas are freshwater birds that live and breed in forested borders of lakes, lagoons, or rivers, feeding exclusively on fish. All members of this family are foot-propelled swimmers, with very good diving capabilities. They dart their prey underwater with the beak. This mode of seizing prey is related to an adaptive specialization of the vertebral anatomy of their long necks. Despite its uncertain taxonomic status, there is no reason to suppose that *L. hesternus* was biologically different from extant species of darters when one considers the striking resemblance between the morphology of their vertebrae.

Ciconiiformes *Protibis cnemialis* could be classified as a wading bird owing to its affinities with plataleids (spoonbills) and its long and slender tibiotarsus. It is known that spoonbills feed mainly on aquatic insects and larvae, but also eat molluscs, crustaceans, small fish, tadpoles, and frogs. Although inferring the diet of *P. cnemialis* in detail without knowing its skull morphology is highly speculative, it is probable that *P. cnemialis* was a carnivorous predator on small prey items, as are most wading birds.

Falconiformes *Thegornis musculosus* and *Thegornis debilis* are medium-sized herpetotherines (Fig. 9.6), closely related to the living Laughing Falcon (*Herpetotheres cachinnans*), a snake-specialist predator (Skutch, 1999; DuVal *et al.*, 2006). Stout tarsi and short toes, typical features of snake-eating raptors (Bierregard, 1994), are present in *T. musculosus*. Given the anatomical similarities between *H. cachinnans* and *T. musculosus*, we might infer that the latter also could have fed preferentially upon snakes. Morphotype resemblance and a close phylogenetic relationship lead us to suspect a similar biology for both species. In the case of *Badiostes patagonicus*, a small falcon with a generalized raptorial morphotype, it is not possible to infer precise dietary habits.

Cariamiformes Living seriemas range over a variety of semi-open and dry landscapes in South America, such as the thorny scrub and other semi-arid woodland areas of the Brazilian "caatinga," the grassy savanna-like "cerrados," and the "monte" and "chaco" forests in Bolivia, Paraguay, and Argentina (Gonzaga, 1996). They are capable of performing occasional, but not sustained flights. Seriemas have diets consisting of a variety of invertebrates and small vertebrates. The skull, leg bone morphology, and body mass of *Cariama santacrucensis* (Fig. 9.6) are very similar to that of the extant forms, which would have allowed them to step on similar prey items while tearing them into pieces with the bill (Noriega *et al.*, 2009).

Phorusrhacids, together with marsupial carnivores, have been considered as the top predators of the Santacrucian fauna (Alvarenga and Höfling, 2003; Blanco and Jones, 2005; Chiappe and Bertelli, 2006; Bertelli *et al.*, 2007), dominating the Tertiary stages of South America in the absence of placental mammalian carnivores. This conclusion was reached primarily on the basis of skull morphology (especially the shape of the beak), large body size, reduced forelimbs, and slender hindlimbs.

Their bizarre skull morphology is characterized by the presence of a very developed lacrimal bone; the occipital region is very expanded and vertically oriented, providing an insertion area for strong neck muscles; the temporal fossae are large, deep, and approach the mid-sagittal plane of the skull, producing a sagittal crest. However, the most outstanding feature is their huge beak. Contrary to the vast majority of extant birds, the beaks of phorusrhacids are long, high, very mediolaterally compressed, completely hollow (Degrange *et al.*, 2010a), and endowed with a projecting hook as in raptors. The lower jaws of phorusrhacids are relatively narrow and slender, describing a slight sigmoid wave longitudinally, and with a very well-developed medial process. Phorusrhacids are characterized by the loss of cranial kinesis, an interpretation based on the absence (by secondary loss) of permissive linkages, or an intracranial *zonae flexoria* (Degrange *et al.*, 2010a). The absence of zones of flexibility indicates that the palate was immobile in all the subfamilies. These features, together with the intimate contact between a descending branch of the lacrimal bone and the jugal, convert the skull of the larger phorusrhacids into a very rigid structure (Degrange *et al.*, 2010a, b).

Traditionally, two functional groups of phorusrhacids have been recognized: non-volant cursorial predators (Phorusrhacinae, Mesembriornithinae, and Patagornithinae), and predators with some flying capability (Psilopterinae) (Tonni, 1977; Tambussi, 1997).

In the gigantic phorusrhacines and the slimmer medium-sized patagornithines (Fig. 9.6), the tarsometatarsus is gracile and its length reaches 70% of the tibiotarsus length (Alvarenga and Höfling, 2003). The elongated tibiotarsus and tarsometatarsus (relative to body size and femur length), the highly developed *cristae cnemialis*, and toes of moderate to short length are all features related to cursoriality (Engels, 1938; Storer, 1971; Coombs, 1978) present in the terror birds. They seem to indicate a great capability for running at high velocities and with high maneuverability, as proposed for *Andalgalornis steulleti* (Kraglievich, 1931) and *Patagornis marshi* which are very similar to those abilities observed in ostriches (Struthionidae) (Tambussi, 1997; Blanco and Jones, 2005). Moreover, the pelvis of patagornithines is narrow and has very elongated post- and preacetabular portions which give a mechanical advantage associated with a more effective origin and stronger development of the femoral retractor and protractor muscles. In addition, their ungual phalanges are robust, allowing them to restrain struggling prey with their feet (Jones, 2010). Collectively, the morphology of the leg and pelvis in phorusrhacids can be interpreted as adaptations for chasing and subduing prey.

The psilopterines are the smallest and most gracile phorusrhacids (Fig. 9.6). *Psilopterus lemoinei* only reached 90 cm in height and *P. bachmanni* 70 cm (Alvarenga and Höfling, 2003), and their hind limbs are more gracile than other phorusrhacids (Tambussi and Noriega, 1996). *Psilopterus lemoinei* had a skull with a deep maxilla, whereas that of *P. bachmanni* is shallower. Notably, the long and slim tarsometatarsus reaches 70 to 75% of the tibiotarsal length. Compared with the pelvis of *Patagornis marshi*, the pelvis of psilopterines are relatively wide, suggesting that they were somewhat less adapted for sustained and fast running than were the patagornithines.

Two predatory styles are traditionally proposed for the phorusrhacids. Some phorusrhacids might have used only their beak to subdue prey whereas others might have used their hindlimbs to also destabilize their prey during capture, killing, and dismemberment, or to gain access to brains and bone marrow (Blanco and Jones, 2005).

Regarding the manipulation of prey with the beak, previous reconstructions and biomechanical analyses of the jaw adductor musculature of *Andalgalornis steulleti* (Degrange, 2007, 2008) showed that the jaw apparatus of phorusrhacids was optimized for strength at the expense of speed. This is consistent with the loss of cranial kinesis, increased skull rigidity, and the increased area of attachment for the nucal muscles that phorusrhacids experienced during their evolutionary history (Degrange *et al.*, 2010a). Bite force will be increased with increasing stiffness of the jaw apparatus because muscle force is not attenuated at flexible areas (Wroe *et al.*, 2007, 2008). Other factors being equal, any bird without cranial kinesis can bite with 1.3 times the force of those with cranial flexion (Bout and Zweers, 2001). However, a further consequence of the loss of flexibility is an increased potential for catastrophic failure when an organism bites into unexpectedly resistant materials (Wroe *et al.*, 2007). In an akinetic skull, more bone, or bone of greater density, will be needed to maintain effective safety margins. In akinetic phorusrhacids such as patagornithines and phorusrhacines, the requisite increased bone mass was presumably tolerated because they were flightless and hence the advantages of ultra-light structures were lost (Degrange *et al.*, 2010a). However, the tall, narrow beak of phorusrhacids would have been incapable of resisting great stress if they were to shake their head from side to side. A finite elements analysis of *Andalgalornis steulleti* showed that the skull was optimized to resist rostrocaudally and dorsoventrally directed loads, but it was less able to resist laterally directed loadings without failure (Degrange *et al.*, 2010a). Taking into account that the morphology of the beak is very similar in the majority of phorusrhacids, it is probable that none of them could resist laterally directed loadings, especially the psilopterines with their very thin beak walls (Degrange and

Tambussi, 2008, 2011). This fact is in agreement with the lower bite force values calculated for the group (Table 9.1) because with higher bite forces the beak is more sensitive to fracture. It seems reasonable to expect that if phorusrhacids used their beaks to dispatch relatively large prey, then the strike must have been applied with considerable precision in order to avoid sustaining high lateral loads. *Patagornis marshi* probably would have consumed prey of small size, but larger phorusrhacids such as *Phorusrhacos longissimus*, with their stronger bite force, probably were able to kill and ingest larger prey using a technique of multiple well-targeted strikes in a repetitive attack-and-retreat strategy. On the other hand, the psilopterines, with the lowest values for bite force and more fragile beaks, were probably restricted in their carnivorous diet to small vertebrates and invertebrates, as in living cariamids.

It has been proposed that the forelimb reduction and the lack of the *processus acrocoracoideus* and *procoracoideus* of the coracoid in phorusrhacids (Alvarenga and Höfling, 2003) are related to the loss of flight (Alvarenga *et al.*, 2011), whereas the species belonging to *Psilopterus* were able to fly (Tonni and Tambussi, 1988; F. Degrange and C. Tambussi, unpublished data, 2011; Degrange, 2012). The flight of the latter birds was probably brief and clumsy, as in extant seriemas.

9.5.2 Paleoenvironmental inferences

The Santacrucian fossil record of birds is relatively meager, which makes it difficult to reconstruct the avifauna from a paleoecological point of view and then to infer the paleoenvironment (Tambussi, 2011). However, we might suggest that the paleoenvironment was not unlike that found in the humid Chaco today. Vegetated open habitat dominated by shrubs or grasslands harbored cursorial predators (phorusrhacids and seriemas) and grazers (rheas and tinamous), while limpkins and spoonbills frequented flooded areas. Ecotonal margins between grasslands with seasonal water bodies and forest edges, ponds or marshes within the closed forest areas, as well as gallery forest along riversides, are appropriate habitats for anhingids.

The Laughing Falcon, *Herpetotheres cachinnans*, the living herpetotherine species considered to be the sister-taxon of the Santacrucian *Thegornis musculosus*, is common in tropical and subtropical zones of South America at forest edges, open forests, and mixed palm savanna and forest habitats, always being observed near large clearings if nesting in closed primary forests (Noriega *et al.*, 2011). Taking into account the habitat preferences mentioned above and the close phylogenetic relationship between both taxa, it is plausible to think that *Thegornis musculosus* was probably a forest dweller at the ecotonal margins of the gradually vanishing wet and humid Santacrucian forests that have been

Fig. 9.7. Life restoration of the Santacrucian birds. 1, *Liptornis hesternus*; 2, *Eutelornis patagonicus*; 3, *Protibis cnemialis*; 4, *Anisolornis excavatus*; 5, *Psilopterus bachmanni*; 6, *Psilopterus lemoinei*; 7, *Thegornis musculosus*; 8, Tinamidae; 9, *Badiostes patagonicus*; 10, *Cariama santacrucensis*; 11, *Brontornis burmeisteri*; 12, *Phorusrhacos longissimus*; 13, *Patagornis marshi*; 14, *Opistodactylus patagonicus*; 15, *Eoneornis australis*. Drawings by F. J. Degrange.

inferred from paleobotanical data (Barreda y Palazzesi, 2007; Noriega *et al*., 2011; Brea *et al*., Chapter 7).

In sum, the paleoenvironmental interpretation of a mixture of open and relatively closed vegetation in relatively dry conditions proposed by Tauber (1997a, b), and reinforced by Vizcaíno *et al*. (2006) based on the diversity of the Santacrucian armadillos, is consistent with the avian fossil record (Fig. 9.7).

9.6   Conclusions

1. In the absence of placental carnivores, a number of birds represented by four species of phorusrhacids and a seriema, along with several groups of marsupials, occupied the large-predator niche in the Santacrucian paleocommunity. The avian small predator niche was occupied by falcons, darters, limpkins, and spoonbills. The first are

classified as terrestrial raptors, darters are exclusively freshwater predators, and the latter are included among the wading birds.

2. The omnivorous niche is represented by the rheid *Opisthodactylus*. Tinamous occupied the herbivorous niche, probably together with *Brontornis*.

3. Medium- to large-sized phorusrhacids (Phorusrhacinae and Patagornithinae) are thought to have preferred open habitats because of their cursorial mode of locomotion.

4. Waterfowl, limpkins, spoonbills, and darters indicate the presence of seasonally flooded areas, marshes, or permanent water bodies in more forested areas.

5. Habitat preferences of living seriemas, rheas, some tinamous, and *Herpetotheres*, all analogs of Santacrucian taxa discussed here, are consistent with the hypothesis of a Chacoan-like scenario for the Santacrucian

paleoenvironment, the latter being characterized by open vegetation alternating with bushy or wooded areas.

ACKNOWLEDGMENTS
We thank Sandra Chapman for photographs of the materials from BMNH, and we thank Jim Holstein (FMNH), Carl Mehling (AMNH), Christopher Norris, and Dan Brinkman (YPMPU) for access to collections in their care. We thank the editors for the invitation to participate. The first author thanks Claudia Tambussi for her constant support and advice. Thanks to the anonymous reviewers and the editors whose observations improved the manuscript. Part of this study was supported by FONCYT-PICT grant 32617 and PICT 2007–392. This is a contribution to the grants PICT 26219, 0143 to Sergio Vizcaíno and NSF 0851272, 0824546 to Richard Kay.

Appendix 9.1   List of the material studied

Anseriformes

Incertae familiae

Brontornis burmeisteri

MLP 20–88, left femur and tibiotarsus. Horizon and locality: Santa Cruz Formation, Lago Argentino, Santa Cruz Province.

MLP 20–91, left tarsometatarsus. Horizon and locality: Santa Cruz Formation, Lago Argentino, Santa Cruz Province.

Eoneornis australis

BMNH-A595, distal end of radius. Horizon and locality: Santa Cruz Formation, Santa Cruz Province.

Eutelornis patagonicus

BMNH-A596, distal end of humerus and a proximal tibiotarsus. Horizon and locality: Santa Cruz Formation, Santa Cruz Province.

Cariamiformes

Cariamidae

Cariama santacrucensis

MPM-PV 3511, incomplete cranium. Horizon and locality: Santa Cruz Formation, Estancia La Costa Member, Fossiliferous Level 5.3, Puesto Estancia La Costa, Santa Cruz Province.

Cariaminae indet.

MPM-PV 3510 and 3512, two unassociated fragments of tibiotarsi. Horizon and locality: Santa Cruz Formation, Estancia La Costa Member, Fossiliferous Levels 6 (MPM-PV 3510) and 5.3 (MPM-PV 3512), Puesto Estancia La Costa, Santa Cruz Province.

Phorusrhacidae

Patagornis marshi

BMNH-A516, skull, mandible, right coracoid, right scapula, distal end of left humerus, right ulna, part of the right radius, both carpometacarpi, pelvis, both femur, right tibiotarsus, and right tarsometatarsus. Horizon and locality: Santa Cruz Formation, Santa Cruz Province; MPM-PV 4242, sternum. Horizon and locality: Santa Cruz Formation, Estancia La Costa Member, Fossiliferous Level 6, Puesto Estancia La Costa, Santa Cruz Province.

Phorusrhacos longissimus

MPM-PV 4241, mandible, left femur, tibiotarsus, and both fibulae. Horizon and locality: Santa Cruz Formation, Estancia La Costa Member, Campo Barranca, Santa Cruz Province.

Psilopterus bachmanni

MPM-PV 4243, furcula and fragments of the left fore and hindlimb. Horizon and locality: Santa Cruz Formation, Estancia La Costa Member, Fossiliferous Level 5.3, Puesto Estancia La Costa, Santa Cruz Province; YPM-VPPU 15904, an almost complete skeleton. Horizon and locality: Santa Cruz Formation, Lago Pueyrredón, Santa Cruz Province.

Psilopterus lemoinei

AMNH 9257, skull, humerus, vertebral column, pelvis and left hindlimb. Horizon and locality: Santa Cruz Formation, Estancia Halliday, Santa Cruz Province; MPM-PV 4240, left tibiotarsus. Horizon and locality: Santa Cruz Formation, Estancia La Costa Member, Anfiteatro, Santa Cruz Province.

Ciconiiformes

Incertae familiae

Protibis cnemialis

BMNH-A 598, distal end of right tibiotarsus. Horizon and locality: Santa Cruz Formation, Santa Cruz Province.

Falconiformes

Falconidae

Badiostes patagonicus

BMNH-A 602, very damaged and fragmentary specimen including ulna, femur and tarsometatarsus. Horizon and locality: Santa Cruz Formation, La Cueva, Santa Cruz Province.

Thegornis debilis

BMNH-A 601, distal end of a fragmentary right tarsometatarsus. Horizon and locality: Santa Cruz Formation, Corriguen Aike, Santa Cruz Province.

Thegornis musculosus

BMNH-A 600, fragmentary right tarsometatarsus. Horizon and locality: Santa Cruz Formation, Yegua Quemada, Santa Cruz Province; MPM-PV 3433, an almost complete skeleton. Horizon and locality: Santa Cruz Formation, Estancia La Costa member, Fossiliferous Level 6, Puesto Estancia La Costa, Santa Cruz Province.

"Gruiformes"

Incertae familiae

Anisolornis excavatus

BMNH-A594, distal end of left tarsometatarsus. Horizon and locality: Santa Cruz Formation, Santa Cruz Province.

Pelecaniformes

Anhingidae

Liptornis hesternus

BMNH-A599, cervical vertebra. Horizon and locality: Santa Cruz Formation, Santa Cruz Province; FMNH-PA 22, cast of BMNH-A599.

Rheiformes

Rheidae

Opisthodactylus patagonicus

BMNH-A586, distal ends of both tarsometatarsi and phalanges. Horizon and locality: Santa Cruz Formation, Santa Cruz Province; BMNH-A587, mandibular symphysis, distal end of left tibiotarsus, fragmentary distal end of right tarsometatarsus. Horizon and locality: Santa Cruz Formation, Santa Cruz Province; BMNH-A588, fragments of distal tarsometatarsus. Horizon and locality: Santa Cruz Formation, Santa Cruz Province.

REFERENCES

Agnolín, F. L. (2006a). Posición sistemática de algunas aves fororracoideas (Ralliformes; Cariamae) Argentinas. *Revista del Museo de Ciencias Naturales*, **8**, 27–33.

Agnolín, F. L. (2006b). Notas sobre el registro de Accipitridae (Aves, Accipitriformes) fósiles argentinos. *Studia Geologica Salmanticensia*, **42**, 67–80.

Agnolín, F. L. (2007). *Brontornis burmeisteri* Moreno y Mercerat, un Anseriformes (Aves) gigante del Mioceno Medio de Patagonia, Argentina. *Revista del Museo Argentino de Ciencias Naturales*, **9**, 15–25.

Agnolín, F. L. (2009). *Sistemática y filogenia de las aves fororracoideas (Gruiformes: Cariamae)*. Buenos Aires: Monografías Fundación Azara/Adrián Giacchino.

Alvarenga, H. M. F. (1985). Um novo Psilopteridae (Aves: Gruiformes) dos sedimentos Terciários de Itaboraí, Rio de Janeiro, Brasil. *Anais do Congresso Brasileiro de Paleontologia, 8, Série Geologia*, **27**, 17–20.

Alvarenga, H. M. F. (1995). A large and probably flightless anhinga from the Miocene of Chile. *Courier Forschungsinstitut Senckenberg*, **181**, 149–161.

Alvarenga, H. M. F. and Guilherme, E. (2003). The anhingas (Aves: Anhingidae) from the Upper Tertiary (Miocene–Pliocene) of southwestern Amazonia. *Journal of Vertebrate Paleontology*, **23**, 614–621.

Alvarenga, H. M. F. and Höfling, E. (2003). Systematic revision of the Phorusrhacidae (Aves: Ralliformes). *Papeis Avulsos de Zoologia, Museu de Zoologia da Universidade de Sao Paulo*, **43**, 55–91.

Alvarenga, H. M. F., Jones, W. W. and Rinderknecht, A. (2010). The youngest record of phorusrhacid birds (Aves, Phorusrhacidae) from the late Pleistocene of Uruguay.

Neues Jahrbuch für Geologie und Paläontologie, Abhandlungen, **256**, 229–234.

Alvarenga, H. M. F., Bertelli, S. and Chiappe, L. M. (2011). Phorusrhacids: the terror birds. In *Living Dinosaurs: The Evolutionary History of Modern Birds*, ed. G. Dyke and G. Kaiser. New York: John Wiley and Sons, pp. 187–208.

Ameghino, F. (1887). Enumeración sistemática de las espécies de mamíferos fósiles coleccionados por Carlos Ameghino en los terrenos Eocenos de la Patagonia austral y depositados en el Museo de La Plata. *Boletín del Museo de La Plata*, **1**, 1–26.

Ameghino, F. (1889). Contribución al conocimiento de los mamíferos fósiles de la República Argentina. *Actas Academia Nacional de Ciencias de Córdoba*, **6**, 1–1028.

Ameghino, F. (1895). Sobre las aves fósiles de Patagonia. *Boletín del Instituto Geográfico de Argentina*, **15**, 501–602.

Ameghino, F. (1906). Les formations sédimentaires du Crétacé superieur et du Tertiaire de Patagonie, avec un parallèle entre leurs faunes mammalogiques et celles de l'ancient continent. *Anales del Museo Nacional de Historia Natural de Buenos Aires*, **3**, 1–568.

Andors, V. A. (1988). Giant groundbirds of North America (Aves, Diatrymidae). Unpublished Ph.D. thesis, Columbia University, New York.

Andors, V. A. (1992). Reappraisal of the Eocene groundbird *Diatryma* (Aves, Anserimorphae). *Science Series Natural History Museum of Los Angeles County*, **36**, 109–126.

Andrews, C. (1899). On the extinct birds of Patagonia. I. The skull and skeleton of *Phororhacos inflatus* Ameghino. *Transactions of the Zoological Society of London*, **15**, 55–86.

Areta, J. I., Noriega, J. I. and Agnolín, F. L. (2007). A giant darter (Pelecaniformes: Anhingidae) from the Upper Miocene of Argentina and weight calculation of fossil Anhingidae. *Neues Jahrbuch für Mineralogie, Geologie und Paleontologie*, **243**, 343–350.

Barreda, V. and Palazzesi, L. (2007). Patagonian vegetation turnovers during the Paleogene–Early Neogene: origin of arid adapted floras. *The Botanical Review*, **73**, 31–50.

Barrio, R. E., Scillato-Yane, G. and Bown, M. (1984). La Formación Santa Cruz en el borde occidental del macizo del Deseado (Provincia de Santa Cruz) y su contenido paleontológico. *Actas del Congreso Geológico Argentino*, **1**, 539–556.

Bertelli, S. and Chiappe, L. M. (2005). Earliest Tinamous (Aves: Palaeognathae) from the Miocene of Argentina and their phylogenetic position. *Contributions in Science*, **502**, 1–20.

Bertelli, S., Chiappe, L. M. and Tambussi, C. (2007). A new phorusrhacid (Aves, Cariamae) from the middle Miocene of Patagonia, Argentina. *Journal of Vertebrate Paleontology*, **27**, 409–419.

Bierregard, R. O. (1994). Family Accipitridae (Hawks and Eagles). In *Handbook of the Birds of the World*. Vol. 2, ed. J. D. Del Hoyo, A. Elliott and J. Sargatal. Barcelona: Lynx Edicions, pp. 52–205.

Blanco, R. E. and Jones, W. W. (2005). Terror birds on the run: a mechanical model to estimate its maximum running speed. *Proceedings of Royal Society B*, **272**, 1769–1773.

Bout, R. G. and Zweers, G. A. (2001). The role of cranial kinesis in birds. *Comparative Biochemistry and Physiology A*, **131**, 197–205.

Brodkorb, P. (1963). Catalogue of fossil birds. Part I. Archaeopterygiformes through Ardeiformes). *Bulletin of Florida State Museum*, **7**, 179–293.

Brodkorb, P. (1964). Catalogue of fossil birds. Part II. Anseriformes through Galliformes. *Bulletin of Florida State Museum*, **8**, 195–335.

Brodkorb, P. (1967). Catalogue of fossil birds. Part III. Ralliformes, Ichthyornithiformes, Charadriiformes. *Bulletin of Florida State Museum*, **2** (3), 99–220.

Bryan, D. C. (1996). Family Aramidae (Limpkin). In *Handbook of the Birds of the World*. Vol. 3, ed. J. D. Del Hoyo, A. Elliott and J. Sargatal. Barcelona: Lynx Edicions, pp. 90–95.

Cabot, J. (1992). Order Tinamiformes. In *Handbook of the Birds of the World*. Vol. 1, ed. J. D. Del Hoyo, A. Elliott and J. Sargatal. Barcelona: Lynx Edicions, pp. 112–138.

Campbell, K. E. Jr. (1996). A new species of giant Anhinga (Aves: Pelecaniformes: Anhingidae) from the upper Miocene (Huayquerian) of Amazonian Perú. *Contributions in Science*, **460**, 1–9.

Campbell, K. E. Jr. and Marcus, L. (1992). The relationship of hindlimb bone dimensions to body weight in birds. *Natural History Museum of Los Angeles County, Science Series*, **36**, 395–412.

Campbell, K. E. Jr. and Tonni, E. P. (1983). Size and locomotion in teratorns (Aves, Theratornithidae). *The Auk*, **100**, 390–403.

Cenizo, M. M. and Agnolín, F. L. (2010). The southernmost records of Anhingidae and a new basal species of Anatidae (Aves) from the lower–middle Miocene of Patagonia, Argentina. *Alcheringa*, **34**, 493–514.

Chiappe, L. M. (1991). Fossil birds from the Miocene Pinturas Formation of Southern Argentina. *Journal of Vertebrate Paleontology*, **11**, 21–22A.

Chiappe, L. M. and Bertelli, S. (2006). Skull morphology of giant terror birds. *Nature*, **443**, 929.

Coombs, W. P. (1978). Theoretical aspects of cursorial adaptations in Dinosaurs. *The Quarterly Review of Biology*, **53**, 393–418.

Cracraft, J. (1973). The systematics and evolution of the Gruiformes (Class Aves). 3. Phylogeny of the suborder Grues. *Bulletin American Museum Natural History*, **151**, 1–127.

Davies, S. J. (2002). *Ratites and Tinamous*. New York: Oxford University Press.

Degrange, F. J. (2007). Mandible muscle reconstruction in Phorusrhacid birds (Aves, Gruiformes, Cariamae). *Ciencias Morfológicas*, **9**, 36.

Degrange, F. J. (2008). M. adductor mandibulae externus de *Andalgalornis steulleti* (Aves, Phorusrhacidae): Reconstrucción y biomecánica. *III Congreso Latinoamericano de Paleontología de Vertebrados, Actas*, 76.

Degrange, F. J. (2012) Morfología del cráneo y complejo apendicular posterior de aves fororracoideas: implicancias en la dieta y modo de vida. Unpublished Ph.D. thesis, Universidad de La Plata, Argentina, pp. 1–390.

Degrange, F. J. and Tambussi, C. P. (2008). Estructura interna del pico de los Psilopterinae (Aves, Gruiformes, Phorusrhacidae). *Journal of Morphology*, **26** (3), 746.

Degrange, F. J. and Tambussi, C. P. (2011). Re-examination of *Psilopterus lemoinei* (Moreno and Mercerat 1891), a late early Miocene little terror bird from Patagonia (Argentina). *Journal of Vertebrate Paleontology*, **31**, 1080–1092.

Degrange, F. J., Tambussi, C. P., Moreno, K., Witmer, L. W. and Wroe, S. (2010a). Mechanical analysis of feeding behavior in the extinct "terror bird" *Andalgalornis steulleti* (Gruiformes: Phorusrhacidae). *PLoS ONE*, **5**, e11856.

Degrange, F. J., Tambussi, C. P., Jones, W. W. and Blanco, E. R. (2010b). Fororracos (Aves, Paleoceno–Pleistoceno): pérdida de quinesis craneana e implicancias funcionales. *X Congreso Argentino de Paleontología y Bioestratigrafía y VII Congreso Latinoamericano de Paleontología, Actas*, 156–157.

Dolgopol de Sáez, M. (1927). Las aves corredoras fósiles del Santacrucense. *Anales de la Sociedad Científica Argentina*, **103**, 145–64.

DuVal, E. H., Greene, H. W. and Manno, K. L. (2006). Laughing Falcon (*Herpetotheres cachinnans*) predation on coral snakes (*Micrurus nigrocinctus*). *Biotropica*, **38**, 566–568.

Engels, W. L. (1938) Cursorial adaptations in birds. Limb proportions in the skeleton of *Geococcyx*. *Journal of Morphology*, **63**, 207–217.

Feduccia, A. (1996). *The origin and evolution of birds*. New Haven and London: Yale University Press.

Fleagle, J. G., Bown, T. M., Swisher, C. C. III and Buckley, G. A. (1995). Age of the Pinturas and Santa Cruz Formations. *VI Congreso Argentino de Paleontología y Bioestratigrafia, Actas*, 129–135.

Folch, A. (1992). Family Rheidae (Rheas). In *Handbook of the Birds of the World*. Vol. 1, ed. J. D. Del Hoyo, A. Elliott and J. Sargatal. Barcelona: Lynx Edicions, pp. 84–89.

Frenguelli, J. (1931). Nomenclatura estratigráfica Patagónica. *Anales Sociedad Ciencias Santa Fe*, **3**, 1–115.

Gonzaga, L. P. (1996). Family Cariamidae (Seriemas). In *Handbook of the Birds of the World*. Vol. 3, ed. J. D. Del Hoyo, A. Elliott and J. Sargatal. Barcelona: Lynx Edicions, pp. 234–239.

Hatcher, J. B. (1903). Narrative of the expedition. In *Reports of the Princeton University Expeditions to Patagonia, 1896–1899*. Vol. 1, *Narrative and Geography*, ed. W. B. Scott. Princeton: Princeton University Press, pp. 1–314.

Herrel, A., Van Damme, R., Vanhooydonck, B. and De Vree, F. (2001). The implications of bite performance for diet in two species of lacertid lizards. *Canadian Journal of Zoology*, **79**, 662–670.

Herrel, A., O'Reilly, J. C. and Richmond, A. M. (2002). Evolution of bite force in turtles. *Journal of Evolutionary Biology*, **15**, 1083–1094.

Herrel, A., Podos, J., Huber, S. K. and Hendry, A. P. (2005a). Bite performance and morphology in a population of Darwin's finches: implications for the evolution of beak shape. *Functional Ecology*, **19**, 43–48.

Herrel, A., Podos, J., Huber, S. K. and Hendry, A. P. (2005b). Evolution of bite force in Darwin's finches: a key role for head width. *Journal of Evolutionary Biology*, **18**, 669–675.

Jones, W. W. (2010). Nuevos aportes sobre la paleobiología de los fororrácidos (Aves: Phorusrhacidae) basados en el análisis de estructuras biológicas. Unpublished Ph.D. thesis, PEDECIBA – Universidad de la República, Uruguay.

Jutglar, F. (1992). Family Cariamidae (Seriemas). In *Handbook of the Birds of the World*. Vol. 1, ed. J. D. Del Hoyo, A. Elliott and J. Sargatal. Barcelona: Lynx Edicions, pp. 234–239.

Marshall, L. G. (1976). Fossil localities for Santacrucian (early Miocene) Mammals, Santa Cruz Province, Southern Patagonia, Argentina. *Journal of Paleontology*, **50**, 1129–1142.

Moreno, F. P. and Mercerat, A. (1891). Catálogo de los pájaros fósiles de la República Argentina conservados en el Museo de La Plata. *Anales del Museo de La Plata*, **1**, 7–71.

Mourer-Chauviré, C. (1981). Première indication de la présence de phorusrhacidés, famille d'oiseaux géants d'Amérique du Sud, dans le Tertiare européen: *Ameghinornis* nov. gen. (Aves, Ralliformes) des phosphorites du Quercy, France. *Geobios*, **14**, 637–647.

Noriega, J. I. (1995). The avifauna from the "Mesopotamian" (Ituzaingó Formation; Upper Miocene) of Entre Rios Province, Argentina. *Courier Forschungsinstitut Senckenberg*, **181**, 141–148.

Noriega, J. I. (2002). Additional material of *Macranhinga paranensis* Noriega 1992 (Aves: Pelecaniformes: Anhingidae) from the "Mesopotamian" (Ituzaingó Formation; Upper Miocene) of Entre Rios Province, Argentina. In *Proceedings of the 5th International Meeting of the Society of Avian Paoleontology and Evolution*, ed. Z. Zhou and F. Zhang. Beijing: China Science Press, 51–61.

Noriega, J. I. and Agnolín, F. L. (2008). El registro paleontológico de las aves del 'Mesopotamiense' (Formación Ituzaingó; Mioceno tardío-Plioceno) de la provincia de Entre Ríos, Argentina. *INSUGEO Serie Correlación Geológica*, **17**, 123–142.

Noriega, J. and Chiappe, L. M. (1993). An early passeriform from Argentina. *The Auk*, **114**, 936–938.

Noriega, J. I., Areta, J. I., Vizcaíno, S. F. and Bargo, M. S. (2008). Reassessment of *Thegornis musculosus* Ameghino 1894 (Aves: Falconidae) based on new material recovered from Santacrucian (Early–Middle Miocene) beds of Patagonia. *Ameghiniana*, **45** (4, Suppl.), 30R.

Noriega, J. I., Vizcaíno, S. F. and Bargo, M. S. (2009). First record and a new species of seriema (Aves: Ralliformes: Cariamidae) from Santacrucian (Early–Middle Miocene) beds of Patagonia. *Journal of Vertebrate Paleontology*, **29**, 620–626.

Noriega, J. I., Areta, J. I., Vizcaíno, S. F. and Bargo, M. S. (2011). Phylogeny and taxonomy of the Patagonian Miocene Falcon *Thegornis musculosus* Ameghino, 1895 (Aves: Falconidae). *Journal of Paleontology*, **85**, 1089–1104.

Olson, S. (1981). The generic allocation of *Ibis pagana* Milne-Edwards, with a review of fossil ibises (Aves: Threskiornithidae). *Journal of Vertebrate Paleontology*, **1**, 165–170.

Olson, S. (1985). The fossil record of birds. In *Avian Biology*. Vol. 8, ed. D. Farner, J. King and K. Parkes. New York: Academic Press, pp. 79–252.

Oppenheimer, M. and Silveira, L. F. (2009). A taxonomic review of the Dark-winged Trumpeter *Psophia viridis* (Aves: Gruiformes: Psophiidae). *Papeis Avulsos de Zoologia, São Paulo*, **49**, 547–555.

Owre, O. T. (1967). Adaptations for locomotion and feeding in the anhinga and the double-crested cormorant. *Ornithological Monographs*, **6**, 1–138.

Pascual, R. and Odreman-Rivas, O. E. (1971). Evolución de las comunidades de vertebrados del Terciario Argentino, los aspectos paleozoogeográficos y paleoclimáticos relacionados. *Ameghiniana*, **8**, 372–412.

Patterson, B. and Kraglievich, L. (1960). Sistemática y nomenclatura de las aves fororracoideas del Plioceno Argentino. *Publicación del Museo Municipal Ciencias Naturales y Tradicionales de Mar del Plata*, **1**, 1–51.

Rasmussen, D. T. and Kay, R. F. (1992). A Miocene *Anhinga* from Colombia, and comments on the zoogeographic relationships of South America's Tertiary avifanua. In *Avian Paleontology*, ed. K. E. Campell Jr. Los Angeles: Natural History Museum of Los Angeles County, pp. 225–230.

Ré, G. H., Bellosi, E. S., Heizler, M. *et al.* (2010). A geochronolgy for the Sarmiento Formation at Gran Barranca. In *The Paleontology of Gran Barranca: Evolutional and Environmental Change through the Middle Cenozoic of Patagonia*, ed. R. H. Madden, A. A. Carlini, M. G. Vucetich and R. F. Kay. New York: Cambridge University Press, pp. 46–58.

Rinderknecht, A. and Noriega, J. I. (2002). Un Nuevo género de Anhingidae (Aves: Pelecaniformes) del Plioceno– Pleistoceno de Uruguay (Formación San José). *Ameghiniana*, **39**, 183–192.

Sherman, P. T. (1996). Family Psophiidae (Trumpeters). In *Handbook of the Birds of the World*. Vol. 3, ed. J. D. Del Hoyo, A. Elliott and J. Sargatal. Barcelona: Lynx Edicions, pp. 96–107.

Sinclair, W. and Farr, M. (1932). Aves of the Santa Cruz beds. In *Reports of the Princeton University Expeditions to Patagonia (1896–1899)*. Vol. 7, *Paleontology IV*, ed. W. B. Scott. Princeton: Princeton University Press, pp. 157–191.

Skutch, A. F. (1999). *Trogons, Laughing Falcons, and Other Neotropical Birds*. College Station, Texas: Texas A&M University Press.

Storer, R. W. (1971). Adaptative radiation in birds. In *Avian Biology*, ed. D. S. Farner and J. R. King. New York: Academic Press, pp. 149–188.

Sustaita, D. and Hertel, F. (2010). *In vivo* bite and grip forces, morphology and prey-killing behavior of North American accipiters (Accipitridae) and falcons (Falconidae). *The Journal of Experimental Biology*, **213**, 2617–2628.

Tambussi, C. P. (1995). The fossil Rheiformes from Argentina. *Courier Forchungs-institut Senckenberg*, **181**, 121–130.

Tambussi, C. P. (1997). Algunos aspectos biomecánicos de la locomoción de los fororracos (Gruiformes). *Ameghiniana*, **34**, 541.

Tambussi, C. P. (2011). Paleoenvironmental and faunal inferences based upon the avian fossil record of Patagonia and Pampa: what works and what does not. *Biological Journal of the Linnean Society*, **103**, 458–474.

Tambussi, C. P. and Noriega, J. I. (1996). Summary of the avian fossil record from southern South America. In *Contributions of the southern South America to vertebrate paleontology*, ed. G. Arratia. Munich: Müncher Geowissenschaftliche Abhandlungen, pp. 245–264.

Tauber, A. A. (1997a). Bioestratigrafía de la Formación Santa Cruz (Mioceno inferior) en el extremo sudeste de la Patagonia. *Ameghiniana*, **34**, 413–426.

Tauber, A. A. (1997b). Paleoecología de la Formación Santa Cruz (Mioceno inferior) en el extremo sudeste de la Patagonia. *Ameghiniana*, **34**, 517–529.

Tonni, E. P. (1977). El rol ecológico de algunas aves fororracoideas. *Ameghiniana*, **14**, 316.

Tonni, E. P. (1980). The present state of knowledge of the cenozoic birds of Argentina. *Contributions in Science. Natural History Museum of Los Angeles County*, **330**, 105–114.

Tonni, E. P. and Tambussi, C. P. (1986). Las aves del cenozoico de la República Argentina. *IV Congreso Argentino de Paleontología y Bioestratigrafía, Actas* **2**, 131–142.

Tonni, E. P. and Tambussi, C. P. (1988). Un nuevo Psilopterinae (Aves: Ralliformes) del Mioceno tardío de la provincia de Buenos Aires, República Argentina. *Ameghiniana*, **25**, 155–160.

van der Meij, M. A. A. and Bout, R. G. (2004). Scaling of jaw muscle size and maximal bite force in finches. *Journal of Experimental Biology*, **207**, 2745–2753.

van der Meij, M. A. A. and Bout, R. G. (2006). Seed husking time and maximal bite forces in finches. *Journal of Experimental Biology*, **209**, 3329–3335.

Vizcaíno, S. F., Bargo, M. S., Kay, R. F. and Milne, N. (2006). The armadillos (Mammalia, Xenarthra, Dasypodidae) of the Santa Cruz Formation (early–middle Miocene): An approach to their paleobiology. *Palaeogeography, Palaeoclimatology, Palaeoecology*, **237**, 255–269.

Vizcaíno, S. F., Bargo, M. S., Kay, R. F. *et al.* (2010). A baseline paleoecological study for the Santa Cruz Formation (late-early Miocene) at the atlantic coast of Patagonia, Argentina. *Palaeogeography, Palaeoclimatology, Palaeoecology*, **292**, 507–519.

Watson, G. E. (1976). . . .And birds took wing. In *Our Continent, A Natural History of North America*, ed. E. H Colbert and S. L. Fishbein. Washington, D.C.: National Geographic Society, pp. 98–107.

Worthy, T. H. (2008). Tertiary fossil waterfowl (Aves: Anseriformes) from Australia and New Zealand. Unpublished Ph.D. thesis, School of Earth and Environmental Sciences, University of Adelaide, Australia, pp. 1–407.

Wroe, S., Moreno, K., Clausen, P. D. and McHenry, C. R. (2007). High-resolution computer simulation of hominid cranial mechanics. *The Anatomical Record*, **290**, 1248–1255.

Wroe, S., Huber, D. R., Lowry, M. *et al.* (2008). Three-dimensional computer analysis of white shark jaw mechanics: how hard can a great white bite? *Journal of Zoology*, **276**, 336–342.

10 Paleoecology of the Paucituberculata and Microbiotheria (Mammalia, Marsupialia) from the late Early Miocene of Patagonia

María Alejandra Abello, Edgardo Ortiz-Jaureguizar, and Adriana M. Candela

Abstract

This chapter presents a paleoecological analysis of non-carnivorous Santacrucian marsupials of the orders Paucituberculata and Microbiotheria. Different ecological niches are inferred from estimates of body mass, diet, and locomotor behavior. Body masses were estimated using a regression analysis based on living marsupials. Possible dietary preferences were explored by an analysis of the development of molar shearing crests. Inferences about locomotor behaviors of some species with well-preserved postcranial skeletal remains were derived from a published morphofunctional analysis. From the wide range of estimated body masses and diet several niches were inferred: small- to medium-sized insectivores, small- to medium-sized insectivore–frugivores, and medium- to large-sized frugivores. According to our results, Paucituberculata and Microbiotheria of the Santa Cruz Formation constitute an ecologically diverse assemblage that inhabited forested habitats, developed under warm temperatures and seasonal rainfall. These forested habitats could have supported several non-carnivorous marsupial niches, offering diverse resources both in the spatial dimensions and in the trophic ones.

Resumen

En este capítulo se presenta un análisis paleoecológico de los marsupiales no carnívoros del Santacrucense pertenecientes a los órdenes Paucituberculata y Microbiotheria. Para establecer los distintos nichos ecológicos se analizaron las masas corporales, dietas y estrategias locomotoras. El tamaño corporal fue estimado a partir de una recta de regresión linear obtenida a partir de marsupiales actuales. Las posibles preferencias dietarias fueron exploradas mediante un análisis del grado de desarrollo de las crestas cortantes de los molares. Las estrategias locomotoras, en el caso de aquellas especies que poseen un esqueleto postcraneano bien preservado, fueron inferidas a partir de un análisis morfofuncional. A partir del

análisis de los resultados de las masas corporales y la dietas, se reconocieron diversos nichos ecológicos: pequeños a medianos insectívoros, pequeños a medianos insectívoros-frugívoros y medianos a grandes frugívoros. Nuestros resultados nos permiten concluir que los Paucituberculata y los Microbiotheria de la Formación Santa Cruz constituyen un agregado ecológicamente diverso que habitó zonas boscosas, desarrolladas bajo un clima cálido y con lluvias estacionales. Estas zonas habrían ofrecido una amplia diversidad de recursos tanto espaciales como tróficos para los diversos nichos ecológicos de los marsupiales no carnívoros. Nuestra reconstrucción paleoambiental es compatible con la existencia de heterogeneidad ambiental durante el Santacrucense, inferencia esta derivada de otros indicadores climático-ambientales.

10.1 Introduction

During the Cenozoic, a high diversity of metatherians occupied a broad range of ecological niches in South America. This array included medium to large carnivorous and carnivorous–omnivorous (i.e. Sparassodonta; Prevosti *et al.*, Chapter 11), small granivorous (e.g. polydolopimorphian Argyrolagoidea; Goin *et al.*, in press), and several small to medium insectivorous and insectivorous–frugivorous species (e.g. didelphimorphian Didelphidae, Paucituberculata; Goin *et al.*, in press, Dumont *et al.*, 2000). All of the larger taxa are now extinct, and a few of the small- to medium-sized insectivorous, insectivorous–frugivorous, and carnivorous taxa (microbiotherians, paucituberculatans, and didelphids) survive to the present.

Paucituberculata and Microbiotheria are two major Marsupialia clades (Fig. 10.1). As a result of all current phylogenies (e.g. Amrine-Madsen *et al.*, 2003; Horovitz and Sánchez-Villagra, 2003; Phillips *et al.*, 2006; Asher *et al.*, 2004; Nilsson *et al.*, 2004; Beck, 2008; Meredith *et al.*, 2008), there is consensus on the close affinities between the extant microbiotheriid *Dromiciops gliroides* and the Australasian marsupials (Australidelphia; Szalay, 1982, 1994). However, the position of *D. gliroides* within Australidelphia is still controversial (Nilson *et al.*, 2010). On the other hand, most

Early Miocene Paleobiology in Patagonia: High-Latitude Paleocommunities of the Santa Cruz Formation, ed. Sergio F. Vizcaíno, Richard F. Kay and M. Susana Bargo. Published by Cambridge University Press. © Cambridge University Press 2012.

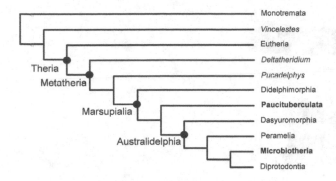

Fig. 10.1. Phylogenetic tree showing the relationships of living marsupial orders (modified from Horovitz and Sánchez-Villagra, 2003).

phylogenetic studies based on molecular or combined data (e.g., Nilson *et al.*, 2004; Asher *et al.*, 2004; Beck, 2008; Meredith *et al.*, 2008), as well as some morphological studies based on cranial, postcranial, and soft tissue anatomy (Horovitz and Sánchez-Villagra, 2003), indicate that Paucituberculata is the sister group of Australidelphia.

Microbiotherians and paucituberculatans are poorly represented in present-day ecosystems (Flores, 2006a, b; Patterson, 2007; Patterson and Rogers, 2007). The only extant microbiotherian is the so-called "monito del monte" (*Dromiciops gliroides*), a small insectivorous marsupial endemic to the temperate forests of southern Chile and Argentina, associated with the southern beech forests (*Nothofagus*) and South American mountain bamboos (*Chusquea*) (Hershkovitz, 1999). *Dromiciops gliroides* is the only South American marsupial reported to exhibit deep torpor or hibernation (Greer, 1966; Bozinovic *et al.*, 2004). In the summer season this species is active during the night, being a common mammal of the understory stratum (Rodríguez-Cabal *et al.*, 2008).

The living Paucituberculata include five species that are grouped in the genera *Caenolestes*, *Lestoros*, and *Rhyncholestes*, all belonging to Caenolestidae ("shrew opossums"). This clade has a disjunct Andean distribution that ranges from Venezuela to northern Peru (*Caenolestes*; Albuja and Patterson, 1996), central Peru and Bolivia (*Lestoros inca*; Anderson, 1997; Ramirez *et al.*, 2007) and southern Chile and Argentina (*Rhyncholestes*; Patterson and Gallardo, 1987; Birney *et al.*, 1996). Caenolestids have a wide latitudinal and altitudinal (up to 4000 meters above sea level) distribution, spanning several biomes including Páramo, Montane forest, and Valdivian forest. Extant caenolestids are small shrew-sized marsupials, which inhabit moist and dense vegetated microhabitats (Kirsh and Waller, 1979; Kelt *et al.*, 1994). The scanty ecological data reported in the literature indicate that caenolestids have a cursorial-leaping locomotion, and a primarily insectivorous diet (Kirsh and Waller, 1979; Barkley and Whitaker, 1984, Patterson and Gallardo, 1987; Patterson, 2007).

The fossil record of Paucituberculata and Microbiotheria indicates that these marsupials had a wider geographic distribution and higher taxonomic diversity than those of the present (Abello, 2007; Goin *et al.*, in press). The oldest-known Paucituberculata and Microbiotheria date from the Paleocene and include forms such as the paucituberculatan *Bardalestes* Goin, Candela, Abello and Oliveira, 2009 (Itaboraian Age, Argentina; Goin *et al.*, in press) and the microbiotherian *Mirandatherium* (Paula Couto, 1952) from Brazil (Itaboraian Age; Goin *et al.*, in press). Both groups achieved their highest taxonomic diversity in the Early Miocene Colhuehuapian and Santacrucian Ages, but the inferred cladogenetic events that gave rise to the Miocene forms seem to have occurred during the Oligocene (Abello, 2007; Goin *et al.*, 2010). By the Early Miocene, microbiotherians are represented by nine species belonging to Microbiotheriidae. At the same time, paucituberculatans are represented for 23 species grouped among Caenolestidae, Pichipilidae, Palaeothentidae, and Abderitidae.

Despite the abundant representation of small marsupials (particularly paucituberculatans) in the Early Miocene (Bown and Fleagle, 1993; Abello, 2007), they are mainly known by mandibular and maxillary remains and isolated teeth. Consequently, the reconstructions of certain paleoecological aspects (e.g. body size, diet) have been derived from the study of dental remains.

Several ecological niches have been identified among Paucituberculata (Dumont *et al.*, 2000): small insectivores (Caenolestidae and Pichipilidae), small- to medium-size insectivore–frugivores (Palaeothentidae), and small- to medium-size frugivores (Abderitidae). As yet only two specimens including postcranial and cranial remains are reported for Paucituberculata (Abello and Candela, 2010). These were referred to two palaeothentid species, *Palaeothentes minutus* Ameghino, 1887 and *Palaeothentes lemoinei* Ameghino, 1887, from the late Early Miocene (Santa Cruz Formation). Curso-saltatorial locomotor strategies were inferred for both species (Abello and Candela, 2010).

In this chapter we summarize previous paleoecological studies of Santacrucian Paucituberculata, and present the results of a new paleoecological analysis of Santacrucian Microbiotheria. Additionally, we evaluate the paleoenvironmental significance of non-carnivorous Miocene marsupials.

10.2 Santacrucian paucituberculatans and microbiotherians

During the South American Miocene, paucituberculatans and microbiotherians coexisted with other metatherians such as Sparassodonta, polydolopimorphian Argyrolagoidea, and Didelphimorphia. Paucituberculatans occur in most Miocene assemblages, but Argyrolagidae and Didelphimorphia are rare or absent as is the case for the late Early Miocene fauna of Santa

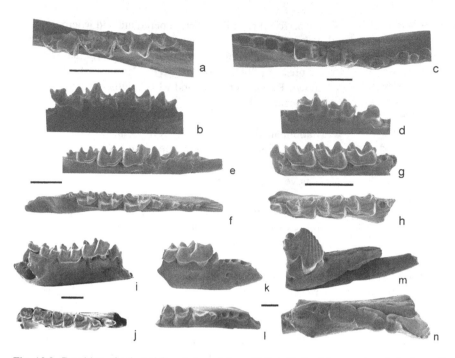

Fig. 10.2. Dentition of selected Paucituberculata and Microbiotheriidae species of the Santa Cruz Formation. a, b, *Microbiotherium acicula* (MACN-A 5727) left mandibular fragment with m1–4 in labial (a) and occlusal (b) views. c, d, *Microbiotherium gallegosense* (type AMNH 9591) right mandibular fragment with p3–m3 in labial (c) and occlusal (d) views. e, f, *Stilotherium dissimile* (type MACN-A 8464) right mandibular fragment with i2, i3 alveolous, three one-rooted teeth and p2–m4 in labial (e) and occlusal (f) views. g, h, *Phonocdromus gracilis* (type MACN-A 8457), left mandibular fragment with p3–m3 in labial (g) and occlusal (h) views. i, j, *Palaeothentes minutus* (MACN-A 5591–5518a), right mandibular fragment with p3–m4 in labial (i) and occlusal (j) views. k, l, *Acdestis owenii* (type MACN-A 1379) mandibular fragment with i2, four alveoli of one-rooted teeth, complete p3–m1 and m2 trigonid in labial (k) and occlusal (l) views. m, n, *Abderites meridionalis*, (m) (MLP 55-XII-13–145) mandibular fragment with i1, four alveolous of one-rooted teeth and p3–m1 in labial view, (n) (type MACN-A 12) mandibular fragment with m1–3 and m4 alveolous in occlusal view. Scale bar, 2 mm.

Cruz beds. To date, microbiotherians (Fig. 10.2a–d) and paucituberculatans (Figs. 10.2e–n and 10.3) are the only non-carnivorous metatherians recorded in the Santa Cruz Formation. Metatherians of this formation were first described by Ameghino (1887, 1891). Sinclair (1906) and Marshall (1976a, 1980, 1982) made significant contributions to the knowledge of the diversity of Santacrucian paucituberculatans and microbiotheriids. Recent revisions include those of Bown and Fleagle (1993), Tauber (1997), Abello (2007), and Abello and Rubilar-Rogers (in press). An updated taxonomic list and the records of paucituberculatans and microbiotheriids from localities of the Santa Cruz Formation are presented in Appendix 10.1.

10.3 Materials and methods

The material studied is listed in Appendix 10.2.

10.3.1 Abbreviations
The generalized metatherian dental formula is: I/i 5/4, C/c 1/1, P/p 3/3, M/m 4/4.

Institutional abbreviations
MACN, Museo Argentino de Ciencias Naturales "B. Rivadavia," Buenos Aires, Argentina. MACN-A, Colección Nacional Ameghino.

MLP, Museo de La Plata, La Plata, Argentina.

AMNH, American Museum of Natural History, New York, EEUU.

MPM-PV, Museo Regional Provincial Padre Manuel Jesús Molina, Río Gallegos, Argentina.

10.3.2 Methodology
To estimate body masses and infer dietary preferences of Santacrucian non-carnivorous marsupials, we follow the methodologies of Dumont *et al.* (2000). In the context of an analysis of dietary preferences of Paucituberculata, Dumont *et al.* (2000) studied all available Santacrucian species. Here we summarize the results of these dietary reconstructions, and apply the same methodology to the Santacrucian Microbiotheriidae. In contrast to Dumont *et al.* (2000), we follow Abello and Rubilar-Rogers (in press) in considering *A. meridionalis* Ameghino, 1887 to be a different species from the Colhuehuapian *Abderites crispus* Ameghino, 1902.

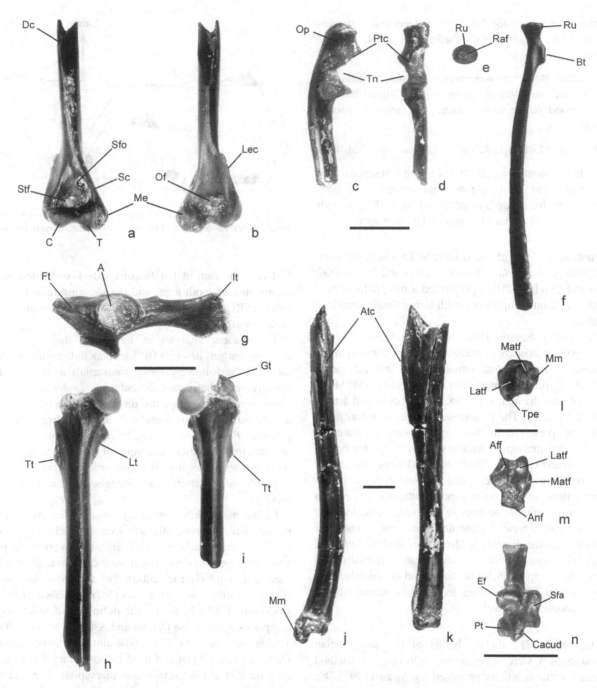

Fig. 10.3. Postcranial skeleton of *Palaeothentes* species. *Palaeothentes minutus*: a, b, right humerus (MACN-A 5619–5639c) in anterior (a) and posterior (b) views; c, d, right ulna (MACN-A 5619–5639e) in lateral (c) and anterior (d) views. e, f, Right radius (MACN-A 5619–5639d) in proximal (e) and medial (f) views. g, Pelvis (MACN-A 5619–5639f) in lateral view; h, right femur (MACN-A 5619–5639g) in anterior view; i, left femur (MACN-A 5619–5639h) in anterior view. *Palaeothentes lemoinei* (MPM-PV 3494): j, k, left tibia in anterior (j) and lateral (k) views; l, right tibia in distal view; m, right astragalus in dorsal view. *Palaeothentes minutus* (MACN-A 5619–5639i); n, right calcaneum in dorsal view. Scale bar, 4 mm. Abbreviations: A, acetabulum; Aff, astragalofibular facet; Anf, astragalonavicular facet; Atc, anterior tibial crest; Bt, bicipital tuberosity; C, capitulum; Cacud, cuboid facet distal half; Dc, deltopectoral crest; Ef, ectal facet; Ft, femoral tubercle; Gt, great trochanter; It, ischial tuberosity; Latf, lateral astragalotibial facet; Lec, lateral epicondylar crest; Lt, lesser trochanter; Matf, medial astragalotibial facet; Me, medial epicondyle; Mm, medial malleolus; Of, olecranon fossa; Op, olecranon process; Pt, peroneal tubercle; Ptc, proximal trochlear crest; Raf, radial articular facet for the capitulum; Ru, radio-ulnar facet; Sc, supracondyloid crest; Sfa, sustentacular facet; Sfo, supracondyloid foramen; Stf, supratrochlear foramen; T, trochlea; Tn, trochlear notch; Tpe, tibial posterior extension; Tt, third trochanter.

Regarding the locomotor habits of *Palaeothentes* species, we present here a synthesis of Abello and Candela's (2010) main results.

Body mass Body masses were estimated from the occlusal area (mesiodistal length × labiolingual breadth) of the second lower molar, using the linear regression equation

$$\ln \text{body mass} = 2.419 + (1.727 \times \ln \text{m2 area}) \qquad \text{(Eq. 10.1)}$$

derived by Dumont *et al.* (2000) for living marsupials. This equation included 27 extant marsupial species whose body masses range from approximately 10 to 1500 g, a wide range encompassing that of Santacrucian marsupials.

Locomotion and use of the substrate To assess the locomotor strategies of *Palaeothentes minutus* and *P. lemoinei*, Abello and Candela (2010) performed a morphofunctional analysis based on comparisons with living South American marsupials.

Palaeothentes minutus (MACN-A 5619–5639a–i) is the most important specimen available, being represented by the humerus, radius, ulna, femur, pelvis, and calcaneum (Fig. 10.3a–i, n). The postcranium of *P. lemoinei* (MPM-PV 3494) preserves the humerus, ulna, pelvis, tibiae, and astragalus (Fig. 10.3j–m). The postcranial features with functional significance, preserved in these fossils, were compared to those of extant marsupials having various locomotor behaviors. Abello and Candela (2010) evaluated traits functionally related to different movements (e.g. flexion/extension, pronation/supination). In this context, special attention was paid in the analysis of articular surfaces (e.g. elbow, cruroastragalar, and hip joints) and areas of muscular attachment. *Caenolestes fuliginosus* and the didelphids *Metachirus nudicaudatus* and *Monodelphis dimidiata* were considered as terrestrial model species. The didelphid *Didelphis aurita* and the microbiotheriid *Dromiciops gliroides* were included as scansorial and arboreal models, respectively.

Feeding habits The dietary habits of the Santacrucian microbiotheriids were investigated following the method of dietary reconstructions proposed by Strait (1993a, b). This approach relies on the assumption that the development of shearing crests on the molars should reflect the physical properties of the foods that must be comminuted (Strait, 1993b; Kay and Hylander, 1978).

Two molar shearing ratios are calculated to discriminate frugivorous from faunivorous small-bodied taxa in a comparative framework with extant mammals whose dietary preferences are well known. Both ratios consider the sum of six shearing crests (crests 1–6 of Kay and Hiiemae, 1974; Fig. 10.4), and two size surrogates: m2 length and the square root of m2 area. The shearing ratio based on area

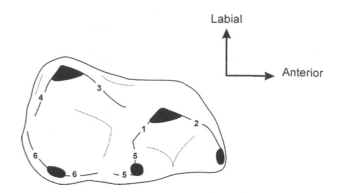

Fig. 10.4. Occlusal view of left lower second molar showing the shearing crests 1–6 (Kay and Hiiemae, 1974) measured in this study.

(SRA) is the sum of lengths of crests 1–6 divided by the square root of tooth area, and the shearing ratio based on length (SRL) is the sum of lengths of crests 1–6 divided by molar length.

The present analysis is based on those performed by Dumont *et al.* (2000) for paucituberculatans, who included the following living marsupials as models: four insectivores (the paucituberculatans *Lestoros inca* and *Caenolestes caniventer*, the dasyurids *Antechinus stuartii* and *Sminthopsis crassicaudata*), and two frugivores (the petaurids *Petaurus breviceps* and *Acrobates pygmaeus*). To broaden the comparative sample of extant marsupials, we included here the insectivorous microbiotheriid *Dromiciops gliroides* (Patterson and Rogers, 2007, and references therein).

Living marsupials, especially those of the Australasian region, have a broad dietary diversity including a wide range of food, such as vertebrate and invertebrate prey, vegetative and reproductive tissues of plants, plant exudates, and fungi (Hume, 2006). For the living marsupials several feeding categories may be established (Lee and Cockburn, 1985); however, the definition of strict trophic categories is imprecise (Vieira and Astúa de Moraes, 2003). Taking into account this constraint, we have followed Dumont *et al.* (2000) and used broad dietary categories to infer the diet of the Santacrucian marsupials. These authors used two categories, faunivores and frugivores, but we have preferred to use "insectivores" instead of "faunivores" for two reasons: (1) faunivores is an equivocal term, because it has been used to refer to those carnivorous mammals that eat vertebrates (e.g. Hume, 2003), as well as to those that eat invertebrates (e.g. Heesy, 2008); and (2) all the living species used by Dumont *et al.* (2000) to represent "faunivores" are actually insectivores (= insectivore/omnivore of Lee and Cockburn, 1985) because they feed primarily on arthropods (which may be consumed as adults or juvenile stages) but their diet may include, in some cases, vegetable matter and fungi. On the other hand, the marsupials used by Dumont *et al.* (2000) to represent frugivores feed mainly on plant materials, such as nectar, pollen, exudates (gum, sap, and sugar encrustations produced by sap-sucking insects and known as "manna") and fruits, but their diet may be complemented with insects when seasonally available. In this context, this category includes those exudate feeders/insectivores and frugivores/omnivores of Lee and Cockburn (1985). Finally, Dumont *et al.* (2000) also found a third feeding category (i.e. "mixed feeders" or "frugivorous–faunivorous"), formed by those marsupials that emerge as frugivorous or "faunivorous" depending on shearing ratio (i.e. the caenolestid *Stilotherium dissimile* Amegino, 1887, the abderitid *Pitheculites minimus* Ameghino, 1902, and the palaeothentid *Acdestis lemairei* Bown and Fleagle, 1993). We prefer to use insectivore–frugivore because both "mixed feeders" and "frugivorous–faunivorous" are ambiguous terms. The former is more frequently used to designate those herbivorous mammals that are intermediate between grazers and browsers, eating both browse and grass (e.g. the red deer *Cervus elaphus*, the impala *Aepyceros melampus*, the nyala *Tragelaphus angasi*; Staver *et al.*, 2009). The

difficulty with the latter is the ambiguity in the usage of faunivore, as explained above.

Measurements of lower second molar maximum length, maximum width, and the lengths of six of its shearing crests were made on a Carl Zeiss Microscope using a measurement module (Axiovision 4.2 and 4.6).

10.4 Results

10.4.1 Body mass

The estimated body masses were 20 g for *Microbiotherium acicula*, 40 g for *M. patagonicum*, 61 g for *M. tehuelchum*, 147 g for *M. gallegosense*, and 487 g for *Abderites meridionalis*. Table 10.1 shows the current body mass estimations for microbiotherians and *A. meridionalis*, and those previously presented by Dumont *et al.* (2000) for paucituberculatans, with the exception of *A. meridionalis* (see Methodology).

10.4.2 Locomotion and use of substrate

The results of comparative analysis with extant marsupials showed that several postcranial features of *Palaeothentes* species resemble those of modern curso-saltatorial forms

Table 10.1. *Body mass estimates and diet of Santacrucian Microbiotheriidae and Paucituberculata*

	Sample size	m2 area (mm^2)	Body mass	Diet
Microbiotheria				
Microbiotheriidae				
Microbiotherium acicula	1	1.38	20 g	Insectivorous
Microbiotherium patagonicum	1	2.07	40 g	Insectivorous
Microbiotherium tehuelchum	1	2.66	61 g	Insectivorous
Microbiotherium gallegosense	1	4.44	147 g	Insectivorous–frugivorous
Paucituberculata				
Caenolestidae				
Stilotherium dissimile	3	1.48	37 g	Insectivorous–frugivorous
Pichipilidae				
Phonocdromus gracilis	2	1.04	28 g	Insectivorous
Palaeothentidae				
Palaeothentes aratae	13	11.34	860 g	Frugivorous
Palaeothentes minutus	112	3.06	82 g	Insectivorous
Palaeothentes intermedius	30	4.67	192 g	Insectivorous
Palaeothentes lemoinei	28	7.18	425 g	Insectivorous
Palaeothentes pascuali	6	2.25	38 g	Insectivorous
Acdestis owenii	113	6.71	344 g	Frugivorous
Acdestis lemairei	18	5.13	256 g	Insectivorous–frugivorous
Abderitidae				
Abderites meridionalis	15	8.86	487 g	Frugivorous

The sample size and mean m2 area (mm^2) from which the body mass was reconstructed are presented. Inferred diet (Diet) is based on shearing ratio and body mass data.

such as *Caenolestes fuliginosus* and *Metachirus nudicaudatus* (Abello and Candela, 2010). Characteristics of the forelimbs include, among other features, a deep and high humeral trochlea (Fig. 10.3a), a deep olecranon fossa (Fig. 10.3b), and mediolaterally broad proximal trochlear crest (Fig. 10.3c, d). These features indicate a well-stabilized elbow joint. The short lateral epicondylar crest (Fig. 10.3b) and the suboval radial head (Fig. 10.3e) suggest that pronation–supination was limited (Abello and Candela, 2010: 1520). The pelvis, better preserved in *P. minutus*, shows a restrictive acetabulum (i.e. tightly articulating; Szalay and Sargis, 2001), lengthened ischium, and prominent femoral tubercle as well as ischial tuberosity (Fig. 10.3g). The morphology of the pelvis indicates a high stability at the ilio-femoral joint and an increased mechanical advantage of the flexors and extensors of the hip (i.e. *rectus femoris* and hamstring muscles), which are well developed in saltatorial species (Argot, 2003a). The greater trochanter of the femur projects beyond the proximal surface of the femoral head (Fig. 10.3i), indicating that the gluteal muscles were favorably positioned to produce fast extension at the hip joint, as occurs in agile and leaping species. Additionally, the hind limbs of *Palaeothentes* species, like those of curso-saltatorial forms, have features associated with marked stability for flexion and extension in the parasagittal plane. Among these functionally significant traits are the right angle between the medial and lateral astragalotibial facets at the cruroastragalar joint (Fig. 10.3l, m), and the conformation of the transverse tarsal joint, characterized by the distal and proximal calcaneo-cuboid facet forming a right angle (Fig. 10.3n).

10.4.3 Feeding habits

Plots of SRL and SRA show the separation between extinct and extant frugivorous and insectivorous species (Fig. 10.5). Both ratios indicate that most extinct microbiotheriids were insectivorous. *Microbiotherium acicula* is the microbiotheriid with the highest SRA and SRL values, equaling the living insectivorous *Caenolestes caniventer* and *Sminthopsis crassicaudata* in SRA ratio. The SRL value of *Microbiotherium patagonicum* is close to that estimated for *Dromiciops gliroides*, while *M. tehuelchum* and *M. gallegosense* present lesser SRL values than *D. gliroides* and *M. acicula*, indicating a less shearing development in these species. *Microbiotherium gallegosense* has an SRA value comparable to those of the living Australian frugivores *Petaurus breviceps* and *Acrobates pygmaeus*. However, *M. gallegosense* clusters with the modern insectivores based on SRL ratio (Fig. 10.5b).

Among paucituberculatans, three species (*Abderites meridionalis*, *Acdestis owenii* Ameghino, 1887, and *Palaeothentes aratae* Ameghino, 1887) cluster with modern frugivores on the basis of SRA and SRL ratios. Conversely, *Phonocdromus gracilis* Ameghino, 1894 (Fig. 10.2g, h) and the palaeothentids *Palaeothentes pascuali* Bown and Fleagle, 1993, *P. minutus*, *P. intermedius* Ameghino, 1887, and *P. lemoinei* cluster with the living insectivores. Finally, the caenolestid *Stilotherium dissimile* (Fig. 10.2e, f) and the palaeothentid *Acdestis lemairei* are classified as frugivorous or insectivorous depending on shearing ratio (i.e. as insectivore–frugivore, Fig. 10.5a, b).

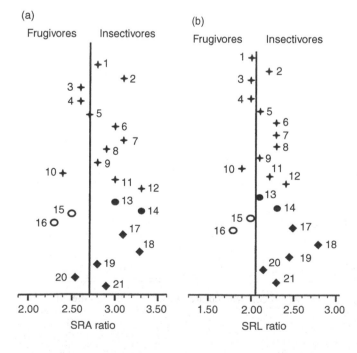

(a) Frugivores Insectivores — SRA ratio 2.00 2.50 3.00 3.50

(b) Frugivores Insectivores — SRL ratio 1.50 2.00 2.50 3.00

Fig. 10.5. Partition between frugivorous and insectivorous taxa based on SRA (a) and SRL (b) ratios. Paucituberculatans (crosses), petaurids (filled circles), dasyurids (open circles), microbiotherids (diamonds). Species are: 1, *Stilotherium dissimile*; 2, *Phonocdromus gracilis*; 3, *Abderites meridionalis*; 4, *Acdestis owenii*; 5, *Acdestis lemairei*; 6, *Palaeothentes minutus*; 7, *Palaeothentes pascuali*; 8, *Palaeothentes intermedius*; 9, *Palaeothentes lemoinei*; 10, *Palaeothentes aratae*; 11, *Lestoros inca*; 12, *Caenolestes caniventer*; 13, *Antechinus stuartii*; 14, *Sminthopsis crassicaudata*; 15, *Petaurus breviceps*; 16, *Acrobates pygmaeus*; 17, *Dromiciops gliroides*; 18, *Microbiotherium acicula*; 19, *Microbiotherium patagonicum*; 20, *Microbiotherium gallegosense*; and 21, *Microbiotherium tehuelchum*.

10.5 Discussion

10.5.1 Body mass

In a study of extant Neotropical marsupials, Birney and Monjeau (2003) analyzed the distribution of body size among species and established three size categories: (a) small-sized (less than 100 g); (b) medium-sized (from 100 to 499 g); and (c) large-sized (500 g or more). Taking into account these size categories, we found that Santacrucian microbiotheriids were mainly small-sized, with estimated body masses ranging from 20 g in *Microbiotherium acicula* to 147 g in *M. gallegosense* (Fig. 10.2c, d). Santacrucian paucituberculatans show a much greater body-mass range, extending from 13 g in *Phonocdromus gracilis* to 800 g in *Palaeothentes aratae*, and with more than half of the species being small- to medium-sized (Table 10.1).

Throughout the Cenozoic, most South American metatherians were small- to medium-sized (see Goin, 2003). In extant faunas, small- to medium-sized marsupials represent about 90% of the marsupial fauna of the Americas (see Birney and Monjeau, 2003), and they are also abundant in the marsupial fauna of Australasia (Dickman and Vieira, 2006). Our results are in agreement with this pattern, because 52% of the Santacrucian marsupial species are small- to medium-sized (nine paucituberculatans plus four microbiotherian species; see Table 10.1), and 48% are large (one paucituberculatan species plus 11 sparassodont species; see Table 10.1 and Prevosti *et al.*, Chapter 11).

Santacrucian paucituberculatans and microbiotheriids show a body mass overlap in the small and medium size categories (Fig. 10.6a, c). Nevertheless, the differential use of space and food resources may have minimized or avoided competition among species of similar body mass, as in extant marsupial assemblages (Charles-Dominique *et al.*, 1981; Charles-Dominique, 1983; Vieira and Monteiro-Filho, 2003).

10.5.2 Locomotion and use of the substrate

Locomotor strategies constitute an important dimension of a species' niche. For extinct mammals, locomotion may be inferred from fossil postcranial remains, and have implications for species paleoecology. Unfortunately, only two skeletons of non-carnivorous Santacrucian marsupials are known. Both skeletons (pertaining to two different palaeothentid species) were recently analyzed by Abello and Candela (2010) who paid particular attention to the reconstruction of locomotor pattern. As noted earlier, in terms of functionally significant features, the major similarities were found with the extant marsupials *Caenolestes fuliginosus* and *Metachirus nudicaudatus*. The latter is the most terrestrial didelphid (Miles *et al.*, 1981; Delciellos and Vieira, 2006, 2009). This marsupial inhabits a wide range of forest habitats (Miranda *et al.*, 2009), from open scrub

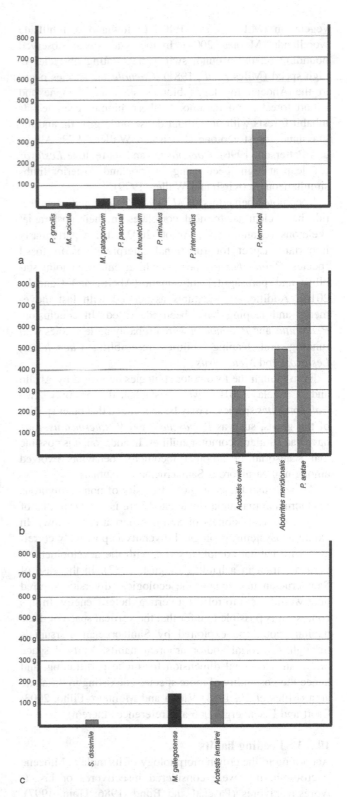

Fig. 10.6. Body mass averages of insectivorous (a), frugivorous (b), and insectivorous–frugivorous (c) non-carnivorous marsupials of Santa Cruz Formation. Gray bars: paucituberculatans; black bars: microbiotheriids. Values from Table 10.1.

vegetation (Miles *et al.*, 1981) to lowland and hillside woodlands (Moraes, 2004). In the wild it was observed bounding on the ground, swiftly negotiating obstacles at high speed (Miles *et al.*, 1981). *Caenolestes* species occur in the Andean highlands biomes such as Montane and Cloud forests, and Páramo. In these biomes caenolestids inhabit forests with or without dense undergrowth, and the grassland–forest ecotone (Kirsh and Waller, 1979; Albuja and Patterson, 1996). *Caenolestes* and its relative *Lestoros* can leap at high speed using anterior and posterior limbs simultaneously (Kirsh and Waller, 1979).

Locomotor behaviors of *M. nudicaudatus* and caenolestids have clear anatomical correlates in their postcranial skeletons (Grand, 1983; Argot, 2003a, b), a particularly important aspect for functional interpretations in fossil species. *Palaeothentes* species exhibit enhanced joint stability and parasagittal movement (Abello and Candela, 2010). Additionally, features associated with fast movements and leaping have been identified. In conclusion, *P. minutus* and *P. lemoinei* were probably agile species with running and leaping abilities resembling *Caenolestes*, *Lestoros*, and *Metachirus*.

Even though the locomotor strategies described by Abello and Candela (2010) were established for only two *Palaeothentes* species, it may be speculated that other species of the genus, such as *P. pascuali* and *P. intermedius*, may have had similar locomotor abilities. In addition, it is possible that scansorial or arboreal locomotor behaviors evolved among non-carnivorous Santacrucian marsupials.

We note that the ecological diversity of non-carnivorous Santacrucian marsupials evaluated here is similar to that of current tropical habitats of South America (see below). In extant ecosystems, ecological diversity is positively correlated with habitat complexity (i.e. with the development of vertical strata in a habitat; August, 1983). In the case of Santacrucian marsupials, the ecological diversity inferred here would seem to reflect a vertical heterogeneity. In this context, it is possible to infer that the vertical space of these habitats could be exploited by Santacrucian marsupials through scansorial and/or arboreal habits. Vertical space offers an additional dimension for niche partitioning, and hence the possibility of more species coexisting in the same area (Miles *et al.*, 1981; Vieira and Monteiro-Filho, 2003; Croft and Eisenberg, 2006 and references therein).

10.5.3   Feeding habits
According to the crown morphology of its molars, Miocene microbiotheriids were considered insectivores or insectivores/frugivores (Pascual and Bond, 1986; Goin, 1997). Based on an analysis of molar wear facets, Goin *et al.* (in press) concluded that Early Oligocene Patagonian members of this family (including *Microbiotherium*, a genus that we analyzed in this study) were insectivores.

According to our results, Miocene microbiotheriids also were mainly insectivores. Only one species, *M. gallegosense*, has shearing ratios that indicate a more limited shearing component, suggesting an insectivorous–frugivorous diet (Fig. 10.5a, b). *Microbiotherium acicula* is the most extreme insectivorous species, showing the highest SRL and SRA values. *Microbiotherium patagonicum* is closer to the extant *Dromiciops gliroides* in SRL values, and it achieved a similar degree of molar crest development as in the living species.

In our analysis, *D. gliroides* is grouped with insectivorous species (Fig. 10.5a, b). This result is consistent with the information of its main dietary preferences in the wild. Analysis of stomach contents indicated that this species feeds primarily on arthropods and other invertebrates (Mann, 1955; Meserve *et al.*, 1988). However, it also feeds seasonally on fruits (Amico *et al.*, 2009). This strategy could be extrapolated to Santacrucian microbiotheriids, as insectivores may consume plant material, either regularly or seasonally, when prey species are scarce or unavailable (Hume, 2003).

According to molar crown morphology, Miocene paucituberculatans were considered insectivores, insectivores–frugivores, and insectivores–phytophages (e.g. Ortiz-Jaureguizar, 2003). Our analysis corroborates this inference, showing that insectivores are more diverse (five species) than insectivore–frugivores and frugivores (including the insectivorous–phytophagous category of Ortiz-Jaureguizar, 2003; see above) (Table 10.1). As Dumont *et al.* (2000) pointed out, folivory seems not to have evolved in paucituberculatans, as no taxon with high shearing ratios has an estimated body mass higher than 600 g (Smith and Lee, 1984).

Most insectivorous paucituberculatans differ markedly in body mass (e.g. *Palaeothentes minutus*, *P. intermedius*, and *P. lemoinei*), suggesting that ecological separation among them may have been achieved by differences in dietary composition and/or consumption of prey of different body size. In living insectivores, such as dasyurid marsupials and soricid placentals, there is a positive correlation between body size of predators and their prey (Fisher and Dickman, 1993; Churchfield *et al.*, 1999). Even so, both small and large dasyurids can exploit a relatively large range of prey sizes, and there is no physical constraint on the size of prey consumed. Consequently, dasyurid species maximized their rates of energy intake by feeding on prey of a selected size (Fisher and Dickman, 1993). Thus, larger dasyurids prefer larger invertebrates because of greater energy return per prey item; conversely, for the smaller dasyurids prey consumption requires more chewing time, and this could result in a fall in the rate of energy intake from each prey (Fisher and Dickman, 1993). Despite small differences in body size and overall ecology, niche overlap is minimized in shrews as they differ in terms of the

Fig. 10.7. Life reconstructions. a, *Abderites meridionalis* and b, *Palaeothentes minutus*. Drawing by Pablo Motta.

percentage of several invertebrate prey in the diet (Churchfield *et al.*, 1999). We expect that dietary composition (as it covaries with body size) would have produced niche separation among extinct insectivorous paucituberculatans.

As mentioned above, the body mass of Santacrucian microbiotheriids and paucituberculatans overlaps to some degree in the small to medium size range (Fig. 10.6a), initially suggesting that some niche overlap may have occurred. However, differences in body size and molar morphology suggest that niche overlap was, in fact, minimal, and it is also possible that differences occurred not only in diet composition, but also in foraging mode (Sanson, 1985; Churchfield and Sheftel, 1994; Churchfield *et al.*, 1999).

Frugivorous and insectivorous–frugivorous marsupials analyzed here do not exhibit body mass overlap (compare Fig. 10.6b with 10.6c). Frugivorous species are paucituberculatans belonging to Palaeothentidae (*Palaeothentes aratae* and *Acdestis owenii*, Fig. 10.2k, l) and Abderitidae (*Abderites meridionalis*, Figs. 10.2m, n and 10.7a), which are quite distinct in dental morphology (Abello, 2007). The dentition of Abderitidae is characterized by the presence of well-developed lophs and a plagiaulacoid complex (Simpson, 1933; Ortiz-Jaureguizar, 2003). The plagiaulacoid complex of *Abderites* was considered by Dumont *et al.* (2000) as a dietary indicator in addition to body mass and extent of molar shearing. According to their analysis, occlusion between P3 and m1 in *Abderites* was a shearing complex that operated in a manner similar to that of living Australian phalangerids such as *Phalanger*, but dissimilar to those of other mammals with plagiaulacoid dentition such as certain Cretaceous to Early Eocene multituberculates (e.g. the cimolodontan *Ptilodus* Cope, 1881) and North American Paleocene

plesiadapiform primates (e.g. *Carpodaptes* Matthew and Granger, 1921). In the multituberculates and plesiadapiforms, the shearing teeth mainly exhibit apical wear, because the lower cutting teeth become worn when the food is ground against a cuspate upper tooth. In contrast, in *Phalanger*, and probably also in *Abderites*, the food items are cut between the upper and lower teeth as they shear across one another in a scissor-like fashion (Dumont *et al.*, 2000). A shearing complex similar to that of the abderitids occurs in many extant Diprotodonta, including the *Phalanger* already mentioned as well as *Burramys* and the Potoroidae *Hypsiprymnodon* and *Bettongia*. Related to their function, these dental modifications are adequate to break open food items with a hard covering (Dimpel and Calaby, 1972; Parker, 1973).

On the other hand, *Palaeothentes aratae* and *Acdestis owenii* have a strong shearing crest (paracrista) on m1 (Fig. 10.2k, l), but they lack a plagiaulacoid dentition and lophs. These differences suggest that frugivorous paucituberculatans were distinct not only in body mass but in their diets as well (Ortiz-Jaureguizar, 2003).

10.5.4 Paleoenvironmental implications

From the wide range of estimated body masses and diet of non-carnivorous Santacrucian marsupials, several ecologic niches were reconstructed: small- to medium-sized insectivores, and small to large frugivores and insectivores–frugivores. In regard to a paleoecological reconstruction, this ecological diversity suggests the existence of forested habitats that could have supported the diverse marsupial niches. In modern ecosystems, and particularly in tropical forests, high values of mammalian species richness arise from habitat heterogeneity, mainly owing to the partitioned vertical space (August, 1983; Bakker and Kelt, 2000).

Species richness also tracks plant productivity, with higher productivity leading to more species (Kay *et al.*, 1997).

Paleoenvironmental inferences can also be advanced by analyzing the trophic guild structure of Santacrucian marsupials. In a macroecological study of Neotropical marsupials, Birney and Monjeau (2003) evaluated the latitudinal variation among several biological characters, such as trophic guild structure, habitat use, and body size. Taking into account that latitude is a surrogate for physical environmental variables such as temperature and precipitation, the trophic guild structure of Santacrucian marsupials may offer clues to the Santacrucian paleoenvironments. According to Birney and Monjeau (2003), three guilds were considered: frugivores, carnivores, and insectivores. Current carnivorous marsupials include, among others, the didelphids *Didelphis*, *Lutreolina*, and *Philander*, all of them having body masses less than 2 kg. In the Santacrucian assemblage, carnivores with body masses less than or equal to 2 kg include the sparassodonts *Sipalocyon gracilis* Ameghino, *S. obusta* Ameghino, *Pseudonotictis pusillus* Ameghino, and *Perathereutes pungens* Ameghino (see Prevosti *et al.*, Chapter 11). Thus, small carnivores represent 22% of the Santacrucian marsupial species richness, insectivore–frugivores plus frugivores 33%, and insectivores 45%. Comparing the trophic guild structure of Santacrucian marsupials with those of the modern Neotropics, we find that in the Santacrucian fauna there is a relatively high percentage of frugivores in relation to carnivores, a ratio that occurs at low latitudes in the extant marsupial faunas of South America. Birney and Monjeau (2003) concluded that the thermal range (i.e. the difference between mean maximum extreme temperatures and mean minimum extreme temperatures) is the best predictor of the proportion of frugivores and carnivores. Closer to the equator, the narrower thermal range allows year-long availability of fruits, and the percentage of frugivores is higher than at high latitudes. Additionally, differences in precipitation (i.e. seasonal rainfall) seem to be another important factor influencing the abundance of frugivores (Birney and Monjeau, 2003). If we consider this last factor, the sharp differences between dry and wet seasons in the tropics would have allowed more diversification in the trophic niche than more constant precipitation. On the basis on these ecological similarities, we can infer that Santacrucian marsupials lived in relatively warm climates, but with seasonal rainfall. However, it should be kept in mind that taphonomic bias and time averaging are common factors that affect the composition of fossil assemblages. Consequently, in the absence of a precise evaluation of the incidence of these biases, the inferences derived from a fossil assemblage have a less heuristic value that those based on living assemblages.

In summary, the ecological characteristics of non-carnivorous marsupials of the Santacrucian suggest that they lived in forested habitats developed under warm temperatures and rain seasonality. According to the fossil record of vascular plants, the distribution of megathermal and mesothermal angiosperms expanded their distribution at middle–high latitudes of Patagonia during the Late Oligocene to Early Miocene (Barreda and Palazzesi, 2007; Brea *et al.*, Chapter 7). Nevertheless, the first records of some shrubby and herbaceous angiosperms in southern South America suggest that the vegetation acquired a more complex physiognomy than that of the Early Oligocene (Barreda and Palazzesi, 2007). During the Early Miocene, xerophytic (or mesophytic) vascular plants were dominant in Patagonia, and the development of all these specialized communities was probably related to a water deficit in open forest regions or marginal marine areas. During the late Early Miocene, rainforest trees may have formed riparian or gallery forests in central Patagonia, while drier conditions would have prevailed in lowland areas (Barreda and Palazzesi, 2007). Finally, during the latest Early Miocene, xerophytic elements suffered a geographic retraction in Patagonia, with an increase of megathermals and a dominance of aquatic herbs and hydrophytes. Forests persisted across extra-Andean Patagonia until about the Middle Miocene (Barreda, 2002; Barreda and Palazzesi, 2007; Palazzesi and Barreda, 2007).

If we consider the ecological information provided by all the Santacrucian mammals (Prevosti *et al.*, Chapter 11; Vizcaíno *et al.*, Chapter 12; Bargo *et al.*, Chapter 13; Cassini *et al.*, Chapter 14; Candela *et al.*, Chapter 15; Kay *et al.*, Chapter 16), this fauna was dominated by grazers and mixed-feeders, with frugivores and browsers remaining highly diverse. This distribution of trophic niches indicates that during Santacrucian times a balance existed between grasslands and woodlands, probably represented by a park savanna (Webb, 1978; Pascual and Ortiz-Jaureguizar, 1990; Pascual *et al.*, 1996; Ortiz-Jaureguizar and Cladera, 2006, and references therein). From a climatic point of view, the presence of primates and other warm climate-sensitive vertebrates that have been recorded as far south as 51°S suggests that warm and forested habitats were well developed in Patagonia. Nevertheless, together with these indicators of warm and forested habitats, there were also other mammals such as some rodents that indicate the existence of open habitats encroaching on areas of wet forests in Patagonia (Pascual *et al.*, 1996; Ortiz-Jaureguizar and Cladera, 2006; Pérez, 2010).

Considering the evidence provided by vascular plants and mammals, we may conclude that the ecological diversity of the Santacrucian marsupials is compatible with the existence of habitat patchiness during Santacrucian times, with a balance between closed and open habitats represented by a park savanna. In this scenario, the non-carnivorous marsupial species would have occupied the more forested areas (Fig. 10.7).

10.6 Conclusions

Microbiotheria and Paucituberculata of the Santa Cruz Formation constitute an ecologically diverse assemblage of non-carnivorous marsupials. Among microbiotheriids we recognize small insectivores such as *Microbiotherium acicula* and *M. tehuelchum*, and a medium-sized insectivore–frugivore, *M. gallegosense*. Compared with microbiotheriids, paucituberculatans are taxonomically and ecologically more diverse, having a wider range of body mass and diet. For this group we identify small insectivores such as *Phonocdromus gracilis*, medium-sized curso-saltatorial insectivores such as *Palaeothentes minutus* (Fig. 10.7b) and *P. lemoinei*, the small- to medium-sized insectivores–frugivores *Stilotherium dissimile* and *Acdestis owenii*, the medium-sized frugivore *Abderites meridionalis* (Fig. 10.7a), and the large frugivore *Palaeothentes aratae*.

Regarding the marsupials of the insectivore guild, a minimal niche overlap was inferred from differences in body mass and its relations to prey size and diet composition, as is observed in extant insectivorous mammals. In addition to the inferred terrestrial, curso-saltatorial locomotion of *Palaeothentes minutus* and *P. lemoinei*, scansorial or arboreal locomotor behaviors could have evolved among non-carnivorous Santacrucian marsupials that allowed their exploitation of resources in the vertical space.

From the present study, the ecological diversity of the non-carnivorous Santacrucian marsupial indicates that they lived in forested habitats, under warm temperatures and rain seasonality. Forested habitats could have supported the varied previously mentioned marsupial niches, offering diverse resources in the spatial and trophic dimensions. Considering also the paleoecological information provided by vascular plants and all the mammalian species, we conclude that the ecological diversity of non-carnivorous marsupials is compatible with a patchy environment, with a balance between vegetation typical of closed and open habitats, represented by a park savanna.

ACKNOWLEDGMENTS
We thank D. Flores (MACN), A. Kramarz (MACN), and M. Reguero (MLP) for facilitating access to specimens of marsupials in their care, and F. Goin (MLP) for allowing us to study specimens in his personal collection. Special thanks to the editors of this volume, S. F. Vizcaíno, R. F. Kay, and M. S. Bargo, for inviting us to make this contribution, and to P. Posadas (LASBE-MLP) and C. Morgan for their help with the English version. This is a contribution to the projects PICT 0143 to Sergio F. Vizcaíno and NSF 0851272, 0824546 to Richard F. Kay.

Appendix 10.1 Paucituberculata and Microbiotheria of the Santa Cruz Formation

Microbiotheria Ameghino, 1889

Microbiotheriidae Ameghino, 1887
Microbiotherium acicula (Ameghino, 1891)
Microbiotherium patagonicum Ameghino, 1887
Microbiotherium tehuelchum Ameghino, 1887
Microbiotherium gallegosense Sinclair, 1906

Paucituberculata Ameghino, 1894

Caenolestidae Trouessart, 1898
Stilotherium dissimile Ameghino, 1887
Pichipilidae (Marshall, 1980)
Phonocdromus gracilis Ameghino, 1894
Palaeothentidae Sinclair, 1906
Palaeothentes aratae Ameghino, 1887
Palaeothentes minutus Ameghino, 1887
Palaeothentes intermedius Ameghino, 1887
Palaeothentes lemoinei Ameghino, 1887
Palaeothentes pascuali Bown and Fleagle, 1993
Acdestis owenii Ameghino, 1887
Acdestis lemairei Bown and Fleagle, 1993
Abderitidae (Ameghino, 1889)
Abderites meridionalis Ameghino, 1887

List of Paucituberculata and Microbiotheria from localities of the Santa Cruz Formation

For details of the specimens, localities, data of collection and collectors (Col.) see Marshall (1980, 1982), Bown and Fleagle (1993) and Vizcaíno *et al.* (Chapter 1: Figs. 1.1 and 1.2, and Appendix 1.1).

Gobernador Gregores: *Acdestis owenii.*

Lago Cardiel: *Acdestis owenii.*

Río Chalía – Ea. Viven Aike – (Col. Bown y Fleagle): *Palaeothentes minutus, Palaeothentes intermedius, Acdestis owenii, Acdestis lemairei.*

Monte León (Col. Bown and Fleagle): *Palaeothentes aratae, Palaeothentes minutus, Palaeothentes intermedius, Palaeothentes lemoinei, Acdestis owenii, Acdestis lemairei.*
La Cueva (Col. Ameghino): *Phonocdromus gracilis, Palaeothentes minutus, Palaeothentes intermedius, Palaeothentes lemoinei, Stilotherium dissimile, Abderites meridionalis, Microbiotherium patagonicum, Microbiotherium tehuelchum.*

Yegua Quemada (Col. Ameghino): *Palaeothentes intermedius, Palaeothentes lemoinei, Microbiotherium tehuelchum.*

Santa Cruz (Col. Ameghino): *Palaeothentes aratae, Palaeothentes minutus, Palaeothentes lemoinei, Microbiotherium patagonicum.*

Monte Observación (= Cerro Observatorio, see Marshall, 1976b, and Vizcaíno *et al.*, Chapter 1; Col. Ameghino, Bown and Fleagle): *Palaeothentes aratae, Palaeothentes minutus, Palaeothentes pascuali, Palaeothentes intermedius, Palaeothentes lemoinei, Stilotherium dissimile, Abderites meridionalis, Acdestis owenii, Acdestis lemairei, Microbiotherium acicula, Microbiotherium patagonicum, Microbiotherium tehuelchum.*

Puesto Estancia La Costa (= Corriguen-Aike; Col. MLP-Duke University): *Palaeothentes lemoinei.*
Corriguen-Kaik (Col. Ameghino): *Palaeothentes lemoinei, Microbiotherium tehuelchum.*

Killik-Aike (Col. H. Felton): *Palaeothentes minutus, Microbiotherium tehuelchum.*

Río Gallegos (Col. B. Brown): *Palaeothentes aratae, Palaeothentes minutus.*

Near Felton's Estancia, along the north bank of the Río Gallegos (Col. Barnum Brown): *Phonocdromus gracilis, Microbiotherium acicula, Microbiotherium gallegosense.*

Sehuen (= Río Chalía; Col. Ameghino): *Palaeothentes aratae, Palaeothentes lemoinei, Palaeothentes intermedius, Stilotherium dissimile, Abderites meridionalis, Microbiotherium patagonicum.*

Appendix 10.2 Material studied

Microbiotheriidae

Microbiotherium acicula (Ameghino, 1891),
MACN-A 5727, a left mandibular ramus with p3–m4.
Horizon and locality: Santa Cruz Formation, Monte Obser-
vación, Santa Cruz Province.

Microbiotherium patagonicum Ameghino, 1887
MLP 11–30, a right mandibular ramus with m1–4. Horizon
and locality: Santa Cruz Formation, collected from "*las
barrancas del río Santa Cruz*" (Ameghino, 1889: 264),
Santa Cruz Province.

Microbiotherium tehuelchum Ameghino, 1887
MLP 11–36, a right mandibular ramus with p2–m4. Horizon and locality: Santa Cruz Formation, without locality
data (see Marshall, 1982), Santa Cruz Province.

Microbiotherium gallegosense Sinclair, 1906
AMNH 9591, a right mandibular ramus with p3–m2. Hori-
zon and locality: Santa Cruz Formation, Estancia Felton,
Santa Cruz Province.

Dromiciops gliroides Thomas, 1894
MACN 19142, MACN 22918, MACN 22919, and MACN
13038.

Paucituberculata

Abderites meridionalis Ameghino, 1887
MACN-A 5542, left mandibular ramus with m1–4 and
MACN-A 2037, left mandibular ramus with p3–m4. Horizon and locality: Santa Cruz Formation, Monte Observación, Santa Cruz Province; MACN-A 5541, left mandibular
ramus with p3–m4 and PU 15079, left mandibular ramus with
p3–m3. Horizon and locality: Santa Cruz Formation, Río
Chalía, Santa Cruz Province; MACN-A 8248, left mandibular
ramus with p3–m3. Horizon and locality: Santa Cruz Formation, La Cueva, Santa Cruz Province; MACN-A 2031, left
mandibular ramus with m2; MACN-A 2032, right mandibular
ramus with m2–4; MACN-A 2033, right mandibular ramus
with m2–4; MACN-A 2034, right mandibular ramus with
m1–3; MLP 11–109, right mandibular ramus with m1–2;
MLP 11–133, left mandibular ramus with m2–3; MLP
55-XII-13–144, right mandibular ramus with m2; and MACN
11651, right mandibular ramus with p3-m4. Horizon and
locality: unknown, Santa Cruz Province.

Palaeothentes minutus
MACN-A 5619–5639a–I, right and left mandibular rami
with m3–4, distal portion of right humerus, proximal
portion of right ulna, partially complete pelvis, proximal
portion of right femur, proximal portion of left femur, and
right calcaneum. Horizon and locality: Santa Cruz Formation, Killik-Aike, Santa Cruz Province.

Palaeothentes lemoinei
MPM-PV 3494, right mandibular fragment with m2–4 and
left edentulous mandibular fragment, left humerus, left
ulna, left fragment of pelvis, fragment of right tibia, fragment of left tibia, and right astragalus. Horizon and locality:
Santa Cruz Formation, Estancia La Costa Member, Fossiliferous Level 5.3, Puesto Estancia La Costa, Santa Cruz
Province.

REFERENCES

Abello, M. A. (2007). Sistemática y bioestratigrafía de los
    Paucituberculata (Mammalia, Marsupialia) del Cenozoico
    de América del Sur. Unpublished Ph.D. thesis, Universidad
    Nacional de La Plata, Argentina.
Abello, M. A. and Candela, A. M. (2010). Postcranial skeleton
    of the Miocene marsupial *Palaeothentes* (Paucituberculata,
    Palaeothentidae): paleobiology and phylogeny. *Journal of
    Vertebrate Paleontology*, **30**, 1515–1527.
Abello, M. A. and Rubilar-Rogers, D. (in press). Revisión del
    género *Abderites* Ameghino, 1887 (Marsupialia,
    Paucituberculata). *Ameghiniana*.
Albuja, L. and Patterson, B. D. (1996). A new species of northern
    shrew-opossum (Paucituberculata: Caenolestidae) from the
    Cordillera del Cóndor, Ecuador. *Journal of Mammalogy*,
    77, 41–53.
Ameghino, F. (1887). Enumeración sistemática de las especies de
    mamíferos fósiles coleccionados por Carlos Ameghino en
    los terrenos eocenos de la Patagonia austral. *Boletín del
    Museo de La Plata*, **1**, 1–26.
Ameghino, F. (1889). Contribución al conocimiento de los
    mamíferos fósiles de la República Argentina. *Actas de
    la Academia Nacional de Ciencias de Córdoba*, **6**,
    XXXIII-1027.
Ameghino, F. (1891). Nuevos restos de mamíferos fósiles
    descubiertos por Carlos Ameghino en el Eoceno inferior
    de Patagonia austral. Especies nuevas, adiciones y

correcciones. *Revista Argentina de Historia Natural*, **1**, 289–328.

Amico, G. C., Rodríguez-Cabal, M. A. and Aizen, M. A. (2009). The potential key seed-dispersing role of the arboreal marsupial *Dromiciops gliroides*. *Acta Oecologica*, **35**, 8–13.

Amrine-Madsen, H., Scally, M., Westerman, M. *et al.* (2003). Nuclear gene sequences provide evidence for the monophyly of australidelphian marsupials. *Molecular Phylogenetics and Evolution*, **28**, 186–196.

Anderson, S. (1997). Mammals of Bolivia, taxonomy and distribution. *Bulletin of the American Museum of Natural History*, **231**, 1–652.

Argot, C. (2003a). Postcranial functional adaptations in the South American Miocene borhyaenoids (Mammalia, Metatheria): *Cladosictis*, *Pseudonotictis* and *Sipalocyon*. *Alcheringa*, **27**, 303–356.

Argot, C. (2003b). Functional adaptations of the postcranial skeleton of two Miocene borhyaenoids (Mammalia, Metatheria), *Borhyaena* and *Prothylacinus*, from South America. *Palaeontology*, **46**, 1213–1267.

Asher, R. J., Horovitz, I. and Sánchez-Villagra, M. R. (2004). First combined cladistic analysis of marsupial mammal interrelationships. *Molecular Phylogenetics and Evolution*, **33**, 240–250.

August, P. V. (1983). The role of habitat complexity and heterogeneity in structuring tropical mammal communities. *Ecology*, **64** (6), 1495–1507.

Bakker, V. J. and Kelt, D. A. (2000). Scale-dependent patterns in body size distributions of Neotropical mammals. *Ecology*, **81**, 3530–3547.

Barkley, L. J. and Whitaker, J. O. Jr. (1984). Confirmation of *Caenolestes* in Peru with information on diet. *Journal of Mammalogy*, **65**, 328–330.

Barreda, V. D. (2002). Palinofloras cenozoicas. In *Geología y Recursos Naturales de Santa Cruz, Relatorio del XV Congreso Geológico Argentino*, ed. M. J. Haller. Buenos Aires: Asociación Geológica Argentina, pp. 545–567.

Barreda, V. and Palazzesi, L. (2007). Patagonian vegetation turnovers during the Paleogene–early Neogene: origin of arid-adapted floras. *The Botanical Review*, **73**, 31–50.

Beck, R. M. D. (2008). A dated phylogeny of marsupials using a molecular supermatrix and multiple fossil constraints. *Journal of Mammalogy*, **89**, 175–189.

Birney, E. C. and Monjeau, J. A. (2003). Latitudinal variation in South American marsupial biology. In *Predators with pouches: the biology of carnivorous marsupials*, ed. M. Jones, C. Dickman and M. Acher. Hobart, Australia: CSIRO Publishing, pp. 297–317.

Birney, E. C., Sikes, R. S., Monjeau, J. A., Guthmann, N. and Carleton, J. P. (1996). Comments on Patagonian marsupials of Argentina. In *Contribution in Mammalogy, A Memorial Volume Honoring Dr. J. Knox Jones Jr.*, ed. H. H. Genoways and R. J. Baker. Museum of Texas Tech University Press, pp. 149–154.

Bown, T. M. and Fleagle, J. G. (1993). Systematics, biostratigraphy, and dental evolution of the Palaeothentidae, later Oligocene to Early–Middle Miocene (Deseadan–Santacrucian) Caenolestoid Marsupials of South America. *Journal of Palaeontology Memoir* 29, **67**, 1–76.

Bozinovic, F., Ruiz, G. and Rosenmann, M. (2004). Energetics and torpor of a South American "living fossil", the microbiotheriid *Dromiciops gliroides*. *Journal of Comparative Physiology B*, **174**, 293–297.

Charles-Dominique, P. (1983). Ecology and social adaptation in didelphid marsupials: comparisons with eutherians of similar ecology. In *Advances in the Study of Mammalian Behavior*, ed. J. F. Eisenberg. Shippensburg: American Society of Mammalogist, pp. 305–422.

Charles-Dominique, P., Atramentowicz, M., Charles-Dominique, M. *et al.* (1981). Les mammiferes frugivores arboricoles nocturnes d'une foret guyanaise: inter-relations plantes–animaux. *Revue d'Ecologie (Terre Vie)* **35**, 341–436.

Churchfield, S. and Sheftel, B. I. (1994). Food niche overlap and ecological separation in a multi-species community of shrews in the Siberian taiga. *Journal of Zoology*, **234**, 105–124.

Churchfield, S., Nesterenko V. A. and Shvarts, E. A. (1999). Food niche overlap and ecological separation amongst six species of coexisting forest shrews (Insectivora: Soricidae) in the Russian Far East. *Journal of Zoology*, **248**, 349–359.

Croft, D. B. and Eisenberg, J. F. (2006). Behaviour. In *Marsupials*, ed. P. J Armati, C. R, Dickman and I. D Hume. Cambridge, UK: Cambridge University Press, pp. 229–298.

Delciellos, A. C. and Vieira, M. V. (2006). Arboreal walking performance in seven didelphid marsupials as an aspect of their fundamental niche. *Austral Ecology*, **31**, 449–457.

Delciellos, A. C. and Vieira, M. V. (2009). Jumping ability in the arboreal locomotion of didelphid marsupials. *Mastozoología Neotropical*, **16**, 299–307.

Dickman, C. R. and Vieira, E. (2006). Ecology and life histories. In *Marsupials*, ed. J. P. Armati, C. R. Dickman and I. D. Hume. Cambridge: Cambridge University Press, pp. 199–228.

Dimpel, H. and Calaby, J. H. (1972). Further observations on the mountain pygmy-possum (*Burramys parvus*). *Victorian Naturalist*, **89**, 101–106.

Dumont, E. R., Strait, S. G. and Friscia, A. R. (2000). Abderitid marsupials from the Miocene of Patagonia: an assessment of form, function, and evolution. *Journal of Paleontology*, **74**, 1161–1172.

Fisher, D. O. and Dickman, C. R. (1993). Body size–prey size relationships in insectivorous marsupials: tests of three hypotheses. *Ecology*, **74**, 1871–1883.

Flores, D. A. (2006a). Orden Paucituberculata Ameghino, 1894. In *Mamíferos de Argentina. Sistemática y Distribución*, ed. R. M. Bárquez, M. M. Díaz and R. A. Ojeda. Tucumán: SAREM, p. 45.

Flores, D. A. (2006b). Orden Microbiotheria Ameghino, 1897. In *Mamíferos de Argentina. Sistemática y Distribución*, ed. R. M. Bárquez, M. M. Díaz and R. A. Ojeda. Tucumán: SAREM, p. 46.

Goin, F. J. (1997). New clues for understanding Neogene marsupial radiations. In *A History of the Neotropical Fauna. Vertebrate Paleobiology of the Miocene in Colombia*, ed. R. F. Kay, R. H. Madden, R. L. Cifelli and J. J. Flynn. Washington: Smithsonian Institution Press, pp. 185–204.

Goin, F. J. (2003). Early marsupial radiations in South America. In *Predators with Pouches: The Biology of Carnivorous Marsupials*, ed. M. Jones, C. Dickman and M. Acher. Hobart, Australia: CSIRO Publishing, pp. 30–42.

Goin, F. J., Candela, A. M., Abello, A. and Oliveira, E. O. (2009). Earliest South American Paucituberculatans and their significance in the understanding of "pseudodiprotodont" marsupial radiations. *Zoological Journal of the Linnean Society*, **155**, 867–884.

Goin, F. J., Abello, M. A. and Chornogubsky, L. (2010). Middle Tertiary marsupials from central Patagonia (early Oligocene of Gran Barranca): understanding South America's *grande coupure*. In *The Paleontology of Gran Barranca: Evolution and Environmental Change through the Middle Cenozoic of Patagonia*, ed. R. H. Madden, A. A. Carlini, M. G. Vucetich and R. F. Kay. Cambridge: Cambridge University Press, pp. 69–105.

Goin, F. J., Zimicz, A. N., Forasiepi, A. M., Chornogubsky, L. and Abello, M. A. (in press). The rise and fall of South American Metatherians: contexts, adaptations, radiations, and extinctions. In *Origins and Evolution of Cenozoic South American Mammals*, ed. A. L. Rosenberger and M. F. Tejedor. Vertebrate Paleobiology and Paleoanthropology series. Dordrecht: Springer.

Grand, T. I. (1983). Body weight: its relationship to tissue composition, segmental distribution of mass and motor function. III. The Didelphidae of French Guyana. *Australian Journal of Zoology*, **31**, 299–312.

Greer, J. K. (1966). Mammals of Malleco Province, Chile. *Publications of the Museum of the Michigan State University (Biological Series)*, **3**, 49–152.

Heesy, C. P. (2008). Ecomorphology of orbit orientation and the adaptive significance of binocular vision in primates and other mammals. *Brain, Behavior and Evolution*, **71**, 54–67.

Hershkovitz, P. (1999). *Dromiciops gliroides* Thomas, 1894, last of the Microbiotheria (Marsupialia), with a review of the family Microbiotheriidae. *Fieldiana Zoology* (new series), **93**, 1–60.

Horovitz, I. and Sánchez-Villagra, M. R. (2003). A morphological analysis of marsupial mammal higher-level phylogenetic relationships. *Cladistics*, **19**, 181–212.

Hume, I. D. (2003). Nutrition of carnivorous marsupials. In *Predators with pouches: the biology of carnivorous marsupials*, ed. M. Jones, C. Dickman and M. Acher. Hobart, Australia: CSIRO Publishing, pp. 221–228.

Hume, I. D. (2006). Nutrition and digestion. In *Marsupials*, ed. J. P. Armati, C. R. Dickman and I. D. Hume. Cambridge: Cambridge University Press, pp. 137–158.

Kay, R. F. and Hiiemae, K. (1974). Mastication in *Galago crassicaudatus*: a cineflourographic and occlusal study. In *Prosimian biology*, ed. R. D. Martin, G. A. Doyle, and A. C. Walker. Pittsburgh: University of Pittsburgh Press, pp. 501–530.

Kay, R. F. and Hylander, W. L. (1978). The dental structure of mammalian folivores with special reference to primates and Phalangeroidea (Marsupialia). In *The Ecology of Arboreal Folivores*, ed. G. G. Montgomery. Washington D.C.: Smithsonian Institution Press, pp. 173–191.

Kay, R. F., Madden, R. H., Van Schaik, C. and Higdon, D. (1997). Primate species richness is determined by plant productivity: Implications for conservation. *Proceedings of the National Academy of Sciences, USA*, **94**, 13023–13027.

Kelt, D. A., Meserve, P. L. and Lang, B. K. (1994). Quantitative habitat associations of small mammals in a temperate rainforest in southern Chile: empirical patterns and the importance of ecological scale. *Journal of Mammalogy*, **75**, 890–904.

Kirsch, J. A. W. and Waller, P. F. (1979). Notes on the trapping and behavior of the Caenolestidae (Marsupialia). *Journal of Mammalogy*, **60**, 390–395.

Lee, A. K. and Cockburn, A. (1985). *Evolutionary Ecology of Marsupials*. Cambridge: Cambridge University Press.

Mann, G. (1955). Monito del monte *Dromiciops australis* Philippi. *Investigaciones Zoologicas Chilenas*, **2**, 159–166.

Marshall, L. G. (1976a). Revision of the South American Fossil Marsupial subfamily Abderitinae (Mammalia, Caenolestidae). *Publicaciones del Museo Municipal Ciencias Naturales "Lorenzo Scaglia,"* **2**, 57–90.

Marshall, L. G. (1976b). Fossil localities for Santacrucian (early Miocene) mammals, Santa Cruz Province, southern Patagonia, Argentina. *Journal of Paleontology*, **50**, 1129–1142.

Marshall, L. G. (1980). Systematics of the South American marsupial family Caenolestidae. *Fieldiana Geology* (new series), **5**, 1–145.

Marshall, L. G. (1982). Systematics of the South American marsupial family Microbiotheriidae. *Fieldiana Geology*, **10**, 1–75.

Meredith, R. W., Westerman, M., Case, J. A. and Springer, M. S. (2008). A phylogeny and timescale for marsupial evolution based on sequences for five nuclear genes. *Journal of Mammalian Evolution*, **15**, 1–26.

Meserve, P. L., Lang, B. K. and Patterson, B. D. (1988). Trophic relationships of small mammals in a Chilean temperate rain forest. *Journal of Mammalogy*, **69**, 721–730.

Miles, M. A., Sousa, A. A. and Póvoa, M. M. (1981). Mammal tracking and nest location in Brazilian forest with an improved spool-and-line device. *Journal of Zoology*, **195**, 331–347.

Miranda, C. L., Rossi, R. V., Silva Júnior, J. S., Lima, M. G. M. and Santos, M. P. D. (2009). Mammalia, Didelphimorphia, Didelphidae, *Metachirus nudicaudatus*, Municipality of José de Freitas, State of Piauí, Northeastern Brazil: Distribution extension. *Check List*, **5**, 360–363.

Moraes, E. A. Jr. (2004). Radio tracking of one *Metachirus nudicaudatus* (Desmarest, 1817) individual in Atlantic Forest of Southeastern Brasil. *Boletim do Museu de Biologia Mello Leitão*, **17**, 57–64.

Nilsson, M., Arnason, U., Spencer, P. B. S. and Janke, A. (2004). Marsupial relationships and a timeline for marsupial radiation in South Gondwana. *Gene*, **340**, 189–196.

Nilsson, M. A., Churakov, G., Sommer, M. *et al.* (2010). Tracking Marsupial evolution using archaic genomic retroposon insertions. *Plos Biology*, **8**, e1000436.

Ortiz-Jaureguizar, E. (2003). Relaciones de similitud, paleoecología y extinción de los Abderitidae (Marsupialia, Paucituberculata, Caenolestoidea). *Coloquios de Paleontología* (Vol. Extr.), **1**, 475–498.

Ortiz-Jaureguizar, E. and Cladera, G. (2006). Paleoenvironmental evolution of southern South America during the Cenozoic. *Journal of Arid Environments*, **66**, 489–532.

Palazzesi, L. and Barreda, V. (2007). Major vegetation trends in the Tertiary of Patagonia (Argentina): a qualitative paleoclimatic approach based on palynological evidence. *Flora*, **202**, 328–337.

Parker, S. A. (1973). An annotated checklist of the mammals of the Northern Territory. *Records of the South Australian Museum*, **16**, 1–57.

Pascual, R. and Bond, M. (1986). Evolución de los marsupiales cenozoicos de Argentina. *IV Congreso Argentino de Paleontología y Bioestratigrafía*, Mendoza, Actas **2**, 143–150.

Pascual, R. and Ortiz-Jaureguizar, E. (1990). Evolving climates and mammal faunas in Cenozoic South America. *Journal of Human Evolution*, **19**, 23–60.

Pascual, R., Ortiz-Jaureguizar, E. and Prado, J. L. (1996). Land mammals: Paradigm of Cenozoic South American geobiotic evolution. In *Contribution of Southern South America to Vertebrate Paleontology*, ed. G. Arratia. München: Müncher Geowissenscaftliche Abhandlungen (A) Verlag Dr. Fiedrich Pfeil, **30**, pp. 265–319.

Patterson, B. D. (2007). Order Paucituberculata Ameghino, 1894. In *Mammals of South America*, Vol. 1, *Marsupials, Xenarthrans, Shrews, and Bats*, ed. A. L. Gardner. Chicago and London: University of Chicago Press, pp. 119–127.

Patterson, B. D. and Gallardo, M. H. (1987). *Rhyncholestes raphanurus*. *Mammalian Species*, **286**, 1–5.

Patterson, B. D. and Rogers, M. A. (2007). Order Microbiotheria Ameghino, 1889. In *Mammals of South America*. Vol. 1, *Marsupials, Xenarthrans, Shrews, and Bats*, ed. A. L. Gardner. Chicago and London: University of Chicago Press, pp. 117–119.

Pérez, M. A. (2010). Sistemática, ecología y bioestratigrafía de Eocardiidae (Rodentia, Hystricognathi, *Cavioidea) del Mioceno temprano y medio de Patagonia*. Unpublished Doctoral thesis, Universidad Nacional de La Plata.

Phillips, M. J., McLenachan, P. A., Down, C., Gibb, G. C. and Penny, D. (2006). Combined mitochondrial and nuclear DNA sequences resolve the interrelations of the major Australasian marsupial radiations. *Systematic Biology*, **55**, 122–137.

Ramirez, O., Arana, M., Bazán, E., Ramirez, A. and Cano, A. (2007). Assemblages of birds and mammals communities in two major ecological units of the Andean highland plateau of southern Peru. *Ecología Aplicada*, **6**, 139–148.

Rodríguez-Cabal, M. A., Amico, G. C., Novaro, A. J. and Aizen, M. A. (2008). Population characteristics of *Dromiciops gliroides* (Philippi, 1893), an endemic marsupial of the temperate forest of Patagonia. *Mammalian Biology*, **73**, 74–76.

Sanson, G. D. (1985). Functional dental morphology and diet selection in dasyurids (Marsupialia: Dasyuridae). *Australian Mammalogy*, **8**, 239–247.

Simpson, G. G. (1933). The "Plagiaulacoid" type of mammalian dentition. *Journal of Mammalogy*, **14**, 97–107.

Sinclair, W. J. (1906). Marsupialia of the Santa Cruz beds. In *Reports of the Princeton University Expeditions of Patagonia 1896–1899*. Vol. 4. *Palaeontology I*. Part 3, ed. W. B. Scott. Princeton: Princeton University Press, pp. 333–460.

Smith, A. P. and Lee, A. (1984). The evolution of strategies for survival and reproduction in possums and gliders. In *Possums and Gliders*, ed. A. Smith and I. Hume. Chipping Norton, Australia: Surrey Beatty and Sons, pp. 17–34.

Staver, C. A., Bond, W. J., Stock, W. D., van Rensburg, S. J. and Waldram, M. S. (2009). Browsing and fire interact to suppress tree density in an African savanna. *Ecological Applications*, **19** (7), 1909–1919.

Strait, S. G. (1993a). Differences in occlusal morphology and molar size in frugivores and faunivores. *Journal of Human Evolution*, **25**, 471–484.

Strait, S. G. (1993b). Molar morphology and food texture among small bodied insectivorous mammals. *Journal of Mammalogy*, **74**, 391–402.

Szalay, F. S. (1982). A new appraisal of marsupial phylogeny and classification. In *Carnivorous Marsupials*, ed. M. Archer. Sydney, Australia: Royal Zoological Society of New South Wales, pp. 621–640.

Szalay, F. S. (1994). *Evolutionary History of the Marsupials and an Analysis of Osteological Characters*. New York: Cambridge University Press.

Szalay, F. S. and Sargis, E. J. (2001). Model-based analysis of postcranial osteology of marsupials from the Paleocene of Itaboraí (Brazil) and the phylogenetics and biogeography of Metatheria. *Geodiversitas*, **23**, 139–302.

Tauber, A. A. (1997). Paleoecología de la Formación Santa Cruz (Mioceno Inferior) en el extremo sudeste de la Patagonia. *Ameghiniana*, **34**, 517–529.

Vieira, E. M. and Astúa de Moraes, D. (2003). Carnivory and insectivory in Neotropical marsupials. In *Predators with pouches: the biology of carnivorous marsupials*, ed. M. Jones, C. Dickman and M. Acher. Hobart, Australia: CSIRO Publishing, pp. 267–280.

Vieira, E. M. and Monteiro-Filho, E. L. A. (2003). Vertical stratification of small mammals in the Atlantic rain forest of south-eastern Brazil. *Journal of Tropical Ecology*, **19**, 501–507.

Webb, S. D. (1978). A history of savanna vertebrates in the New World. Part II: South America and the Great Interchange. *Annual Review of Ecology and Systematics*, **9**, 393–426.

11 Paleoecology of the mammalian carnivores (Metatheria, Sparassodonta) of the Santa Cruz Formation (late Early Miocene)

Francisco J. Prevosti, Analía M. Forasiepi, Marcos D. Ercoli, and Guillermo F. Turazzini

Abstract

South America had an endemic mammalian fauna for much of the Cenozoic, largely evolved during its long isolation. The predator guild was mainly occupied by metatherians (Sparassodonta), as well as large terrestrial birds (Phorusrhacidae), agile terrestrial crocodiles (Sebecidae), and giant snakes (Madtsoiidae). Sparassodonta was a diverse clade, recorded from the Paleocene to the Middle Pliocene, with its acme in the late Early Miocene (Santacrucian Age). In this chapter, we review the paleoecology of the sparassodonts known from the Santa Cruz Formation and include new results obtained by geometric morphometric analyses. The Santa Cruz Formation contains 11 sparassodont species: six Hathliacynidae (*Acyon tricuspidatus*, *Cladosictis patagonica*, *Sipalocyon gracilis*, *Sipalocyon obusta*, *Pseudonotictis pusillus*, *Perathereutes pungens*) and five Borhyaenoidea (*Prothylacynus patagonicus*, *Lycopsis torresi*, and three Borhyaenidae, *Borhyaena tuberata*, *Acrocyon sectorius*, and *Arctodictis munizi*). These sparassodonts were mainly hypercarnivores exhibiting different locomotor abilities (from scansorial to terrestrial), and a wide range of body masses (from 1 kg to more than 50 kg). The reconstruction of the Santacrucian predator guild suggests that there was good ecological separation within the sparassodonts, determined by particular combinations of body size, locomotion, and diet. The diversity of sparassodonts recorded in the Santa Cruz Formation (11 species) and in the Estancia La Costa Member (seven species), is similar to that observed in present and past placental hypercarnivore communities.

Resumen

Durante la mayor parte del Cenozoico, cuando se encontraba aislada de otros continentes, América del Sur poseía una fauna de mamíferos endémica. El gremio de los depredadores estaba ocupado principalmente por mamíferos metaterios (Sparassodonta), así como por grandes aves terrestres (Phorusrhacidae), cocodrilos terrestres (Sebecidae) y víboras gigantes (Madtsoiidae). Los Sparassodonta constituyen un clado diverso con registros desde el Paleoceno al Plioceno medio, siendo el Mioceno Temprano tardío (Edad Santacrucense) el momento de mayor diversidad del grupo. En este capítulo, se revisa la paleoecología y la diversidad de los esparasodontes de la Formación Santa Cruz y se presentan nuevos resultados obtenidos con análisis de morfometría geométrica. La Formación Santa Cruz posee una rica diversidad de esparasodontes representada por once especies: seis Hathliacynidae (*Acyon tricuspidatus*, *Cladosictis patagonica*, *Sipalocyon gracilis*, *Sipalocyon obusta*, *Pseudonotictis pusillus* y *Perathereutes pungens*), cinco Borhyaenoidea (*Prothylacynus patagonicus*, *Lycopsis torresi* y tres Borhyaenidae, *Borhyaena tuberata*, *Acrocyon sectorius* y *Arctodictis munizi*). Estos esparasodontes fueron mayormente hipercarnívoros, con diferentes habilidades locomotoras (desde trepadores a terrestres) y un amplio rango de tamaños corporales (entre 1 kg y más de 50 kg). La reconstrucción del gremio de los depredadores del Santacrucense sugiere que había una buena separación entre los esparasodontes, dada por una combinación de diferencias en los tamaños corporales, los hábitos locomotores y la dieta. La diversidad de esparasodontes registrados en la Formación Santa Cruz (once especies) y en el Miembro Estancia La Costa (siete especies) es similar a la observada en comunidades de hipercarnívoros placentarios actuales y del pasado.

11.1 Introduction

During the Cenozoic, owing to its geographic isolation, South America developed a particular continental fauna with strange groups in comparison with other continents: xenarthrans (Cingulata and Pilosa), metatherians (opossums and fossil relatives), and native ungulates (e.g. Notoungulata, Astrapotheria, Litopterna, Pyrotheria), among others (Simpson, 1950, 1980; Patterson and Pascual, 1972; Reig, 1981; Pascual, 2006). The predator guild was mainly occupied by metatherians (Mammalia, Sparassodonta) and by

Early Miocene Paleobiology in Patagonia: High-Latitude Paleocommunities of the Santa Cruz Formation, ed. Sergio F. Vizcaíno, Richard F. Kay and M. Susana Bargo. Published by Cambridge University Press. © Cambridge University Press 2012.

"terror birds" (Phorusrhacidae), large, agile, terrestrial crocodiles (Sebecidae), and giant snakes (Madtsoiidae) (Simpson, 1950, 1980; Patterson and Pascual, 1972; Reig, 1981; Gasparini, 1984; Albino, 1996; Pascual, 2006; Riff et al., 2010). In contrast, the predatory guilds of the Old World and North America were mainly filled by two groups of placental mammals (Creodonta and Carnivora) (Savage, 1977; Martin, 1989; Hunt, 1996; Werdelin, 1996; Flynn and Wesley-Hunt, 2005).

Known sparassodonts had a wide range of body sizes (from approximately 1 kg opossum-like species to large forms of nearly 100 kg; Argot, 2003a, b, 2004a, b, c), diverse locomotor habits (from scansorial-semi-arboreal to terrestrial-cursorial; Argot, 2004a, b, c, vide infra), and relatively limited variation within the carnivorous diet (mostly strict hypercarnivorous and a few omnivorous taxa). Some species, such as the saber-tooth *Thylacosmilus atrox* Riggs, 1933, developed extreme morphotypes, with large hypertrophied upper canines that resemble those of saber-tooth felids (Felidae, Machairodontinae; e.g. Simpson, 1980). The oldest record of Sparassodonta is from the Paleocene (probably Tiupampan Age, ~64–62 Ma), while the youngest fossils come from the Middle Pliocene (Chapadmalalan Age, ~3.3 Ma). The acme of the group was reached during the late Early Miocene (Santacrucian Age, 18–16 Ma) when their diversity increased up to 11 species (Figs. 11.1, 11.2, and 11.3; Sinclair, 1906; Argot, 2004a; Forasiepi, 2009; Prevosti et al., 2009; Prevosti et al., in press). Although most of the known sparassodonts are from Argentina, some fossil material comes from localities in Bolivia, Brazil, Chile, Colombia, and Uruguay (Forasiepi, 2009). In South American ecosystems, sparassodonts were gradually replaced by the placental carnivores which immigrated during the Late Miocene (8 Ma) to Pleistocene. Sparassodonts disappear from the fossil record in the middle Pliocene (see Goin, 1989; Forasiepi et al., 2007; Prevosti et al., 2009).

In this chapter, we review the available paleoecological information for the sparassodonts of the Santa Cruz Formation (late Early Miocene, Santa Cruz Province, Argentina), including some new studies using geometric morphometrics and other statistical techniques.

11.2   Systematics and phylogeny

Carnivorous South American metatherians are a monophyletic group, Sparassodonta, consisting of more than 50 species, 11 of which are currently recognized from the Santa Cruz Formation (Table 11.1) (e.g. Cabrera, 1927; Simpson, 1945; Marshall, 1976, 1978, 1979, 1981; Villarroel and Marshall, 1982, 1983; Hoffstetter and Peter, 1983; Goin and Pascual, 1987; Marshall et al., 1990; Goin, 1997; Babot et al., 2002; Forasiepi et al., 2003, 2006; Shockey and Anaya, 2008; Forasiepi and Carlini, 2010). Sparassodonta is the clade that includes hathliacynids, borhyaenoids, and all the taxa that are more closely related to

hathliacynids and borhyaenoids than to living marsupials (Forasiepi, 2009). In this sense it is a sister taxon of the crown group Marsupialia (Fig. 11.4). Several sparassodont stem taxa represent successive sister taxa of hathliacynids plus borhyaenoids (e.g. *Patene* Simpson, 1935, *Nemolestes* Ameghino, 1902, *Stylocynus* Mercerat, 1917, *Hondadelphys* Marshall, 1976; Forasiepi, 2009). The Santacrucian sparassodonts are arranged within Hathliacynidae and Borhyaenoidea.

Hathliacynidae includes small- to medium-sized sparassodonts, with long slender, skulls and dentaries. Their anatomy is relatively more uniform than among borhyaenoids. Hathliacynidae includes the common ancestor of *Sipalocyon* Ameghino, 1887 and *Cladosictis* Ameghino, 1887 and all the taxa that are more closely related to them than to borhyaenoids (Forasiepi, 2009; Fig. 11.4). Six hathliacynids are recognized in the Santacrucian beds (Table 11.1).

Acyon tricuspidatus was originally described by Ameghino (1887) on the basis of a fragmentary dentary. Later, Mercerat (1891) regarded this species as belonging to the genus *Hathliacynus*, whereas Cabrera (1927) synonymized it with *Cladosictis*. Marshall (1981) considered *Anatherium* Ameghino, 1887 a valid genus and included the type specimen of *Ac. tricuspidatus* within the hypodygm of *Anatherium defossus* Ameghino, 1887. More recently, Forasiepi et al. (2006) returned to Ameghino's concept of the genus *Acyon* based on new material from the Middle Miocene of Bolivia. *Acyon* species are among the largest known hathliacynids and are morphologically similar to *Cladosictis*, which represents its sister taxon (Forasiepi et al., 2006).

Sipalocyon is represented by two species in the Santacrucian beds: *Sipalocyon gracilis* Ameghino, 1887 (Figs. 11.1a, b, 11.3a–c) and *Sipalocyon obusta* Ameghino, 1891. The former, which is one of the most frequently found in the Santa Cruz Formation, was based on the material collected during the first expedition to Patagonia led by Carlos Ameghino. Several specimens were later referred to this species and, similarly to other well-known sparassodonts, its synonym list is large and includes 16 other different names (Marshall, 1981). In turn, *S. obusta* is known by a single specimen, which is "*virtually identical to ... S. gracilis*" (Marshall, 1981: 60), but with a shallower and more slender dentary and more reduced m4 talonid. Based on the intraspecific variability observed in *S. gracilis* and other sparassodont species, *S. obusta* may represent a synonym of the former. However, we follow Marshall's view pending taxonomic revision.

Perathereutes pungens Ameghino, 1891 is a small hathliacynid, slightly smaller than *Sipalocyon* spp., and is known by a few fragmentary specimens described by Ameghino (1891) (Fig. 11.1c). The synonymy list of the species includes *Abderites altiramis* Ameghino, 1894 (previously referred to *Epanorthus aratae* Ameghino, 1889) which was first considered as a member of the Paucituberculata and

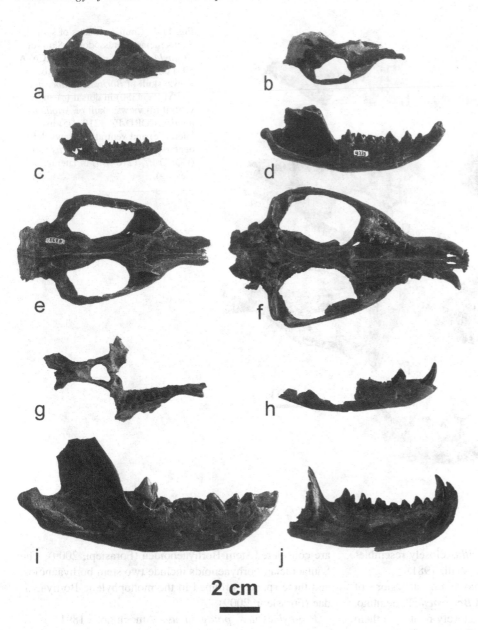

Fig. 11.1. New sparassodont specimens collected in the Santa Cruz Formation. Skull of *Sipalocyon gracilis* (MPM-PV 4316) in dorsal (a) and ventral (b) views; right dentary of *Perathereutes pungens*? (MPM-PV 4322) in lateral view (c); right dentary of *Cladosictis patagonica* (MPM-PV 4333) in lateral view (d); skull of *Cladosictis patagonica* (MPM-PV 4333) in dorsal view (e); skull of *Cladosictis patagonica* (MPM-PV 4326) in ventral view (f); fragment of skull of *Lycopsis torresi* (MPM-PV 4328) in ventral view (g); right fragmentary dentary of *Lycopsis torresi* (MPM-PV 4328) in lateral view (h); right dentary of *Borhyaena tuberata* (MPM-PV 4330) in lateral view (i); left dentary (reflected) of *Prothylacynus patagonicus* (MPM-PV 4318) in lateral view (j).

2 cm

later identified by Marshall (1981) as *P. pungens* and, therefore, as a Sparassodonta.

Cladosictis patagonica Ameghino (1887) (Figs. 11.1d–f, 11.3d–j) is the most abundant hathliacynid recovered from the Santa Cruz Formation. This species is fox-like in size and shape. Like other sparassodonts known by a large amount of abundant material, the specimens assigned to *C. patagonica* encompass considerable intraspecific variability: some specimens are larger and more robust, with deeper dentaries and more crowded dentition than others. The taxonomy of the species is not exempt from controversy, mainly because of

the strong variability observed among the specimens assigned. More than 21 other different names have been applied to this taxon (see synonymy list in Marshall, 1981).

Pseudonotictis pusillus (Ameghino, 1891) is the smallest sparassodont species recovered in outcrops of the Santa Cruz Formation. Unfortunately, this taxon is known only by limited and fragmentary specimens. The genus was created by Marshall (1981) on the basis of Ameghino's species *Sipalocyon pusillus* (Ameghino, 1891) and the species *Hathliacynus kobyi* Mercerat, 1891 (considered as *Perathereuthes kobyi* by Cabrera, 1927). According to

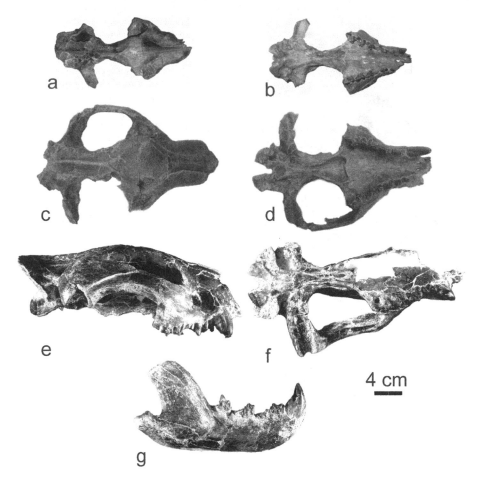

Fig. 11.2. Cranial remains of some Santacrucian sparassodonts. Skull of *Prothylacynus patagonicus* (MACN-A 5931) in dorsal (a) and ventral (b) views; skull of *Borhyaena tuberata* (MPM-PV 4380) in dorsal (c) and ventral (d) views; skull of *Artodictis munizi* (CORD-PZ-1210–1/5) in lateral (e) and ventral (f) views; left dentary (reflected) of *Artodictis munizi* (CORD-PZ-1210–1/5) in lateral view (g).

4 cm

Marshall's view, *Pseudonotictis pusillus* closely resembles the Huayquerian genus *Notictis* (Marshall, 1981).

Borhyaenoidea includes the common ancestor of *Prothylacynus* Ameghino, 1891 and *Borhyaena* Ameghino, 1887 and all the taxa that are more closely related to them than to hathliacynids (Forasiepi, 2009; Fig. 11.4). Borhyaenoidea are mostly medium- to large-sized sparassodonts. Some taxa have relatively slender jaws and teeth whereas others are massive and robust, with deep dentaries sometimes fused at the symphysis. The taxa in this group are more diverse than among hathliacynids, with differences in the architecture of the cranium, dentition, and postcranial skeleton. Borhyaenoidea includes Proborhyaenidae (the monophyly of which has been recently questioned; Babot *et al.*, 2002), Thylacosmilidae, Borhyaenidae, and successive sister taxa placed at the base of the clade (Forasiepi, 2009). The classically recognized family Prothylacynidae (e.g. Ameghino, 1894; Marshall, 1979; Marshall *et al.*, 1990) is not recognized because the characters used to diagnose the group are plesiomorphies. "Prothylacynids"

are considered stem Borhyaenoidea (Forasiepi, 2009). The Santacrucian borhyaenoids include two stem borhyaenoids, and three species grouped in the monophyletic Borhyaenidae (Forasiepi, 2009).

Prothylacynus patagonicus Ameghino, 1891 is a common medium-sized sparassodont of the Santacrucian (Figs. 11.1j, 11.2a, b, 11.3k–m). This species was described by Ameghino (1891), who compared it and allied it with the Australian marsupial carnivore *Thylacynus cynocephalus* (Harris 1808), as others did later (e.g. Sinclair, 1906). *Prothylacynus* is a stem borhyaenoid (Forasiepi, 2009). *Prothylacynus patagonicus* is represented by several specimens that include the dentitions, skulls, and postcranial skeletons. As among other sparassodonts, the variability is significant, and thus its synonymy list is long, including five different names (Marshall, 1979).

Lycopsis torresi Cabrera, 1927 is a medium-sized sparassodont that is rare in the Santacrucian in comparison with other borhyaenoids (Fig. 11.1g, h). The genus also includes "*Lycopsis*" *longirostrus* Marshall, 1977 (Laventan Age,

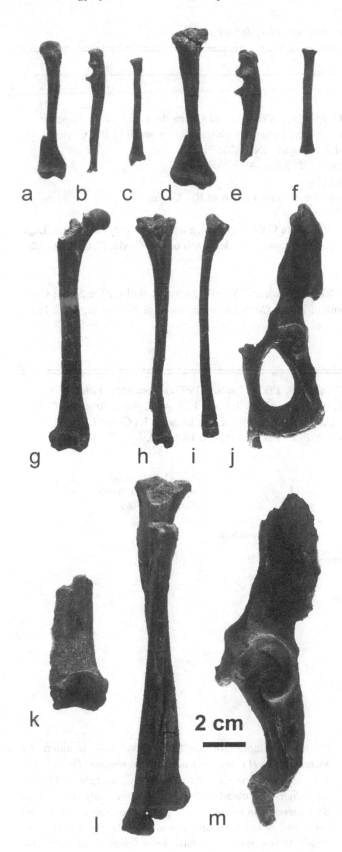

Middle Miocene beds of Colombia (Marshall, 1977a), and "*Lycopsis*" *viverensis* Forasiepi, Goin, and Di Martino, 2003 (Chasicoan Age, Late Miocene beds of Argentina; Forasiepi *et al.*, 2003). The monophyletic nature of the genus is far from being certain because the characters used in the diagnosis are plesiomorphies (Forasiepi, 2009); nonetheless, at least provisionally, the species names are maintained as they were originally described. Based on the phylogenetic position of "*L.*" *longirostrus* (Forasiepi, 2009), other species currently grouped in the genus would represent basal stem taxa of Borhyaenoidea (although they probably do not form a monophyletic group).

Borhyaena tuberata Ameghino, 1887 is the most common large sparassodont found in the Santacrucian (Figs. 11.1i, 11.2c, d; *vide infra*). This species was named by Ameghino (1887) on the basis of an isolated P3 collected during the first trip to southern Patagonia led by Carlos Ameghino during the austral summer of 1887. Today, several other specimens are available that include cranial and postcranial skeletons. The variability among specimens is considerable: some are larger, more massive, and more robust than others; the development of crests in the skull as well as the disposition of the teeth and the development of the canines also are somewhat variable. These differences are interpreted as intraspecific variability (similar to that currently observed among large carnivorous eutherians; Kurtén, 1967, 1973) attributed to sex or age, or even to small differences in the stratigraphic level from which the specimens were found (Marshall, 1978). At least 14 other names were applied to *B. tuberata* (synonymy list in Marshall, 1978).

Acrocyon sectorius Ameghino, 1887 is a borhyaenid very similar in size and morphology to *Borhyaena tuberata*. The species was erected by Ameghino (1887) and widely accepted (e.g. Cabrera, 1927; Marshall, 1978). However, a recent revision of the specimens assigned to this genus led to doubt about the validity of the generic name (Forasiepi, 2009). *Borhyaena tuberata* is represented by

Fig. 11.3. Postcranial remains of the new specimens of Sparassodonta collected in the Santa Cruz Formation. Right humerus *Sipalocyon gracilis* (MPM-PV 4316) in anterior view (a); left ulna and radius of *Sipalocyon gracilis* (MPM-PV 4316) in lateral view (b and c, respectively); right humerus of *Cladosictis patagonica* (MPM-PV 4329) in anterior view (d); left ulna and radius of *Cladosictis patagonica* (MPM-PV 4329) in lateral and anteriorposterior views (e and f, respectively); right femur and tibia of *Cladosictis patagonica* (MPM-PV 4333) in anterior view (g and h, respectively); right fibula of *Cladosictis patagonica* (MPM-PV 4333) in posterior view (i); pelvis of *Cladosictis patagonica* (MPM-PV 4333) in left-lateral view (j); distal fragment of left femur and left tibia + fibula of *Prothylacynus patagonicus* (MPM-PV 4318) in anterior view (k and l, respectively); pelvis of *Prothylacynus patagonicus* (MPM-PV 4318) in left-lateral view (m).

Table 11.1. *Species of Sparassodonta from the Santa Cruz Formation, including localities*

Species	Localities
Borhyaenidae	
Arctodictis munizi	ML; PLC; CT
Borhyaena tuberata	ML; MO; Karaiken; LC; Río Santa Cruz; Yegua Quemada; Estancia La Angelina; 12 miles north of Cape Fairthweather; La Costa, 8 miles south of Río Coyle; 8, 10, and 12 miles south of Río Coyle; ANF; CB; PLC
Prothylacynus patagonicus	MO; CdV; Estancia Felton; CT; Killik Aike Norte; PLC
Lycopsis torresi	Río Santa Cruz; PLC; ELC
Acrocyon sectorius	Río Santa Cruz; Shehuén; about 5 miles south of Río Coyle
Hathliacynidae	
Cladosictis patagonica	ML; MO; LC; Shehuén; Río Santa Cruz; Río Gallegos; Jack Harvey; Karaiken; Lago Pueyrredón; Río Coyle; CdV; about 10 miles north of Río Coyle; PLC; ANF; CS; CB; ELC
Acyon tricuspidatus	Río Santa Cruz; Shehuén
Sipalocyon gracilis	Río Santa Cruz?; ML; MO; LC; Shehuén; Yegua Quemada; Killik Aike Norte; CdV; La Costa, 7 miles south of Puerto Coyle; 10 miles south of Río Coyle; ELC; PLC.
Sipalocyon obusta	MO
Pseudonotictis pusillus	ML; MO
Perathereutes pungens	MO; La Cueva; PLC

Locality data from Sinclair (1906), Cabrera (1927), Marshall (1978, 1979, 1981), Tauber (1997a), Forasiepi *et al.* (2004), Kay *et al.* (2008), and Vizcaíno *et al.* (2010) and Vizcaíno *et al.*, (Chapter 1: Figs. 1.1 and 1.2). ANF: Anfiteatro; CB: Campo Barranca; CdV: Cañadón de Las Vacas, CS: Cañadón Silva; CT: Cañadón las Totoras; ELC: Estancia La Costa; LC: La Cueva; ML: Monte León; MO: Monte Observación (= Cerro Observatorio); PLC: Puesto Estancia La Costa.

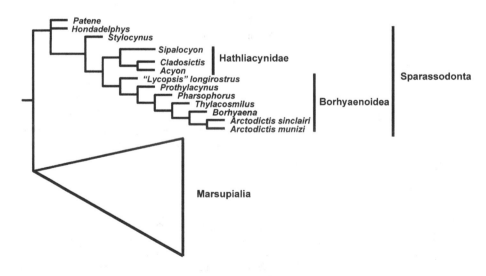

Fig. 11.4. Phylogenetic tree of Sparassodonta (modified from Forasiepi, 2009).

several specimens with a notable intraspecific variation. *Acrocyon* Ameghino, 1887 may be a synonym of *Borhyaena*. A detailed revision of all the specimens assigned to *A. sectorius* and *B. tuberata* is required to evaluate the taxonomy of these taxa.

Arctodictis munizi Mercerat, 1891 (Fig. 11.2e–g) is the largest Santacrucian borhyaenid and the largest post-Deseadan sparassodont species known (Forasiepi *et al.*,

2004; Forasiepi, 2009). The taxon was founded by Mercerat (1891) and subsumes the species *Dynamyctis fera* and *Borhyaena fera* named by Ameghino (1891). *Arctodictis* is clearly a borhyaenid closely related to *Borhyaena* (Forasiepi *et al.*, 2004; Forasiepi, 2009). As compared with *Borhyaena* and despite its larger size, relatively few specimens have been found in the Santacrucian outcrops.

Table 11.2. *Body mass and relative grinding area of lower last molar of Santacrucian sparassodonts*

Species	BM Argot, 2003a, b; 2004b, c	BM Wroe *et al.*, 2004	BM Vizcaíno *et al.*, 2010	BM This chapter	RGA
Borhyaenidae					
Arctodictis munizi		51.60	37.00	>37.00[a]	0
Borhyaena tuberata	24.15	21.40	23.31	36.40[b]	0
Prothylacynus patagonicus	32.05	26.80	13.83	31.79[b]	0.17
"Lycopsis" longirostrus	16.85	12.80		29.77[b]	
Lycopsis torresi		19.40		20.07[a]	0.30
Acrocyon sectorius		28.70		11.49[a]	0
Hathliacynidae					
Cladosictis patagonica	6.05	4.00	3.7	6.60[b]	0.17
Acyon tricuspidatus		8.00		4.30[a]	0.30
Sipalocyon gracilis	3.00		1.93	2.11[b]	0.33
Sipalocyon obusta				2.06[a]	0.27
Pseudonotictis pusillus				1.17[b]	0.30
Perathereutes pungens		2.5		1.00[a]	0.34

BM: body mass (mean figure in kg); **RGA**: relative grinding area.
[a] Estimation based on dental measurements.
[b] Estimation based on centroid size.

11.3 Previous paleobiological studies

Previous studies inferred body mass, locomotion, and diet of sparassodonts based mainly on descriptive and qualitative approaches, as well as on linear measurements, and comparison of their anatomy with living placental and marsupial carnivores (e.g. Sinclair, 1906; Marshall, 1977b, 1978, 1979, 1981; Muizon, 1998; Forasiepi *et al.*, 2004; Forasiepi, 2009).

Marshall (1977b, 1978) separated the taxa into large, medium, and small, according to the absolute size of the dentition and long bones. More recent studies estimated body mass of sparassodonts using allometric equations. These equations are based on a variety of mammals that are not phylogenetically close to sparassodonts (Argot, 2003a, b, 2004b, c; Wroe *et al.*, 2004; Vizcaíno *et al.*, 2010; *vide infra*). Argot (2003a, b; 2004b, c) used the allometric equations published by Anyonge (1993) based on the area of the distal condyles of the femur, femur length, and circumference of the femur at midshaft of living carnivores. Wroe *et al.* (2004) used different equations, such as that of Anderson *et al.* (1985; based on the minimum circumference of the femur shaft of a wide sample of living mammal) and Myers (2001; based on upper and lower molar row lengths of living marsupials). Vizcaíno *et al.* (2010) estimate the body size of Santacrucian sparassodonts through the equations of Van Valkenburgh (1990; based on the length of the lower carnassial of a sample of living Carnivora), and Gordon (2003; based on the length of the third and fourth lower molars of living marsupials). With the exception of the estimation by Argot (2003a, 2004a) for *B. tuberata*, all the estimates of Vizcaíno *et al.* (2010) are lower than those of previous authors (see Table 11.2).

Sinclair (1906) was one of the first authors who discussed the ecology of Sparassodonta, mostly based on the fine skeletons collected from the Santacrucian outcrops. He suggested that *S. gracilis* and *C. patagonica* were plantigrade or semi-plantigrade with arboreal or semi-arboreal habits. *Prothylacynus patagonicus* and *B. tuberata* were described as having plantigrade feet and terrestrial habits, the former being more agile than the latter (Sinclair, 1906). Marshall (1977b, 1978) extended Sinclair's inferences about the terrestrial locomotion to *A. munizi* and largely agreed with Sinclair's assessments about the locomotor abilities of the other Santacrucian taxa. Argot and Muizon (Muizon, 1998; Argot, 2003a, b, 2004a, b, c; Muizon and Argot, 2003) undertook more detailed studies of sparassodont postcrania, based on comparisons and reconstruction of morphological structures and morphometric indices (e.g. crural index: tibia/femur; brachial index: radius/humerus). In general, the conclusions were as follows: *Ps. pusillus* probably had arboreal abilities (Argot, 2003b), *S. gracilis* was probably scansorial (Argot, 2003b, 2004b), *C. patago-nica* was a scansorial ambush predator with plantigrade forefeet and semi-digitigrade hindfeet (Argot, 2003b, 2004b), *P. patagonicus* was a scansorial ambush predator with its forefoot capable of supination and grasping and a flexible vertebral column (Argot, 2003a, 2004b), "*L.*" *longirostrus* was interpreted as a terrestrially adapted form with grasping capabilities and semi-digitigrade forefoot (Argot, 2004a, b), *A. sinclairi* was a terrestrial (Argot, 2004a, b), and *B. tuberata* was terrestrial with cursorial tendencies including digitigrade fore- and hindlimbs that would have allowed hunting over some distance (Argot, 2003a, 2004b). Forasiepi (1999, 2009) suggested that *A. sinclairi* was a plantigrade terrestrial predator and that digitigrady in *B. tuberata* was possible, but not as yet reliably demonstrated. She also suggested that *B. tuberata* is the most cursorial sparassodont, but its morphology is still not to the extreme seen in living cursorial placental carnivores.

The diet of sparassodonts was strictly carnivorous as suggested by the morphology of the dentition (e.g. Muizon and Lange-Badré, 1997). However, there are subtle variations in the tooth architecture that suggest differences in the type of food ingested. Previous authors made inferences on the basis of comparison with the dentition of living marsupial and placental carnivores. Hathliacynids and basal borhyaenoids (*P. patagonicus* and *L. torresi*) were considered omnivores, or mainly so (Marshall, 1977b, 1978, 1979, 1981). Because of their size, hathliacynids were viewed as analogs of didelphids, mustelids, or canids (Marshall, 1977b, 1978), preying upon small vertebrates (e.g. small mammals, reptiles, birds, and amphibians), invertebrates, and eggs (Argot, 2004a). Despite large differences in the architecture of the dentition and skeleton, *P. patagonicus* and *L. torresi* were compared in earlier studies with ursids and procyonids (Marshall, 1977b, 1978). The morphology of the dentition and the preservation of rodent remains (*Scleromys colombianus* Fields, 1957) in the body cavity of "*Lycopsis*" *longirostrus* from Colombia led Marshall (1977a) to conclude that the consumption of meat was important. On the other hand, the architecture of the postcranial skeleton of *P. patagonicus* is more in agreement with an active predator than that of the placentals (ursids and procyonids) mentioned above, preying upon small mammals and birds, and with the capability for dragging and transporting large prey (Argot, 2003a, 2004a). Large borhyaenids (i.e. *B. tuberata*, *A. sectorius*, and *A. munizi*) have been considered as specialized carnivores and compared with canids and felids (Marshall, 1977b, 1978). The robust nature of the teeth, bulbous roots, and robust, deep, and sometimes fused dentaries suggest that probably at least *A. munizi* and *B. tuberata* had some capacity to break hard elements like bones, as currently occurs among scavengers (Argot, 2004a; Forasiepi *et al.*, 2004), but probably less so

than in specialist bone-cracking extant species such as living hyaenids.

In the Results section of this chapter, we present the inferences that we have recently obtained about body mass, locomotion, and diet, using geometric morphometric analyses and multivariate studies combined with morphometric indices.

11.4 Materials and methods

In this contribution, we included a broad sample of Sparassodonta collected from the nineteenth century up to the present (see Appendix 11.1), including material collected by researchers from the Museo de La Plata/Duke University expeditions between 2003 and 2011 at the coastal localities of Santa Cruz Province (see Vizcaíno *et al.*, Chapter 1, Figs. 1.1 and 1.2, and Appendix 1.1). Where possible, we analyzed the Santacrucian taxa, although in some cases we based our conclusions on closely related and better known taxa of similar morphology from other ages. For inferences on locomotion, we included *Arctodictis sinclairi* from Early Miocene (Colhuehuapian) levels at Gran Barranca (Argentina) and "*Lycopsis*" *longirostrus* from Middle Miocene (Laventan) levels of La Venta (Colombia), because there are either limited or no postcranial remains for the Santacrucian species. Because there is no evidence supporting the monophyly of the *Lycopsis* Cabrera, 1927 species (Forasiepi, 2009; see above), we stress that the extrapolations from "*L.*" *longirostrus* to *L. torresi* are only tentative.

Body mass was estimated by allometric equations using the centroid size from geometric morphometric analyses of the postcranial elements (Ercoli, 2010; Ercoli and Prevosti, 2012). For some species that do not have preserved postcranial remains or for which the remains are very fragmentary (e.g. *Lycopsis torresi*, *Acrocyon sectorius*, *Perathereutes pungens*, *Acyon tricuspidatus*, *Sipalocyon obusta*, and *Arctodictis munizi*), we used the estimations published by Vizcaíno *et al.* (2010) based on the dentition, and provided estimates for the species that were not included in that paper using the same equations (see above). Phylogenetic eigenvector regression was used to correct the bias imposed by the phylogenetic patterns (Diniz-Filho *et al.*, 1998).

We use the geometric morphometric approach (following Zelditch *et al.*, 2004) to infer locomotion. Landmarks and semi-landmarks were defined and placed on the distal portion of the humerus in anterior view, the proximal part of the ulna in lateral view, and the proximal articular surface of the tibia (Ercoli, 2010; Ercoli and Prevosti, 2012; Ercoli *et al.*, 2012; Fig. 11.5). We used these to explore the relationship between form and locomotor habits, based on a wide sample, including over 200 specimens, of more than 100 species of living Carnivora (Canidae, Ursidae, Felidae,

Mustelidae, Mephitidae, Viverridae, and Herpestidae) and Marsupialia (Didelphimorphia and Dasyuroidea). The landmarks and semi-landmarks were placed using the tpsDig 2.14 software (Rohlf, 2009), while the Procrustes superimposition and the Relative Warp analysis were performed by using the tpsRelw software (Rohlf, 2008). Several analyses were carried out separately for metatherians and eutherians and with differing landmark configurations to explore possible phylogenetic "constraints." The centroid size was retained and used in allometric studies. The Relative Warp scores obtained for each specimen were used in discriminant analyses (same probability for each locomotor group) to classify the locomotor habits in Sparassodonta. Living species are classified as arboreal (those that live mostly in trees and only rarely descend to the ground), scansorial (those that move on the ground and in the trees and have good climbing skills, but carry out a large proportion of their life on the ground), terrestrial (those that move mostly on the ground and either never or only occasionally climb or dig), cursorial (those that move almost exclusively on the ground, can perform long and fast runs and occasionally dig), semi-aquatic (those that swim either frequently or always), and semi-fossorial (those terrestrial mammals that frequently dig) (modified from Van Valkenburgh, 1987, and Polly, 2007).

To infer diet, we used geometric morphometric analysis and a morphometric index (see below). In the geometric morphometric analysis, landmarks and semi-landmarks were placed on the dentary in the lateral view of the dentary (Fig. 11.6). Pictures from a large sample of living placental and marsupial carnivores (see above) were taken in a standardized way (Prevosti *et al.*, 2012; Turazzini *et al.*, 2010, and unpublished data). Living carnivores were classified as hypercarnivores (those which feed mainly on meat from other vertebrates), mesocarnivores (those which feed mainly on other vertebrates, usually species smaller than themselves, but also on plants and invertebrates), omnivores (those which feed mainly on plants and invertebrates), herbivores (those which feed mostly on plant material), and insectivores (those which feed mostly on insects) (modified from Van Valkenburgh, 1989; Van Valkenburgh and Koepfli, 1993). These categories were used in discriminant and canonical variate analyses (same probability for each diet group), using the scores of the Relative Warps, to infer the diet of Santacrucian sparassodonts (Prevosti *et al.*, 2012; Turazzini *et al.*, 2010, and unpublished data). Lastly, we used a morphometric index based on the last lower molar (m4):

Relative grinding area RGA = square root of talonid area/ trigonid length

(modified from Van Valkenburgh, 1989), to represent a measurement of the dietary information taken from the dentition (Prevosti *et al.*, 2009, Table 11.2). In the case of extinct species, a taxon is considered hypercarnivorous when the RGA index is lower than 0.48; mesocarnivorous when the

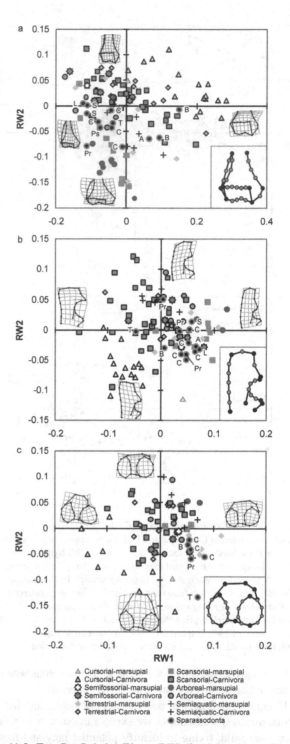

Fig. 11.5. Two first Relative Warps (RW), for the analyses of the anterior view of humerus (a), lateral view of ulna (b), and proximal view of tibia (c). Symbols are marked in the key. A: *Arctodictis sinclairi*; B: *Borhyaena tuberata*; C: *Cladosictis patagonica*; L: "*Lycopsis*" *longirostrus*; Pr: *Prothylacynus patagonicus*; Ps: *Pseudonotictis pusillus*; S: *Sipalocyon gracilis*; and T: cf. *Thylacosmilus/T. atrox*. The figure at the right lower corner of each graph represents the consensus shape with the landmarks (black dots) and semi-landmarks (gray dots) used.

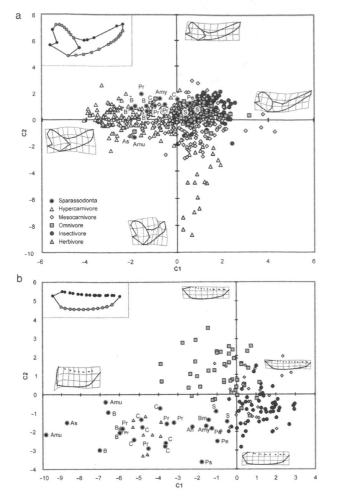

Fig. 11.6. Two first canonical axes (C), for the analyses of mandibular shape in Carnivora and in the marsupial sample (a), and horizontal ramus shape of the marsupial sample (b). Symbols are marked in the key. Ah: *Acyon herrerae*; Amu: *Arctodictis munizi*; Amy: *Acyon myctoderos*; As: *Arctodictis sinclairi*; B: *Borhyaena tuberata*; Bm: *Borhyaenidium musteloides*; C: *Cladosictis patagonica*; Pr: *Prothylacynus patagonicus*; Pe: *Perathereutes pungens*; Ps: *Pseudonotictis pusillus*; S: *Sipalocyon gracilis*. The figure at the left upper corner of each graph represents the consensus shape with the landmarks (black dots) and semi-landmarks (gray dots) used.

index ranges between 0.48 and 0.54; and omnivorous when it is larger than 0.54 (Prevosti *et al.*, 2009, in press).

Using the inferences of body size, locomotion, and diet of Santacrucian sparassodonts, we explore the structure of this carnivore guild, trying to identify potential prey and intra-guild competition.

11.5 Results

11.5.1 Body size

New estimations were obtained using the centroid size of the humerus, ulna, and tibia, and analyzed by geometric

morphometrics, in which the phylogenetic pattern and other statistical biases were corrected (see Table 11.2; Ercoli, 2010; Ercoli and Prevosti, 2012). Masses more than 10 kg higher than previous estimates were found for *Borhyaena tuberata* (~36 kg) and "*Lycopsis*" *longirostrus* (~30 kg), while our estimates for *Prothylacynus patagonicus* (~32 kg), *Cladosictis patagonica* (~7 kg), and *Sipalocyon gracilis* (~2 kg) were within the range of previous estimations. The body size of *Pseudonotictis pusillus* is below the estimation for *Sipalocyon gracilis* (Table 11.2). For Santacrucian taxa for which the postcranial skeleton is either unknown or extremely fragmentary, the mass was not estimated by this method. For these taxa, estimates were based on the dentition following the methodology previously applied for the group by Vizcaíno *et al.* (2010; Table 11.2).

Comparison between the masses obtained from the size centroid of the postcranial elements obtained here and those based on dental measurements by Vizcaíno *et al.* (2010) suggests that dental equations underestimate the body size of the sparassodonts (Table 11.2). This is clearly exemplified by *A. munizi*, which has a skull comparable in size to that of the largest jaguar (*Panthera onca*, ~120 kg), but whose dental measurements give an estimate of 37 kg, similar to the mass of the gray wolf (*Canis lupus*, ~40 kg) and to that estimated for *B. tuberata* with the centroid size, which has a skull at least one-third smaller than *A. munizi*. According to linear measurements, the body mass of this giant sparassodont was above 50 kg.

11.5.2 Locomotion

The shapes of some bones of the postcrania of Santacrucian sparassodonts were plotted together with extant placentals and marsupials in morphogeometric and multivariate studies, based on limb-bone anatomy (Ercoli, 2010; Ercoli and Prevosti, 2012; Ercoli *et al.*, 2012). In living species, there is a gradation in the morphospace from arboreal through scansorial to cursorial taxa (Fig. 11.5). Semi-fossorial and semi-aquatic species widely overlap the scansorial space in intermediate areas of the graph. Terrestrial taxa are closer to cursorial species, but overlapped with some scansorial taxa. According to the diagram resulting from the two first Relative Warps of the humerus and ulna (Fig. 11.5a, b), a gradient is found in the Santacrucian taxa. At one extreme, *P. patagonicus* has a humerus with a sharp ectepicondylar crest and robust medial epicondyle, as well as wide areas on the ulna for the origins of the digital flexors and extensors (reflected by the convex distal border in the ulna; Fig. 11.5a, b). At the other extreme, reduction of these structures and area is evident in *B. tuberata* and *A. sinclairi* (Fig. 11.5a, b). The morphology of the humerus and ulna suggests that *P. patagonicus* likely had better forelimb supination capabilities than *Arctodictis* (*A. sinclairi*) and *B. tuberata*, in which movements were likely more restricted to the sagittal plane. *C. patagonica*, *S. gracilis*, and "*L.*" *longirostrus* fall in the middle of the sparassodont morphospace (Fig. 11.5a, b).

The tibial platforms of *B. tuberata*, *C. patagonica*, and *P. patagonicus* are roughly similar, with a robust and transversely narrowed tibial tuberosity, and nearly circular articular condyles (Fig. 11.5c). These sparassodonts fall in a region of the morphospace dominated by terrestrial mammals, although *Chironectes*, a semi-aquatic living marsupial, *Meles*, a semi-fossorial placental, and *Paguma*, an arboreal placental taxon, also lie near the fossil species (Fig. 11.5).

The discriminant analysis presented here correctly reclassified between 65.5% and 70.1% of the extant taxa. *P. patagonicus*, *C. patagonica*, *L. torresi*, *S. gracilis*, and *Ps. pusillus* are classified as scansorial, while *A. sinclairi* (and by extension *A. munizi*) and *B. tuberata* are classified as terrestrial. Taking into account these locomotor habits and the morphological analyses using Relative Warps, the Santacrucian sparassodonts would have been diverse in their locomotor capabilities and habitat utilization. Some species (e.g. *P. patagonicus*, *C. patagonica*, *S. gracilis*, and *Ps. pusillus*) would have had substantial limb joint mobility, especially for forelimb supination, resembling extant climbers such as procyonids and viverrids (e.g. *Nasua*). *P. patagonicus* would have had grasping capabilities in the anterior limbs which were even better developed than in *C. patagonica*, *S. gracilis*, and *Ps. pusillus*. Nevertheless, the morphology of the tibia of *P. patagonicus* suggests more restricted movements than the fully arboreal taxa (e.g. *Arctictis binturong*; Muizon, 1998). The limb morphology of *B. tuberata* and *A. sinclairi* is compatible with strictly terrestrial taxa because of the small medial epicondyle and a larger humeral trochlea.

11.5.3 Diet
Using the RGA ratio, which characterizes the relative development of the talonid (RGA) in the last lower molar (m4), all Santacrucian sparassodonts are comparable to living marsupial and placental hypercarnivores, most of which have RGAs less than 0.48 (Table 11.2). Hypercarnivores are characterized by possessing deep jaws, large canines and incisors, longer shearing blades on carnassials, and reduced molar grinding areas (Van Valkenburgh and Koepfli, 1993; Van Valkenburgh et al., 2004). Like living felids, *A. munizi*, *B. tuberata*, and *Acr. sectorius* have virtually no talonid (RGA = 0). *Prothylacynus patagonicus* and *C. patagonica* have better developed talonids, while other species (e.g. most hathliacynids) have slightly larger talonids with distinct basins, with RGA values that are comparable to canids and mustelids (e.g. RGA between 0.27 and 0.34).

Based on the geometric morphometric analysis (Fig. 11.6; Turazzini et al., 2010, and unpublished data),

two separate groups are identified (Fig. 11.6b). One group consists of *B. tuberata*, *A. munizi*, *C. patagonica*, and *P. patagonicus*, grouped by the presence of short and deep dentaries, and an anteriorly displaced masseteric fossa. The other group includes *Pe. pungens*, *S. gracilis*, probably *Ac. tricuspidatus* by extension from both *Acyon herrerae* (Marshall, 1981) and *Acyon myctoderos* Forasiepi et al., 2006, and *Ps. pusillus*, which have longer and more shallow dentaries. The first group overlaps with hypercarnivores, while the second overlaps with omnivores, mesocarnivores, and insectivores. This is clear using the sample of metatherians (Fig. 11.6b), but in the analysis including Carnivora, one specimen of *Cladosictis* is placed near *Sipalocyon* in an area dominated by non-hypercarnivores, while the others are placed with hypercarnivores. Discriminant functions, with percentages of correct reclassification between 69.59% and 79.39%, classified the members of the first group as hypercarnivores and partially classified those of the second group as omnivores or insectivores. The shape of the dentary is the most important discriminative parameter in these functions, and therefore explains why sparassodonts with a hypercarnivorous dentition but more generalized dentaries are not classified as hypercarnivores. Our results agree with the hypothesis that *A. munizi* and probably *B. tuberata* (Marshall, 1977b, 1978; Argot, 2004a; Forasiepi et al., 2004) may have been scavengers at least opportunistically, an inference supported by their robust, extremely deep, and short dentaries (Fig. 11.6b).

11.6 Discussion
Although the range of locomotion and dietary specialization of the Santacrucian sparassodonts is smaller than that observed among living placental carnivores (see Van Valkenburgh, 1985, 1999; Prevosti and Vizcaíno, 2006; Croft et al., 2010), there is a clear niche partitioning among the Santacrucian species. At one extreme are the strongly specialized hypercarnivorous terrestrial species, such as the borhyaenids *A. munizi* and *B. tuberata*, and at the opposite extreme are scansorial species, including most hathliacynids, with slightly more generalized dental morphology. Body size spans a broad range of at least 50 kg (Argot, 2004a), with a good separation of the species across the entire range. *Arctodictis munizi* is the largest species, followed by *Borhyaena tuberata*, and then by *Prothylacynus patagonicus*, *Lycopsi torresi*, *Acrocyon sectorius*, *Cladosictis patagonica*, *Acyon tricuspidatus*, *Sipalocyon* spp., *Pseudonotictis pusillus*, and *Perathereutes pungens* (Table 11.2). Larger species are clearly separated by size though this is less evident among smaller species (*Sipalocyon* species are similar sized, and the same is observed between *P. pungens* and *P. pusillus*). It is worth mentioning that according to our results (and without

considering *A. munizi*, whose size was underestimated; see above), we find no mammalian predators with a body mass above 100 kg in the Santacrucian. This absence contrasts with most living and extinct placental communities (e.g. Van Valkenburgh, 1985, 1999; Fariña, 1996; Prevosti and Vizcaíno, 2006), but it is similar to some present South American ecosystems (e.g. Patagonia).

Body mass, locomotor, and dietary variations, in addition to environmental factors (see below), suggest that there was a good ecological separation of niches amongst the sparassodonts of the Santa Cruz Formation that could have allowed the coexistence of the 11 species (e.g. Rosenzweig, 1966; Van Valkenburgh, 1985; 1991; Van Valkenburgh and Hertel, 1998; Palomares and Caro, 1999; Donadio and Buskirk, 2006; Meiri *et al.*, 2007). Although size clearly separated most of the species, locomotion and diet also contributed to the discrimination. The larger taxa *B. tuberata* and *A. munizi* (extrapolating the locomotion habits of *A. sinclairi*) were the most terrestrial species. The dietary range of *B. tuberata* and *A. munizi* was limited to hypercarnivory (RGA = 0) and according to the robustness of the dentary and morphology of the teeth, it is probable that at least the large *A. munizi* may have had the ability to crush hard elements, such as bones, at least opportunistically (Forasiepi *et al.*, 2004). The largest Santacrucian sparassodonts were probably ambush predators that stalked and waited for their prey and then made a short, fast dash for the kill (Argot, 2004a). *Prothylacynus patagonicus* was scansorial and had a more generalized hypercarnivorous diet (RGA = 0.17) than the similar-sized *Borhyaena tuberata*. *Lycopsis torresi* was scansorial but smaller than *P. patagonicus* and more generalized in its hypercarnivory (RGA = 0.30). *Acrocyon sectorius* was smaller in body mass than *L. torresi* and highly specialized to carnivory (RGA = 0), as in other borhyaenids. If *Acr. sectorius* is a valid taxon, then it represents the smallest sparassodont of the Santacrucian assemblage with an extreme hypercarnivorous dental morphology. *Cladosictis patagonica* seems to have had scansorial capabilities and an RGA of 0.17, similar to *Prothylacynus*, but with a considerably smaller body mass (Table 11.2). The remaining five hathliacynids have similar RGAs (~0.30) and therefore, they would have had similar diets. The trophic segregation in these smaller sparassodont species could be more subtle than between larger species, as is the case in modern mustelid guilds (e.g. Dayan *et al.*, 1989; Meiri *et al.*, 2007). Size clearly segregates *Acyon tricuspidatus* from the remaining hathliacynids. Species of *Sipalocyon* are similar-sized, and almost no size difference is observed between *Pe. pungens* and *Ps. pusillus* (Table 11.2). Postcranial remains are known only in *S. gracilis* and *Ps. pusillus* among the last five hathliacynids, and their morphology suggests that both species would have had

scansorial habits (see above). Possible segregation among the hathliacynid species based on locomotor capabilities cannot be tested for want of material.

Despite the variations commented upon above and the inferred segregation of the Santacrucian sparassodonts within the predatory guild, the morphological diversity is limited with respect to that observed today in placental carnivore communities. Strict cursorial, arboreal, semi-aquatic, and semi-fossorial taxa are absent in the known sparassodont assemblages of the Santa Cruz Formation, indeed in pre-Pliocene South America Cenozoic assemblages as a whole.

Using the paleoautecological characterization of sparassodonts, and considering the associated mammalian fauna and their body sizes (see Tauber, 1997a, b; Vizcaíno *et al.*, 2010, and Chapter 12; Bargo *et al.*, Chapter 13; Cassini *et al.*, Chapter 14), we propose the following trophic relationships for Santacrucian sparassodonts (see also Argot, 2004a; Fig. 11.7). *Arctodictis munizi* and *B. tuberata* would have hunted mammals spanning a wide range of size, from rodents (e.g. *Neoreomys* Ameghino, 1887: about 4.22 kg), small xenarthrans (e.g. *Peltephilus* Ameghino, 1887: between 8.3 and 11 kg), small notoungulates (Interatheriidae: 2–8 kg), and the juveniles of the largest Santacrucian mammals (e.g. *Astrapotherium magnum* (Owen, 1853) with an adult mass above 900 kg). It is likely that the most accessible prey would have occupied a range of body size that ranged between 20 and 150 kg, including sloths (e.g. *Hapalops* Ameghino, 1887 and *Eucholoeops* Ameghino, 1887: 35–85 kg), toxodonts (e.g. *Adinotherium* Ameghino, 1887: 100 kg), and litopterns (e.g. *Thoatherium* Ameghino, 1887 and *Diadiaphorus* Ameghino, 1887: 24–82 kg). *Prothylacynus patagonicus* and *L. torresi* may have had similar diets to *A. munizi* and *B. tuberata*, but adult toxodonts and astrapotheres were probably beyond their hunting capabilities, and the most probable common prey would have been sloths, armadillos (e.g. *Proeutatus* Ameghino, 1891: 15 kg), litopterns, and hegetotheriids (e.g. *Hegetotherium* Ameghino, 1887: 8 kg). The potential prey of *Acr. sectorius* would have included small litopterns (e.g. *Licaphrium* Ameghino, 1887 and *Thoatherium*: 18–24 kg), hegetotheriids, armadillos, rodents, interatheriids, small armadillos (e.g. *Stenotatus* Ameghino, 1891 and *Prozaedyus* Ameghino, 1891: 1–4 kg), and primates (*Homunculus* Ameghino, 1891: 2.7 kg). The smaller sparassodonts probably concentrated their diet on small rodents (e.g. *Eocardia* Ameghino, 1887 and *Stichomys* Ameghino, 1887: 0.16–0.79 kg), interatheriids, and small marsupials (e.g. *Palaeothentes* Ameghino 1887: 0.08–0.36 kg). In particular, the larger hathliacynids, *C. patagonica* and *Ac. tricuspidatus*, would have preyed more frequently on small armadillos, larger rodents (e.g., *Neoreomys*), and primates (*Homunculus*).

1. *Arctodictis munizi*
2. *Borhyaena tuberata*
3. *Cladosictis patagonica*
4. *Lycopsis torresi*
5. *Prothylacynus patagonicus*
6. *Pseudonotictis pusillus*
7. *Sipalocyon gracilis*

Fig. 11.7. Reconstruction of the Santacrucian sparassodont guild.

The known sparassodonts from the Santa Cruz Formation represent the richest assemblage of mammalian carnivores in the South American Cenozoic. Santacrucian sparassodonts have different morphologies and body sizes which would have enabled guild differentiation. Although sparassodonts displayed a considerable range of habits and morphotypes, the ranges of their inferred behavior and anatomy were smaller than observed in extant placental carnivore guilds, in terms of inferred diet and locomotion. One interesting point is that sparassodonts have a generalized postcranial skeleton (in comparison to Carnivora) combined with an extremely specialized dentition suggesting hypercarnivory (Argot, 2004a; Forasiepi *et al.*, 2004; Wroe *et al.*, 2004). This combination is not common among living carnivores, except in some clades of mustelids (e.g. *Mustela, Galictis, Gulo*; Argot, 2004a). Overall, sparassodont niche breadth more nearly resembled that of the extinct placental carnivorous clade Creodonta, a group of mainly Paleogene predators that was distributed through Eurasia, North America, and Africa (Van Valkenburgh, 1999, 2007).

Sparassodont diversity in the Santa Cruz Formation reached 10 genera and 11 species. This number of taxa is high and comparable to that of other fossil and living faunas (e.g. Van Valkenburgh, 1985; Croft, 2001, 2006; Prevosti and Vizcaíno, 2006), and contrasts with the claims of other authors, who have suggested that the richness of the carnivorous mammals in the Cenozoic of South America was low (Croft, 2001, 2006; Wroe *et al.*, 2004; Vizcaíno *et al.*, 2010). The difference of opinion may be a matter of the geologic time and geographic scale of the analysis, and body size limits imposed in the different approaches. For example, Vizcaíno *et al.* (2010) included carnivores above 10 kg at some levels and localities of the Estancia La Costa Member; Croft (2001, 2006) considered separately in his analyses the faunal lists of Estancia La Costa and Estancia Angelina Members of the Santa Cruz Formation and included all the species then known, while Wroe *et al.* (2004) used a continental scale and a longer time span (7 Ma). Not surprisingly, Wroe *et al.* (2004) reported that the richness of hypercarnivores was higher than that estimated by Croft (2001, 2006), a difference that can be related to greater time averaging, which conflates alpha with beta species richness. It is also worth mentioning that the coexistence of the 11 species at the same time and in the same space would be overestimated because the Santa Cruz Formation is several hundred meters thick and may comprise a time span of about 1.5 Ma (Perkins *et al.*, Chapter 2; Fleagle *et al.*, Chapter 3), if, as suggested by Fleagle *et al.* (1995), this unit is restricted to one single polarity interval (see also Tauber, 1997a, b; Kay *et al.*, 2008). The time averaging is reduced by analyzing the members of the Santa Cruz Formation as units (e.g. Croft, 2001, 2006), or even better if the analysis is limited and the spatial and temporal averaging constrained, as in Vizcaíno *et al.* (2010). The latter authors found a low richness and abundance of sparassodonts in the localities of Puesto Estancia La Costa (= Corriguen Aike) and Campo Barranca. This conclusion is congruent with the assertion of Croft (2006), who suggested that the low abundance of sparassodonts in paleontological localities is a biological signal related to the low density of predators in the past and not a mere taphonomic artifact. However, we believe that the taphonomic bias should not be ignored, especially given that predators usually have lower densities than other mammals and would be less likely to appear as fossils for this reason alone (e.g. Sunquist and Sunquist, 2001). Despite the strong sampling effort made by the MLP/Duke University team during the past decade, 1128 fossil specimens were collected in Puesto Estancia La Costa and Campo Barranca (Vizcaíno *et al.*, Chapter 1) and only 14 (1.24%) of them were sparassodonts. Additionally, only 26 sparassodont specimens were collected in all the localities studied during these field seasons: *Cladosictis patagonica*: 10 specimens; *Borhyaena tuberata*: 8 specimens; *Sipalocyon gracilis*: 2 specimens; *Sipalocyon* sp.: 2 specimens; *Prothylacynus patagonicus*: 2 specimens; *Lycopsis torresi*: 1 specimen; cf. *Perathereutes pungens*: 1 specimen. The relative abundance of specimens known for each species is similar to that collected by previous field groups in other areas (see Marshall, 1978, 1979, 1981; Tauber, 1997a; Forasiepi *et al.*, 2004): *C. patagonica*: 66; *S. gracilis*: 42; *B. tuberata*: 35; *P. patagonicus*: 14; *Arctodictis munizi*: 7; *L. torresi*: 2; *Acyon tricuspidatus*: 5; *Acrocyon sectorius*: 3; *Sipalocyon obusta*: 2; *Pe. pungens*: 2; *Pseudonotictis pusillus*: 2. The rank abundances of both samples are positively and significantly correlated (Spearman R correlation index: 0.68; *p* = 0.014). The most frequent taxon (*C. patagonica*) was also the most ubiquitous in the localities studied in the recent fieldwork, and sometimes the only sparassodont recovered (e.g. at Cañadón Silva). These data suggest that in the latter localities, we are recovering the most common species, and that the absence of some rare taxa may be due to low sample size.

Taking the best sampled levels exposed at the Puesto Estancia La Costa, there are at least five sparassodonts

(*Cladosictis patagonica*, *Sipalocyon gracilis*, *Borhyaena tuberata*, *Prothylacynus patagonicus*, and cf. *Perathereutes pungens*), and it is probable that an old specimen of *Arctodictis munizi* was collected from the same levels (Tauber, 1997a; Forasiepi *et al.*, 2004; but see Vizcaíno *et al.*, 2010). Extending the analysis to the Estancia La Costa Member, we add to the previously mentioned species a new record of *Lycopsis torresi* (Fig. 11.1), *Acrocyon sectorius* (found about 8 km south of Río Coyle; Marshall, 1981), and *A. munizi* (found at Puesto Estancia La Costa; Marshall, 1978; Tauber, 1997a; Forasiepi *et al.*, 2004). These records increase the number of predator species for the Estancia La Costa Member from two (in Croft, 2001, 2006) to seven. In addition, the predator/prey index increases from 0.02 to 0.18 for the Estancia La Costa Member (changing the numerator from two to seven in the index calculated by Croft, 2006), falling within the observed range of placental faunas. If we consider that these sparassodonts are only hypercarnivores, and if we included only hypercarnivore placentals in the comparison, this fauna would be even more similar to placental faunas. Admittedly, it is possible that this richness of the Santacrucian sparassodonts is overestimated owing to systematic inaccuracies (taxonomic inflation). Even taking into account that *Sipalocyon gracilis* and *Sipalocyon obusta*, and *B. tuberata* and *A. sectorius* may be synonyms (although this must be justified; see above), the predator richness and predator index remain similar.

11.7 Conclusions

The Santa Cruz Formation contains a rich assemblage of 11 Sparassodonta. Six species (*Sipalocyon obusta*, *Sipalocyon gracilis*, *Cladosictis patagonica*, *Acyon tricuspidatus*, *Pseudonotictis pusillus*, and *Perathereutes pungens*) belong to Hathliacynidae; two (*Prothylacynus patagonicus* and *Lycopsis torresi*) are basal stem Borhyaenoidea; and three (*Acrocyon sectorius*, *Borhyaena tuberata*, and *Arctodictis munizi*) are grouped among Borhyaenidae.

These Santacrucian sparassodonts had a wide variation in body size, from very small species of ~1 kg (e.g. *Pe. pungens*) to large ones that exceeded 50 kg (e.g. *A. munizi*), and a clear separation exists among species within the range.

Sparassodont postcranial anatomy points to the existence of locomotor differentiation among the Santacrucian sparassodonts. Some taxa (e.g. *S. gracilis*, *Ps. pusillus*, *C. patagonica*, and *P. patagonicus*) had more generalized limbs and were probably scansorial, while others (e.g. *Borhyaena tuberata* and *Arctodictis munizi*, inferred from *A. sinclairi*) were more terrestrial.

The Santacrucian guild of metatherian predators was formed by hypercarnivorous taxa according to the dental morphology (RGA ranges from 0.34 in *Pe. pungens* to 0.00 in borhyaenids). However, several morphological special- izations are recognizable with suggested implications for dietary separation. Most hathliacynids (e.g. *S. gracilis*, *Pe. pungens*, and *Ps. pusillus*) have a long and shallow dentary body and better-developed lower molar talonids, indicating a less hypercarnivorous diet than most borhyaenoids. At the other extreme, larger borhyaenids (e.g. *B. tuberata* and *A. munizi*) have more robust dentaries (i.e. short and deep hori- zontal rami) and a lower dentition virtually lacking talonids, which is related to extreme hypercarnivory and the ability to scavenge, at least opportunistically. The analogy of sparasso- donts with some placental carnivores (e.g. canids, ursids, and mephitids) is not supported.

The paleoautecological reconstruction of the Santacrucian sparassodonts suggests that there was a good ecological separation within the guild, imposed by the combination of body size, locomotion, and diet. The absolute and relative numbers of sparassodont species recorded in the Santa Cruz Formation (11 species) and in the Estancia La Costa Member (seven species) are comparable to those observed in present and past placental communities of hypercarnivores.

ACKNOWLEDGMENTS
We especially acknowledge Sergio Vizcaíno, Susana Bargo, and Richard Kay for their invitation to participate in this book. We also extend our thanks to the curators who helped us during our visits to the different collections: M. Reguero, L. Pomi, A. Kramarz, W. Simpson, M. Carrano, J. Gadkin, J. Flynn, W. Joyce, D. Brinkman, M. Palacios, and Patricia Holroyd; those who participated in the recent fieldwork at the Santa Cruz Formation, led by S. Vizcaíno and R. Kay, in which several of the sparassodonts studied here were col- lected: S. Bargo, J. C. Fernicola, J. Perry, A. Weil, N. Toledo, L. Pomi, R. Madden, J. Moly, M. Malinzak, and A. Tauber; the editors (S. Vizcaíno, S. Bargo, and R. Kay), reviewers (B. Van Valkenburgh and D. Croft) for their corrections and comments; and the colleagues who are always available for discussing ideas (F. J. Goin). We also appreciate the "study collection grants" given to F.J.P. by the American Museum of Natural History and the Field Museum of Natural History, which allowed the study of the sparassodonts deposited in these institutions. This is a contribution to the projects PIP 1054, and to PICT 26219 and 0143 to Sergio Vizcaíno, and NSF 0851272, and 0824546 to Richard Kay.

Appendix 11.1   Specimens of Sparassodonta studied

See Vizcaíno *et al.*, Chapter 1, Figs. 1.1 and 1.2 and Appendix 1.1 for details about localities, and Matheos and Raigemborn, Chapter 4, Figs. 4.3 and 4.4 for information about geology.

Institutional abbreviations

FMNH: Field Museum of Natural History, Chicago, USA.

CORD-PZ: Museo de Paleontología, Facultad de Ciencias Exactas, Físicas y Naturales de la Universidad Nacional de Córdoba, Córdoba, Argentina.

MACN-A: Museo Argentino de Ciencias Naturales "Bernardino Rivadavia," Colección Nacional Ameghino, Buenos Aires, Argentina.

MLP: Museo de La Plata, La Plata, Argentina.

MNHN-Bol: Museo Nacional de Historia Natural, La Paz, Bolivia.

MPM-PV: Museo Regional Provincial Padre M. Jesús Molina, Río Gallegos, Argentina.

UCMP: University of California Museum of Paleontology, Berkeley, USA.

USNM: National Museum of Natural History; Smithsonian Institution, Washington DC, USA.

YPM-VPPU: Yale Peabody Museum, Princeton University Collection, New Haven, USA.

Hathliacynidae

Pseudonotictis pusillus

MLP 11–26, incomplete left and right dentary, fragment of right maxilla, incomplete right humerus, radius, and ulna. Horizon and locality: Santa Cruz Formation; Monte León, Santa Cruz Province, Argentina.

Sipalocyon sp.

MPM-PV 4316, incomplete skull and skeleton. Horizon and locality: Santa Cruz Formation, Estancia La Costa Member, Fossiliferous Level 3; Estancia La Costa, Santa Cruz Province, Argentina. MPM-PV 4319, rostrum. Horizon and locality: Santa Cruz Formation, Estancia La Costa Member Fossiliferous Level 5.3; Puesto Estancia La Costa, Santa Cruz Province, Argentina.

Sipalocyon gracilis

YPM-VPPU 15029, rostral part of the skull and both dentaries. Horizon and locality: Santa Cruz Formation; 15 km south of Río Coyle, Santa Cruz Province, Argentina. YPM-VPPU 15373, skull, dentaries, a patella, phalanges, and a metapodial bone. Horizon and locality: Santa Cruz Formation; Killik Aike, Santa Cruz Province, Argentina. MPM-PV 4332, dentary. Horizon and locality: Santa Cruz Formation, Estancia La Costa Member, Fossiliferous Level 3; Estancia La Costa, Santa Cruz Province, Argentina. MPM-PV 4317, incomplete dentary. Horizon and locality: Santa Cruz Formation, Estancia La Costa Member, Fossiliferous Level 5.3; Puesto Estancia La Costa, Santa Cruz Province, Argentina. MACN-A 5938–5949, left and right dentaries, fragment of maxillae, and postcranial elements. Horizon and locality: Santa Cruz Formation; La Cueva, Santa Cruz Province, Argentina.

Sipalocyon obusta

MACN-A 686, fragment of left dentary. Horizon and locality: Santa Cruz Formation; Monte Observación (= Cerro Observatorio), Santa Cruz Province, Argentina. MACN-A 687, fragment of right dentary. Horizon and locality: Santa Cruz Formation; Monte Observación (= Cerro Observatorio), Santa Cruz Province, Argentina.

Acyon tricuspidatus

MLP 11–64, right dentary. Horizon and locality: Santa Cruz Formation; Monte León, Santa Cruz Province, Argentina. MACN-A 669, right dentary. Horizon and locality: Santa Cruz Formation, Monte Observación (= Cerro Observatorio), Santa Cruz Province, Argentina. MACN-A 9, fragment of left dentary. Horizon and locality: Santa Cruz Formation; Santa Cruz Province, Argentina.

Cladosictis patagonica

MACN-A 5927, skull, right dentary, one cervical vertebra, two vertebral bodies, and a fragment of the right pelvic girdle. Horizon and locality: Santa Cruz Formation; Puesto Estancia La Costa, Santa Cruz Province, Argentina. MACN-A 6288–6298, both dentaries, fragment of left maxilla, two vertebrae, proximal part of the scapula, distal part of right humerus, radius, and fragment of both ulnae.

Horizon and locality: Santa Cruz Formation; Santa Cruz Province, Argentina. YPM-VPPU 15046, skull and dentaries, fragment of limb bones. Horizon and locality: Santa Cruz Formation; Río Coyle, Santa Cruz Province, Argentina. YPM-VPPU 15170, skull, both dentaries, and postcranial elements. Horizon and locality: Santa Cruz Formation; Río Coyle, Santa Cruz Province, Argentina. YPM-VPPU 15702, rostral part of the skull and postcranial elements. Horizon and locality: Santa Cruz Formation; Río Coyle, Santa Cruz Province, Argentina. MPM-PV 3584, incomplete left dentary and fragment of right humerus. Horizon and locality: Santa Cruz Formation, Estancia La Costa Member, Fossiliferous Level 5.1; Cañadón Silva, Santa Cruz Province, Argentina. MPM-PV 4253, incomplete skull, dentaries, and cervical vertebrae. Horizon and locality: Santa Cruz Formation, Estancia La Costa Member, Fossiliferous Level 5.3; Puesto Estancia La Costa, Santa Cruz Province, Argentina. MPM-PV 3645, skull and postcranial elements. Horizon and locality: Santa Cruz Formation, Estancia La Costa Member; Anfiteatro, Santa Cruz Province, Argentina. MPM-PV 3646, fragment of left dentary. Horizon and locality: Santa Cruz Formation, Estancia La Costa Member, Fossiliferous Level 7; Puesto Estancia La Costa, Santa Cruz Province, Argentina. MPM-PV 4326, skull, dentaries, and postcranial bones. Horizon and locality: Santa Cruz Formation, Estancia La Costa Member; Anfiteatro, Santa Cruz Province, Argentina. MPM-PV 4325, fragment of left dentary. Horizon and locality: Santa Cruz Formation, Estancia La Costa Member; Anfiteatro, Santa Cruz Province, Argentina. MPM-PV 4327, fragment of right maxilla. Horizon and locality: Santa Cruz Formation, Estancia La Costa Member; Anfiteatro, Santa Cruz Province, Argentina. MPM-PV 4333, skull, dentaries, and postcranial elements. Horizon and locality: Santa Cruz Formation, Estancia La Costa Member; Campo Barranca, Santa Cruz Province, Argentina. MPM-PV 4323, skull and left dentary. Horizon and locality: Santa Cruz Formation, Estancia La Costa Member, Fossiliferous Level 3; Estancia La Costa, Santa Cruz Province, Argentina. MPM-PV 4329, partial skull, dentary, and incomplete postcrania. Horizon and locality: Santa Cruz Formation, Estancia La Costa Member; Anfiteatro, Santa Cruz Province, Argentina.

Perathereutes pungens

MACN-A 684, left dentary. Horizon and locality: Santa Cruz Formation; Monte Observación (= Cerro Observatorio), Santa Cruz Province, Argentina.

Perathereutes pungens?

MPM-PV 4322, partial skull and dentary. Horizon and locality: Santa Cruz Formation, Estancia La Costa Member, Fossiliferous Level 5.3. Puesto Estancia La Costa, Santa Cruz Province, Argentina.

Acyon herrerae

FMNH P13521, associated left and right dentaries. Horizon and locality: Colhuehuapian Age (Early Miocene); Lago Colhué-Huapi, Chubut Province, Argentina.

Acyon myctoderos

MNHN Bol V-003668, skull and dentaries, and postcranial elementes. Horizon and locality: Laventan Age (Early Miocene); Papachacra, Quebrada Honda, Bolivia.

Borhyaenidium musteloides Pascual and Bocchino 1963

MLP 57-X-10–153, partial skull, left dentary, fragment of right dentary, some postcranial elements. Horizon and locality: Huayquerian Age (Late Miocene); Salinas Grandes de Hidalgo, La Pampa, Argentina.

Borhyaenoidea

Lycopsis torresi

MPM-PV 4328, partial skull and right dentary. Horizon and locality: Santa Cruz Formation, Estancia La Costa Member, Fossiliferous Level 3; Estancia La Costa, Santa Cruz Province, Argentina.

Prothylacynus patagonicus

MACN-A 706–720, almost complete left dentary, left maxilla, and postcranial elements. Horizon and locality: Santa Cruz Formation; Monte Observación (= Cerro Observatorio), Santa Cruz Province, Argentina. MACN-A 5931–5937, complete skull, part of right dentary, and fragmentary postcranial elements. Horizon and locality: Santa Cruz Formation; Monte Observación (= Cerro Observatorio), Santa Cruz Province, Argentina. YPM-VPPU 15700, partial skull, dentary, and partial skeleton. Horizon and locality: Santa Cruz Formation; Santa Cruz Province, Argentina. MPM-PV 4318, incomplete dentaries and postcranial elements. Horizon and locality: Santa Cruz Formation, Estancia La Costa Member, Fossiliferous Level 5.3; Puesto Estancia La Costa, Santa Cruz Province, Argentina. MPM-PV 4331, incomplete right dentary. Horizon and locality: Santa Cruz Formation, Estancia La Costa Member, Fossiliferous Level 5.3; Puesto Estancia La Costa, Santa Cruz Province, Argentina.

Acrocyon sectorius

MLP 11–70, fragment of right dentary. Horizon and locality: Santa Cruz Formation; Río Santa Cruz, Santa Cruz Province, Argentina. MACN-A 9364–9385, large part of an associated upper and lower dentition. Horizon and locality: Santa Cruz Formation; Shehuén, Santa Cruz Province, Argentina.

Borhyaena tuberata

FMNH 13253, partial skull and dentary. Horizon and locality: Santa Cruz Formation; Río Coyle, Santa Cruz Province, Argentina. FMNH 13266, partial skull and dentary. Horizon and locality: Río Coyle, Santa Cruz Province, Argentina. USNM 15701, skull dentary and partial skeleton. Horizon and locality: Santa Cruz Formation; Río Coyle, Santa Cruz Province, Argentina. MPM-PV 3625, skull and thoracic vertebra. Horizon and locality: Santa Cruz Formation, Estancia La Costa Member, Fossiliferous Level 5.3; Puesto Estancia La Costa, Santa Cruz Province, Argentina. MPM-PV 4254, left scapula and maxillae. Horizon and locality: Santa Cruz Formation, Estancia La Costa Member, Fossiliferous Level 5.3; Puesto Estancia La Costa, Santa Cruz Province, Argentina. MPM-PV 3554, incomplete dentaries. Horizon and locality: Santa Cruz Formation, Estancia La Costa Member; Campo Barranca, Santa Cruz Province, Argentina. MPM-PV 4324, fragment of left dentary. Horizon and locality: Santa Cruz Formation, Estancia La Costa Member; Anfiteatro, Santa Cruz Province, Argentina. MPM-PV 4321, fragment of left dentary. Horizon and locality: Santa Cruz Formation, Estancia La Costa Member, Fossiliferous Level 5.3; Puesto Estancia La Costa, Santa Cruz Province, Argentina. MPM-PV 4320, fragment of right dentary. Horizon and locality: Santa Cruz Formation, Estancia La Costa Member, Fossiliferous Level 5.3; Puesto Estancia La Costa, Santa Cruz Province, Argentina. MPM-PV 4330, complete dentary. Horizon and locality: Santa Cruz Formation, Estancia La Costa Member, Fossiliferous Level 7; Puesto Estancia La Costa, Santa Cruz Province, Argentina. MPM-PV 4380, skull and left incomplete dentary. Horizon and locality: Santa Cruz Formation, Estancia La Costa Member, Fossiliferous Level 7; Puesto Estancia La Costa, Santa Cruz Province, Argentina. MACN-A 2076–2078, distal fragment of humerus and radius, partial calcaneus. Horizon and locality: Santa Cruz Formation; Karaiken, Santa Cruz Province, Argentina. MACN-A 6203–6365, part of both dentaries and maxillae, and postcranial elements. Horizon and locality: Santa Cruz Formation; La Cueva, Santa Cruz Province, Argentina. MACN-A 9343, proximal portion of left humerus and tibia, navicular, left calcaneus, proximal fragment of a metapodial. Horizon and locality: Santa Cruz Formation; Corriguen Aike (= Puesto Estancia La Costa), Santa Cruz Province, Argentina. YPM-VPPU 15701, skull, dentaries, and postcranial elements. Horizon and locality: Santa Cruz Formation; South of Río Coyle, Santa Cruz Province, Argentina.

Arctodictis munizi

MLP 11–65, skull and dentary. Horizon and locality: Santa Cruz Formation; Santa Cruz, Santa Cruz Province, Argentina. CORD-PZ-1210–1/5, skull, both dentaries, fragment

of right humerus and tibia, two vertebrae, and fragments of ribs. Horizon and locality: Santa Cruz Formation; Cañadón Las Totoras, Santa Cruz Province, Argentina.

Arctodictis sinclairi

MLP 85-VIII-3–1, nearly complete skeleton. Horizon and locality: Colhuehuapian Age; Gran Barranca, Chubut, Argentina.

"Lycopsis" longirostrus

UCMP 38061, nearly complete skeleton. Horizon and locality: Laventan Age (Middle Miocene); La Venta, Colombia.

Thylacosmylus atrox

FMNH 14344, basicranium, left dentary, and several postcranial elements. Horizon and locality: Huayquerian Age (Late Miocene); Chiquimil, Catamarca, Argentina. FMNH 14531, nearly complete skull, and postcranial elements. Horizon and locality: Chapadmalalan Age (Late Pliocene); Chiquimil, Catamarca, Argentina.

Thylacosmilidae cf. Thylacosmylus

MACN-A 10956, left ulna. Horizon and locality: Chapadmalalan Age (Late Pliocene); Chapadmalal, Buenos Aires Province, Argentina.

REFERENCES

Albino, A. (1996). The South American fossil squamata (Reptilia: Lepidosauria). In *Contributions of Southern South America to Vertebrate Paleontology*, ed. G. Arratia. München: Münchner Geowissenschaftliche Abhandlungen, pp. 9–72.

Ameghino, F. (1887). Enumeración sistemática de las especies de mamíferos fósiles coleccionados por Carlos Ameghino en los terrenos eocenos de la Patagonia Austral y depositados en el Museo de La Plata. *Boletín del Museo de La Plata*, **1**, 1–26.

Ameghino, F. (1891). Nuevos restos de mamíferos fósiles descubiertos por Carlos Ameghino en el Eoceno Inferior de la Patagonia austral. Especies nuevas, adiciones y correcciones. *Revista Argentina de Historia Natural*, **1**, 289–328.

Ameghino, F. (1894). Enumération synoptique des espèces de mammifères fossiles des formations éocènes de Patagonie. *Boletín de la Academia Nacional de Ciencias de Córdoba*, **13**, 259–452.

Anderson, J. F., Hall-Martin, A. and Russell, D. A. (1985). Longbone circumference and weight in mammals, birds, and dinosaurs. *Journal of Zoology, London*, **207**, 53–61.

Anyonge, W. (1993). Body mass in large extant and extinct carnivores. *Journal of the Zoological Society of London*, **231**, 339–350.

Argot, C. (2003a). Functional adaptation of the postcranial skeleton of two Miocene borhyaenoids (Mammalia,

Metatheria), *Borhyaena* and *Prothylacinus*, from South America. *Paleontology*, **46**, 1213–1267.

Argot, C. (2003b). Postcranial functional adaptations in the South American Miocene borhyaenoids (Mammalia, Metatheria): *Cladosictis, Pseudonotictis* and *Sipalocyon*. *Alcheringa*, **27**, 303–356.

Argot, C. (2004a). Evolution of South American mammalian predator (Borhyaenoidea): anatomical and paleobiological implications. *Zoological Journal of the Linnean Society*, **140**, 487–521.

Argot, C. (2004b). Functional-adaptive features and paleobiologic implications of the postcranial skeleton of the late Miocene sabretooth borhyaenoid *Thylacosmilus atrox* (Metatheria). *Alcheringa*, **28**, 229–266.

Argot, C. (2004c). Functional-adaptive analysis of the postcranial skeleton of a Laventan borhyaenoid, *Lycopsis longirostris* (Marsupialia, Mammalia). *Journal of Vertebrate Paleontology*, **24**, 689–708.

Babot, M. J., Powell, J. E. and Muizon, C. de (2002). *Callistoe vincei*, a new Proborhyaenidae (Borhyaenoidea, Metatheria, Mammalia) from the Early Eocene of Argentina. *Geobios*, **35**, 615–629.

Cabrera, A. (1927). Datos para el conocimiento de los dasyuroideos fósiles argentinos. *Revista del Museo de La Plata*, **30**, 271–315.

Croft, D. A. (2001). Cenozoic environmental change in South America as indicated by mammalian body size distributions (cenograms). *Diversity and Distributions*, **7**, 271–287.

Croft, D. A. (2006). Do marsupials make good predators? Insights from predator–prey diversity ratios. *Evolutionary Ecology Research*, **8**, 1192–1214.

Croft, D., Dolgushina T. and Wesley-Hunt, G. (2010). Morphological diversity in extinct South American sparassodonts (Mammalia: Metatheria). *Journal of Vertebrate Paleontology*, **30**, 76A.

Dayan, T., Simberloff, D., Tchernov, E. and Yom-Tov, Y. (1989). Inter- and intra-specific character displacement in mustelids. *Ecology*, **70**, 1526–1539.

Diniz-Filho, J. A. F., Ramos de Sant'Ana, C. E. and Bini, L. M. (1998). An eigenvector method for estimating phylogenetic inertia. *Evolution*, **52**, 1247–1262.

Donadio, E. and Buskirk, S. W. (2006). Diet, morphology, and interspecific killing in Carnivora. *The American Naturalist*, **167**, 524–536.

Ercoli, M. D. (2010). Estudio de los hábitos locomotores en los Borhyaenoidea (Marsupialia, Sparassodonta) de la Formación Santa Cruz (Mioceno Inferior de la provincia de Santa Cruz) a partir de la diferenciación morfológica en depredadores vivientes. Unpublished Lic. thesis, Universidad de Buenos Aires, Argentina.

Ercoli, M. D. and Prevosti, F. J. (2012). Estimación de masa de las especies de Sparassodonta (Metatheria, Mammalia) de la Edad Santacrucense (Mioceno Temprano) a partir de tamaños de centroide de elementos apendiculares: inferencias paleoecológicas. *Ameghiniana*, **48**, 462–479.

Ercoli, M. D., Prevosti, F. J. and Alvarez, A. (2012). Form and function within a phylogenetic framework: locomotory

habits of extant predators and some Miocene sparassodonta (Metatheria). *Zoological Journal of the Linnean Society*, **165**, 224–251.

Fariña, R. A. (1996). Trophic relationships among Lujanian mammals. *Evolutionary Theory*, **11**, 125–134.

Fleagle, J. G., Bown, T. M., Swisher, C. C. III and Buckley, G. A. (1995). Age of the Pinturas and Santa Cruz formations. *VI Congreso Argentino de Paleontologia y Bioestratigrafia, Actas*, 129–135.

Flynn, J. J. and Wesley-Hunt, G. D. (2005). Carnivora. In *The Rise of Placental Mammals: Origins and Relationships of the Major Extant Clades*, ed. D. Archibald, D. and K. Rose. Baltimore: Johns Hopkins University Press, pp. 175–198.

Forasiepi, A. M. (1999). Plantigradismo en un Borhyaenidae (Metatheria) del Mioceno Temprano de Patagonia. *Ameghiniana*, **36**, 11R.

Forasiepi, A. M. (2009). Osteology of *Arctodictis sinclairi* (Mammalia, Metatheria, Sparassodonta) and phylogeny of Cenozoic metatherian carnivores from South America. *Monografías del Museo Argentino de Ciencias Naturales*, **6**, 1–174.

Forasiepi, A. M., Goin, F. J. and Di Martino, V. (2003). Una nueva especie de *Lycopsis* (Metatheria, Prothylacyninae) de la Formación Arroyo Chasicó (Mioceno Tardío), de la Provincia de Buenos Aires. *Ameghiniana*, **40**, 249–253.

Forasiepi, A. M., Goin, F. J. and Tauber, A. A. (2004). Las especies de *Arctodictis* Mercerat 1891 (Metatheria, Borhyaenidae), grandes carnívoros del Mioceno de América del Sur. *Revista Española de Paleontología*, **19**, 1–22.

Forasiepi, A. M., Sánchez-Villagra, M. R., Goin, F. J. and Takai, M. (2006). A new species of Hathliacynidae (Metatheria, Sparassodonta) from the Middle Miocene of Quebrada Honda, Bolivia. *Journal of Vertebrate Paleontology*, **26**, 670–684.

Forasiepi, A. M., Martinelli, A. G. and Goin, F. J. (2007). Revisión taxonómica de *Parahyaenodon argentinus* Ameghino y sus implicancias en el conocimiento de los grandes mamíferos carnívoros del Mio-Plioceno de América del Sur. *Ameghiniana*, **44**, 143–159.

Forasiepi, A. M. and Carlini, A. A. (2010). A new thylacosmilid (Mammalia, Metatheria, Sparassodonta) from the Miocene of Patagonia, Argentina. *Zootaxa*, **2552**, 55–68.

Gasparini, Z. (1984). New tertiary Sebecosuchia (Crocodylia: Mesosuchia) from Argentina. *Journal of Vertebrate Paleontology*, **4**, 85–95.

Goin, F. J. (1989). Late Cenozoic South American marsupial and placental carnivores: changes in predator–prey evolution. *5° International Theriological Congress, Abstracts*, **1**, 271–272.

Goin, F. J. (1997). New clues for understanding Neogene marsupial radiations. In *Vertebrate Paleontology in the Neotropics: the Miocene Fauna of La Venta, Colombia*, ed. R. F. Kay, R. H. Madden, R. L. Cifelli and J. J. Flynn. Washington DC: Smithsonian Institution Press, pp. 187–206.

Goin, F. J. and Pascual, R. (1987). News on the biology and taxonomy of the marsupials Thylacosmilidae (late Tertiary of Argentina). *Anales de la Academia Nacional de Ciencias Exactas, Físicas y Naturales*, **39**, 219–256.

Gordon, C. L. (2003). A first look at estimating body size in dentally conservative marsupials. *Journal of Mammalian Evolution*, **10**, 1–21.

Hoffstetter, R. and Petter, G. (1983). *Paraborhyaena boliviana* et *Andinogale sallensis*, deux marsupiaux (Borhyaenidae) nouveaux du Déséadien (Oligocène inférieur) de Salla (Bolivie). *Comptes Rendus de l'Académie des Sciences, Paris, Série II*, **296**, 205–208.

Hunt Jr., R. M. (1996). Biogeography of the order Carnivora. In *Carnivore Behavior, Ecology and Evolution*, ed. J. L. Gittleman. New York: Cornell University Press, pp. 451–485.

Kay, R. F., Vizcaíno, S. F., Bargo, M. S. *et al.* (2008). Two new fossil vertebrate localities in the Santa Cruz Formation (late early–early middle Miocene, Argentina), ~51° south latitude. *Journal of South American Earth Sciences*, **25**, 187–195.

Kurtén, B. (1967). Pleistocene bears of North America, Part 2. Genus *Arctodus*, short-faced bears. *Acta Zoologica Fennica*, **117**, 1–60.

Kurtén, B. (1973). Geographic variation in size in the puma (*Felis concolor*). *Commentationes Biologicae*, **63**, 1–8.

Marshall, L. G. (1976). Evolution of the Thylacosmilidae, extinct saber-tooth marsupials of South America. *PaleoBios*, **23**, 1–30.

Marshall, L. G. (1977a). A new species of *Lycopsis* (Borhyaenidae, Marsupialia) from the La Venta Fauna (Miocene) of Colombia, South America. *Journal of Paleontology*, **51**, 633–642.

Marshall, L. G. (1977b). Evolution of the carnivorous adaptative zone in South America. In *Major Patterns in Vertebrate Evolution*, ed. M. K. Hecht, P. C. Goody and B. M. Hecht. New York: Plenum Press, pp. 709–722.

Marshall, L. G. (1978). Evolution of the Borhyaenidae, extinct South American predaceous marsupials. *University of California Publications in Geological Sciences*, **117**, 1–89.

Marshall, L. G. (1979). Review of the Prothylacyninae, an extinct subfamily of South American 'dog-like' marsupials. *Fieldiana Geology N. S.*, **3**, 1–50.

Marshall, L. G. (1981). Review of the Hathlyacyninae, an extinct subfamily of South American 'dog-like' marsupials. *Fieldiana Geology N. S.*, **7**, 1–120.

Marshall, L. G., Case, J. A. and Woodburne, M. O. (1990). Phylogenetic relationships of the families of marsupials. *Current Mammalogy*, **2**, 433–502.

Martin, L. D. (1989). Fossil history of the terrestrial Carnivora. In *Carnivore Behaviour, Ecology and Evolution*. Vol. 1, ed. J. L. Gittleman. London: Chapman & Hall, pp. 382–409.

Meiri, S., Dayan, T. and Simberloff, D. (2007). Guild composition and mustelid morphology – character displacement but no character release. *Journal of Biogeography*, **34**, 2148–2158.

Mercerat, A. (1891). Caracteres diagnósticos de algunas especies de Creodonta conservadas en el Museo de La Plata. *Revista del Museo de La Plata*, **2**, 51–52.

Muizon, C. de (1998). *Mayulestes ferox*, a borhyaenoid (Metatheria, Mammalia) from the early Palaeocene of Bolivia. Phylogenetic and palaeobiologic implications. *Geodiversitas*, **20**, 19–142.

Muizon, C. de and Argot, C. (2003). Comparative anatomy of the Tiupampa didelphimorphs; an approach to locomotory habits of early marsupials. In *Predators with Pouches. The Biology of Carnivorous Marsupials*, ed. M. Jones, C. Dickman and M. Archer. Collingood: CSIRO, pp. 43–62.

Muizon, C. de and Lange-Badré, B. (1997). Carnivorous dental adaptations in tribosphenic mammals and phylogenetic reconstruction. *Lethaia*, **30**, 351–366.

Myers, T. J. (2001). Marsupial body mass prediction. *Australian Journal of Zoology*, **49**, 99–118.

Palomares, F. and Caro, T. M. (1999). Interspecific killing among mammalian carnivores. *The American Naturalist*, **153**, 492–508.

Pascual, R. (2006). Evolution and geography: the biogeographic history of South American land mammals. *Annals of the Missouri Botanical Garden*, **93**, 209–230.

Patterson, B. and Pascual, R. (1972). The fossil mammal fauna of South America. In *Evolution, Mammals and Southern Continents*, ed. A. Keast, F. C. Erk and B. Glass. New York: State University of New York Press, pp. 274–309.

Polly, D. (2007). Limbs in mammalian evolution. In *Fins into Limbs Evolution, Development and Transformation*, ed. B. K. Hall. Chicago and London: University of Chicago Press, pp. 245–268.

Prevosti, F. J. and Vizcaíno, S. (2006). The carnivore guild of the late Pleistocene of Argentina: paleoecology and carnivore richness. *Acta Paleontologica Polonica*, **51**, 407–422.

Prevosti, J. F., Forasiepi, A., Soibelzon, L. and Zimicz, N. (2009). Sparassodonta vs. Carnivora: Ecological relationships between carnivorous mammals in South America. *10th International Mammalogical Congress*, **1**, 61–62.

Prevosti, F. J., Turazzini, G. F., Ercoli, M. D. and Hingst-Zaher, E. (2012). Mandible shape in marsupial and placental carnivorous mammals: morphological comparative study using geometric morphometry. *Zoological Journal of the Linnean Society*, **164**, 836–855.

Prevosti, F. J., Forasiepi, A. M. and Zimicz, N. (in press). The evolution of the Cenozoic terrestrial mammalian predator guild in South America: competition or replacement? *Journal of Mammalian Evolution*, DOI 10.1007/s10914–011–9175–9.

Reig, O. A. (1981). Teoría del origen y desarrollo de la fauna de mamíferos de América del Sur. *Monographie Naturae, Museo Municipal de Ciencias Naturales Lorenzo Scaglia, Mar del Plata*, 1–162.

Riff, D., Seyferth, P. R. R., Ribeiro Oliveira, G. and Aguilera, O. A. (2010). Neogene crocodile and turtle fauna in northern South America. In *Amazonia, Landscape and Species Evolution: A Look into the Past*, ed. C. Hoorn and F. P. Wesselingh. Oxford: Wiley-Blackwell Publishing, pp. 259–280.

Rohlf, F. J. (2008). tpsRelw. Version 1.46. Suny at Stony Brook. http://life.bio.sunysb.edu/morph/index.html.

Rohlf, F. K. (2009). tpsDig. Version 2.14. Suny at Stony Brook. http://life.bio.sunysb.edu/morph/index.html

Rosenzweig, M. L. (1966). Community structure in sympatric Carnivora. *Journal of Mammalogy*, **47**, 602–612.

Savage, R. J. G. (1977). Evolution of carnivorous mammals. *Palaeontology*, **20**, 237–271.

Shockey, B. J. and Anaya, F. (2008). Postcranial osteology of mammals of Salla, Bolivia (late Oligocene): form, function, and phylogenetic. In *Mammalian Evolutionary Morphology, A tribute to Frederick S. Szalay*, ed. E. J. Sargis and D. Dagosto. Dordrecht: Springer, pp. 135–157.

Simpson, G. G. (1945). The principles of classification and a classification of Mammals. *Bulletin of the American Museum of Natural History*, **85**, 1–350.

Simpson, G. G. (1950). History of the fauna of Latin America. *American Scientist*, **38**, 361–389.

Simpson, G. G. (1980). *Splendid Isolation. The Curious History of South American Mammals.* New Haven: Yale University Press.

Sinclair, W. J. (1906). Marsupialia of the Santa Cruz beds. In *Reports of the Princeton University Expeditions to Patagonia.* Vol. 4, *Palaeontology* I, Part 3, ed. W. B. Scott. Princeton: Princeton University Press, pp. 333–460.

Sunquist, M. E. and Sunquist, F. (2001). Changing landscapes consequences for carnivores. In *Carnivore Conservation*, ed. J. L. Gittleman, J. M. Funk and D. W. MacDonald. London: Cambridge University Press, pp. 399–418.

Tauber, A. A. (1997a). Bioestratigrafía de la Formación Santa Cruz (Mioceno inferior) en el extremo sudeste de la Patagonia. *Ameghiniana*, **34**, 413–426.

Tauber, A. A. (1997b). Paleoecología de la Formación Santa Cruz (Mioceno inferior) en el extremo sudeste de la Patagonia. *Ameghiniana*, **34**, 517–529.

Turazzini, G. F., Ercoli, M. D. and Prevosti, F. J. (2010). Estudio de la morfología de la mandíbula y la dieta en mamíferos depredadores vivientes y representantes de la Superfamilia Borhyaenoidea (Marsupialia, Sparassodonta). *X Congreso Argentino de Paleontología y Bioestratigrafía y VII Congreso Latinoamericano de Paleontología, Actas*, 218–219.

Van Valkenburgh, B. (1985). Locomotor diversity within past and present guilds of large predatory mammals. *Paleobiology*, **11**, 406–428.

Van Valkenburgh, B. (1987). Skeletal indicators of locomotor behavior in living and extinct carnivores. *Journal of Vertebrate Paleontology*, **7**, 162–182.

Van Valkenburgh, B. (1989). Carnivore dental adaptations and diet: a study of trophic diversity within guilds. In *Carnivore Behaviour, Ecology and Evolution*, ed. J. L. Gittleman. London: Chapman and Hall, pp. 410–436.

Van Valkenburgh, B. (1990). Skeletal and dental predictors of body mass in carnivores. In *Mammalian Paleobiology: Estimation and Biological Implications*, ed. J. Damuth and B. J. MacFadden. New York: Cambridge University Press, pp. 181–205.

Van Valkenburgh, B. (1991). Iterative evolution of hypercarnivory in canids (Mammalia: Carnivore): evolutionary interactions among sympatric predators. *Paleobiology*, **17**, 340–362.

Van Valkenburgh, B. (1999). Major patterns in the history of carnivorous mammals. *Annual Review Earth and Planetary Science*, **27**, 463–493.

Van Valkenburgh, B. (2007). Déjà vu: the evolution of feeding morphologies in the Carnivora. *Integrative and Comparative Biology*, **47**, 147–163.

Van Valkenburgh, B. and Hertel, F. (1998). The decline of North American predators during the late Pleistocene. *Illinois State Museum Scientific Papers*, **27**, 357–374.

Van Valkenburgh, B. and Koepfli, K. P. (1993). Cranial and dental adaptation to predation in canids. In *Mammals as Predators*, ed. N. Dunstone and M. Gorman. L. Symposia of the Zoologica Society of London, **65**, 15–37.

Van Valkenburgh, B., Wang, X. and Damuth, J. (2004). Cope's rule, hypercarnivory, and extinction in North American canids. *Science*, **306**, 101–104.

Villarroel, C. and Marshall, L. G. (1982). Geology of the Deseadan (early Oligocene) age Estratos Salla in the Salla-Luribay Basin, Bolivia with description of new Marsupialia. *Geobios, Memoire Special*, **6**, 201–211.

Villarroel, C. and Marshall, L. G. (1983). Two new late Tertiary marsupials (Hathlyacyninae and Sparassocyninae) from the Bolivian Altiplano. *Journal of Paleontology*, **57**, 1061–1066.

Vizcaíno, S. F., Bargo, M. S., Kay, R. F. *et al.* (2010). A baseline paleoecological study for the Santa Cruz Formation (late–early Miocene) at the Atlantic coast of Patagonia, Argentina. *Palaeogeography, Palaeoclimatology, Palaeoecology*, **292**, 507–519.

Werdelin, L. (1996). Carnivoran ecomorphology: A phylogenetic perspective. In *Carnivore Behavior, Ecology, and Evolution*. Vol. 2, ed. J. L. Gittleman. Ithaca and London: Cornell University Press, pp. 582–624.

Wroe, S., Argot, C. and Dickman, C. (2004). On the rarity of big fierce carnivores and primacy of isolation and area: tracking large mammalian carnivore diversity on two isolated continents. *Proceedings of the Royal Society London*, **271**, 1203–1211.

Zelditch, M. L., Swinderski, D. L., Sheets, H. D. and Fink, W. L. (2004). *Geometric Morphometrics for Biologists: A Primer*. San Diego: Elsevier Academic Press.

12 Paleobiology of Santacrucian glyptodonts and armadillos (Xenarthra, Cingulata)

Sergio F. Vizcaíno, Juan C. Fernicola, and M. Susana Bargo

Abstract

This chapter describes the paleobiology of Santacrucian cingulates (armadillos and glyptodonts). At least five genera of armadillos and four genera of glyptodonts were sympatric in the Santa Cruz Formation. Body masses were calculated based on a variety of scaling models, allometric equations, multiple regressions, and geometric similarity. The locomotor habits were inferred from indices previously modeled in living dasypodids, using morphogeometric analyses and the application of a strength indicator. Feeding habits were inferred from jaw biomechanics, and the shape, arrangement, and wear patterns on teeth, and from ecomorphological analyses. All armadillos fall in the range of medium-sized living armadillos, and all glyptodonts are larger than any living armadillo, slightly greater than 100 kg, but are smaller than Middle Miocene–Pleistocene glyptodonts. All Santacrucian armadillos were good diggers but none reached the degree of fossoriality found in some specialized living taxa; all glyptodonts were ambulatory. The variation in the masticatory apparatus of the armadillos exceeds that in the living species, denoting a broader range of specializations and strong niche partitioning among the fossil species. The degree of variation in the masticatory apparatus of glyptodonts suggests differences in the selectivity of feeding and habitat preference. The taxonomic richness of armadillos is similar to that recorded today in the Chaqueña biogeographic province, supporting the environmental interpretation of a mixture of open and relatively closed vegetation in relatively dry conditions.

Resumen

En este capítulo se trata la paleobiología de los cingulados santacrucenses (armadillos y gliptodontes). Al menos cinco géneros de armadillos y cuatro géneros de gliptodontes fueron simpátricos en la Formación Santa Cruz. Las masas corporales se calcularon usando modelos a escala, ecuaciones alométricas, regresiones múltiples y la similitud geométrica. Los hábitos locomotores se infirieron a partir de índices generados previamente a partir de dasipódidos vivientes, análisis morfogeométricos y la

aplicación de un indicador de fuerza. Los hábitos de alimentación fueron interpretados a partir de la mecánica de la mandíbula, la forma, disposición y patrones de desgaste en los dientes y análisis ecomorfológicos. Todos los armadillos están en el rango de los armadillos vivientes de tamaño mediano, mientras que los gliptodontes son más grandes que cualquier armadillo viviente, apenas superando los 100 kg, pero menos que los gliptodontes del Mioceno Medio-Pleistoceno. Todos los armadillos extintos eran buenos cavadores y ninguno alcanzó el grado de fosorialidad que se observa en algunas formas vivientes especializadas; los gliptodontes fueron ambulatorios. La mayor variación en el aparato masticatorio de los armadillos que en las especies vivientes denota una amplia gama de especializaciones y una importante partición de nicho en los fósiles. El grado de variación en el aparato masticatorio de los gliptodontes sugiere diferencias en el grado de alimentación selectiva y la preferencia de hábitat de alimentación. La gran riqueza taxonómica de los armadillos es similar a la registrada hoy en la provincia biogeográfica chaqueña, apoyando una interpretación ambiental de una mezcla de vegetación abierta y relativamente cerrada en condiciones relativamente secas.

12.1 Introduction

Cingulates (armadillos *s.l.*, and glyptodonts) are grouped together with anteaters and ground and tree sloths as Xenarthra, an assemblage that initially radiated in South America and one of the four major clades of placental mammals (Delsuc and Douzery, 2008; Gaudin and McDonald, 2008; and references therein). Their most conspicuous feature is their armor, composed of dermal scutes covered with epidermal scales and divided into shields that protect the head and body, with a sheath for the tail.

Living cingulates are represented only by armadillos, which constitute two-thirds of the diversity of living xenarthrans (Wetzel, 1985; Aguiar and Fonseca, 2008). With over 65 genera and 150 species (Mones, 1986; McKenna and Bell, 1997), fossil cingulates represent a much greater diversity than their living representatives and were common elements of the Santacrucian Age faunas. Fossil cingulates include typical armadillos (Dasypodidae), the horned peltephilines (Peltephilidae), the pampatheres (Pampatheriidae),

Early Miocene Paleobiology in Patagonia: High-Latitude Paleocommunities of the Santa Cruz Formation, ed. Sergio F. Vizcaíno, Richard F. Kay and M. Susana Bargo. Published by Cambridge University Press. © Cambridge University Press 2012.

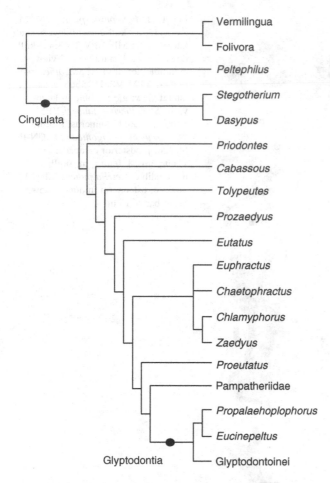

Fig. 12.1. Phylogeny of Cingulata, modified from Gaudin and Wible (2006), Fernicola (2008) and Porpino *et al.* (2010). *Propalaehoplophorus* and *Eucinepeltus* are representatives of the subfamily Propalaehoplophorinae.

and the glyptodonts (Glyptodontia). According to Gaudin and Wible (2006), peltephilids are the sister group of the remaining cingulates, dasypodids are paraphyletic, and within the latter neither the Eutatini tribe nor the Euphractini tribe of McKenna and Bell (1997) is monophyletic. Pampatheres + glyptodonts constitute a monophyletic group. Pampatheres have not been recorded for the Santacrucian Age, and hence they are not further discussed in this chapter. The monophyly and internal relationships of glyptodonts were recently examined by Fernicola (2008) and Porpino *et al.* (2010) who support a basal dichotomy between Propalaehoplophorinae – the only group represented during Santacrucian times – and Glyptodontoinei, grouping the remaining glyptodonts (Fig. 12.1). In the following section we synthesize the latest data on cingulate diversity as recorded during the Santacrucian Age.

Armadillos usually have flexible armor, formed by a variable number of imbricating bands joining the immobile anterior and posterior bucklers (Figs. 12.2, 12.3, and 12.9b–d),

and are mostly specialized digging animals, with limbs well-designed for that activity. They have well-developed claws, proximally and distally fused tibia and fibula, large tuberosities for strong muscular attachments, and long lever arms such as the elongated olecranon process on the ulna for the line of action of the principal muscles (Vizcaíno *et al.*, 1999; Vizcaíno and Milne, 2002). The masticatory apparatus of armadillos is very peculiar, and its adaptive interpretation presents an interesting challenge because of the strong phylogenetic constraints identified among armadillos (Vizcaíno and De Iuliis, 2003). The dentition is greatly simplified; nine or ten rather small and mostly peglike teeth are present in each jaw quadrant (except in *Priodontes*). Enamel is absent (except in the Late Eocene *Utaetus* Ameghino, 1902 and juveniles of *Dasypus*; Kalthoff, 2011), and all extant and possibly all extinct xenarthrans lack deciduous teeth, except for *Dasypus*. The cuspal patterns present in other mammals are also absent. In most armadillos the teeth are composed mainly of orthodentine; there may be a thin outer cementum layer and sometimes orthodentine occurs together with vasodentine or poorly developed osteodentine in the center (Kalthoff, 2011). Eutatines feature osteodentine as an inner layer, and an orthodentine outer layer, with the outermost portion standing out as a rim. Teeth are always hypselodont, and may be lobate, but are usually simple and separated by short diastemata. These peculiarities of the dentition must have imposed severe functional and biomechanical constraints as lineages adapted to different diets. Although living armadillos are mainly animalivorous or omnivorous (Redford, 1985; Vizcaíno *et al.*, 2004), a variety of dietary habits have been inferred for the fossils, including herbivory (Vizcaíno *et al.*, 2004, Vizcaíno, 2009 and references therein).

Glyptodonts traditionally are considered functionally unsuited to digging because the relatively rigid carapace (Figs. 12.5 and 12.9a), which lacks the movable bands of armadillos, is fused to the pelvic girdle, there is extensive fusion among dorsal and lumbar vertebrae to form a tube, and the other skull and limb structures typical of burrowing mammals are absent (Kraglievich, 1934; Quintana, 1992). The morphology of the skull, and the masticatory apparatus in particular, is even more bizarre than the carapace. Compared with that of armadillos, the snout is shortened, with reduced premaxillae, and the skull has undergone a very peculiar process of telescoping, in which the braincase lies above the posterior half of the tooth row. Another conspicuous feature is the huge anteroposteriorly compressed descending process of the zygomatic arch. The jaw has the distinctive feature of a highly developed angle. The teeth are hyselodont (hypsodont and ever-growing) and lobated, with an internal structure like that of eutatines, i.e. inner osteodentine and outer orthodentine (Kalthoff, 2011). Preliminary observations by Fariña and Vizcaíno (2001) indicate that the stout architecture of the masticatory apparatus and the very hypsodont teeth with lobes

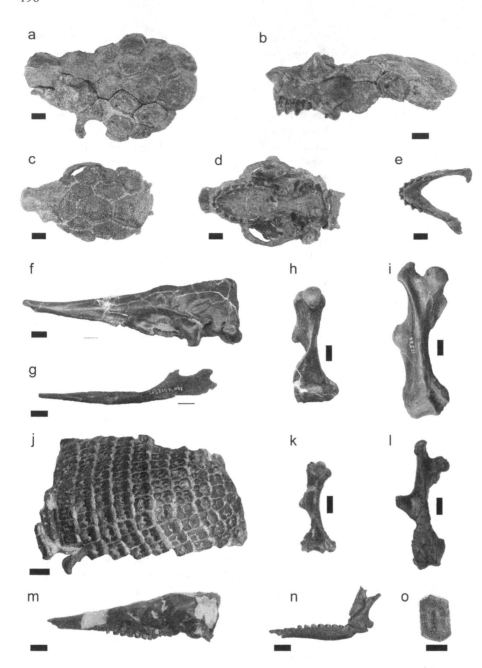

Fig. 12.2. *Peltephilus pumilus* MPM-PV 3433: a, skull dorsal view; b, skull lateral view; MPM-PV 3643: c, skull dorsal view; d, skull ventral view; e, mandible dorsal view. *Stegotherium tauberi* YPM-VPPU 15565: f, skull in lateral view; g, mandible in lateral view. *Stegotherium tauberi* YPM-VPPU 15566: h, humerus; i, femur. *Stenotatus* cf. *patagonicus* MACN-Pv SC406: j, posterior carapace; k, humerus; l, femur; m, skull; n, mandible. *Vetelia puncta* MPM-PV 3652: o, osteoderm in dorsal view. Scale bar = 1 cm.

and rims of hard dentine on the occlusal surface suggest that glyptodonts in general were probably grazers.

Recent morphofunctional, biomechanical, and ecomorphological studies provide the basis for understanding cingulate paleobiology (Bargo, 2003, and Vizcaíno *et al.*, 2008, and references therein; Vizcaíno, 2009; Milne *et al.*, 2009; Vizcaíno *et al.*, 2011a; Vizcaíno *et al.*, 2011b, and Vizcaíno *et al.*, 2012). In this chapter we provide a comprehensive approach to the paleobiology of Santacrucian cingulates based on an overview of this recent literature. We also evaluate their diversity in comparison with that of living cingulates and offer some paleoenvironmental hypotheses.

12.2 Diversity of Santacrucian cingulates

Florentino Ameghino (1887) provided the first descriptions of Santacrucian cingulates, including 11 species of armadillos that are currently in the genera *Peltephilus* Ameghino, 1887 (Fig. 12.2a–e), *Stegotherium* Ameghino, 1887 (Fig. 12.2f–i), *Prozaedyus* Ameghino, 1891 (Fig. 12.3a–e), *Vetelia* Ameghino, 1891 (Fig. 12.2o), *Proeutatus* Ameghino, 1891 (Fig. 12.4a–g), and *Stenotatus* Ameghino, 1891 (Fig. 12.2j–n). He also named two species of the glyptodont *Propalaehoplophorus* Ameghino, 1887 (Fig. 12.5c and e). In the years following the original

Fig. 12.3. *Prozaedyus proximus* MPMP-PV 3617, articulated skeleton: a, ventral; b, dorsal view; MPM-PV 3506: c, complete articulated skeleton and carapace in matrix; MPM-PV 3423: d, skull and cervical vertebrae in ventral view; e, skull and cephalic shield in dorsal view. Scale bar = 1 cm.

descriptions, he named three new genera of armadillos, *Anantiosodon* Ameghino, 1891, *Peltecoelus* Ameghino, 1902, and *Eodasypus* Ameghino, 1894, and added four glyptodonts, *Asterostemma* Ameghino, 1889, *Cochlops* Ameghino, 1889 (Fig. 12.5f–h), *Eucinepeltus* Ameghino, 1891 (Fig. 12.5a, b) and *Metopotoxus* Ameghino, 1898 (Ameghino, 1889, 1891, 1894, 1898, and 1902). Lydekker (1894) synonymized most of the Santacrucian taxa that Ameghino had proposed up to 1894. However, in his exhaustive taxonomic revision of the cingulates, Scott (1903–04) revalidated almost all taxa erected by Ameghino. Subsequent taxonomic revisions have been limited to peltephilids (Bordas, 1936, 1938), and the genus *Stegotherium* (Fernicola and Vizcaíno, 2008).

Among armadillos, the Peltephilidae are certainly represented by *Peltephilus* (Fig. 12.2a–e), with four species -

Pe. nanus Ameghino, 1898, from La Cueva locality; *Pe. ferox* Ameghino, 1891, collected in the Río Sehuen (= Shehuen), at the mouth of the Río Coyle and at Monte Observación (= Cerro Observatorio); *Pe. strepens* Ameghino, 1887 and *Pe. pumilus* Ameghino, 1887, both collected along the Río Santa Cruz, La Cueva, Monte Observación (= Cerro Observatorio), and Río Sehuen (for information on the localities see Vizcaíno *et al.*, Chapter 1: Figs. 1.1 and 1.2, and Appendix 1.1). The other two taxa mentioned for the Santacrucian, *Anantiosodon rarus* Ameghino, 1891 collected at Monte Observación (= Cerro Observatorio), and *Peltecoleus grandis* Moreno y Mercerat, 1891 found at Río Sehuen, at La Cueva and Monte Observación (= Cerro Observatorio), were assigned by Scott (1903–04) to *Peltephilus*. Bordas (1938) did not accept this taxonomic decision, but our preliminary studies of these

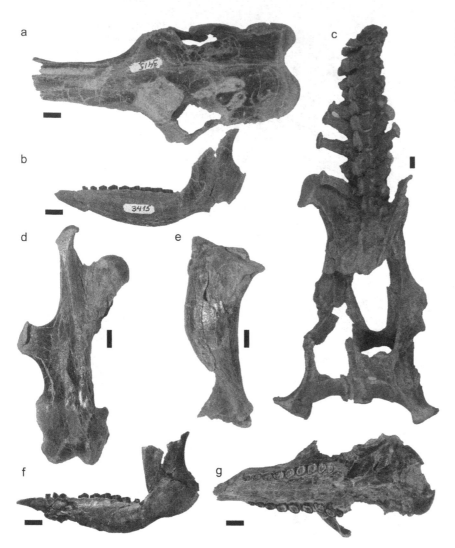

Fig. 12.4. *Proeutatus oenophorus* MPM-PV 3415: a, skull in dorsal view; b, mandible in lateral view; c, pelvis and part of the vertebral column; d, femur; e, tibia. *Proeutatus oenophorus* MPM-PV 3430: f, mandible in lateral view; g, skull in palatal view. Scale bar = 1 cm.

taxa agree with Scott (1903–04) in considering them as synonyms of *Peltephilus*. Studies in progress allow assigning to *Pe. pumilus* the specimens collected by us in the lower member of the Santa Cruz Formation, south of the Río Coyle.

One genus of Stegotheriini, *Stegotherium*, is represented in the Santacrucian. It includes three species: *Stegotherium tessellatum* Ameghino, 1887 was collected along the Río Santa Cruz and coast of Santa Cruz Province; *S. notohippidensis* González and Scillato-Yané, 2009, was recovered near Lago Argentino; and *S. tauberi* González and Scillato-Yané, 2008 (Fig. 12.2f–i) was collected in Estancia La Angelina and Killik Aike Norte and probably Coy Inlet.

Two genera of "Euphractini," *Prozaedyus* and *Vetelia*, were reported from the Santacrucian by different authors (Ameghino, 1887, 1889, 1902; Marshall *et al.*, 1983; Tauber,

1999; Fernicola *et al.*, 2009). Scott (1903–04) accepted two species within *Prozaedyus* (with reservations): *P. proximus* Ameghino, 1887 and *P. exilis* Ameghino, 1887; he considered that there were only minor and perhaps inconstant differences between the two. Two well-preserved specimens collected by us from the coastal beds of Santa Cruz Province support the validity of these two species. Both species are represented in collections along the banks of the Río Santa Cruz, and along the coast of Santa Cruz Province between Monte Observación and Cabo Buen Tiempo (Cape Fairweather). The specimens collected by us from the coastal beds of Santa Cruz Formation, including two very well-preserved individuals, are assigned to *P. proximus*. The other "Euphractini," *Vetelia puncta* Ameghino, 1891 (Fig. 12.2o), was first reported from the north bank of the Río Santa Cruz,

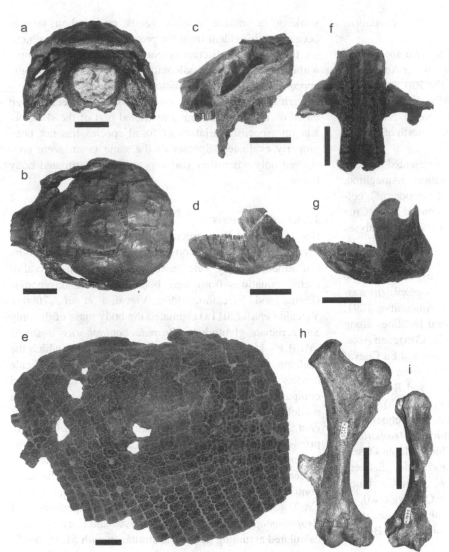

Fig. 12.5. *Eucinepeltus petesatus* MACN-A 4758 (Holotype): a, skull in anterior view; b, skull in dorsal view. *Propalaehoplophorus australis* MPM-PV 3422: c, skull in lateral view; d, mandible in lateral view; e, carapace in dorsal view. *Cochlops muricatus* MPM-PV 3432: f, skull in palatal view; g, mandible in lateral view; *Cochlops muricatus* MACN-A 2120: h, femur. Propalaehoplophoridae indet. MACN-A 7693: i, humerus. Scale bar = 5 cm.

in outcrops assigned by Ameghino (1902) to the "Notohippidian" (the oldest Santacrucian faunal assemblage according to Marshall *et al.*, 1983). Recently, Fernicola *et al.* (2009) reported the same species from the coast of Santa Cruz Province, 3km south of the mouth of the Rio Coyle. A third genus, *Eodasypus* Ameghino, 1894, was considered *incertae sedis* by Scott (1903–04) and Scillato-Yané (1980), and will not be considered here.

Among "Eutatini", *Proeutatus*, *Stenotatus* (Fig. 12.2j–n) and *Pareutatus* Scott 1903 have been recorded in Santacrucian beds. Following Scott (1903–04), we recognize five species of *Proeutatus*: *Pr. oenophorus* Ameghino, 1887, *Pr. lagena* Ameghino, 1887, *Pr. carinatus* Ameghino, 1891, *Pr. deleo* Ameghino, 1891, and *Pr. robustus* Scott, 1903.

Proeutatus carinatus was collected at Monte Observación (= Cerro Observatorio) (Ameghinos catalog), *Pr. lagena* and *Pr. oenophorus* are from along the Río Santa Cruz and Río Sehuen, from Monte Observación (= Cerro Observatorio), Corriguen Aike, the mouth of Río Coyle, and from Cabo Buen Tiempo (Cape Fairweather, Catalogs of Ameghino and Yale Peabody Museum collections; Tauber, 1999; Ameghino, 1902; Scott, 1903–04). The specimens of *Pr. robustus* were found at the mouth of the Rio Coyle and at Cabo Buen Tiempo (Yale Peabody Museum catalog). A preliminary revision, including new material recently collected by us, suggests that the anatomical differences among *Pr. oenophorus*, *Pr. deleo*, *Pr. lagena*, and *Pr. robustus* that were noted by Ameghino (1891) and Scott (1903–04) likely represent intraspecific variation within *Pr. oenophorus* (J. C. Fernicola, unpublished data).

The genus *Stenotatus* is represented by two species: *St. hesternus* Ameghino, 1889 and *St. patagonicus* Ameghino, 1887. The first species is known from the Río Santa Cruz and the coastal outcrops between Monte Observación (= Cerro Observatorio) and Killik-Aike Norte. *Stenotatus patagonicus*, was collected along the coast, north of the Río Coyle and up to Monte Observación.

Finally, *Pareutatus* Scott, 1903 is represented by one poorly known species, *Pareutatus distans* (Ameghino, 1894) (based on a fragmentary skull and osteoderms), collected along the Río Santa Cruz and the coastal Santa Cruz Formation. We do not include this taxon in our analyses because of its fragmentary state.

Scott (1903–04) accepted the five glyptodont genera named by Ameghino (mentioned above) and also the following Santacrucian species: *Propalaehoplophorus australis* Ameghino, 1887 and *Pro. minor* Ameghino, 1891, both from the Río Santa Cruz and several localities along the Atlantic coast such as Yegua Quemada, Corriguen Aike, Monte Observación (= Cerro Observatorio) and La Cueva; *Asterostemma depressa* Ameghino, 1889 from Canadón Jack (Yack-Harvey), Monte Observación, and Río Chico; *Cochlops muricatus* Ameghino, 1889 (Fig. 12.5f–h), and *C. debilis* Ameghino, 1891 from Río Chico, Río Sehuen, and Monte Observación (= Cerro Observatorio); *Metopotoxus laevatus* (Ameghino, 1889) from Río Chico, Monte Observación (= Cerro Observatorio), and Río Sehuen; *Eucinepeltus petesatus* Ameghino, 1891 from Corriguen Aike and Monte Observación (= Cerro Observatorio), *E. crassus* Scott, 1903, represented by only one specimen collected at Killik-Aike Norte, and *E. complicatus* Brown, 1903 from Río Gallegos.

A preliminary study of the recently collected material from along the coast of Santa Cruz Province, and of the type specimens housed in the Ameghino Collection at Museo Argentino de Ciencias Naturales "B. Rivadavia," led us to recognize the following taxa: cf. *Asterostemma depressa*, *Propalaehoplophorus australis*, *Cochlops muricatus*, *Eucinepeltus petesatus* (Fig. 12.5a, b), *E. crassus*, *E. complicatus*. We consider *Metopotoxus* to be a synonym of *Cochlops* (J. C. Fernicola, unpublished data).

12.3 Materials and methods

The abbreviations and a list of the materials are provided in Appendix 12.1. The specimens studied include those collected during the nineteenth century and the material collected by the Museo de La Plata/Duke University expeditions over the past nine years (2003–2011). This recently collected material permits a larger and more varied database, consequently producing more robust results. The

working taxonomic unit is genus rather than species because, as is evident from the preceding taxonomic synopsis, the genera are discrete taxonomic units accepted almost without debate by most paleontologists. Also, paleobiologic approaches based on morphology usually are not sensitive enough for discriminating among congeneric species when based on a few traits from a restricted part of the skeleton, and intraspecific variation of fossil species has not been properly evaluated. Species of the same genus were considered only when they had very different estimated body masses.

12.3.1 Body mass

Body masses of Santacrucian cingulates were estimated in several ways, depending on the peculiarities of each group. For armadillos they were based on adult interspecific allometric equations from long bones of living dasypodids (Fariña and Vizcaíno, 1997; Vizcaíno et al., 2006a). Vizcaíno et al. (2011a) estimated the body mass of the only Santacrucian glyptodont, *Propalaehoplophorus australis* (MLP 16–15, Plate XXII of Scott 1903–04), for which the skeleton is almost entirely preserved. They used scale models and computer-generated geometric models, and compared them with previous results obtained from scale models and from allometric equations (Fariña, 1995; Fariña et al., 1998). For the same specimen, Vizcaíno et al. (2011b) provided a body mass estimate as an average of multiple regressions calculated from measurements of the postcranial skeleton (De Esteban-Trivigno et al., 2008). In the same contribution, the body mass of *Eucinepeltus* was taken from Croft (2000) and the masses of other specimens of *Propalaehoplophorus*, cf. *Asterostemma* and *Cochlops* were calculated assuming geometric similarity with MLP 16–15.

12.3.2 Locomotion and use of substrate

Vizcaíno et al. (2006a) characterized the habits of Santacrucian armadillos by calculating indices on proportions of the proximal and middle segments of the limb bones. These indices were the shoulder moment index (SMI); brachial index (BI); index of fossorial ability (IFA); hip moment index (HMI); leg robusticity index (LRI); crural index (CI); and intermembral index (IMI). Vizcaíno and Milne (2002) and Vizcaíno et al. (2003) provide descriptions of the measurements involved in the calculation of each index (Fig. 12.6) and their individual functional interpretation. Vizcaíno et al. (2006a) carried out a principal components analysis (PCA) to explore how the Santacrucian armadillos fit the patterns of structure and function observed in the limbs of living armadillos by Vizcaíno and Milne (2002). Vizcaíno et al. (2011a) calculated the same indices for glyptodonts of different geologic ages, including the Santacrucian *Propalaehoplophorus australis*.

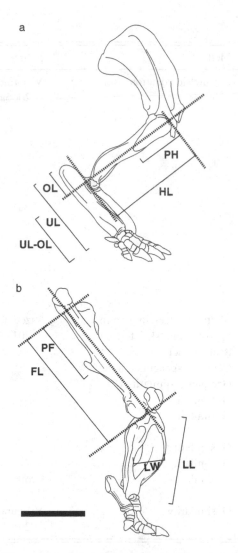

Fig. 12.6. Measurements of cingulate limbs (based on the specimen MLP 16–15 *Propalaehoplophorus australis*) used to calculate the indices (modified from Vizcaíno *et al.* 2011a). a, Forelimb: **HL**, humeral length; **PH**, proximal humeral length; **UL**, ulna total length; **OL**, olecranon length; **UL–OL**, ulna total length less olecranon length. b, Hindlimb: **FL**, femur length; **PF**, proximal femoral length; **LL**, leg length; **LW**, mid-leg width. Scale bar: 10 cm.

12.3.3 Mastication and feeding

Vizcaíno *et al.* (2004, 2006a) summarized previous masticatory and dietary interpretations for Santacrucian armadillos using several approaches, mainly a geometric model of jaw mechanics. The mandible is viewed as a lever, with its fulcrum at the temporomandibular joint, an input (muscular) force generated by the masticatory musculature, and an output (masticatory) force exerted by the teeth on food. The moment arms of the *m. temporalis* and *m. massetericus* muscles are estimated so that the effective power of each muscle and the corresponding bite force may be estimated.

For a complete explanation see Vizcaíno *et al.* (1998, 2004). This approach could not be applied to *Stegotherium* because its peculiar morphology implies that its mandible did not act as a lever (see below). Occlusal patterns and mandibular movements were determined through analysis of the craniomandibular joint (including manipulation), the shape and arrangement of teeth (including occlusal wear patterns), and the shape and arrangement of the symphysis.

So far there are no comparable studies for glyptodonts, in part because they pose different problems given the design of their masticatory apparatus (Fariña, 1985, 1988; Fariña and Vizcaíno, 2001) which is not completely understood. Here we provide some preliminary observations on masticatory movements (Fariña and Vizcaíno, 2001).

Vizcaíno *et al.* (2011b) performed an ecomorphological analysis on glyptodonts from different ages, evaluating morphological variables previously quantified in other xenarthrans (Bargo *et al.*, 2006a, b; Vizcaíno *et al.*, 2006b; Bargo and Vizcaíno, 2008). The taxa studied include the Santacrucian *Propalaehoplophorus*, *Cochlops*, and *Eucinepeltus* and the Pleistocene *Neuryurus* Ameghino, 1889, *Neosclerocalyptus* Paula Couto, 1957, *Panochthus* Burmeister, 1866, *Doedicurus* Burmeister, 1874, and *Glyptodon* Ower, 1839. The variables analyzed were -relative muzzle width (RMW), hypsodonty index (HI), and dental occlusal surface area (OSA).

RMW was calculated following the original procedure of Janis and Ehrhardt (1988) for ungulates, as palatal width (PW) divided by maximum muzzle width (MMW). According to these authors, the width of the palate could signal the rate of food ingestion, with high RMW values indicating selective feeders with narrow muzzles. Because the premaxillae are reduced in glyptodonts, MMW generally occurs along the maxilla and was thus measured following Bargo *et al.* (2006b) as the distance between labial alveolar margins of the first molariforms. PW was measured between the lingual alveolar margins of the central lobe of the fourth molariform, where the palate is narrowest in glyptodonts. The values obtained cannot be compared directly with those for ungulates, but they provide a framework for comparison among glyptodonts. HI was calculated as the depth of the mandible (DM) divided by the length of the molariform tooth row (LTR), as used by Bargo *et al.* (2006a) for ground sloths. OSA was evaluated following Vizcaíno *et al.* (2006b) as the total occlusal surface area of the cheek teeth, including the infolded contour of each tooth. While HI and RMW are size-independent, the allometric relationship between OSA and BM was assessed by calculating the least-squares regression lines, using log transformed OSA as the dependent variable and log transformed BM as the independent variable. Residuals from the regression (size-independent OSA) were plotted against both indices (HI and RMW).

Table 12.1. *Body mass estimates for Santacrucian cingulates*

Catalog number	Taxa	Body mass (kg)	Method	Source
MACN-A 1063	*Proeutatus* sp.	16.48	Allometric equations from long	Vizcaíno *et al.*,
MLP 69-IX-8–11	*Proeutatus* sp.	20.65	bones	2006a
YPM-VPPU 15957	*Proeutatus robustus*	13.45		
YPM-VPPU 15214	*Pr. robustus*	19.07		
YPM-VPPU 15389	*Pr. robustus*	10.46		
YPM-VPPU 15613	*Pr. lagena*	7.58		
FMNH P 13197	*Pr. oenophorus*	15.96		
MLP 69-IX-5–11	*Stenotatus* sp.	3.91		
YPM-VPPU 15863	*Stenotatus patagonicus*	3.67		
YPM-VPPU 15579	*Prozaedyus exilis*	0.75		
YPM-VPPU 15604	*Prozaedyus* sp.	1.86		
YPM-VPPU 15566	*Stegotherium tauberi*	11.47		
MACN-A 7910 /40	*Peltephilus ferox*	11.07		
YPM-VPPU 15390	*Pe. strepens*	8.22		
MLP 16–15	*Propalaehoplophorus australis*	81.10	Multiple regressions from postcranial skeleton	Vizcaíno *et al.*, 2011b
		85.00	Scale model	
		67.00	Computer-generated model	
MACN-A 4754	*Pro. australis*	93.48	Geometric similarity (*)	
MACN-A 7655	*Pro. incisivus*	74.49	G. similarity	
MACN-A 7656	*Propalaehoplophorus* sp.	77.98	G. similarity	
MPM-PV 3420	*Cochlops muricatus*	79.27	G. similarity	
MPM-PV 3432	*Cochlops muricatus*	86.71	G. similarity	
MACN-A 4758	*Eucinepeltus petesatus*	115.00	G. similarity	Croft, 2000
MACN-A 7663	*cf. Asterostemma depressa*	78.13	G. similarity	Vizcaíno *et al.*, 2006b

(*) Geometric similarity with a specimen with body mass estimated by other methods.

12.4   Results

12.4.1   Body mass

Body mass estimates for Santacrucian cingulates are provided in Table 12.1.

Mean estimated body masses (rounded to the nearest kilogram) for the armadillos are 15 kg for *Proeutatus*, 4 kg for *Stenotatus*, 1 kg for *Prozaedyus*, and 11 kg for *Stegotherium* and *Peltephilus* (Vizcaíno *et al.*, 2006a).

Using plastic scale models and computer-generated models, Vizcaíno *et al.* (2011a) obtained body mass estimates of 85 kg and 67 kg, respectively, for the glyptodont *Propalaehoplophorus australis* MLP 16–15. Using the multiple regression equations provided by De Esteban-Trivigno *et al.* (2008), Vizcaíno *et al.* (2011b) obtained a value of 81 kg for the same individual. Scaling it geometrically, Vizcaíno *et al.* (2011b) obtained values between 74 and 93 kg for the several specimens of *Propalaehoplophorus*, 78 kg for one

specimen of cf. *Asterostemma*, and 72 and 87 kg for two specimens of *Cochlops muricatus*. Croft (2000) provided an estimate of 115 kg for *Eucinepeltus petesatus*.

12.4.2   Locomotion and use of substrate

Table 12.2 summarizes the indices of the five Santacrucian genera of armadillos and the glyptodont *Propalaehoplophorus australis* as provided in Vizcaíno *et al.* (2006a) and Vizcaíno *et al.* (2011a) respectively, in comparison with those of living forms (Vizcaíno and Milne, 2002).

In the forelimb, the highest IFA among Santacrucian taxa occurs in *Proeutatus*, followed by *Peltephilus*, *Prozaedyus*, and *Stenotatus*; we do not have data for *Stegotherium*. In *Proeutatus* and *Peltephilus*, IFA is higher than in extant *Tolypeutes matacus*, dasypodines, and euphractines, but lower than in priodontines and *Chlamyphorus truncatus*. *Prozaedyus* is close to the average for extant euphractines

Table 12.2. *Indices of the Santacrucian armadillos and the glyptodont* Propalaehoplophorus australis, *in comparison with the mean values of extant forms*

Catalog number	Taxa	IFA	SMI	BI	HMI	LRI	CI	IMI
MACN-A 1063	*Proeutatus* sp.	–	47.29	–	–	–	–	–
MLP 69-IX-8-11	*Proeutatus* sp.	–	49.74	–	–	–	–	–
YPM-VPPU 15957	*Proeutatus robustus*	78.50	–	–	–	–	–	–
YPM-VPPU 15214	*Pr. robustus*	75.65	55.14	61.0	–	–	–	–
YPM-VPPU 15389	*Pr. robustus*	–	–	–	47.65	37.59	75.77	–
YPM-VPPU 15613	*Pr. lagena*	–	–	–	50.00	–	–	–
FMNH P 13197	*Pr. oenophorus*	–	53.52	–	48.12	39.94	91.77	–
YPM-VPPU 15863	*Stenotatus patagonicus*	62.16	53.14	68.51	–	–	–	–
YPM-VPPU 15579	*Prozaedyus exilis*	69.67	52.63	64.21	38.46	31.89	94.87	82.10
YPM-VPPU 15604	*Prozaedyus* sp.	–	49.0	–	–	–	–	–
YPM-VPPU 15566	*Stegotherium tauberi*	–	53.33	–	53.3	30.1	97.17	–
MACN-A 7910 /40	*Peltephilus ferox*	74.49	51.97	59.45	–	–	–	–
MLP 16–15	*Propalaehoplophorus australis*	68.75	48.48	48.48	43.47	51.16	53.48	74.24
Extant armadillos	*Tolypeutes matacus*	58.2	58.78	80.17	46.05	18.91	101.5	68.29
	Dasypodini	66.85	48.44	69.29	54.49	36.75	81.93	70.93
	"Euphractini"	69.44	56.07	63.21	50.40	36.48	85.48	84.71
	"Priodontini"	92.86	68.67	55.92	51.08	42.35	81.22	77.11
	Chlamyphorus truncatus	112.68	77.40	61.54	42.24	22.13	105.02	70.60

Note: data for fossil armadillos and the glyptodont from Vizcaíno *et al.* (2006a) and Vizcaíno *et al.* (2011a) respectively; data for extant armadillos from Vizcaíno and Milne (2002). Abbreviations: **SMI**: shoulder moment index; **BI**: brachial index; **IFA**: index of fossorial ability; **HMI**: hip moment index; **LRI**: leg robusticity index; **CI**: crural index; **IMI**: intermembral index.

and within the range of extant dasypodines. In *Stenotatus* the value lies within the upper maximum range of *T. matacus*, and lower than the rest of the living forms. SMI values for the five Santacrucian genera are within the range of euphractines and dasypodines. BI was calculated for *Proeutatus*, *Stenotatus*, *Prozaedyus*, and *Peltephilus*. The first three have values within the range of euphractines and dasypodines, while *Peltephilus* has a lower value – between the priodontines and *C. truncatus*.

In the hindlimb, the HMI values for *Proeutatus*, *Stegotherium*, and *Peltephilus* are lower than for dasypodines, falling within the range of *T. matacus*, euphractines, and priodontines. *Prozaedyus* shows the lowest HMI, close to the lower minimum range of *C. truncatus*. LRI and CI were estimated for *Proeutatus*, *Prozaedyus*, and *Stegotherium*. *Proeutatus* LRI is higher than in dasypodines and euphractines and within the lower range of priodontines; the other two have lower values than in dasypodines, euphractines, and priodontines, but higher than in *T. matacus* and *C. truncatus*. CI values in *Prozaedyus* and *Stegotherium* are higher than in dasypodines, euphractines, and priodontines, but lower than in *T. matacus* and *C. truncatus*. *Proeutatus* falls within the range of dasypodines, euphractines, and priodontines. IMI is known only

for *Prozaedyus*, and it is within the range of euphractines and priodontines.

In the first two principal components of the PCA analysis performed by Vizcaíno *et al.* (2006a), PC1 factor loadings are interpreted as indicating digging ability and PC2 correlates with size (Fig. 12.7). Despite some missing data, the results suggest that all the Santacrucian genera had limb proportions similar to the extant digging armadillos, i.e. euphractines and dasypodines. *Prozaedyus* was about as fossorial as the extant *Euphractus*. *Stenotatus* appears to have been as good a digger as the dasypodines, and probably more cursorial than *Prozaedyus*. *Peltephilus* was apparently a better digger than euphractines, but not as powerful as the priodontines. The same applies to *Proeutatus* and *Stegotherium*, although with a lower degree of confidence.

Comparing the limb proportion indices of glyptodonts of different sizes and geologic ages with those of the living armadillos (Vizcaíno *et al.*, 2011a), it appears that the glyptodonts have IFA and SMI values similar to those of the living Dasypodini and Euphractini (generalized diggers). The BI and CI of glyptodonts are much lower than those of living armadillos, and seem to be more related to their larger body masses. Similarly, the hindlimb indices HMI and LRI

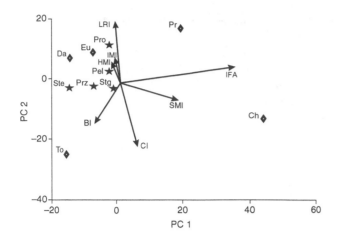

Fig. 12.7. Bivariate plot of the first and second axes of a principal components analysis of armadillo limbs (modified from Vizcaíno *et al.*, 2006a). Extant taxa (diamonds): Ch, *Chlamyphorus truncatus*; Da, dasypodines; Eu, euphractines; Pr, priodontines; To, *Tolypeutes matacus*. Santacrucian taxa (stars): Pel, *Peltephilus*; Pro, *Proeutatus*; Prz, *Prozaedyus*; Ste, *Stenotatus*; Stg, *Stegotherium*.

of the glyptodonts are much higher than in living armadillos, and this also seems to be related to their larger body masses and to the fact that glyptodonts carry more of their body weight on their hindlimbs (Fariña, 1995; Vizcaíno *et al.*, 2011a). It is noteworthy that the forelimb indices correlate strongly with body mass among glyptodonts (Vizcaíno *et al.*, 2011a), but not among living armadillos (Vizcaíno and Milne, 2002) where IFA is strongly related to digging behavior. In *Propalaehoplophorus* the strength indicator is higher for the femur than for the humerus.

12.4.3 Mastication and feeding

The masticatory apparatus of peltephilines is distinct from that of other armadillos (Vizcaíno and Fariña, 1997). It bears a peculiar morphology (Fig. 12.2b, d and e) suggesting that the main bite force must have been exerted by the larger front teeth, which implies a poor mechanical advantage in biting, as the output force has a longer lever arm. The moment arms of the bite points along the tooth row are long compared with jaw lengths as a consequence of the shortness of the tooth row. However, the ratios of the muscle to bite points are moderate, owing to the short moment arm of the masseter muscle. This may have been partially compensated by enlarged masticatory musculature. Vizcaíno and Fariña (1997) postulated that a compromise is reached in regard to the tooth row length between stronger bite and a larger gape, the latter presumably related to the relatively large size of food items.

The skull of *Stegotherium* (Fig. 12.2f, g) resembles that of the living dasypodines of the genus *Dasypus*, although it is more specialized in some respects (Vizcaíno, 1994). The rostrum is more elongated and the minute teeth are

concentrated more posteriorly. The glenoid cavity is a narrow, longitudinally elongated groove and the mandibular condyle is very low, longitudinal and scroll-shaped. The descending process of the zygomatic arch and the elongation of the masseteric fossa combine with the shapes and locations of the pterygoid and angular processes in such a way as to emphasize the anteroposterior component of the line of action of the masseter and pterygoid muscles.

A morphofunctional evaluation of the masticatory apparatus of *Prozaedyus* (Vizcaíno and Fariña, 1994) identified a resemblance to the living euphractine *Zaedyus pichiy*. Although the rostrum of *Prozaedyus* (Fig. 12.3c–e) is more elongated, the coronoid process of the mandible is more slender, the symphysis shorter, the masseteric fossa deeper, the angular process more hook-like, and the teeth comparatively smaller.

Vizcaíno and Bargo (1998) analyzed the masticatory apparatus of some eutatines, including the Santacrucian *Proeutatus* (Fig. 12.4a, g) and *Stenotatus* (Fig. 12.2m, n). The general shape of the mandible and teeth of *Proeutatus* resembles those of some ungulates. The mandibular condyle is elevated above the tooth row improving the mechanical advantage of the masseter muscle. Compared with living armadillos, the moment arms of the *m. temporalis* and *m. massetericus* in *Proeutatus* are relatively large compared with the moment arms of the bite forces at all points on the tooth row. This is also true for the combined moment arm of both muscles. The concave-to-flat morphology of the condyle allows considerable lateral and anteroposterior mandibular excursion. The unfused symphysis may be related to transverse chewing movements, produced by the pterygoid muscles, as in artiodactyls. Greaves (1978) suggested that with an unfused symphysis, the independent movement of the jaws allows the balancing-side jaw to tilt, causing a more advantageous orientation (less laterally directed) for the vector of the balancing-side medial pterygoid muscle. The teeth are better suited for cutting or shearing than those of living armadillos. Santacrucian eutatines display flat grinding surfaces on the posterior two-thirds of the tooth row. The outer and the inner hard dentine layers, elevated above the level of softer medial dentine, are almost parallel to the long axis of the tooth, which implies a strong lateral component in mastication. In *Stenotatus* the palate and the rostrum are more elongated and the teeth are more euphractine-like, i.e. chisel-shaped, and the different layers of dentine are not very evident. The general shape of the masticatory apparatus suggests that *Stenotatus* could not bite as powerfully as *Proeutatus* or living euphractines of similar size, such as the hairy armadillo *Chaetophractus villosus* (Vizcaíno and Bargo, 1998).

The masticatory mechanics of glyptodonts are not completely understood, owing to the peculiar morphology of the skull (Fig. 12.5). As mentioned above, glyptodont skulls have undergone a very peculiar process of telescoping, in

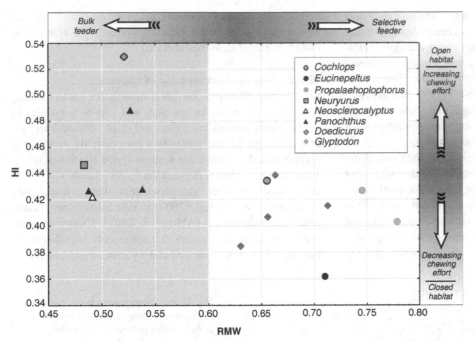

Fig. 12.8. Ecomorphology of Santacrucian and Pleistocene glyptodonts. Plot of relative muzzle width index (RMW) against hypsodonty index (HI). Modified from Vizcaíno *et al.* (2011b).

which the braincase is placed above the posterior half of the tooth row, the zygomatic arch is anteriorly displaced, both the condylar and coronoid processes are very markedly displaced anteriorly, and the tooth row is displaced distally (Vizcaíno *et al.*, 1998; Fariña and Vizcaíno, 2001). Fariña and Vizcaíno (2001) addressed a number of differences with regard to the masticatory mechanics that resulted from this derived condition. The articular surfaces of the jaw joint are shallowly convex and vertical. If the jaw joint acts as a hinge, each tooth in the lower jaw must move along an arc centered at the joint as the mouth opens and closes. This implies that the front tooth would rise almost vertically to close against the front tooth of the upper jaw, but the most distal tooth would slide forward almost horizontally to close against its opposite. Any crushing action of the hind teeth would depend on the jaw joint's capacity to slide up and down. The vertical articular surfaces of the craniomandibular joint in glyptodonts would allow some side-to-side movement of the jaw, as well as up and down sliding. Fariña and Vizcaíno (2001) were unable to recognize striae in glyptodont teeth and, while wear facets were not observable on the first tooth, different regional variations are seen in the remaining individual teeth, and no leading edge is clearly identifiable as observed in ungulates by Greaves (1973). This suggests little lateral movement in the power stroke. Preliminary observations of *Propalaehoplophorus* teeth reveal the presence of cupping wear that suggests a preferred direction of movement, which would appear to be posterior; that is, the lower jaw was moved backwards while the teeth were in or near occlusion (R. F. Kay, personal communication).

The ecomorphological approach by Vizcaíno *et al.* (2011b) provided some interesting results. When RMW and HI are taken together (Fig. 12.8), a general relationship becomes apparent that suggests selective-feeding taxa have narrower muzzles and lower hypsodonty whereas bulk-feeding taxa have wider muzzles and higher hypsodonty. Santacrucian propalaehoplophorines would all have been selective feeders, and among them the higher RMW values indicate that in *Eucinepeltus* and *Propalaehoplophorus*, selective behavior would be even more emphasized than in *Cochlops*, while the higher HI of the last two genera indicates that they probably preferred more open environments than *Eucinepeltus*.

12.5 Discussion

12.5.1 Body mass

With the probable exception of *Eucinepeltus*, Santacrucian cingulates were all smaller than 100 kg (Table 12.1), and only glyptodonts qualified as "large mammals," i.e. larger than 44 kg, following a definition of "large mammal" commonly used in archeology and paleontology (see Martin and Steadman, 1999; Johnson, 2002; Cione *et al.*, 2003, 2009; Vizcaíno *et al.*, 2012). The range of body sizes of Santacrucian armadillos falls between the extremes of the most common and diverse living armadillos. Vizcaíno *et al.* (2006a) remarked that none of them reached the extremes recorded today, which range from 85 grams for the tiny *Chlamyphorus truncatus* to 50 kilograms for the giant armadillo *Priodontes maximus*. *Prozaedyus* was the smallest, being slightly smaller than the small euphractine Patagonian pichy, *Zaedyus pichyi*. *Stenotatus* was closely comparable to the common long-nosed armadillo *Dasypus novemcinctus*. *Proeutatus*, *Peltephilus*, and *Stegotherium* were the largest of the Santacrucian armadillos (Fig. 12.9b, c, d); on average, they were equivalent to the size of the second-largest living armadillo, the extant greater long-nosed armadillo *Dasypus kappleri*.

All Santacrucian glyptodonts were larger than any living armadillo, with the lowest mass estimate about 20 kg higher than the highest figure recorded for the extant giant armadillo *Priodontes maximus*, but none attained the size of the Plio-Pleistocene glyptodonts (see Vizcaíno *et al.*, 2011a and b, 2012). Body mass estimates obtained (67–90 kg) also reflect a high degree of size homogeneity among

Propalaehoplophorus, *Cochlops* (Fig. 12.9a), and cf. *Asterostemma*, with the sole exception of *Eucinepeltus* (which is larger than the others).

12.5.2 Locomotion and use of substrate

Living armadillos provide good models for understanding limb specializations, locomotion, and use of substrates in Santacrucian armadillos (Vizcaíno *et al.*, 2003; Vizcaíno *et al.*, 2006a). Living armadillos are primarily specialized diggers and their limbs are well designed for that activity. Vizcaíno *et al.* (1999) divided the fossorial habits of living armadillos into three categories that were subsequently modified by Milne *et al.* (2009) to take into account recent behavioral information (Abba *et al.*, 2005). These categories are: (1) non-diggers, species that are mainly ambulatory (in this contribution we use ambulatory instead of cursorial as

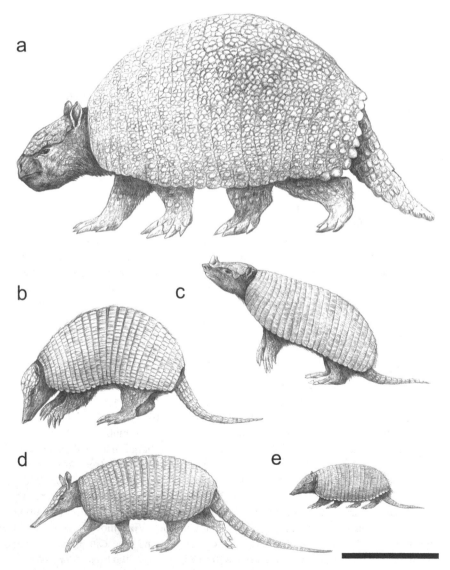

Fig. 12.9. Life reconstructions. a, *Cochlops muricatus*; b, *Proeutatus oenophorus*; c, *Peltephilus pumilus*; d, *Stegotherium tessellatum*; and e, *Prozaedyus proximus*. Scale bar = 25cm.

Milne *et al.* (2009) did, given that the latter implies running; see Stein and Casinos, (1997) for a discussion on the use of the term cursorial); (2) generalized diggers, species that dig short burrows for protection or in search of food and that feed on the surface or just below it by making "food probes" (Abba *et al.*, 2005); (3) specialized diggers, species that are either burrowers or those that feed on underground termites or ants. Non-diggers are represented by the three-banded armadillo *Tolypeutes matacus*, which is the most ambulatory within the family and does not seem to dig burrows (Nowak, 1999). Generalized diggers include members of Dasypodini and Euphractini, which display the typical fossorial behaviors associated with this group. Specialized diggers include the naked tailed armadillo *Cabassous* spp. and the pygmy armadillo *Chlamyphorus truncatus*, both of which demonstrate extreme fossorial habits, and the giant armadillo *Priodontes maximus*, which is considered to be a powerful and rapid digger that shelters in burrows of its own construction (Nowak, 1999).

The Santacrucian armadillos have a more restricted range of limb features than seen in living armadillos and all appear to be well adapted for digging. The forelimb proportions are most similar to those of the living dasypodines and euphractines (Category 2 of Vizcaíno *et al.*, 1999). Only in *Proeutatus* (IFA) and *Peltephilus* (IFA, BI) do some values begin to approach those observed in the most specialized extant digging priodontines and *Chlamyphorus truncatus*. This result calls into question the conclusions of Hoffstetter (1958) and Scillato-Yané (1977) who interpreted *Peltephilus* as having cursorial habits. In *Stenotatus* IFA values fall within the range of *Tolypeutes matacus*, which would indicate a lesser degree of specialization for digging. Unfortunately, the forelimb of *Stegotherium* is not well enough known to calculate IFA and BI; however, SMI is very similar to that in the other Santacrucian species.

The structure of the hindlimbs of the Santacrucian armadillos is more difficult to interpret. *Proeutatus*, *Prozaedyus*, and *Stegotherium* have higher CIs than the living dasypodines, euphractines and priodontines, but lower CIs than *Tolypeutes matacus* and *Chlamyphorus truncatus*. This suggests a relatively longer distal part of the hindlimb and some degree of ambulatory ability in the Santacrucian species.

In summary, although there is some diversity in limb bone proportions in Santacrucian armadillos, the range is far less than in the living species. None of the Santacrucian taxa attained the degree of cursoriality of *Tolypeutes matacus*, the fossoriality of *Priodontes maximus*, or the subterranean habits of *Chlamyphorus truncatus*.

As mentioned above, glyptodonts have forelimb IFA and SMI values similar to those of the living Dasypodini and Euphractini (generalized diggers) while other proportions of the fore- and hindlimbs are, respectively, much lower or much higher than in living armadillos. It is noteworthy that these two forelimb indices correlate positively with body mass among glyptodonts, but not among living armadillos (Vizcaíno and Milne, 2002), in which IFA is strongly related to digging behavior. Considering the caveats about this inference mentioned in the introduction, the high values of IFA and SMI do not necessarily reflect functional specialization, and can be considered a phylogenetic constraint given that both the morphological and molecular phylogenies support the interpretation that generalized digging habits are ancestral for all cingulates, and that the marked ambulatory habits of the armadillo *Tolypeutes matacus* are derived from digging ancestors (Milne *et al.*, 2009). Vizcaíno *et al.* (2011a) suggested that these features seem to be more closely tied to the larger body masses of glyptodonts and to the fact that glyptodonts carry most of their body weight on their hindlimbs – indeed glyptodonts were probably capable of adopting bipedal stances to perform strenuous activities, as suggested by Fariña (1995). These findings raise the possibility that the structure of the forelimb in glyptodonts is directly related to body support, locomotion, and maneuvering, and not to digging behavior (Vizcaíno *et al.*, 2011a). The ability to stand erect upon a tripod formed by the hindlimbs and tail has been mentioned as one of the traits that unify Xenarthra by Wetzel (1982), as this posture is a common response for defense, observation, and often for feeding in the anteaters, tree sloths, and armadillos, and is assumed to have occured in extinct ground sloths (Casinos, 1996; Blanco and Czerwonogora, 2003). Hence, it may be that the ability to adopt a bipedal, though not fully erect, stance is a retained primitive character enhanced in glyptodonts because they bear most of the weight on their hindlimbs.

12.5.3 Mastication and feeding
Comparisons with living armadillos provided a reasonable framework for interpreting the masticatory specializations and diet of Santacrucian armadillos. All living armadillos consume a considerable amount of animal material (Redford, 1985, and references therein). Redford (1985: 429) noted that living armadillos "show a range of trophic specialization from the generalized carnivore–omnivore through the generalist insectivore to the specialist insectivore."

Peltephilines were traditionally regarded as specialized carnivores (Ameghino, 1910; Winge, 1941; Hoffstetter, 1958; Scillato-Yané, 1977). However, Vizcaíno and Fariña (1997) concluded that the peculiar peltephiline morphology would have rendered them ill-designed for strict carnivory. Instead, these authors suggested that peltephilines fed on soft plant material of an underground origin (roots or tubers), although scavenging habits cannot be excluded as a possibility for peltephiline diet, since such habits are commonplace in armadillo biology (Redford, 1985).

The great reduction in the number and size of the teeth of *Stegotherium* and the extreme elongation of the rostrum clearly suggest a myrmecophagous diet (Scott, 1913; Patterson and Pascual, 1968; Vizcaíno, 1994).

Vizcaíno and Fariña (1994) observed a functional differentiation along the tooth row of *Proeutatus*, where the anterior part of the tooth row is euphractine-like (carnivore–omnivore) and the posterior is herbivore-like (as in ungulates and rodents). Consequently, they postulated that although *Proeutatus* also may have fed on animal items, it showed a clear tendency towards herbivory. Later Vizcaíno and Bargo (1998) confirmed that *Proeutatus* exhibits the most specialized morphology known for an herbivore with an armadillo-like skull pattern. The other Santacrucian eutatine, *Stenotatus*, has a relatively weaker masticatory apparatus that, in addition to having elongation of the palate and the rostrum, suggests some specialization towards insectivory. Thus, it may have been an omnivore morphologically and ecologically intermediate between living euphractines and the living long-nosed armadillo *Dasypus* (Table 1 of Redford, 1985).

The similarity of *Prozaedyus* to the living euphractine *Zaedyus pichy* and to *Chaetophractus vellerosus* may indicate a grossly comparable diet, i.e. omnivorous with a preference for rotten flesh as well as larvae and cocoons of ants and/or termites. Vizcaíno and Fariña (1994) pointed out that the differences mentioned above suggest a higher degree of insectivory.

Noting the relatively limited variation in locomotion (compared with the range observed today), Vizcaíno et al. (2006a) proposed that the differences in feeding styles could be the best clue to understanding niche partitioning among the Santacrucian armadillos. Some dietary specializations not present in living armadillos can be detected readily among the Santacrucian armadillos. Herbivory, in particular, must have been better developed than today in *Proeutatus* and *Peltephilus*. Thus the dietary range of Santacrucian armadillos surpasses that of living armadillos and might explain the probable sympatry of so many genera in a single area, as it would have potentially diminished the degree of competition for food resources (Vizcaíno et al., 2006a).

So far there are no comparable functional studies of glyptodonts. Fariña and Vizcaíno (2001) proposed that glyptodonts were not highly efficient at chewing, because of the relatively small grinding area of their dentition, as evidenced by the fact that dental measurements underestimate body mass (Fariña et al., 1998). Recent analysis of the relation between occlusal surface area and body mass (Vizcaíno et al., 2006a; Vizcaíno et al., 2011b) confirmed that glyptodonts, like other xenarthrans, have less OSA available for triturating food than other placentals of similar size. The idea of incomplete oral processing of food is congruent with McNab's (1985) suggestion of a low metabolic rate for extinct xenarthrans (Vizcaíno et al., 2006b; Vizcaíno et al., 2011b).

The ecomorphological analysis by Vizcaíno et al. (2011b) suggested that, among glyptodonts, *Eucinepeltus* would have been a highly selective feeder in relatively closed environments, *Propalaehoplophorus* a highly selective feeder in moderately open habitats, and *Cochlops* a less selective feeder in moderately open habitats.

12.5.4 Paleoenvironment

Five Santacrucian genera (*Prozaedyus*, *Proeutatus*, *Stenotatus*, *Stegotherium*, and *Peltephilus*) with up to 16 species of armadillos, and four genera (cf. *Asterostemma*, *Propalaehoplophorus*, *Cochlops*, and *Eucinepeltus*) with up to eight species of glyptodonts potentially were sympatric in the lower levels (Estancia La Costa Member) of the Santa Cruz Formation along the present coast of Southern Patagonia between Monte León and Cabo Buen Tiempo (= Cape Fairweather) (Vizcaíno et al., Chapter 1: Figs. 1.1 and 1.2).

Following Vizcaíno et al. (2006a), an analysis of the distribution of the living armadillos demonstrated that the highest diversity of armadillos today is present in a more or less restricted region east of the Andes in central South America, between the latitudes 12° and 32° S. This area extensively overlaps the Chaqueña Province of the biogeographic division of the Neotropical Region (figure 5 of Vizcaíno et al., 2006a:) proposed by Cabrera and Willink (1980). Depending on the accuracy of the distribution maps, marginal areas of the Amazonia, Yungas, Cerrado, Paranense, Puneña, Monte, and Espinal Provinces are also involved. The first four belong to the Amazonic Domain, which is characterized by dense vegetational cover, and abundant and diverse flora and fauna living in a warm and humid climate with limited seasonality in rainfall. On the other hand, the Puneña, Monte, and Espinal Provinces together with the Chaqueña Province belong to the Chaqueño Domain, with predominantly xerophytic vegetation in a continental climate with low to moderate rainfall and with cool winters and hot summers. In this province, seven genera and 11 species of armadillos belonging to the five tribes recognized by Wetzel (1985) are recorded. They represent the complete range of size and digging capacities of living armadillos, including the ambulatory tolypeutines, and the digging euphractines, dasypodines, and priodontines, as well as the subterranean chlamyphorines. Also, all the categories of feeding styles of living armadillos are present (Redford, 1985). They include carnivore–omnivores (euphractines), generalist-terrestrial insectivores (dasypodines), generalist-fossorial insectivores (chlamyphorines), and ant and/or termite specialists (tolypeutines and priodontines).

Vizcaíno et al. (2006a) proposed that, if the metabolic requirements and other related biologic parameters of extinct armadillos such as population density were comparable to those of the living faunas, the generic richness recorded in the Santacrucian is consistent with the environmental interpretation of Tauber (1997; 1999): that is, open vegetation in relatively dry conditions, with marked seasonality for the upper levels of the lower Estancia La Costa Member and at least the lower levels of the upper Estancia La Angelina Member. However, the information recovered from the ecomorphology of glyptodonts suggests the presence of closed habitats, perhaps bushlands or dry forests, still consistent with the Chaqueña Province scenario.

12.6 Conclusions

In Santacrucian times there is clear niche differentiation in size, feeding, and locomotion between armadillos and glyptodonts.

All Santacrucian armadillos fall in the range of living medium-sized armadillos (Fig. 12.9b–d). Despite some variation in limb proportions, all were good diggers, although none reached the degree of fossoriality found in some specialized living forms. In contrast, morphological and functional diversity in the masticatory apparatus exceeds that of the living species. This denotes a broader range of specializations, which includes herbivory and strict myrmecophagy. This diversification of feeding styles suggests strong niche partitioning: *Proeutatus* spp. were omnivorous, showing the most specialized morphology known for a cingulate herbivore with an armadillo-like skull pattern; *Stenotatus* spp. may have been omnivorous, morphologically and ecologically intermediate between living euphractines and dasypodines, with some specialization towards insectivory; *Prozaedyus* spp. was perhaps omnivorous with a preference for rotting flesh and larvae as well as cocoons of ants and/or termites – a diet comparable to that of small living euphractines; *Stegotherium* was clearly more specialized for a myrmecophagous diet than any living armadillo; and *Peltephilus* may have specialized on roots or tubers, although scavenging habits cannot be ruled out.

Santacrucian glyptodonts also are conservative in terms of body sizes and locomotor features, being moderately large sized (Fig. 12.9a) and ambulatory. Although their masticatory apparatus does not show the range of morphological diversity present in contemporaneous armadillos, some morphological traits allowed us to identify some ecological partitioning within the category of selective feeding in relatively closed to very closed habitats. At one extreme, *Eucinepeltus* spp. would have been highly selective feeders in relatively closed environments; on the other hand, *Propalaehoplophorus* spp. would have been highly selective feeders in moderately open habitats; and *Cochlops* might have been a less selective feeder in moderately open habitats.

Considering the taxonomic richness of Santacrucian armadillos compared with the distribution of living species and the ecomorphological diversity of glyptodonts, their environment is interpreted as a mixture of open and relatively closed vegetation in relatively dry conditions, perhaps bushlands or dry forests, agreeing in general with the modern Chaqueña biogeographic province.

ACKNOWLEDGMENTS
We thank A. Kramarz (MACN), W. Joyce and D. Brinkman (YPM), and J. Flynn (AMNH) who kindly gave access to the collections in their care. We also express our gratitude to Dirección de Patrimonio Cultural, and Museo Regional Provincial "Padre M. J. Molina" (Río Gallegos, Santa Cruz Province), and the Battini family for its hospitality during the field work. R.F. Kay and J.M. Perry critically reviewed early versions of the manuscript. We are especially indebted to Marcelo Canevari for the life reconstructions. We thank the reviewers G. McDonald and R. Fariña for their valuable comments and suggestions. This is a contribution to the projects PICT 26219 and 0143, UNLP N647 and PIP-CONICET 1054, and National Geographic Society grants 8131–06 to S. F. Vizcaíno, and NSF 0851272 and 0824546 to Richard F. Kay.

Appendix 12.1　Acronyms of the institutions, abbreviations, and list of the material studied

Acronyms of the institutions

AMNH: American Museum of Natural History, New York, USA.

FMNH: Field Museum of Natural History, Chicago, USA.

MACN: Museo Argentino de Ciencias Naturales "B. Rivadavia", Buenos Aires, Argentina. MACN-A: Colección Nacional Ameghino MACN-PV: Colección Paleovertebrados

MLP: Museo de La Plata, La Plata, Argentina.

MPM-PV: Museo Regional Provincial Padre M.J. Molina, Río Gallegos, Santa Cruz, Argentina.

YPM-VPPU: Yale Peabody Museum, Vertebrate Paleontology, Princeton University, New Haven, USA.

FL: Fossiliferous Level.

Materials

All specimens come from Santa Cruz Province, Argentina. See Vizcaíno *et al.* (Chapter 1: Figs. 1.1 and 1.2) and Appendix 1.1 for information on the localities, and Matheos and Raigemborn (Chapter 4: Figs. 4.3 and 4.4) for information on the stratigraphy.

Armadillos

Anantiosodon rarus
MACN-A 5119. Mandible. Horizon and locality: Santa Cruz Formation; Monte Observación (= Cerro Observatorio).

Pareutatus distans
MACN-A 7272–7274. Fragment of skull, mandible, and osteoderms. Horizon and locality: Santa Cruz Formation; La Cueva.

Peltephilus sp.
MACN-A 7910–40. Fragment of skull, mandible, cervical vertebrae, humerus, femur, osteoderms. Horizon and locality: unknown.

Peltephilus ferox
MACN-A 4901. Skull. Horizon and locality: Santa Cruz Formation; Monte Observación (= Cerro Observatorio). MACN-A 4902–4918. Skull and osteoderms. Horizon and locality: Santa Cruz Formation; Sehuen.

Peltephilus giganteus
MACN-A 4891–4900. Osteoderms of casqued shield and carapace. Horizon and locality: Santa Cruz Formation; Sehuen.

Peltephilus nanus
MACN-A 7958–7959. Femur and two osteoderms. Horizon and locality: Santa Cruz Formation; La Cueva.

Peltephilus pumilus
MACN-A 866–870. Osteoderms. Horizon and Locality: Santa Cruz Formation; Río Santa Cruz. MACN-A 7784–91. Skull with casqued shield, mandible, fragment of humerus, fragment of femur, vertebrae and calcaneum. Horizon and locality: Santa Cruz Formation; unknown. MPM-PV 3414. Skull and fragment of pelvis. Horizon and locality: Santa Cruz Formation, Estancia La Costa Member, FL 5.3; Puesto Estancia La Costa (= Corriguen Aike). MPM-PV 3427. Skull. Horizon and locality: Santa Cruz Formation, Estancia La Costa Member, FL 3; Estancia La Costa. MPM-PV 3431. Osteoderms. Horizon and locality: Santa Cruz Formation, Estancia La Costa Member, FL 5.3; Puesto Estancia La Costa (= Corriguen Aike). MPM-PV 3433. Skull and partial postskeleton. Horizon and locality: Santa Cruz Formation; Anfiteatro. MPM-PV 3643. Skull with casqued shield and mandible. Horizon and locality: Santa Cruz Formation; Anfiteatro. YPM-VPPU 15391. Skull. Horizon and locality: Santa Cruz Formation; Coy Inlet.

Peltephilus strepens
MACN-A 771. Osteoderms. Horizon and locality: Santa Cruz Formation; Río Santa Cruz. MACN-A 7784–91. Skull with casqued shield, mandible, fragment of humerus, fragment of femur, vertebrae and calcaneum. Horizon and locality: unknown. YPM-VPPU 15390. Postskeleton, including fore- and hindfoot, fragment of limb bones and osteoderms. Horizon and locality: Santa Cruz Formation; 10 miles south of Coy Inlet.

Proeutatus sp.
MACN-A1063. Humerus. Horizon and locality: Santa Cruz Formation; unknown. MLP 69-IX-8–11. Almost complete skeleton: humeri, ulna and radius, distal end of femora, both tibiae, both astragali and calcaneous, numerous phalanges. Horizon and locality: Santa Cruz Formation; unknown.

Proeutatus carinatus
MACN-A 8041–8042. Fragment of carapace and isolated osteoderms. Horizon and locality: Santa Cruz Formation; Monte Observación (= Cerro Observatorio).

Proeutatus deleo
MACN-A 4800–4802. Mandible, three osteoderms, and a molariform. Horizon and locality: Santa Cruz Formation; Monte Observación (= Cerro Observatorio).

Proeutatus lagena
YPM-VPPU 15613. Skull, mandible, and postskeleton. Horizon and locality: Santa Cruz Formation; near Coy Inlet.

Proeutatus oenophorus
FMNH P 13197. Skull and mandible, left humerus, pelvis, right femur, left tibia-fibula, fragments of carapace with articulated vertebrates and limb bones. Horizon and locality: Santa Cruz Formation; La Angelina Ranch, 12 miles north of Cape Fairweather. MPM-PV 3413. Skull, fragment of carapace with vertebrae, isolated osteoderms. Horizon and locality: Santa Cruz Formation, Estancia La Costa Member, FL 2; Estancia La Costa. MPM-PV 3415. Skull, mandible, pelvis and part of the vertebral column, femora, tibiae, and foot bones; osteoderms. Horizon and locality: Santa Cruz Formation, Estancia La Costa Member, FL 5.3; Puesto Estancia La Costa (= Corriguen Aike). MPM-PV 3425. Fragment of skull. Horizon and locality: Santa Cruz Formation, Estancia La Costa Member, FL 5.3; Puesto Estancia La Costa (= Corriguen Aike). MPM-PV 3430. Skull, mandible, fragment of femur and vertebrae, osteoderms. Horizon and locality: Santa Cruz Formation, Estancia La Costa Member, FL 5.3; Puesto Estancia La Costa (= Corriguen Aike). MPM-PV 3435. Osteoderms. Horizon and locality: Santa Cruz Formation, Estancia La Costa Member; Anfiteatro. MPM-PV 3438. Osteoderms. Horizon and locality: Santa Cruz Formation, Estancia La Costa Member; Campo Barranca. MPM-PV 3439. Partial skull. Horizon and locality: Santa Cruz Formation, Estancia La Costa Member; Campo Barranca. MPM-PV 3440. Fragment of carapace. Horizon and locality: Santa Cruz Formation, Estancia La Costa Member; Campo Barranca. MPM-PV 3442. Osteoderms. Horizon and locality: Santa Cruz Formation, Estancia La Costa Member; Campo Barranca. MPM-PV 3489. Fragment of carapace with fragment of femur and vertebrae, isolated osteoderms. Horizon and locality: Santa Cruz Formation, Estancia La Costa Member, below FL 1; Estancia La Costa. MPM-PV 4256. Skull. Horizon and locality: Santa Cruz Formation, Estancia La Costa Member, FL 5.3; Puesto Estancia La Costa (= Corriguen Aike).

Proeutatus robustus
YPM-VPPU 15214. Skull and postskeleton (Type). Horizon and locality: Santa Cruz Formation; 10 miles south of Coy Inlet. YPM-VPPU 15389. Femur, tibia and pes. Horizon and locality: Santa Cruz Formation; Cape Fairweather. YPM-VPPU 15957. Associated skeleton in a nodule, and ulna. Horizon and locality: Santa Cruz Formation; 10 miles south of Coy Inlet.

Stenotatus sp.
MLP 69-IX-5–11. Femur. Horizon and locality: Santa Cruz Formation; unknown. MLP 55-III-13–310. Skull and mandible articulated. Horizon and locality: Santa Cruz Formation; Monte León.

Stenotatus hesternus
MACN-A 4873. Fragment of carapace and fragment of bones. Horizon and locality: Santa Cruz Formation; Monte Observación (= Cerro Observatorio). MACN-A 871–878. Osteoderms. Horizon and locality: Santa Cruz Formation; unknown.

Stenotatus patagonicus
MACN-A 898–900. Osteoderms. Horizon and locality: Santa Cruz Formation; unknown. MACN A 4874–4875. Mandible and osteoderms. Horizon and locality: Santa Cruz Formation; Monte Observación (= Cerro Observatorio). MACN-A 7800. Osteoderms. Horizon and locality: Santa Cruz Formation; La Cueva. MACN-A 7801–7802. Fragment of mandible and osteoderms. Horizon and locality: Santa Cruz Formation; Monte Observación (= Cerro Observatorio). MACN-A 8235. Fragment of carapace and postskeleton. Horizon and locality: Santa Cruz Formation; La Cueva.

Stenotatus cf. *patagonicus*
MACN-Pv SC406. Skull, mandible, femur, humerus, pelvic shield. Horizon and locality: Santa Cruz Formation; Monte Observación (= Cerro Observatorio). MPM-PV 3437. Osteoderms. Horizon and locality: Santa Cruz Formation, Estancia La Costa Member; Campo Barranca. MPM-PV 3429. Mandible and osteoderms. Horizon and locality: Santa Cruz Formation, Estancia La Costa Member, FL 5.3; Puesto Estancia La Costa (= Corriguen Aike).

Prozaedyus sp.
YPM-VPPU 15604. Skull and humerus. Horizon and locality: Santa Cruz Formation; Coy Inlet.

Prozaedyus exilis
MACN-A 4814–4863. Mandible, fragment of carapace, postskeleton. Horizon and locality: Santa Cruz Formation; Sehuen. YPM-VPPU 15579. Skull and postskeleton. Horizon and locality: Santa Cruz Formation; Killik Aike.

Prozaedyus proximus
MACN-A 2201. Partial carapace. Horizon and locality: Santa Cruz Formation; unknown. MACN-A 4865. Partial carapace. Horizon and locality: Santa Cruz Formation; Sehuen. MACN-A 7806–1828. Mandible, vertebrae, fragment of pelvis, fragment of femur, astragalus, calcaneus, and osteoderms. Horizon and locality: Santa Cruz Formation; Yegua Quemada. MPM-PV 3418. Fragment of skull. Horizon and locality: Santa Cruz Formation, Estancia La Costa Member, FL 5.3; Puesto Estancia La Costa (= Corriguen Aike).

MPM-PV 3423. Skull with cephalic shield, hyoid bones, and cervical vertebrae articulated. Horizon and locality: Santa Cruz Formation, Estancia La Costa Member, FL 5.3; Puesto Estancia La Costa (= Corriguen Aike). MPM-PV 3428. Incomplete mandible, carapace, fragment of femur. Horizon and locality: Santa Cruz Formation, Estancia La Costa Member; Cañadón Silva locality. MPM-PV 3436. Osteoderms. Horizon and locality: Santa Cruz Formation, Estancia La Costa Member; Campo Barranca. MPM-PV 3441. Fragment of carapace. Horizon and locality: Santa Cruz Formation, Estancia La Costa Member; Campo Barranca. MPM-PV 3506. Complete skeleton and carapace articulated in matrix. Horizon and locality: Santa Cruz Formation, Estancia La Costa Member, FL 5.3; Puesto Estancia La Costa (= Corriguen Aike). MPM-PV 3617. Skeleton and carapace articulated. Horizon and locality: Santa Cruz Formation, Estancia La Costa Member, FL 5.3; Puesto Estancia La Costa (= Corriguen Aike). MPM-PV 4182. Skull, mandible, and osteoderms. Horizon and locality: Santa Cruz Formation, Estancia La Costa Member, FL 5.3; Puesto Estancia La Costa (= Corriguen Aike). YPM-VPPU 15567. Skull and pes. Horizon and locality: Santa Cruz Formation; Coy Inlet.

Stegotherium tauberi
YPM-VPPU 15565. Skull, cervical and thoracic vertebrae, pes, and fragment of carapace. Horizon and locality: Santa Cruz Formation; Killik Aike. YPM-VPPU 15566. Skull, postskeleton (pelvis, femora and tibiae, left humerus and left scapula), and fragment of carapace. Horizon and locality: Santa Cruz Formation; Coy Inlet?

Stegotherium tessellatum
MACN-A 781–785. Osteoderms. Horizon and locality: Santa Cruz Formation; Río Santa Cruz.

Vetelia puncta
MACN-A 2139. Three osteoderms. Horizon and locality: Santa Cruz Formation; near Lago Argentino. MPM-PV 3652. Osteoderm. Horizon and Locality: Santa Cruz Formation, Estancia La Costa Member; Punta Sur.

Glyptodonts

Propalaehoplophoridae indet.
MPM-PV 3417. Fragment of carapace. Horizon and locality: Santa Cruz Formation, Estancia La Costa Member, FL 6; Puesto Estancia La Costa (= Corriguen Aike). MACN-A 7693. Humerus. Horizon and locality: Santa Cruz Formation, La Cueva.

Propalaehoplophoridae cf. *Asterostemma depressa*.
MACN-A 7663. Partial skull. Horizon and locality: Santa Cruz Formation; Cañadón Jack Harvey.

Cochlops muricatus
MPM-PV 3420. Skull, partial skeleton, and carapace. Horizon and locality: Santa Cruz Formation, Estancia La Costa

Member, FL 5.3; Puesto Estancia La Costa (= Corriguen Aike). MPM-PV 3432. Partial cranium including maxilla with dentition, and left mandible; fragment of carapace. Horizon and locality: Santa Cruz Formation, Estancia La Costa Member; Anfiteatro. MACN-A 2120. Mandible and right femur. Humerus. Horizon and locality: Santa Cruz Formation; unknown.

Eucinepeltus petesatus
MACN-A 4758. Complete cranium and mandible. Horizon and locality: Santa Cruz Formation; Monte Observación (= Cerro Observatorio). MACN-A 4760. Mandible. Horizon and locality: Santa Cruz Formation; Monte Observación (= Cerro Observatorio). MACN-A 4761. Mandible. Horizon and locality: Santa Cruz Formation; Monte Observación (= Cerro Observatorio).

Metopotoxus laevatus
MACN-A 1042. Fragment of carapace. Horizon and locality: Santa Cruz Formation; unknown.

Propalaehoplophorus sp.
MACN-A 4754. Skull. Horizon and locality: Santa Cruz Formation; Corriguen Aike.

Propalaehoplophorus australis
MLP 16–15. Skull, mandible, and skeleton. Horizon and locality: Santa Cruz Formation; unknown. Figured in Lydekker (1894, **3**, Pl. 32). MPM-PV 3419. Mandible and partial carapace. Horizon and locality: Santa Cruz Formation, Estancia La Costa Member, FL 6; Puesto Estancia La Costa (= Corriguen Aike). MPM-PV 3422. Partial skull, mandible; humerus, partial scapula, and other fragments of skeleton; fragment of carapace. Horizon and locality: Santa Cruz Formation, Estancia La Costa Member, FL 5.3; Puesto Estancia La Costa (= Corriguen Aike).

Propalaehoplophorus incisivus
MACN-A 7655. Skull. Horizon and locality: Santa Cruz Formation; Corriguen Aike. MACN-A 7656. Skull. Horizon and locality: Santa Cruz Formation; Corriguen Aike.

Propalaehoplophorus minus
MACN-A 4757. Mandible. Horizon and locality: Santa Cruz Formation; Corriguen Aike.

REFERENCES
Abba, A. M., Udrizar Sauthier, D. E. and Vizcaíno, S. F. (2005). Use and distribution of burrows and tunnels built by *Chaetophractus villosus* (Mammalia, Xenarthra) in the eastern Argentinian Pampas. *Acta Theriologica*, **50**, 115–124.
Aguiar, J. M. and Fonseca, G. A. B. (2008). Conservation status of Xenarthra. In *The Biology of the Xenarthra*, ed. S. F. Vizcaíno and W. J. Loughry. Gainesville, FL: University Press of Florida, pp. 215–231.

Ameghino, F. (1887). Enumeración sistemática de las especies de mamíferos fósiles coleccionados por Carlos Ameghino en los terrenos eocenos de Patagonia Austral y depositados en el Museo de La Plata. *Boletín del Museo de La Plata*, **1**, 1–26.

Ameghino, F. (1889). Contribución al conocimiento de los mamíferos fósiles de la República Argentina. *Actas de la Academia Nacional de Ciencias*, **6**, 1–1027.

Ameghino, F. (1891). Nuevos restos de mamíferos fósiles descubiertos por Carlos Ameghino en el Eoceno inferior de la Patagonia austral. Especies nuevas, adiciones y correcciones. *Revista Argentina de Historia Natural*, **1**, 289–328.

Ameghino, F. (1894). Enumération synoptique des espéces de Mammifères fossiles des formations Eocènes de Patagonie. *Boletín de la Academia Nacional de Ciencias de Córdoba*, **13**, 259–455.

Ameghino, F. (1898). Primera sinopsis geológico-paleontológica. *2° Censo de la República Argentina*, **1**, 111–255.

Ameghino, F. (1902). Notices préliminaires sur des mammifères nouveaux des terrains crétacés de Patagonie. *Boletín de la Academia Nacional de Ciencias*, **17**, 5–70.

Ameghino, F. (1910). Geología, paleogeografía, paleontología y antropología de la República Argentina. *Obras Completas* (1934), **18**, 297–335.

Bargo, M. S. (2003). Biomechanics and Palaeobiology of the Xenarthra: state of the art. *Senckenbergiana biologica*, **83**, 41–50.

Bargo, M. S. and Vizcaíno, S. F. (2008). Paleobiology of Pleistocene ground sloths (Xenarthra, Tardigrada): biomechanics, morphogeometry and ecomorphology applied to the masticatory apparatus. *Ameghiniana*, **45**, 175–196.

Bargo, M. S., De Iuliis, G. and Vizcaíno, S. F. (2006a). Hypsodonty in Pleistocene ground sloths. *Acta Paleontologica Polonica*, **51**, 53–61.

Bargo, M. S., Toledo, N. and Vizcaíno, S. F. (2006b). Muzzle of South American ground sloths (Xenarthra, Tardigrada). *Journal of Morphology*, **267**, 248–263.

Blanco, R. E. and Czerwonogora, A. (2003). The gait of *Megatherium* Cuvier 1796 (Mammalia, Xenarthra, Megatheriidae). *Senckenbergiana Biologica*, **83**, 61–68.

Bordas, A. (1936). Los "Peltateloidea" de la colección Ameghino. *Physis*, **41**, 1–18.

Bordas, A. F. (1938). Sobre un nuevo "Peltephiloda" del Trelewense. *Physis*, **12**, 267–277.

Cabrera, A. L. and Willink, A. (1980). *Biogeografía de América Latina*. Organización de los Estados Americanos (OEA): Serie Biología.

Casinos, A. (1996). Bipedalism and quadrupedalism in *Megatherium*: an attempt at biomechanical reconstruction. *Lethaia*, **29**, 87–96.

Cione, A. L, Tonni, E. P. and Soibelzon, L. H. (2003). The broken zig-zag: late Cenozoic large mammal and turtle extinction in South America. *Revista del Museo Argentino de Ciencias Naturales "Bernardino Rivadavia,"* **5**, 1–19.

Cione, A. L., Tonni, E. P. and Soibelzon, L. H. (2009). Did humans cause the Late Pleistocene-Early Holocene mammalian extinctions in South America in a context of shrinking open areas? In *American Megafaunal Extinctions at the End of the Pleistocene*, ed. G. Haynes. Vertebrate Paleobiology and Paleoanthropology Series. New York: Springer, pp. 125–144.

Croft, D. (2000). Archaeohyracidae (Mammalia: Notoungulata) from the Tinguiririca Fauna, central Chile, and the evolution and paleoecology of South American mammalian herbivores. Unpublished Ph.D. thesis, University of Chicago.

De Esteban-Trivigno, S., Mendoza, M. and De Renzi, M. (2008). Body mass estimation in Xenarthra: a predictive equation suitable for all quadrupedal terrestrial placentals? *Journal of Morphology*, **269**, 1276–1293.

Delsuc, F. and Douzery, E. J. P. (2008). Recent advances and future prospects in xenarthran molecular phylogenetics. In *The Biology of the Xenarthra*, ed. S. F. Vizcaíno and W. J. Loughry. Gainesville: University Press of Florida, pp. 11–23.

Fariña, R. A. (1985). Some functional aspects of mastication in Glyptodontidae (Mammalia). *Fortschritte der Zoologie*, **30**, 277–280.

Fariña, R. A. (1988). Observaciones adicionales sobre la biomecánica masticatoria en Glyptodontidae (Mammalia, Edentata). *Boletín de la Sociedad Zoológica*, **4**, 5–9.

Fariña, R. A. (1995). Limb bone strength and habits in large glyptodonts. *Lethaia*, **28**, 189–196.

Fariña, R. A., and Vizcaíno, S. F. (1997). Allometry of the leg bones in armadillos (Mammalia, Dasypodidae). A comparisson with other mammals. *Zeitschrift für Säugetierkunde*, **62**, 65–70.

Fariña, R. A. and Vizcaíno, S. F. (2001). Carved teeth and strange jaws: How glyptodonts masticated. *Acta Paleontologica Polonica*, **46**, 87–102.

Fariña, R. A., Vizcaíno, S. F. and Bargo, M. S. (1998). Body size estimations in Lujanian (Late Pleistocene-Early Holocene of South America) mammal megafauna. *Mastozoología Neotropical*, **5**, 87–108.

Fernicola, J. C. (2008). Nuevos aportes para la sistemática de los Glyptodontia Ameghino 1889 (Mammalia, Xenarthra, Cingulata). *Ameghiniana*, **45**, 553–574.

Fernicola, J. C. and Vizcaíno, S. F. (2008). Revisión del género *Stegotherium* Ameghino, 1887 (Mammalia, Xenarthra, Dasypodidae). *Ameghiniana*, **45**, 321–332.

Fernicola, J. C., Vizcaíno, S. F. and Bargo, M. S. (2009). Primer registro de *Vetelia puncta* Ameghino (Xenarthra, Cingulata) en la Formación Santa Cruz (Mioceno temprano) de la costa atlántica de la provincia de Santa Cruz, Argentina. *Ameghiniana*, **46** Suppl.: 77R.

Gaudin, T. J. and McDonald, G. H. (2008). Morphology-based investigations of the phylogenetic relationships among extant and fossil xenarthrans. In *The Biology of the Xenarthra*, ed. S. F. Vizcaíno and W. J. Loughry. Gainesville: University Press of Florida, pp 24–36.

Gaudin, T. J. and Wible, R. W. (2006). The phylogeny of living and extinct armadillos (Mammalia, Xenarthra, Cingulata): a craniodental analysis. In *Amniote Paleobiology:*

Perspectives on the Evolution of Mammals, Birds and Reptiles, ed. M. T. Carrano, T. J. Gaudin, R. W. Blob and J. R. Wible. Chicago: University of Chicago Press, pp. 153–198.

Greaves, W. S. (1973). The inference of jaw motion from tooth wear facets. *Journal of Paleontology*, **47**, 1000–1001.

Greaves, W. S. (1978). The jaw lever system in ungulates: a new model. *Journal of Zoology,* **184**, 271–285.

Hoffstetter, R. (1958). Xenarthra. In *Traité de Paléontologie* ed. J. Piveteau. Vol. 6 (2). Paris: Masson et Cie, pp. 535–636.

Janis, C. M. and Ehrhardt, D. (1988). Correlation of the muzzle width and relative incisor width with dietary preference in ungulates. *Zoological Journal of the Linnean Society*, **92**, 267–284.

Johnson, C. N. (2002). Determinants of loss of mammal species during the Late Quaternary 'megafauna' extinctions: life history and ecology, but not body size. *Proceedings of the Royal Society London B: Biological Sciences*, **269**, 2221–2227.

Kalthoff, D. C. (2011). Microstructure of dental hard tissues in fossil and recent xenarthrans (Mammalia, Folivora and Cingulata). *Journal of Morphology*, **272**, 641–661.

Kraglievich, L. (1934). La antigüedad Pliocena de las faunas de Monte Hermoso y Chapadmalal deducidas de su comparación con las que le precedieron y le sucedieron. *Imprenta El Siglo Ilustrado*, **983**, 1–36.

Lydekker, R. (1894). Contributions to a knowledge of the fossil vertebrates of Argentina. Part II. *Anales del Museo La Plata (Paleontología Argentina III)*. pp. 1–118; 68 Plates.

Marshall, L. G., Hoffstetter, R. and Pascual, R. (1983). Mammals and Stratigraphy: Geochronology of the Continental Mammal-bearing Tertiary of South America. *Paleovertebrata, Mémoire Extraordinaire*, 1–93.

Martin, P. S. and Steadman, D. W. (1999). Prehistoric extinctions on islands and continents. In *Extinctions in Near Time: Causes, Contexts and Consequences*, ed. R. D. E. MacPhee. New York: Kluwer/Plenum, pp. 17–56.

McKenna, M. C. and Bell, S. K. (1997). *Classification of Mammals Above the Species Level*. New York: Columbia University Press.

McNab, B. K. (1985). Energetics, population biology, and distribution of Xenarthra, living and extinct. In *Evolution and Ecology of Armadillos, Sloths and Vermilinguas*, ed. G. G. Montgomery. Washington and London: Smithsonian Institution Press, pp. 219–232.

Milne, N., Vizcaíno, S. F. and Fernicola, J. C. (2009). A 3D geometric morphometric analysis of digging ability in the extant and fossil cingulate humerus. *Journal of Zoology*, **278**, 48–56.

Mones, A. (1986). Palaeovertebrata Sudamericana. Catálogo sistemático de los vertebrados fósiles de América del Sur. Parte I. Lista Preliminar y Bibliografía. *Courier Forschungsinstitut Senckenberg*, **82**, 1–625.

Nowak, R. M. (1999). *Walker's Mammals of the World*, 6th Edn. Baltimore: Johns Hopkins University Press.

Patterson, B. and Pascual, R. (1968). Evolution of mammals on southern continents. *Quarterly Review of Biology*, **43**, 409–451.

Porpino, K. O., Fernicola, J. C. and Bergqvist, L. P. (2010). Revisiting the intertropical Brazilian species *Hoplophorus euphractus* (Cingulata, Glyptodontoidea) and the phylogenetic affinities of *Hoplophorus*. *Journal of Vertebrate Paleontology*, **30**, 911–927.

Quintana, C. A. (1992). Estructura interna de una paleocueva, posiblemente de un Dasypodidae (Mammalia, Edentata) del Pleistoceno de Mar del Plata (Provincia de Buenos Aires, Argentina). *Ameghiniana*, **29**, 87–91.

Redford, K. H. (1985). Food habits of Armadillos (Xenarthra: Dasypodidae). In *The Evolution and Ecology of Armadillos, Sloths, and Vermilinguas*, ed. G. G. Montgomery. Washington: Smithsonian Institution Press, pp. 429–437.

Scillato-Yané, G. J. (1977). Notas sobre los Dasypodidae (Mammalia, Edentata) del Plioceno del territorio argentino. I. Los restos de Edad Chasiquense (Plioceno inferior) del Sur de la Provincia de Buenos Aires. *Ameghiniana*, **14**, 133–144.

Scillato-Yané, G. J. (1980). Catálogo de los Dasypodidae fósiles (Mammalia, Edentata) de la República Argentina. *II Congreso Argentino de Paleontología y Bioestratigrafía y I Congreso Latinoamericano de Paleontología*, **3**, 7–36.

Scott, W. B. (1903–04). Mammalia of the Santa Cruz beds. I. Edentata. In *Reports of the Princeton University Expeditions to Patagonia 1896–1899*. Vol. 5, Paleontology II, ed. W. B. Scott. Princeton: Princeton University Press, pp. 1–364.

Scott, W. B. (1913). *A History of Land Mammals in the Western Hemisphere*. New York: MacMillan.

Stein, B. R. and Casinos, A. (1997). What is a cursorial mammal? *Journal of Zoology London*, **242**, 185–192.

Tauber, A. A. (1997). Paleoecología de la Formación Santa Cruz (Mioceno inferior) en el extremo sudeste de la Patagonia. *Ameghiniana*, **34**, 517–529.

Tauber, A. A. (1999). Los vertebrados de la Formación Santa Cruz (Mioceno inferior-medio) en el extremo sureste de la Patagonia y su significado paleoecológico. *Revista Española de Paleontología*, **14**, 173–182.

Vizcaíno, S. F. (1994). Mecánica masticatoria de *Stegotherium tessellatum* Ameghino (Mammalia, Xenarthra) del Mioceno temprano de Santa Cruz (Argentina). Algunos aspectos paleoecológicos relacionados. *Ameghiniana*, **31**, 283–290.

Vizcaíno, S. F. (2009). The teeth of the "toothless". Novelties and key innovations in the evolution of xenarthrans (Mammalia, Xenarthra). *Paleobiology*, **35**, 343–366.

Vizcaíno, S. F. and Bargo, M. S. (1998). The masticatory apparatus of *Eutatus* (Mammalia, Cingulata) and some allied genera. Evolution and paleobiology. *Paleobiology*, **24**, 371–383.

Vizcaíno, S. F. and De Iuliis, G. (2003). Evidence for advanced carnivory in fossil armadillos (Mammalia: Xenarthra: Dasypodidae). *Paleobiology*, **29**, 123–138.

Vizcaíno, S. F. and Fariña, R. A. (1994). Caracterización trófica de los armadillos (Mammalia, Xenarthra, Dasypodidae) de Edad Santacrucense (Mioceno temprano) de Patagonia (Argentina). *Acta Geologica Leopoldensia*, **39**, 191–200.

Vizcaíno, S. F. and Fariña, R. A. (1997). Diet and locomotion of the armadillo *Peltephilus*: a new view. *Lethaia*, **30**, 79–86.

Vizcaíno, S. F. and Milne, N. (2002). Structure and function in armadillo limbs (Mammalia: Xenarthra: Dasypodidae). *Journal of Zoology London*, **257**, 117–127.

Vizcaíno, S. F., De Iuliis, G. and Bargo, M. S. (1998). Skull shape, masticatory apparatus, and diet of *Vassallia* and *Holmesina* (Mammalia: Xenarthra: Pampatheriidae). When anatomy constrains destiny. *Journal of Mammalian Evolution*, **5**, 291–322.

Vizcaíno, S. F., Fariña, R. A. and Mazzetta, G. (1999). Ulnar dimensions and fossoriality in armadillos and other South American mammals. *Acta Theriologica*, **44**, 309–320.

Vizcaíno, S. F., Milne, N. and Bargo, M. S. (2003). Limb reconstruction of *Eutatus seguini* (Mammalia: Dasypodidae). Paleobiological implications. *Ameghiniana*, **40**, 89–101.

Vizcaíno, S. F., Bargo, M. S., Kay, R. F. and Milne, N. (2006a). The armadillos (Mammalia, Xenarthra) of the Santa Cruz Formation (early–middle Miocene). An approach to their paleobiology. *Palaeogeography, Palaeoclimatology, Palaeoecology*, **237**, 255–269.

Vizcaíno, S. F., Fariña, R. A., Bargo, M. S. and De Iuliis, G. (2004). Functional and phylogenetical assessment of the masticatory adaptations in Cingulata (Mammalia, Xenarthra). *Ameghiniana*, **41**, 651–664.

Vizcaíno, S. F., Bargo, M. S. and Cassini, G. H. (2006b). Dental occlusal surface area in relation to body mass, food habits and other biologic features in fossil Xenarthrans. *Ameghiniana*, **43**, 11–26.

Vizcaíno, S. F., Bargo, M. S. and Fariña, R. A. (2008). Form, Function and Paleobiology in Xenarthrans. In *The Biology of the Xenarthra*, ed. S. F. Vizcaíno and W. J. Loughry. Gainsville: University Press of Florida, pp. 86–99.

Vizcaíno, S. F., Blanco, R. E., Bender, J. B. and Milne, N. (2011a). Proportions and function of the limbs of glyptodonts (Mammalia, Xenarthra). *Lethaia*, **44**, 93–101.

Vizcaíno, S. F., Cassini, G. H., Fernicola, J. C. and Bargo, M. S. (2011b). Evaluating habitats and feeding habits through ecomorphological features in glyptodonts (Mammalia, Xenarthra). *Ameghiniana*, **48**, 305–319.

Vizcaíno, S. F., Cassini, G. H., Toledo, N. and Bargo, M. S. (2012). On the evolution of large size in mammalian herbivores of Cenozoic faunas of Southern South America. In *Bones, Clones and Biomes: The History and Geography of Recent Neotropical Mammals*, ed. B. Patterson and L. P. Costa. Chicago: University of Chicago Press, pp. 76–101.

Wetzel, R. M. (1982). Systematics, distribution, ecology, and conservation of South American Edentates. In *Mammalian Biology in South America*, ed. M. A. Mares and H. H. Genoways. Pymatuning Laboratory of Ecology, University of Pittsburgh, *Special Publication Series* **6**, 345–375.

Wetzel, R. M. (1985). Taxonomy and distribution of armadillos, Dasypodidae. In *The Evolution and Ecology of Armadillos, Sloths, and Vermilinguas*, ed. G. G. Montgomery. Washington: Smithsonian Institution Press, pp. 23–46.

Winge, H. (1941). Edentates (Edentata). In *The Interrelationships of the Mammalia Genera*, ed. S. Jensen, R. Spärck and H. Volsoe. Copenhagen: Reitzels Forlag, pp. 319–341.

13 Paleobiology of the Santacrucian sloths and anteaters (Xenarthra, Pilosa)

M. Susana Bargo, Néstor Toledo, and Sergio F. Vizcaíno

Abstract

This chapter reviews the paleobiology of pilosans (anteaters and sloths) from the Santa Cruz Formation, which comprise at least one genus of vermilinguan and 11 genera of sloths. Paleobiological studies performed on these xenarthrans include: estimation of body mass (through multivariate regression); analysis of the limbs so as to infer the locomotor habits and substrate use (through morphometric and qualitative-comparative morphofunctional analyses), and studies of the masticatory apparatus to infer probable feeding habits (mainly through analysis of tooth morphology and wear facets, plus the anatomy of the masticatory apparatus as a whole). Santacrucian anteaters were small animals, about 6 kg, well suited for climbing and for scratch-digging the substrate in searching for their preferred food, social insects. Various sloths were moderately large-sized forms, the largest reaching about 100 kg, with a locomotor pattern distinct from that of living sloths, resembling more that of vermilinguans and pangolins. The results suggest well-developed digging capabilities, but semi-arboreal habits cannot be ruled out. Megatherioid sloths were most likely leaf eaters, and the primary method of food reduction must have been by shearing or cutting (with a predominance of orthal movements). Mylodontid masticatory movements included a larger transverse component, and food reduction must have been by crushing and grinding, which suggests they fed on more compact, three-dimensional, and fibrous food items such as the underground storage organs of plants. The semi-arboreal habits suggested for anteaters and, probably, sloths indicate they lived in forests or that forested areas were present nearby. The specialized feeding habits of vermilinguans are indicative of subtropical and warm temperate environments because they would have depended on a year-round availability of social insects.

Resumen

Este capítulo trata la paleobiología de los Pilosa (osos hormigueros y perezosos) de la Formación Santa Cruz que incluyen al menos un género de Vermilingua y 11 géneros de perezosos. Los estudios paleobiológicos realizados sobre estos xenartros incluyen: estimaciones de masa corporal (a través de regresiones multivariadas), análisis de los miembros (a través de análisis morfométricos y morfofuncionales comparativos cualitativos), con el fin de deducir hábitos locomotores y uso de sustrato, y estudios del aparato masticatorio (principalmente a través del análisis de la morfología de los dientes y facetas de desgaste, además de otras evidencias anatómicas del aparato masticatorio), para inferir probables hábitos de alimentación. Los osos hormigueros santacrucenses eran animales pequeños, cerca de seis kg, bien preparados para trepar y escarbar el sustrato en busca de insectos sociales para alimentarse. Los perezosos eran formas de tamaño moderadamente grande (los más grandes alcanzarían los 100 kg) con un patrón de locomoción diferente a los perezosos actuales, más similar al de los vermilinguas y pangolines. Poseían una buena capacidad de excavación, pero no se pueden descartar hábitos semiarborícolas. Los perezosos megaterioideos se alimentaban de hojas y el método principal de trituración habría sido por corte (predominio de movimientos ortales). Los movimientos masticatorios de los milodóntidos incluían una importante componente lateral y la reducción de los alimentos habría sido por trituración y molienda, lo que sugiere que ingerían alimentos más compactos, turgentes (bulbos subterráneos) y fibrosos. Los hábitos semiarborícolas de los osos hormigueros y, probablemente de los perezosos, indican la presencia de áreas boscosas. Los hábitos alimenticios especializados de los vermilinguas sugieren ambientes tropicales y templados debido a su dependencia a la disponibilidad de insectos a lo largo del año.

13.1   Introduction

Pilosa includes Vermilingua (anteaters) and Folivora (ground and tree sloths; also called Tardigrada or Phyllophaga; see Delsuc *et al.*, 2001; Fariña and Vizcaíno, 2003; Vizcaíno and Loughry, 2008; Shockey and Anaya, 2011) that together with cingulates (armadillos and glyptodonts, see Vizcaíno, Fernicola and Bargo, Chapter 12) constitute Xenarthra, one of the four major clades of pre-interchange South American placental mammals (see Delsuc and Douzery, 2008; Gaudin and McDonald, 2008).

Early Miocene Paleobiology in Patagonia: High-Latitude Paleocommunities of the Santa Cruz Formation, ed. Sergio F. Vizcaíno, Richard F. Kay and M. Susana Bargo. Published by Cambridge University Press. © Cambridge University Press 2012.

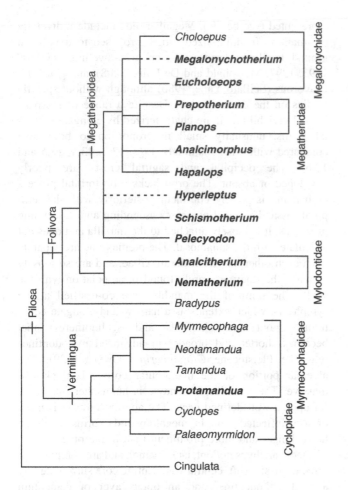

Fig. 13.1. Phylogeny of Pilosa depicting the Santacrucian genera in bold (modified from Gaudin and Branham, 1998, and Gaudin, 2004).

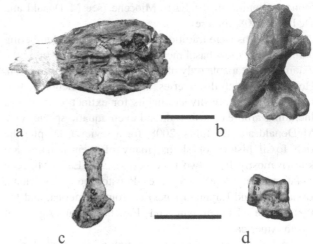

Fig. 13.2. *Protamandua rothi* YPM-VPPU 15267: a, skull in ventral view; Myrmecophagidae indet., MLP 69-IX-8–8a; b, left humerus. *Protamandua rothi* MACN-A 10901b: c, right calcaneum. MACN-A 11530: d, right astragalus. Scale bar = 2.5 cm.

Living anteaters are dietary specialists feeding on social insects, such as termites and ants. They include all members, fossil and living, of the two extant clades, Myrmecophagidae and Cyclopedidae. The fossil record of vermilinguans is not very abundant and comes mainly from the Miocene of Argentina and Colombia, and from the Pleistocene of Uruguay, Brazil, and Mexico (see McDonald *et al.*, 2008 for a review). The earliest certain fossil Vermiligua is from Colhuehuapian Age sediments (Early Miocene) from the Atlantic coast of Chubut Province, Argentina (Carlini *et al.*, 1992). This yet-to-be described specimen is nearly complete, and similar in size to *Tamandua*.

We follow a phylogenetic arrangement of Vermilingua proposed by Gaudin and Branham (1998). Along with the three extant genera (*Myrmecophaga*, *Tamandua*, and *Cyclopes*), they included only three extinct genera in their analysis: *Protamandua* Ameghino, 1904 (Early Miocene), *Palaeomyrmidon* Rovereto, 1914 (Late Miocene), and *Neotamandua* Rovereto, 1914 (Late Miocene–Pliocene). As proposed by previous workers (Hirschfeld, 1976; Engelmann, 1985), Gaudin and Branham (1998) confirm that

Palaeomyrmidon is the sister group of *Cyclopes*, *Protamandua* is the sister group to the other Vermilingua taxa, and *Neotamandua* is the sister group to *Myrmecophaga* (Fig. 13.1).

The skull of living anteaters has an elongated rostrum and hard palate, with complete loss of the dentition. The rostral modifications are no doubt functionally related to the slender, elongated, sticky tongue used as a means of capturing prey. Although the preserved skulls of the Santacrucian anteater *Protamandua* are incomplete (Fig. 13.2a), they can be characterized as having an elongated skull, an edentulous maxilla, and a jugal reduced to a small process attached to the zygomatic process of the maxilla, as in living forms. Patterson *et al.* (1992) indicated that the auditory region of *Protamandua* clearly shows the charateristics of Myrmecophagidae.

Living sloths are represented by only two genera, the tree sloths *Bradypus* and *Choloepus*, which inhabit tropical forests, and are mainly folivorous and almost completely arboreal (Chiarello, 2008). In contrast, the sloth fossil record is extraordinarily rich (nearly 100 sloth genera are listed by McKenna and Bell, 1997), and together with cingulates, sloths are among the largest and most dominant and distinctive groups of the Cenozoic mammalian fauna of South America. The oldest unquestionable sloths come from the Deseadan Age (Late Oligocene) of Patagonia (Hoffstetter, 1982) and Bolivia (Engelmann, 1987). Recently, the Tinguirirican (Early Oligocene) *Pseudoglyptodon* sp. was reported as a very basal sloth by McKenna *et al.* (2006), so it may be considered the oldest record of the clade. Sloths become conspicuous elements of the South American fossil fauna beginning in the Early Miocene (see McDonald and De Iuliis, 2008, for a review).

Fossil sloths were traditionally viewed as large lumbering animals with a low basal metabolism (a characteristic of the extant taxa), capable only of slow, quadrupedal locomotion. However, recent discoveries and morphofunctional analyses suggest more diverse habits for extinct xenarthrans, including arboreal, fossorial, and even aquatic species (see McDonald and De Iuliis, 2008, for a review). Despite the rich fossil history of sloths, many of their remains are known mostly from two time periods, the Early Miocene (Santacrucian Age) and the Pleistocene (Ensenadan, Bonaerian, and Lujanian Ages) of South America, and the Pleistocene (Irvingtonian and Rancholabrean Ages) of North America.

The latest comprehensive morphology-based phylogenetic analysis of sloths is that of Gaudin (2004), who strongly supports the hypothesis that living sloths (*Bradypus* and *Choloepus*) evolved their distinctive suspensory arboreal specializations independently from different "ground" sloths (Fig. 13.1). Gaudin proposed *Bradypus* as the sister-taxon to all other sloths, in contrast to Patterson and Pascual (1972) and Webb (1985), who considered it closely allied with the megatheres, and to the views of Carlini and Scillato-Yané (2004) who proposed that *Bradypus* differentiated through a process of heterochrony within Megalonychidae. *Choloepus* was placed by Gaudin (2004) within Megalonychidae, a clade that includes the extinct Antillean sloths, results that are congruent with those of Patterson and Pascual, Webb, and Carlini and Scillato-Yané. Although the relationships among tree sloths and ground sloths proposed by Gaudin (2004) differ from those promoted in other morphological analyses (White and MacPhee, 2001; Pujos *et al.*, 2007), it is notable that all arrangements support the diphyly of tree sloths. Gaudin's (2004) phylogeny also corroborates the monophyly of the four clades of sloths, Mylodontidae, Megatheriidae, Nothrotheriidae, and Megalonychidae. Megatheriids, nothrotheriids, and megalonychids are joined in a monophyletic clade Megatherioidea, within which Megatheriidae and Nothrotheriidae form a monophyletic group, Megatheria. Within Mylodontidae, two major subdivisions are recognized, the scelidotheres and the mylodontines. Pujos *et al.* (2007) supports Megalonychidae and Megatherioidea as monophyletic clades, and considered *Hapalops* (Ameghino, 1887) to be a stem folivoran, but the monophyly of Mylodontidae is not supported. Although the proposed relationships among Folivora are different in both studies, it is remarkable that both analyses support the diphyly of tree sloths.

Santacrucian sloths fall mainly within the megatherioid radiation. Nothrotheriids have not been recorded from the Santa Cruz Formation, but mylodontids are also represented (see below). Megatherioids include a diversity of small- to medium-sized forms, proposed to have been ground-dwelling, scansorial, or tree-dwelling (White, 1993, 1997; McDonald and De Iuliis, 2008), and generally folivorous (Scillato-Yané, 1986) although without specific studies on the dietary habits. Despite a range of morphological variability, all are characterized by elongated, tubular skulls, unusually small in proportion to body size compared with other mammals (Figs. 13.3, 13.4, 13.5 and 13.6). The occipital and sagittal crests are poorly developed or absent. The orbit lacks a postorbital process or bar and is widely confluent posteriorly with the temporal fossa. The jugal possesses ascending and descending processes. It is loosely attached to the maxilla and does not articulate with the squamosal. The premaxillae are separate from each other and from the maxillae, and are so loosely attached that they are rarely found in association with the skull. The rami of the mandible are co-ossified at the symphysis, which extends as a narrow and elongate edentulous spout (megatherioids and scelidotheres) that becomes shorter and broader in advanced mylodontines (e.g. the Pleistocene *Glossotherium* Owen, 1839). The alveolar portion of the mandibular corpus is thick and massive. The mandibular condyle is hemispherical and is set upon a well-defined neck. The angular process is more or less inflected, and in megalonychids forms a curved hook, longer and more prominent than in the other clades. The teeth are hypselodont, lack enamel, and are composed of three tissues: a soft, central vasodentine core surrounded by a hard orthodentine, and an outer layer of cementum (Kalthoff, 2011). The dental formula is 5/4, all upper teeth are set in the maxilla and separated by diastemata. In megalonychids and basal megatherioids the first upper and lower teeth are caniniforms, usually cylindrical or triangular in section, and variable in size (Figs. 13.3 and 13.4). The upper caniniform is positioned at the rostral end of the maxilla adjacent to the contact with the premaxilla and in occlusion it passes in front of the lower caniniform. Molariform cheek teeth are transversely oval to rectangular in megatherioids, with two transverse crests of hard dentine (orthodentine) separated by a deep valley. In mylodontids, molariforms are oval or lobate with flat or shallowly concave occlusal surfaces.

The appendicular skeleton of Santacrucian sloths is very conservative when compared with that of their living relatives. Their overall features are more similar to the limb elements of vermilinguans and cingulates than to those of extant sloths, especially the highly derived *Bradypus*. Except for the scapula, the general robustness and diaphyseal proportions of the limb elements, the great development of tuberosities, ridges, and crests for muscular attachment, and the morphologies of articular surfaces in most cases resemble the anteaters.

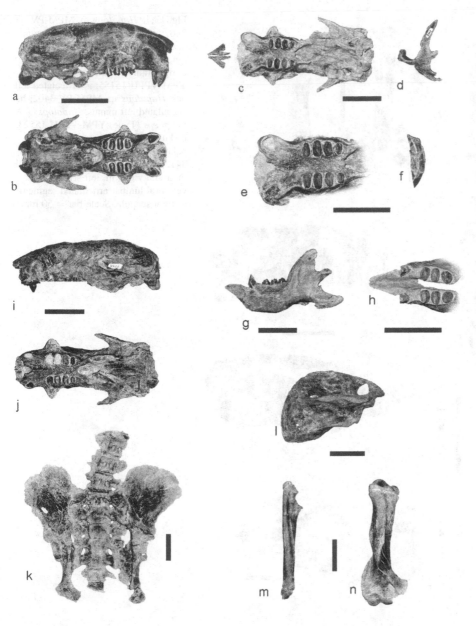

Fig. 13.3. *Eucholoeops ingens* MPM-PV 3451: a, skull in lateral view; b, skull in palatal view. *Eucholoeops ingens* MPM-PV 3401: c, skull in palatal view and premaxillae; d, left zygoma; e, close-up of the palate; f, left upper caniniform; g, mandible in lateral view; h, occlusal view of lower teeth. *Eucholoeops* cf. *E. fronto* MPM-PV 3403: i, skull in lateral view; j, skull in palatal view; k, pelvis in ventral view; l, right scapula; m, right ulna; n, right humerus. Scale bar = 50 mm.

13.2   Overview of the diversity and systematics of the group

McDonald *et al.* (2008) provided the most recent and complete overview of the fossil record of Vermilingua. Several taxa from the Santacrucian coastal localities (i.e. Corriguen Aike and Ameghino's Monte Observación, now Cerro Observatorio; Vizcaíno *et al.*, Chapter 1: Figs. 1.1 and 1.2 and Appendix 1.1) have been referred to this group: *Adiastaltus habilis* Ameghino, 1893, *Adiastaltus procerus* Ameghino, 1894, *Plagiocoelus obliquus* Ameghino, 1894, *Anathitus revelator* Ameghino, 1894, *Protamandua rothi*

Ameghino, 1904 (Fig. 13.2a, c, d), *Promyrmephagus dolichoarthrus* Ameghino, 1904, *Promyrmephagus euryarthrus* Ameghino, 1904, *Argyromanis patagonica* Ameghino, 1904, and *Orthoarthrus mixtus* Ameghino, 1904.

Ameghino (1893, 1894) placed *Adiastaltus* and *Plagiocoelus* into his new family Adiastaltidae, and *Anathitus* in his new family Anathitidae, and allied them with monotremes. More recently, Mones (1986) considered *A. habilis*, *A. procerus*, *An. revelator*, *Ar. patagonica*, and *O. mixtus*, as Xenarthra *incertae sedis*, and *Plagiocoelus obliquus* as Mammalia *incertae sedis*, reflecting the need for a critical re-examination of this material. *Protamandua*

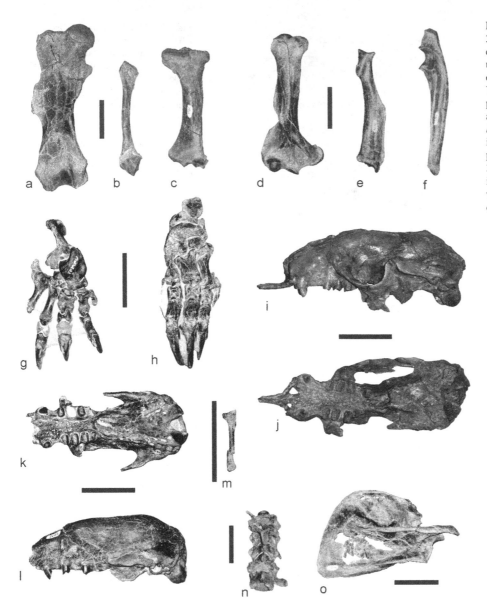

Fig. 13.4. *Hapalops* sp. MPM-PV 3467: a, right femur; b, right fibula; c, right tibia; d, left humerus; f, left ulna. *Hapalops* sp. MPM-PV 3404: e, right radius. *Hapalops elongatus* YPM-VPPU 15155: g, articulated left pes. *Hapalops* sp. MPM-PV 3402: h, articulated left manus. *Hapalops longiceps* (Type) YPM-VPPU 15523: i, skull in lateral view; j, skull in palatal view. *Hapalops* sp. MPM-PV 3412: k, skull in palatal view; l, skull in lateral view; m, left stylohyal; n, vertebral lumbar articulated segment; o, right scapula. Scale bar = 50 mm.

and *Promyrmephagus* have been referred to as myrmeco-phagids since their original descriptions (Ameghino, 1904). *Promyrmephagus* was placed in synonymy with *Protaman-dua* (Hirschfeld, 1976; Carlini *et al.*, 1993). More recently, Vizcaíno *et al.* (2004) assigned to Myrmecophagidae indet. a left humerus (MLP 69-IX-8–8a; Fig. 13.2b) originally attributed to the armadillo *Peltephilus* Ameghino, 1887 (Peltephilidae) by Lyddeker (1894).

Following McDonald *et al.* (2008), the large number of taxa named reflects the generally fragmentary nature of the types, and that each new discovery was described as a new taxon. Moreover, because different skeletal elements often were utilized as a type, it is not possible to make comparisons between taxa and, consequently, to produce a taxonomic revision that would permit their synonymy. McDonald *et al.* (2008) concluded, as did Gaudin and Branham (1998), that *Protamandua rothi* is the only species from the Santa Cruz Formation that can be confidently referred to Vermilingua (Myrmecophagidae). Based on the morphology of the distal humerus, McDonald *et al.* (2008) provisionally considered Adiastaltidae and Anathitidae to be junior synonyms of Myrmecophagidae, and all members to be poorly known vermilinguans. Further examination of *Argyromanis patagonica* and *Orthoarthrus mixtus* is needed to determine their validity, possible synonymies, and systematic relationships.

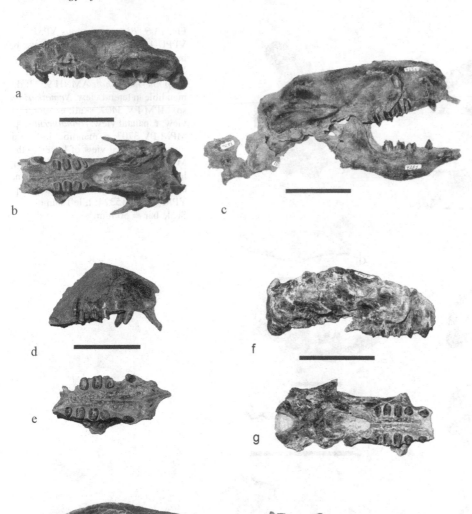

Fig. 13.5. *Hapalops* sp. MPM-PV 3467 skull: a, lateral view; b, palatal view. *Hyperleptus* sp. MPM-PV 4251: c, articulated skull, mandible, and first two cervical vertebrae. cf. *Hyperleptus* MPM-PV 3410, anterior skull: d, lateral view; e, palatal view. *Pelecyodon cristatus* MPM-PV 3409 skull: f, lateral view; g, palatal view. *Pelecyodon cristatus* YPM-VPPU 15049 skull: h, lateral view; i, palatal view.

Although Santacrucian sloths represent the first major sloth radiation, and could provide a wealth of information on sloth evolution and diversity, their taxonomy remains poorly understood, and this has hindered higher-level systematic studies (McDonald and De Iuliis, 2008). The plethora of genera and species erected by earlier workers were based in large part on fragmentary remains, even though most Santacrucian genera are clearly identifiable taxonomic units (De Iuliis, 1994; Gaudin, 1995, 2004; Gaudin and McDonald, 2008; McDonald and De Iuliis, 2008). According to McKenna and Bell (1997) the following genera are recorded in the Santa Cruz Formation: *Eucholoeops* Ameghino, 1887 (Fig. 13.3), and *Megalonychotherium* Scott, 1904 (Megalonychidae); *Hapalops* (Figs. 13.4 and 13.5a,b), *Hyperleptus* Ameghino, 1891 (Fig. 13.5c, d), *Pelecyodon*

Ameghino, 1891 (Fig. 13.5f–i), *Analcimorphus* Ameghino, 1891, and *Schismotherium* Ameghino, 1887 (Megatheriidae, Schismotheriinae; although, as noted by McDonald and De Iuliis, 2008: 46, no phylogenetic analyses exist that support the monophyly of Schismotheriinae); *Planops* Ameghino, 1887 and *Prepotherium* Ameghino, 1891 (Megatheriidae, Megatheriinae) (Fig. 13.6a–d), and *Nematherium* Ameghino, 1887 (Fig. 13.6e–h), and *Analcitherium* Ameghino, 1891 (Scelidotheriidae, Scelidotheriinae).

We follow Gaudin (2004) in accepting *Eucholoeops* as a basal Megalonychidae, *Schismotherium* and *Pelecyodon* as basal megatherioids, and *Hapalops* and *Analcimorphus* as basal members of either megalonychids or megatherioids (Gaudin, 2004, suggested referring to

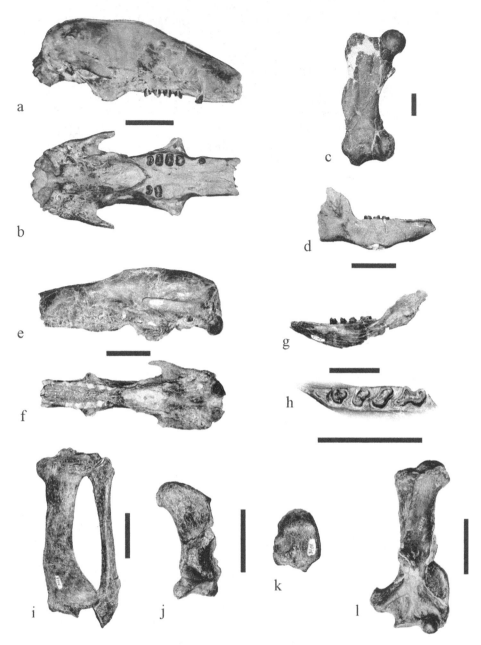

Fig. 13.6. *Planops magnus* YPM-VPPU 15346, skull: a, lateral view; b, palatal view. *Prepotherium potens* YPM-VPPU 15345: c, right femur. *Prepotherium filholi* AMNH 9573: d, mandible in lateral view. *Nematherium* sp. MPM-PV 3407, skull: e, lateral view; f, palatal view. *Nematherium* sp. MPM-PV 3703: g, mandible in lateral view; h, occlusal view of lower teeth. Mylodontidae indet. MPM-PV 3406: i, left tibia and fibula; j, left calcaneum; and k, left astragalus. *Nematherium* sp. YPM-VPPU 15374: l, left humerus. Scale bar = 50 mm.

these last four genera as basal megatherioids). Gaudin (2004) did not include *Hyperleptus* in his analysis, and we consider it here a basal megatherioid, essentially following McKenna and Bell (1997). Studies in progress based on newly recovered material will allow a better understanding of its relationships to other members of the group. *Planops* and *Prepotherium* are regarded as Megatheriidae (in accordance with De Iuliis, 1994; Gaudin, 2004, also assigned *Planops* to this taxon), and *Nematherium*

and *Analcitherium* are placed as a sister group to all remaining mylodontids (contra other less comprehensive recent phylogenetic studies, e.g. McDonald and Perea, 2002) (Fig. 13.1).

As already noted by Scott (1903–04), in Santacrucian beds the greatest number of taxa (and also specimens) belong to Scott's Megalonychidae (i.e. megalonychids and basal megatherioids *sensu* Gaudin, 2004) whereas megatheriids are much less common, and mylodontids are very rare. He also mentioned that, within "megalonychids," specimens of *Eucholoeops* are much less frequently encountered than those of *Hapalops*. Although Mones (1986) listed some 12 Miocene species of *Eucholoeops,* only the four recognized by Scott (1903–04) have generally been considered valid: *E. ingens* Ameghino, 1887, *E. externus* Ameghino, 1891, *E. fronto* Ameghino, 1891, and *E. curtus* Ameghino, 1894. In the case of *Hapalops,* 26 species are still formally recognized (Mones, 1986), even though most are based on partial or fragmentary specimens from the same region or even locality. A revision is in progress (G. De Iuliis and F. Pujos, unpublished data), and it appears that few of these species will prove to be valid, particularly in view of the wide range of intraspecific variation established recently for other sloth species by several authors.

Recent field work in the coastal exposures of the Santa Cruz Formation (Santa Cruz Province; Vizcaíno *et al.*, Chapter 1: Figs. 1.1 and 1.2) has yielded several new specimens, including skulls and mandibles, some with associated postcranial remains, particularly of *Eucholoeops* (Fig. 13.3) and *Hapalops* (Fig. 13.4) and, in less proportion, *Pelecyodon, Hyperleptus* (Fig. 13.5) and *Nematherium* (Fig. 13.6). Preliminary studies of this material (De Iuliis and Pujos, 2006; De Iuliis *et al.*, 2009), together with the type specimens available in older collections, provide a basis for beginning to unravel the chaotic taxonomy of some of the Santacrucian sloths, suggesting that most named species are likely invalid. De Iuliis and Pujos (2006) proposed the existence of only four main size classes and morphotypes of *Hapalops,* based on the skull and dentition. De Iuliis *et al.* (2009) considered that there are two morphotypes (i.e. species) of *Eucholoeops,* both with relatively large triangular caniniform teeth, the size of the canines being the main distinction between the two. This pattern was recognized by Scott (1903–04), who also suggested that within these morphotypes the caniniforms teeth exhibited sexual dimorphism, with males having the larger caniniforms. This last assertion, however, is not clearly supported by the available evidence. The name of the species with larger canines is *E. ingens* (as already recognized by Bargo *et al.*, 2009; Fig. 13.3a–h). The name of the second species, regarded as *E. fronto* by Scott (1903–04), is still under consideration, but here we tentatively assign the remains to *Eucholoeops* cf. *E. fronto* (Fig. 13.3i–n). The mylodontids *Nematherium* and *Analcitherium* are currently under revision (T. Gaudin and G. De Iuliis, unpublished data). Mones (1986) listed up to six Miocene species of *Nematherium,* and only one for *Analcitherium.*

Since ongoing taxonomic study leaves the species-level taxonomy in flux, and because Santacrucian sloth genera are well-identifiable and more stable taxonomic units, we analyse the morphology and habits of the genera. The following taxa are recognized in this chapter: the megalonychids *Eucholoeops* and *Megalonychotherium,* the megatheriids *Planops* and *Prepotherium,* the basal megatherioids *Hapalops, Hyperleptus, Analcimorphus, Schismotherium,* and *Pelecyodon,* and the mylodontids *Nematherium* and *Analcitherium.*

13.3 Materials and methods

Abbreviations for the institutions are given in Appendix 13.1, together with the list of materials. The fossil specimens studied include those collected by the expeditions carried out during the nineteenth and early twentieth centuries (housed at MLP, MACN, YPM-VPPU, AMNH, and FMNH) and the material collected by the Museo de La Plata/Duke University expeditions (housed at MPM-PV) over the past 9 years (2003–2011).

The Santacrucian vermilinguans from the older collections include two incomplete skulls, one of them with a partial skeleton (FMNH 13134), of *Protamandua rothi.* Unfortunately, the skull of the specimen FMNH 13134 is missing, so the only cranium available for study was the one from YPM (YPM-VPPU 15267; Fig. 13.2a). No new material of Vermilingua was found during the recent field expeditions.

The abundant new sloth material, which includes unusually complete specimens, complements the samples analyzed by previous workers on locomotion (e.g. White, 1997) and enables us to undertake new studies on mastication (e.g. Bargo *et al.*, 2009). As stated for the cingulates (Vizcaíno *et al.*, Chapter 12), paleobiologic approaches based on morphology are not usually sensitive enough to discriminate between congeneric species. Consequently, and following the taxonomic overview, the taxonomic unit chosen for investigation is genus. As most of the recently collected specimens are largely megalonychids and basal megatherioids, mylodontids and megatheriids being very rare, our analysis is based mostly on the better-preserved specimens of *Eucholoeops, Hapalops, Nematherium,* and *Planops.*

For the morphometric comparisons we used a database of 126 specimens (31 species) of living mammals representing a diversity of marsupials, xenarthrans, pangolins, rodents, primates, and carnivorans (Appendix 13.1).

13.3.1 Body mass
For the body mass estimations a multivariate regression analysis of living xenarthrans (36 specimens; Appendix 13.1) was carried out, using log-transformed data (imparting normality to the data) of linear measurements of the forelimb as independent variables and logarithm of body mass as the dependant variable. Measurements used are shown in Fig. 13.7, most of which follow traditional linear

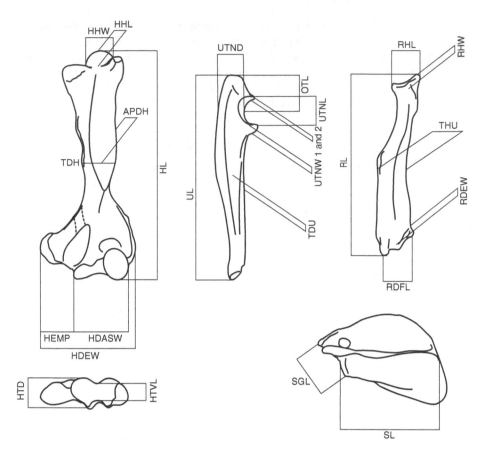

Fig. 13.7. Measurements used in morphometric analysis. For simplicity, some measurements are not depicted. For abbreviations and explanations see Appendix 13.2.

measurements seen in literature (e.g. Sargis, 2002a; Elissamburu and Vizcaíno, 2004; Candela and Picasso, 2008). The body masses of living mammals were taken from Nowak (1991). The regression analysis was performed only on living xenarthrans in an attempt to avoid bias produced by phylogenetic context, as might occur if other mammals were used. Because most fossil specimens do not have all the forelimb elements preserved, separate multiple regression analyses were performed for scapula, humerus, ulna, and radius. The regression coefficients obtained were used to estimate body mass in fossils. In vermilinguans, only the variables of the single humerus available (MLP 69-IX-8–8a) could be used. In sloths, different estimates were obtained for each forelimb element, and the single mean of these different estimates was chosen for those fossil specimens that preserve more than one forelimb element.

13.3.2 Locomotion and use of substrate
The rare and fragmentary nature of the remains of fossil vermilinguans precludes the application of statistical approaches for the interpretation of locomotor abilities. However, some information can be obtained by comparing the position of muscle insertion areas on the humerus of

the specimen identified as Myrmecophagidae indet. (MLP 69-IX-8–8a) (Fig. 13.2b) with that of living vermilinguans.

A principal components analysis (PCA) was performed for the forelimbs of sloths (Toledo *et al.*, 2012). PCA was chosen for robustness, simplicity, ease of interpretation and, most importantly, obviating the need for assumptions of normality of the data, variance homogeneity, and allocation of specimens to taxonomic groups. The analysis was performed on logarithms of the raw measurements, because otherwise the large variability in body size in the sample, from *Cyclopes* (0.5 kg) to *Gorilla* (80 kg), would swamp other morphological variation. A correlation matrix was used. Missing measurements (especially among fossils) were replaced by global mean values. Additionally, a qualitative-comparative morphofunctional analysis was carried out both for the forelimbs and for the hindlimbs of the sloths. Specimens were visually examined, with functionally relevant traits, including muscle scars (attachment areas) such as tuberosities and crests, joint surfaces, and other structures, described and compared with homologous features in living xenarthrans.

Often, it is more informative to refer to substrate preference (arboreal, semi-arboreal, or terrestrial), or to substrate use (digging) than to locomotor mode. A discrete and unimodal locomotor categorization of mammals would be impossible to propose because each animal exhibits a wide spectrum of locomotor modes. In order to avoid misconceptions, in this contribution we refer to locomotor categories, substrate preference categories, and substrate use when needed.

13.3.3 Mastication and feeding behavior

The skull of *Protamandua* (YPM-VPPU 15267) preserves the glenoid fossa of the temporal, allowing several conclusions about the shape of the mandibular condyle and possible jaw movements (Fig. 13.2a). These were complemented with the examination of a cast of the mandible of a Colhuehuapian (Early Miocene) specimen reported by Carlini *et al.* (1992).

Bargo *et al.* (2009) undertook a detailed morphofunctional analysis of the jaw apparatus of the megalonychid sloth *Eucholoeops*. The results provide the basis for comments concerning contemporaneous megatherioid sloths (i.e. *Hapalops, Pelecyodon, Planops*). A comprehensive description of the teeth of *Eucholoeops* introduced a new nomenclature for the crown features in order to infer jaw movements from the occlusal pattern, in a manner analogous to that applied to other mammals with clearly tribosphenic teeth. Having done this, the occlusal movements were complemented through analysis of the wear facets following Naples (1982) and the manipulation of casts of the upper and lower tooth series. In her study of mastication of living sloths, Naples (1982) pointed out that it is possible to discern movement direction on the basis of differential wear in the hard, cortical orthodentine and softer core of vasodentine, by analogy with the wear patterns of mammals with an outer hard layer of enamel and a softer dentine core (Rensberger, 1973; Costa and Greaves, 1981). This produces a characteristic "cupping" wear in front of the trailing edge but not behind the leading edge.

13.4 Results

13.4.1 Body mass

Anteaters The myrmecophagid MLP 69-IX-8–8a is a left humerus lacking the proximal epiphysis. The available measurements for calculating body mass were the anteroposterior diameter at the midshaft, the width at epicondyles, and the estimated humeral length. They provided an average body mass of 5.9 kg (Table 13.1).

Sloths All the forelimb elements are tightly correlated with body size, with 93% and 97% of the variance explained by scapular dimensions and ulnar dimensions respectively (Table 13.1). The stem megatherioid *Hapalops* measurements provided a wide range of estimates of body mass (between 11 kg and 85 kg) with an average of 38 kg. Estimates for the megalonychid *Eucholoeops* averaged 77 kg (with a range

between 63 kg and 90 kg), while for the mylodontid *Nematherium* only one estimate of 95 kg was obtained.

13.4.2 Locomotion and use of substrate

Anteaters The humerus of the myrmecophagid MLP 69-IX-8–8a is, in overall shape, shorter and more robust (Fig. 13.2b) than that of *Tamandua, Myrmecophaga*, and the Colloncuran (Middle Miocene) *Neotamandua? australis* (MLP 91-IX-6–5), both laterally and anteroposteriorly. The humeral head is missing. As in the other myrmecophagids, the distal end of the humerus is very broad with a large medial epicondyle. On the anterior surface, the pectoral tuberosity is much more proximally placed and stouter than in *Tamandua* and *Myrmecophaga*. On the lateral side the deltoid tuberosity is much more proximal and proportionately much larger than in *Tamandua* and *Myrmecophaga*. The proximal end of the supracondyloid ridge is broken, but clearly suggests the presence of a stout deltoepicondylar ligament as in *Tamandua* (Taylor, 1978, 1985). On the medial side, the muscle scar for the insertion of the *m. teres major* is a prominent ridge that extends much farther down the humeral shaft (almost to the distal end) than in the living species. The epitrochlear foramen is relatively smaller. On the posterior side, the olecranon fossa (which receives the olecranon process of the ulna in elbow extension) is proportionately larger and deeper than in the living species. The condyle for the radius (capitulum) is similar to that of *Cyclopes*, smaller and flatter than in *Tamandua* (spherical in *Neotamandua? australis*). The trochlea is flatter and separated from the epitrochlea by a prominent groove for the medial head of the triceps. These features indicate enhanced mechanical advantage of the antebrachial extensors and brachial retractor and, judging from the development of the ridges, tuberosities, and areas available for muscular attachment, the muscles must have been comparatively larger than in living myrmecophagids.

Based on the morphology of the caudal vertebrae (i.e. unslanted, transversely widened centra with double transverse processes, as in *Cyclopes* and *Tamandua*, but unlike *Myrmecophaga*), Hirschfeld (1976) and Gaudin and Branham (1998) concluded that *Protamandua* possessed a prehensile tail and hence was almost certainly arboreal. The astragalar morphology of *P. rothi* resembles that of *Tamandua*, especially in the shape of the astragalar trochlea, while the calcaneum is more similar to that of *Myrmecophaga* (Fig. 13.2c, d).

Sloths The first three principal components account for 90.3% of the total variance (Table 13.2 and Figs. 13.8 and 13.9). Uniform loadings indicate that PC1 roughly represents the overall size of the forelimb elements (not exactly body size); *Cyclopes* and *Gorilla* are at opposite extremes in the point-cloud. With respect to PC2, negative values indicate relatively short and massive forelimb elements, with traits that improve force (leverage) rather than speed in flexion or extension of the proximal and medium segments of the forelimb, such as a long olecranon process, increased diameter of the humerus, and a wide medial epicondyle (e.g. vermilinguans, the giant armadillo *Priodontes*, pangolins, the aardvark, the wombat, and the African porcupine). Species with high positive scores have slender, gracile bones with less development of such features (e.g. the gibbon and extant sloths). Positive values of PC3 represent a wide distal humeral epiphysis and a large and medially protruding humeral entepicondyle (again, grouping vermilinguans, the giant armadillo, pangolins, the aardvark, the wombat, and the African porcupine), while negative values represent a narrower distal humeral epiphysis and less developed entepicondyle (primates and extant sloths, with *Hylobates* being the most extreme taxon).

In PC1, Santacrucian sloths occupy a position between *Gorilla* and the giant panda *Ailuropoda* (the largest extant taxa analyzed) on the one hand, and the giant armadillo *Priodontes* on the other, showing similar loading to the giant anteater *Myrmecophaga*, *Papio*, and the cheetah *Acynonix*. In both PC2 and PC3 (Fig. 13.9), Santacrucian sloths share a common morphospace with vermilinguans, the giant armadillo, the African porcupine, pangolins, the

Table 13.1. *Multiple regression coefficients (above) and body mass estimations for Santacrucian pilosans*

	Multiple R	adjusted R^2	F	df
Scapula	0.96975765	0.93274343	122.349	4.31
Humerus	0.97883264	0.9527086	177.273	4.31
Ulna	0.98855076	0.97252211	207.459	6.29
Radius	0.98480748	0.96701882	343.071	3.32

df: degree of freedom; F: F-test

Taxon	Catalog number	Mass estimation in kg (scapula)	Mass estimation in kg (humerus)	Mass estimation in kg (ulna)	Mass estimation in kg (radius)	Mean in kg
Myrmecophagidae indet.	MLP 69-IX-8–8a		5.87			5.87
Eucholoeops ingens	MPM-PV 3401		90.70			90.70
Eucholoeops* cf. *E. fronto	**MPM-PV 3403**	53.07	102.92			78.00
Eucholoeops ingens	MPM-PV 3451				63.07	63.07
***Eucholoeops* sp.**	**MPM-PV 3651**		114.00	20.92	64.96	66.63
Megalonychidae indet.	**AMNH 9249**		32.71		18.35	25.53
Megalonychidae indet.	**AMNH 94754**		45.53	13.73	30.82	30.03
Hapalops angustipalatus	**YPM-VPPU 15562**			14.44	38.39	26.41
Hapalops elongatus	YPM-VPPU 15155	29.82				29.82
Hapalops elongatus	YPM-VPPU 15160		37.87	13.35		25.61
Hapalops longiceps	**YPM-VPPU 15523**			32.14	60.43	46.29
Hapalops platycephalus	YPM-VPPU 15564		51.42			51.42
Hapalops ponderosus	YPM-VPPU 15520		85.17			85.17
Hapalops rectangularis	AMNH 9222	16.16			14.92	15.54
Hapalops sp.	YPM-VPPU 15414	12.03				12.03
Hapalops sp.	AMNH 9252	10.96				10.96
Hapalops sp.	YPM-VPPU 15005	32.88			23.24	28.06
Hapalops sp.	YPM-VPPU 15183	57.99				57.99
Hapalops sp.	MPM-PV 3412	61.17				61.17
***Hapalops* sp.**	**MPM-PV 3467**		79.60	20.36		49.98
cf. *Hapalops*	**MPM-PV 3404**				44.96	44.96
cf. *Hapalops*	MPM-PV 3462	17.80				17.80
Nematherium sp.	YPM-VPPU 15374		95.02			95.02

Specimens in bold font are those included in the PCA.

Table 13.2. *PCs eigenvalues (above) and contribution of each variable*

	Eigenvalue	% Total	Cumulative
PC1	25.97861	81.18317	81.1832
PC2	2.09377	6.54302	87.7262
PC3	0.81625	2.55077	90.2770

	PC1	PC2	PC3
SL	−0.930442	0.003902	−0.289461
SGL	−0.952918	0.032938	−0.040272
SSE	−0.932492	0.002944	−0.146045
SGW	−0.961047	0.034031	−0.106401
HTVL	−0.796821	0.444804	−0.014989
TDH	−0.841357	−0.354592	0.239456
HL	−0.835287	0.518578	0.048266
APDH	−0.944376	−0.156533	−0.081334
HHL	−0.963397	0.131864	−0.019285
HDASW	−0.968221	0.028455	0.072726
HTVD	−0.946037	0.163341	−0.062402
HTD	−0.938200	0.053726	−0.054242
HEMP	−0.770528	−0.366348	0.450885
HDEW	−0.915589	−0.176736	0.303859
HGT	−0.915279	−0.145645	−0.084992
HLT	−0.667517	−0.308469	−0.060413
ItuW	−0.912570	−0.211122	−0.049628
HHW	−0.963332	0.072121	−0.105189
UL	−0.814796	0.522300	0.036411
THU	−0.912362	−0.268184	−0.082550
THUn	−0.959333	−0.183607	−0.027465
OTL	−0.804607	−0.473229	−0.251516
UTNL	−0.953847	0.037502	0.010598
TDU	−0.935652	−0.101415	−0.100237
UTNW1	−0.939840	−0.117323	−0.070329
UTNW2	−0.882801	0.033611	−0.260355
RL	−0.712701	0.640953	0.103997
RH	−0.877545	−0.054230	0.284538
RDFL	−0.939546	0.000006	0.141520
RHL	−0.961483	0.030768	0.081082
RHW	−0.945321	0.226844	0.132721
RDEW	−0.933618	−0.002607	0.090075

Abbreviations as in Fig. 13.7 and explained in Appendix 13.2.

aardvark, and the wombat, and are clearly separated from their living suspensory arboreal relatives.

The morphofunctional comparative analysis indicates that the forelimb exhibits a fairly uniform morphology among all Santacrucian sloths. The scapula is slender, triangular, and very similar to that of extant sloths (Figs. 13.3l and 13.4o). The scapular acromion, when preserved, does not project as far beyond the glenoid fossa as in vermilinguans and cingulates, indicating that the *m. deltoideus*, the main shoulder joint stabilizer in extant vermilinguans and cingulates, was not as well developed in the Santacrucian sloths. The humerus of Santacrucian sloths is wider and more robust than those of extant sloths, and exhibits better-developed crests for muscular attachments (Figs. 13.3n, 13.4d, and 13.6l). The humeral tuberosities are lower than the humeral head, allowing great shoulder mobility. The deltopectoral ridge is strongly raised, at its distalmost portion, as a shelf from the diaphysis, thus increasing the muscle leverage for abduction, adduction, and retraction at the shoulder. The entepicondyle is well developed, providing enlarged origin sites for the carpal and digital flexor musculature, as in extant vermilinguans. A conspicuous entepicondylar foramen is present in all forms, although in *Eucholoeops* it is slightly smaller. The humeral trochlea is wide and low, as in *Myrmecophaga*, and the capitulum is rounded, as in *Choloepus*. The epicondylar ridge is well developed, providing an extensive origin attachment site for powerful forearm flexor musculature. In the mylodontids *Nematherium* and *Analcitherium*, the humerus is proportionately shorter and more robust and the deltopectoral crest is more expanded medially and laterally than in the megatherioids and megalonychids (Fig. 13.6l). Santacrucian sloths resemble vermilinguans in having a well-developed olecranon process (Figs. 13.3m and 13.4f), indicating powerful forearm extension and/or humeral retraction when the elbow is flexed (Argot, 2001). The radial diaphysis is bent, similarly to that of *Choloepus*, but is much shorter and more robust, indicating strong forearm flexor musculature, particularly in *Nematherium*. The radial head is oval (though not to the degree that occurs in vermilinguans), limiting the capacity for pronation and supination. The hand is pentadactylous, with the three central digits being the most developed (Fig. 13.4h). The third metacarpal is more robust and shorter than the second, and the ungual phalanges are compressed and curved, suggesting the existence of enlarged claws.

Although the overall morphology of the hindlimb of Santacrucian sloths roughly resembles that of vermilinguans, a more detailed examination reveals a mix of vermilinguan and extant sloth features. The pelvis is intermediate in width between living sloths (wider), and vermilinguans (narrower) (Fig. 13.3k). The acetabulum is large and laterally oriented, suggesting considerable femoral abduction capabilities. The femur is wide and flattened, increasing the area for thigh adductor musculature, with a well-developed third trochanter positioned approximately at midshaft (Figs. 13.4a and 13.6c). The greater trochanter is low and, as in other pilosans, does not project far above the femoral head, allowing enhanced abduction at the hip. The femoral head is hemispherical and comparatively large. The lesser

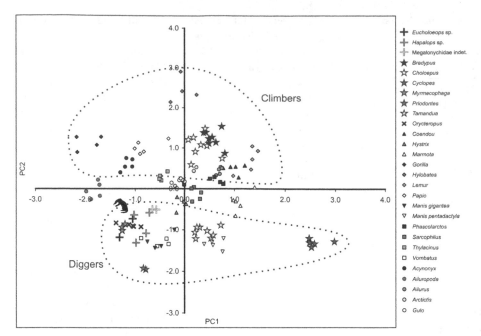

Fig. 13.8. Plots of PC 1 and PC 2. Crosses indicating fossil sloths, and small *Hapalops* silhouette showing position of *H. longiceps* (YPM-VPPU 15523).

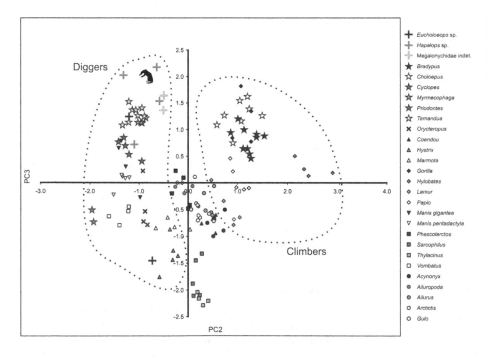

Fig. 13.9. Plots of PC 2 and PC 3. As in Figure 13.8, crosses indicating fossil sloths, and small *Hapalops* silhouette showing position of *H. longiceps* (YPM-VPPU 15523).

trochanter is conspicuous and medially projecting, increasing leverage for lateral thigh rotator musculature. The distal epiphysis is wide and anteroposteriorly compressed, a trait related to knee-flexed postures (Sargis, 2002b; Candela and Picasso, 2008). Both femoral condylar articular surfaces are continuous with a shallow and wide patellar groove. The condyles are asymmetric: the medial condyle is anteroposteriorly deeper than the lateral, the reverse of the condition in extant sloths, and more nearly resembling that in cingulates and the ground-dwelling *Myrmecophaga*. The tibia and fibula are unfused as in extant sloths. The tibia is short and slightly mediolaterally bent, especially in *Hapalops*

(Fig. 13.4c) and *Pelecyodon*, with condylar facets disposed at different levels. The proximal medial facet for the medial femoral condyle is concave and distal, while the lateral facet is convex and proximal. The astragalar articular facet is narrow and triangular, and the tibial malleolus is strongly reduced (as in other xenarthrans). The fibula is short and straight, and its distal epiphysis is very robust, showing a flat and very well-marked articular facet for the astragalus (Figs. 13.4b and 13.6i). The astragalus is dorsoventrally deep with a well-defined but asymmetric trochlea (Fig. 13.6k). The medial astragalar condyle is smaller than the lateral one, presenting a small odontoid process (Pujos *et al.*, 2007). The contact between the ectal facet of astragalus and the astragalar facet of the fibula is very precise, suggesting limited mobility in the mediolateral plane. The astragalar head is big, concave, and sessile. The calcaneum possesses a remarkably large and mediolaterally expanded tuber calcanei (Fig. 13.6j), wing-shaped in most forms, with an inner process that projects far medially, particularly in *Hapalops*. This particular morphology increases the lever arm for foot extensor muscles. The foot is pentadactylous, and the three middle digits are largest (Fig. 13.4g). The third metatarsal is shorter than the second one, and very robust. The fifth metatarsal exhibits an expanded and flattened lateral border (metatarsal tuberosity of humans).

13.4.3 Mastication and feeding behavior

Anteaters The poorly preserved skull of *Protamandua* permits us to say only a few things about functional design. YPM-VPPU 15267 shows a smooth, anteroposteriorly elongated glenoid fossa, which suggests a convex mandibular condyle (Fig. 13.2a). A mandible of a Colhuehuapian myrmecophagid mentioned by Carlini *et al.* (1992) suggests that the face and jaws were already elongated as in *Cyclopes* and non-xenarthran myrmecophages such as that of pangolins. It retains a prominent coronoid process more like *Cyclopes* than *Tamandua* or *Myrmecophaga*. Although broken, the angular process also resembles that of *Cyclopes*. The mandibular symphysis is unfused as in the living xenarthran anteaters, and the condyle is relatively flat.

Sloths Despite a range of morphological variability, Santacrucian sloths are characterized by elongated, tubular skulls, curiously small in proportion to body size (Figs. 13.3a–c, 13.4i–l and 13.5). Bargo *et al.* (2009) summarized the main features of the cranium of *Eucholoeops* that were relevant for the analysis of occlusion and masticatory movements. These features are shared by the Santacrucian basal megatherioids and megatheriids in general, as will be discussed below. The temporomandibular joint is located slightly dorsal to the occlusal plane. The glenoid fossa is a shallow depression that allows the mandibular condyle great freedom of anteroposterior motion. The most anterior teeth are large caniniforms, usually sharply pointed by wear, and placed externally to the line of the cheek teeth in both the upper and lower jaw. The upper caniniform has a distally facing facet on its distal surface for occlusion with a mesially facing facet on the lower caniniform (Fig. 13.3a, b, f). Caniniforms are separated from the first molariforms by relatively long diastemata. *Eucholoeops* has four maxillary and three mandibular molariforms (M1–M4 and m1–m3 respectively; Fig. 13.3b, e, g, h) which are isognathic, but occlusion appears to have occurred on one side of the mandible at a time, as in living sloths (Naples, 1982). Molariforms are short mesiodistally and broad vestibulolingually and occlude alternately (as in living sloths, e.g. *Choloepus*): m1 occludes between and against the M1 and M2; m2 and m3 have the same relationship with M2–M3 and M3–M4 respectively. The occlusal details and proportions of the teeth vary from m1 to m3 (and in the upper teeth) but the basic pattern of occlusion and the inferred occlusal relationships remain essentially the same. Bargo *et al.* (2009) described the morphology and occlusion between m2 and the corresponding M2–M3 as representative of molariform occlusion as a whole, and generated a nomenclature for describing the inferred occlusal pattern analogous to that applied to other mammals. Upper molariforms have three cusps (A, B, and C) and two transverse lophs (mesial and distal) composed of orthodentine. Lower molariforms are almost a mirror image of the uppers: they also have three cuspids (A, B, and C) and two transverse lophids. A basin or sulcus, excavated in the vasodentine, lies between the mesial and distal lophs (and lophids) (see Bargo *et al.*, 2009, for a comprehensive description; see also their Figure 4).

Based on scale drawings and examination and mapping of occlusal wear facets Bargo *et al.* (2009) reconstructed two main jaw movements during the power stroke, one of the three components or strokes of the chewing cycle of eutherian and metatherian mammals (Crompton and Hiiemae, 1969a, 1969b; Hiiemae and Kay, 1972, 1973; Kay and Hiiemae, 1974; Hiiemae, 1978; Hiiemae and Crompton, 1985). Movement A corresponds to the basic therian pattern equivalent to Phase I (Hiiemae and Kay, 1972) of the power stroke: the working side dentary is moved dorsally, mainly orthally but also anteriorly and slightly medially; the result is puncturing, tearing, and shearing of food. Movement B is a distinct and unrelated movement of the working side dentary dorsally, mainly orthally, but also posteriorly and slightly medially; the dominant result is again to produce shearing of food. Additionally, the analysis of the matching wear surfaces on the basis of differential wear in the hard cortical orthodentine and softer core of vasodentine (pattern of "cupping" wear) in *Eucholoeops* indicates that the jaw movement during the power stroke is mainly orthal, as indicated by Naples (1982) for extant sloths (see Discussion).

The basal megatherioids *Hapalops* and *Pelecyodon* (Figs. 13.4 and 13.5) and the large megatheriids *Planops* (Fig. 13.6a) and *Prepotherium* show the same general dental morphology as *Eucholoeops*: the first tooth is a caniniform with a variable diastema separating it from the first molariform, and the molariforms are transversely oval to rectangular, with the typical two transverse crests separated by a valley. Even though the occlusal patterns and mandibular movements of these taxa were not described in such detail as in *Eucholoeops*, wear facets and cusps are clearly visible in all of them, suggesting the predominance of the same two distinct mainly orthal movements.

The Santacrucian mylodontid sloths, *Nematherium* and *Analcitherium*, have a very elongated and tubular skull (Fig. 13.6e–h) resembling those of the Pleistocene scelidotheres (e.g. *Catonyx* Ameghino, 1891 and *Scelidotherium* Owen, 1840). The mandibular condyle is positioned lower than in megatherioids. The dentition lacks caniniforms and the typical two transverse shearing crests (lophs) of the megatherioids. Instead, the teeth are semi-oval or lobed, with relatively flat occlusal surfaces and a shallow basin excavated in the core of soft dentine (vasodentine) (Fig. 13.6h), a morphology also observed – with minor variations – in *Scelidotherium*, as well as in Pleistocene mylodontines, i.e. *Glossotherium* Owen, 1839 and *Mylodon* Owen, 1839 (Bargo and Vizcaíno, 2008). The analysis of the occlusal wear facets of *Nematherium* suggests a predominance of lateral masticatory movement, instead of the orthal movements proposed for megatherioids, and also suggests that grinding was emphasized over shearing.

13.5 Discussion

13.5.1 Body mass

Anteaters A qualitative comparison of the skull of *Protamandua rothi* suggests that it is a small animal, intermediate in size between *Cyclopes* (0.270 kg; Nowak, 1991) and *Tamandua* (about 6 kg; Nowak, 1991). The multiple regression of the myrmecophagid humerus MLP 69-IX-8–8a based on total length and two transverse diameters indicates a body mass comparable to that of the living *Tamandua* albeit with different proportions. Although the bone lacks the proximal epiphysis, its length is about half that of adult *Tamandua* while width at the epicondyles is approximately 90% that of *Tamandua*.

Sloths Santacrucian sloths range in body mass from 40 to 90 kg, surpassing by an order of magnitude that of extant sloths. Most would be classified as large mammals (*sensu* Martin and Steadman, 1999; Johnson, 2002; Cione *et al.*,

2003, 2009; Vizcaíno *et al.*, 2012). Not surprisingly, in PC1 of our analysis (interpreted as forelimb elements size) the Santacrucian sloths show loadings in the range between a giant armadillo and a gorilla (Fig. 13.8). It is noteworthy that the ulna, the best predictor for body mass for living mammals, produced lower estimates than other elements such as the humerus in fossil sloths. Since very few specimens preserved at least three limb elements, the comparative power of these body size estimations is relatively weak, and they must be considered as preliminary values until regression analyses using all hindlimb variables are completed (N. Toledo, unpublished data).

13.5.2 Locomotion and use of substrate

Anteaters As described above, although the overall morphology of the limb bones of Santacrucian vermilinguans resembles that of the living ones, they show some peculiarities (e.g. shorter and stouter humeri). Among the fossils, differences in the morphology of the humerus (e.g. capitulum rounded in *Adiastaltus* and *Plagiocoelus*, and more ovoid in the myrmecophagid MLP 69-IX-8–8a) parallel those between the extant forms, making *Adiastaltus* and *Plagiocoelus* more similar to *Tamandua*, and MLP 69-IX-8–8a more similar to *Cyclopes*. This may suggest functional differences within the Santacrucian vermilinguans. At first sight, a compressed capitulum would indicate reduced freedom of movements of the antebrachium so that it is more restricted to the sagittal plane. The morphology of the astragalus of *Protamandua* resembles that of *Tamandua*, while the calcaneum bears a resemblance to *Myrmecophaga*. *Protamandua* differs from *Tamandua* and *Myrmecophaga* in the morphology of the caudal vertebrae, indicating it had a prehensile tail (Hirschfeld, 1976; Gaudin and Branham, 1998).

Taylor (1978, 1985) stated that the major specializations in the forelimb of living myrmecophagines (*Tamandua* and *Myrmecophaga*; Gaudin and Branham, 1998) imply powerful flexion of enlarged central claws of the manus, and flexion and axial rotation of the humerus. The morphological features of the humerus that correlate with such specialization are already present, and even emphasized, in the MLP 69-IX-8–8a specimen. The medial head of the triceps would have passed through the groove separating the trochlea from the epitrochlea to insert in a common tendon with the *flexor digitorum profundus,* improving the claw flexion even when the antebrachium was fully extended (Taylor, 1978, 1985). The slightly proximal position of the deltoid tuberosity implies a longer delto-epicondilar ligament with a consequent increase in the mechanical advantage of the *m. brachioradialis, m. extensors carpi radialis,* and *carpi ulnaris, m. extensor digitorum,* and *m. supinator* in elbow flexion. The lateral expansion of the deltoid tuberosity displaces the insertion of the *m. spinodeltoideus*, augmenting lateral axial rotation of the humerus in vermilinguans. The insertion of the long head of the *teres major* muscle extends much farther distally than in the living species, imparting increased mechanical advantage to this powerful retractor (extensor) of the humerus. As mentioned above, not only is the mechanical advantage enhanced, but the muscles must also have been comparatively larger. Taylor (1978, 1985) clearly explained how these features relate functionally to the ability of *Tamandua* to tear apart hard materials with its claws, in the efficient harvesting of ants or termites that nest inside wood, or break apart hard carton nests whilst keeping its body as far as possible from stings or bites of the insects. The strong, prehensile hands of *Tamandua* are also obviously important for climbing, as *Tamandua* is able to move, feed, and rest in trees as well as on the ground (Montgomery, 1985). The forelimb of *Myrmecophaga* functionally resembles that of *Tamandua*, although the former uses its forelimb for digging or tearing wood in searching for ants or termites and defense from predators, but does not climb. Taylor (1985) hypothesized that such similarities are holdovers from the arboreal ancestry of *Myrmecophaga*, bearing witness to the strong phylogenetic signals that constrain xenarthran evolution (Vizcaíno and De Iuliis, 2003).

The obvious overall functional similarities of Santacrucian myrmecophagids with living vermilinguans indicate that the Miocene taxa were capable of clawing substrates in search of ants and termites. Interestingly, Tauber (1994: 52) reported the presence of biogenetic structures that he interpreted as probable ant or termite nests in the lower levels of the Santa Cruz Formation, at the coastal locality of Cañadón Silva (see Krapovickas, Chapter 6). The functional similarities with living anteaters also suggest that the Santacrucian myrmecophagids could climb trees, and hence that arboreal habits were already established in the clade by the Early Miocene. The presence of a prehensile tail like that of *Cyclopes* and *Tamandua* in the Santacrucian *Protamandua*, plus the fact that two of the three modern genera have arboreal adaptations, led Gaudin and Branham (1998) to infer that arboreality is likely the primitive condition for vermilinguans. In their brief report, Carlini *et al.* (1992) indicated that the myrmecophagine from the Early Miocene Gaiman Formation is larger than *Tamandua* and that certain features of the postcranium indicate arboreal life habits.

Sloths Principal components analysis (PCA) shows that while extant sloths share a common forelimb morphospace with suspensory arboreal primates like gibbons, both groups having elongated, slender, gracile forelimbs, Santacrucian sloths are clearly unlike their living relatives in these respects and instead share a morphospace with anteaters

and other capable digging forms, such as pangolins, aardvarks, and wombats. This morphometric similarity is an expression of relatively short forelimb elements as well as other functionally interesting variables, such as long olecranon processes, involved in mechanical advantage for the *m. triceps*, and broad humeral diaphyses, related to strengthening the cortical bone when loaded. Hence, Santacrucian sloths have a different functional pattern than extant sloths, probably more similar to that of living anteaters and pangolins, animals able to dig or at least to rip the substrate to obtain their food. Several of them are scansorial or arboreal, such as the lesser pangolin and the lesser and pigmy anteaters, *Tamandua* and *Cyclopes*. The very slender and gracile forms, such as the extant sloths and the gibbon *Hylobates* at the opposite extreme of PC2 axis, are arboreal and suspensory. Between the extreme morphologies there are several ground-dwelling, scansorial, and arboreal mammals without highly derived morphologies for fossoriality and/or arboreality. The Santacrucian sloths also show positive values of PC3, which represent variables that are apparently related to powerful hand prehension, such as wide distal humeral epiphysis and a large, medially protruding humeral entepicondyle. Both traits imply available area for well-developed carpal and digital flexor musculature, involved both in digging and in climbing activities.

The results of PCA of Santacrucian sloths seem to reflect to a great extent the way the animals may have used the substrate, obscuring inferences in locomotor pattern *per se*. As mentioned above, the morphospace including fossil sloths comprises most specialized digging forms. All of them have robust and massive forelimb elements with wide humeri, and protruding entepicondyles, long olecranon processes, and strong and relatively short radii.

Bargo *et al.* (2000) analyzed the limb proportions and resistance to bending forces in the Pleistocene mylodontids *Scelidotherium*, *Glossotherium*, and *Lestodon* Gervais, 1855 to infer their locomotor styles. The conclusion was that the forelimbs of these ground sloths were well suited for activities such as digging where force is enhanced over velocity. It is noteworthy that *Nematherium* exhibits the most robust and massive humerus (Fig. 13.6l) among Santacrucian sloths, suggesting that mylodontids had a long phylogenetic history of digging abilities.

An extensive description of the morphological traits related to climbing habits of the Lujanian sloth *Diabolotherium* was made by Pujos *et al.* (2007). These authors considered the increased shoulder mobility, the curved radius, a relatively short olecranon, and highly mobile hip and ankle joints, among other features, as indicating climbing or arboreal habits.

It is interesting to note that many morphological features are associated with the mechanical requirements of both digging and climbing: powerful humeral adduction and retraction capabilities, as suggested by the great development of the deltopectoral ridge; powerful (rather than fast) antebrachial extension, indicated by the relatively long olecranon process; and capacity for powerful carpal and digital flexion. Other traits, such as the increased shoulder mobility indicated by the low humeral tuberosities, are more related to mechanical properties advantageous only for climbing (Pujos *et al.*, 2007). Finally, other features, such as a limited capacity for supination at the elbow suggested by the radial head morphology, seem to be better adapted for digging alone. In short, forelimb mechanical requirements for digging and climbing are similar in some ways, so their morphological patterns overlap (Toledo *et al.*, 2012).

The hindlimbs of Santacrucian sloths are characterized by a mix of morphological features. Several traits could be related to climbing, especially those involved with increased abduction and adduction capabilities at the hip, the habitual knee-flexed postures, and powerful extension (plantar flexion) of feet.

A laterally oriented acetabulum allows great femoral abduction, and a greater trochanter lower than the femoral head increases the mobility of the hip joint (Pujos *et al.*, 2007). A conspicuous and medially oriented lesser trochanter improves the mechanical advantage for external rotation of the femur. Flattening of the femoral diaphysis allows powerful adduction of the femur, assisting the grasping strength of the feet. A shallow and wide patellar groove suggests slow movements in a knee-flexed posture and a more mobile knee joint. Finally, a large and expanded tuber calcanei suggests powerful extension (plantar flexion) of a plantigrade foot.

The asymmetric astragalar trochlea of Santacrucian sloths suggest increased astragalocrural mobility, a trait commonly shared among most climbing mammals (Argot, 2002; Candela and Picasso, 2008; Croft and Anderson 2008). The same capability was mentioned by Pujos *et al.* (2007) for the Lujanian climbing sloth *Diabolotherium*. However, other ankle joint features of Santacrucian sloths seems to indicate more constrained mediolateral, i.e. less ability for inversion and eversion at the ankle joint, suggesting terrestrial locomotion.

Overall, the hind limbs of Santacrucian sloths are characterized by plantigrady, restricted lateral joint mobility at the ankle, a slightly flexed knee, and considerable lateral excursion (abduction) at the hip.

13.5.3 Mastication and feeding behavior

Anteaters The overall morphology of the mandible of the Colhuehuapian Vermilingua indicates that the specialized morphology related to social-insect eating was attained before the Santacrucian. Ants and termites are small enough to swallow without biting or cutting (Montgomery, 1985), and food acquisition efficiency in anteaters is not limited by mastication as it is in other mammals, but rather by ingestion rate (Naples, 1999). The mandibular rami are joined at the midline by a ligament and other connective tissues; the mandible depresses little, but rotates about its long axis as the anteater protracts and retracts its tongue (Naples, 1999). The shape of the glenoid fossa of *Protamandua rothi* and the mandibular condyle and the symphysis of the Colhuehuapian specimen indicate that the same kind of movements were present in the Miocene myrmecophagids.

The ant and termite feeding specialization of the Santacrucian myrmecophagids allowed them to take advantage of a food source not available to other Santacrucian mammals, thus avoiding competition from other mammals of similar size, except perhaps for the armadillos of the genus *Stegotherium* (Vizcaíno *et al.*, Chapter 12). However, *Stegotherium* was certainly ground-dwelling and fossorial, whereas *Protamandua* may have been ground-dwelling in part but certainly was also a capable arborealist.

Sloths The masticatory apparatus of the extant sloths, *Bradypus* and *Choloepus*, was studied in detail by Naples (1982). Their occlusion pattern shows some similarities to that observed in *Eucholoeops*, and megatherioids in general, which provides a framework for inferring masticatory specializations and diet in Santacrucian sloths. In *Bradypus*, leaves constitute 94–99% of diet, fruit about 1%, and flowers also 1% (Chiarello, 2008). There are no published dietary studies on *Choloepus* in its natural habitat, but an unpublished report by Alvarez *et al.* (2004) that analyzed fecal samples indicates that more than 90% of the diet is constituted by leaves. Naples (1982) indicated that during the power stroke the jaw movement is mainly orthal, with smaller anterior and medial components, and involves contact between the distally facing hard dentine edges of the upper molariforms and the mesially facing durodentine of the lower molariforms. Bargo *et al.* (2009) proposed that the tooth contacts described in Santacrucian megatherioid sloths are homologous to those of the living sloths, and the primary method of food reduction must have been by simple shearing or cutting; some grinding in the basins (between the shearing lophs) could have been accomplished through a slight mortar and pestle effect. Therefore, *Eucholoeops*, and probably nearly all other Miocene megatherioids, were mostly leaf eaters. The anteriorly edentulous muzzle of these sloths suggests a well-developed upper lip to complement oral food acquisition, as was inferred also for some Pleistocene ground sloths (Bargo *et al.*, 2006).

Muizon *et al.* (2004) studied occlusion in five species of other megatherioid sloths, the nothrotheriids of the genus *Thalassocnus* from the marine Pisco Formation (Late Miocene and Pliocene) in Peru. *Thalassocnus* species, which lack caniniforms and possess four upper and three lower molariforms, are considered to have had aquatic or semi-aquatic habits. Muizon *et al.* (2004) analysed the wear facets and identified the two sets of tooth contacts described for *Eucholoeops*. Although these authors had a different interpretation of their significance (see Bargo *et al.*, 2009, for further discussion), they concluded that the main masticatory movement of these nothrotheriids was predominatly orthal. Muizon *et al.* (2004) also reconstructed the areas of origin and insertion of the masticatory muscles, *m. temporalis* and *m. massetericus,* of *Thalassocnus*. They concluded that the large area of origin of the *m. temporalis* indicates that the *m. temporalis* was the most powerful muscle used during the power stroke and responsible for the posterodorsal movement inferred from tooth morphology. In Santacrucian megatherioid sloths, the great development of the temporal fossa clearly indicates the importance of the *m. temporalis* in adduction and retraction during orthal movements (Bargo *et al.*, 2009).

Bargo *et al.* (2009) also noted that *Eucholoeops* (and sloths generally) exhibits a complete symphyseal fusion. This condition has been implicated as especially important for the recruitment of contralateral muscles that assist in transverse, as well as orthal, movement (Hylander *et al.*, 1998). Interestingly, however, the transverse component is not emphasized in sloths, which may explain the peculiar modifications of the zygomatic arch to allow more advantageous positioning of the fibers of the *m. massetericus* when compared with the emphasis on posterior fibers of the *m. temporalis* in ungulates that have substantial transverse movement.

Finally, the combination of dental morphology and occlusal pattern, and other anatomical features of the skull of Santacrucian megatherioid sloths, foreshadows the predominance of orthal masticatory movements for other megatherioids, as proposed at least for the giant Pleistocene megatheriid *Megatherium americanum* Cuvier, 1796 (Bargo, 2001).

The overall morphology of the skull of Santacrucian mylodontids *Nematherium* and *Analcitherium* resembles that of the Pleistocene scelidotheres. Instead of the mainly orthal component proposed for megatherioids, the analysis of the tooth wear facets suggest that the jaw movements during the power stroke were mainly lateral (mesiolingual), as has been described for the Pleistocene mylodontid sloths *Paramylodon* Brown, 1903 (Naples, 1989) and *Scelidotherium*, *Glossotherium*, and *Mylodon* (Bargo and Vizcaíno, 2008). In *Nematherium*, and probably all mylodontids, food reduction must have been achieved by crushing and grinding (more three-dimensional, turgid, and fibrous food items like tubers or fruits), the concave occlusal surfaces acting as a mortar and pestle system. This suggests that Santacrucian mylodontids were better suited for processing different food items than sympatric megatherioids.

13.5.4 Paleoenvironment

A paleobiological interpretation based on locomotor and feeding habits of the Santacrucian pilosans permits several inferences about the environments they may have inhabited. The semi-arboreal habits suggested for anteaters and probably sloths indicate the presence of forested areas. Also, the combination of specialized feeding habits and possible low metabolism suggests preference for tropical and warm temperate environments. Ant/termite eating in mammals, including vermilinguans, is limited to tropical and warm temperate environments and these mammals have among the lowest basal metabolic rates (McNab, 1985, 2008). Thus, the specialized locomotory and feeding habits of Santacrucian vermilinguans, in combination with inferred low metabolism, are indicative of forests developed in tropical and warm temperate environments. The inferred feeding habits of sloths are also consistent with forests, although they do not preclude the existence of open environments: most megatherioids were leaf eaters and mylodontids could have fed on tubers or fruits, although the latter may also have grazed to some extent.

13.6 Conclusions

Santacrucian Vermilingua are represented by one unquestionable species, *Protamandua rothi*. Vermilinguans were small forms, of the size of the living *Tamandua*, well adapted for climbing, i.e. probably semi-arboreal, and for scratch-digging the substrate in feeding on ants or termites.

Santacrucian sloths include a great variety of forms, represented by 11 genera (the megatherioids *Eucholoeops*, *Megalonychotherium*, *Hapalops*, *Hyperleptus*, *Analcimorphus*, *Schismotherium*, *Pelecyodon*, *Planops*, and *Prepotherium*, and the mylodontids *Nematherium* and *Analcitherium*). They were moderately large-sized forms with locomotor habits different from those of living sloths (Fig. 13.10), and more similar to those of living vermilinguans and pangolins. They were capable diggers, but a mixture of terrestrial and scansorial activities was likely. Megatherioid sloths were most likely folivorous, and the predominant method of food reduction was orthal shearing or cutting. Mylodontids exhibit evidence for more transverse masticatory movements; food

Fig. 13.10. Reconstructions of the external appearance of the head. a, *Eucholoeops*; b, *Hapalops*. c, Life reconstruction of *Hapalops*. Drawings by Néstor Toledo.

reduction must have involved more crushing and grinding, which suggests they fed probably on fibrous three-dimensional foods such as fruits and tubers.

The specialized locomotory and feeding habits of Santacrucian vermilinguans are indicative of forests in tropical and warm temperate environments. The locomotory habits of sloths would also be indicative of woodlands, which is consistent with the feeding habits proposed for some of them (leaf eaters), although the existence of open environments cannot be ruled out based on the feeding habits of mylodontids. The paleobiological interpretations on locomotory and feeding habits of the Santacrucian pilosans is overall more consistent with the presence of a forested environment.

ACKNOWLEDGMENTS

We thank A. Kramarz (MACN), W. Joyce and D. Brinkman (YPM), J. Flynn, N. B. Simmons, and C. Norris (AMNH), and W. F. Simpson and K. D. Angielczyk (FMNH), who kindly provided access to the collections in their care. We also express our gratitude to Dirección de Patrimonio Cultural, and Museo Regional Provincial "Padre M. J. Molina" (Río Gallegos, Santa Cruz Province), and the Battini family for its hospitality during the fieldwork. We thank the reviewers G. DeIuliis and F. Pujos for their valuable suggestions and comments. This is a contribution to the projects PICT 26219 and 0143, UNLP N647, PIP-CONICET 1054, and National Geographic Society 8131–06 to SFV; NSF 0851272, and 0824546 to Richard F. Kay.

Appendix 13.1 Institutional abbreviations and list of the material studied

Acronyms of the institutions

AMNH: American Museum of Natural History, New York, USA.

FMNH: Field Museum of Natural History, Chicago, USA.

MACN-A: Museo Argentino de Ciencias Naturales "Bernardino Rivadavia," Colección Nacional Ameghino, Buenos Aires, Argentina.

MLP: Museo de La Plata, La Plata, Argentina.

MPM-PV: Museo Regional Provincial Padre M. Jesús Molina, Río Gallegos, Argentina.

YPM-VPPU: Yale Peabody Museum, Vertebrate Paleontology, Princeton University Collection, New Haven, USA.

FL: Fossiliferous Level.

Fossil specimens

All specimens come from Santa Cruz Province, Argentina (see Vizcaíno *et al.*, Chapter 1, Figs. 1.1 and 1.2 and Appendix 1.1 for localities, and Matheos and Raigemborn, Chapter 4 for stratigraphy, Figs. 4.3 and 4.4).

Vermilingua

Myrmecophagidae indet.

MLP 69-IX-8–8a. Left humerus. Horizon and locality: Santa Cruz Formation; unknown locality.

Protamandua rothi

FMNH 13134. Astragalus and fragments of ribs and vertebrae. Horizon and locality: Santa Cruz Formation, 10 miles south of Coy Inlet. MACN–A 11530. Right astragalus (Type of *Promyrmephagus euryarthrus*). Horizon and locality: Santa Cruz Formation; unknown locality. MACN–A 10901b- Right calcaneum. Horizon and locality: Santa Cruz Formation; unknown locality. YPM-VPPU 15267. Partial skull. Horizon and locality: lower levels Santa Cruz Formation; 10 miles south of Coy Inlet.

Folivora

Eucholoeops sp.

MPM-PV 3402. Incomplete postcranial skeleton. Horizon and locality: Santa Cruz Formation, Estancia La Costa Member, FL 6; Puesto Estancia La Costa. MPM-PV 3651. Skull and incomplete postcranial skeleton. Horizon and

locality: Santa Cruz Formation, Estancia La Costa Member; Campo Barranca.

Eucholoeops ingens

MACN-A 11614. Left half of skull and left dentary. Horizon and locality: Santa Cruz Formation; unknown locality. MPM-PV 3401. Skull, mandibles, and partial postcranial skeleton. Horizon and locality: Santa Cruz Formation, Estancia La Costa Member, FL 7.2; Puesto Estancia La Costa. MPM-PV 3451. Skull and incomplete postcranial elements. Horizon and locality: Santa Cruz Formation, Estancia La Costa Member, FL 5.3; Puesto Estancia La Costa. MPM-PV 3452. Skull. Horizon and locality: Santa Cruz Formation, Estancia La Costa Member. Campo Barranca.

Eucholoeops fronto

YPM-VPPU 15314. Mandible. Horizon and locality: Santa Cruz Formation; 10 miles south of Coy Inlet.

Eucholoeops cf. *E. fronto*

MPM-PV 3403. Skull, mandible, and postcranial skeleton. Horizon and locality: Santa Cruz Formation, Estancia La Costa Member, FL 6; Puesto Estancia La Costa.

cf. *Hapalops*

MPM-PV 3404. Incomplete skull and postcranial skeleton. Horizon and locality: Santa Cruz Formation, Estancia La Costa Member, FL 6; Puesto Estancia La Costa. MPM-PV 3462. Incomplete skull and postcranial skeleton. Horizon and locality: Santa Cruz Formation, Estancia La Costa Member, FL 5.3; Puesto Estancia La Costa.

Hapalops sp.

AMNH 9252. Incomplete postcranial skeleton. Horizon and locality: Santa Cruz Formation; Estancia Felton. MPM-PV 3400. Skeleton. Horizon and locality: Santa Cruz Formation, Estancia La Costa Member, FL 3; Estancia La Costa. MPM-PV 3412. Skull and incomplete postcranial skeleton. Horizon and locality: Santa Cruz Formation, Estancia La Costa Member, FL 5.3; Puesto Estancia La Costa. MPM-PV 3467. Skull and postcranial skeleton. Horizon and locality: Santa Cruz Formation, Estancia La Costa Member, FL 5.3; Puesto Estancia La Costa. YPM-VPPU 15005. Incomplete postcranial skeleton. Horizon and locality: Santa Cruz Formation; Coy Inlet. YPM-VPPU 15183. Scapula. Horizon and locality: Santa Cruz Formation; Coy Inlet. YPM-VPPU 15264. Mandible and incomplete postcranial skeleton. Horizon and locality: Santa Cruz Formation; Coy Inlet. YPM-VPPU 15535. Incomplete postcranial skeleton. Horizon and locality: Santa Cruz Formation; Arroyo Aike. YPM-VPPU 15618. Incomplete postcranial skeleton. Horizon and locality: Santa Cruz Formation; Killik Aike. YPM-VPPU 15836. Incomplete postcranial skeleton. Horizon and locality: Santa Cruz Formation; Monte Observación.

Hapalops longiceps

YPM-VPPU 15523. Holotype. Skull, mandible, and postcranial skeleton. Horizon and locality: Santa Cruz Formation; 8 miles south of Coy Inlet.

Hapalops elongatus

YPM-VPPU 15597. Skull and mandible. Horizon and locality: Santa Cruz Formation; Killik Aike. YPM-VPPU 15155. Incomplete postcranial skeleton. Horizon and locality: lower levels Santa Cruz Formation; 10 miles south of Coy Inlet. YPM-VPPU 15160. Skull and incomplete postcranial skeleton. Horizon and locality: lower levels Santa Cruz Formation; 10 miles south of Coy Inlet.

Hapalops indifferens

YPM-VPPU 15110. Skull, mandible and incomplete postcranial skeleton. Horizon and locality: Santa Cruz Formation; 10 miles south of Coy Inlet.

Hapalops angustipalatus

YPM-VPPU 15562. Holotype. Skull and incomplete postcranial skeleton. Horizon and locality: Santa Cruz Formation; 10 miles south of Coy Inlet.

Hapalops platycephalus

YPM-VPPU 15536. Postcranial skeleton. Horizon and locality: Santa Cruz Formation; Lago Pueyrredón. YPM-VPPU 15564. Skull and incomplete postcranial skeleton. Horizon and locality: Santa Cruz Formation; Lago Pueyrredón.

Hapalops ponderosus

YPM-VPPU 15520. Holotype. Skull and postcranial skeleton. Horizon and locality: Santa Cruz Formation; 10 miles south of Coy Inlet.

Hapalops rectangularis

AMNH 9222. Holotype. Mandible and postcranial skeleton. Horizon and locality: Santa Cruz Formation; Río Gallegos.

Pelecyodon cristatus

MPM-PV 3409. Skull. Horizon and locality: Santa Cruz Formation, Estancia La Costa Member, FL 3; Estancia La Costa. YPM-VPPU 15049. Skull. Horizon and locality: Santa Cruz Formation; 10 miles south of Coy Inlet.

Hyperleptus sp.

MPM-PV 4251. Articulated skull and mandible. Horizon and locality: Santa Cruz Formation, Estancia La Costa Member, FL 7; Puesto Estancia La Costa.

cf. Hyperleptus

MPM-PV 3410. Skull and some postcranial skeleton. Horizon and locality: Santa Cruz Formation, Estancia La Costa Member, FL 5.3; Puesto Estancia La Costa.

Mylodontidae indet.

MPM-PV 3406. Incomplete postcranial skeleton. Horizon and locality: Santa Cruz Formation, Estancia La Costa Member, FL 3; Estancia La Costa.

Nematherium sp.

MPM-PV 3407. Skull. Horizon and locality: Santa Cruz Formation, Estancia La Costa Member, FL 3; Estancia La Costa. MPM-PV 3703. Left mandible. Horizon and locality: Santa Cruz Formation, Estancia La Costa Member, FL 5.3; Puesto Estancia La Costa. YPM-VPPU 18009. Skull, mandible, and incomplete postcranial skeleton. Horizon and locality: Santa Cruz Formation; Santa Cruz. YPM-VPPU 15374. Incomplete postcranial skeleton. Horizon and locality: Santa Cruz Formation; Güer Aike.

Prepotherium potens

YPM-VPPU 15345. Incomplete postcranial skeleton. Horizon and locality: Santa Cruz Formation; Killik Aike.

Prepotherium filholi

AMNH 9573. Type. Mandible. Horizon and locality: Santa Cruz Formation; Estancia Felton.

Planops magnus

YPM-VPPU 15346. Type. Skull. Horizon and locality: Santa Cruz Formation; Güer Aike Department.

Megalonychidae indet.

AMNH 9249. Incomplete postcranial skeleton. Horizon and locality: Santa Cruz Formation; Estancia Felton, north of Río Gallegos, Killik Aike, Güer Aike. AMNH 94754. Incomplete postcranial skeleton. Horizon and locality: Santa Cruz Formation; Río Gallegos.

Living specimens

Xenarthra, Folivora

Bradypodidae. *Bradypus variegatus*: AMNH 209940; 261304; 42838; 135474; *B. tridactylus*: AMNH 42454; 74136; 74137; 97315; 133437; *B. boliviensis*: AMNH 211663.

Megalonychidae. *Choloepus* sp.: AMNH 35483; *C. hoffmanni*: AMNH 16873; 70440; 90269; 139772; 139773; 209941; *C. didactylus*: AMNH 265952.

Xenarthra, Vermilingua

Myrmecophagidae. *Myrmecophaga tridactyla*: AMNH 1020; 100068; 100139.

Tamandua mexicana: AMNH 23432; 23436; 23437; 23565; 23567; *T. tetradactyla*: AMNH 96258; 211659; 21660.

Cyclopidae. *Cyclopes didactylus*: AMNH 4780; 167845; 171297; 204662; 213188.

Xenarthra, Cingulata

Dasypodidae. *Priodontes giganteus*: AMNH 130387; *P. maximus*: AMNH 208104.

Tubulidentata

Orycteropodidae. *Orycteropus afer*: AMNH 51370; 51374; 51905; 51909; 65540.

Rodentia

Erethizontidae. *Coendou prehensilis*: AMNH 80045; 100097; 100119; 134073; 212611.

Hystricidae. *Hystrix cristata*: AMNH 51735; 87220; 87222; 119506.

Sciuridae. *Marmota monax*: AMNH 70338; 97386; 179934; 180314; 235648.

Primates

Cercopithecidae. *Papio ursinus*: AMNH 80771; 80774; 216247; 216251; *P. hamadryas*: AMNH 120388.

Hominidae. *Gorilla gorilla*: AMNH 54089; 54090; 54091; 54092.

Hylobatidae. *Hylobates syndactylus*: AMNH 90268; 102463; 106581; 106584.

Lemuridae. *Lemur catta*: AMNH 22912; 150039; 170739; 170740; *Lemur* sp. AMNH 35396.

Pholidota

Manidae. *Manis gigantea*: AMNH 53847; 53851; 53857; 53858. *Manis pentadactyla*: AMNH 60004; 60006; 60007; 172147; 184959.

Diprotodontia

Phascolarctidae. *Phascolarctos cinereus*: AMNH 65607; 65608; 65609; 65610; 107805.

Vombatidae. *Vombatus ursinus*: AMNH 42997; 65619; 65622; 70209; 146850.

Dasyuromorphia

Dasyuridae. *Sarcophilus laniarius*: AMNH 65670; 65672; 65673; 70406; 150211.

Thylacinidae. *Thylacinus cynocephalus*: AMNH 35244; 35504; 35866; 42259.

Carnivora

Ailuridae. *Ailurus fulgens*: AMNH 35433; 119474; 146682; 146778; 185346.

Felidae. *Acinonyx jubatus*: AMNH 36426; 119654; 119655; 119656; 119657.

Mustelidae. *Gulo gulo*: AMNH 35054; 35081; 149692; 165766.

Ursidae. *Ailuropoda melanoleuca*: AMNH 87242; 89028; 110453; 110454; 147746.

Viverridae. *Arctictis binturong*: AMNH 181; 22906; 35469; 80163; 119600.

Appendix 13.2

Abbreviations of measurements used for the analyses and shown in Figure 13.7 and Table 13.2.

	Measurement definition	Abbrev.	Source
Scapula	Length from supraglenoid apophysis to ventral border of glenoid fossa (lateral aspect)	SGL	Sargis, 2002a
	Maximum width of glenoid fossa in ventral view	SGW	Sargis, 2002a
	Length from supraglenoid apophysis to posterior border of scapula (at level of spine)	SL	Toledo et al., 2012
	Length from supraglenoid apophysis to lateralmost border of the spine or acromion (in ventral view)	SSE	N. Toledo, unpublished data
Humerus	Anteroposterior diaphyseal width at midshaft	APDH	Sargis, 2002a; Elissamburu and Vizcaíno, 2004; Candela and Picasso, 2008
	Width from medialmost border of trochlea to lateralmost border of capitulum	HDASW	Sargis, 2002a; Candela and Picasso, 2008
	Width from medialmost border of the epitrochlea to lateralmost border of epicondyle	HDEW	Sargis, 2002a; Elissamburu and Vizcaíno, 2004; Candela and Picasso, 2008
	Medial protrusion of the epitrochlea	HEMP	Toledo et al., 2012
	Width from bicipital groove to greater tuberosity	HGT	N. Toledo, unpublished data
	Length from anteriormost edge of head to the posteriormost border	HHL	Candela and Picasso, 2008
	Head width between posteriormost edges of both tuberosities	HHW	Candela and Picasso, 2008
	Length from proximalmost border of head to distalmost border of the trochlea	HL	Elissamburu and Vizcaíno, 2004; Candela and Picasso, 2008
	Width from bicipital groove to lesser tuberosity	HLT	N. Toledo, unpublished data
	Width from anteriormost border of the trochlea to posteriormost border of the epitrochlea	HTD	Candela and Picasso, 2008
	Anteroposterior width of trochlear valley	HTVD	N. Toledo, unpublished data
	Height of distal articular surface between trochlea and capitulum	HTVL	N. Toledo, unpublished data
	Maximum width between both tuberosities	ItuW	Toledo et al., 2012
	Transversal diaphyseal width at midshaft	TDH	Sargis, 2002a; Elissamburu and Vizcaíno, 2004; Candela and Picasso, 2008
Ulna	Length from semilunar notch midpoint to the posteriormost end of olecranon	OTL	Elissamburu and Vizcaíno, 2004; Candela and Picasso, 2008
	Diaphyseal width at midshaft	TDU	Elissamburu and Vizcaíno, 2004; Candela and Picasso, 2008
	Diaphyseal height at midshaft	THU	Toledo et al., 2012
	Diaphyseal width at semilunar notch midpoint	THUn	N. Toledo, unpublished data

(*cont.*)

	Measurement definition	Abbrev.	Source
	Total length	UL	Sargis, 2002a; Elissamburu and Vizcaíno, 2004; Candela and Picasso, 2008
	Anteroposterior length of semilunar notch	UTNL	Sargis, 2002a; Candela and Picasso, 2008
	Semilunar notch width at anteriormost border	UTNW1	N. Toledo, unpublished data
	Semilunar notch width at posteriormost border	UTNW2	Sargis, 2002a; Candela and Picasso, 2008
Radius	Minimum diameter of carpal fossa	RDEW	Sargis, 2002a
	Width of carpal fossa	RDFL	Sargis, 2002a; Candela and Picasso, 2008
	Diaphyseal width at midshaft	RH	Sargis, 2002a; Candela and Picasso, 2008
	Maximum diameter of capitular fossa	RHL	Toledo *et al.*, 2012
	Minimum diameter capitular fossa	RHW	Sargis, 2002a
	Total length	RL	Sargis, 2002a; Candela and Picasso, 2008

REFERENCES

Alvarez, S. J., Sanchez, A. and Carmona, M. M. (2004). Density, diet and habitat preference of the two-toed sloth *Choloepus hoffmanni in an andean forest of Colombia*. Unpublished Final Report of the The Ruffor Small Grants for Nature Conservation.

Ameghino, F. (1893). Nouvelles découvertes paléontologiques dans la Patagonie australe. *Revue Scientifique*, **51**, 731pp.

Ameghino, F. (1894). Enumération synoptique des espèces de Mammifères fossiles des formations éocènes de Patagonie. *Boletín de la Academia Nacional de Ciencias en Córdoba*, **13**, 259–455.

Ameghino, F. (1904). Nuevas especies de Mamíferos cretáceos y terciarios de la República Argentina. *Anales de la Sociedad Científica Argentina*, **57**, 162–175; 327–341.

Argot, C. (2001). Functional-adaptive anatomy of the forelimb in the Didelphidae, and the paleobiology of the Paleocene marsupials *Mayulestes ferox* and *Pucadelphys andinus*. *Journal of Morphology*, **247**, 51–79.

Argot, C. (2002) Functional-adaptative anatomy of the hindlimb of the Didelphidae, and the paleobiology of the Paleocene marsupials *Mayulestes ferox* and *Pucadelphys andinus*. *Journal of Morphology*, **253**, 76–108.

Bargo, M. S. (2001). The ground sloth *Megatherium americanum*: skull shape, bite forces, and diet. In *Biomechanics and Paleobiology of Vertebrates*, ed. S. F. Vizcaíno, R. A. Fariña and C. Janis. *Acta Paleontologica Polonica*, **46**, 41–60.

Bargo, M. S. and Vizcaíno, S. F. (2008). Paleobiology of Pleistocene ground sloths (Xenarthra, Tardigrada): biomechanics, morphogeometry and ecomorphology applied to the masticatory apparatus. *Ameghiniana*, **45**, 175–196.

Bargo, M. S., Vizcaíno, S. F., Archuby, F. M. and Blanco, R. E. (2000). Limb bone proportions, strength and digging in some Lujanian (Late Pleistocene–Early Holocene) mylodontid ground sloths (Mammalia, Xenarthra). *Journal of Vertebrate Paleontology*, **20**, 601–610.

Bargo, M. S., Toledo, N. and Vizcaíno, S. F. (2006). Muzzle of South American ground sloths (Xenarthra, Tardigrada). *Journal of Morphology*, **267**, 248–263.

Bargo, M. S., Vizcaíno, S. F. and Kay, R. F. (2009). Predominance of orthal masticatory movements in the early Miocene *Eucholaeops* (Mammalia, Xenarthra, Tardigrada, Megalonychidae) and other megatherioid sloths. *Journal of Vertebrate Paleontology*, **29**, 870–880.

Candela, A. M. and Picasso, M. B. J. (2008). Functional anatomy of the limbs of Erethizontidae (Rodentia, Caviomorpha): indicators of locomotor behaviour in Miocene porcupines. *Journal of Morphology*, **269**, 552–593.

Carlini, A. A. and Scillato-Yané, G. J. (2004). The oldest Megalonychidae (Xenarthra: Tardigrada): phylogenetic relationships and an emended diagnosis of the family. *Neues Jahrbuch für Geologie und Paläontologie Abhandlungen*, **233**, 423–443.

Carlini, A. A., Scillato-Yané, G. J., Vizcaíno, S. F. and Dozo, M. T. (1992). Un singular Myrmecophagidae (Xenarthra, Vermilingua) de Edad Colhuehuapense (Oligoceno tardío, Mioceno temprano) de Patagonia, Argentina. *Ameghiniana*, **29**, 176.

Carlini, A. A., Scillato-Yané, G. J. and Vizcaíno, S. F. (1993). Un Myrmecophagidae (Vermilingua) del Mioceno de Cerro Boleadoras (Santa Cruz, Argentina). *Ameghinana*, **30**, 102.

Chiarello, A. G. (2008). Sloth ecology: an overview of field studies. In *The Biology of the Xenarthra*, ed. S. F. Vizcaíno and W. J. Loughry. Gainesville: University Press of Florida, pp. 269–280.

Cione, A. L., Tonni, E. P. and Soibelzon, L. H. (2003). The broken zig-zag: late Cenozoic large mammal and turtle extinction in South America. *Revista del Museo Argentino de Ciencias Naturales "Bernardino Rivadavia,"* **5**, 1–19.

Cione, A. L., Tonni, E. P. and Soibelzon, L. H. (2009). Did humans cause the late Pleistocene–early Holocene mammalian extinctions in South America in a context of shrinking open areas? In *American Megafaunal Extinctions at the end of the Pleistocene*, ed. G. Haynes. Vertebrate

Paleobiology and Paleoanthropology Series, Springer, pp. 125–144.

Costa, R. L. Jr. and Greaves, W. S. (1981). Experimentally produced tooth wear facets and the direction of jaw motion. *Journal of Paleontology*, **55**, 635–638.

Croft, D. A. and Anderson, L. C. (2008). Locomotion in the extinct notoungulate *Protypotherium*. *Paleontologia Electronica*, **11**, 1–20.

Crompton, A. W. and Hiiemae, K. M. (1969a). Functional occlusion in tribosphenic molars. *Nature*, **222**, 678–679.

Crompton, A. W. and Hiiemae, K. M. (1969b). How mammalian molar teeth work. *Discovery*, **5**, 23–34.

De Iuliis, G. (1994). Relationships of the Megatheriinae, Nothrotheriinae, and Planopsinae: some skeletal characteristics and their importance for phylogeny. *Journal of Vertebrate Paleontology*, **14**, 577–591.

De Iuliis, G. and Pujos, F. (2006). On the systematics of *Hapalops* (Xenarthra: Megatherioidea). *Journal of Vertebrate Paleontology* **26**(3), 55A.

De Iuliis, G., Pujos, F., Bargo, M. S., Toledo, N. and Vizcaíno, S. F. (2009). *Eucholoeops* (Xenarthra, Tardigrada) remains from the Santa Cruz Formation (early Miocene), Patagonia, Argentina. *10th International Mammalogical Congress*. Mendoza: SAREM (Sociedad Argentina para el Estudio de los Maníferos), pp. 342.

Delsuc, F. and Douzery, E. J. P. (2008). Recent advances and future prospects in xenarthran molecular phylogenetics. In *The Biology of the Xenarthra*, ed. S. F. Vizcaíno and W. J. Loughry. Gainesville: University Press of Florida, pp. 11–23.

Delsuc, F., Catzeflis, F. M., Stanhope, M. J. and Douzery, E. J. P. (2001). The evolution of armadillos, anteaters and sloths depicted by nuclear and mitochondrial phylogenies: implications for the status of the enigmatic fossil *Eurotamandua*. *Proceedings of the Royal Society of London B*, **268**, 1605–1615.

Elissamburu, A. and Vizcaíno, S. F. (2004). Limb proportions and adaptations in caviomorph rodents (Rodentia: Caviomorpha). *Journal of Zoology*, **262**, 145–159.

Engelmann, G. (1985). The phylogeny of the Xenarthra. In *The Evolution and Ecology of Armadillos, Sloths, and Vermilinguas*, ed. G. G. Montgomery. Washington, D.C.: Smithsonian Institution Press, pp. 51–64.

Engelmann, G. F. (1987). A new Deseadan sloth (Mammalia: Xenarthra) from Salla, Bolivia, and its implications for the primitive condition of the dentition in edentates. *Journal of Vertebrate Paleontology*, **7**, 217–223.

Fariña, R. A. and Vizcaíno, S. F. (2003). Slow moving or browsers? A note on nomenclature. In *Morphological Studies in Fossil and Extant Xenarthra (Mammalia)*, ed. R. A. Fariña, S. F. Vizcaíno and G. Storch. *Senckenbergiana Biologica*, **83**, 3–4.

Gaudin, T. J. (1995). The ear region of Edentates and the phylogeny of the Tardigrada (Mammalia, Xenarthra). *Journal of Vertebrate Paleontology*, **15**, 672–705.

Gaudin, T. J. (2004). Phylogenetic relationships among sloths (Mammalia, Xenarthra, Tardigrada): the craniodental

evidence. *Zoological Journal of the Linnean Society*, **140**, 255–305.

Gaudin, T. J. and Branham, D. G. (1998). The phylogeny of the Myrmecophagidae (Mammalia, Xenarthra, Vermilingua) and the relationship of *Eurotamandua* to the Vermilingua. *Journal of Mammalian Evolution*, **5**, 237–265.

Gaudin, T. and McDonald, H. G. (2008). Morphology-based investigations of the phylogenetic relationships among extant and fossil xenarthrans. In *The Biology of the Xenarthra*, ed. S. F. Vizcaíno and W. J. Loughry. Gainsville: University Press of Florida, pp. 24–36.

Hiiemae, K. M. (1978). Mammalian mastication: a review of the activity of the jaw muscles and movements they produce in chewing. In *Development, Function and Evolution of Teeth*, ed. P. M. Butler and K. A. Joysey. New York: Academic Press, pp. 359–398.

Hiiemae, K. M. and Crompton, A. W. (1985). Mastication, food transport, and swallowing. In *Functional Vertebrate Morphology*, ed. M. Hildebrand, D. M. Bramble, K. F. Liem and D. B. Wake. Cambridge and London: Harvard University Press, pp. 262–290.

Hiiemae, K. M. and Kay, R. F. (1972). Trends in the evolution of primate mastication. *Nature*, **240**, 486–487.

Hiiemae, K. M. and Kay, R. F. (1973). Evolutionary trends in the dynamics of primate mastication. *IVth International Congress of Primatology, Symposium Craniofacial Biology of Primates*, **3**, 28–64.

Hirschfeld, S. E. (1976). A new fossil anteater (Edentata, Mammalia) from Colombia, S.A., and evolution of the Vermilingua. *Journal of Paleontology*, **50**, 419–432.

Hoffstetter, R. (1982). Les édentés xénarthres, un groupe singulier de la faune néotropicale (origines, affinités, radiation adpatative, migrations et extinctions). In *Proceedings of the First International Meeting on "Paleontology, Essential of Historical Geology,"* ed. E. Montanaro Gallitelli. Modena: STEM Mucchi, pp. 385–443.

Hylander, W. L., Ravosa, M. J., Ross, C. F. and Johnson, K. R. (1998). Mandibular corpus strain in primates: Further evidence for a functional link between symphyseal fusion and jaw-adductor muscle force. *American Journal of Physical Anthropology*, **107**, 257–271.

Johnson, C. N. (2002). Determinants of loss of mammal species during the Late Quaternary 'megafauna' extinctions: life history and ecology, but not body size. *Proceedings Royal Society London B: Biological Sciences*, **269**, 2221–2227.

Kalthoff, D. (2011). Microstructure of dental hard tissues in fossil and recent Xenarthrans (Mammalia: Folivora and Cingulata). *Journal of Morphology*, **272**, 641–661.

Kay, R. F. and Hiiemae, K. M. (1974). Mastication in *Galago crassicaudatus*, a cinefluorographic and occlusal study. In *Prosimian Biology*, ed. R. D. Martin, G. A. Doyle and A. C. Walker. London: Duckworth, pp. 501–530.

Lydekker, R. (1894). Contributions to knowledge of the fossil vertebrates of Argentina. Part II: The extinct edentates of Argentina. *Anales del Museo de La Plata (Paleontología Argentina)*, **3**, 1–118.

Martin, P. S. and Steadman, D. W. (1999). Prehistoric extinctions on islands and continents. In *Extinctions in Near Time: Causes, Contexts and Consequences*, ed. R. D. E. MacPhee. New York: Kluwer/Plenum, pp. 17–56.

McDonald, H. G. and De Iuliis, G. (2008). Fossil history of sloths. In *The Biology of the Xenarthra*, ed. S. F. Vizcaíno and W. J. Loughry. Gainsville: University Press of Florida, pp. 39–55.

McDonald, H. G. and Perea, D. (2002). The large scelidothere *Catonyx tarijensis* (Xenarthra, Mylodontidae) from the Pleistocene of Uruguay. *Journal of Vertebrate Paleontology*, **22**, 677–683.

McDonald, H. G., Vizcaíno, S. F. and Bargo, M. S. (2008). Skeletal anatomy and the fossil history of the Vermilingua. In *The Biology of the Xenarthra*, ed. S. F. Vizcaíno and W. J. Loughry. Gainesville: University Press of Florida, pp. 64–78.

McKenna, M. C. and Bell, S. K. (1997). *Classification of Mammals Above the Species Level*. New York: Columbia University Press.

McKenna, M. C., Wyss, A. R. and Flynn, J. J. (2006). Paleogene Pseudoglyptodont Xenarthrans from Central Chile and Argentine Patagonia. *American Museum Novitates*, **3536**, 1–18.

McNab, B. K. (1985). Energetics, population biology, and distribution of xenarthrans, living and extinct. In *Evolution and Ecology of Armadillos, Sloths, and Vermilinguas*, ed. G. G. Montgomery. Washington, D.C.: Smithsonian Institution Press, pp. 219–232.

McNab, B. K. (2008). An analysis of the factors that influence the level and scaling of mammalian BMR. *Comparative Biochemistry and Physiology*, **151**, 5–28.

Mones, A. (1986). Palaeovertebrata Sudamericana. Catálogo sistemático de los vertebrados fósiles de América del Sur. Parte I. Lista preliminar y bibliografía. *Courier Forschungsinstitut Senckenberg*, **82**, 1–625.

Montgomery, G. G. (1985). Impact of vermilinguas (*Cyclopes, Tamandua*: Xenarthra = Edentata) on arboreal ant populations. In *The Evolution and Ecology of Armadillos, Sloths and Vermilinguas*, ed. G. G. Montgomery. Washington D.C.: Smithsonian Institution Press, pp. 351–363.

Muizon, C. de, McDonald, H. G., Salas, R. and Urbina, M. (2004). The evolution of feeding adaptations of the aquatic sloth *Thalassocnus*. *Journal of Vertebrate Paleontology*, **24**, 401–414.

Naples, V. L. (1982). Cranial osteology and function in the tree sloths, *Bradypus* and *Choloepus*. *American Museum Novitates*, **2739**, 1–41.

Naples, V. L. (1989). The feeding mechanism in the Pleistocene ground sloth, *Glossotherium*. *Contributions in Science, Los Angeles County Museum of Natural History*, **425**, 1–23.

Naples, V. L. (1999). Morphology, evolution and function of feeding in the giant anteater (*Myrmecophaga tridactyla*). *Journal of Zoology, London*, **249**, 19–41.

Nowak, R. M. (1991). *Walker's Mammals of the World* 5th Edn. Baltimore: Johns Hopkins University Press.

Patterson, B. and Pascual, R. (1972). The fossil mammal fauna of South America. In *Evolution, Mammals, and Southern Continents*, ed. A. Keast, F. C. Erk and B. Glass. Albany: State University of New York Press, pp. 246–309.

Patterson, B., Turnbull, W. D., Segall, W. and Gaudin, T. J. (1992). The ear region in Xenarthrans (= Edentata: Mammalia). Part II. Pilosa (sloths, anteaters), Palaeanodonts, and a miscellany. *Fieldiana Geology*, **1438**, 1–78.

Pujos, F., De Iuliis, G., Argot, C. and Werdelin, L. (2007). A peculiar climbing Megalonychidae from the Pleistocene of Peru, and its implications for sloth history. *Zoological Journal of the Linnean Society*, **149**, 179–235.

Rensberger, J. M. (1973). An occlusion model for mastication and dental wear in herbivorous mammals. *Journal of Paleontology*, **47**, 515–528.

Sargis, E. J. (2002a). Functional morphology of the forelimbs of tupaiids (Mammalia, Scandentia) and its phylogenetic implications. *Journal of Morphology*, **253**, 10–42.

Sargis, E. J. (2002b). Functional morphology of the hindlimbs of tupaiids (Mammalia, Scandentia) and its phylogenetic implications. *Journal of Morphology*, **254**, 149–185.

Scillato-Yané, G. J. (1986). Los Xenarthra fósiles de Argentina (Mammalia, Edentata). *IV Congreso Argentino de Paleontología y Bioestratigrafía*, **2**, 151–155.

Scott, W. B. (1903–04). Mammalia of the Santa Cruz beds. I. Edentata. In *Reports of the Princeton University Expeditions to Patagonia 1896–1899*. Vol. 5, *Paleontology II*, ed. W. B. Scott. Princeton: Princeton University Press, pp. 1–364.

Shockey, B. J. and Anaya, F. (2011). Grazing in a new late Oligocene mylodontid sloth and a mylodontid radiation as a component of the Eocene-Oligocene faunal turnover and the early spread of grasslands/savannas in South America. *Journal of Mammalian Evolution*, **18**, 101–115.

Tauber, A. A. (1994). *Estratigrafía y vertebrados fósiles de la Formación Santa Cruz (Mioceno inferior) en la costa atlántica entre las rías del Coyle y Río Gallegos, Provincia de Santa Cruz, República Argentina*. Unpublished thesis, Facultad de Ciencias Exactas, Físicas y Naturales, Universidad Nacional de Córdoba, República Argentina.

Taylor, B. K. (1978). The anatomy of the forelimb in the anteater (*Tamandua*) and its functional implications. *Journal of Morphology*, **157**, 347–368.

Taylor, B. K. (1985). Functional anatomy of the forelimb in vermilinguas (anteaters). In *The Evolution and Ecology of Armadillos, Sloths, and Vermilinguas*, ed. G. G. Montgomery. Washington, D.C.: Smithsonian Institution Press, pp. 163–171.

Toledo, N., Bargo, M. S., Cassini, G. H. and Vizcaíno, S. F. (2012). The forelimb of Early Miocene sloths (Mammalia, Xenarthra, Folivora): morphometrics and functional implications for substrate preferences. *Journal of Mammalian Evolution* Online 19 Jan 2012/DOI: 10.1007/s10914-012-9189-y.

Vizcaíno, S. F. and De Iuliis, G. (2003). Evidence for advanced carnivory in fossil armadillos. *Paleobiology*, **29**, 123–138.

Vizcaíno, S. F. and Loughry, W. J. (2008). Xenarthran biology: past, present and future. In *The Biology of the Xenarthra*, ed. S. F. Vizcaíno and W. J. Loughry. University Press of Florida, pp. 1–7.

Vizcaíno, S. F., Bargo, M. S., Tauber, A. A. and Kay, R. F. (2004). Myrmecophagidae (Mammalia, Xenarthra) de edad Santacrucense (Mioceno temprano-medio). *Ameghiniana*, **41**, 67R.

Vizcaíno, S. F., Cassini, G. H., Toledo, N. and Bargo, M. S. (2012). On the evolution of large size in mammalian herbivores of Cenozoic faunas of South America. In *Bones, Clones, and Biomes: History and Geography of Recent Neotropical Mammals*, ed. B. Patterson and L. Costa. Chicago University Press, pp. 76–101.

Webb, S. D. (1985). The interrelationships of tree sloths and ground sloths. In *The Evolution and Ecology of Armadillos,*

Sloths, and Vermilinguas, ed. G. G. Montgomery. Washington D.C.: Smithsonian Institution Press, pp. 105–112.

White, J. L. (1993). Indicators of locomotor habits in Xenarthrans: evidence for locomotor heterogeneity among fossil sloths. *Journal of Vertebrate Paleontology*, **13**, 230–242.

White, J. L. (1997). Locomotor adaptations in Miocene Xenarthrans. In *Vertebrate Paleontology in the Neotropics. The Miocene Fauna of La Venta, Colombia*, ed. R. F. Kay, R. H. Madden, R. L. Cifelli and J. J. Flynn. Washington D.C.: Smithsonian Institution Press, pp. 246–264.

White, J. L. and MacPhee, R. D. (2001). The sloths of the West Indies: a systematic and phylogenetic review. In *Biogeography of the West Indies: Patterns and Perspectives*, ed. C. A. Woods. New York: CRC Press, pp. 201–235.

14 Paleobiology of Santacrucian native ungulates (Meridiungulata: Astrapotheria, Litopterna and Notoungulata)

Guillermo H. Cassini, Esperanza Cerdeño, Amalia L. Villafañe, and Nahuel A. Muñoz

Abstract

A paleobiological study of Santacrucian native ungulates is presented in this chapter. Seven families are recorded: Hegetotheriidae, Interatheriidae, Toxodontidae, and Homalodotheriidae (Notoungulata); Proterotheriidae and Macraucheniidae (Litopterna); and Astrapotheriidae (Astrapotheria); however, a detailed systematic revision is still pending. A broad body size range is recorded. Typotheres (Hegetotheriidae + Interatheriidae) vary from 2 to 10 kg, proterotheriids vary from 20 to 100 kg, toxodonts and macraucheniids surpass 100 kg, and astrapotheres reach 1000 kg. The highest taxonomic richness corresponds to the range between 20 and 100 kg. Locomotor behavior is interpreted as mostly cursorial. Typotheres might have engaged in occasional digging, but this group also includes the most agile and fastest forms, more so than proterotheriids. Only two taxa in the sample (*Interatherium* and *Astrapotherium*) show evidence of swimming capabilities and potential aquatic habits. In contrast, feeding behavior presents less variation, which agrees with previous interpretations: notoungulates inhabited open habitats and fed mainly on grass, while litopterns and astrapotheres inhabited closed habitats and were mainly browsers. We infer that notoungulates had exceptional chemical digestion capabilities whereas litopterns may have relied on long periods of chewing to process their food.

Resumen

En este capítulo se aborda el estudio de la paleobiología de los ungulados nativos santacrucenses, representados por diferentes familias: Hegetotheriidae, Interatheriidae, Toxodontidae y Homalodotheriidae (Notoungulata), Proterotheriidae y Macraucheniidae (Litopterna) y Astrapotheriidae (Astrapotheria); aún falta una revisión sistemática detallada. Se registra un amplio rango de tamaño corporal. Los tipoterios varían entre 2 y 10 kg, los proterotéridos entre 20 y 100 kg, los toxodontes y los

macrauquénidos sobrepasan los 100 kg y sólo los Astrapotheria alcanzarían una masa corporal de 1000 kg. La mayor riqueza taxonómica corresponde al rango entre 20 y 100 kg. Los comportamientos locomotores inferidos son principalmente cursoriales; los tipoterios podrían ser cavadores ocasionales, pero también incluyen a las formas más ágiles y rápidas, tal vez más que los proterotéridos; solo dos taxones (*Interatherium* y *Astrapotherium*) tendrían capacidad de nadar y potenciales hábitos acuáticos. Por el contrario, los comportamientos alimentarios inferidos presentan menor variación, en concordancia con interpretaciones previas: los notoungulados como pastadores de ambientes abiertos y los litopternos y astrapoterios como ramoneadores de ambientes cerrados. Los Notoungulata y Astrapotheria tendrían una buena capacidad digestiva, mientras que los Litopterna requerirían mayor tiempo de procesamiento en la cavidad oral.

14.1 Introduction

There are five groups of endemic extinct South American ungulates: astrapotheres, pyrotheres, notoungulates, litopterns, and xenungulates (Patterson and Pascual, 1968; Simpson, 1980; Bond *et al.*, 1995). They all have been traditionally considered mostly herbivores. The phylogenetic relationships of these groups are unclear (Cifelli, 1985, 1993; Gelfo *et al.*, 2008; Billet, 2010). They were all once united in a single taxon, Meridiungulata, originally founded on the idea that pre-Interchange South American ungulates were monophyletic (McKenna, 1975). So far, Horovitz (2004) has published the only cladistic analysis that includes South American ungulates and representatives of most extant mammal orders and "condylarths" (Fig. 14.1a). This author postulates a polyphyletic origin for Holarctic condylarths, as well as South American (including Antarctic) ungulates. However, whether meridiungulates are monophyletic or not is still a subject of debate. Therefore, in this chapter, the use of the term "Meridiungulata" does not imply that we are accepting the monophyly of the group. Similarly, the term "ungulate" as used here does

Early Miocene Paleobiology in Patagonia: High-Latitude Paleocommunities of the Santa Cruz Formation, ed. Sergio F. Vizcaíno, Richard F. Kay and M. Susana Bargo. Published by Cambridge University Press. © Cambridge University Press 2012.

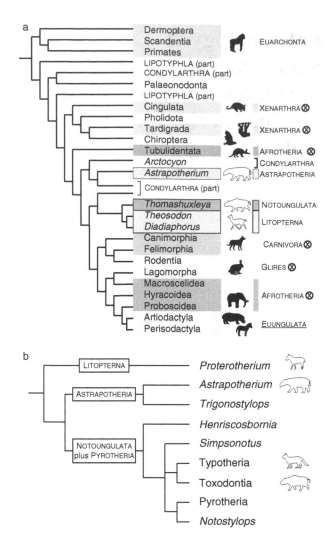

Fig. 14.1. Hypotheses of South American ungulate phylogenetic relationships. a, Modified from Horovitz (2004). Monophyletic supraordinal clade supported by morphological studies is underlined. Crossed symbols: unrecovered monophyletic supraordinal clades supported by molecular studies. b, Modified from Billet (2010).

not imply that endemic South American groups and modern ungulates share a most recent common ancestor exclusive of other mammals or even form a single clade among themselves. Modern ungulates include Perissodactyla and Cetartiodactyla, which are currently placed in the Euungulata clade (Asher and Helgen, 2010). Of the five aforementioned groups, only astrapotheres, notoungulates, and litopterns survived until Santacrucian times or later. For a summary of South American ungulates see Croft (1999).

Astrapotheres include the most bizarre mammals among native terrestrial fauna from the Tertiary of South America (Kramarz, 2009). They are recorded from the Paleocene (Soria and Powell, 1981) to the Middle Miocene (Johnson

and Madden, 1997). They attained their maximum taxonomic richness during the Early Miocene Colhuehuapian and Santacrucian ages (see below; Marshall and Cifelli, 1989; Johnson and Madden, 1997). According to Cifelli (1993), Astrapotheria are a monophyletic clade and following Billet (2010) they constitute the sister group of Notoungulata (Fig. 14.1b). In the coastal exposures of the Santa Cruz Formation they are represented by a single genus, *Astrapotherium* Burmeister, 1879 (Fig. 14.2a), which together with *Astrapothericulus* Ameghino, 1902 constitutes the most derived taxa (Astrapotheriinae) among astrapotheres (Kramarz, 2009). *Astrapotherium* species are large (i.e. above 44 kg, *sensu* Martin and Steadman, 1999) to very large mammals (including strict megamammals, i.e. 1000 kg or more *sensu* Owen-Smith, 1988). They are described as morphologically intermediate between a tapir and an elephant, having large canine tusks, brachydont cheek teeth, and a large diastema due to the loss of all upper incisors and some upper and lower premolars. Based on enamel structure, Rensberger and Pfretzschner (1992) concluded that astrapothere cheek teeth have similar functional and mechanical demands to the teeth of rhinoceroses. The nasals are retracted and premaxillae reduced, suggesting they had a tapir-like proboscis (Scott, 1913, 1928; Riggs, 1935; Croft, 1999; Kramarz, 2009). Since Scott (1913), astrapotheres have been considered inhabitants of riparian or meadow habitats who fed upon vegetation with a high water content (Riggs, 1935; Scott, 1937a). Marshall *et al.* (1990) considered *Astrapotherium* to be a good indicator for lowland continental environments.

Notoungulates constitute the most abundant and diverse clade of endemic South American ungulates, both taxonomically and morphologically (Simpson, 1936; Patterson and Pascual, 1968; Cifelli, 1993; Croft, 1999). Following Billet (2010), the clade comprises two main monophyletic groups, Toxodontia and Typotheria (Fig. 14.1b). Toxodonts include large to very large animals. Among them, the toxodontids (Fig. 14.2c, d) are the most abundant and are sometimes compared to hippos or rhinos because of their inferred general appearance as well as their molar crown patterns, which suggest a grinding masticatory action (Ameghino, 1907; Scott, 1912; Bond, 1999). Although homalodotheriid remains are not very common, most of the skeleton is known, and they are characterized by their long and clawed forelimbs (Fig. 14.2b), a convergent feature with Holarctic perissodactyl chalicotheres (Scott, 1930). On the other hand, typotheres are small- to medium-sized mammals, mostly described as rodent-like in overall form (Fig. 14.3a–d), although different families resemble extant capybaras (Mesotheriidae), hares (Hegetotheriidae), or hyraxes (Interatheriidae; but see discussion) (Ameghino, 1889; Sinclair, 1909; Bond *et al.*, 1995; Croft, 1999; Reguero *et al.*, 2007). In both toxodonts and typotheres there is an

a

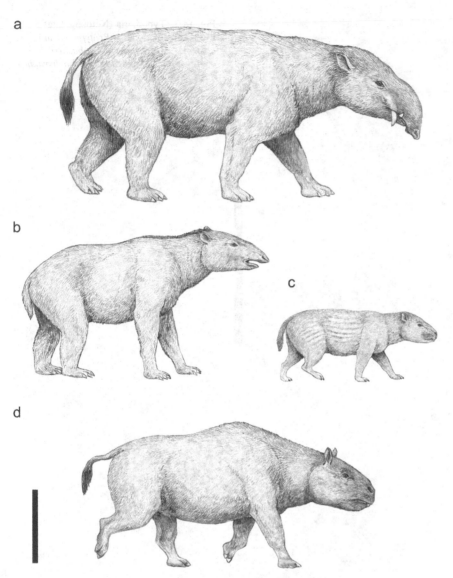

Fig. 14.2. Astrapotheriidae (Astrapotheria) and Toxodontia (Notoungulata) life reconstructions. a, *Astrapotherium magnum*; b, *Homalodotherium cunninghami*; c, *Adinotherium ovinum*; and d, *Nesodon imbricatus*. Scale bar = 0.5 m.

b

c

d

apparent tendency to evolve from a generalized masticatory apparatus with a complete dentition, no diastema, and brachydont cheek teeth, to a very specialized one with, for instance, hypertrophied incisors, simplified crown patterns, and high-crowned, ever-growing (euhypsodonty *sensu* Mones, 1982) cheek teeth (Ameghino, 1887, 1894; Sinclair, 1909; Simpson, 1967; Cifelli, 1985).

Following Scott (1937a) and based primarily on their high-crowned cheek teeth, notoungulates have been considered inhabitants of open plains, eating mostly grasses and/or other open habitat vegetation (Patterson and Pascual, 1968; Cifelli, 1985; Billet *et al.*, 2009). Bond (1986) and Tauber (1997b) also supported the grazing hypothesis based

on incisor morphology and the degree of hypsodonty. Madden (1997) proposed that the most complex nesodontine molar crowns would have provided great shearing ability, allowing them to break down grasses. However, Townsend and Croft (2008), using a microwear approach, concluded that *Nesodon imbricatus* Owen, 1846 was a leaf browser that focused more on hard browse, potentially including bark; *Adinotherium ovinum* (Owen, 1853) was a pure leaf browser; and *Protypotherium* Ameghino, 1887 was a browser that ate both soft browse and soft fruits. More recently, Cassini *et al.* (2011) applied machine learning techniques of knowledge discovery (i.e. decision trees and discriminant analyses; see Methods) to identify new

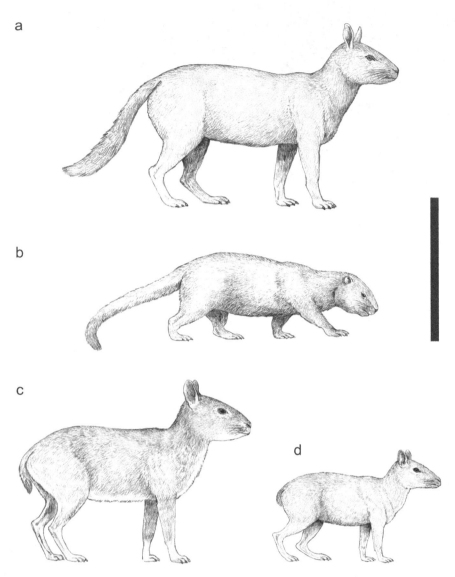

a

b

c

d

Fig. 14.3. Typotheria (Notoungulata) life reconstructions. a, *Protypotherium australe*; b, *Interatherium robustum*; c, *Hegetotherium mirabile*; d, *Pachyrukhos moyani*. Scale bar = 0.25 m.

morphological patterns and infer the habitat and feeding behavior of the notoungulates from the Santacrucian mammal assemblage. Their results, which will be further described and discussed below, suggest that all Santacrucian notoungulates present morphologies that also characterize extant ungulates living in open habitats. Furthermore, while the Toxodontia exhibit the same morphological patterns seen in extant mixed-feeders and grazers, the Typotheria possess exaggerated traits of ungulates that are specialized grazers.

Litopterns are the second most diverse and abundant clade of endemic South American ungulates (Pascual *et al.*, 1996). They are recorded throughout the Cenozoic, from the Early Paleocene (Bonaparte and Morales, 1997) to the Late Pleistocene (Bondesio, 1986; Bond, 1999). They reached their greatest generic richness during the Early Miocene (Santacrucian; Villafañe *et al.*, 2006) and gradually diminished throughout the Pliocene as forms became progressively more specialized, until they went extinct by the Late Pleistocene to Early Holocene (Bond *et al.*, 1995). Cifelli and Soria (1983) and Cifelli (1993) regarded them as a clade, and de Muizon and Cifelli (2000) postulated that they are a clade derived from Panameriungulata. The most abundant litoptern families are Macraucheniidae and Proterotheriidae. Macraucheniids include large to very large animals characterized by the progressive retraction of the nasals (suggesting a putative proboscis) and uniformly sharp-pointed incisors and canines. Additionally, the canal

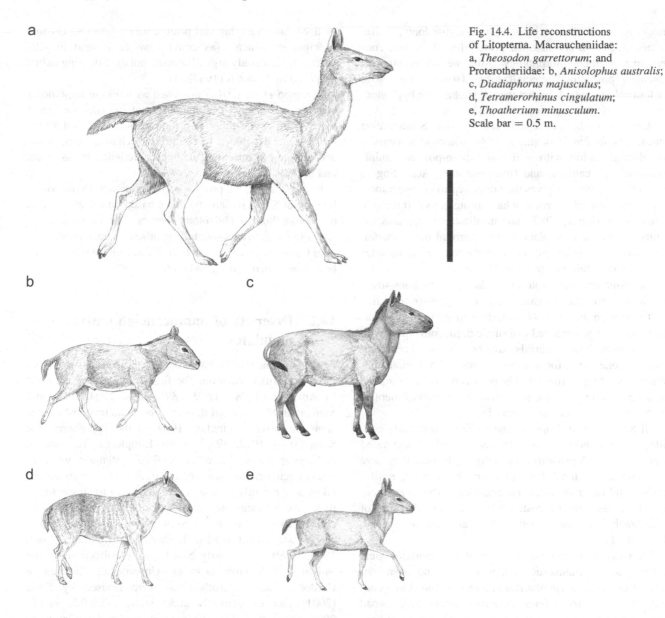

Fig. 14.4. Life reconstructions of Litopterna. Macraucheniidae: a, *Theosodon garrettorum*; and Proterotheriidae: b, *Anisolophus australis*; c, *Diadiaphorus majusculus*; d, *Tetramerorhinus cingulatum*; e, *Thoatherium minusculum*. Scale bar = 0.5 m.

for the vertebral artery passes through the neural arch of most cervical vertebrae in macraucheniids instead of being in the normal position, perforating the transverse processes (see details in Scott, 1910). This, and the possession of a long neck are features convergent with Tylopoda (Fig. 14.4a) (Scott, 1913; Bond, 1999). Proterotheriids are medium- to large-sized mammals mostly described as resembling primitive Holarctic horses owing to their body size and overall cranial shape, whereas digit reduction and mesaxonic limbs are features convergent with modern horses (Fig. 14.4b–e). However, their tooth crown morphology resembles that of artiodactyls like deer and camels (Bond *et al.*, 1995; Bond

et al., 2001). Most proterotheriids present a masticatory apparatus with an incomplete dentition, a diastema, and persistently brachydont buno-selenodont to selenodont cheek teeth of finite growth (brachyodonty, *sensu* Mones, 1982). A very few proterotheriids have simplified crown patterns and somewhat higher tooth crowns.

In macraucheniids there is an apparent tendency to evolve from more generalized forms, with craniad nares, a masticatory apparatus with a complete dentition with no diastema and brachydont cheek teeth, to very specialized forms with dorsally oriented and retracted nares – indicating the presence of a putative proboscis – and higher-crowned cheek teeth of finite growth (protohypsodonty, *sensu* Mones, 1982). In contrast, the proterotheriids remain conservative in their morphology. Miocene forms attained advanced postcranial specializations (some genera with anatomical monodactyly) but they did not evolve hypsodont teeth as in modern equids.

Macraucheniids are represented by one Santacrucian genus, *Theosodon* Ameghino, 1887. *Theosodon* bears a complete dentition with uniformly sharp-pointed caniniform incisors, canines, and first premolars. According to Scott (1913: 248), this gives the nose a reptilian appearance. The cheek teeth of *Theosodon* have finite growth (brachyodonty *sensu* Mones, 1982), and no diastema separates the canines from the premolars. The postorbital bar is incomplete, and the attachment areas of the temporalis muscles are larger than those of the masseter.

The Santacrucian Proterotheriidae are medium-sized, primitive horse-like animals. They show perissodactyl-like digit reduction, but only *Thoatherium* Ameghino, 1887 has the manus and pes reduced to a single digit, convergent with extant horses. Proterotheriids have brachydont cheek teeth and only one caniniform upper incisor, with a small diastema separating it from P1. The postorbital bar is complete as in horses and there are large areas for the attachment of the temporal and masseteric muscles.

All Santacrucian litopterns have been traditionally considered mainly herbivorous. Since Scott (1937a), and based primarily on limb proportions and digit reduction, they have been considered inhabitants of open plains eating mostly grasses and other open-habitat vegetation. However, more recent studies have suggested browsing habits because of their brachydont cheek teeth (Cifelli and Guerrero, 1997; Soria, 2001).

The paleobiological implications of this morphological diversity among Santacrucian ungulates have not been considered in an ecomorphological and morphofunctional context. Also, dietary inferences have mostly been broad generalizations based on the direct application of indices developed for extant analogs (see below). In addition to the earlier literature, some recent papers have permitted the identification of working hypotheses related to locomotion and substrate use. Santacrucian ungulates have been hypothesized to span a wide range of body sizes and functional types: small burrowing forms (e.g. some Hegetotheriidae; see Elissamburu, 2004), medium-sized runners (e.g. Proterotheriidae Scott, 1913; Soria, 2001), incipiently cursorial species, with arboreal/semi-fossorial characteristics (e.g. *Protypotherium*, see Croft and Anderson, 2008), and large, probably semi-aquatic forms (e.g. Nesodontinae and Astrapotheriinae; Scott, 1937a; Marshall *et al.*, 1990; Avilla and Vizcaíno, 2005). A recent biomechanical study of *Homalodotherium* Flower, 1873 (Elissamburu, 2010) describes this taxon as a browser on tree leaves with the

ability to assume a bipedal posture and a forelimb capable of rapid movements (as could have been used in self-defense). That study argued against putative digging habits proposed by Coombs (1983).

Vizcaíno *et al.* (2010) proposed as working hypotheses that the coexistence of such different lineages implies a partitioning of vegetation as a food resource and that different lineages developed masticatory mechanics comparable and convergent on extant ungulates, rodents, lagomorphs, and hyraxes.

In this chapter, we provide a first approach to the paleobiology of Santacrucian ungulates based on the application of ecomorphology and biomechanics. We examine some previous paleoenvironmental hypotheses and provide new ones based on the reconstruction of three biological aspects: body mass, locomotion, and diet.

14.2 Diversity of Santacrucian native ungulates

Few taxonomic studies have focused on the ungulates from the Santa Cruz Formation. The first descriptions were done by Ameghino (1887, 1889, 1891, 1894, 1904, 1907) and Mercerat (1893a, b), but the most complete taxonomic revisions are those by Sinclair (1909) of the Typotheria, and Scott (1910, 1912, 1928) of the Litopterna, Toxodontia, and Astrapotheria. No other detailed systematic work on native ungulates from the Santa Cruz Formation were undertaken until recently, although repeated visits to classic localities have supplemented the fossil material, and the older collections have been revisited (see Vizcaíno *et al.*, Chapter 1). Among recent works, Tauber's contributions (1994, 1996, 1999) were mostly based on the collections of the Museo de Paleontología of the Universidad Nacional de Córdoba. The Proterotheriidae were revised by Soria (2001), and are currently under study (Villafañe *et al.*, 2006; Schmidt, 2011). We present here a compilation of updated information on the recorded orders and families, whilst a full systematic revision of Santacrucian native ungulates remains to be completed.

The order Astrapotheria has been formerly divided into two families, Trigonostylopidae and Astrapotheriidae, although the former is likely a paraphyletic stem group relative to the latter. Patagonian astrapotheres are united in the subfamily Astrapotheriinae, while extra-Patagonian species form the subfamily Uruguaytheriinae, which is recorded in Uruguay, Colombia, Venezuela, Brazil, Peru, and Bolivia (Kramarz and Bond, 2009). Among Astrapotheriinae, the well-known *Astrapotherium magnum* (Owen, 1853) (Fig. 14.5a–e) is typical of the Santacrucian Formation and is well represented by both juvenile and adult specimens (Scott, 1928). Besides

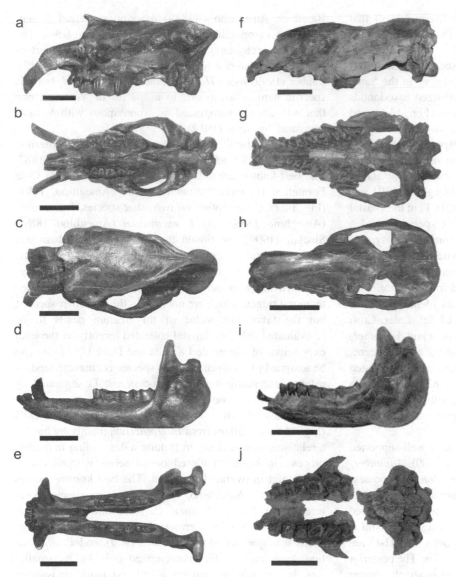

Fig. 14.5. *Astrapotherium magnum* (AMNH 9278) cranium: a, lateral; b, palatal; c, dorsal. Mandible: d, lateral and e, occlusal. *Nesodon imbricatus* cranium: f, lateral (MPM-PV 4377); g, palatal (YPM-VPPU 15256); h, dorsal (YPM-VPPU 15256); i, Mandible (MPM-PV 3659). j, *Homalodotherium cunninghami* (MPM-PV 3706). Scale bar = 100 mm.

A. magnum, Scott (1928) also recognized the Santacrucian species *A. giganteum* and *A. nanum*, which are differentiated from *A. magnum* by their greater and lesser body sizes, respectively.

Kramarz (2009) considered the genus *Astrapothericulus* to be characteristic of the Pinturas Formation. Its remains have been recovered mainly from the lower and middle sections of this formation, but some specimens also come from the upper beds that crop out at the locality of La Cañada (Santa Cruz Province), which is correlated with the lower levels of the Santa Cruz Formation (Kramarz, 2009). Nevertheless, no specimens of *Astrapothericulus* have been reported from the coastal outcrops of Santa Cruz Formation (Tauber, 1997a, 1999; Kay *et al.*, 2008).

Among native ungulates of the Santa Cruz Formation, Notoungulata and Litopterna are diverse and abundant. Notoungulates are represented by three genera of Toxodontia (two toxodontids and one homalodotheriid; Scott, 1912) and four genera of Typotheria (two interatheriids and two hegetotheriids; Sinclair, 1909). The family Notohippidae, with the genus *Notohippus*, has been also mentioned among the Santacrucian fauna (i.e. Scott, 1912; Marshall *et al.*, 1983). However, the latter taxon comes from the "Notohippidense" (Ameghino, 1900–1902), which corresponds to the lower part of the Santa Cruz Formation (Marshall and

Pascual, 1977, 1978), a different fossiliferous level that outcrops in the west of the Santa Cruz Province, not in the coastal beds herein studied.

Toxodontids have a high skull, hypsodont dentition, I2/i3 developed as tusks, and a robust skeleton. In the Santa Cruz Formation, two small- to medium-sized toxodontids are well known, *Adinotherium ovinum* (Fig. 14.6) and *Nesodon imbricatus* (Fig. 14.5f–i), each well described and illustrated by Scott (1912). Other species of the same genera are also mentioned by Scott, but most of them are poorly characterized and of questionable taxonomic status. Among them, *Adinotherium robustum* Ameghino, 1891 was recognized by Tauber (Tauber, 1999: table 1) in the coastal levels of the Santa Cruz Formation. The species *Nesodon cornutus* Scott, 1912 is known only from its holotype, a nearly complete skull from Lago Pueyrredón (Santa Cruz Province). Croft *et al.* (2003) recognized *N. conspurcatus* Ameghino 1887, which is differentiated from *N. imbricatus* based on its smaller size and also was identified in the Santacrucian Cura Mallín Formation (Chile). Later, Croft *et al.* (2004) recognized *Nesodon imbricatus* and *Adinotherium* sp. in the Chucal Formation (late Early Miocene, Chile). Marshall *et al.* (1983) and Madden (1990) listed *Hyperoxotodon speciosus* (Ameghino, 1887) for the Santa Cruz Formation, but this species is typical of the younger Collón Cura Formation (Colloncuran Age) and has not been encountered among our specimens.

Adinotherium and *Nesodon* constitute a well-supported clade within Toxodontidae (Nasif *et al.*, 2000). *Adinotherium* is smaller and more slender than *Nesodon*. Both possess upper teeth with a typical Y-shaped central valley/fossette and a posterior fossette. The I3, C, and P1 are more reduced in *Adinotherium*, there is a marked narrowing of the skull at this level (premaxillary–maxillary junction), and I1-I2 are more transversely positioned than in *Nesodon*. The posterior part of the skull and the sagittal crest are relatively shorter in *Nesodon*.

The family Homalodotheriidae is represented in the Santa Cruz Formation by *Homalodotherium* Flower, 1873 (Scott, 1912) (Fig. 14.5j). Ameghino (1891) described the genus and species *Diorotherium egregium*, but its taxonomic validity was questioned by Scott (1912), and its original locality is unknown. Originally, four species of this genus were recognized, mainly based on their different size: *H. cunninghami* Flower, 1873, *H. segoviae* Ameghino, 1891, *H. excursum* Ameghino, 1894, and *H. crassum* Ameghino, 1894. Scott (1912) validated *H. cunninghami* and *H. segoviae*, differentiating them by the narrower muzzle, smaller incisors, and reduced P1 of *H. segoviae*. A nice cranial fragment was recently recovered from Puesto Estancia La Costa locality (Fig. 14.5j) that preserves the posterior part of the cranium and the palate with right C and P2–M3 as well as left P2–M3.

Based on Ameghino's (1889) description, dental dimensions are comparable to those of *H. cunninghami*. The zygomatic arch starts at the level of the anterior part of M2, slightly anterior to that of *H. segoviae* (Scott, 1912). Tauber (1999) listed *Homalodotherium rutimeyeri*, but this specific name corresponds to a species of *Nesodon*, one that was already recognized as synonymous with *N. conspurcatus* by Scott (1912).

The interatheriids are represented by *Interatherium* Ameghino, 1887 and *Protypotherium* Ameghino, 1887. The best-known interatheriid species from the Santa Cruz Formation is *Interatherium robustum* (Ameghino, 1891) (Fig. 14.7). Commenting on two other species, *I. extensum* (Ameghino 1889) and *I. excavatum* (Ameghino 1889), Sinclair (1909) questioned the validity of the characters used to separate *I. extensum* from *I. robustum* such as the smaller I3 and the straight fronto-nasal suture. *Interatherium excavatum*, in turn, is distinguished by its lyre-shaped temporal ridges, which are not seen in the two other species, but the taxonomic value of this feature needs to be re-evaluated. All the material collected recently by the joint expeditions of Museo de La Plata and Duke University can be assigned to *I. robustum*. This species is characterized by a long descending maxillary process and I3 separated by small diastemata from I2 and C.

The interatheriid *Protypotherium* Ameghino, 1885 (Fig. 14.8a–g) differs from *Interatherium* mainly by having a relatively longer skull, in lacking a descending maxillary process, in having a closed dental series without a diastema, and in overall body form. The best known species is *P. australe* Ameghino, 1887. Two other species were referred to *Protypotherium*: *P. praerutilum* Ameghino, 1887, with relatively narrower molars, whose taxonomic status was questioned by Sinclair (1909), and *P. attenuatum* Ameghino, 1887, characterized only by its smaller size. In our opinion, pending a full taxonomic revision of the whole *Protypotherium* material from Santa Cruz, these differences are not enough to justify recognition of two species.

In addition to these interatheriid taxa, Tauber (1999) indicated the presence of *Cochilius* nov. sp. in Estancia La Costa locality without further comments or later published data, which prevents any discussion on this record. So far, the genus *Cochilius* is known only from Deseadan and Colhuehuapian levels (Marshall *et al.*, 1983; Reguero, 1999).

Hegetotheriids, by comparison with the interatheriids, are characterized by a wider skull, more laterally expanded zygomatic arches, longer postorbital apophyses, procumbent incisors with a very well developed I1, and cheek teeth with a convex lingual wall. The recognized taxa are *Hegetotherium mirabile* Ameghino, 1887 (Fig. 14.8h–k) and *Pachyrukhos moyani* Ameghino, 1885 (Fig. 14.8l–o). The former is larger, with a higher

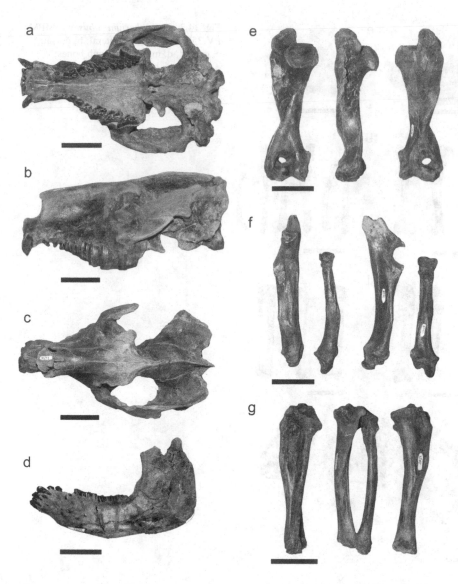

Fig. 14.6. *Adinotherium ovinum* cranium: a, palatal (MPM-PV 3668); b, lateral (MPM-PV 3667); c, dorsal (MPM-PV 3532); d, mandible (MPM-PV 3666). Limb bones (MPM-PV 3542): e, humerus; f, ulna and radio; g, tibia and fibula. Scale bar = 50 mm.

rostrum, and more robust zygomatic arch. *Pachyrukhos moyani* is the smallest notoungulate in the Santa Cruz Formation, with a more rodent-like skull, larger orbits, greater development of the mastoid bullae, and a long diastema between I1/i2 and P2/p2.

According to the faunal list in Marshall *et al.* (1983), the litopterns from the Santa Cruz Formation include representatives of the families Proterotheriidae (*Diadiaphorus* Ameghino, 1887; *Thoatherium* Ameghino, 1887; *Licaphrium* Ameghino, 1887; *Licaphrops* Ameghino, 1904; and *Proterotherium* Ameghino, 1883) and Macraucheniidae (*Theosodon*). *Adianthus* Ameghino, 1891 (Family Adianthidae) was also included among the Litopterna from Santa Cruz, but the original locality was not specified (Scott,

1910). Following Cifelli and Soria (1983), it probably comes from the Corriguen Aike locality.

Scott (1910) published a diagnosis of the genera *Proterotherium*, *Licaphrium*, *Thoatherium*, and *Diadiaphorus*, providing the first postcranial descriptions and illustrations. However, he did not revise the specimens directly, and perpetuated some previous errors. The Proterotheriidae were reviewed by Soria (2001) who tried to resolve the chaotic systematic situation of the whole family and recognized the following Santacrucian taxa: (1) *Tetramerorhinus lucarius* Ameghino, 1894, one of the better represented taxa, previously recognized mainly as several species of *Proterotherium* Ameghino, 1883; (2) *Tetramerorhinus cingulatum* (Ameghino, 1891), also

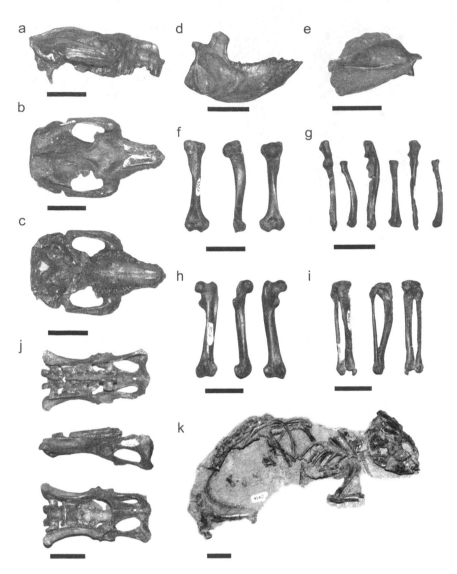

Fig. 14.7. *Interatherium robustum* (MPM-PV 3527) cranium: a, lateral; b, dorsal; c, palatal; d, mandible. Limb bones: e, scapula; f, humerus; g, radio and ulna; h, femur; i, tibia and fibula; j, pelvis. k, Articulated skeleton (MPM-PV 4263). Scale bar = 25 mm.

previously included in *Proterotherium*, as *P. cingulatum* Ameghino, 1891, and *P. principale* Ameghino, 1892. Soria (2001) differentiated *T. cingulatum* into two subspecies: *T. c. cingulatum* (Ameghino, 1891) and *T. c. fleaglei* Soria, 2001, but Kramarz and Bond (2005) ranked them at species level, restricting *T. fleaglei* to the Pinturas Formation (Early Miocene); (3) *Tetramerorhinus mixtum* (Ameghino, 1894) including as synonyms five *Proterotherium* species described by Ameghino (1894) and Scott (1910); (4) *Thoatherium minusculum* Ameghino, 1887, a small monodactyl species; (5) *Diadiaphorus majusculus* Ameghino, 1887, a large species; (6) *Anisolophus australis* (Burmeister, 1879); (7) *Anisolophus floweri* (Ameghino, 1887); and (8) *Anisolophus minusculus* (Roth,

1899), recognized also in the Collón Cura Formation (Río Negro Province).

Soria also recognized two other taxa, *Lambdaconus lacerum* (Ameghino, 1902) and *Tretramerorhinus prosistens* (Ameghino, 1899), from the Río Pinturas (Pinturas Formation).

Tauber (1999: table 1) identified five proterotheriid taxa from the lower member of the Santa Cruz Formation at the coastal beds: "*Proterotherium*" *cavum* now *Tetramerorhinus lucarius* after Soria (2001); *Licaphrium floweri* (= *Anisolophus floweri* after Soria, 2001); *Diadiaphorus robustus* (= *D. majusculus* after Soria, 2001); *Thoatherium minusculum*; and "*Proterotherium*" *intermedium* (= *Anisolophus australis* after Soria, 2001).

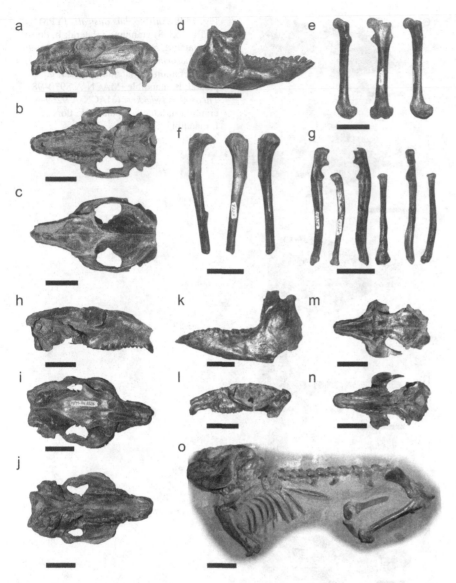

Fig. 14.8. *Protypotherium australe* (AMNH 9565) cranium: a, lateral; b, palatal; c, dorsal; d, mandible. Limb bones: e, femur; f, tibia; g, radio and ulna. *Hegetotherium mirabile* (MPM-PV 3526) cranium: h, lateral; i, dorsal; j, palatal; k, mandible (MPM-PV 4316). *Pachyrukhos moyani* (YPM-VPPU 15743) cranium: l, lateral; m, dorsal; n, palatal; o, articulated skeleton (AMNH 9283). Scale bar = 25 mm.

In addition, he described *Licaphrium* sp. and *Diadiaphorus* sp. The proterotheriid remains recovered during the past few years in the Santa Cruz Formation have been identified as *Anisolophus australis* (Fig. 14.9a–d), *Tetramerorhinus cingulatum* (Fig. 14.9e–h), *Thoatherium minusculum* (Fig. 14.10a–j), and *Diadiaphorus majusculus* (Fig. 14.10k–s).

Following Scott (1910), the only macraucheniid from Santa Cruz Formation is the genus *Theosodon*, being a common element of the fauna. Scott recognized the species *Theosodon lydekkeri* Ameghino, 1887 (Fig. 14.9i–l), *T. lallemanti* Mercerat, 1891, the most common species, *T. garrettorum* Scott, 1910, *T. fontanae* Ameghino, 1891,

T. gracilis Ameghino, 1891, *T. patagonicum* Ameghino, 1891, and *T. karaikensis* Ameghino, 1904. However, the validity of these taxa is still pending a full taxonomic revision. Tauber (1999) recorded *Theosodon lallemanti* at lower levels of the Santa Cruz Formation and Croft *et al.* (2004) reported the presence of *Theosodon* sp. in the Santacrucian Chucal Formation (Chile).

14.3 Materials and methods

Instituional abbreviations and the list of materials are provided in Appendix 14.1. The specimens studied include those collected by the expeditions carried out during the

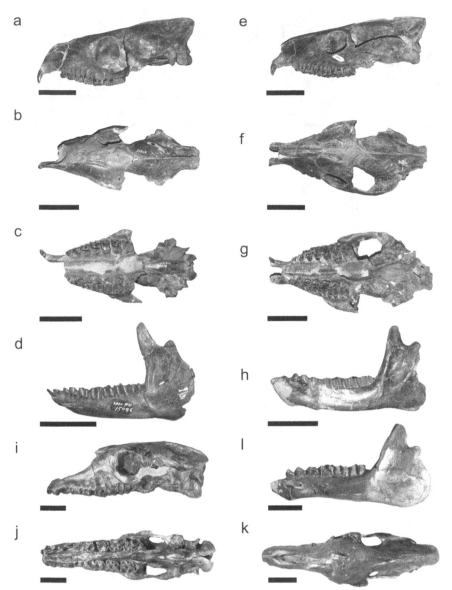

Fig. 14.9. *Anisolophus australis* (YPM-VPPU 15368) cranium: a, lateral; b, dorsal; c, palatal; d, mandible (YPM-VPPU 15996). *Tetramerorhinus cingulatum* (MPM-PV 3493) cranium: e, lateral; f, dorsal; g, palatal; h, mandible (MACN-A 8970–98). *Theosodon lydekkeri* (MACN-A 9269–88) cranium: i, lateral; j, palatal; k, dorsal; l, mandible. Scale bar = 50 mm.

late nineteenth and early twentieth centuries and the material collected by joint Museo de La Plata/Duke University expeditions during the period 2003–2011 (see Vizcaíno *et al.*, Chapter 1).

For the purposes of this book both new and old material was included in the analyses, enlarging the samples analyzed and consequently improving our confidence in the results obtained. In general, the genus was chosen as the working taxonomic level. As in other chapters (see Bargo *et al.*, Chapter 13; Vizcaíno *et al.*, Chapter 12), we prefer to use genera rather than species because, as is evident from the taxonomic overview above, they are more stable taxonomic units accepted by most paleontologists and, particularly among meridiungulates, a full taxonomic revision of species is still needed (but beyond the objectives of this chapter). Moreover, paleobiological approaches based on morphology often are not sensitive enough to discriminate among species of the same genus when they are only based on a few traits in a restricted part of the skeleton, and intraspecific variation in fossils is not properly evaluated. However, species of the same genus were considered separately where they attained different body masses.

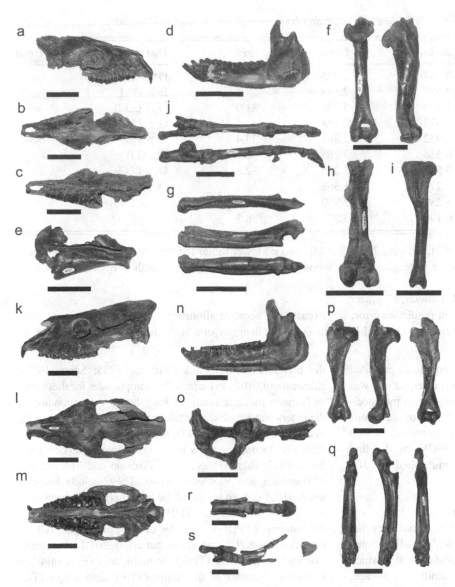

Fig. 14.10. *Thoatherium minusculum* (MPM-PV 3529) cranium: a, lateral; b, dorsal; c, palatal; d, mandible. Limb bones: e, scapula; f, humerus; g, radio and ulna; h, femur; i, tibia; j, feet. *Diadiaphorus majusculus* (MPM-PV 3397) cranium: k, lateral; l, dorsal; m, palatal; n, mandible; o, pelvis. Limb bones: p, humerus; q, radio and ulna; r, manus; s, feet. Scale bar = 50 mm.

14.3.1 Body mass

Body masses of Santacrucian ungulates were estimated using craniodental interspecific allometric equations of extant ungulates (Artiodactyla and Perissodactyla) proposed by Janis (1990) and Mendoza *et al.* (2006), which are based on the same dataset but use a different framework (bivariate and multivariate equations, respectively). Janis (1990) provided independent equations for families of Perissodactyla and Artiodactyla as well as a general equation for all ungulates. None of the orders considered in this chapter (Astrapotheria, Notoungulata, and Litopterna) include living representatives and their phylogenetic relationships with extant ungulates are not entirely clear (see Horovitz, 2004). Therefore, we chose equations based on traits available in the fossil record that are present in all extant ungulates rather than on artiodactyls or perissodactyls alone. We selected seven equations for "*all ungulates*" from Janis (1990) based on the value of the percentage of variance explained (R^2), the percent predictive error (%PE), and availability of the variable involved in the equation among the fossil specimens (Table 14.1).

With respect to the equations proposed by Mendoza *et al.* (2006) in a multivariate context, when using adjusted weights for the equations to minimize the taxonomic bias

Table 14.1. *Janis (1990)* "all ungulates" *selected equations to predict body mass from craniodental variables*

Variable	R^2	Intercept	Slope	%PE	Diet	Trend
TSL	0.950	−2.344	2.975	30.5	O* > I.B*	iso
OCH*	0.948	−0.457	2.873	28.1	B > G*.I	iso
SLML	0.944	1.130	3.201	31.9	O > G.I.B	−
PSL	0.942	−0.973	2.758	33.4	G > I*.B	iso
TJL	0.942	−1.952	2.884	33.4	O > I.B*	iso
LMRL*	0.941	−0.536	3.265	31.9	I > G.B	−
LMRL	0.940	−0.552	3.285	32.8	O > G.I.B	−
SUMA	0.939	1.277	1.568	32.7	O > I*	−
MFL	0.938	−1.289	2.950	35.0	G* > I> B	−
PAW*	0.917	−0.196	3.270	38.2		−

Abbreviations: TSL, total skull length; OCH, occipital height; SLML, second lower molar length; PSL, posterior skull length; TJL, total jaw length; LMRL, lower molar row length; SUMA, second upper molar area; MFL, length of masseteric fossa; PAW, palatal width.

O, omnivores; I, intermediate feeders; B, browser, G, grazer.

R^2 determination coefficient; %PE, percent prediction error; iso, isometry; −, negative allometry.

* Variable considered without suines in data set; O > G.I.B means O greater than any one of the following feeding types.

of the sample, the authors argue that the equations provided are free of taxonomic bias. The selection of equations was based on the availability of all measurements in the specimens (Table 14.2). The mean of each set of equations (single, multivariate) per specimen was calculated, as well as the percent coefficient of variation (%CV) by dividing the standard deviation by the mean, multiplied by 100 (Christiansen and Harris, 2005).

14.3.2 Locomotion and substrate use

In order to analyze functional capabilities of Santacrucian ungulate limbs, several indices were calculated. Functional indices are informative about the mechanical context in which limb elements function (Howell, 1944; Smith and Savage, 1955; Hildebrand, 1988; Samuels and Van Valkenburgh, 2008). Such a mechanical context (the "average biomechanical situation" of Oxnard, 1984) can suggest useful hypotheses about the spectrum of activities that the limb could perform (digging, climbing, and running, among others) without being limited to fixed locomotor categories. When needed, we refer to "cursorial" mammals in the sense of the term as defined by Stein and Casinos (1997): that is, "cursorial mammals are those terrestrial quadrupeds that possess vertically-oriented limbs which move in a mainly parasagittal plane, regardless of the gait being employed." When alternative definitions for "cursorial" from previous works are mentioned, the source and meaning are cited.

Forty-two specimens of notoungulates (six genera) and 10 litopterns (five genera) were analyzed (see Appendix 14.1). Additionally, for one specimen of *Astrapotherium* and three specimens of *Homalodotherium* (Notoungulata), values from the literature were included (Riggs, 1935; Scott, 1937a; Elissamburu, 2010) to increase the sample sizes for these taxa.

Thirteen measurements of long bones corresponding to diameters and lengths were taken with digital calipers to the nearest 0.01 mm (Fig. 14.11), following previous works that deal with locomotor traits based on proportions (Lessa and Stein, 1992; Biknevicius, 1993; Vizcaíno and Milne, 2002; Elissamburu and Vizcaíno, 2004). They include humerus length (HL), deltoid length of the humerus (DLH), transverse diameter of the humerus (TDH), anteroposterior diameter of the humerus (APDH), diameter of the epicondyles (DEH), total ulna length (UL), ulnar olecranon length (OL), transverse diameter of the ulna (TDU), femur length (FL), midshaft transverse diameter of the femur (TDF), tibia length (TL), proximal tibia length (PT), and midshaft transverse diameter of the tibia (TDT) (see details in Elissamburu, 2004, 2010).

These measurements were combined into 10 functional indices that represent attributes of bones and the mechanical efficiency of principal muscles related to limb function as proposed in previous works (Vizcaíno and Milne, 2002; Elissamburu and Vizcaíno, 2004; Samuels and Van Valkenburgh, 2008). Ratios are commonly used in biological studies because they reflect functional and easily understandable features of organisms. We recognize that using ratios can create problems in statistical analyses because of the non-independence of the two variables, as well as the possible violation of assumptions of normality and homoscedasticity included in parametric tests (Sokal and Rohlf, 1995). Nevertheless, as many studies have found the use of ratios in statistical analyses to be robust (Corruccini, 1987; Van Valkenburgh and Koepfli, 1993; Croft and Anderson, 2008;

Table 14.2. *Mendoza* et al.'*s (2006) selected equations to estimate body mass from craniodental variables*

Alg.	R^2	Equation	Range %PE
1.1	0.995	1.276*SLML + 1.268*JMA + 0.493*MZW + 0.442*PAW − 0.580*SD + 1.047	4.5 to 7.5%
2.1	0.986	−1.602*LMRL + 2.791*SLML + 0.576*JLB + 1.005*JMA + 2.402	13.5 to 15%
2.2	0.986	−1.352*LMRL + 2.434*SLML + 0.587*JLB + 0.866*JMA + 0.263*JMC + 1.890	
2.3	0.987	−1.366*LMRL + 2.421*SLML + 0.542*JLB + 1.017*JMA + 0.716*JMC -0.509*JMB + 2.006	
3.1	0.978	1.119*LMRL + 0.210*LPRL + 0.730*JMA + 0.637*JMC + 0.181*JD − 0.619	21 to 25%
3.2	0.978	1.086*LMRL + 0.176*LPRL + 0.823*JMA + 0.968*JMC + 0.167*JD − 0.331*JMB − 0.573	
5.1	0.988	0.593*MZW + 0.515*PAW + 0.996*SA + 0.601*SB + 0.384*BL − 0.266*SD + 0.313*SE − 1.55	17.5 to 21.5%

Abbreviations: Alg., algorithm number; R^2, determination coefficient; %PE, percent prediction. Variables (names in parenthesis from Janis, 1990): BL, basicranial length; JD, length of the coronoid process; JLB, anterior jaw length (AJL); JMA, posterior jaw length (PJL); JMB, depth of mandibular angle (DMA); JMC, maximum width of the mandibular angle (WMA); LMRL, lower molar row length; LPRL, lower premolar row length; MZW, muzzle width; PAW, palatal width; SA, length of the ridge for the attachment of masseter origin (MFL); SB, occipital height (OCH); SD, depth of the face under the orbit; SE, length of the paraoccipital process; SLML, second lower molar length.

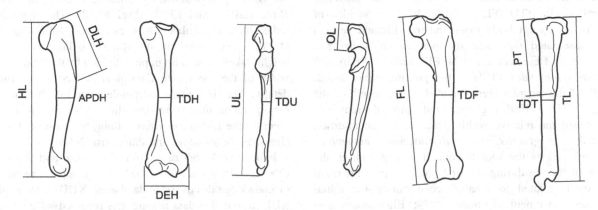

Fig. 14.11. Measurements of long bones based on *Interatherium*. HL, functional humerus length; DLH, deltoid length of the humerus; TDH, transverse diameter of the humerus; APDH, anteroposterior diameter of the humerus; DEH, diameter of the epicondyles; UL, total ulna length; OL, olecranon length; TDU, transverse diameter of the ulna; FL, functional femur length; TDF, transverse diameter of the femur; TL, tibia length; PT, proximal tibial length; TDT, transverse diameter of the tibia. See details in Elissamburu and Vizcaíno (2004).

Samuels and Van Valkenburgh, 2008), we consider them useful for inferring locomotor habits.

The indices used, their definition, and functional meanings are as follows (see also references for each index and bibliography included therein):

Shoulder moment index (SMI) is the deltoid length of the humerus divided by the functional length of the humerus [(DLH/HL) × 100]. This index is an indication of the mechanical advantage of the *deltoideus* and *pectoralis major* muscles acting across the shoulder joint. This index increases from generalized forms to specialized forms that emphasize more powerful shoulder movements (Vizcaíno and Milne, 2002).

Humerus robustness index (HRI) is the transverse diameter of the humerus divided by functional length of the humerus [(TDH/HL) × 100]. It gives an indication of the robustness of the humerus (Elissamburu and Vizcaíno, 2004).

Index of fossorial ability (IFA) is the length of the ulnar olecranon process divided by the functional ulna length [(OL/(UL − OL)) × 100]. This index gives a measure of the mechanical advantage of the *triceps* and *dorsoepitrochlearis* muscles in elbow extension. It is considered a good indicator of fossoriality (Vizcaíno and Milne, 2002; Elissamburu and Vizcaíno, 2004).

Epicondyle index (EI) is the epicondylar width of the
humerus divided by functional length of the humerus
[(DEH/HL) × 100]. It gives an indication of the relative
space available for the origin of the flexor, pronator, and
supinator muscles of the forearm. It is also considered a
good indicator of fossoriality (Hildebrand, 1988; Lessa and
Stein, 1992).

Brachial index (BI) is the functional forearm length (difference between ulnar length and olecranon length) divided
by the humerus length [((UL − OL)/HL) × 100]. It gives an
indication of the extent to which the forelimb is adapted for
fast movement (Howell, 1944; Fleagle, 1979; Vizcaíno and
Milne, 2002).

Ulna robustness index (URI) is the transverse diameter of
the ulna taken at midshaft divided by the functional ulna
length [(TDU/FUL) × 100]. It gives an indication of the
robustness of the forearm and the relative width available
for the insertion of muscles involved with pronation and
supination of the forearm and flexion of the manus and
digits (Elissamburu, 2010).

Femur robustness index (FRI) is the transverse diameter
of the femur taken at midshaft divided by the functional
femoral length [(TDF/FL) × 100]. It gives an idea of
capacity to support body mass and to withstand vertical
forces associated with velocity increase (Biewener and
Taylor, 1986; Demes *et al.*, 1994; Elissamburu, 2010).

Tibial spine index (TSI) is the proximal tibia length
(length of the cnemial crest) divided by the length of the
tibia [(PT/TL) × 100]. It gives an indication of strength of
the leg and the relative width available for the insertion
site of the *mm. gracilis*, *semitendinosus*, and *semimembra-
nosus* as well as the foot flexors. It is important in the
flexion of the leg during the first phase of gait. Proximal
insertion is related to greater speed during the initial
propulsor movement (Elftman, 1929; Elissamburu and
Vizcaíno, 2004).

Tibia robustness index (TRI) is the transverse diameter of
the tibia taken at midshaft divided by the tibial length
[(TDT/TL) × 100]. It gives an indication of the strength
of the leg and the relative width available for the origins of
the muscles acting across the ankle (Elissamburu and Vizcaíno, 2004).

Crural index (CI) is the tibia length divided by the
functional femur length [(TL/FL) × 100]. It gives a measure
of the extent to which the hindlimb is built for speed
(Howell, 1944; Fleagle, 1979; Bond *et al.*, 1995; Vizcaíno
and Milne, 2002).

14.3.3 Feeding behavior
Only specimens with little or no apparent deformation were
measured. Forty-seven specimens of notoungulates (six
genera), 19 litopterns (seven genera), and three astrapotheres (one genus) were analyzed (see Appendix 14.1).

Data on 119 extant species of artiodactyls and perissodactyls, distributed among 13 families, taken from Mendoza
et al. (2002), were used as learning sample to analyze the
relationship between the craniodental morphology of ungulates and their ecology (i.e. habitat and feeding behavior).

Fourteen craniodental measurements and the hypsodonty
index (HI) were used in the analyses (Fig. 14.12). They
include: length of the masseteric fossa (SA), occipital
height (SB), length of the posterior portion of the skull
(SC), depth of the face under the orbit (SD), length of the
paraoccipital process (SE), muzzle width (MZW), palatal
width (PAW), posterior jaw length (JMA), depth of mandibular angle (JMB), maximum width of the mandibular
angle (JMC), length of the coronoid process (JD), lower
molar tooth row length (LMRL), and lower premolar tooth
row length (LPRL). The length of diastema (JLB) was
measured from the base of the third incisor to the first
premolar (see table 2 in Mendoza *et al.*, 2002) in extant
and fossil ungulates when a diastema is present. This
includes Astrapotheriidae, Proterotheriidae, and only
Pachyrukhos among Typotheria. The total length of the
jaw (JAW) was obtained by summing JLB (when present),
JMA, LMRL, and LPRL (Fig. 14.12). The hypsodonty
index of notoungulates was calculated following Janis
(1988), i.e. crown height ratio of m3 calculated as the crown
height (taken from radiographs) divided by the labio-lingual
width of the tooth (see Cassini *et al.*, 2011, for further
details). The HI is a size-independent variable, but the other
measurements used were size-adjusted by dividing each of
them by the LMRL, measured along the base of the teeth
(Janis, 1990; Mendoza and Palmqvist, 2006).

Following Mendoza (2007), Mendoza and Palmqvist
(2008), and Cassini *et al.* (2011), two computer techniques
of knowledge discovery in databases (KDD) were applied.
KDD, also called data mining, has received wide attention
from practitioners and researchers (Fayyad *et al.*, 1996).
There are several techniques from machine learning, statistics, and databases that can be conveniently combined to
obtain useful methods and systems for KDD (Mannila and
Toivonen, 1997). A typical task of knowledge discovery
and data mining is to find "rules" that explain a huge set of
examples well (Domingo *et al.*, 2002). In this contribution,
we use discriminant analysis together with classification
trees as a technique of KDD (Cios *et al.*, 1998) to find these
rules and apply them to infer fossil feeding behavior (see
Mendoza, 2007, for an extensive discussion). Discriminant
analysis is a classical technique of multivariate statistics
that allows new samples to be classified within predefined
groups using the discriminant functions adjusted to maxi-
mize the between-groups to within-groups ratio of variance.
Classification trees, developed by Breiman *et al.* (1984), are
a KDD technique (Larose, 2004) that stems from the realm
of machine learning (Michie *et al.*, 1994). Classification

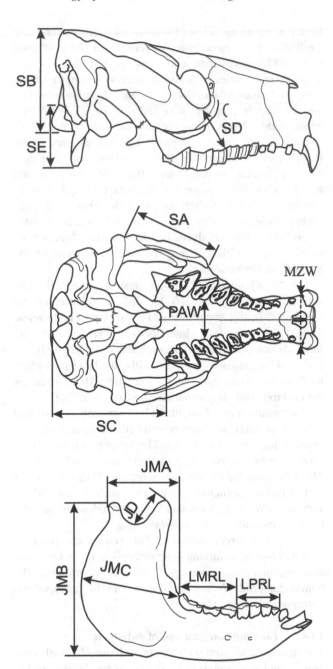

Fig. 14.12. Skull of *Adinotherium* with selected measurements used in this study. SA, length of the masseteric fossae; SB, occipital height; SC, length of the posterior portion of the skull; SD, depth of the face under the orbit; SE, length of the paraoccipital process; MZW, muzzle width; PAW, palatal width; JMA, posterior jaw length; JMB, depth of mandibular angle; JMC, maximum width of the mandibular angle; JD, length of the coronoid process; LMRL, lower molar tooth row length; LPRL, lower premolar tooth row length (see details in Mendoza *et al.*, 2002; reproduced from Cassini *et al.*, 2011).

trees are non-parametric models. Some of their advantages are that they look at variables hierarchically rather than simultaneously, are easy to interpret, and do not assume

that the dependent variable follows any given distribution. The approach followed in this contribution is a hierarchical analysis identifying the correlation between morphology and habitat types, and then feeding behavior between certain habitats, versus Spencer (1995), who evaluated habitat preference within diet categories.

Inferences about habitat For the first set of analyses, performed to characterize types of habitat, extant species were classified among the following three categories (Mendoza *et al.*, 2005): (1) open habitats (i.e. treeless or scarcely wooded savannas, grasslands, dry deserts, and semidesert steppes); (2) mixed habitats (i.e. wooded savannas, bush land, open forests, and also including species dwelling both in closed and open habitats); (3) closed habitats (i.e. closed woodlands, riverine, moist deciduous, and evergreen forests).

Fifty-eight out of the 119 extant ungulate species included in this study were categorized as living in open habitats, 30 in mixed habitats, and 31 in closed habitats (see Mendoza *et al.*, 2002: appendix, and Mendoza and Palmqvist, 2008: supplementary material). Mendoza and Palmqvist (2008) interpreted JLB to be the best mandibular measurement, together with HI, for discriminating species from open and mixed habitats. As mentioned above, most Santacrucian notoungulates do not have a diastema. Applying the same KDD techniques as Mendoza and Palmqvist (2008), Cassini *et al.* (2011) found that another measurement shows a similar pattern: the length of the jaw (JAW). Using JAW and HI, Cassini *et al.*, (2011) defined a habitat preference index (HPI) that facilitates inferring whether an ungulate species is more likely to dwell in open or mixed habitats. Using that index it is possible to characterize the extant species that forage in open, mixed, and closed habitats. Only selective browsers from open habitats, and those species dwelling in open but water-associated habitats, cannot be distinguished from those dwelling in mixed habitats.

Inferences about diet For the second set of analyses, performed to characterize feeding behavior, we subdivided each habitat category by feeding behavior as in Mendoza and Palmqvist (2006, 2008).

Extant open-habitat species were classified into the following three feeding subcategories: (1) grazers, feeding mainly on grasses; (2) mixed feeders, including those species that consume grass and leaves depending on availability; and (3) open habitat browsers, feeding predominantly on dicotyledonous plants.

Closed-habitat species were classified among the following four feeding categories: (1) omnivores, eating vegetation, fungus, and occasionally animal tissues; (2) frugivores, feeding predominantly on fruits and other non-fibrous soft material; (3) closed habitat browsers, feeding predominantly on dicotyledonous plants; and (4) closed-habitat mixed feeders, including those species that consume grass and leaves.

Starting from 14 measurements, there are 588 possible combinations involving three or fewer measurements. The KDD techniques were then used to identify those combinations of three or fewer variables that allow a better characterization of the ecological groups. In addition, these results can be graphically represented in order to visualize the position of each extinct species in the subspaces depicted by these combinations of variables, optimized for the task of ecological characterization (feeding behavior).

Inferences about digestive capabilities Vizcaíno et al. (2006) proposed that evaluating how occlusal surface area (OSA) scales with body mass (BM) and comparing the normalization constant (intercepts) would allow inferences to be made about digestive capabilities.

The OSA was calculated, following Vizcaíno et al. (2006), as the total cheek-tooth occlusal surface area. This measure considers the infolding contour of the tooth. The allometric relationship (scaling) of OSA with BM was assessed by the standardized major axis (SMA = reduced major axis) method using the base-10 logarithmic transformation of two variables. Here, the purpose of line-fitting is not to predict Y from X (ordinary least squares), but to summarize the relationship between two variables; therefore, SMA is more appropriate for dealing with allometric approaches (see Warton et al., 2006 for further discussion on use of these terms and methodological procedures). Deviations from isometry were assessed by comparing the allometric coefficient with that expected under geometric similarity (Alexander, 1985). Expected coefficients under isometry are equal to 0.67 (2/3) for variables involved, because they are all measurements of surface (OSA) vs volume (BM). Consequently, we performed F-tests with the null coefficient set at 2/3 to assess significant deviations from isometry (Warton and Weber, 2002). We obtained some new equations from Vizcaíno et al. (2006), using their extant mammals sample as a reference. The residuals of each group were compared by means of the Mann–Whitney U-test, which is the most powerful (or sensitive) non-parametric alternative to the t-test for independent samples.

14.4 Results

14.4.1 Body mass
The BM averages of native ungulates obtained are listed in Table 14.3. For several species, in particular the less-abundant ones, the value for a single specimen is presented.

Bivariate equations allowed a more inclusive sample, and the BM values obtained were in general smaller than those estimated by multivariate equations (Table 14.3).

Only some specimens of *Astrapotherium* reaches 1000 kg. Among Notoungulata, *Nesodon* and *Homalodotherium* are the largest (approx. 640 and 400 kg, respectively), while both *Adinotherium* species are over 100 kg. All Typotheria are smaller than 10 kg. Among typotheres, *Protypotherium australe* and *Hegetotherium mirabile* are the largest (~8 kg) and *Interatherium robustum* and *Pachyrukhos moyani* the smallest (~2 kg). Among Litopterna, all Proterotheriidae fall in the 20 to 50 kg range except *Diadiaphorus majusculus*, for which a body mass around 80 kg was obtained. In addition, only Macraucheniidae exceed 120 kg, being the largest litoptern of the Santa Cruz Formation.

The "small ungulates" group presents estimated mass values between 1 and 10 kg and is represented by typotheres. In this group some taxa overlap in body mass. Thus, the smallest interatheriids, represented by *Interatherium*, overlap in size with the small hegetotheriid *Pachyrukhos* (around 2 kg). Similarly, the large interatheriid *Protypotherium australe*, not exceeding 10 kg, overlaps the hegetotheriid *Hegetotherium mirabile* (~7.7 kg).

The "medium-sized ungulates" group (between 10 and 100 kg) includes the proterotheriid litopterns and the toxodontid *Adinotherium ovinum*. The lower part of the range is composed exclusively of proterotheriids, *Anisolophus* and *Thoatherium* being the smallest genera (~20 kg). The upper part of this range includes forms that reach almost 100 kg, such as *Adinotherium ovinum* among toxodontids, and the largest proterotheriid, *Diadiaphorus*.

Finally, the "larger ungulates" (between 100 and 1000 kg) include *Theosodon* among macraucheniids, the toxodontids *Adinotherium robustum* and *Nesodon imbricatus*, the homalodotheriid *Homalodotherium*, and the astrapothere *Astrapotherium magnum*.

14.4.2 Locomotion and use of substrate
The mechanical context ("average biomechanical situation") and interpretation of indices were defined in the Methods section. The indices obtained for native ungulate genera are listed in Table 14.4. For several genera, the value for a single specimen is presented. The median, 25th percentile, 75th percentile (box), minimum, and maximum values are shown in the box and whiskers plot (Fig. 14.13).

Some indices show very similar values within clades. For example, among litopterns, Proterotheriidae and Macraucheniidae have very similar index values for the forelimb, with the exception of brachial index (BI). These groups have different values for the hindlimbs, but among proterotheriids the values are very homogeneous (Fig. 14.13 and

Table 14.3. *Summarized mean body-mass estimates for native Santacrucian ungulates based on Janis (1990) and Mendoza* et al. *(2006) equations*

Family	Species	Janis (1990)			Mendoza *et al.* (2006)			Mean body mass of species (kg)
		Mean	s.d.	*n*	Mean	s.d.	*n*	
Astrapotheriidae	*Astrapotherium magnum*	933.74	*357.10*	4	908.90	*223.07*	4	**921.32**
Macraucheniidae	*Theosodon garrettorum*	142.74		1	173.34		1	**158.04**
	Theosodon gracilis	112.44	*60.09*	2	130.66	*83.20*	2	**121.55**
	Theosodon lydekkeri	105.53		1	156.32		1	**130.93**
Proterotheriidae	*Anisolophus australis*	36.61		1				**36.61**
	Diadiaphorus majusculus	72.03	*14.85*	9	92.08	*22.28*	7	**82.05**
	Tetramerorhinus cingulatum	33.09	*4.87*	3	50.34		1	**41.71**
	Tetramerorhinus lucarius	29.50		1				**29.50**
	Tetramerorhinus mixtum	26.29		1	43.82		1	**35.06**
	Thoatherium minusculum	20.55	*4.75*	6	27.84	*1.09*	2	**24.20**
Homalodotheriidae	*Homalodotherium* sp.	405.08		1				**405.08**
Toxodontidae	*Adinotherium ovinum*	100.29	*9.18*	8				**100.29**
	Adinotherium robustum	126.24	*20.22*	2				**126.24**
	Nesodon imbricatus	644.85	*203.26*	12	630.17	*112.09*	10	**637.51**
Interatheriidae	*Interatherium robustum*	1.80	*0.45*	13	2.96	*0.58*	12	**2.38**
	Protypotherium attenuatum	3.01	*0.79*	4	4.59	*1.23*	3	**3.80**
	Protypotherium australe	5.12	*0.45*	4	10.35	*1.24*	3	**7.73**
	Protypotherium praerutilum	3.10	*0.29*	3	5.92	*1.55*	3	**4.51**
Hegetotheriidae	*Hegetotherium mirabile*	7.20	*2.66*	5	8.21	*1.78*	2	**7.71**
	Pachyrukhos moyani	1.62	*0.16*	4	2.64	*0.23*	2	**2.13**

Abbreviations: s.d., standard deviation; *n*, number of specimens.

Table 14.4). Among notoungulates, the Toxodontia present similar values, particularly within Nesodontinae, while the Typotheria (excepting *Interatherium*) have similar values for forelimb indices and shows more variability in the hindlimb indices (Fig. 14.13).

The Toxodontia have the highest robustness indices of both fore- and hindlimbs (HRI, SMI, URI, and FRI). *Homalodotherium* in particular has exceptional femur and tibial robustness as indicated by the FRI and TRI, reaching the highest values among the Santacrucian ungulates assemblage (Fig. 14.13 and Table 14.4). Litopterns and typotheres have low limb robustness indices.

The IFA index, originally considered as indicative of digging capabilities (but see discussion) shows the highest values in nesodontines. All typotheres and *Homalodotherium* have the lowest values, with *Interatherium* and *Pachyrukhos* having the highest and lowest values among typotheres, respectively.

In the hindlimb, the typotheres have the highest crural index (CI) values among Santacrucian ungulates, with *Pachyrukhos* and *Protypotherium* presenting the highest values and *Interatherium* and *Hegetotherium* the lowest. *Hegetotherium* and *Pachyrukhos* are the more gracile typotheres, having the lowest values of TSI, FRI, and TRI. On the other hand, *Interatherium* has significant differences between the fore- and hindlimb, showing the lowest values of SMI and BI and the highest values of IFA, FRI, and TSI among typotheres.

Table 14.4. *Locomotor indices for Santacrucian ungulates*

Catalog	Genera	IFA	SMI	BI	HRI	EI	URI	FRI	TSI	TRI	CI
FMNH 14251	*Astrapotherium*	28.86	49.99	67.09	12.29	21.05	14.01	8.95	52.35	8.60	71.51
YPM-VPPU 15164	*Theosodon*	31.39	45.31	116.49	10.73	28.78	10.86	12.25	32.22		79.26
YPM-VPPU 15216	*Theosodon*		56.83			29.64				13.50	
MACN-A 2545–57	*Theosodon*								39.37	12.95	
MACN-A 9252–53	*Theosodon*								39.20	12.38	
MACN-A 2832	*Anisolophus*								37.36	13.37	
YPM-VPPU 15711	*Anisolophus*							11.21	36.54	10.86	86.96
YPM-VPPU 15799	*Diadiaphorus*	34.20	48.60	67.89	11.37	23.76	12.26	10.20			
YPM-VPPU 15107	*Tetramerorhinus*	32.34	49.31	86.05	11.24	24.37	10.05				
MACN-A 8970–98	*Tetramerorhinus*						13.79	9.86	40.89	10.59	86.00
MPM-PV 3529	*Thoatherium*		46.27		10.64	24.39			41.23	10.03	
FMNH 13092*	*Homalodotherium*			102.00				21.85		16.53	53.00
YPM-VPPU 15747	*Homalodotherium*	27.08					7.87				
AMHN 9229	*Adinotherium*		66.51		13.02	33.01		16.97	53.44	8.01	83.16
YPM-VPPU 15004	*Adinotherium*	48.85	62.36	79.73	13.86	29.93	11.72				
YPM-VPPU 15127	*Adinotherium*							16.42			
YPM-VPPU 15131	*Adinotherium*	45.49	68.49	77.51	15.98	31.29	13.52	16.92	44.70	10.47	78.36
YPM-VPPU 15480	*Adinotherium*								45.85	9.53	
YPM-VPPU 15966	*Adinotherium*				18.44	34.43					
MPM-PV 3542	*Adinotherium*		65.45		16.39	32.03					
AMHN 9192	*Nesodon*	51.87	61.54	79.02	13.48	36.44	13.23				
AMHN 9553	*Nesodon*	45.14					17.84		49.11	10.74	
YPM-VPPU 15132	*Nesodon*	42.80					13.33				
YPM-VPPU 15132	*Nesodon*	48.00					15.37				
YPM-VPPU 15256	*Nesodon*	46.05	67.25	79.44	18.57	37.11	14.82				
YPM-VPPU 15967	*Nesodon*							18.41	49.14	10.52	83.90
YPM-VPPU 15041	*Interatherium*							12.51	48.40	7.86	108.18
YPM-VPPU 15108	*Interatherium*	23.77		74.35	9.05	27.33	7.06		45.93	7.55	

Table 14.4. (*cont.*)

Catalog	Genera	IFA	SMI	BI	HRI	EI	URI	FRI	TSI	TRI	CI
YPM-VPPU 15202	*Interatherium*								49.51	7.67	
YPM-VPPU 15401	*Interatherium*	26.56	33.50	73.64	10.97	27.55	7.62	13.24	52.26	9.05	101.36
MPM-PV 3469	*Interatherium*	27.36					8.61				
MPM-PV 3471	*Interatherium*		33.26		8.65	27.71					
MPM-PV 3527	*Interatherium*	31.03	31.60	73.07	9.51	28.78	8.35	11.58	50.08	6.97	103.60
MPM-PV 3539	*Interatherium*								50.07	8.59	
AMHN 9149	*Protypotherium*	26.22	48.35	86.37	9.37	25.85	8.12	11.23	37.09	8.28	113.42
YPM-VPPU 15161	*Protypotherium*									6.76	115.79
YPM-VPPU 15340	*Protypotherium*							12.22	38.84	9.39	114.90
YPM-VPPU 15341	*Protypotherium*				7.82	22.03		10.43		7.04	114.29
YPM-VPPU 15386	*Protypotherium*				9.29	23.46		10.97		9.15	116.49
YPM-VPPU 15659	*Protypotherium*							11.21			
YPM-VPPU 15828	*Protypotherium*	21.24	45.33	88.34	9.63	26.67	8.04				
YPM-VPPU 15892	*Protypotherium*				9.40	24.47					
MACN-A 9657–63	*Protypotherium*		43.33		9.69	25.45					
YPM-VPPU 15176	*Hegetotherium*		53.21		8.66	27.73					
YPM-VPPU 15298	*Hegetotherium*								18.84	6.48	
YPM-VPPU 15395	*Hegetotherium*							10.01	21.49	6.91	99.07
YPM-VPPU 15431	*Hegetotherium*							11.08	18.14		103.82
YPM-VPPU 15542	*Hegetotherium*	25.65					6.25				
AMHN 9285	*Pachyrukhos*	20.87						9.31			
AMHN 9481	*Pachyrukhos*	23.14	48.67	87.18							
AMHN 9242	*Pachyrukhos*							8.83	26.58	4.79	128.57

* Values from Elissamburu (2010).

14.4.3 Feeding behavior

Inferences about habitat Among all of the craniodental variables that could be used to discriminate habitat preference, the relative length of the jaw (JAW) and HI are the best at characterizing species that forage in open and mixed habitats. The HI, which permits open-habitat species to be distinguished from those of mixed habitats, scales with JAW (Fig. 14.14a). In addition, applying the habitat preference index (HPI = HI/1.4 JAW − 2.4) developed by Cassini *et al.* (2011), we inferred the habitat preferences of the Santacrucian ungulates.

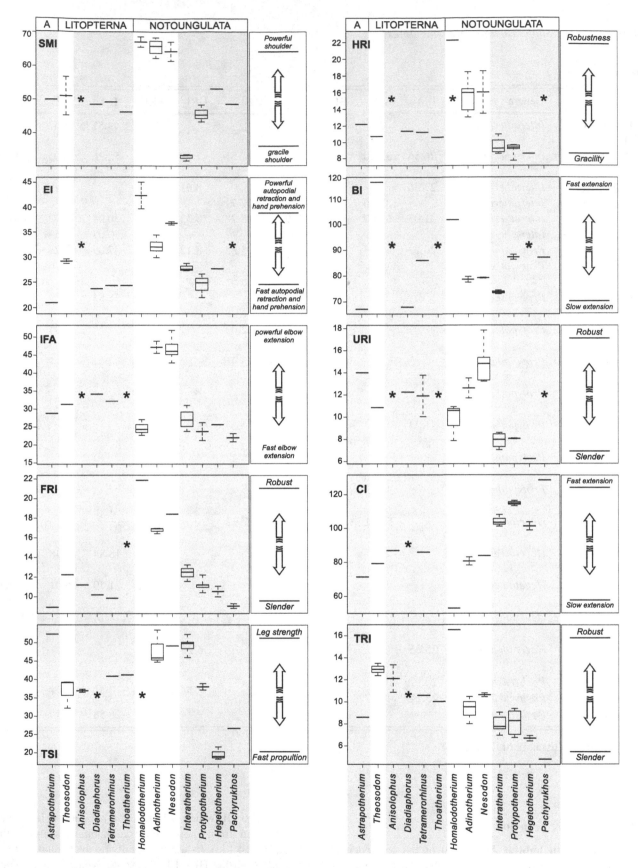

Fig. 14.13. Biomechanical indices for each genus indicating median (middle bar), 25th percentile, 75th percentile (inferior and superior edges of boxes), minimum, and maximum values. Abbreviations: SMI, shoulder moment index; HRI, humerus robustness index; EI, epicondylar index; BI, brachial index; IFA, index of fossorial ability; URI, ulnar robustness index; FRI, femur robustness index; CI, crural index; TSI, tibial spine index and TRI, tibia robustness index. Asterisks represent missing data.

Astrapotheres and litopterns have similar HPI to the extant ungulates that inhabit closed habitats and, coincidently, share the morphospace suggested by JAW and HI (Fig. 14.14a). HI for *Astrapotherium* ranges from 1.1 to 1.3 (Table 14.5), while HPI ranges from 0.3 to 0.5. Among litopterns, the HI of the macraucheniid *Theosodon* varies between 0.8 and 1.2, and HPI is comparable to that of *Astrapotherium*. Among the Proterotheriidae, *Thoatherium* and *Anisolophus* have the lowest values of hypsodonty (HI ~0.78) and HPI (from 0.31 to 0.35). In contrast, specimens of *Diadiaphorus* have the highest HI (~1.33) and HPI of all litopterns (0.5 to 0.65), although they are still classified as closed habitat inhabitants. The specimens of *Tetramerorhinus* have HI and HPI values intermediate between the last two taxa (HI from 0.9 to 1.16 and HPI from 0.37 to 0.49; Table 14.5).

In contrast, notoungulates as a whole have a combination of HI and JAW characteristic of extant ungulates that live in open habitats. Among typotheres, *Protypotherium* specimens show somewhat higher hypsodonty (HI ~4; Table 14.5), but most of them are also placed near camelids in the bottom-left corner of the morphospace depicted by HI and JAW (Fig. 14.14a). Two *Protypotherium* specimens, however, have longer jaws, showing a combination of jaw length and hypsodonty more similar to many extant ungulates from open habitats. *Interatherium* specimens have an intermediate degree of hypsodonty (HI ~3.3; Table 14.5) but they show a wide range in the relative length of their jaws, which results in a wide range of HPI values (0.8–1.3). Most of them show combinations of jaw length and hypsodonty similar to extant ungulates from both open and mixed habitats. Finally, the most hypsodont are two specimens of *Hegetotherium* (HI ~7.8; Table 14.5), which are comparable to *Equus asinus* (HI ~8.7). Their HPI values are very high (2.92 and 3.18), so their HPI values are much higher than the threshold that separates open-habitat species from those of mixed habitats. However, as there are no extant species with such a short jaw combined with such a high degree of hypsodonty, they alone occupy the upper-left region in the morphospace depicted by HI and JAW.

Among toxodontids, the specimens of *Adinotherium* are very hypsodont (HI ~4.9; Table 14.5), and their jaw lengths are variable, so their HPI values are also very different (1.1, 1.8, and 2.2). All of them, however, show combinations of jaw length and hypsodonty similar to extant ungulates from open habitats. *Nesodon* specimens show HI values around 3.2 (Table 14.5) and their HPI ranges from 1.5 to 2.2, so they exceed the limits of HI and HPI that separate species dwelling in open habitats from those dwelling in mixed or closed habitats. They lie in the bottom-left region of the morphospace depicted by HI and JAW, without overlapping with any extant ungulates, but surrounded by open-habitat camelids.

Summarizing, all Litopterna (*Anisolophus, Tetramerorhinus, Diadiaphorus,* and *Thoatherium*) and Astrapotheria (*Astrapotherium*) lived and/or foraged in closed habitats. In turn, all Notoungulata (*Adinotherium, Nesodon, Protypotherium, Interatherium,* and probably *Hegetotherium* and *Pachyrukhos*) lived and/or foraged in open habitats.

Inferences about diet We performed the analyses about feeding behavior considering astrapotheres and litopterns as inhabitants of closed habitats and all the notoungulates as from open or mixed habitats. Species from closed habitats can be classified as omnivores, frugivores, browsers, or mixed feeders, while species of open or mixed habitats can be classified as grazers, mixed feeders, or browsers.

Among the closed-habitat species, the variables SA, LPRL, and JMA allow us to distinguish omnivores from non-omnivores. The omnivores have relatively lower values of these three variables than non-omnivores. The position of Santacrucian litopterns in the morphospace depicted by LPRL and JMA allows us to reject omnivory for them. This is reinforced by the fact that none of the studied litopterns have bunodont cheek teeth. Instead, they exhibit a combination of traits similar to those of browsing perissodactyls (Fig. 14.14b). The specimens of *Astrapotherium* fall within a morphospace close to the extant omnivore *Hylochoerus meinertzhageni*, which include a high percentage of vegetation in its diet. Owing to the peculiar morphology and the reduced premolars of astrapotheres (see Discussion), we prefer to treat the specimens of *Astrapotherium* as putative non-omnivorous ungulates.

Among non-omnivorous ungulates it is difficult to obtain a clear pattern that combines only a few variables. However, the morphospace depicted by JMC and JMB allows us to discriminate browsers from frugivores-plus-mixed-feeders, even though the cephalophine bovids invade the lower portion of browser morphospace (Fig. 14.14c). All litopterns and *Astrapotherium* have a combination of these two variables intermediate between browsing perissodactyls and artiodactyls. While one specimen of Macraucheniidae (*Theosodon garrettorum*) lies next to the mixed-feeding moschid *Moschus moschiferus* and the cervid *Elaphodus cephalophus*, the others lie in the morphospace of browsers.

In addition, the variables SC and MZW permit us to characterize the mixed feeders to the exclusion of frugivores and browsers (Fig. 14.15a). While browsers have higher values of SC than forms with other diets, and similar values of MZW, the mixed feeders have a more well-defined morphospace despite the fact that Cephalophinae lie on the boundary between frugivores and mixed feeders, as defined by MZW. The specimens of Macraucheniidae have a combination of SC and MZW characteristic of extant mixed feeders.

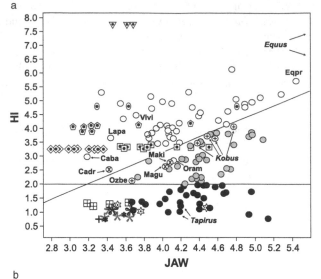

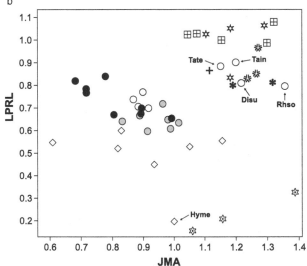

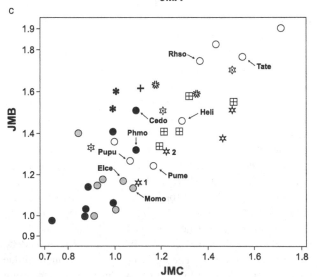

Among open habitat ungulates, Mendoza and Palmqvist (2008), using KDD, arrived at the same conclusion as Janis and Ehrhardt (1988), who showed that grazers have a wider muzzle than non-grazers. The morphospace depicted by the relative width of the muzzle (MZW) and the maximum width of mandibular angle (JMC) allows a very good characterization of grazers versus non-grazers (see Fig. 14.15b). Non-grazers among artiodactyls and perissodactyls show a common and relatively homogeneous pattern, characterized by both narrow muzzles and narrow mandibular angles. Santacrucian notoungulates share a wide mandibular angle with equids and a wide muzzle with bovids. In fact, the highest value of MZW corresponds to a specimen of *Interatherium* (see Fig. 14.15b). According to its muzzle and mandibular angle width, *Nesodon* was the least specialized genus for the consumption of grass, although some specimens have

Fig. 14.14. a, Distribution of 119 extant ungulates, 3 astrapotheres, 13 litopterns and 26 notoungulates in the morphospace depicted by the hypsodonty index (HI) and the relative length of the jaw (JAW), which allows the characterization of the craniodental morphology of species from open habitats (white symbols), mixed habitats (gray symbols), and closed habitats (black symbols); water-associated species (symbols with a plus sign) and browsers from open habitats (symbols with a cross). b, Distribution of 30 extant ungulates, astrapotheres, and litopterns in the morphospace depicted by posterior jaw length (JMA) and lower premolar tooth row length (LPRL), which allows the morphological characterization of omnivores (diamonds) from non-omnivores, i.e. frugivores (black circles), browsers (white circles), and mixed feeders (gray circles). c, Distribution of 30 extant ungulates, astrapotheres, and litopterns in the morphospace depicted by maximum width of the mandibular angle (JMC) and depth of mandibular angle (JMB), which allows the morphological characterization of browsers (white circles) from mixed feeders (gray circles) and frugivores (black circles). Extant ungulates: *Camelus bactrianus* (Caba), *Camelus dromedarius* (Cadr), *Cephalophus dorsalis* (Cedo), *Dicerorhinus sumatrensis* (Disu), *Elaphodus cephalophus* (Elce), *Hexaprotodon liberiensis* (Heli), *Hylochoerus meinertzhageni* (Hyme), *Lama pacos* (Lapa), *Lama guanicoe* (Lagu), *Madoqua guentheri* (Magu), *Madoqua kirkii* (Maki), *Moschus moschiferus* (Momo), *Oreamnos americanus* (Oram), *Ozotoceros bezoarticus* (Ozbe), *Philantomba monticola* (Phmo), *Pudu mephistophiles* (Pume), *Pudu puda* (Pupu), *Rhinoceros sondaicus* (Rhso), *Tapirus indicus* (Tain), *Tapirus terrestris* (Tate), *Vicugna vicugna* (Vivi). Santacrucian ungulates: Astrapotheria: *Astrapotherium* (dotted six-tipped star), Litopterna: *Anisolophus* (cross), *Diadiaphorus* (square with a cross), *Tetramerorhinus* (white five-tipped star), *Thoatherium* (asterisks), *Theosodon* (black six-tipped star), Notoungulata (dotted symbols): *Interatherium* (squares), *Nesodon* (diamonds), *Protypotherium* (pentagons), *Adinotherium* (ovoids), *Hegetotherium* (inverse triangles).

Table 14.5. *Summary statistics [mean ± 1 standard deviation (n)] of craniodental measurements from Santacrucian ungulates database (in millimeters)*

Genera	HI	SA	SB	SC	SD	SE	MZW	LPRL	LMRL	JMA	JMB	JMC	JD
Adinotherium	4.92 ± 0.42 (2)	105.29 ± 6.35 (4)	90.03 ± 2.45 (3)	112.64 ± 1.29 (4)	52.92 ± 4.60 (4)	49.83 ± 0.07 (2)	39.20 ± 5.56 (3)	37.47 ± 10.38 (7)	55.97 ± 17.16 (7)	95.72 ± 19.70 (3)	115.66 ± 26.89 (3)	82.08 ± 17.82 (7)	33.70 ± 8.42 (4)
Nesodon	3.24 (1)	177.91 ± 8.91 (5)	139.75 ± 4.54 (3)	195.82 ± 10.43 (5)	92.15 ± 10.83 (7)	64.92 ± 12.13 (5)	70.64 ± 5.58 (6)	64.24 ± 8.32 (12)	123.53 ± 9.45 (13)	170.51 ± 14.54 (10)	224.09 ± 16.47 (9)	159.87 ± 23.82 (13)	55.15 ± 6.69 (8)
Interatherium *	3.33 ± 0.05 (3)	36.96 ± 3.66 (5)	22.74 ± 2.31 (6)	31.81 ± 2.95 (5)	15.10 ± 0.79 (6)	13.28 ± 2.07 (6)	12.12 ± 1.57 (7)	11.72 ± 1.20 (10)	15.46 ± 1.50 (10)	25.31 ± 2.54 (10)	35.32 ± 2.54 (9)	29.56 ± 2.43 (10)	11.94 ± 2.02 (10)
Protypotherium *	4.05 ± 0.13 (4)	42.41 ± 4.51 (6)	26.19 ± 4.64 (4)	38.68 ± 3.60 (6)	15.30 ± 1.53 (7)	16.71 ± 4.27 (5)	16.69 ± 2.08 (6)	13.54 ± 2.14 (10)	21.28 ± 2.75 (10)	31.39 ± 8.00 (7)	40.45 ± 4.74 (8)	36.18 ± 7.88 (7)	15.63 ± 6.05 (7)
Hegetotherium *	7.76 (1)	57.37 ± 0.18 (2)	31.23 ± 1.16 (2)	41.69 ± 1.47 (2)	27.65 ± 4.39 (2)	13.18 ± 0.00 (1)	19.00 ± 0.00 (1)	16.59 ± 1.46 (3)	24.02 ± 0.83 (3)	40.29 ± 1.88 (2)	56.91 ± 0.80 (2)	41.05 ± 11.70 (3)	
Pachyrukhos *		37.32 (1)	17.83 ± 1.75 (2)	29.63 ± 0.14 (2)	19.63 ± 1.27 (3)	14.40 ± 3.07 (3)	11.64 (1)	10.80 ± 0.23 (3)	14.66 ± 0.63 (3)	27.32 ± 3.26 (3)	39.24 (1)	26.36 ± 3.61 (3)	4.43 ± 0.86 (2)
Anisolophus	0.78 (1)							33.02 (1)	38.05 (1)	42.42 (1)	61.43 (1)	42.22 (1)	25.78 (1)
Diadiaphorus	1.33 ± 0.06 (7)	81.66 ± 2.92 (5)	64.69 ± 5.20 (5)	96.96 ± 2.97 (5)	41.36 ± 0.98 (6)	26.38 (1)	27.98 ± 6.14 (5)	59.95 ± 2.40 (7)	58.55 ± 2.61 (7)	68.70 ± 4.19 (5)	85.11 ± 3.26 (5)	75.97 ± 3.49 (6)	32.06 ± 1.06 (4)
Tetramerorhinus	1.07 ± 0.12 (4)	69.31 (1)	47.12 (1)	73.74 (1)	27.05 (1)		17.95 (1)	37.94 ± 2.68 (4)	43.81 ± 1.52 (4)	55.72 ± 1.33 (3)	69.31 ± 3.44 (3)	55.36 ± 3.42 (2)	24.84 (1)
Thoatherium	0.94 ± 0.21 (3)							31.77 ± 0.93 (3)	38.48 ± 1.23 (3)	48.68 ± 1.67 (2)	60.62 ± 0.12 (2)	34.76 ± 7.12 (3)	19.65 ± 5.78 (2)
Theosodon	0.94 ± 0.15 (4)	86.19 ± 3.15 (3)		139.37 ± 8.23 (3)	63.85 ± 1.96 (3)	37.50 (1)	52.18 ± 1.74 (3)	69.42 ± 7.14 (4)	69.64 ± 1.07 (4)	82.89 ± 4.23 (4)	93.28 ± 9.10 (4)	92.10 ± 12.22 (4)	41.69 ± 1.47 (2)
Astrapotherium	1.23 ± 0.11 (3)	169.48 ± 22.16 (2)	160.86 ± 31.29 (2)	190.01 ± 12.08 (2)	84.89 ± 4.53 (2)	62.06 ± 16.88 (2)	68.07 ± 52.53 (2)	29.07 ± 4.75 (3)	132.65 ± 30.28 (3)	148.44 ± 19.87 (3)	193.62 ± 28.20 (3)	154.87 ± 21.60 (3)	66.36 ± 10.66 (3)

Taxa tagged by an asterisk (*) have high-crowned ever-growing cheek teeth (i.e. euhypsodonty *sensu* Mones, 1982). Means are given in italics.

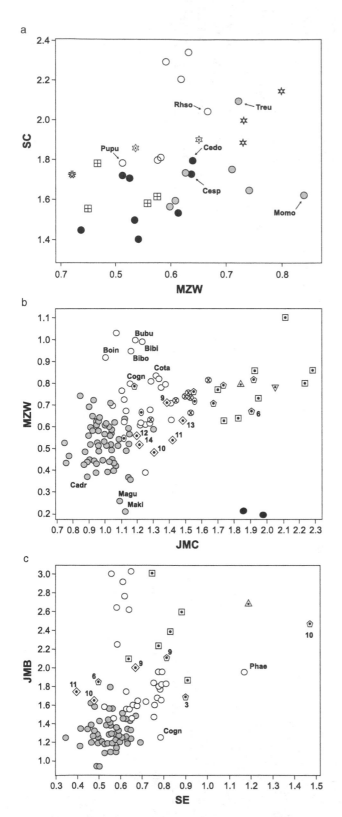

both the muzzle and the mandibular angle as wide as in some equids (see Fig. 14.15b).

Other traits also seem to be involved in the morphological patterns that make ungulates appropriate for feeding mainly on grasses. As shown in Fig. 14.15c, the morphospace depicted by the depth of the mandibular angle (JMB) and the length of the paraoccipital process (SE) also allows us to distinguish grazers from non-grazers. Again, grazers from the families Bovidae and Equidae have a deeper mandibular angle (JMB), but this aspect is much more pronounced among bovids, and only bovids show a longer paraoccipital process (SE, see Fig. 14.15c). Thus, non-grazing open-habitat ungulates are again the category that shows a common pattern, characterized by both a shallower mandibular angle and a shorter paraoccipital process than grazers. These features are merely the primitive condition for artiodactyl and perissodactyl ungulates (not something specifically indicating browsing). Notoungulates share the traits of grazers, having a deep mandibular angle (JMB) and a long paraoccipital process (SE, see Fig. 14.15c). The *Pachyrukhos* specimen examined also has an extreme SE value, although its paraoccipital process is not much longer than that of the only grazer species of the family Suidae (Fig. 14.15c). Although all suids have a long paraoccipital process, it is proportionally longer in the grazer *Phacochoerus* (SE = 1.17) than in the omnivorous forms (SE from 0.87 to 1.02). The feeding behavior of *Nesodon* is not very clear, but also taking into account its location in the morphospace depicted by JMB and SE (Fig. 14.15c), it seems more probable that it was a mixed feeder.

Summarizing, all proterotheriids (*Anisolophus*, *Tetramerorhinus*, *Diadiaphorus*, and *Thoatherium*) fed on dicotyledonous plants, while the notoungulates (*Adinotherium*, *Protypotherium*, *Interatherium Hegetotherium*,

morphological characterization of mixed feeders (gray circles) from browsers (white circles) and frugivores (black circles).
b, Distribution of the 89 extant ungulates (circles) inhabitants of open and mixed habitats and 24 notoungulates (dotted symbols) in the morphospace depicted by MZW and JMC, which allows the morphological characterization of grazers (white circles) and non-grazers (gray circles). c, Distribution of the 89 extant ungulates (circles) inhabitants of open and mixed habitats and 24 notoungulates (dotted symbols) in the morphospace depicted by JMB and SE, which allows the morphological characterization of grazers (white circles) and non-grazers (gray circles). Extant ungulates: references as in Fig. 14.14 plus *Bison bison* (Bibi), *Bison bonasus* (Bibo), *Bos indicus* (Boin), *Bubalus bubalis* (Bubu), *Cephalophus spadix* (Cesp), *Connochaetes gnou* (Cogn), *Connochaetes taurinus* (Cota), *Equus* species (crossed circles), *Syncerus caffer* (Syca), *Tragelaphus eurycerus* (Treu). Santacrucian ungulates: references as in Figure 14.14 (b and c reproduced from Cassini *et al.*, 2011).

Fig. 14.15. a, Distribution of 30 extant ungulates, astrapotheres, and litopterns in the morphospace depicted by muzzle width (MZW) and length of the posterior portion of the skull (SC), which allows the

Table 14.6. *Regression results for log-transformed cheek tooth occlusal surface area* vs. *body mass*

Taxon	Number of spp.	Range of W	n	R^2	$F_{(1, n-2)}$	Intercept$_{SMA}$[a]	Slope$_{SMA}$	Trend	F iso$_{(1, n-2)}$	Residuals
Mammals	51	0.084–4637	125	**0.958**	2826.86	**1.448**	**0.716**	+	15.152	X < G, Af, P
Xenarthra	7	1.175–8.41	26	**0.613**	37.99	**1.277**	**0.705**	iso	0.195	
Afrotheria	6	2.29–4637	18	**0.982**	876.26	**1.581**	**0.684**	iso	0.617	
Glires	8	0.084–50.07	22	**0.924**	243.15	**1.566**	**0.656**	iso	0.070	
Euungulata	30	12.00–3729	59	**0.845**	311.29	**1.518**	**0.691**	iso	0.476	A < P
†Meridiungulata	17	1.67–922	39	**0.953**	743.27	**1.453**	**0.795**	+	24.514	N < L
Perissodactyla	13	210–1637	22	**0.632**	34.31	**2.112**	**0.516**	iso	3.652	Tp < Eq
Artiodactyla	17	12.00–727	37	**0.937**	518.30	**1.594**	**0.621**	iso	2.730	R = Hp < Tl
Paenungulata	5	2.29–4637	17	**0.990**	1540.47	**1.604**	**0.684**	iso	0.990	
†Litopterna	7	14.74–158	12	**0.694**	22.70	**2.074**	**0.526**	iso	1.872	
†Notoungulata	9	1.67–738.6	25	**0.991**	2552.42	**1.407**	**0.788**	+	72.952	Tx = Ty
†Proterotheriidae	4	14.74–105.8	9	**0.665**	13.89	**2.084**	**0.521**	iso	1.286	
†Toxodontidae	2	85.78–38.6	8	**0.971**	197.86	**1.393**	**0.793**	iso	6.187	
†Typotheria	6	1.67–11.66	16	**0.909**	139.48	**1.502**	**0.626**	iso	0.601	In = Hg

Results of standardized major axis (SMA) regression for each group. Except for "Meridiungulata," all extant mammals data came from Vizcaíno *et al.* (2006). The more common parameters are marked in bold. Non-parametrical Mann–Whitney U analyses of residual values of OSA are indicated when the differences between taxa were statistically significant (p < 0.01).
[a] SMA intercept at body mass = 1 kg.
† indicates extinct taxon.
Abbreviations: (iso) isometric trend, F_{iso}-test no significant differences from expected value of 0.67 at $\alpha = 0.05$ level; (+) positive allometric trend, F_{iso}-tests for slope were significantly different from the expected value of 0.67 at $\alpha = 0.05$ level. A = Artiodactyla, Af = Afrotheria, C = Cingulata, Eq = Equidae, G = Glires, Hg = Hegetotheriidae, Hp = Hippopotamidae, In = Interatheriidae, L = Litopterna, N = Notoungulata, P = Perissodactyla, R = Ruminantia, Tl = Tylopoda, Tp = Tapiridae, Tx = Toxodontia, Ty = Typotheria, X = Xenarthra.

and *Pachyrukhos*) fed on monocotyledonous plants. Macraucheniids, *Nesodon* and astrapotheres were more likely mixed-feeders.

Inferences about digestive capabilities All regressions of occlusal surface area (OSA) versus body mass are highly correlated, most of them with values of R^2 above 0.9 (Table 14.6). Positive allometric trends were found for the all-mammal sample (omitting "Meridiungulata"), as well as for the "Meridiungulata" regression and the Notoungulata regression, while all other regressions were isometric (Table 14.6). Analysis of residuals shows that, among "Meridiungulata," the Litopterna have a greater OSA available for triturating food than the Notoungulata; this pattern is also found within Euungulata between Perissodactyla and Artiodactyla irrespective of diet. Among extant ungulates, monogastric forms such as zebras (Equidae) and tapirs (Tapiridae) are above the mammalian regression line, while ruminants (Cervidae and Bovidae) lie below it. This coincides with the feeding strategies in relation to

differences in the physiology of digestion between hindgut fermenter and foregut fermenters (Vizcaíno *et al.*, 2006; but see Janis *et al.*, 2010).

The specimens of *Astrapotherium* show relative OSA values (i.e. residuals from a least-squares regression) similar to that of Proboscidea, which are placed very close to the mammalian regression line (Fig. 14.16a).

The Litopterna, especially the Proterotheriidae, have relative OSA values comparable to those of Equidae but greater than those of Cervidae of similar BM (Fig. 14.16a). *Thoatherium* specimens show the highest regression residual values, *Diadiaphorus* and *Tetramerorhinus* are intermediate, and *Anisolophus* has the lowest residuals (Fig. 14.16b).

Among Notoungulata we find two different patterns: the Toxodontia lie above the mammalian regression line (Fig. 14.16a) and all Typotheria tend to be on or below the mammalian regression line, but never above it. Among toxodontians, *Adinotherium* is close to the regression line, and *Homalodotherium* and three *Nesodon* specimens lie far above it. Among typotheres, the Hegetotheriidae are just

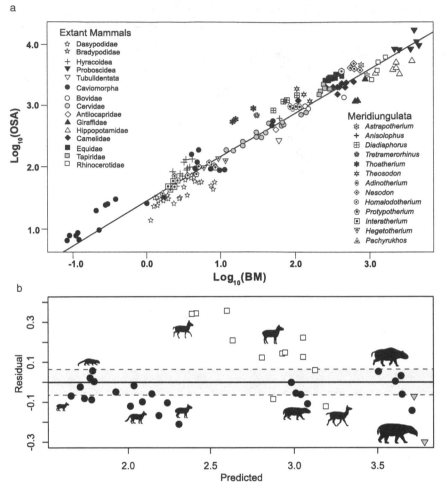

Fig. 14.16. a, Regression of OSA against body mass for extant mammals ($n = 125$) and Meridiungulata. Symbols: cross, Hyracoidea; squares, Perissodactyla (black, Equidae; gray, Tapiridae; white, Rhinocerotidae); white triangles, Hippopotamidae; black triangles, Giraffidae; black inverted triangles, Elephantidae; white inverted triangle, Tubulidentata; white rhombus, Antilocapridae; black rhombus, Camelidae; circles, Bovidae; gray disks, Cervidae; black disks, Rodentia; black five-tipped star, Bradypodidae; white five tipped stars, Dasypodidae. b, Graphical distribution of residuals from Meridiungulata regression line of OSA *vs.* body mass (see Table 14.6). Symbols: Notoungulata (black circles); Litopterna (white squares); Astrapotheria (inverted triangles).

below the line, with *Pachyrukhos* placed very close to the similar-sized caviomorph *Pediolagus* (Fig. 14.16a).

14.5 Discussion

14.5.1 Body mass

Previously reported body masses for Santacrucian genera or species are listed in Table 14.7 together with those obtained here. The body mass estimates in the present study are higher than those based only on simple regressions of m1 measurements previously reported by Croft (2000). Instead, they are closer to those based on postcranial remains given by Elissamburu (2004, 2007, 2011). The major exceptions are *Homalodotherium* with a far lower value (405 *vs.* 1171.6 kg), and *Nesodon* with an estimate ~100 kg greater (637 *vs.* 587.9 kg). The body masses reported by Vizcaíno *et al.* (2010) based on cranial remains and using only simple regressions are somewhat higher than those obtained here for Notoungulata (with the exceptions of *Homalodotherium*

and *Nesodon*, which are lower) and Astrapotheria, but lower for the Litopterna. Our values for the species of *Protypotherium* are very close to those reported by Scarano *et al.* (2011), being equal in the case of *P. praerutilum* (4.5 kg) and slightly higher for *P. australe* and *P. attenuatum* (7.73 *vs.* 5.86 kg and 3.8 *vs.* 2.85 kg, respectively). The values calculated here for the latter two species are very similar to the maximum values (7.4 kg and 3.1 kg, respectively) obtained by Scarano *et al.* (2011). Among proterotheriids, our estimates are all lower than those reported by Villafañe (2005). This is the first time that body masses for the litopterns *Theosodon garrettorum* and *T. gracilis* are reported and, hence, no contrast can be made for these species.

Santacrucian ungulates can be divided into three size ranges that can be characterized into base-10 logarithmic scale ranges, where each log unit corresponds to one order of magnitude (Fig. 14.17). There is an overlap in size of taxa belonging to different orders (see below) at around 100 kg. These taxa include the macraucheniid litoptern

Table 14.7. *Comparison of mean body mass (kg) obtained here to estimates reported previously*

Family	Species	Croft (2000)	Other sources	Vizcaíno *et al.* (2010)	This chapter
Astrapotheriidae	*Astrapotherium magnum*	503.8		1021.63	**921.32**
Macraucheniidae	*Theosodon garrettorum*				**158.04**
	Theosodon gracilis				**121.55**
	Theosodon lydekkeri	46.04			**130.93**
Proterotheriidae	*Anisolophus australis*	8.84*	62.63(a)	18.14	**36.61**
	Diadiaphorus majusculus	26.48	190.11(a)	70.25	**82.05**
	Tetramerorhinus cingulatum		82.76(a)		**41.71**
	Tetramerorhinus lucarius		53.52(a)		**29.50**
	Tetramerorhinus mixtum		65.58(a)		**35.06**
	Thoatherium minusculum	6.25	45.00(a)	21.0	**24.20**
Homalodotheriidae	*Homalodotherium*	300.547*	1171.6(b)*	340.0*	**405.08**
Toxodontidae	*Adinotherium ovinum*	33.75	119.45(b)*	121.0*	**100.29**
	Adinotherium robustum	24.32			**126.24**
	Nesodon imbricatus	293.66	587.9(b)*	554.0*	**637.51**
Interatheriidae	*Interatherium robustum*	0.399	3.33(b)*	3.5	**2.38**
	Protypotherium attenuatum	0.906	2.85(c)	4.4	**3.80**
	Protypotherium australe	2.817	5.86(c)/6.74(b)*	7.8	**7.73**
	Protypotherium praerutilum	0.976	4.57(c)		**4.51**
Hegetotheriidae	*Hegetotherium mirabile*	2.190	9.69(b)*	14.23	**7.71**
	Pachyrukhos moyani		1.77(b)*		**2.13**

Asterisk (*), when mass was reported only for the genus.
(a) Villafañe (2005); (b) Elissamburu (2011); (c) Scarano *et al.* (2011).

Theosodon, the proterotheriid litoptern *Diadiaphorus*, and the toxodontid notoungulate *Adinotherium*. This pattern implies that there are representatives of three families of ungulates around 100 kg in size living in roughly the same environment.

14.5.2 Locomotion and substrate use

Riggs (1935: 176) proposed that *Astrapotherium* inhabited forests or riverine grassland plains (meadowland) and frequented lagoons and river banks. This inference was based on features of the dentition, its inferred diet, its putatively padded feet, and sedimentological evidence. Marshall *et al.* (1990) found depositional evidence that supported Riggs's paleohabitat interpretations. Scott (1937b: 325) pointed out that *Astrapotherium* had remarkably short and weak limbs, which suggested an aquatic habit. Recently, Avilla and Vizcaíno (2005), using the same indices applied here, concluded that "*A. magnum* was cursorial, similar to large ungulates" and "had a pelvic limb-dominated semi-aquatic locomotor pattern, using the four limbs alternatively (similar to hippos, *Hippopotamus*)." In our analysis, the HRI is

similar to values for the Indian rhinoceros *Rhinoceros* (~12.7; Table 14.8) and higher than that of the African elephant *Loxodonta* (~9.7), forms that reach larger body masses (2500 and 5000 kg respectively; Nowak, 1991) than that estimated for *Astrapotherium* (around 1000 kg; see Table 14.3). This indicates that the forelimb of *Astrapotherium* was capable of withstanding high mechanical loads, even more than its own body mass. The femur is slender, with a FRI value between *Loxodonta* and *Rhinoceros* (~8 and ~10, respectively; Table 14.8). The tibia is more gracile than in *Loxodonta* and *Rhinoceros* (TRI ~14). Considering the geometry of the limbs, Biewener (1983) showed that large mammals avoid putting excessive stress on their bones by keeping their legs straighter than those of small mammals. There are, however, differences among large mammals between the rather straight legs of elephants and the somewhat flexed (but also much shorter) legs of the rhinoceroses and hippopotamus. In this sense, it is noteworthy that the limbs of *Astrapotherium* more resemble those of rhinos than elephants. Our results do not support Scott's (1937b) idea of weak limbs. In sum, the body size and

Table 14.8. *Reference values of limb indices for extant mammals*

Catalog	Species	Mass (kg)	IFA	BI	HRI	FRI	TRI	CI
MLP 186	*Myocastor coypus*	4.2	21.29	114.15	12.92	10.71	10.00	111.76
MLP 1079	*Dolichotis patagonum*	12	20.00	130.43	9.56	8.92	6.18	130.77
MLP 1012	*Pteronura brasiliensis*	5.8	28.47	76.80	10.00	9.59	9.41	116.44
MLP 1123	*Loxodonta africana*	2500	31.15	70.93	9.77	8.10	13.41	55.14
MLP 1125	*Rhinoceros unicornis*	1300	42.42	76.74	12.79	9.80	14.64	70.59
MLP 1124	*Hippopotamus amphibius*	1200	50.00	57.89	13.18	12.05	24.63	61.36
MLP 1130	*Camelus dromedarius*	326	17.02	123.68	11.42	8.38	7.53	91.49
MLP w/n	*Cervus elaphus*	250	30.00	100.00	14.47	9.50	7.53	105.26
MLP 1068	*Sus scrofa*	100	42.86	70.00	11.95	10.48	10.67	85.71
MLP w/n	*Madoqua kirkii*	4.89			9.16	7.98	5.68	124.37

Mass values estimated for the specimens come from Bargo *et al.* (2000).

w/n, without catalog number available.

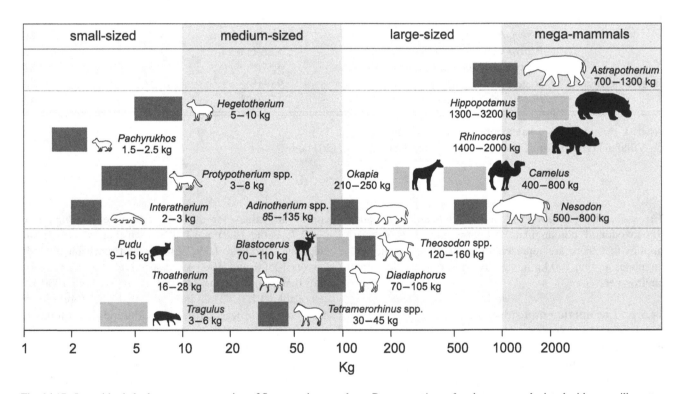

Fig. 14.17. Logarithmic body mass representation of Santacrucian ungulates. Representatives of each group are depicted with open silhouettes looking left. Extant references (artiodactyls and perissodactyls) are in black silhouette looking right. Horizontal length of rectangles indicates the range of body masses obtained.

limb proportions (due to the combination of all indices) of *Astrapotherium* seem to be comparable to those of the Indian rhinoceros, which besides being a good runner is also an excellent swimmer (Nowak, 1991; Neuschulz and Meister, 1998; Konwar *et al.*, 2009). Considering that most mammals are able to swim, the proposal by Avila and Vizcaíno (2005) cannot be discarded but should be re-evaluated.

Litopterns are characterized by slender limbs, but Scott (1910, 1913) emphasized that the Proterotheriidae were more gracile than Macraucheniidae. Based on reduction in the number of digits, *Thoatherium* was considered more gracile and more of a runner than *Diadiaphorus*. Our fore-limb indices are very similar among the macraucheniid *Theosodon* and the proterotheriids. For example, they show similar SMI, HRI, and URI values, indicating approximately the same forelimb robustness. The IFA value is similar to that of proterotheriids and close to that of the elk *Cervus elaphus* (~30; Table 14.8). The EI values suggest good available space for autopodial retractor musculature. In addition, the high BI value indicates that *Theosodon* was able to extend its forelimb faster than any other Santacrucian ungulate – much like the dromedary *Camelus dromedarius* (~123; Table 14.8). All hindlimb robustness indices indicate that proterotheriids were more gracile than *Theosodon*. Among proterotheriids, *Thoatherium* and *Tetramerorhinus* are more gracile than *Diadiaphorus* and *Anisolophus*. Even though speed and reaction indices (BI and CI) are not available for all litopterns and variability cannot be assessed, all available values are similar and particularly homogeneous among proterotheriids, suggesting similar mechanical capabilities for their limbs. Unfortunately, our data set is rather incomplete and we lack a better mechanical understanding that might allow a biomechanically supported hypothesis to contrast with those from classic works. However, these forms are unguligrade mammals with a reduced number of digits (see Introduction), indicating good capabilities for fast locomotor activities like running.

Among notoungulates, typotheres are mostly referred to as rodent-like in overall form (Ameghino, 1889; Sinclair, 1909; Bond *et al.*, 1995; Croft, 1999; Reguero *et al.*, 2007). Sinclair (1909), based on hindfoot morphology, argued that hegetotheriids were jumping animals with saltatory locomotion (particularly *Pachyrukhos*), while the interatheriids had cursorial habits. A locomotor morphofunctional analysis by Elissamburu (2004) compared typotheres with extant caviomorph rodents and revealed that certain indices characterize *Pachyrukhos* and *Protypotherium* as occasional diggers, with *Interatherium* and *Hegetotherium* as frequent diggers. Croft and Anderson (2008) pointed out that *Protypotherium* was most likely a generalized terrestrial mammal tending towards cursoriality (i.e. quadrupedal runner *sensu* Croft and Anderson, 2008), whereas *Interatherium* was more similar to a mustelid and probably more fossorial (*sensu* Elissamburu, 2004), certainly not cursorial. The results of Seckel and Janis' (2008) analyses, based on scapular morphology, led these authors to hypothesize that the small pachyrukine notoungulates may well have had a rabbit-like form of rapid half-bounding locomotion.

In our analysis of the forelimb, typotheres were the most gracile and agile of all Santacrucian ungulates (perhaps even more so than proterotheriids), although this could just

reflect their smaller body size. Comparing absolute values of forelimb indices such as HRI, EI, and IFA (which indicate mechanical capabilities involved in digging) among caviomorphs and typotheres, *Interatherium* and *Hegetotherium* segregate as intermediate between digging and occasionally digging caviomorphs, while *Protypotherium* and *Pachyrukhos* plot with non-digging caviomorphs (see Table 2 from Elissamburu and Vizcaíno, 2004). Although *Interatherium* and *Hegetotherium* may have been good diggers when compared with other Santacrucian typotheres (Elissamburu, 2004), according to our results all Santacrucian typotheres were poor diggers with respect to caviomorph rodents.

The hind limb indices of typotheres support their characterization as gracile as interpreted from forelimb indices. The femur is as gracile as that of protherotheriids, and the tibia is even more gracile. The indices that indicate fast initial propulsion during locomotion (TSI) and higher velocity development (CI) are higher in typotheres (except *Interatherium*) than in proterotheriids, showing values comparable to quadrupedal fast runners and jumpers ("cursorial" and "bounding" *sensu* Croft and Anderson, 2008).

Thus, all Typotheria seem to be capable of developing fast locomotion and to have had poor digging capabilities. For example, while *Pachyrukhos* was the most agile and fastest (like the caviomorph *Dolichotis*, the mara, or the bovid *Madoqua*, the dik dik, according to their index values; Table 14.8), *Interatherium* may have had low initial reactions and developed less final speed, optimizing muscular power. The latter is true also for semi-aquatic forms like the otter and the coypu (see index values for *Pteronura brasiliensis*, Lutrinae, and *Myocastor coypus*, Myocastoridae). As Toledo *et al.* (2012) suggest for the forelimb, mechanical requirements for swimming can be similar to those for climbing and digging activities. Thus, *Interatherium* shows a generalized pattern without any well-marked specialization towards a specific activity. *Hegetotherium* to some degree shows limbs that are well-suited for agile and fast locomotor behavior, intermediate between *Protypotherium* and *Pachyrukhos*.

The Toxodontia have the most robust fore- and hind-limb proximal elements among Santacrucian ungulates. The nesodontines *Adinotherium* and *Nesodon* have the most robust forearm, and *Homalodotherium* the most robust distal hindlimb. The nesodontines show a combination of indices that indicates a strong and specialized shoulder, robust forelimb elements (humerus and ulna) that resist mediolateral tension, with a great mechanical advantage for powerful, but not for fast elbow extension. They have index values of the forearm characteristic of digging mammals (i.e. high EI and IFA values and low BI). Additionally, the hindlimb indices show a robust femur (FRI) for their body mass and a strong leg (TSI and CI of 10.7 and 85, respectively, similar to pigs like *Sus scrofa*) as well adapted for initial acceleration and developing speed as proterotheriids (see above). Although IFA was originally conceived as an indicator of digging ability in armadillos (see Vizcaíno and Milne, 2002) and also works well for caviomorph rodents (Elissamburu and Vizcaíno, 2004), it must be used carefully when applied to different clades. New functional interpretations suggest that in large glyptodonts, high values of IFA (long olecranon process) were related to reasons other than digging (Milne *et al.*, 2009). Moreover, Vizcaíno *et al.* (2011) suggest that high values of IFA in large glyptodonts are directly related to body support and movement and not to digging behavior, such as supporting and maneuvering its massive body mass on flexed elbows. So, taking into account all the index values, these results are consistent with a generalized model of walking cursorial (following the nomenclature of Stein and Casinos, 1997) ungulates of moderate to large size. Further studies are needed for a better characterization.

Homalodotherium is the Santacrucian ungulate that perhaps least resembles the locomotor model of a typical extant ungulate (i.e. quadrupedal with an unguligrade foot posture). Riggs (1937) interpreted its postcranial features as specializations to access arboreal leaves with the manus while most of the weight of the body was supported by the hindlimbs in a bipedal position. The indices reported by Elissamburu (2010) and those calculated here suggest that *Homalodotherium* must have optimized its shoulder structure to emphasize speed of extension rather than power. However, EI suggests well-developed musculature for hand prehension and retraction. Elissamburu's (2010) muscular reconstruction suggests a high degree of habitual elbow supination in contrast to the usual prone forearm of extant ungulates. The FRI also indicates a robust hindlimb able to support a large proportion of the body mass, optimizing muscular power over speed.

14.5.3 Feeding behavior

Inferences about habitat Several studies (Janis and Ehrhardt, 1988; Solounias and Moelleken, 1993; Janis, 1995; Pérez-Barbería and Gordon, 1999; Mendoza *et al.*, 2002) have used the correspondence between craniodental morphology and ecological traits to infer habits of extinct mammals. However, morphological adaptations of the craniodental skeleton for feeding are not easily distinguished from adaptations to habitat. These could be reflecting the fact that food is not distributed randomly with respect to the habitat. In fact, analyses of craniodental measures that take phylogenetic constraints into account, by Williams and Kay (2001) on rodents, primates, and ungulates, and

Pérez-Barbería *et al.* (2001) on extant ungulates, demonstrate that habitat and diet are correlated. Crown height is the morphological trait that has been most frequently considered as indicative of feeding behavior in ungulates in relation to grass consumption and the spread of grassland habitats (for extensive overviews on the subject see: Janis, 1995; MacFadden, 2000; Strömberg, 2006; Reguero *et al.*, 2010). In a recent contribution, Billet *et al.* (2009) summarized the factors that favored increasing hypsodonty as two alternative hypotheses, namely increasing chewing effort and increase of consumed abrasives. Although hypsodonty, as an adaptation for wear tolerance, has been firmly linked to dry climate (Fortelius, 1985; Janis, 1988; Fortelius *et al.*, 2002; Mendoza and Palmqvist, 2008), the mechanisms remain elusive.

Generating hypotheses about habitat preferences was the first step in our analysis. In the analyses of Cassini *et al.* (2011) and Mendoza and Palmqvist (2008), all ungulates (except species with special adaptations, such as high-level browsers like the giraffe) that dwell in open or mixed habitats have a HI greater than 2. The ability of HI to distinguish between ungulates from open habitats and those from mixed habitats increases with the relative length of the jaw (JAW). Using JAW, HI, or HPI, it is possible to characterize species adapted to forage in open, mixed, and closed habitats.

Following the above studies of ungulate morphospace (based primarily on HPI, HI, and JAW), all Santacrucian notoungulates studied (*Homalodotherium* was not included in the analysis, see Methods) foraged in open habitats, although *Interatherium* could have also been from mixed or water-associated open habitats. The Astrapotheria and Litopterna foraged in closed habitats.

Inferences about diet Among Santacrucian closed-habitat ungulates, astrapotheres and litopterns, based on cranial structure, we obtain a good discrimination of omnivorous forms from non-omnivorous forms. For *Astrapotherium* the results are not conclusive, although in the first step they were classified as omnivorous based on premolar row length and masseteric fossa length. This could be interpreted as an artifact due to the loss and reduction in size of premolars (Kramarz and Bond, 2009). Regardless, the position that astrapotheres occupy in the morphospace depicted by the other variables shows some characteristics intermediate between artiodactyl and perissodactyl browsers, particularly in the features of the mandibular angle. This is in agreement with the traditional idea of *Astrapotherium* being a browser mainly based on the putative presence of a proboscis, and supported by the enamel structure (which is shared with rhinocerotids, most of them browsers) and low-magnification enamel microwear analyses (Rensberger and Pfretzschner, 1992; Townsend and Croft, 2005).

Santacrucian litopterns show a combination of variables very different from those seen in extant omnivorous ungulates like some pigs, and none of them were classified as such. Bunodont protherotheriids are absent from the Santa Cruz Formation, unlike the bunodont Miocene proterotheriids of La Venta fauna (Huila, Colombia) i.e. *Megadolodus molariformis* McKenna, 1956 and *Prothoatherium colombianus* (Hoffstetter and Soria, 1986), which may have had a similar dietary preference to omnivorous suoids (Cifelli and Guerrero, 1997). Among Santacrucian litopterns, proterotheriids present a combination of mandibular traits characteristic of extant perissodactyl browsers (i.e. in width and depth of mandibular angle and premolar row length) and narrow muzzles like the extant artiodactyl browsers (e.g. the pudu deer *Pudu* spp.; Figs. 14.14c and 14.15a) and frugivores (e.g. the duikers *Cephalophus* spp; Figs. 14.14c and 14.15a). Soria (2001) proposed that the proterotheriids were browsing herbivores. This was based mainly on their brachydont cheek teeth (here used in combination with other craniodental traits for habitat characterization). In addition, the presence of proterotheriids in the fossil record tends to be correlated with forested habitats (Cifelli and Guerrero, 1997; Bond *et al.*, 2001; Kramarz and Bond, 2005).

Turning to macraucheniids, *Theosodon* presents similarities with both browsers and mixed feeders, as defined above. Kramarz and Bond (2005) argue that the broad geographic and chronologic distribution of macraucheniids as a group precludes the use of them as environmental indicators. However, the largest Santacrucian species, *Theosodon garrettorum*, has a combination of traits closer to the mixed feeders of closed habitats (i.e. in muzzle width and the depth and width of mandibular angle), whereas the smaller species, *T. gracilis* and *T. lydekkeri*, show a combination of mandibular traits similar to the extant closed-habitat browsers and a muzzle width close to that of mixed feeders. In sum, the Santacrucian *Theosodon* would likely have inhabited mainly closed habitats, with the larger species having more mixed-feeding habits than the smaller ones, which would have had more browsing habits.

Among the more open-habitat ungulates, the typotheres *Interatherium*, *Protypotherium*, *Hegetotherium*, and *Pachyrukhos* show some characteristics of extant grazers, i.e. a wide muzzle, an extremely wide and deep mandibular angle, and a large paraoccipital process. Based on incisor morphology and implantation, Tauber (1996) proposed that the larger species of *Protypotherium* (*P. australe*) was adapted to forage on tougher grasses than the other two (*P. praerutilum* and *P. attenuatum*).

Among toxodontids, *Adinotherium* also shows craniodental morphology characteristic of extant grazers, i.e. a wide

muzzle, an extremely wide and deep mandibular angle, and a large paraoccipital process, while the specimens of *Nesodon* occupy a morphospace between grazers and nongrazers, showing no clear discrimination (see Cassini *et al.*, 2011, for detailed discussion).

Inferences about digestive capabilities Allometric coefficients obtained here by SMA regression did not differ from the parameters reported by Vizcaíno *et al.* (2006) for the same taxonomic groups and same data set. Copes and Schwartz (2010) find a negative allometric relationship between the postcanine occlusal surface area and body size in mammals (slope = 0.59). In contrast, the allometric coefficients obtained here do not deviate significantly from isometry (slope not significantly different from 0.67). However, it must be noted that in the Copes and Schwartz (2010) approach, occlusal surface area was approximated as a rectangle and the infolding and fossettes were not considered. Therefore, although both studies attempt to quantify the same property, the level of detail of quantification is different.

Our results, using a wide range of sizes and morphological diversity of herbivorous mammals, from the guinea pigs *Cavia* to the African elephant *Loxodonta*, show that the occlusal surface area (OSA) increases with an exponent of 0.716 (Table 14.6; significantly different from 2/3), a value that barely exceeds that reported by Vizcaíno *et al.* (2006). It shows a positive allometric growth trend, so large herbivorous mammals have proportionately more OSA than small ones.

Gould (1975) suggested that the occlusal area of the mammalian postcanines should maintain a relationship with respect to body mass to meet metabolic requirements. McMahon and Bonner (1983) proposed the elastic similarity rule, whereby the cross-sectional area of any anatomical feature grows in proportion to mass raised to the power 3/4 (0.75). West *et al.* (1997, 1999a, b) proposed that exponents in multiples of 1/4 found in many biological traits in relation to body mass have their origin in the fractal patterns of structure and function of systems (e.g. surfaces exchange in the respiratory system), so the power laws obtained empirically describe mathematically the hierarchical organization and fractal design (dynamic and self-similarity) over a wide range of scales (Brown *et al.*, 2002).

Regardless of the mechanistic origin of allometric coefficients, the reasoning that led Gould (1975) to link the occlusal tooth area with the metabolism was more than a coincidental power of multiples of 1/4. He suggested that this was directly related to the amount of food that an organism can process in the oral cavity. While the coefficient we obtained for the regression of herbivorous mammals differs significantly from the 3/4 value expected for Kleiber's law or basal metabolic law (Kleiber, 1932), it is close enough to potentially lend support to Gould's reasoning. However, as most of the equations do not differ significantly from 0.67 (isometry), the relationship between OSA and metabolism is not really strong enough to support this hypothesis (but see Fortelius, 1988). Isometry of tooth size (usually molar area) with body mass has been identified and reviewed (Fortelius, 1990; Ungar, 1998; Lucas, 2004), as well as other particular aspects of the mammalian masticatory system (e.g. Fortelius, 1985), but see Cassini *et al.* (2012) for discussion on positive allometric scaling of teeth and masticatory apparatus across ontogeny in nesodontines. Ungar's (1998) review on primates states that molar size varies isometrically with body size within diet categories and is positively allometric across them (see also Kay, 1975), and he hypothesized that large mammals eat abundant low-energy foods, whereas small taxa tend to eat foods with a higher energy and protein content; larger primates would therefore require more tooth surface area for processing greater quantities of lower-energy foods. However, in the allometric analyses followed here, we attempt to explain the functional value of OSA independent of body mass.

Therefore, given a pattern of isometry, the functional value of OSA independent of body mass is not explained by the allometric coefficient, but perhaps by the normalization constant (intercept) and the regression residuals (see Vizcaíno *et al.*, 2006). For example, among the ungulates, the artiodactyls (especially ruminants) are distributed below the regression line and have lower values than monogastric forms such as equids. This parallels the observations made by Janis (1988, 1995) that equids have a longer and more molarized premolar series than ruminants. These differences reflect different physiological mechanisms of digestion and food-processing strategies (Janis *et al.*, 2010). They also reflect the time spent chewing food by different dietary groups (grazers *vs.* browsers) and the quality of the diet (Pérez-Barbería and Gordon, 1998). This led Vizcaíno *et al.* (2006) to propose correlations amongst OSA and the ability to process food in the oral cavity, aspects of digestive physiology, and energy metabolism.

The present results indicate that, within the "Meridiungulata," the Santacrucian litopterns have proportionately higher OSA than observed for similar-sized notoungulates. The Litopterna, particularly proterotheriids, have proportionally similar values of OSA to those of extant equids (Fig. 14.16a), reflecting even higher values of the normalization constant (intercepts: 2.11 and 2.08 respectively; Table 14.6). Although proterotheriid crowns are selenodont instead of lophoselenodont as in equids, both share a molarized premolar series. In contrast, the notoungulates, especially typotheres, have proportionally lower values of OSA, similar to those of extant deer, and the intercept is closer to that of artiodactyls (1.5 and 1.59, Table 14.6). In typotheres this could be related to tooth infolding and/or smaller

premolars, whereas Artiodactyla exhibit loss of premolars, but an increase in enamel surface in relation to total occlusal surface.

Residual analysis accounts for these differences too (Meridiungulata regression line; Table 14.6), because the residual values of OSA (i.e. the differences between the observed values of OSA and fitted values) for Litopterna (white squares, Fig. 14.16b) are significantly higher than those for notoungulates (black circles, Fig. 14.16b) of comparable size. Therefore, per unit of body mass, Litopterna have more dental occlusal surface for grinding food than do notoungulates. If the digestive and metabolic physiology can be inferred from craniodental features, such as dental occlusal surface (Gould, 1975; Janis and Fortelius, 1988; Janis and Constable, 1993; Janis, 1995; Vizcaíno *et al.*, 2006), the results suggest that litopterns, especially proterotheriids, used more intraoral food preparation, while notoungulates (especially typotheres) used less. The fact that extant hindgut fermenters are known to have a greater food intake than ruminants (of similar body size and diet), and horses may chew their food more thoroughly than ruminants do on initial ingestion, led us to presume that Litopterna had a gut physiology comparable to hindgut fermenters, and Notoungulata likely had a gut physiology comparable to ruminants. However, this scenario is not in agreement with the conclusion of Fletcher *et al.* (2010). Following these authors, hindgut feeders have jaws which are more "robust" than the relatively more gracile jaws of ruminants. As notoungulates have a more robust mandible than litopterns (even more than perissodactyls; see above), they likely could have had hindgut fermentation. Therefore, at least for notoungulates, two not necessarily conflicting inferences can be made. On the one hand, taking into account the hypothesis of a primarily grazing diet (which implies longer retention time of ingesta; see Clauss *et al.*, 2008) and a reduced cheek tooth surface area by unit of mass, Notoungulata would have had better digestive capabilities than Litopterna (while at the same time not assuming that one group was a foregut fermenter whilst the other was a hindgut fermenter). On the other hand, the presence of a very robust jaw and the greatest HI in notoungulates could be compensating for having a small OSA by unit of mass for intraoral food processing.

14.5.4 Paleoenvironment

At least 13 genera of native ungulates (one Astrapotheria, seven Notoungulata, and five Litopterna) potentially coexisted in the lower levels of the Santa Cruz Formation in the area of the present coast of Southern Patagonia (from Monte Observación (= Cerro Observatorio) to Cabo Buen Tiempo).

Locomotor behavior, habitat preference, and feeding behavior analyses present a coherent picture and suggest a mixture of open and closed habitats. The Typotheria are indicative of the availability of open habitats (i.e. treeless or scarcely wooded savannas, grasslands, or semi-desert steppes), while litopterns (particularly the proterotheriids) were associated with closed habitats (i.e. closed woodlands or moist deciduous and evergreen forests). Additionally, two taxa could be interpreted as generalists, including habitats close to standing water. Habitat preference and locomotor characterization of *Interatherium* indicate edge to open areas with swamps or inundated plains, whereas for *Astrapotherium* locomotor patterns indicate the presence of riverine forests or streams in closed habitats. The coexistence of both closed and open habitats is also supported by sedimentological features, paleoflora indicators (Barreda and Palazzesi, 2007; Brea *et al.*, Chapter 7), and faunal composition (Kay *et al.*, Chapter 17).

14.6 Conclusions

Santacrucian ungulates fall into three size classes: (1) small-sized (1 to 10 kg) including all Typotheria, (2) medium-sized (10 to 100 kg) including all Proterotheriidae and *Adinotherium* (Toxodontidae: Notoungulata), and (3) large-sized (100 to 1000 kg) including *Theosodon* (Macraucheniidae: Litopterna), *Nesodon* and *Homalodotherium* (Toxodontia: Notoungulata), and *Astrapotherium magnum* (Astrapotheria). Most of the taxonomic richness and morphological diversity is composed of the small- to medium-sized ungulates (i.e. those below 100 kg).

Santacrucian ungulates show a range of locomotor strategies, ranging from quadrupedal walking to running. Typotheria were likely the most agile and fastest runners among Santacrucian ungulates, and among them *Hegetotherium* and *Interatherium* may have occasionally engaged in digging behaviors, not discounting for *Interatherium* putative swimming or climbing capabilities. Nesodontines probably were walking cursors, typical of medium-sized to heavier mammals. *Homalodotherium* has a peculiar locomotion not seen among extant ungulates, having the ability to adopt a bipedal posture and use the forelimbs to forage on tree leaves. Among Litopterna, the Proterotheriidae show a homogeneous functional pattern of a walking to running mammal, and they were probably more agile than the macraucheniid *Theosodon*. Finally, *Astrapotherium* resembles quadrupedal walking megamammals, but aquatic locomotor activities cannot be precluded.

According to the ecomorphological analyses, all Notoungulata (*Adinotherium*, *Nesodon*, *Protypotherium*, *Interatherium*, and probably *Hegetotherium* and *Pachyrukhos*) lived and/or foraged in open habitats. All Litopterna (*Anisolophus*, *Tetramerorhinus*, *Diadiaphorus*, and *Thoatherium*) and Astrapotheria (*Astrapotherium*) lived and/or foraged in closed habitats.

The relationship between the area of occlusal tooth wear and body size suggests that the Santacrucian ungulates (Notoungulata, Litopterna, and Astrapotheria) had a high basal metabolic rate, equivalent to extant herbivores like perissodactyls and artiodactyls. Among the Notoungulata, the Typotheria could compensate for a reduced intraoral food preparation with better digestive capabilities. The Litopterna would have been browsers that had predominantly intraoral food preparation and perhaps not as good digestive capabilities as notoungulates.

The herbivore niche partitioning is based on the differential use of environments (open, mixed, and closed) and the differentiation of the diet, mainly reflected in three biological attributes: (1) body size, (2) form and function of the cranial and dental traits, and (3) energy requirements.

Small ungulates (Typotheria) were open habitat grazers. *Pachyrukhos* and *Interatherium* exaggerate the traits attributed to grazing habits, while *Hegetotherium* exaggerates the features indicative of more open environments but not necessarily exclusively grazing.

Among the ungulates of medium size, the proterotheriids (*Anisolophus*, *Tetramerorhinus*, *Diadiaphorus*, and *Thoatherium*) likely were browsers in closed habitats. Gracile species of *Theosodon* likely had intermediate to stricter browser diets in mixed environments. *Adinotherium* likely was a grazer in open environments.

All large-sized ungulates likely had intermediate diets. *Theosodon garrettorum* probably dwelt in mixed environments, *Nesodon* in open environments, and *Astrapotherium* in more closed environments.

ACKNOWLEDGMENTS
We thank the following persons and institutions: the editors of the book, Susana Bargo, Sergio Vizcaíno, and Richard Kay, for inviting us to participate; Dirección de Patrimonio Cultural, and Museo Regional Provincial Padre M. J. Molina (Río Gallegos, Santa Cruz Province) for permission and support for fieldwork; vertebrate paleontological collections managers M. Reguero from MLP, A. Kramarz and J.C. Fernicola from MACN, J. Flynn from AMNH, and W. Joyce from YPM; Nestor Toledo for illustrating Fig. 14.11, for the use of his unpublished database on limb measurements, and for the revision of sections on locomotion; Marcelo Canevari for passionate labor in the life reconstructions of taxa herein studied; Jonathan Perry, Darin Croft and an anonymous reviewer, for their valuable suggestions on the manuscript. The study of the YPM collection was partially funded by the John H. Ostrom Research Fund. This is a contribution to the projects: PICT 26219 and 0143, CONICET-PIP 1054, UNLP N647 and National Geographic Society 8131–06 to Sergio F. Vizcaíno, and NSF 0851272 and 0824546 to Richard F. Kay.

Appendix 14.1   List of the material studied

Institutional abbreviations

AMNH: American Museum of Natural History, New York, USA.

FMNH: Field Museum of Natural History, Chicago, USA.

MACN-A: Museo Argentino de Ciencias Naturales "Bernardino Rivadavia," Colección Nacional Ameghino, Buenos Aires, Argentina.

MLP: Museo de La Plata, La Plata, Argentina.

MPM-PV: Museo Regional Provincial Padre M. Jesús Molina, Río Gallegos, Argentina.

YPM-VPPU: Yale Peabody Museum, Vertebrate Paleontology Princeton University Collection, New Haven, USA.

FL: Fossiliferous Level.

All specimens come from Santa Cruz Province, Argentina. See Vizcaíno *et al.* Chapter 1, Figs. 1.1 and 1.2 and Appendix 1.1) for references on the localities, and Matheos and Raigemborn (Chapter 4, Figs. 4.3 and 4.4) for information on the stratigraphy.

Astrapotheria
Astrapotherium magnum

AMNH 9278. Cranium and mandible. Horizon and locality: Santa Cruz Formation, Río Gallegos. YPM-VPPU 15142. Mandible. Horizon and locality: Santa Cruz Formation, 10 miles south of Coy Inlet. YPM-VPPU 15261. Cranium. Horizon and locality: Santa Cruz Formation, 10 miles South of Coy Inlet. YPM-VPPU 15332. Cranium and mandible. Horizon and locality: Santa Cruz Formation, 10 miles South of Coy Inlet.

Notoungulata
Typotheria
Interatheriidae
Interatherium excavatum

YPM-VPPU 15043. Cranium and mandible. Horizon and locality: Santa Cruz Formation, Coy Inlet.

Interatherium extensum

AMNH 9299. Cranium. Horizon and locality: Santa Cruz Formation, Halliday's Estancia, Río Gallegos.

Interatherium robustum

AMNH 9154. Cranium and mandible. Horizon and locality: Santa Cruz Formation, Cañadón de Las Vacas. AMNH 9284. Cranium and mandible. Horizon and locality: Santa Cruz Formation, Felton's Estancia, Río Gallegos. AMNH 9483. Mandible. Horizon and locality: Santa Cruz Formation, Río Gallegos. MPM-PV 3469. Cranium, mandible, and ulna. Horizon and locality: Santa Cruz Formation, Estancia La Costa Member, Estancia La Costa. MPM-PV 3471. Cranium, mandible, and humerus. Horizon and locality: Santa Cruz Formation, Estancia La Costa Member, Anfiteatro. MPM-PV 3527. Cranium, mandible, humerus, ulna, femur, and tibia. Horizon and locality: Santa Cruz Formation, Estancia La Costa Member, Anfiteatro locality. MPM-PV 3528. Mandible. Horizon and locality: Santa Cruz Formation, Estancia La Costa Member, Anfiteatro locality. MPM-PV 3539. Tibia. Horizon and locality: Santa Cruz Formation, Estancia La Costa Member, Anfiteatro. MPM-PV 4263. Articulated skeleton. Horizon and locality: Santa Cruz Formation, Estancia La Costa Member, FL 7, Puesto Estancia La Costa (= Corriguen Aike). YPM-VPPU 15041. Femur and tibia. Horizon and locality: Santa Cruz Formation, Güer Aike Department. YPM-VPPU 15100. Cranium and mandible. Horizon and locality: Santa Cruz Formation, unknown specific locality. YPM-VPPU 15108. Humerus, ulna, and tibia. Horizon and locality: Santa Cruz Formation, unknown specific locality. YPM-VPPU 15202. Humerus and tibia. Horizon and locality: Santa Cruz Formation, 10 miles south of Coy Inlet. YPM-VPPU 15293. Cranium. Horizon and locality: Santa Cruz Formation, Coy Inlet. YPM-VPPU 15296. Cranium and mandible. Horizon and locality: Santa Cruz Formation, 10 miles south of Coy Inlet. YPM-VPPU 15300. Cranium and mandible. Horizon and locality: Santa Cruz Formation, 10 miles south of Coy Inlet. YPM-VPPU 15401. Cranium, mandible, humerus, ulna, femur, and tibia. Horizon and locality: Santa Cruz Formation, Patagonia. YPM-VPPU 15554. Mandible. Horizon and locality: Santa Cruz Formation, unknown specific locality.

Protypotherium attenuatum

AMNH 9187. Cranium and mandible. Horizon and locality: Santa Cruz Formation, 15 miles south of Monte León. MACN-A 3991. Mandible (Paratype *Protypotherium leptocephalum*). Horizon and locality: Santa Cruz Formation, Corriguen Aike. MPM-PV 3470. Cranium and mandible. Horizon and locality: Santa Cruz Formation, Estancia La Costa Member, FL 5.3, Puesto Estancia La Costa (= Corriguen Aike). YPM-VPPU 15341. Humerus, ulna, femur, and tibia. Horizon and locality: Santa Cruz Formation, Killik Aike, Río Gallegos. YPM-VPPU 15665. Cranium. Horizon and locality: Santa Cruz Formation, Killik Aike, Río Gallegos.

Protypotherium australe

AMNH 9149. Humerus, ulna, femur, and tibia. Horizon and locality: Santa Cruz Formation, unknown specific locality. AMNH 9286. Cranium and mandible. Horizon and locality: Santa Cruz Formation, Felton's Estancia, Río Gallegos. AMNH 9565. Cranium and mandible. Horizon and locality: Santa Cruz Formation, Río Gallegos. MPM-PV 3531. Mandible. Horizon and locality: Santa Cruz Formation, Estancia La Costa Member, Campo Barranca. YPM-VPPU 15340. Femur and tibia. Horizon and locality: Santa Cruz Formation, Killik Aike, Río Gallegos. YPM-VPPU 15659. Femur and tibia. Horizon and locality: Santa Cruz Formation, Killik Aike, Río Gallegos. YPM-VPPU 15828. Cranium, mandible, humerus, and ulna. Horizon and locality: Santa Cruz Formation, Killik Aike, Río Gallegos. YPM-VPPU 15892. Humerus. Horizon and locality: Santa Cruz Formation, Killik Aike, Río Gallegos.

Protypotherium praerutilum

MACN-A 3920–21. Cranium and mandible. Horizon and locality: Santa Cruz Formation, Corriguen Aike. MPM-PV 3530. Mandible. Horizon and locality: Santa Cruz Formation, Estancia La Costa Member, FL 5.3, Puesto Estancia La Costa (= Corriguen Aike). MPM-PV 3659. Cranium and mandible. Horizon and locality: Santa Cruz Formation, Estancia La Costa Member, FL 5.3, Puesto Estancia La Costa (= Corriguen Aike). YPM-VPPU 15386. Cranium, mandible, humerus, femur and tibia. Horizon and locality: Santa Cruz Formation, Coy Inlet. YPM-VPPU 15161. Femur and tibia. Horizon and locality: Santa Cruz Formation, 10 miles south of Coy Inlet.

Protypotherium sp.

MACN-A 9657–63. Humerus. Horizon and locality: Santa Cruz Formation, unknown specific locality.

Hegetotheriidae

Hegetotherium mirabile

AMNH 9159. Cranium and mandible. Horizon and locality: Santa Cruz Formation, unknown specific locality. AMNH 9223. Cranium. Horizon and locality: Santa Cruz Formation, Halliday's Estancia, Río Gallegos. MPM-PV 3526.

Cranium. Horizon and locality: Santa Cruz Formation, Estancia La Costa Member, FL 5.3, Puesto Estancia La Costa (= Corriguen Aike). MPM-PV 4316. Mandible. Horizon and locality: Santa Cruz Formation, Estancia La Costa Member, FL 5.3, Puesto Estancia La Costa (= Corriguen Aike). YPM-VPPU 15176. Humerus and ulna. Horizon and locality: Santa Cruz Formation, Cabo Buen Tiempo. YPM-VPPU 15298. Mandible and tibia. Horizon and locality: Santa Cruz Formation, unknown specific locality. YPM-VPPU 15395. Tibia. Horizon and locality: Santa Cruz Formation, unknown specific locality. YPM-VPPU 15431. Femur and tibia. Horizon and locality: Santa Cruz Formation, Killik Aike, Río Gallegos. YPM-VPPU 15542. Cranium, mandible, humerus, and ulna. Horizon and locality: Santa Cruz Formation, Killik Aike, Río Gallegos.

Pachyrukhos moyani

AMNH 9219. Cranium and mandible. Horizon and locality: Santa Cruz Formation, Halliday's Estancia, Río Gallegos. AMNH 9242. Femur and tibia. Horizon and locality: Santa Cruz Formation, Felton's Estancia, Río Gallegos. AMNH 9283. Cranium and mandible. Horizon and locality: Santa Cruz Formation, Felton's Estancia, Río Gallegos. AMNH 9285. Ulna and femur. Horizon and locality: Santa Cruz Formation, Felton's Estancia, Río Gallegos. AMNH 9481. Humerus and ulna. Horizon and locality: Santa Cruz Formation, Felton's Estancia, Río Gallegos. YPM-VPPU 15743. Cranium. Horizon and locality: Santa Cruz Formation, Killik Aike, Río Gallegos. YPM-VPPU 15744. Cranium and mandible. Horizon and locality: Santa Cruz Formation, Killik Aike, Río Gallegos.

Toxodontia
Homalodotheriidae

Homalodotherium cunninghami

YPM-VPPU 15747. Ulna. Horizon and locality: Santa Cruz Formation, Patagonia. MPM-PV 3706. Fragmentary cranium. Horizon and locality: Santa Cruz Formation, Estancia La Costa Member, FL 7, Puesto Estancia La Costa (= Corriguen Aike).

Toxodontidae

Adinotherium ovinum

AMNH 9571. Cranium. Horizon and locality: Santa Cruz Formation, Río Gallegos. MACN-A 5352–53. Cranium and mandible (Type *Adinotherium ferum*). Horizon and locality: Santa Cruz Formation, Corriguen Aike. MPM-PV 3666 Cranium and mandible. Horizon ad locality: Formación Santa Cruz, Estancia La Costa Member, Campo Barranca. MPM-PV 3667. Cranium and mandible. Horizon and locality: Santa Cruz Formation, Estancia La Costa Member, Campo Barranca. MPM-PV 3668. Cranium and mandible.

Horizon and locality: Santa Cruz Formation, Estancia La Costa Member, arcillas below FL 2, Estancia La Costa. MPM-PV 3532. Cranium. Horizon and locality: Santa Cruz Formation, Estancia La Costa Member, FL 6, Puesto Estancia La Costa (=Corriguen Aike). MPM-PV 3542. Humerus, ulna, and tibia. Horizon and locality: Santa Cruz Formation, Estancia La Costa Member, Anfiteatro. YPM-VPPU 15003. Cranium and mandible. Horizon and locality: Santa Cruz Formation, 20 miles south of Coy Inlet. YPM-VPPU 15004. Humerus and ulna. Horizon and locality: Santa Cruz Formation, 10 miles south of Coy Inlet. YPM-VPPU 15118. Cranium. Horizon and locality: Santa Cruz Formation, south of Coy Inlet. YPM-VPPU 15127. Femur. Horizon and locality: Santa Cruz Formation, 10 miles south of Coy Inlet. YPM-VPPU 15131. Humerus, ulna, femur, and tibia. Horizon and locality: Santa Cruz Formation, Coy Inlet. YPM-VPPU 15136. Mandible. Horizon and locality: Santa Cruz Formation, 10 miles south of Coy Inlet. YPM-VPPU 15480. Tibia. Horizon and locality: Santa Cruz Formation, 15 miles south of Coy Inlet. YPM-VPPU 15966. Humerus. Horizon and locality: Santa Cruz Formation, 10 miles south of Coy Inlet.

Adinotherium robustum

AMNH 9497. Cranium. Horizon and locality: Santa Cruz Formation, Patagonia. AMNH 9532. Cranium. Horizon and locality: Santa Cruz Formation, Cobaredonda, 5 miles north of Coyle Inlet.

Adinotherium sp.

AMNH 9141. Mandible. Horizon and locality: Santa Cruz Formation, Cañadón de Las Vacas. AMNH 9229. Humerus, femur and tibia. Horizon and locality: Santa Cruz Formation, unknown specific locality. MACN SC 4355. Cranium and mandible. Horizon and locality: Santa Cruz Formation, Monte Observación (= Cerro Observatorio).

Nesodon imbricatus

AMNH 9192. Mandible, humerus, and ulna. Horizon and locality: Santa Cruz Formation, unknown specific locality. MACN-A 5145. Mandible. Horizon and locality: Santa Cruz Formation, Corriguen Aike. MACN-A 774–775. Cranium. Horizon and locality: Santa Cruz Formation, Corriguen Aike. MLP 12–250. Mandible (Type *Protoxodon marmoratus*). Horizon and locality: Santa Cruz Formation, Santa Cruz River. MPM-PV 4377. Cranium. Horizon and locality: Santa Cruz Formation, Estancia La Costa Member, FL 5.3, Puesto Estancia La Costa (= Corriguen Aike). MPM-PV 3659. Cranium and mandible. Horizon and locality: Santa Cruz Formation, Estancia La Costa Member, FL 2, Estancia La Costa. YPM-VPPU 15000. Cranium and mandible. Horizon and locality: Santa Cruz Formation, 10 miles south of Coy Inlet. YPM-VPPU 15132. Ulna. Horizon and locality: Santa Cruz Formation, Patagonia. YPM-VPPU

15141. Cranium. Horizon and locality: Santa Cruz Formation, unknown specific locality. YPM-VPPU 15215. Cranium and mandible. Horizon and locality: Santa Cruz Formation, Coy Inlet. YPM-VPPU 15256. Cranium and mandible, humerus, and ulna. Horizon and locality: Santa Cruz Formation, unknown specific locality. YPM-VPPU 15260. Mandible. Horizon and locality: Santa Cruz Formation, unknown specific locality. YPM-VPPU 15336. Cranium and mandible. Horizon and locality: Santa Cruz Formation, 10 miles south of Coy Inlet. YPM-VPPU 15492. Cranium and mandible. Horizon and locality: Santa Cruz Formation, 8 miles south of Coy Inlet. YPM-VPPU 15967. Femur and tibia. Horizon and locality: Santa Cruz Formation, Coy Inlet.

Nesodon sp.

AMNH 9128. Cranium and mandible. Horizon and locality: Santa Cruz Formation, unknown specific locality. AMNH 9168. Cranium and mandible. Horizon and locality: Santa Cruz Formation, Río Gallegos. AMNH 9510. Mandible. Horizon and locality: Santa Cruz Formation, Felton's Estancia, Río Gallegos. AMNH 9553. Ulna and tibia. Horizon and locality: Santa Cruz Formation, Río Gallegos.

Litopterna
Proterotheriidae

Anisolophus australis

MACN-A 2832. Tibia. Horizon and locality: Santa Cruz Formation, Monte Observación (= Cerro Observatorio). YPM-VPPU 15368. Cranium. Horizon and locality: Santa Cruz Formation, Killik Aike, Río Gallegos. YPM-VPPU 15996. Mandible. Horizon and locality: Santa Cruz Formation, 4 miles north of Coy Inlet. YPM-VPPU 15711. Cranium, femur, and tibia. Horizon and locality: Santa Cruz Formation, Lago Pueyrredón.

Diadiaphorus majusculus

AMNH 9196. Mandible. Horizon and locality: Santa Cruz Formation, unknown specific locality. AMNH 9291. Cranium and mandible. Horizon and locality: Santa Cruz Formation, unknown specific locality. MACN-A 2711–12. Cranium and mandible. Horizon and locality: Santa Cruz Formation, Corriguen Aike. MACN-A 9137. Cranium. Horizon and locality: Santa Cruz Formation, Monte Observación (= Cerro Observatorio). MACN-A 9180–82. Cranium and mandible. Horizon and locality: Santa Cruz Formation, Corriguen Aike. MACN-A 9200–08. Cranium and mandible (Type *Diadiaphorus robustus*). Horizon and locality: Santa Cruz Formation, Corriguen Aike. MPM-PV 3397. Cranium and mandible. Horizon and locality: Formación Santa Cruz, Estancia La Costa Member, Campo

Barranca. YPM-VPPU 15799. Mandible. Horizon and locality: Santa Cruz Formation, Coy Inlet.

Tetramerorhinus cingulatum

MACN-A 5971. Cranium (Type *Tetramerorhinus fortis*). Horizon and locality: Santa Cruz Formation, unknown specific locality. MPM-PV 3493. Cranium. Horizon and locality: Santa Cruz Formation, Estancia La Costa Member, FL 5.3, Puesto Estancia La Costa (= Corriguen Aike). YPM-VPPU 15436. Mandible. Horizon and locality: Santa Cruz Formation, Killik Aike, Río Gallegos. YPM-VPPU 15732. Mandible. Horizon and locality: Santa Cruz Formation, Río Gallegos.

Tetramerorhinus lucarius

YPM-VPPU 15722. Mandible. Horizon and locality: Santa Cruz Formation, Killik Aike, Río Gallegos.

Tetramerorhinus mixtum

MACN-A 8970–71. Cranium and mandible, humerus, ulna, femur, and tibia (Type). Horizon and locality: Santa Cruz Formation, Corriguen Aike. YPM-VPPU 15107. Humerus and ulna. Horizon and locality: Santa Cruz Formation, Coy Inlet.

Thoatherium minusculum

AMNH 9245. Cranium. Horizon and locality: Santa Cruz Formation, Felton's Estancia, Río Gallegos. MPM-PV 3529. Cranium, femur, ulna, and tibia. Horizon and locality: Santa Cruz Formation, Estancia La Costa Member, FL 5.3, Puesto Estancia La Costa (= Corriguen Aike). MACN-A 2958. Mandible. Horizon and locality: Santa Cruz Formation, Monte Observación (= Cerro Observatorio). YPM-VPPU 15240. Cranium. Horizon and locality: Santa Cruz Formation, Coy Inlet. YPM-VPPU 15714. Mandible. Horizon and locality: Santa Cruz Formation, Coy Inlet. YPM-VPPU 15719. Mandible. Horizon and locality: Santa Cruz Formation, Coy Inlet. YPM-VPPU 15724. Cranium. Horizon and locality: Santa Cruz Formation, Coy Inlet.

Macraucheniidae

Theosodon garrettorum

YPM-VPPU 15164. Mandible, humerus, ulna, femur, and tibia (Holotype). Horizon and locality: Santa Cruz Formation, Güer Aike Department.

Theosodon gracilis

AMNH 9230. Cranium. Horizon and locality: Santa Cruz Formation, Halliday's Estancia, Río Gallegos. MACN-A 9297. Mandible. Horizon and locality: Santa Cruz Formation, Corriguen Aike.

Theosodon lallemanti

YPM-VPPU 15216. Humerus and tibia. Horizon and locality: Santa Cruz Formation, Güer Aike Department.

Theosodon lydekkeri

MACN-A 2487–90. Cranium and mandible (Type). Horizon and locality: Santa Cruz Formation, Corriguen Aike. MACN-A 2545–57. Tibia. Horizon and locality: Santa Cruz Formation, Monte Observación (= Cerro Observatorio). MACN-A 9252–53. Tibia. Horizon and locality: Santa Cruz Formation, Monte Observación (= Cerro Observatorio). MACN-A 9269–88. Cranium and mandible. Horizon and locality: Santa Cruz Formation, Corriguen Aike.

REFERENCES

Alexander, R. M. (1985). Body support, scaling and allometry. In *Functional Vertebrate Morphology*, ed. M. Hildebrand and D. B. Wake. Cambridge, Massachusetts: Belknap Press of Harvard University Press, pp. 27–37.

Ameghino, F. (1887). Enumeración sistemática de las especies de mamíferos fósiles coleccionados por Carlos Ameghino en los terrenos eocenos de la Patagonia austral y depositados en el Museo La Plata. *Boletín del Museo La Plata*, **1**, 1–26.

Ameghino, F. (1889). Contribución al conocimiento de los mamíferos fósiles de la República Argentina. *Actas de la Academia Nacional de Ciencias de Córdoba*, **6**, 1–1027.

Ameghino, F. (1891). Nuevos restos de mamíferos fósiles descubiertos por Carlos Ameghino en el Eoceno inferior de la Patagonia Austral. Especies nuevas, adiciones y correcciones. *Revista Argentina de Historia Natural*, **1**, 286–328.

Ameghino, F. (1894). Enumération synoptique des espèces de mammifères fossiles des formations éocènes de Patagonie. *Boletín de la Academia de Ciencias de Córdoba*, **13**, 259–452.

Ameghino, F. (1900–1902). L'âge des formations sédimentaires de Patagonie. *Anales de la Sociedad Científica Argentina*. **50**, 109–130; 145–165; 209–229 (1900); **51**, 20–39, 65–91 (1901); **52**, 189–197, 244–250 (1901); **54**, 161–180, 220–249, 283–342 (1902).

Ameghino, F. (1904). Nuevas especies de mamíferos cretáceos y terciarios de la República Argentina. *Anales de la Sociedad Científica Argentina*, **56**, 193–208.

Ameghino, F. (1907). Les toxodontes à cornes. *Anales del Museo Nacional de Historia Natural de Buenos Aires*, **16**, 49–91.

Asher, R. J. and Helgen, K. M. (2010). Nomenclature and placental mammal phylogeny. *BMC Evolutionary Biology*, **10**, 1–9.

Avilla, L. D. S. and Vizcaíno, S. F. (2005). Locomotory pattern of *Astrapotherium mangun* (Owen) (Mammalia: Astrapotheria) from the Neomiocene (Colhuehuapian–Santacrucian) of Argentina. *II Congresso Latino-Americano de Paleontologia de Vertebrados, Boletim de Resumos*, p. 44.

Bargo, M. S., Vizcaíno, S. F., Archuby, F. M. and Blanco, R. E. (2000). Limb bone proportions, strength and digging in some Lujanian (late Pleistocene–early Holocene) mylodontid ground sloths (Mammalia, Xenarthra). *Journal of Vertebrate Paleontology*, **20**, 601–610.

Barreda, V. and Palazzesi, L. (2007). Patagonian vegetation turnovers during the Paleogene–early Neogene: Origin of arid-adapted floras. *Botanical Review*, **73**, 31–50.

Biewener, A. A. (1983). Locomotory stresses in the limb bones of two small mammals: the ground squirrel and chipmunk. *Journal of Experimental Biology*, **103**, 135–154.

Biewener, A. A. and Taylor, C. R. (1986). Bone strain: a determinant of gait and speed? *Journal of Experimental Biology*, **123**, 383–400.

Biknevicius, A. R., McFarlane, D. A. and MacPhee, R. D. E. (1993) Body size in *Amblyrhiza inundata* (Rodentia: Caviomorpha), an extinct megafaunal rodent from the Anguilla Bank, West Indies: Estimates and implications. *American Museum Novitates*, **3079**, 1–25.

Billet, G. (2010). New observations on the skull of *Pyrotherium* (Pyrotheria, Mammalia) and new phylogenetic hypotheses on South American ungulates. *Journal of Mammalian Evolution*, **17**, 21–59.

Billet, G., Blondel, C. and Muizon, C. de (2009). Dental microwear analysis of notoungulates (Mammalia) from Salla (Late Oligocene, Bolivia) and discussion on their precocious hypsodonty. *Palaeogeography, Palaeoclimatology, Palaeoecology*, **274**, 114–124.

Bonaparte, J. F. and Morales, J. (1997). Un primitivo Notonychopidae (Litopterna) del Paleoceno inferior de Punta Peligro, Chubut, Argentina. *Estudios Geologicos*, **53**, 263–274.

Bond, M. (1986). Los ungulados fósiles de Argentina: evolución y paleoambientes. In *Simposio "Evolución de los Vertebrados Cenozoicos", IV Congreso Argentino de Paleontología y Bioestratigrafía Actas* **2**, 187–190.

Bond, M. (1999). Quaternary native ungulates of Southern South America. A synthesis. In *Quaternary of South America and Antarctic Peninsula*, ed. J. Rabassa and M. Salemme. Ushuaia, Tierra de Fuego: Centro Austral de Investigaciones Científicas and Universidad Nacional de la Patagonia, pp. 177–205.

Bond, M., Cerdeño, E. and López, G. (1995). Los ungulados nativos de América del Sur. In *Evolución biológica y climática de la región Pampeana durante los últimos cinco millones de años. Un ensayo de correlación con el Mediterráneo occidental*, ed. M. T. Alberdi, G. Leone and E. P. Tonni. Madrid: Monografías del MNCN, CSIC, pp. 259–275.

Bond, M., Perea, D., Ubilla, M. and Tauber, A. A. (2001). *Neolicaphrium recens* Frenguelli, 1921, the only surviving Proterotheriidae (Litopterna, Mammalia) into the South American Pleistocene. *Palaeovertebrata*, **30**, 37–50.

Bondesio, P. (1986). Lista sistemática de los vertebrados terrestres del Cenozoico de Argentina. In *Simposio "Evolución de los Vertebrados Cenozoicos", IV congreso Argentino de paleontología y bioestratigrafía Actas*, **2**, 187–190.

Breiman, L., Friedman, J. H., Olshen, R. A. and Stone, C. J. (1984). *Classification and Regression Trees*, Florida: CRC Press LLC.

Brown, J. H., Gupta, V. K., Li, B.-L. *et al.* (2002). The fractal nature of nature: power laws, ecological complexity and biodiversity. *Philosophical Transactions: Biological Sciences*, **357**, 619–626.

Cassini, G. H., Mendoza, M., Vizcaíno, S. F. and Bargo, M. S. (2011). Inferring habitat and feeding behaviour of early Miocene notoungulates from Patagonia. *Lethaia*, **44**, 153–165.

Cassini, G. H., Flores, D. A. and Vizcaíno, S. F. (in press). Postnatal ontogenetic scaling of Nesodontine (Notoungulata, Toxodontidae) cranial morphology. *Acta Zoologica*, doi:10.1111/j.1463–6395.2011.00501.x

Christiansen, P. and Harris, J. M. (2005). Body size of *Smilodon* (Mammalia: Felidae). *Journal of Morphology*, **266**, 369–384.

Cifelli, R. L. (1985). South American ungulate evolution and extinction. In *The Great American Biotic Interchange*, ed. F. G. Stehli and S. D. Webb. New York: Plenum Press, pp. 249–266.

Cifelli, R. L. (1993). The phylogeny of the native South American ungulates. In *Mammals Phylogeny: Placentals*, ed. F. S. Szalay, M. J. Novacek and M. C. Mckenna. New York and London: Springer-Verlag, pp. 195–216.

Cifelli, R. L. and Guerrero, J. G. (1997). Litopterns. In *Vertebrate Paleontology in the Neotropics: The Miocene Fauna of La Venta, Colombia*, ed. R. F. Kay, R. H. Madden, R. L. Cifelli and J. J. Flynn. Washington, DC: Smithsonian Institution Press, pp. 289–302.

Cifelli, R. L. and Soria, M. F. (1983). Systematics of the Adianthidae (Litopterna, Mammalia). *American Museum Novitates*, **2771**, 1–25.

Cios, K., Pedrycz, W. and Swiniarski, R. (1998). *Data Mining Methods for Knowledge Discovery*. Boston: Kluwer Publishers.

Clauss, M., Kaiser, T. and Hummel, J. (2008). The Morphophysiological adaptations of browsing and grazing mammals. In *The Ecology of Browsing and Grazer*, ed. I. J. Gordon and H. H. T. Prins. Berlin Heidelberg: Springer-Verlag.

Coombs, M. C. (1983). Large mammalian clawed herbivores: a comparative study. *Transactions of the American Philosophical Society*, **73**, 1–96.

Copes, L. E. and Schwartz, G. T. (2010). The scale of it all: postcanine tooth size, the taxon-level effect, and the universality of Gould's scaling law. *Paleobiology*, **36**, 188–203.

Corruccini, R. S. (1987). Shape in morphometrics: comparative analyses. *American Journal of Physical Anthropology*, **73**, 289–303.

Croft, D. A. (1999). Placentals: endemic South American ungulates. In *The Encyclopedia of Paleontology*, ed. R. Singer. Chicago: Fitzroy-Dearborn.

Croft, D. A. (2000). Archaeohyracidae (Mammalia: Notoungulata) from the Tinguiririca Fauna, central Chile, and the evolution and paleoecology of South American mammalian herbivores. Unpublished thesis, University of Chicago.

Croft, D. A. and Anderson, L. C. (2008). Locomotion in the extinct notoungulate *Protypotherium*. *Palaeontologia Electronica*, **11**, 1–20.

Croft, D. A., Radic, J. P., Zurita, E. *et al.* (2003). A Miocene toxodontid (Mammalia: Notoungulata) from the sedimentary series of the Cura-Mallín Formation, Loquimay, Chile. *Revista Geológica de Chile*, **30**, 285–298.

Croft, D. A., Flynn, J. J. and Wyss, A. R. (2004). Notoungulata and Litopterna of the early Miocene Chucal Fauna, northern Chile. *Fieldiana: Geology (New Series)*, **50**, 1–49.

Demes, B., Larson, S. G., Stern, J. T. J. *et al.* (1994). The kinetics of primate quadrupedalism: 'hindlimb drive' reconsidered. *Journal of Human Evolution*, **26**, 353–374.

Domingo, C., Gavaldà, R. and Watanabe, O. (2002). Adaptive sampling methods for scaling up knowledge discovery algorithms. *Data Mining and Knowledge Discovery*, **6**, 131–152.

Elftman, H. O. (1929). Functional adaptations of the pelvis of marsupials. *Bulletin of the American Museum of Natural History*, **63**, 189–232.

Elissamburu, A. (2004). Análisis morfométrico y morfofuncional del esqueleto apendicular de *Paedotherium* (Mammalia, Notoungulata). *Ameghiniana*, **41**, 363–380.

Elissamburu, A. (2007). Estudio biomecánico del aparato locomotor de ungulados nativos Sudamericanos (Notoungulata). Unpublished doctoral thesis, Universidad Nacional de La Plata, Argentina.

Elissamburu, A. (2010). Estudio biomecánico y morfofuncional del esqueleto apendicular de *Homalodotherium* Flower 1873 (Mammalia, Notoungulata). *Ameghiniana*, **47**, 25–43.

Elissamburu, A. (2011). Estimación de la masa corporal en géneros del Orden Notoungulata. *Estudios Geológicos*, doi:10.3989/egeol.40336.133.

Elissamburu, A. and Vizcaíno, S. F. (2004). Limb proportions and adaptations in caviomorph rodents (Rodentia: Caviomorpha). *Journal of Zoology*, **262**, 145–159.

Fayyad, U., Piatetsky-Shapiro, G. and Smyth, P. (1996). From data mining to knowledge discovery in databases. *Artificial Intelligence Magazine*, **17**, 37–54.

Fleagle, J. G. (1979). *Primate Adaptation and Evolution*. New York: Academic Press.

Fletcher, T. M., Janis, C. M. and Rayfield, E. J. (2010). Finite element analysis of ungulate jaws: can mode of digestive physiology be determined? *Palaeontologia Electronica*, **13**, 21A.

Fortelius, M. (1985). Ungulate cheek teeth: developmental, functional, and evolutionary interrelations. *Acta Zoologica Fennica*, **180**, 1–76.

Fortelius, M. (1988) Isometric scaling of mammalian cheek teeth is also true metabolic scaling. In *Proceedings of the VIIth International Symposium on Dental Morphology, Paris 1986*, ed. D. E. Russell, J. P. Santoro and D. Sigogneau-Russell. Paris: Mémoires du muséum national d'histoire naturelle (série C), pp. 459–462.

Fortelius, M. (1990). The mammalian dentition: a "tangle" view. *Netherland Journal of Zoology*, **40**, 312–328.

Fortelius, M., Eronen, J. T., Jernvall, J. *et al.* (2002). Fossil mammals resolve regional patterns of Eurasian climate change during 20 million years. *Evolutionary Ecology Research*, **4**, 1005–1016.

Gelfo, J. N., López, G. M. and Bond, M. (2008) A new Xenungulata (Mammalia) from the Paleocene of Patagonia, Argentina. *Journal of Paleontology*, **82**, 329.

Gould, S. J. (1975). On the scaling of tooth size in mammals. *American Zoologist*, **15**, 351–362.

Hildebrand, M. (1988). *Analysis of Vertebrate Structure*, USA: John Wiley & Sons, Inc.

Horovitz, I. (2004). Eutherian mammal systematics and the origins of South American ungulates as base on postcranial osteology. In *Fanfare for an Uncommon Paleontologist: Papers in Honor of Malcolm C. McKenna*, ed. M. R. Dawson and J. A. Lillengraven. *Bulletin of Carnegie Museum of Natural History*, **36**, pp. 63–79.

Howell, B. A. (1944). *Speed in Animals. Their Specialization for Running and Leaping*, Chicago: University of Chicago Press.

Janis, C. M. (1988). An estimation of tooth volume and hypsodonty indices in ungulate mammals, and the correlation of these factors with dietary preference. In *Teeth Revisited: Proceedings of the VII International Symposium on Dental Morphology, Paris 1986*, ed. D. E. Russell, J. P. Santoro and D. Sigoneau-Russell. Paris: Mémoires du muséum national d'histoire naturelle (série C), pp. 367–387.

Janis, C. M. (1990). Correlation of cranial and dental variables with body size in ungulates and macropodoids. In *Body Size in Mammalian Paleobiology: Estimation and Biological Implications*, ed. J. Damuth and B. J. MacFadden. Cambridge: Cambridge University Press, pp. 255–300.

Janis, C. M. (1995). Correlations between craniodental morphology and feeding behavior in ungulates: Reciprocal illumination between living and fossil taxa. In *Functional Morphology in Vertebrate Paleontology*, ed. J. J. Thomason. Cambridge: Cambridge University Press, pp. 76–98.

Janis, C. M. (2008). An evolutionary history of browsing and grazing ungulates. In *The Ecology of Browsers and Grazers*, ed. I. J. Gordon and H. H. T. Prins. Berlin: Springer-Verlag, pp. 21–45.

Janis, C. M. and Constable, E. (1993). Can ungulate craniodental features determine digestive physiology? *Journal of Vertebrate Paleontology*, **13**, 43A.

Janis, C. M. and Ehrhardt, D. (1988). Correlation of relative muzzle width and relative incisor width with dietary preference in ungulates. *Zoological Journal of the Linnean Society*, **92**, 267–284.

Janis, C. M. and Fortelius, M. (1988). On the means whereby mammals achieve increased functional durability of their dentitions, with special reference to limiting factors. *Biological Reviews*, **63**, 197–230.

Janis, C. M., Constable, E. C., Houpt, K. A., Streich, W. J. and Clauss, M. (2010). Comparative ingestive mastication in domestic horses and cattle: a pilot investigation. *Journal of Animal Physiology and Animal Nutrition*, **94**, 402–409.

Johnson, S. C. and Madden, R. H. (1997). Uruguaytheriine astrapotheres of tropical South America. In *Vertebrate*

Paleontology in the Neotropics: The Miocene Fauna of La Venta, Colombia, ed. R. F. Kay, R. H. Madden, R. L. Cifelli and J. J. Flynn. Washington, DC: Smithsonian Institution Press, pp. 355–381.

Kay, R. F. (1975). Allometry in early hominids (letter). *Science*, **189**, 63.

Kay, R. F., Vizcaíno, S. F., Bargo, M. S. *et al.* (2008). Two new fossil vertebrate localities in the Santa Cruz Formation (late early Miocene, Argentina), ∼51 degrees south latitude. *Journal of South American Earth Sciences*, **25**, 187–195.

Kleiber, M. (1932). Body size and metabolism. *Hilgardia*, **6**, 315–353.

Konwar, P., Saikia, M. K. and Saikia, P. K. (2009). Abundance of food plant species and food habits of *Rhinoceros unicornis* Linn. in Pobitora Wildlife Sanctuary, Assam, India. *Journal of Threatened Taxa*, **1**, 457–460.

Kramarz, A. G. (2009). Adiciones al conocimiento de *Astrapothericulus* (Mammalia, Astrapotheria): anatomía cráneo-dentaria, diversidad y distribución. *Revista Brasileira de Paleontologia*, **12**, 55–66.

Kramarz, A. G. and Bond, M. (2005). Los Litopterna (Mammalia) de la Formación Pinturas, Mioceno Temprano-Medio de Patagonia. *Ameghiniana*, **42**, 611–625.

Kramarz, A. G. and Bond, M. (2009). A new Oligocene astrapothere (Mammalia, Meridiungulata) from Patagonia and a new appraisal of astrapothere phylogeny. *Journal of Systematic Palaeontology*, **7**, 117–128.

Larose, D. T. (2004). *Discovering Knowledge in Data: An Introduction to Data Mining*. New York: Wiley.

Lessa, E. P. and Stein, B. R. (1992). Morphological constraints in the digging apparatus of pocket gophers (Mammalia: Geomyidae). *Biological Journal of the Linnean Society*, **47**, 439–453.

Lucas, P. W. (2004). *Dental Functional Morphology. How Teeth Work*. Cambridge: Cambridge University Press.

MacFadden, B. J. (2000). Cenozoic mammalian herbivores from the Americas: reconstructing ancient diets and terrestrial communities. *Annual Review of Ecology and Systematics*, **31**, 33–59.

Madden, R. H. (1990). Miocene Toxodontidae (Notoungulata, Mammalia) from Colombia, Ecuador and Chile. Unpublished doctoral thesis, Duke University.

Madden, R. H. (1997). A new toxodontid notoungulate. In *Vertebrate Paleontology in the Neotropics: The Miocene Fauna of La Venta*, ed. R. F. Kay, R. H. Madden, R. L. Cifelli and J. J. Flynn. Washington D.C.: Smithsonian Institution Press, pp. 355–381.

Mannila, H. and Toivonen, H. (1997). Levelwise search and borders of theories in knowledge discovery. *Data Mining and Knowledge Discovery*, **1**, 241–258.

Marshall, L. G. and Cifelli, R. L. (1989). Analysis of changing diversity patterns in Cenozoic. Land mammal age faunas, South America. *Paleovertebrata*, **19**, 169–210.

Marshall, L. G. and Pascual, R. (1977). Nuevos marsupiales Caenolestidae del "Piso Notohipidense" (SW de Santa Cruz, Patagonia) de Ameghino. Sus aportaciones a la cronología y evolución de las comunidades de mamíferos

sudamericanos. *Publicaciones del Museo Municipal de Ciencias Naturales "Lorenzo Scaglia,"* **2**, 91–122.

Marshall, L. G. and Pascual, R. (1978). Una escala temporal radiométrica preliminar de las edades-mamífero del Cenozoico medio y tardío sudamericano. *Obra del Centenario del Museo de La Plata*, **5**, 11–28.

Marshall, L. G., Hoffstetter, R. and Pascual, R. (1983). Mammals and stratigraphy: Geochronology of the continental mammal-bearing Tertiary of South America. *Paleovertebrata, Mémoire Extraordinaire*, 1–93.

Marshall, L. G., Salinas, P. and Manuel, S. (1990). *Astrapotherium* sp. (Mammalia Astrapotheriidae) from Miocene strata along the Quepuca river, central Chile. *Revista Geológica de Chile*, **17**, 215–223.

Martin, P. S. and Steadman, D. W. (1999). Prehistoric extinctions on islands and continents. In *Extinctions in Near Time: Causes, Contexts and Consequences*, ed. R. D. E. MacPhee. New York: Kluwer/Plenum, pp. 17–56.

McKenna, M. C. (1975). Towards a phylogenetic classification of mammals. In *Phylogeny of the Primates*, ed. W. P. Lucket and F. S. Szalay. New York: Plenum, pp. 21–46.

McMahon, T. A. and Bonner, J. T. (eds.) (1983). *On Size and Life*. New York: Scientific American Library distributed by W.H. Freeman.

Mendoza, M. (2007). Decision trees: a machine learning methodology for characterizing morphological patterns resulting from ecological adaptations. In *Automated Recognition of Biological Objects*, ed. N. Macleod. UK Systematics Association's Special Volume Series, pp. 261–276.

Mendoza, M. and Palmqvist, P. (2006). Characterizing adaptive morphological patterns related to diet in Bovidae (Mammalia: Artiodactyla). *Acta Zoologica Sinica*, **52**, 988–1008.

Mendoza, M. and Palmqvist, P. (2008). Hypsodonty in ungulates: An adaptation for grass consumption or for foraging in open habitat? *Journal of Zoology*, **274**, 134–142.

Mendoza, M., Janis, C. M. and Palmqvist, P. (2002). Characterizing complex craniodental patterns related to feeding behaviour in ungulates: A multivariate approach. *Journal of Zoology*, **258**, 223–246.

Mendoza, M., Janis, C. M. and Palmqvist, P. (2005). Ecological patterns in the trophic-size structure of large mammal communities: a 'taxon-free' characterization. *Evolutionary Ecology Research*, **7**, 505–530.

Mendoza, M., Janis, C. M. and Palmqvist, P. (2006). Estimating the body mass of extinct ungulates: a study on the use of multiple regression. *Journal of Zoology*, **270**, 90–101.

Mercerat, A. (1893a). Contribución a la geología de la Patagonia. *Annales de la Sociedad Científica Argentina*, **36**, 65–103.

Mercerat, A. (1893b). Un viaje de exploración en la Patagonia austral. *Boletín del Instituto Geográfico Argentino*, **14**, 267–291.

Michie, D., Spiegelhalter, D. J. and Taylor, C. C. (1994). *Machine Learning, Neural and Statistical Classification*. New York: Ellis Horwood.

Milne, N., Vizcaíno, S. F. and Fernicola, J. C. (2009). A 3D geometric morphometric analysis of digging ability in the extant and fossil cingulate humerus. *Journal of Zoology*, **278**, 48–56.

Mones, A. (1982). An equivocal nomenclature: What means hypsodonty? *Paläontologische Zeitschrift*, **56**, 107–111.

Muizon, C. de and Cifelli, R. L. (2000). The "condylarths" (archaic Ungulata, Mammalia) from the early Palaeocene of Tiupampa (Bolivia): implications on the origin of the South American ungulates. *Geodiversitas*, **22**, 47–150.

Nasif, N. L., Musalem, S. and Cerdeño, E. (2000). A new toxodont from the late Miocene of Catamarca, Argentina, and a phylogenetic analysis of the Toxodontidae. *Journal of Vertebrate Paleontology*, **20**, 591–600.

Neuschulz, N. and Meister, J. (1998). *Nashoernern auf der Spur: Leben und uberleben einer stark bedrohten Tierfamilie.* Erfurt: Verein der Zooparkfreunde in Erfurt.

Nowak, R. M. (1991). *Walker's Mammals of the World*. Baltimore, Maryland: Johns Hopkins Press.

Owen-Smith, N. (1988). *Megaherbivores. The Influence of Very Large Body Size on Ecology*, Cambridge: Cambridge University Press.

Oxnard, C. (1984). *The Order of Man*. Hong Kong: Hong Kong University Press.

Pascual, R., Ortiz-Jaureguizar, E. and Prado, J. L. (1996). Land mammals: paradigm for Cenozoic South American geobiotic evolution. In *Contribution of Southern South America to Vertebrate Paleontology*, ed. G. Arratia. Münich: Münchner Geowissenshaftliche Abhandlungen, pp. 265–319.

Patterson, B. and Pascual, R. (1968). The fossil mammal fauna of South America. *The Quarterly Review of Biology*, **43**, 409–451.

Pérez-Barbería, F. J. and Gordon, I. J. (1998). Factors affecting food comminution during mastication in herbivorous mammals: a review. *Biological Journal of the Linnean Society*, **63**, 233–256.

Pérez-Barbería, F. J. and Gordon, I. J. (1999). The functional relationship between feeding type and jaw and cranial morphology in ungulates. *Oecologia*, **118**, 157–165.

Pérez-Barbería, F. J., Gordon, I. J. and Nores, C. (2001). Evolutionary transitions among feeding styles and habitats in ungulates. *Evolutionary Ecology Research*, **3**, 221–230.

Reguero, M. A. (1999). El problema de las relaciones sistemáticas y filogenéticas de los Typotheria y Hegetotheria (Mammalia, Notoungulata): análisis de los taxones de Patagonia de la edad-mamífero Deseadense (Oligoceno). Unpublished doctoral thesis, Universidad de Buenos Aires.

Reguero, M. A., Dozo, M. T. and Cerdeño, E. (2007). A poorly known rodentlike mammal (Pachyrukhinae, Hegetotheriidae, Notoungulata) from the Deseadan (Late Oligocene) of Argentina. Paleoecology, biogeography, and radiation of the rodentlike ungulates in South America. *Journal of Paleontology*, **81**, 1301–1307.

Reguero, M. A., Candela, A. M. and Cassini, G. H. (2010). Hypsodonty and body size in rodent-like notoungulates.

In *The Paleontology of Gran Barranca: Evolution and Environmental Change through the Middle Cenozoic of Patagonia*, eds. R. H. Madden, A. A. Carlini, M. H. Vucetich and R. F. Kay. Cambridge: Cambridge University Press, pp. 358–367.

Rensberger, J. M. and Pfretzschner, H. U. (1992). Enamel structure in astrapotheres and its functional implications. *Scanning Microscopy*, **6**, 495–510.

Riggs, E. S. (1935). A skeleton of *Astrapotherium*. *Geological Series of Field Museum of Natural History*, **6**, 167–177.

Riggs, E. S. (1937). Mounted skeleton of *Homalodotherium*. *Geological Series of Field Museum of Natural History*, **6**, 233–243.

Samuels, J. X. and Van Valkenburgh, B. (2008). Skeletal indicators of locomotor adaptations in living and extinct rodents. *Journal of Morphology*, **269**, 1387–1411.

Scarano, A. C., Carlini, A. A. and Illius, A. W. (2011). Interatheriidae (Typotheria: Notoungulata), body size and paleoecology characterization. *Mammalian Biology*, **76**, 109–114.

Schmidt, G. I. (2011). Los Proterotheriidae (Litopterna) de Entre Ríos (Argentina): Consideraciones nomenclaturales e implicancias sistemáticas. *Ameghiniana*, **48**, 605–620.

Scott, W. B. (1910). Mammalia of the Santa Cruz beds. Part I. Litopterna. In *Reports of the Princeton University Expeditions to Patagonia, 1896–1899*. Vol. 7, *Paleontology IV*, ed. W. B. Scott, Princeton: Princeton University Press, pp. 1–156.

Scott, W. B. (1912). Mammalia of the Santa Cruz beds. Part II. Toxodonta. Part III. Entelonychia. In *Reports of the Princeton University Expeditions to Patagonia, 1896–1899*. Vol. 6, *Paleontology III*, ed. W. B. Scott, Princeton: Princeton University Press, pp. 111–300.

Scott, W. B. (1913). *A History of Land Mammals in the Western Hemisphere*. New York: MacMillan.

Scott, W. B. (1928). Mammalia of the Santa Cruz beds. Part IV. Astrapotheria. In *Reports of the Princeton University Expeditions to Patagonia, 1896–1899*. Vol. 6, *Paleontology III*, ed. W. B. Scott, Princeton, Princeton: University Press, pp. 301–352.

Scott, W. B. (1930). A partial skeleton of *Homalodontotherium* from the Santa Cruz beds of Patagonia. *Field Museum of Natural History Geology Memoirs*, **1**, 7–34.

Scott, W. B. (1937a). *A History of Land Mammals in the Western Hemisphere*, New York: Macmillan.

Scott, W. B. (1937b). The Astrapotheria. *Proceedings of the American Philosophical Society*, **77**, 309–393.

Seckel, L. and Janis, C. M. (2008). Convergences in scapula morphology among small cursorial mammals: an osteological correlate for locomotory specialization. *Journal of Mammalian Evolution*, **15**, 261–279.

Simpson, G. G. (1936) Structure of a primitive Notoungulate cranium. *American Museum Novitates*, **824**, 1–31.

Simpson, G. G. (1967). The beginning of the age of mammals in South America. Part II. *Bulletin of the American Museum of Natural History*, **137**, 1–260.

Simpson, G. G. (1980). *Splendid Isolation: The Curious History of South American Mammals*. New Haven, Connecticut: Yale University Press.

Sinclair, W. J. (1909). Mammalia of the Santa Cruz beds. Part I. Typotheria. In *Reports of the Princeton University Expeditions to Patagonia, 1896–1899*. Vol. 6, *Paleontology III*, ed. W. B. Scott. Princeton: Princeton University Press, pp. 1–110.

Smith, M. J. and Savage, R. J. G. (1955). Some locomotory adaptations in mammals. *Zoological Journal of the Linnean Society*, **42**, 603–622.

Sokal, R. R. and Rohlf, F. J. (1995). *Biometry: The Principles and Practice of Statistics in Biological Research*, 3rd Edn. New York: W. H. Freeman and Co.

Solounias, N. and Moelleken, S. M. C. (1993). Dietary adaptation of some extinct ruminants determined by premaxillary shape. *Journal of Mammalogy*, **74**, 1059–1071.

Soria, M. F. (2001). Los Proterotheriidae (Litopterna, Mammalia), sitemática, origen y filogenia. *Monografías del Museo Argentino de Ciencias Naturales*, **1**, 1–167.

Soria, M. F. and Powell, J. E. (1981). Un primitivo Astrapotheria (Mammalia) y la Edad de la Formación Río Loro, provincia de Tucumán, República Argentina. *Ameghiniana*, **18**, 155–168.

Spencer, L. M. (1995). Morphological correlates of dietary resource partitioning in the African Bovidae. *Journal of Mammalogy*, **76**, 448–471.

Stein, B. R. and Casinos, A. (1997). What is a cursorial mammal? *Journal of Zoology*, **242**, 185–192.

Strömberg, C. A. E. (2006). Evolution of hypsodonty in equids: Testing a hypothesis of adaptation. *Paleobiology*, **32**, 236–258.

Tauber, A. A. (1994). Estratigrafía y vertebrados fósiles de la Formación Santa Cruz (Mioceno inferior) en la costa atlántica entre las rías del Coyle y Río Gallegos, Provincia de Santa Cruz, República Argentina. Unpublished doctoral thesis, Universidad Nacional de Córdoba.

Tauber, A. A. (1996). Los representantes del género *Protypotherium* (Mammalia, Notoungulata, Interatheridae) del Mioceno Temprano del sudoeste de la Provincia de Santa Cruz, República Argentina. *Academia Nacional de Ciencias (Córdoba)*, **95**, 1–29.

Tauber, A. A. (1997a). Bioestratigrafia de la Formación Santa Cruz (Mioceno inferior) en el extremo sudeste de la Patagonia. *Ameghiniana*, **34**, 413–426.

Tauber, A. A. (1997b). Paleoecología de la Formación Santa Cruz (Mioceno inferior) en el extremo sudeste de la Patagonia. *Ameghiniana*, **34**, 517–529.

Tauber, A. A. (1999). Los vertebrados de la Formación Santa cruz (Mioceno inferior-medio) en el extremo sureste de la Patagonia y su significado paleoecológico. *Revista Española de Paleontología*, **14**, 173–182.

Toledo, N., Bargo, M. S., Cassini, G. H. and Vizcaíno, S. F. (2012). The forelimb of early Miocene sloths (Mammalia, Xenarthra, Folivora): morphometry and functional implications for substrate preferences. *Journal of Mammalian Evolution* DOI 10.1007/s10914-012-9185-2.

Townsend, K. E. B. and Croft, D. A. (2005). Low-magnification microwear analyses of South American endemic herbivores. *Journal of Vertebrate Paleontology*, **25**, 123A.

Townsend, K. E. B. and Croft, D. A. (2008). Diets of notoungulates from the Santa Cruz Formation, Argentina: New evidence from enamel microwear. *Journal of Vertebrate Paleontology*, **28**, 217–230.

Ungar, P. S. (1998). Dental allometry, morphology, and wear as evidence for diet in fossil primates. *Evolutionary Anthropology*, **6**, 205–217.

Van Valkenburgh, B. and Koepfli, K. (1993). Cranial and dental adaptations to predation in canids. *Symposia of the Zoological Society of London*, **65**, 15–37.

Villafañe, A. L. (2005). Paleoecología de los Proterotheriidae (Mammalia: Litopterna): un estudio basado en los cambios en la masa corporal. Unpublished Licentiate thesis, Universidad Nacional de La Patagonia San Juan Bosco.

Villafañe, A. L., Ortiz-Jaureguizar, E. and Bond, M. (2006). Cambios en la riqueza taxonómica y en las tasas de primera y última aparición de los Proterotheriidae (Mammalia, Litopterna) durante el Cenozoico. *Estudios Geológicos*, **62**, 155–166.

Vizcaíno, S. F. and Milne, N. (2002). Structure and function in armadillo limbs (Mammalia: Xenarthra: Dasypodidae). *Journal of Zoology*, **257**, 117–127.

Vizcaíno, S. F., Bargo, M. S. and Cassini, G. H. (2006). Dental occlusal surface area in relation to body mass, food habits and other biological features in fossil xenarthrans. *Ameghiniana*, **43**, 11–26.

Vizcaíno, S. F., Bargo, M. S., Kay, R. F. *et al.* (2010). A baseline paleoecological study for the Santa Cruz Formation (late–early Miocene) at the Atlantic coast of Patagonia, Argentina. *Palaeogeography, Palaeoclimatology, Palaeoecology*, **292**, 507–519.

Vizcaíno S. F., Blanco, E. R., Bender, B. J. and Milne, N. (2011). Proportions and function of the limbs of glyptodonts. *Lethaia*, **44**, 93–101.

Warton, D. I. and Weber, N. C. (2002). Common slope tests for bivariate structural relationships. *Biometrical Journal*, **44**, 161–174.

Warton, D. I., Wright, I. J., Falster, D. S. and Westoby, M. (2006). Bivariate line-fitting methods for allometry. *Biological Reviews of the Cambridge Philosophical Society*, **81**, 259–291.

West, G. B., Brown, J. H. and Enquist, B. J. (1997). A general model for the origin of allometric scaling laws in biology. *Science*, **276**, 122–126.

West, G. B., Brown, J. H. and Enquist, B. J. (1999a). The fourth dimension of life: fractal geometry and allometric scaling of organisms. *Science*, **284**, 167–169.

West, G. B., Brown, J. H. and Enquist, B. J. (1999b). A general model for the structure of plant vascular systems. *Nature*, **400**, 664–667.

Williams, S. H. and Kay, R. F. (2001). A comparative test of adaptive explanations for hypsodonty in ungulates and rodents. *Journal of Mammalian Evolution*, **8**, 207–229.

15 Paleobiology of Santacrucian caviomorph rodents: a morphofunctional approach

Adriana M. Candela, Luciano L. Rasia, and María E. Pérez

Abstract

Caviomorphs constitute a monophyletic group, representative of the oldest rodent radiation in South America. Since the Oligocene they have contributed importantly to the fossil communities of South American mammals. Today, caviomorphs display a diverse array of locomotor and feeding behaviors, occupying a wide range of habitats in which they consume a variety of food items. In this contribution, we analyze species of Santacrucian caviomorphs from a morphofunctional perspective in the context of extant caviomorph behavioral diversity, in order to evaluate their paleobiology and its paleoenvironmental significance. Body masses based on Santacrucian specimens were estimated by applying published equations. Reconstructions were made of areas of origin and insertion of the main masticatory muscles following the muscular pattern observed in living caviomorphs. On this basis, functional analyses of the masticatory apparatus were performed. Major postcranial features in extant caviomorphs identified as functionally associated with different locomotor habits were used as a model to evaluate the functional significance of the postcranial features in Santacrucian species. Santacrucian caviomorphs examined show the range of body mass from small- to relatively large-sized living caviomorphs. The porcupine *Steiromys duplicatus* and the dasyproctid *Neoreomys australis* present features that indicate a diet of harder and more abrasive food items than those consumed by living Neotropical phylogenetically related species. Features of the masticatory apparatus of the eocardiid *Eocardia fissa* are compatible with a wide feeding behavior, including abrasive food. *Steiromys duplicatus* was a semi-arboreal form, which did not achieve the degree of specialization that is seen in the Neotropical extant *Coendou*. *Eocardia fissa* was a cursorial agile rodent, similar to living *Pediolagus salinicola*. *Neoreomys australis* was a cursorial species, more similar to living *Agouti paca* than to *Dasyprocta* species. Santacrucian caviomorphs could have lived in an environment like the Cerrado biome, showing differential utilization of food resources and locomotor behaviors.

Resumen

Los caviomorfos constituyen un clado representativo de la más antigua radiación de roedores en América del Sur. Desde el Oligoceno fueron un importante componente de las comunidades de mamíferos fósiles de América del Sur. Actualmente los caviomorfos muestran una gran diversidad de comportamientos de locomoción y alimentación, ocupando diferentes hábitats en los cuales consumen variados tipos de alimentos. En esta contribución se analizan especies de caviomorfos Santacrucenses desde una perspectiva morfofuncional, en el contexto de la diversidad de comportamientos de caviomorfos vivientes, con el objeto de evaluar su paleobiología y significado paleoambiental. Las masas corporales de los especímenes analizados fueron estimadas aplicando ecuaciones publicadas. Las reconstrucciones de las áreas de origen e inserción de los principales músculos masticatorios fueron realizadas siguiendo el patrón observado en caviomorfos vivientes. Sobre esta base, se realizó el análisis funcional del aparato masticatorio en las especies fósiles examinadas. Los rasgos postcranianos funcionalmente asociados con diferentes hábitos locomotores en caviomorfos vivientes fueron usados como modelo para evaluar el significado funcional de los presentes en las especies extintas. Éstas muestran el rango de masa corporal comprendido entre las más pequeñas a las relativamente más grandes especies de caviomorfos vivientes. El puercoespín *Steiromys duplicatus* y el dasipróctido *Neoreomys australis* presentan rasgos que indican una dieta basada en alimentos más duros y abrasivos que aquellos consumidos por las especies neotropicales vivientes filogenéticamente relacionadas. Rasgos del aparato masticatorio del eocárdido *Eocardia fissa* son compatibles con un comportamiento alimenticio relativamente generalista, incluyendo alimentos abrasivos. *Steiromys duplicatus* habría sido una forma semi-arborícola que no alcanzó el grado de especialización observado en el género neotropical viviente *Coendou*. *Eocardia fissa* fue un roedor cursorial ágil, similar al cávido viviente *Pediolagus salinicola*. *Neoreomys australis* fue una especie cursorial más similar a la especie viviente *Agouti paca* que a *Dasyprocta* spp. Los caviomorfos Santacrucenses habrían habitado un ambiente similar al Bioma del Cerrado, mostrando diferentes patrones de utilización de recursos alimenticios y comportamientos de locomoción.

Early Miocene Paleobiology in Patagonia: High-Latitude Paleocommunities of the Santa Cruz Formation, ed. Sergio F. Vizcaíno, Richard F. Kay and M. Susana Bargo. Published by Cambridge University Press. © Cambridge University Press 2012.

15.1 Introduction

Caviomorphs constitute a very diverse and ancient mono-phyletic group of rodents (e.g. Huchon and Douzery, 2001; Churakov *et al.*, 2010), endemic to Central and South America. Recorded since the Late Eocene? – Early Oligo-cene (Wyss *et al.*, 1993; Frailey and Campbell, 2004; Vucetich *et al.*, 2010), caviomorphs evolved in isolation during most of the Cenozoic, forming part of all of the fossil communities of South American mammals, particu-larly of those recorded in Patagonia. During their long evolutionary history they developed a significant taxonom-ical richness (e.g. McKenna and Bell, 1997; Vucetich *et al.*, 1999). With about 59 genera and 250 extant species grouped in 12 families (Woods and Kilpatrick, 2005; Honeycutt *et al.*, 2007), caviomorphs comprise a notable component of the faunas of living Neotropical mammals. They show a wide range of body size, from about 200 g, as in some echimyids, to 50–60 kg, as in the capybara. In terms of ecology and lifestyles, caviomorphs display a diverse array of locomotor behaviors and use of substrate, ranging from cursorial, saltatorial, semi-fossorial, and fossorial, to arboreal and swimming forms, occupying a wide range of habitats (e.g. Redford and Eisenberg, 1992; Elissamburu and Vizcaíno, 2004). Although basically herbivorous, living species consume varied food items, such as fruits, seeds, leaves, and grasses, and display diverse dietary preferences (e.g. Nowak, 1991; Redford and Eisenberg, 1992; Emmons and Feer, 1997). Among Santacrucian (late Early Miocene) mammals, caviomorphs are represented by numerous species (e.g. Ameghino, 1887; Scott, 1905; Candela, 2000; Kramarz, 2006; Pérez, 2010a), many of them known by well-preserved remains repre-senting all the major clades recognized within this group, i.e. Cavioidea, Erethizontoidea, Chinchilloidea, and Octo-dontoidea (Fig. 15.1). Therefore, comparisons between Santacrucian and living caviomorphs can be made within the same phylogenetic levels. In spite of the important diversity and in many cases the exceptional preservation of the caviomorphs from the Santa Cruz Formation (late Early Miocene), inferences on their possible feeding habits and substrate preferences were based mainly on allied extant taxa (with some exceptions; see Candela and Vizcaíno, 2007; Candela and Picasso, 2008). Unlike other mammalian lineages (e.g. Vizcaíno *et al.*, 2010), morphofunctional, ecomorphological, and biomechanical studies as a basis for understanding paleobiological aspects in Santacrucian caviomorphs are limited. Some species, such as those of the cavioid clade, represent the earliest caviomorphs with known postcranial remains. In this context, analyses of these remains will lead to an improved understanding of the early history of the locomotor behaviors in different lineages of Caviomorpha.

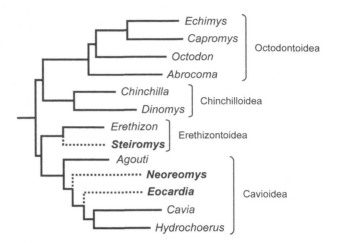

Fig. 15.1. Cladogram of Caviomorpha (taken from Blanga-Kanfi *et al.*, 2009). For the phylogenetic position of *Eocardia* and *Neoreomys* we follow Pérez (2010b). Santacrucian genera in bold.

In this contribution we analyze exceptionally well-preserved osteological remains of species of Santacrucian caviomorphs belonging to Erethizontidae, Dasyproctidae, and "Eocardiidae" from a morphofunctional perspective, in order to expand our knowledge of their paleobiology. On this basis, we evaluate the paleoenvironmental signifi-cance of Santacrucian caviomorphs.

15.2 Santacrucian rodents in the context of Caviomorpha

Erethizontidae, the New World porcupines, are a distinctive clade of caviomorph rodents, known in South America since the Late Eocene–Late Oligocene (see Wood and Patterson, 1959; Candela, 2000; Frailey and Campbell, 2004). One characteristic that makes them easily recogniz-able among rodents is the presence of sharp quills. Living Neotropical porcupines represent one of the several radi-ations of mammals adapted to a strictly arboreal life and include approximately 15 species, extending from northern Uruguay and Argentina to southern Mexico (Emmons and Feer, 1997). They live mainly in different types of forests (e.g. rainforest, Atlantic forest, semi-deciduous forest, gallery forest; Emmons and Feer, 1997) of the Brazilian Subregion. With the exception of *Chaetomys subspinosus*, species of *Coendou* comprise the totality of Neotropical porcupines (Handley and Pine, 1992). Erethizontids are the only caviomorphs that survived in North America after their ingression during the Great American Biotic Inter-change. The semi-arboreal *Erethizon dorsatum*, the only extant species in North America, is distributed from north-ern Mexico to Alaska and Canada. It lives principally in warm forests and tolerates some climatic seasonal variation

(Roze, 1989). Fossil erethizontids were particularly abundant in Patagonia from the Late Oligocene to the Middle Miocene, during which they reached a considerable taxonomic diversity (e.g. Wood and Patterson, 1959; Candela, 1999, 2000, 2003; Kramarz, 2004). Santacrucian porcupines are grouped in the genus *Steiromys* Ameghino, 1887 (the most diverse of the extinct Erethizontidae and the first described by Ameghino as a member of this clade) which includes four (*S. duplicatus* Ameghino, 1887, *S. detentus* Ameghino, 1887, *S. principalis* Ameghino, 1901, *S. annectens* Ameghino, 1901) and possibly five (*S. intermedius* Scott, 1905; see Candela, 2000) species. *Steiromys duplicatus* is one of the most abundant Miocene porcupines from Patagonia.

In the extant Cavioidea, Dasyproctidae is represented by *Dasyprocta* (agoutis) with 11 species and *Myoprocta* (acouchis) with two species. All are extremely agile cursorial forms, inhabiting tropical and subtropical forested habitats and scrublands of the Brazilian Subregion (Nowak, 1991; Redford and Eisenberg, 1992). Dasyproctids are recognized since the Late Eocene? of Peru (Wyss *et al.*, 1993; Flynn *et al.*, 2003; Frailey and Campbell, 2004). Several genera from the Late Oligocene–Middle Miocene of Bolivia, Argentina, and Colombia have been assigned to this clade (e.g. Hoffstetter and Lavocat, 1970; Patterson and Wood, 1982; Kramarz, 1998, 2006; Fields, 1957; Walton, 1997). However, the phylogenetic relationships of the extinct taxa with *Dasyprocta* and *Myoprocta* have not been clearly established (Vucetich, 1984; Patterson and Wood, 1982; Walton, 1997). *Neoreomys* Ameghino, 1887 is one of the most abundant extinct caviomorphs of the Miocene of Patagonia, and is recognized from the "Pinturan" (late Early Miocene; Kramarz, 2006) to the "Colloncuran" (Middle Miocene; Vucetich *et al.*, 1993). *Neoreomys australis* Ameghino, 1887 is the only species of the genus recognized in the Santa Cruz Formation (Kramarz, 2006).

Eocardiids include small- to medium-sized caviomorphs, recognized from the Late Oligocene to the Middle Miocene (e.g. Ameghino, 1887; Scott, 1905; Wood and Patterson, 1959; Patterson and Wood, 1982; Vucetich, 1984; Kramarz, 2006; Pérez *et al.*, 2010a, b) mainly in Patagonia, but have recently been recorded from San Juan, Argentina (López *et al.*, 2011), Pampa Castillo and Curá Mallín, Chile (Flynn *et al.*, 2002; Flynn *et al.*, 2008), and Quebrada Honda, Bolivia (Croft, 2007).These rodents were traditionally related to the origin of Caviidae (cavies and maras) and Hydrochoeridae (the large capybaras) (e.g. Scott, 1905; Kraglievich, 1932; Patterson and Wood, 1982; Kramarz, 2006). However, recent phylogenetic studies (Pérez, 2010a, b; Pérez and Vucetich, 2011) of Cavioidea *sensu stricto* (*sensu* Patterson and Wood, 1982: 510) supported the paraphyly of "Eocardiidae" and, therefore, quotation marks are used for this assemblage of basal cavioids. The crown group of Cavioidea *s.s.* is formed by Caviidae + Hydrochoeridae, excluding the basal forms (i.e. "Eocardiidae"). According to Pérez (2010a) the second large diversification in the evolutionary history of cavioids *s.s.* occurred during the Santacrucian, represented by several Patagonian species of "Eocardiidae", such as *Eocardia montana* Ameghino, 1887, *E. excavata* Ameghino, 1894, *E. fissa* Ameghino, 1891, *Luantus toldensis* Kramarz, 2006, *Phanomys mixtus* Ameghino, 1887, *Ph. vetulus* Ameghino, 1891, *Schistomys erro* Ameghino, 1887, and *S. rollinsii* Scott, 1905 (Pérez, 2010a). *Eocardia fissa* is one of the most common caviomorph rodents recovered from the Santa Cruz Formation (Pérez, 2010a).

Octodontoids (not discussed in this chapter) constitute the most diverse clade of caviomorphs (Nowak, 1991). They were well represented in Patagonia during the Santacrucian (Ameghino, 1887; Scott, 1905), with six genera (i.e. *Sciamys* Ameghino, 1887, *Acaremys* Ameghino, 1887, *Spaniomys* Ameghino, 1887, *Stichomys* Ameghino, 1887, *Adelphomys* Ameghino, 1887, and *Acarechimys* Patterson, 1965) and more than 20 recognized species (Ameghino, 1887; Scott, 1905) that require detailed systematic revision. Available information indicates that Santacrucian octodontoids were small rodents with a body size apparently no larger than 1 kg (*Spaniomys riparius* Ameghino, 1887, 649 g; *Stichomys regularis* Ameghino, 1887, 492 g; Croft, 2000).

Chinchilloids (not treated in this chapter) include Dinomyidae, Neopiblemidae, and Chinchillidae (McKenna and Bell, 1997). The extinct *Perimys* Ameghino, 1887 (Neopiblemidae) is the only Neopiblemidae identified during the Santacrucian age, with several recognized species (Scott, 1905; Kramarz, 2002), which show a relatively wide range of body size (*Perimys* cf. *erutus* 585 g, *Perimys* sp. 326 g; Croft, 2000). Among Santacrucian chinchillids, nine species are included in the genera *Pliolagostomus* Ameghino, 1887 and *Prolagostomus* Ameghino, 1887, all smaller than the living *Lagostomus maximus* (*Prolagostomus oblicuidens* Scott, 1905, 2591 g; Croft, 2000).

15.3 Materials and methods

15.3.1 Materials

Institutional abbreviations Specimens of extinct caviomorphs studied here are housed in the Colección Nacional Ameghino of the Museo Argentino de Ciencias Naturales "Bernardino Rivadavia" (MACN-A), Buenos Aires, Argentina and the Museo Regional Provincial "Padre M. J. Molina" (MPM-PV), Rio Gallegos, Santa Cruz Province, Argentina.

Specimens of living caviomorphs examined are housed in the Museo de La Plata, La Plata, Argentina (MLP); Museo Municipal de Ciencias Naturales "Lorenzo Scaglia," Mar del Plata, Argentina (MMP); Museo Argentino de Ciencias

Naturales "Bernardino Rivadavia," Buenos Aires, Argentina (MACN); Museu Nacional, Rio de Janeiro, Brazil (MN); Fundação Zoobotânica do Rio Grande do Sul, (FZB/RS); Museu de Ciências Naturais, Porto Alegre, Brazil (MCN), and Centro Nacional Patagónico, Puerto Madryn, Argentina (CENPAT).

We studied those Santacrucian caviomorphs that are known by associated and well-preserved cranial and post-cranial remains. These include the cavioids *Eocardia fissa* ("Eocardiidae") and *Neoreomys australis* (Dasyproctidae), and the porcupine *Steiromys duplicatus* (Erethizontidae). Scott (1905) provided a mainly descriptive treatment of the species. Although chinchilloids and octodontoids are not analyzed in this study, they are noted where relevant.

The specimens of extinct caviomorphs studied herein are listed in Appendix 15.1. Fossil materials housed in MPM-PV were recovered between 2003 and 2011 by a joint team from the Museo de La Plata (Argentina) and Duke University (USA). All fossil specimens are from the Santa Cruz Formation (late Early Miocene, Santa Cruz Province, Argentina).

15.3.2 Methods

Body mass Body masses of the Santacrucian rodents analyzed here were estimated from allometric equations obtained by Bikneviciu *et al.* (1993), based on the anteroposterior diameters of both the femur and humerus, and by Croft (2000), based on the length of the upper (M1) and lower (m1) first molars. For functional interpretations we use the body mass estimates obtained by applying the equations of Bikneviciu *et al.* (1993), because they resulted in better estimators than the equations of Croft (2000), as explained below in Results. The body masses so obtained were averaged for subsequent analyses.

Feeding habits The identification of the areas of origin and insertion of the masticatory muscles in extinct species is based on muscle scars and fossae on the skull and the lower jaw, following the pattern observed by us and from the literature (Woods, 1972) in living caviomorphs, and information on other living mammals (Maynard Smith and Savage, 1959; Turnbull, 1970; Woods and Howland, 1979; Woods and Mackeen, 1989; Greaves, 1991; Samuels, 2009). Qualitative features that have been identified as functionally significant for interpreting the feeding habits of other rodents (e.g. Turnbull, 1970; Samuels, 2009) are also considered in the evaluation of the extinct species. Dental features and degree of hypsodonty are qualitatively evaluated in each case by comparison with the characteristic patterns of the extant related forms.

Locomotion Locomotor habits of *Steiromys duplicatus* were recently evaluated by Candela and Picasso (2008). Here we provide a synopsis of the main postcranial features

that, according to these authors, are functionally significant for inferring the locomotor behavior of this species. Postcranial remains of *Neoreomys australis* are compared especially with those of the living cavioids *Dasyprocta* (Dasyproctidae) and *Agouti paca* (Agoutidae), and analyzed from a morphofunctional perspective. *Dasyprocta* was selected as it is a member of the same family in which *Neoreomys* is traditionally placed. Extensive comparisons were also made between *Neoreomys* and *Agouti paca* because of the postcranial resemblances between them. The main postcranial features functionally associated with different locomotor habits (including the capacity to climb, run, and leap) identified in caviomorphs and in other rodents by Elissamburu and Vizcaíno (2004), Samuels and Van Valkenburgh (2008), Candela and Picasso (2008), and Candela and Vizcaíno (2007) were used as a model to infer their functional significance in Santacrucian species. We especially discussed the muscular and associated osteological features that are relevant for functional interpretations. In this sense, the areas of articulation and muscular attachment were emphasized in our analysis. We mainly followed the myological nomenclature and system arrangement of McEvoy (1982) and García Esponda and Candela (2010). Terminology of the postcranial features of caviomorphs mainly follows Candela and Picasso (2008).

15.4 Results

15.4.1 Body mass
The results of the body mass estimations are shown in Table 15.1. At least in the case of one specimen of *Pediolagus salinicola* with known body mass, femoral and humeral measurements produced better estimates than molar lengths. Average body masses of the Santacrucian species, obtained from long bones, are 14.17 kg for *Steiromys duplicatus*, 2.83 kg for *Eocardia fissa*, and 7.12 kg for *Neoreomys australis*. Body masses obtained from equations based on molar lengths (Croft, 2000) are shown for comparative purposes.

15.4.2 Feeding habits
Results on feeding habits are preliminary because the analysis of masticatory apparatus in Santacrucian caviomorphs included here is in progress.

The masticatory apparatus of *Steiromys duplicatus* (Fig. 15.2) shows some differences from those of extant porcupines. The sagittal and temporal crests (origin of the main part of the temporal muscle; Woods, 1972) and the posterior zygomatic root (origin of the orbital part of the temporal muscle; see Fig. 15.2a) are more developed than in living porcupines, suggesting that the temporal muscle was better developed in this fossil form (see Turnbull, 1970; Samuels, 2009). The anterior part of the

Table 15.1. *Body mass estimations of the studied specimens*

| | Body mass estimates | | | | | | | |
	Humerus	Femur	Mean	M1	m1	Mean	Body mass (kg)	Sources
Dasyproctidae								
Neoreomys australis	6.13±1.76 (2)	8.9±1.54 (5)	7.52	4.09±1.35 (15)	6.2±1.93 (7)	5.14		
Dasyprocta							1.3–4	a
Myoprocta							0.6–1.3	a
Agoutidae								
Agouti paca							7.5	b
"Eocardiidae"								
Eocardia fissa		2.838	2.838	0.8	1.33	1.06		
Caviidae								
Pediolagus salinicola[*]	2.158	2.267		0.697	1.998		2	b
Dolichotis patagonum							8	c
Erethizontidae								
Steiromys duplicatus	16.27	12.07	14.17	5.72				d
Erethizon dorsatum							10–12	e
Coendou prehensilis							4.7	b
Chaetomys subspinosus							1.3	a
"Sphiggurus"							1.5–2.5	b

Note: Estimates based on anteroposterior diameters of humerus and femur (equations from Biknevicius *et al.*, 1993) and first molar length (equations from Croft, 2000). a, Nowak, 1991; b, Redford and Eisenberg, 1992; c, Campos *et al.*, 2001; d, Candela and Picasso, 2008; e, Miles *et al.*, 1985.

* indicates known body mass of specimens. Species in bold face type indicate extinct genera analyzed in this chapter.

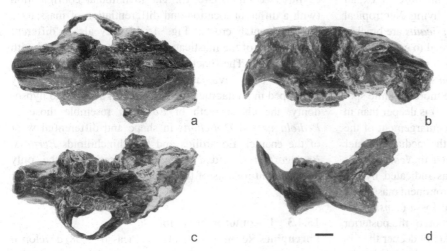

Fig. 15.2. *Steiromys duplicatus* MACN-A 10053–54; a–c, skull in dorsal (a), lateral (b), and ventral (c) views; d, right dentary in lateral view (reversed). Scale bar = 1 cm.

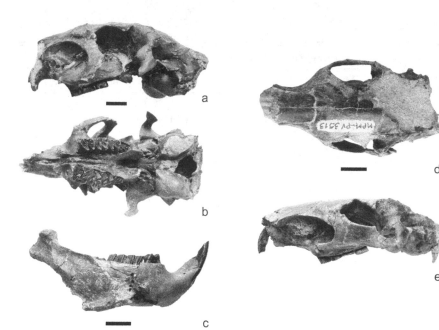

Fig. 15.3. *Neoreomys australis*. MPM-PV 3515, a–c, skull in lateral (a), and ventral (b) views, (c) right dentary in lateral view. MPM-PV 3513, d, e, skull in dorsal (d) and lateral (e) views. Scale bar = 1 cm.

medial masseter muscle also seems to have been more developed in *S. duplicatus* than in the extant species, as indicated by the larger rostral fossa of the zygomatic (*sensu* Candela, 2000) and deeper masseteric fossa of the mandible. The tuberosity on the ventral zygomatic root is very prominent, indicating a strong superficial masseter muscle, and the well-developed masseteric crest indicates also a strong lateral masseter muscle. The mandibular condyle (Fig. 15.2d) is relatively more dorsal with respect to the cheek teeth than in extant genera. In addition, the incisors in *S. duplicatus* (see Fig. 15.2b–d) are more robust, broader, and anteroposteriorly deeper, and more deeply implanted than those of extant porcupines (see Candela, 2000). Like living Neotropical species, the cheek teeth of *Steiromys duplicatus* are brachydont, but with thicker enamel, and the trend to early closure of the flexi (see Candela, 2000). The lower molars have five lophids (one more than those of *Coendou* spp.).

In *Neoreomys australis*, even in those more juvenile specimens, the rostral fossa (see Fig. 15.3a, b) is deeper than in *Dasyprocta*, a condition that reflects an enlargement of the area of the origin of the anterior part of the medial masseter (Samuels, 2009). The superficial masseter in *Neoreomys* is more developed than in *Dasyprocta*, as indicated by the wider ventral zygomatic root and more prominent masseteric tuberosity. On the mandible, the medial fossa (*sensu* Candela, 2000; see Fig. 15.3c), for the insertion of the posterior part of the medial masseter, is also markedly deeper than in *Dasyprocta*, indicating that this muscle was relatively larger. The coronoid process is much more prominent than in

Dasyprocta, indicating a stronger temporal muscle. One of the more distinctive features in *Neoreomys* is the relatively larger size and higher crowns of the cheek teeth (see Figs. 15.3a, b, 15.6a, b, and 15.7a) compared with *Dasyprocta*. In addition, *Neoreomys* has noticeably larger incisors than *Dasyprocta*. Unlike *Dasyprocta* (with sub-parallel tooth rows), the divergent tooth rows in *Neoreomys* reflect a different modality of mastication, but more information is necessary to evaluate the significance of this feature.

Preliminary observations indicate that the incisors in *Eocardia fissa* are proportionately larger than in living caviids (see Fig. 15.4b, d). The mandibular configuration (with a different location and differentiation of masseteric and horizontal crests; Fig. 15.4d) reveals a different arrangement of the masticatory musculature compared with living caviids. The functional significance of these features has not been evaluated yet. Euhypsodonty is fully developed in Santacrucian eocardiids. In addition to hypsodonty, the cheek teeth of *Eocardia* resemble those of *Pediolagus* and *Dolichotis* in shape and differential wear of the enamel. Eocardiids and the chinchilloids *Perimys*, *Prolagostomus*, and *Pliolagostomus* represent the only euhypsodont rodents of the Santacrucian Age.

15.4.3 Locomotor behavior

Porcupines Recently, Candela and Picasso (2008) developed a functional analysis of the postcranium of the Santacrucian erethizontid *Steiromys duplicatu*s (see Fig. 15.5). As in living

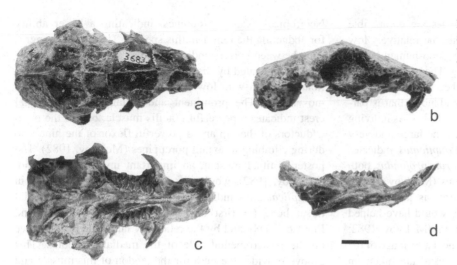

Fig. 15.4. *Eocardia fissa* MPM-PV 3683; a–c, skull in dorsal (a), lateral (b), and ventral (c) views; d, left dentary in lateral view (reversed). Scale bar = 1 cm.

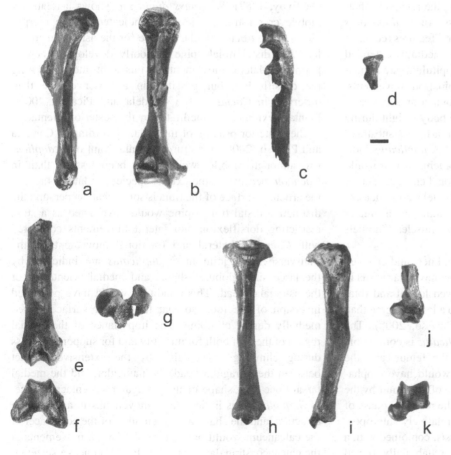

Fig. 15.5. *Steiromys duplicatus* MACN-A 10055–78; a, b, right humerus in medial and anterior views; c, left ulna in lateral view; d, proximal left radius in anterior view; e, f, distal right femur in anterior and distal views; g, proximal portion of the left femur in anterior view; h, i, left tibia in anterior and lateral views; j, right calcaneum in dorsal view; k, left astragalus in dorsal view. Scale bar = 1 cm.

porcupines, several features of this species are compatible with the ability to climb. In the forelimb, the relatively low humeral tuberosities allow rotation at the gleno-humeral joint, a movement that is required for climbing. The prominent and relatively distally extended deltopectoral crest is indicative of a large *m. pectoralis major*, a powerful adductor that is particularly active during climbing (McEvoy, 1982). As in living porcupines and other arboreal mammals, the large and very protruding entepicondyle of *Steiromys duplicatus* indicates large *mm. pronator teres* and *flexor digitorum profundus*, both acting in climbing and grasping functions (McEvoy, 1982). Large *mm. brachioradialis* and *supinator*, as indicated by the prominent lateral epicondylar ridge, would have helped maintain stability at the radio-humeral joint (McEvoy, 1982; Candela and Picasso, 2008). The mechanical advantage of the *m. biceps brachii* is enhanced by the more distal attachment on the bicipital tuberosity, which reflects the powerful action of the flexor muscles (*mm. flexors biceps brachii* and *brachialis*) of the antebrachium during climbing at the beginning of the propulsive phase (McEvoy, 1982). As in living porcupines and other arboreal mammals (e.g. Szalay and Dagosto, 1980; Argot, 2001; Sargis, 2002), and differing from cursorial caviomorphs (see Candela and Picasso, 2008), the relatively shallow and non-perforated olecranon fossa of *S. duplicatus* reflects rotation at the elbow joint. Several features (concave radial notch, convex and very posteromedially extended ulnar facet of the radial head, globular capitulum) are indicative of a great range of pronation–supination movements. The wide trochlea–coronoid process contact area indicates an important function in supporting the body weight during the flexed and abducted forelimb on an arboreal substrate. A short but robust olecranon process of *S. duplicatus* indicates a strong insertion of the triceps brachii, which would act to catch new supports, aiding directional changes, during arboreal locomotion (Argot, 2002; Candela and Picasso, 2008). A short olecranon process also indicates a greater effective range of motion for the triceps muscles (Samuels and Van Valkenburgh, 2008).

In the hindlimb, the shape of the hip, knee, and cruroastragalar, calcaneoastragalar, and astragalonavicular joints in *Steiromys duplicatus* would have allowed lateral and rotational movements, although probably to a lesser degree than in extant porcupines (Candela and Picasso, 2008). The relatively large femoral head of *S. duplicatus* is compatible with a broad range in abduction of the femur, and the medially protruding lesser trochanter would have emphasized the abduction and outward rotation of the femur by the action of the iliopsoas complex during the recovery phase of climbing. As in extant porcupines, the relatively anteroposteriorly narrowed distal femoral epiphysis, combined with a wide femoral trochlea, may indicate a habitually flexed knee, which is characteristic of arboreal locomotion. This condition is less marked in *Steiromys duplicatus* than in

Neotropical extant porcupines, indicating a lesser ability for abducting the femur in this species. The limited congruence between corresponding articular surfaces of the knee joint, as indicated by the shape of the femoral condyles and tibial platform with low tibial spines, facilitates rotational movements. The prominent and distally extended tibial crest indicates a powerful gracilis muscle, one of the main adductors of the hip and a powerful flexor of the hindlimb during climbing in extant porcupines (McEvoy, 1982). The posterior tibial muscle, an important invertor of the foot (McEvoy, 1982), would also have been well developed in *S. duplicatus*, as indicated by the great size of the medial tarsal bone (= first sesamoid; fig. 28 of Candela and Picasso, 2008), and by the relatively large and deep groove on the posteromedial side of the medial malleolus. This groove provides the path for the tendon of this muscle and for the *flexor digitorum tibialis*, an important plantar and digital flexor (McEvoy, 1982). The space between the tibia and fibula in *S. duplicatus* reveals a large origin area for the *m. flexor digitorum fibularis*, another important plantar and digital flexor, particularly active in forcing the claws into the substrate and in grasping branches during climbing (McEvoy, 1982). *Steiromys duplicatus* shows a relatively mobile cruroastragalar joint, as indicated by the morphology of the tibial articular surface for the astragalar trochlea. The distal tibial spine is poorly developed, thereby permitting lateral movements at this joint and facilitating the dorsiflexion, but possibly in a lesser degree than observed in *Coendou* (see Candela and Picasso, 2008). Plantar flexion is also facilitated by the posterior orientation of the posterior process of the tibia. According to Candela and Picasso (2008), the cruroastragalar joint of *S. duplicatus*, although mobile, would have been less so than in *Coendou*, because in the extinct species the lateral facet of the articular surface of the tibia is somewhat deeper, and the distinctive distal tibial spine would have acted as a stop, restricting dorsiflexion and lateral movements compared with *Coendou*. Lateral and rotational movements at the transverse tarsal joint in *S. duplicatus* are indicated by the large size, globular shape, and medial orientation of the astragalar head. This condition would have permitted inversion of the foot, so that the plantar surface turned medially during climbing. The importance of the medial region of the foot both for mobility and for supporting stress during climbing is revealed by the extensive contact between the astragalar head, the navicular, and the medial tarsal bone. The shape of the astragalar sustentacular facet of *S. duplicatus* is indicative of movements in a transverse direction, and the shape and orientation of the ectal facet of the calcaneum would have permitted lateral movements at the calcaneoastragalar joint. The slightly concave sustentacular facet suggests that these movements would have been less emphasized than in extant Neotropical species. As in extant porcupines, *S. duplicatus* has a relatively long tuber calcanei (in-lever arm) and a very short distal region of the calcaneum as well as a relatively short metatarsal (out-lever arm). This condition is more efficient in generating greater power but not speed, providing great effective force during climbing (see Candela and Picasso, 2008).

Among the most specialized features for grasping and foot inversion in living porcupines are the well-developed medial tarsal bone (= first sesamoid; see fig. 22 of Candela and Picasso, 2008), the hypertrophied medial sesamoid (more developed in *Coendou*), the medially oriented metatarsal I, and very curved claws, which reach their extreme development in *Coendou* (Jones, 1953; Grand and Eisenberg, 1982). In *Coendou*, the hallux is reduced and incorporated into the preaxial pad and the medial sesamoid is very large and particularly active during pedal grasping (Jones, 1953). The foot of *Erethizon dorsatum*, bearing a well-developed hallux and a smaller medial sesamoid, has not achieved the degree of specialization of *Coendou*. As in *E. dorsatum*, in *Steiromys duplicatus* the hallux is well developed, the medial sesamoid is smaller than in *Coendou*, the metatarsal I is not as medially oriented with respect to the remaining metatarsals as in *Coendou*, and the claws are less robust and deep than in the latter (fig. 22 of Candela and Picasso, 2008).

Dasyproctidae Recently, postcranial features of *Neoreomys australis* were identified as functionally related to parasagittal movements, compatible with cursorial habits (Candela and Vizcaíno, 2007). Most features of *N. australis* (Figs. 15.6 and 15.7) are shared with those of living *Agouti paca*. Proportions and relative size of all examined elements are more similar to those of *Agouti paca* than of *Dasyprocta* spp. Osteological features indicate that the muscles of *Neoreomys australis* were comparatively larger than in *Dasyprocta* spp. Humeral tuberosities of *N. australis*, especially the greater tuberosity, are higher than the humeral head, restricting the mobility of the glenohumeral joint, a condition shared with other cavioids (see Candela and Picasso, 2008). The deltopectoral crest, which provides the area of insertion for pectoral and deltoid muscles (see García Esponda and Candela, 2010), is more distally extended than in *Dasyprocta*, a condition similar to that of *Agouti paca*. As in the latter, the entepicondyle, which provides the origin for the flexor muscles of the antebrachium and manus, is more robust than in *Dasyprocta*, suggesting a large pronator teres, which stabilizes the humeroradial joint and supports the weight during locomotion. The reduction of the entepicondyle and the related flexor musculature (Jenkins, 1973) in *Dasyprocta* seem to be cursorial specializations of this more genus. As in *Agouti paca*, the lateral epicondylar ridge is more raised than in *Dasyprocta*. As in the latter, *Agouti*, *Dolichotis*, *Cavia*,

and *Hydrochoerus*, the olecranal fossa is perforated in *Neoreomys australis*. This feature would facilitate anteroposterior movements, allowing the olecranal process to pass through the fossa during complete extension. The elbow joint of *N. australis*, which is relatively wider than that of *Dasyprocta*, has several features indicative of a preponderance of parasagittal movements. The trochlea is deep, although to a lesser degree than in *Dasyprocta*, a condition that increases the articular surface in a proximodistal direction (Candela and Picasso, 2008). As in *Agouti paca* and *Dasyprocta*, *N. australis* has a very distinctive medial trochlear keel, extended flanges, and a markedly developed capitular tail. These features would be effective for resisting mediolateral forces at the elbow, maximizing the correspondence of articular surfaces and stability at this joint. The radial head, which as in other cavioids (see Candela and Picasso, 2008) is anterior with respect to the ulna, is subrectangular (more laterally extended than in *Dasyprocta*), with a flattened and posteriorly located ulnar facet. The capitular eminence is well developed (resembling *Agouti paca*), probably acting in the stabilization of the joint. This configuration restricts movements at the elbow joint and increases stability during flexion/extension. The extended radial head would be more efficient in bearing loads at the humeroradial articulation during the propulsive phase of locomotion. As in *Dasyprocta* and *A. paca*, the greater trochanter of the femur projects farther proximally than the femoral head, which increases the mechanical advantage of the gluteal muscles (Maynard Smith and Savage, 1956). As in *Dasyprocta* and other cavioids, the lesser trochanter is posteriorly located. This position would indicate a more anteroposterior orientation of the fibers of the iliopsoas complex, facilitating parasagittal movements (Candela and Picasso, 2008). The distal femoral epiphysis is relatively deep. A deep knee, with an anteriorly projected tibial tuberosity (a feature less marked in *Neoreomys* than in *Dasyprocta*) increases the mechanical advantage of the quadriceps femoris, making it more powerful for flexion of the hip and extension of the knee. These conditions are consistent with efficient parasagittal propulsion during terrestrial locomotion. Like *Dasyprocta* and *Agouti paca*, *N. australis* has a relatively congruent knee articulation, which enhances stability, thereby restricting rotation of the tibia. As in *A. paca*, the tibia of *N. australis* is relatively shorter than that of *Dasyprocta*. The distally extended tibial crest of *N. australis* reflects an increased mechanical advantage of the hamstring muscles which are required for powerful extension of the hip at the beginning of propulsion and for flexion of the knee during the recovery stroke (Maynard Smith and Savage, 1956; Elftman, 1929). Features of the cruroastragalar joint of *N. australis* indicate restricted movements, because the tibial articular surface possesses a deep and concave lateral facet. The anteriorly projecting

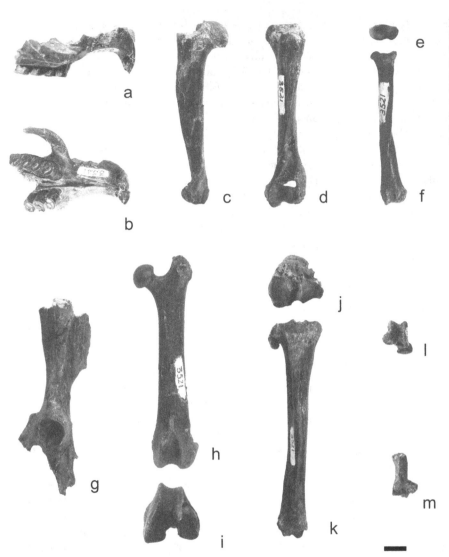

Fig. 15.6. *Neoreomys australis* MPM-PV 3521; a, b, cranial fragment in lateral and ventral views; c, d, left humerus in lateral and anterior views; e, f, left radius in proximal and anterior views; g, left pelvis in lateral view; h, i, left femur in anterior and distal views; j, k, right tibia in proximal and anterior views; MPM-PV 3517; l, right astragalus in dorsal view; m, calcaneum in dorsal view. Scale bar = 1 cm.

posterior process and the prominent anterior distal tibial spine act as stops to the movements at the upper ankle joint. As in *Dasyprocta*, *Agouti*, and caviids (see Candela and Picasso, 2008), the astragalar head in *N. australis* is oriented relatively parallel to the parasagittal plane, a condition that restricts flexion–extension movements. As in other cavioids, the distal portion of the calcaneum is very elongated, a condition that increases speed at the expense of force compared with arboreal forms (see Candela and Picasso, 2008). The shape and relative size of remaining elements of the feet are very similar to those of *Agouti paca* (e.g. the plantar process of the navicular is well developed but to a lesser degree than in *Dasyprocta*; the preserved facet for metatarsal I on metatarsal II is more developed than in

Dasyprocta; metatarsals II–IV are more robust and relatively shorter than in *Dasyprocta*, even more than in *A. paca*).

"Eocardiidae" Forelimb features in *Eocardia fissa* available for this study are represented only by the proximal end of the ulna, for which the articular region is preserved. Even so, it is sufficiently preserved to allow the recognition of elbow joint features that, as in living cavioids, restrict mobility at this joint and increase stability of the joint during flexion and extension (see Candela and Picasso, 2008). In *E. fissa* the coronoid process of the ulna is comparatively narrow with respect to the radial notch, reflecting a narrow anterior contact with the humeral trochlea, as

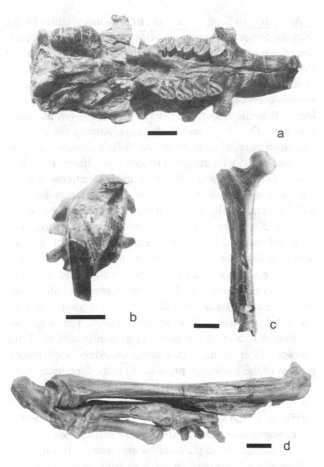

Fig. 15.7. *Neoreomys australis* MPM–PV 4342; a, b, skull in ventral and anterior views, c; right femur in anterior view; d, right tibia and articulated foot in lateral view. Scale bar = 1 cm.

occurs in living caviids (see Candela and Picasso, 2008). This condition indicates that the trochlea–coronoid process contact was anteriorly replaced by the trochlea–radial head contact. The relatively wide radial notch reflects a relatively wide radial head, which, as in living cavioids, was located anteriorly with respect to the ulna. As noted by Scott (1905), this condition prevented any rotational movement at the elbow joint. As in *Dolichotis* or *Cavia* (see fig. 8 of Candela and Picasso, 2008), the radial notch in *Eocardia fissa* is represented by two distinctive articular facets, one anterolaterally located, flattened, and subcircular, and the other extended medially towards the reduced coronoid process. This conformation limits radial supination and increases stability during flexion/extension (Candela and Picasso, 2008). According to Scott (1905: 467), in *Eocardia* there is much more disparity between the lengths of the fore- and hindlimbs than in any other member of the clade. We are unable to corroborate this information, but taking into account the measurements provided by Scott (1905), we note that *Eocardia* possessed relatively longer hindlimbs than those of *Dolichotis*. *Pediolagus* seems to have intermediate limb lengths, but more similar to that of *Dolichotis*. In addition, as was pointed out by Scott (1905: 470–471), *Eocardia* does not exhibit the elongation of the radius/ulna and tibia/fibula present in *Dolichotis*, instead having proportions more similar to those of *Cavia*. The remaining features of the hindlimbs in *Eocardia fissa* are compatible with emphasized parasagittal movements and stabilized joints. The portion preserved of the gluteal dorsal fossa of the ilium (Fig. 15.8) is deeper and wider than that of the living *Dasyprocta* and *Cavia*, being similar to that of

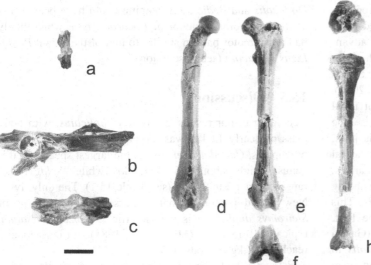

Fig. 15.8. *Eocardia fissa* MPM-PV 3683; a, right proximal fragment of the ulna; b, c, right and left pelvis in lateral view; d, right femur in anterior view; e, f, left femur in anterior and distal views; g, h, right tibia in proximal and anterior views; i, left tibia in anterior view. Scale bar = 1 cm.

Pediolagus salinicola. This condition indicates that the medial gluteus muscle was relatively well developed, in agreement with an improved extension of the hip joint. As in *Pediolagus salinicola*, the tuberosity of the pelvis for the origin of *m. rectus femoris* is prominent, forming a robust protuberance, which is more developed than in the living cavies and *Dasyprocta*. This is in agreement with a strong development of this muscle, which acts as a flexor of the hip and extensor of the knee (García Esponda and Candela, 2010). A rapid extension of the knee is required in species with agile movements (e.g. Szalay and Sargis, 2001; Abello and Candela, 2010). A deep acetabulum limited by raised margins acts to partially restrict rotational movements in living caviids (Candela and Picasso, 2008; García Esponda and Candela, 2010). This configuration provides a relatively elevated congruence at the hip joint necessary to stabilize flexion/extension movements and accentuated parasagittal movements of the femur during locomotion. In the available sample, the degree of proximal extension of the greater trochanter of the femur is unknown, because the base of this structure is partially broken. However, the preserved portion of this trochanter (robust and relatively wide) indicates an extensive insertion area for the deep and medial gluteals, compatible with a powerful extension of the thigh during the propulsive phase of locomotion, a feature typical of cursorial species (e.g. Maynard Smith and Savage, 1956; Candela and Picasso, 2008; Salton and Sargis, 2009). The lesser trochanter is relatively more developed than in examined living cavies and it is caudally located, as in other cursorial caviomorphs (Candela and Picasso, 2008). This condition would improve the function of *m. iliopsoas* as a protractor, rather than as a rotator of the femur (Candela and Picasso, 2008; Salton and Sargis, 2009).

The distal end of the femur and proximal end of the tibia compose a deep knee. As in living caviids, this is characterized by a long and relatively narrow femoral trochlea, delimited by sharp crests, and a cranially projected tibial tuberosity. These features increase the mechanical advantage of *m. quadriceps femoris* in the extension of the knee (Fostowicz-Frelik, 2007; Candela and Picasso, 2008; Salton and Sargis, 2009; García Esponda and Candela, 2010). As in living caviids, the knee joint has prominent tibial spines and concave tibial condyles, which increase the congruence of corresponding articular facets at this joint, restricting mobility and rotation of the tibia. The lateral fossa of the tibia, from which the *m. tibialis cranialis* arises, is deeper than in living cavies and in *Dasyprocta*, being more similar to that of *Pediolagus* and *Dolichotis*. This muscle dorsiflexes the ankle joint and inverts the foot. It is possible that, as was interpreted for *Dasyprocta* (García Esponda and Candela, 2010), the *m. tibialis cranialis* in *Eocardia* acted principally as an extensor rather than as an inverter of the pes.

According to Scott (1905), the tibia is only slightly longer than the femur. In agreement with these data, the comparison of equivalent portions of the tibial shafts of *E. fissa* and *Pediolagus salinicola* indicates that this bone in the extinct species was somewhat longer than the femur but not to the extent observed in living species. Thus, although somewhat longer than the femur, the tibia of *E. fissa* is not as elongated as in *Dolichotis* and *Pediolagus*. Among the cavioids examined here, elongation of the tibia, a feature functionally associated with cursorial or saltatorial habits (Gambaryan, 1974), was only observed in the extreme cursors *Dasyprocta*, *Pediolagus*, and *Dolicthotis* (García Esponda and Candela, 2010). As in living cavioids (see Candela and Picasso, 2008), the distal tibial articular surface possesses deep and concave lateral and medial facets, which indicate restricted lateral movements at the upper ankle joint. The distinctive anterior distal tibial spine would have offered a stop to dorsiflexion and lateral movements at this joint. These features indicate that the upper ankle joint in *E. fissa* was relatively stable, as in living cavioids. The shape and relative size of foot elements (e.g. configuration of the cuboidal facet of the calcaneum, well-developed plantar process of the navicular, presence of three functional metatarsals, rudimentary facet on the metatarsal IV for metatarsal V) indicate that the foot was digitigrade (see Scott, 1905), as is the tarsus of living caviids (A. M. Candela *et al.*, unpublished data). All of these features are compatible with emphasized parasagittal movements. In summary, the postcranial morphology of *Eocardia*, characterized by marked joint stability, a prominent femoral tubercle, and a deep knee, among other features, constitutes an arrangement better suited for flexion–extension movements. Stabilized joints are in agreement with the necessary stability required to run or leap. These features are well suited for a cursorial and agile mode of locomotion. As in living *Dolichotis* and *Pediolagus*, leaping could have been a part of the locomotor behavior of *Eocardia fissa*, which likely had a locomotor pattern similar to that of the extant *Pediolagus salinicola* (see Discussion).

15.5 Discussion

According to our results, *Steiromys duplicatus*, with body mass of nearly 14 kg, was larger than extant Neotropical porcupines. *Coendou prehensilis*, the largest species of this genus, weighs about 4.5 to 4.9 kg, while "*Sphiggurus*" ranges from 1.5 to 2.5 kg (see Table 15.1). The only living New World porcupine that surpasses the body mass of *Steiromys duplicatus* is the Holartic *Erethizon dorsatum*, typically 10 to 12 kg (Miles *et al.*, 1985), but occasionally reaching 18 kg (Woods, 1973).

Estimated body mass for *Eocardia fissa* is similar to that of the caviid *Pediolagus salinicola* ("conejo de los palos").

The latter, with an average body weight of 1.9 to 2.1 kg and a maximum of 2.7 kg, is the smallest living Dolichotinae, but it is larger than any living Caviinae (guinea pig; cavies). *Eocardia fissa* was thus larger than the largest living cavy *Cavia aperea* (0.5 to 1.5 kg; average weight 460 g). The "conejo de los palos" can be a suitable analog for interpreting the paleobiology of *E. fissa*. Average estimated body mass of *Neoreomys australis* is closer to the cursorial rodent *Agouti paca* (6 to 12 kg; average weight 7.5 kg; see Table 15.1) than to the species of *Dasyprocta* (1.3 to 4 kg). *Neoreomys australis* represents the largest Santacrucian cavioid and one of the largest Miocene caviomorphs from Patagonia.

In summary, our results indicate that the caviomorphs from the Santa Cruz Formation examined here exhibit a range of body masses between approximately 1 kg and 14 kg. Small-, medium- and relatively large-sized caviomorphs are recorded from the Santacrucian, with the smallest represented by octodontoids, the largest by the paca-like *Neoreomys australis* and the porcupine *Steiromys duplicatus*, and the medium-sized forms by the eocardiids.

The few studies on the ecology of Neotropical porcupines have noted that their feeding strategy is based mainly on fruits, seeds, buds, and soft leaves (Charles-Dominique *et al.*, 1981; Emmons and Feer, 1997). The semi-arboreal Holartic *Erethizon dorsatum*, mainly inhabiting temperate woodlands, consumes mostly harder food than Neotropical species, such as bark, coniferous leaves, and seeds (Roze, 1989). Osteological features of *Steiromys duplicatus* indicate that the masticatory muscles were stronger than in living Neotropical species. As was demonstrated in other rodents, a more massive skull could serve to accommodate stresses resulting from mastication and to support larger masticatory muscles (Samuels, 2009: 881). This increase of the mass of the adductor muscles in *S. duplicatus* may result from scaling effects related to its larger size with respect to the living species. The relatively anteroposteriorly thick and more deeply implanted incisors in *S. duplicatus* probably better resisted forces resulting from biting harder foods than in Neotropical porcupines (Samuels, 2009). Pentalophodont lower molars of *S. duplicatus* provided an increased number of surfaces available to cut or grind than in living porcupines, which would have favored the grinding action of the occlusal surfaces (Evans *et al.*, 2007; Samuels, 2009).

Consequently, *Steiromys duplicatus* presents features that indicate a diet of harder and more abrasive food items than those consumed by living Neotropical species. Its diet was probably more similar to that of *Erethizon dorsatum*, perhaps a specialization related to living in more open areas than in rainforest (see Section 15.6: Paleoenvironments).

The same situation is found in the dasyproctid *Neoreomys australis*, which exhibits features (e.g. higher crowned molars and relatively larger cheek teeth and incisors) that indicate that this species was capable of consuming harder food items than those consumed by living species of *Dasyprocta*. The feeding strategy of *Dasyprocta* spp. is based mainly on fruits, seeds, buds, and soft leaves (Charles-Dominique *et al.*, 1981; Emmons and Feer, 1997).

Eocardia fissa, with euhypsodont and heart-shaped cheek teeth, similar to those of *Pediolagus salinicola* and *Dolichotis patagonum*, may have favored a diet similar to these extant rodents, both of which consume a relatively high proportion of grasses and herbs, as well as leaves of shrubs and trees (Campos *et al.*, 2001; Chillo, 2007). A recent study (Chillo, 2007) indicated that *Pediolagus salinicola* and *Dolichotis patagonum* consume the same food items (leaves of trees, shrubs, and grasses), but in different proportions (*P. salinicola* prefers leaves of trees, herbs, and grasses while *D. patagonum* prefers grasses). Features of the masticatory apparatus of *E. fissa* are compatible with the ability to consume these food items, and suggest a relatively wide food spectrum, including abrasive foods.

Among Santacrucian caviomorphs different locomotor behaviors were represented, as indicated by the presence of semi-arboreal, relatively large cursorial, and smaller and more agile cursorial forms. Although exhibiting diverse forms of locomotion, none of them achieved the degree of specialization seen in living related taxa. *Steiromys* did not achieve a highly arboreal lifestyle like that of Neotropical porcupines (Candela and Picasso, 2008). *Neoreomys australis* would not have had the capacity to move with the remarkable speed and agility achieved by *Dasyprocta* (Nowak, 1991). In fact, at least some features of *Dasyprocta* may have originated in association with the development of a highly cursorial lifestyle (García Esponda and Candela, 2010). On the other hand, body size and most of the postcranial features of *Neoreomys australis* (e.g. general limb proportions) are more similar to those of *Agouti paca* than to *Dasyprocta*. *Agouti paca*, which inhabits mainly tropical forests, is a cursorial rodent with voluminous body that can run and is capable of swimming with agility (Mondolfi, 1972). It is thus plausible that *Neoreomys* had swimming abilities.

Estimated body mass and functional analysis of postcranial features of *Eocardia fissa* suggest that this species would have had a locomotor behavior similar to that of living *Pediolagus salinicola*. Like the mara, *Dolichotis patagonum*, the smaller *P. salinicola* is characterized by cursorial habits and the ability to leap and run with remarkable speed and agility (Nowak, 1991). Most of the osteological traits observed in *Eocardia fissa* are compatible with such habits, although to a lesser degree than in extant forms. Thus for example, *E. fissa* does not exhibit the marked elongation of the distal limb elements of *Pediolagus* and *Dolichotis*.

Santacrucian caviomorphs indicate ecological niche partitioning: *Steiromys* would have been semi-arboreal and adapted to consume harder food items than those eaten by extant Neotropical species; *Neoreomys* a relatively large cursorial rodent with some ability to swim, and feeding on moderately more abrasive food than extant species; and *Eocardia* an agile cursorial and probably a "mixed feeder" (forbes and grasses) capable of consuming relatively abrasive food items.

15.6 Paleoenvironments

As discussed above, characteristics of *Eocardia fissa* resembled those of living *Pediolagus salinicola*, an agile cursorial rodent adapted to eating relatively abrasive food. *Pediolagus* inhabits the arid and semi-arid bushland and relatively open forested areas of the Chaqueña Province (*sensu* Cabrera and Willink, 1980), and regions with thorn scrubs, characterized by warm climatic conditions and marked seasonality. Relatively open areas, with notable variation in rainfall, may have been the appropriate environment inhabited by *Eocardia fissa*. On the other hand, characteristics of the relatively large porcupine *Steiromys duplicatus* indicate that this species was not as highly arboreal as living Neotropical species, but rather semi-arboreal, resembling the living Holartic *Erethizon dorsatum*. The habitat occupied by *Steiromys duplicatus* was possibly more open than that inhabited by *Coendou*. *Steiromys duplicatus* indicates the presence of forested environments with certain variation in rainfall, unlike the evergreen rainforests inhabited by highly arboreal porcupines. The cursorial paca-like *Neoreomys australis* suggests the presence of relatively warm forested areas, probably associated with water bodies.

Taken together, the composition and paleobiological characteristics of Santacrucian caviomorphs denote the presence of mixed woodland habitats, with probable seasonality in rainfall patterns, similar to the Brazilian Cerrado biome. This is a vast tropical semideciduous savanna composed of a mosaic of subunits, ranging from open grassland to dry forest, with gallery forest, composed of evergreen trees usually with a sparse understory (Eiten, 1972, 1974; Fernandes and Bezerra, 1990). The Cerrado biome includes significant variation in vegetation types, forming a gradient from completely open "cerrado" ("campo limpo," clean field), dominated by grasses and low bushes with a total absence of arboreal cover, to the "cerrado *s.s.*" and the "cerradão," a closed canopy forest characterized by a higher density of trees with a semi-open canopy. The intermediate forms include the dirty field (the "campo sujo") and the "campo cerrado" (savanna), having an increasing density of trees (e.g. Eiten, 1972, 1974; Ferandes and Bezerra, 1990). The climate is hot, semi-humid, with pronounced seasonality marked by a dry winter season (CODEPLAN, 1976).

The wide variability of habitats in the different types of "cerrado" currently supports an enormous diversity of plants and animals species. Redford and Fonseca (1986) noted that 50% of cerrado species are characterized as wide-ranging (found in both forest and savanna biomes), such as armadillos, giant anteater *Myrmecophaga tridactyla*, and the canid *Chrysocyon brachyurus*, which occur in both open and forested areas. For these species, gallery forests may be particularly important during the dry season, as well as in years of decreased rainfall (Redford and Fonseca, 1986). Cerrado also includes several forest-adapted species, such as monkeys and porcupines, and species associated with permanent water bodies, such as *Chironectes minimus*, *Pteronura brasiliensis*, *Lontra longicaudis*, *Hydrochoerus hydrochaeris*, and *Agouti paca* (Redford and Fonseca, 1986; Cáceres *et al.*, 2008). The environment in which the numerous Santacrucian species lived may have been similar to the heterogeneous and diverse landscape of the Cerrado biome. A complex mosaic of open and closed habitats would have been favorable to the coexistence of Santacrucian species with different lifestyles: forest-adapted species, such as primates and porcupines; species associated with water bodies, such as paca-like rodents, and wide-ranging species, which probably lived in both forest and savanna biomes, such as certain armadillos and rodents like "conejo de los palos." Information provided by Santacrucian caviomorphs, in association with other sources of information (e.g. paleoclimate, paleobotanic evidence), will facilitate interpretation of their paleohabitats.

15.7 Conclusions

1. Santacrucian caviomorphs showed a range of body mass, as represented by small- to relatively large-sized living caviomorphs.
2. The range of sizes and morphological disparity in the masticatory apparatus of Santacrucian caviomorphs (including different degree of crown height, development of masticatory muscles) suggest differences in the degree of food selection. The porcupine *Steiromys duplicatus* and the dasyproctid *Neoreomys australis* probably had diets of harder and more abrasive food items than those consumed by related living Neotropical species. The eocardiid *Eocardia fissa* consumed a wide variety of items, including relatively coarse foods.
3. Santacrucian rodents had different locomotor behaviors and use of substrates. *Steiromys duplicatus* was a semi-arboreal form, which did not reach the degree of specialization seen in highly arboreal Neotropical species. *Eocardia fissa* was probably an agile cursorial species, similar to the living *Pediolagus salinicola* ("conejo de los palos"). *Neoreomys australis* was a

cursorial rodent, more similar to the living *Agouti paca* (Agoutidae) than to living species of *Dasyprocta*.

4. These species could have inhabited different areas or microhabitats within a mosaic environment comparable to the Brazilian Cerrado. Santacrucian caviomorphs developed differential resource utilization patterns with marked differentiation of their ecological niches.

ACKNOWLEDGMENTS

We thank C. García Esponda for comments and suggestions that improved the manuscript, and D. Flores (MACN), U. Pardiñas (CENPAT), D. Romero (MMP), and A. Kramarz (MACN), for facilitating access to specimens under their care. We extend special thanks to S. F. Vizcaíno, R. F. Kay, and M. S. Bargo for providing the new specimens collected during Museo de La Plata/Duke University expeditions and for inviting us to take part in this book. We thank A. Gainza for her help with the translation; D. Voglino for the illustrations and information on Cerrado biome; Michelle Arnal for providing information on Santacrucian Octodontoidea. We especially appreciate the editorial support and thank G. De Iuliis and the reviewers X. J. Samuels and M. G. Vucetich for their helpful comments and suggestions that substantially improved the manuscript. This is a contribution to the projects PICT 0143 to Sergio F. Vizcaíno, and NSF 0851272, and 0824546 to Richard F. Kay.

Appendix 15.1 List of examined specimens of extinct caviomorphs

Specimens of extinct caviomorphs are from the Santa Cruz Formation (Santa Cruz Province, Argentina). See Vizcaíno *et al.* (Chapter 1: Figs. 1.1 and 1.2 and Appendix 1.1) for information on the localities, and Matheos and Raigemborn (Chapter 4: Figs. 4.3 and 4.4) for stratigraphy.

Eocardia fissa

MPM-PV 3683, almost complete skull, left and right dentaries, left and right pelvis, left and right femora and tibiae, proximal portion of right ulna. Horizon and locality: Santa Cruz Formation, Estancia La Costa Member, Fossiliferous Level 7; Puesto Estancia La Costa, Santa Cruz Province.

Steiromys duplicatus

MACN-A 10053–54, skull and right dentary. Horizon and locality: Santa Cruz Formation; Corriguen Kaik, Santa Cruz Province. MACN-A 10055–78, maxillary fragment with left P4–M3 and right M1–M2, left dentary with p4–m3, right humerus, distal portion of left humerus, left ulna, proximal portion of left radius, right femur, proximal and distal portions of the left femur, left tibia, distal and proximal portions of right tibia, left astragalus, and portions of the vertebral column. Horizon and locality: Santa Cruz Formation; Corriguen Kaik, Santa Cruz Province. MPM-PV 4273, skull and right mandible. Horizon and locality: Santa Cruz Formation, Estancia La Costa Member, Fossiliferous level 5.3; Puesto Estancia La Costa, Santa Cruz Province.

Steiromys detentus

MPM-PV 4358, right dentary with dp4–m2. Horizon and locality: Santa Cruz Formation, Estancia La Costa Member, Fossiliferous Level 5.3; Puesto Estancia La Costa, Santa Cruz Province. MPM-PV 4271, portion of skull with incisors and right P4 and left P4–M2. Horizon and locality: Santa Cruz Formation, Estancia La Costa Member, Fossiliferous Level 5.3; Puesto Estancia La Costa, Santa Cruz Province.

Neoreomys australis

MPM-PV 3513, almost complete skull, atlas, axis and proximal portion of left femur. Horizon and locality: Santa Cruz Formation, Estancia La Costa Member; Anfiteatro, Santa Cruz Province. MPM-PV 3514, incomplete skull and left femur. Horizon and locality: Santa Cruz Formation, Estancia La Costa Member, Fossiliferous Level 5.3; Puesto Estancia La Costa, Santa Cruz Province. MPM-PV 3515, almost complete skull and right dentary. Horizon and locality: Santa Cruz Formation, Estancia La Costa Member; Anfiteatro, Santa Cruz Province. MPM-PV 3516, proximal portion and shaft of right femur. Horizon and locality: Santa Cruz Formation, Estancia La Costa Member; Anfiteatro, Santa Cruz Province. MPM-PV 3517, partial left dentary, right astragalus, and right calcaneum. Horizon and locality: Santa Cruz Formation, Estancia La Costa Member; Anfiteatro, Santa Cruz Province. MPM-PV 3518, rostrum with incisors, pelvis fragment, proximal fragments of left femur and left tibia. Horizon and locality: Santa Cruz Formation, Estancia La Costa Member; Anfiteatro, Santa Cruz Province. MPM-PV 3519, almost complete skull, left and right dentaries, distal portions of left tibia and fibula, and portion of right scapula. Horizon and locality: Santa Cruz Formation, Estancia La Costa Member; Anfiteatro, Santa Cruz Province. MPM-PV 3520, portion of right dentary, and right humerus. Horizon and locality: Santa Cruz Formation, Estancia La Costa Member; Anfiteatro, Santa Cruz Province. MPM-PV 3521, fragment of skull, left femur and tibia, incomplete pelvis, right humerus, radius and incomplete scapula, two vertebrae. Horizon and locality: Santa Cruz Formation, Estancia La Costa Member; Anfiteatro, Santa Cruz Province. MPM-PV 3522, almost complete skull and left femur. Horizon and locality: Santa Cruz Formation, Estancia La Costa Member; Anfiteatro, Santa Cruz Province. MPM-PV 3523, incomplete skull, proximal portion of left femur. Horizon and locality: Santa Cruz Formation, Estancia La Costa Member; Anfiteatro, Santa Cruz Province. MPM-PV 3674, skull fragment, right dentary fragment, head of left femur, distal portions of left and right tibia. Horizon and locality: Santa Cruz Formation, Estancia La Costa Member, Fossiliferous Level 5.3; Puesto Estancia La Costa, Santa Cruz Province. MPM-PV 3676, incomplete skull. Horizon and locality: Santa Cruz Formation, Estancia La Costa Member, Fossiliferous Level 5.3; Puesto Estancia La Costa, Santa Cruz Province. MPM-PV 4339, almost complete skull. Horizon and locality: Santa Cruz Formation, Estancia La Costa Member, Fossiliferous Level 7; Puesto Estancia La Costa, Santa Cruz Province. MPM-PV 4340, incomplete skull, left dentary, proximal portion of left femur and associated portion of pelvis, right portion of pelvis, distal portion of right tibia. Horizon and locality: Santa Cruz Formation, Estancia La Costa Member,

Fossiliferous Level 5.3; Puesto Estancia La Costa, Santa Cruz Province. MPM-PV 4341, almost complete skull, left and right femora and tibiae, right portion of pelvis, one vertebra. Horizon and locality: Santa Cruz Formation, Estancia La Costa Member, Fossiliferous Level 5.3; Puesto Estancia La Costa, Santa Cruz Province. MPM-PV 4342, almost complete skull, left and right dentaries, right femur, tibia, and almost complete foot. Horizon and locality: Santa Cruz Formation, Estancia La Costa Member, Fossiliferous Level 5.3; Puesto Estancia La Costa, Santa Cruz Province. MPM-PV 4343, almost complete skull and three vertebrae. Horizon and locality: Santa Cruz Formation, Estancia La Costa Member, Fossiliferous Level 5.3; Puesto Estancia La Costa, Santa Cruz Province. MPM-PV 4344, left dentary. Horizon and locality: Santa Cruz Formation, Estancia La Costa Member, Fossiliferous Level 5.3; Puesto Estancia La Costa, Santa Cruz Province. MPM-PV 4345, incomplete skull. Horizon and locality: Santa Cruz Formation, Estancia La Costa Member; Campo Barranca, Santa Cruz Province.

REFERENCES

Abello, M. A. and Candela, A. M. (2010). Postcranial skeleton of the Miocene marsupial *Palaeothentes* (Paucituberculata, Palaeothentidae): paleobiology and phylogeny. *Journal of Vertebrate Paleontology*, **30**, 1515–1527.

Ameghino, F. (1887). Enumeración sistemática de las especies de mamíferos fósiles coleccionados por Carlos Ameghino en los terrenos eocenos de Patagonia austral y depositados en el Museo de La Plata. *Boletín del Museo de La Plata*, **1**, 1–26.

Argot, C. (2001). Functional-adaptative anatomy of the forelimb in the Didelphidae, and the paleobiology of the Paleocene marsupials *Mayulestes ferox* and *Pucadelphis andinus*. *Journal of Morphology*, **217**, 51–79.

Argot, C. (2002). Functional-adaptative analysis of the hindlimb anatomy of extant marsupials and the paleobiology of the Paleocene marsupials *Mayulestes ferox* and *Pucadelphis andinus*. *Journal of Morphology*, **253**, 76–108.

Biknevicius, A. R., McFarlane, D. A. and MacPhee, R. D. E. (1993). Body size in *Amblyrhiza inundata* (Rodentia: Caviomorpha), an extinct megafaunal rodent from the Anguilla Bank, West Indies: estimates and implications. *American Museum Novitates*, **3079**, 1–25.

Blanga-Kanfi, S., Miranda, H., Penn, O. *et al.* (2009). Rodent phylogeny revised: analysis of six nuclear genes from all major rodent clades. *BMC Evolutionary Biology*, **9**, 71.

Cabrera, A. and Willink, A. (1980). *Biogeografía de América Latina*. Monografía No. 13. Serie Biología. Washington, D.C.: Secretaría General de la Organización de los Estados Americanos.

Cáceres, N. C., Carmignotto, A. P., Fischer, E. and Santos, C. F. (2008). Mammals from Mato Grosso do Sul, Brazil. *Check List*, **4**, 321–335.

Campos, C. M., Tognelli, M. F. and Ojeda, R. A. (2001). *Dolichotis patagonum. Mammalian Species*, **652**, 1–5.

Candela, A. M. (1999). The evolution of the molar pattern of the Erethizontidae (Rodentia, Hystricognathi) and the validity of *Parasteiromys* Ameghino 1904. *Palaeovertebrata*, **28**, 53–73.

Candela, A. M. (2000). Los Erethizontidae (Rodentia, Hystricognathi) fósiles de Argentina, Sistemática e historia evolutiva y biogeográfica. Unpublished Ph.D. thesis, Universidad Nacional de La Plata.

Candela, A. M. (2003). A new porcupine (Rodentia, Erethizontidae) from the Early–Middle Miocene of Patagonia. *Ameghiniana*, **40**, 483–494.

Candela, A. M. and Picasso, M. B. J. (2008). Functional anatomy of the limbs of Erethizontidae (Rodentia, Caviomorpha): indicators of locomotor behavior in Miocene porcupines. *Journal of Morphology*, **269**, 552–593.

Candela, A. M. and Vizcaíno, S. F. (2007). Functional analysis of the postcranial skeleton of *Neoreomys australis* (Rodentia, Caviomorpha, Dasyproctidae) from the Miocene of Patagonia. *Journal of Morphology*, **268**, 1055–1056.

Charles-Dominique, P., Atramentowicz, M., Charles-Dominique, M. *et al.* (1981). Les mammifères frugivores arboricoles nocturnes d'une forêt guyanaise: interrelations plantes–animaux. *Revue d'Ecologie (la Terre et la Vie)*, **35**, 341–435.

Chillo, M. V. (2007). Selección de dieta de conejo de los palos (*Pediolagus salinicola*) y mara (*Dolichotis patagonum*) durante la estación seca en un sitio de coexistencia. Unpublished Ph.D. thesis, Universidad Nacional de Córdoba.

Churakov, G., Sadasivuni, M., Rosenbloom, K. *et al.* (2010). Rodent evolution: back to the root. *Molecular Biology and Evolution*, **27**, 1315–1326.

CODEPLAN-Secretaria do Governo do Distrito Federal. (1976). *Diagnostico do Espaço Natural do Distrito Federal*, Brasilia, D. F.

Croft, D. (2000). Archaeohyracidae (Mammalia: Notoungulata) from the Tinguiririca Fauna, central Chile, and the evolution and paleoecology of South American mammalian herbivores. Unpublished Ph.D. thesis, University of Chicago.

Croft, D. A. (2007). The middle Miocene (Laventan) Quebrada Honda Fauna, southern Bolivia, and a description of its notoungulates. *Palaeontology*, **50**, 277–303.

Eiten, G. (1972). The cerrado vegetation of Brazil. *Botanical Review*, **38**, 201–341.

Eiten, G. (1974). An outline of the vegetation of South America. *Proceedings from the Symposia of the 5th Congress of the International Primatological Society*. Nagoya, Japan: Japan Science Press, pp. 529–545.

Elftman, H. O. (1929). Functional adaptations of the pelvis in marsupials. *Bulletin of the American Museum of Natural History*, **58**, 189–232.

Elissamburu, A. and Vizcaíno, S. F. (2004). Limb proportions and adaptations in caviomorph rodents (Rodentia: Caviomorpha). *Journal of Zoology, London*, **262**, 145–159.

Emmons, H. L. and Feer, F. (1997). *Neotropical Rainforest Mammals. A Field Guide*, 2nd Edn. Chicago: University of Chicago.

Evans, A. R., Wilson, G. P., Fortelius, M. and Jernvall, J. (2007). High-level similarity of dentitions in carnivorans and rodents. *Nature*, **445**, 78–81.

Fernandes, A. and Bezerra, P. (1990). *Estudo fitogeográfico do Brasil*. Fortaleza, Brasil: Stylus Comunicações.

Fields, R. W. (1957). Hystricomorph rodents from the Late Miocene of Colombia, South America. *University of California Publications in Geological Sciences*, **32**, 273–404.

Flynn, J. J., Novacekc, M., Dodsond, H. *et al.* (2002). A new fossil mammal assemblage from the southern Chilean Andes: implications for geology, geochronology, and tectonics. *Journal of South American Earth Sciences*, **15**, 285–302.

Flynn, J. J., Wyss, A. R., Croft, D. A. and Charrier, R. (2003). The Tinguiririca Fauna, Chile: biochronology, paleoecology, biogeography, and a new earliest Oligocene South American Land Mammal 'Age'. *Palaeogeography, Palaeoclimatology, Palaeoecology*, **195**, 229–259.

Flynn, J. J., Charrier, R., Croft, D. A. *et al.* (2008). Chronologic implications of new Miocene mammals from the Cura-Mallín and Trapa Trapa formations, Laguna del Laja area, south central Chile. *Journal of South American Earth Sciences*, **26**, 412–423.

Fostowicz-Frelik, Ł. (2007). The hind limb skeleton and cursorial adaptations of the Plio-Pleistocene rabbit *Hypolagus beremendensis*. *Acta Palaeontologica Polonica*, **52**, 447–476.

Frailey, C. D. and Campbell, K. (2004). Paleogene rodents from Amazonian Peru: the Santa Rosa local fauna. In *The Paleogene Mammalian Fauna of Santa Rosa, Amazonian Peru*, ed. K. E. Campbell. Los Angeles: Natural History Museum of Los Angeles County, Science Series 40, pp. 71–130.

Gambaryan, P. P. (1974). *How Mammals Run*. New York: John Wiley and Sons.

García Esponda, C. M. and Candela, A. M. (2010). Anatomy of the hindlimb musculature in the cursorial caviomorph *Dasyprocta azarae* Lichtenstein, 1823 (Rodentia, Dasyproctidae): functional and evolutionary significance. *Mammalia*, **74**, 407–422.

Grand, T. I. and Eisenberg, J. F. (1982). On the affinities of the Dinomyidae. *Sonderdruck aus Saugetierkundliche Mitteilungen*, **2**, 151–157.

Greaves, W. S. (1991). The orientation of the force of the jaw muscles and the length of the mandible in mammals. *Zoological Journal of the Linnean Society*, **102**, 367–374.

Handley, C. and Pine, R. (1992). A new Species of Prehensile-tailed Porcupine, *Coendou* Lacépède, from Brazil. *Mammalia*, **56**, 237–244.

Hoffstetter, R. and Lavocat, R. (1970). Découverte dans le Déséadien de Bolivie des genres pentalophodontes appuyant les affinités africaines des Rongeurs Caviomorphes. *Comptes Rendus de l' Académie des Sciences, Paris*, Série D, **271**, 172–175.

Honeycutt, R. L., Frabotta, L. J. and Rowe, D. L. (2007). Rodent evolution, phylogenetics, and biogeography. In *Rodent Societies: An Ecological and Evolutionary Perspective*, ed. J. O. Wolff and P. Sherman. Chicago: University of Chicago Press, pp. 8–13.

Huchon, D. and Douzery, E. J. P. (2001). From the Old World to the New World: a molecular chronicle of the phylogeny and biogeography of hystricognath rodents. *Molecular Phylogenetics and Evolution*, **18**, 127–135.

Jenkins, F. A. Jr. (1973). The functional anatomy and evolution of the mammalian humero-ulnar articulation. *American Journal of Anatomy*, **137**, 281–298.

Jones, F. W. (1953). Some readaptations of the mammalian pes in response to arboreal habits. *Proceedings of the Zoological Society of London*, **23**, 33–41.

Kraglievich, L. (1932). Diagnosis de nuevos géneros y especies de roedores cávidos y eumegámidos fósiles de la Argentina. *Anales de la Sociedad Científica Argentina*, **114**, 155–181, 211–237.

Kramarz, A. G. (1998). Un nuevo Dasyproctidae (Rodentia, Caviomorpha) del Mioceno temprano de Patagonia. *Ameghiniana*, **35**, 181–192.

Kramarz, A. G. (2002). Roedores chinchilloideos (Hystricognathi) de la Formación Pinturas, Mioceno temprano-medio de la provincia de Santa Cruz, Argentina. *Revista del Museo Argentino de Ciencias Naturales (nueva serie)*, **4**, 167–180.

Kramarz, A. G. (2004). Octodontoids and erethizontoids (Rodentia, Hystricognathi) from the Pinturas Formation, Early-Middle Miocene of Patagonia, Argentina. *Ameghiniana*, **41**, 199–216.

Kramarz, A. G. (2006). *Neoreomys* and *Scleromys* (Rodentia, Hystricognathi) from the Pinturas Formation, late Early Miocene of Patagonia, Argentina. *Revista del Museo Argentino de Ciencias Naturales*, **8**, 53–62.

López, G. M., Vucetich, M. G., Carlini, A. *et al.* (2011). New Miocene mammal assemblages from Neogene Manantiales basin, Cordillera Frontal, San Juan, Argentina. In *Cenozoic Geology of the Central Andes of Argentina*, ed. J. A. Salfity and R. A. Marquillas. Special Paper of the Geological Society of America. Salta: SCS Publisher, pp. 211–226.

Maynard Smith, J. and Savage, R. J. G. (1956). Some locomotory adaptations in mammals. *Zoological Journal of the Linnean Society*, **42**, 603–622.

Maynard Smith, J. and Savage, J. R. G. (1959). The mechanics of mammalian jaws. *School Science Review*, **141**, 289–301.

McEvoy, J. S. (1982). Comparative myology of the pectoral and pelvic appendages of the North American porcupine (*Erethizon dorsatum*) and the prehensile-tailed porcupine (*Coendou prehensilis*). *Bulletin of the American Museum of Natural History*, **173**, 337–421.

McKenna, M. C. and Bell, S. K. (1997). *Classification of Mammals Above the Species Level*. New York: Columbia University Press.

Miles, R., Brand, S. and Maliniak, E. (1985). The biology of captive prehensile-tailed porcupines *Coendou prehensilis*. *Journal of Mammalogy*, **63**, 473–482.

Mondolfi, E. (1972). La lapa o paca. *Defensa de la Naturaleza*, **2**, 4–16.

Nowak, R. M. (1991). *Walker's Mammals of the World*, 5th Edn. Baltimore: The Johns Hopkins University Press.

Patterson, B. and Wood, A. E. (1982). Rodents of the Deseadan Oligocene of Bolivia and the relationships of the Caviomorpha. *Bulletin of the Museum of Comparative Zoology*, **149**, 371–543.

Pérez, M. E. (2010a). Sistemática, ecología y bioestratigráfica de "Eocardiidae" (Rodentia, Hystricognathi, Cavioidea) del Mioceno temprano y medio de Patagonia. Unpublished Ph.D. thesis, Universidad Nacional de La Plata.

Pérez, M. E. (2010b). A new rodent (Hystricognathi, Caviomorpha) from the middle Miocene of Patagonia, the mandibular homologies in Cavioidea *s. s.* and the origin of the serie medialis cavioids. *Journal of Vertebrate Paleontology*, **30**, 1848–1859.

Pérez, M. E. and Vucetich, M. G. (2011). A new extinct genus of Cavioidea (Rodentia, Hystricognathi) from the Miocene of Patagonia (Argentina) and the evolution of Cavioid mandibular morphology. *Journal of Mammalian Evolution* **18**, 163–183.

Pérez, M. E., Vucetich, M. G. and Kramarz, A. G. (2010). The first Eocardiidae (Rodentia) in the Colhuehuapian (Early Miocene) of Bryn Gwyn (northern Chubut, Argentina) and the early evolution of the peculiar cavioid rodents. *Journal of Vertebrate Paleontology*, **30**, 528–534.

Redford, K. H. and Eisenberg, J. F. (1992). *Mammals of the Neotropics. Vol. 2. The Southern Cone: Chile, Argentina, Uruguay, Paraguay*. Chicago: University of Chicago Press.

Redford, K. H. and Fonseca, G. A. B. (1986). The role of gallery forests in the zoogeography of the Cerrado's non-volant mammalian fauna. *Biotropica*, **18**, 126–135.

Roze, U. (1989). *The North American Porcupine*. Washington, DC: Smithsonian Institution Press.

Salton, J. A. and Sargis, E. J. (2009). Evolutionary morphology of the Tenrecoidea (Mammalia) hindlimb skeleton. *Journal of Morphology*, **270**, 367–387.

Samuels, J. X. (2009). Cranial morphology and dietary habits of rodents. *Zoological Journal of the Linnean Society*, **156**, 864–888.

Samuels, J. X. and Van Valkenburgh, B. (2008). Skeletal indicators of locomotor adaptations in living and extinct rodents. *Journal of Morphology*, **269**, 1387–1411.

Sargis, E. J. (2002). Functional morphology of the forelimb of tupaiids (Mammalia. Scandentia) and its phylogenetic implications. *Journal of Morphology*, **253**, 10–42.

Scott, W. B. (1905). Mammalia of the Santa Cruz beds. Glires. In *Reports of the Princeton University Expeditions to Patagonia 1896–1899*, Vol. 5, *Paleontology II*: Part 3, ed. W. B. Scott. Princeton: Princeton University Press, pp. 384–490.

Szalay, F. S. and Dagosto, M. (1980). Locomotor adaptations as reflected on the humerus of Paleogene primates. *Folia Primatologica*, **34**, 1–45.

Szalay, F. S. and Sargis, E. J. (2001). Model-based analysis of postcranial osteology of marsupials from the Palaeocene of Itaboraí (Brazil) and the phylogenetics and biogeography of Metatheria. *Geodiversitas*, **23**, 139–302.

Turnbull, W. D. (1970). Mammalian masticatory apparatus. *Fieldiana Geology*, **18**, 149–356.

Vizcaíno, S. F., Bargo, M. S., Kay, R. F. *et al.* (2010). A baseline paleoecological study for the Santa Cruz Formation (late-early Miocene) at the Atlantic coast of Patagonia, Argentina. *Palaeogeography, Palaeoclimatology, Palaeoecology*, **292**, 507–519.

Vucetich, M. G. (1984). Los roedores de la Edad Friasense (Mioceno medio) de Patagonia. *Revista Museo de La Plata*, Nueva Serie, 8. *Paleontología*, **50**, 47–126.

Vucetich, M. G., Mazzoni, M. A. and Pardiñas, U. F. J. (1993) Los roedores de la Formación Collón Cura (Mioceno Medio), y la Ignimbrita Pilcaniyeu. Cañadón del Tordillo, Neuquén. *Ameghiniana*, **30**, 361–381.

Vucetich, M. G., Verzi, D. H. and Hartenberger, J. L. (1999). Review and analysis of the radiation of the South American Hystricognathi (Mammalia, Rodentia). *Comptes Rendus de l'Académie des Sciences, Paris*, **392**, 763–769.

Vucetich, M. G., Kramarz, A. G. and Candela, A. M. (2010). Colhuehuapian rodents from Gran Barranca and other Patagonian localities: the state of the art. In *The Paleontology of Gran Barranca*, ed. R. Madden, A. A. Carlini, M. G. Vucetich and R. Kay. Cambridge: Cambridge University Press, pp. 206–219.

Walton, A. H. (1997). Rodents. In *Vertebrate Paleontology in the Neotropics*, ed. R. F. Kay, R. H. Madden, R. L. Cifelli and J. J. Flynn, Washington D.C.: Smithsonian Institution Press, pp. 392–409.

Wood, A. E. and Patterson, B. (1959). The rodents of Deseadan Oligocene of Patagonia and the beginnings of South American rodent evolution. *Bulletin of the Museum of Comparative Zoology*, **120**, 280–428.

Woods, C. A. (1972). Comparative myology of jaw, hyoid, and pectoral appendicular regions of New and Old World hystricomorph rodents. *Bulletin of the American Museum of Natural History*, **147**, 117–198.

Woods, C. A. (1973). *Erethizon dorsatum*. *Mammalian Species*, **29**, 1–6.

Woods, C. A. and Howland, E. B. (1979). Adaptative radiation of capromyid rodents: anatomy of the masticatory apparatus. *Journal of Mammalogy*, **60**, 95–116.

Woods, C. A. and Kilpatrick, C. W. (2005). *Hystricognathi*. In *Mammal Species of the World: a Taxonomic and Geographic Reference* (3rd Edn) ed. D. E. Wilson and D. M. Reeder. Washington D.C.: Smithsonian Institution Press, pp. 1538–1600.

Woods, C. A. and Mckeen, B. (1989). Convergence in New World porcupines and West Indian hutías: an analysis of tooth wear, jaw movement, and diet in rodents. In *Advances in Neotropical mammalogy*, ed. K. H. Redford and J. F. Eisenberg. Gainesville, FL: Sandhill Crane Press, pp. 97–124.

Wyss, A. R., Flynn, J. J., Norell, M. A. *et al.* (1993). South America's earliest rodent and recognition of a new interval of mammalian evolution. *Nature*, **365**, 434–437.

16 Paleobiology of Santacrucian primates

Richard F. Kay, Jonathan M. G. Perry, Michael Malinzak, Kari L. Allen, E. Christopher Kirk, J. Michael Plavcan, and John G. Fleagle

Abstract

Over the past century, the Santa Cruz Formation of coastal Argentina (late Early Miocene) has yielded a remarkable collection of platyrrhine primates. With few notable exceptions, most of the specimens have been included in *Homunculus patagonicus* Ameghino, 1891, a stem platyrrhine. *Homunculus patagonicus* was approximately 1.5 to 2.5 kg in body mass, about the size of a living saki monkey (*Pithecia*) or a female *Cebus*. Molar structure indicates that the diet consisted of a mixture of fruit and leaves. A deep jaw, large postcanine tooth roots, large postglenoid processes and moderately large chewing muscle attachments (i.e. massive zygomatic arches, sculpted temporalis origins) suggest that physically resistant foods were key components of the diet. Heavy tooth wear suggests large amounts of ingested silica or exogenous abrasives. Incisor morphology suggests that exudate harvesting may have been part of the behavioral repertoire, although not a specialization. The canines were small, providing no evidence of sclerocarpic foraging. Canines were sexually dimorphic, suggesting that the taxon experienced some intrasexual competition rather than being solitary or pair-bonded. Brain size was small and the frontal cortical region was proportionately small. From the small size and structure of the orbits, the structure of the organ of hearing, the reduced olfactory fossae and the relatively large infraorbital foramina, we infer that *Homunculus* was probably diurnal, with acute vision and hearing, but with a poor sense of smell and little reliance on tactile vibrissae. *Homunculus* was an above-branch arboreal quadruped with leaping abilities. The semicircular canals show evidence of considerable agility, reinforcing the inference of leaping behavior. The overall locomotor repertoire is not unlike that of the forest-dwelling extant saki monkey *Pithecia*. Considered together, the mosaic of dietary and locomotor morphology in *Homunculus* suggests that *Homunculus* inhabited an environment – as compared with earlier Colhuehuapian and Pinturan primate habitats – shifting towards greater seasonality in patchy forests near river courses.

Resumen

Durante el último siglo, los afloramientos de Formación Santa Cruz de la costa de Argentina (Mioceno Temprano tardío) ha provisto una notable colección de primates platirrinos. Excepto unas pocas excepciones, la mayoría de los especímenes han sido incluidos en *Homunculus patagonicus* Ameghino, 1891, un platirrino basal. *Homunculus patagonicus* poseía una masa corporal de 1,5–2,5 kg, el tamaño del mono viviente saki (*Pithecia*) o de una hembra de *Cebus*. La estructura molar indica que la dieta se componía de una mezcla de frutos y hojas. La mandíbula profunda, las raíces de los dientes postcaninos grandes, los procesos postglenoideos grandes y las áreas de inserción de los músculos masticatorios moderadamente grandes (i.e. arcos cigomáticos robustos, áreas de origen del musculo temporal rugosas) sugieren que la dieta estaba compuesta por alimentos físicamente resistentes. El fuerte desgaste de los dientes indica que ingerían una gran cantidad de sílice o abrasivos exógenos. La morfología de los incisivos sugiere que la extracción de exudado puede haber sido parte del comportamiento, pero no una especialización. Los caninos eran pequeños, lo que no aporta ninguna evidencia sobre alimentación esclerocárpica. Los caninos eran sexualmente dimórficos, sugiriendo que esta especie habría experimentado competencia intrasexual en lugar de ser solitaria o vivir en parejas. El cerebro era pequeño y la región cortical frontal era proporcionalmente pequeña. A partir de la estructura y pequeño tamaño de las órbitas, la estructura del órgano de la audición, las fosas olfatorias reducidas y el foramen infraorbital relativamente grande se infiere que *Homunculus* fue probablemente diurno, con los sentidos de visión y audición agudos. El olfato estaba poco desarrollado y no dependían de vibrisas táctiles. *Homunculus* era un arborícola cuadrúpedo con capacidad de saltar de una rama a otra. Los canales semicirculares proveen evidencia de una considerable agilidad, lo que refuerza la conclusión del comportamiento saltador. El repertorio locomotor en general no es diferente del mono saki viviente *Pithecia* que habita en bosques. Considerados en conjunto, la morfología de los aparatos masticatorio y locomotor de *Homunculus* sugiere que – en comparación con los hábitats de los primates más tempranos colhuehuapenses y pinturenses – habitó en un ambiente que cambió hacia una mayor estacionalidad, en bosques en parches cerca de cursos fluviales.

Early Miocene Paleobiology in Patagonia: High-Latitude Paleocommunities of the Santa Cruz Formation, ed. Sergio F. Vizcaíno, Richard F. Kay and M. Susana Bargo. Published by Cambridge University Press. © Cambridge University Press 2012.

16.1 Introduction

Today, the most southerly geographic range of platyrrhine primates is at latitude 29° S, but in the Early Miocene, a time interval frequently referred to as the Mid-Miocene Climatic Optimum (Zachos *et al.*, 2001), several medium-sized platyrrhines lived much further south in Patagonian Argentina. The oldest known taxa come from the Colhue-huapian (~20 Ma) of Chubut Province and include *Tremacebus harringtoni* (Hershkovitz, 1974), *Dolichocebus gaimanensis* Kraglievich, 1951, and *Mazzonicebus almen-drae* Kay, 2010 (Kay *et al.*, 2008; Kay, 2010). Also of Colhuehuapian age, the extra-Patagonian *Chilecebus* Flynn *et al.*, 1995 comes from further north, but still outside the extant platyrrhine geographic range (Flynn *et al.*, 1995). Pinturan localities (~17 to 19 Ma) (Ré *et al.*, 2010) in northern Santa Cruz Province have yielded two genera: *Carlocebus* Fleagle, 1990 and *Soriacebus* Fleagle, 1990 (Fleagle, 1990; Tejedor, 2005). Towards the end of the Early Miocene at 18–16 Ma, platyrrhines reached their southern latitudinal extreme at 51° S. Here, the coastal Santa Cruz Formation yields *Homunculus patagonicus* Ameghino, 1891 and two other species of *Homunculus*: *H. blakei* (= *Killikaike blakei* Tejedor *et al.*, 2006; R. F. Kay *et al.*, unpublished data) and a yet-to-be-named larger species. *Proteropithecia neuquensis* Kay *et al.*, 1999 occurs in the Colloncuran of Neuquén Province at ~15.5 Ma (early Middle Miocene), after which no primate occurs anywhere in Argentina until the Recent. Fossil evidence for most of these Miocene taxa consists mainly of teeth, jaws, and a few broken and unassociated postcranial bones or, in three cases, a single cranium (*Tremacebus harringtoni, Dolichocebus gaimanensis, and Chilecebus carrascoensis*). *Homunculus*, on the other hand, is comparatively well known from crania, mandibles, well-preserved teeth, and a number of postcranial elements. Therefore, it is *Homunculus* that allows the most complete reconstruction of the adaptations of any early platyrrhines.

A full account of the debates about the phyletic position of the above-mentioned Early–Middle Miocene platyrrhines is beyond the scope of this paper, but it is important to note that the two most common current interpretations are very different. Some workers suggest that the above fossil taxa represent two living platyrrhine families: Cebidae (*Tremacebus* (Aotinae); *Chilecebus, Dolichocebus,* and *Killikaike* (Cebinae); and Pitheciidae (*Carlocebus, Homunculus, Mazzonicebus,* and *Soriacebus*)). Other researchers suggest that only one extant platyrrhine family is represented in Patagonia, the pithecine *Proteropithecia* and that the others are stem taxa (for contrasting views, see Kay and Fleagle, 2010; Rosenberger *et al.*, 2011). The two divergent views relate mainly to the question of whether the manifest adaptive similarities between the Early Miocene and modern taxa are the product of convergence or derive from

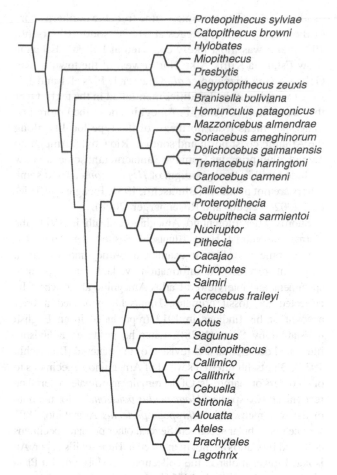

Fig. 16.1. A summary of the phylogeny of early Miocene platyrrhines, based on the data published by Kay *et al.* (2008), as modified by R. F. Kay (unpublished data).

common ancestry. To us, the preponderance of the evidence favors convergence; we take the view that the Patagonian taxa were stem platyrrhines and ecological "vicars" (Eldredge, 1985; Elton, 1927) of some modern platyrrhines (Kay, 2010). Figure 16.1 summarizes the phylogeny of early Miocene platyrrhines, based on the data published by Kay *et al.* (2008), as modified by R. F. Kay (unpublished data).

Whatever the phylogenetic position of these early forms, the available anatomical evidence tells a very interesting story about the paleobiology of Miocene platyrrhines. In this paper, we summarize the anatomical structure of *Homunculus patagonicus* on the basis of which we reconstruct its behavioral profile and pinpoint its place in the community structure of the Santacrucian fauna.

16.2 The Santacrucian species

Homunculus patagonicus, the first known Miocene platyrrhine, was described in 1891 by Florentino Ameghino, based on a mandibular specimen recovered by his brother Carlos. The specimen came from the northern shore of the estuary (Ría) Gallegos at 51° 38′ south (Bluntschli, 1931), and was most likely collected at Felton's Estancia (now Estancia Killik Aike Norte) west of the town of Río Gallegos (see Vizcaíno *et al.*, Chapter 1, Figs. 1.1 and 1.2, for geography of the localities mentioned in the text). Over the succeeding 10 years, F. Ameghino described more primate specimens from northeast of the type locality along the Atlantic coast north and south of Río Coyle. Ameghino named other primates from the Santacrucian; some are now included within the hypodigm of *H. patagonicus* and some others are not primates (Bluntschli, 1931; Fleagle and Tejedor, 2002; Tejedor and Rosenberger, 2008).

Shortly after Florentino Ameghino's death in 1911, the German anatomist Hans Bluntschli visited the Ameghino family home in La Plata to examine the Ameghino collection. Bluntschli recorded the information available for all primate specimens and interviewed Carlos Ameghino about where he collected the fossil primates. Bluntschli published a brief account of his findings in 1913 (reproduced in an English translation by Scott (1928)). Later he produced a lavishly illustrated description and revision of the material (Bluntschli, 1931). Bluntschli referred several of Ameghino's specimens to other orders of mammals. Of the remaining primate material he recognized two species: *Homunculus patagonicus* for the bulk of the specimens and *Anthropops perfectus* Ameghino, 1891 for one mandibular specimen. Several other primate specimens collected by Carlos are not mentioned in Bluntschli's paper. An isolated lower molar in the collections of Museo de La Plata was not cataloged until 1955, but probably belonged to the Ameghino collection (M. Reguero, personal communication). A juvenile mandible in the La Plata collections, the type of *Stilotherium grandis* Ameghino, 1894, thought by Ameghino to be a marsupial, was later shown to be *Homunculus patagonicus*.

In the 1980s, two collecting groups worked to recover fossil mammals in the Santa Cruz Formation. Miguel Soria, John Fleagle, and their colleagues collected extensively in the more northerly exposures near Monte León and Cerro Observatorio (Monte Observación of the older literature). They recovered a large number of specimens, mostly isolated teeth. Working independently at the Universidad de Córdoba, Adán Tauber surveyed localities south of Río Coyle and discovered a partial cranium of *Homunculus patagonicus* (Tauber, 1991). Unknown to either group at the time, North American paleontologists in transit to Antarctica visited Killik Aike Norte on a day trip and discovered and photographed the palate and rostrum of a primate (Judd Case, personal communication). Along with another partial upper dentition, that specimen was later described as a new taxon, *Killikaike blakei* (Tejedor *et al.*, 2006).

In the first decade of this century, joint Museo de La Plata/Duke University expeditions recovered more complete primate specimens from Estancia La Costa (Estancia Montes), Puesto Estancia La Costa, and Estancia Killik Aike Norte, including two mandibles, two relatively complete crania, an adult hemicranium, a juvenile rostrum, and a fairly complete humerus. Thus, the available sample of Santacrucian primates now consists of at least seven crania (including the palate and face of a juvenile), mandibles, many isolated dental specimens, a partial skeleton (distal humerus, ulna, radius, femur) and several isolated humeri. Figure 16.2 illustrates several crania.

The alpha taxonomy of Santacrucian primates is important to the adaptive patterns described here because if the primate bones represent a commingling of more than one species, we would be describing a chimera. After close scrutiny we conclude that with one or two exceptions, all of the known primate material belongs to one genus *Homunculus* and most of it belongs to one species, *H. patagonicus* (R. F. Kay *et al.*, unpublished data). As already mentioned, many taxa named by Ameghino were shown not to be primates and others were brought into synonymy with *Homunculus patagonicus* by Bluntschli (1931). Bluntschli continued to recognize *Anthropops perfectus* (a mandible with one molar) as possibly valid and distinct from *H. patagonicus*, but we follow recent workers who synonymize the two (Tejedor and Rosenberger, 2008). Another new species, *Killikaike blakei*, was described (Tejedor *et al.*, 2006). The type specimen (MPM-PV 5000; see acronyms below) is a palate and rostrum with the complete orbits and part of the frontal with the anterior cranial fossa. Tentatively we consider *Killikaike blakei* as a species of *Homunculus, H. blakei*.

In 2010, we recovered a remarkably complete mandible of *Homunculus* at Puesto Estancia La Costa preserving most of the cheek teeth (see below). While being of similar morphology to *Homunculus*, the teeth are comparatively very large and the jaw is more robust than other examples of *Homunculus* spp. Elsewhere we have assigned this specimen to a new species of *Homunculus* (R. F. Kay and J. M. G. Perry, unpublished).

16.3   Material and methods

Institutional acronyms

CORD-PZ, Museo de Paleontología, Universidad Nacional de Córdoba, Córdoba, Argentina.

MACN, Museo Argentino de Ciencias Naturales "B. Rivadavia," Buenos Aires, Argentina.

MACN-A, Museo Argentino de Ciencias Naturales "B. Rivadavia," Colección Nacional Ameghino, Buenos Aires, Argentina.

MLP, Museo de La Plata, La Plata, Argentina.

MPM-PV, Museo Regional Provincial Padre Manuel Jesús Molina, Río Gallegos, Argentina.

The material analyzed for this study is listed in Appendix 16.1.

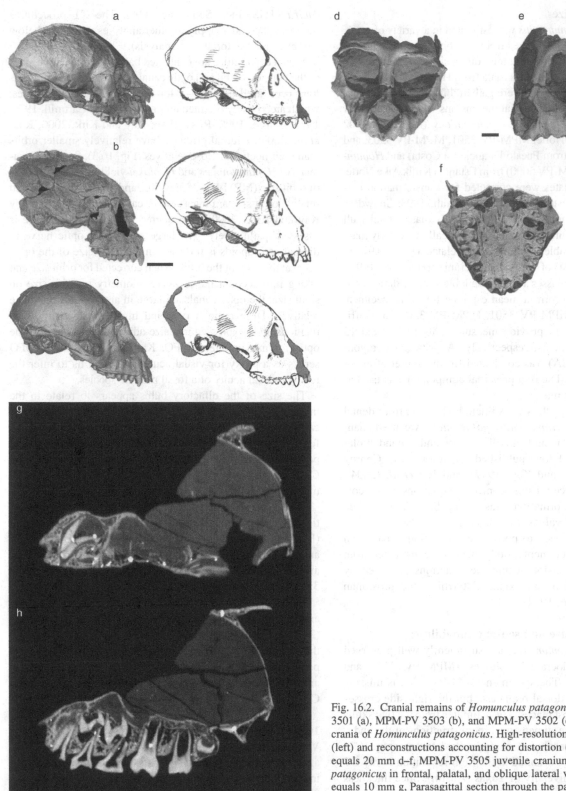

Fig. 16.2. Cranial remains of *Homunculus patagonicus*. MPM-PV 3501 (a), MPM-PV 3503 (b), and MPM-PV 3502 (c) lateral views of crania of *Homunculus patagonicus*. High-resolution CT composites (left) and reconstructions accounting for distortion (right). Scale bar equals 20 mm d–f, MPM-PV 3505 juvenile cranium of *Homunculus patagonicus* in frontal, palatal, and oblique lateral views. Scale bar equals 10 mm g, Parasagittal section through the palate showing the permanent canine tooth bud. h, Parasagittal section through the maxilla showing the fully erupted dP2 (broken), dP3–4. Permanent M1–2 in place; M3 is a tooth bud without roots.

16.3.1 Body size

Body mass of *Homunculus* was estimated in a variety of ways. Cranial estimates are derived from equations based on various combinations of 18 linear cranial dimensions for a sample of 47 extant platyrrhine species (data from Hartwig, 1993). Body masses for extant species were gathered from Smith and Jungers (1997). Preserved cranial dimensions were measured for fossil crania of *Homunculus patagonicus* (MPM-PV 3502 from Killik Aike Norte, MPM-PV 3501, MPM-PV 3503, and CORD-PZ 1130 from Puesto Estancia La Costa) and *Homunculus blakei* (MPM-PV 5000) from Estancia Killik Aike Norte. Body mass estimates were corrected for transformation bias using the "ratio estimator" method (Smith, 1993; Snowdon, 1991). One commonly used scalar for body mass is total skull length. This measurement scales isometrically with body mass and the two variables are highly correlated ($R^2 = 0.97$) in Hartwig's sample of platyrrhine crania (Hartwig, 1993). Estimated body masses for all available cranial dimensions were averaged to form a mean estimate for each specimen.

Two crania (MPM-PV 3501, MPM-PV 3502) are sufficiently complete to provide measurements for 17 and 13 cranial measurements, respectively. A principal components analysis (PCA) was conducted for the subset of available dimensions. The first principal component was used as a proxy for body mass.

A second approach is to estimate body mass from dental dimensions. For *Homunculus patagonicus* we used mandibular first molar and maxillary first and second molar specimens based on published equations by Conroy (1987), Meldrum and Kay (1997), and Egi *et al.* (2004). Estimates produced from Conroy's equations were corrected for transformation bias using Smith's published "ratio estimator" values (Smith, 1993).

The third approach to body size estimation is based on postcranial measurements. Body mass is reconstructed from the femoral dimensions using the equations produced by Runestad, based on an extant platyrrhine and prosimian dataset (Runestad, 1994).

16.3.2 Brain size and sensory capabilities

Two crania of *Homunculus* are sufficiently well preserved to estimate endocranial volumes (MPM-PV 3501 and MPM-PV 3502). The specimen MPM-PV 3502 is missing a part of the left dorsal braincase, but the right side crosses the midline in most places. The estimates from the two skulls are based on three-dimensional reconstructions of the endocast from high-resolution computerized tomography (CT) scans. Skulls were scanned in the high-resolution X-ray CT facility at the University of Texas (Austin). Slice thickness varied between 0.04 mm and 0.1 mm for specimens of different size. Subsequent analysis of stacks of 8-bit jpg files was performed on a Macintosh computer running VGStudio Max 1.2 (Volume Graphics; Heidelberg, Germany) and

Amira (TGS, Inc.; San Diego CA). The CT procedures described here also apply to other analyses described below (for example the tooth root analysis).

Visual capabilities are evaluated based on the relative sizes of the bony eye orbit and optic canal. The size of the orbits has long been used as a proxy for activity pattern in primates, which are visually oriented animals (Kay and Cartmill, 1974, 1977; Martin, 1990; Ross, 1996; Kay and Kirk, 2000; Ross *et al.*, 2007). Diurnal primates have relatively smaller orbits than their nocturnal close relatives (Fig. 16.3). Several specimens of *H. patagonicus* and *H. blakei* yield reliable estimates of orbit size (MPM-PV-3501, 3502, and 5000). Orbit size and relative orbit size are measured or calculated as described by Kay and Kirk (2000). The degree of postorbital closure is assessed qualitatively. The large size of the optic nerve in diurnal anthropoids is reflected in the large size of the optic canal at the apex of the orbit. When corrected for orbit size and taking into account the negative allometry of orbit size on skull size, the optic canals are large in anthropoids that have relatively large optic nerves and high visual acuity. This relationship is expressed in a size-adjusted measurement, the optic foramen quotient (OFQ; Kay and Kirk, 2000). OFQ serves as a proxy for visual acuity, allowing us to infer the relative visual acuity of a fossil primate species.

The size of the olfactory bulbs appears to relate to the importance of olfaction in extant platyrrhine primates, and can be estimated by measuring the breadth of the olfactory fossa in the skull (Kay *et al.*, 2004). The olfactory fossa is preserved in MPM-PV-3501 and 3502 of *Homunculus*, and CT scans allow us to visualize this structure in coronal and midsagittal sections (Fig. 16.4).

Infraorbital foramen size is indicative of the vascularization of the snout and development of the facial vibrissae (Kay and Cartmill, 1977; Muchlinski, 2010). The number and relative size of the infraorbital foramina in *Homunculus* are qualitatively assessed based on specimens MPM-PV-3501, 3502, 5000, and CORD-PZ 1130.

Bony ear morphology provides valuable information about auditory sensitivity (Coleman *et al.*, 2010). Coleman *et al.* (2010) report that auditory sensitivity is highly correlated with the dimensions of several auditory structures, including tympanic membrane area, cochlear length, and stapedial footplate area. These structures were measured by Coleman *et al.* from CT images of *Homunculus* specimen MPM-PV 3502.

16.3.3 Organs of balance

Variation in the semcircular canals of the inner ear in living and, by inference, extinct primates has been attributed to interspecific differences in locomotor behavior (Spoor *et al.*, 2007; Malinzak, 2010; Malinzak *et al.*, in press). When an animal moves, the rotational components of its head movements are sensed by the semicircular canals. A species' locomotor repertoire produces a signature pattern of head

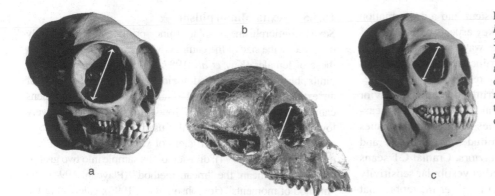

Fig. 16.3. Comparing the orbits of *Homunculus patagonicus* MPM-PV 3502 (b) with diurnal *Callicebus moloch* (c) and nocturnal *Aotus trivirgatus* (a). The white arrow depicts orbit size. Although the images are not precisely to the same scale, they illustrate the differing proportions of orbit size to skull size.

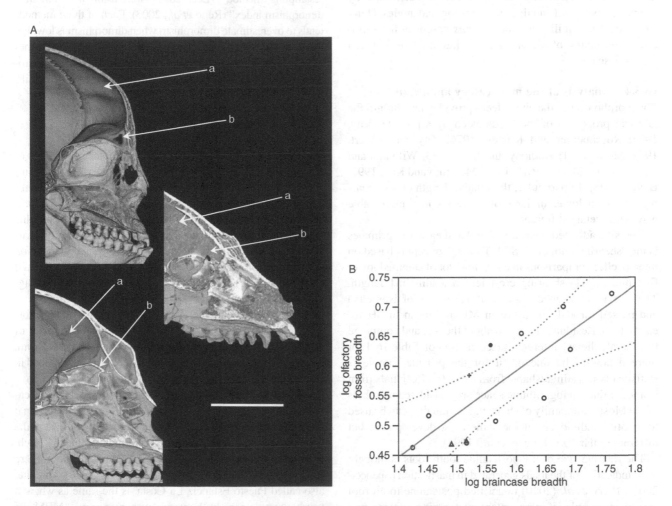

Fig. 16.4. The olfactory fossa of *Homunculus patagonicus*. A, Lateral views in the midsagittal plane of the olfactory fossae of *Saimiri sciureus*, *Homunculus patagonicus* (MPM-PV 3502), and *Callicebus moloch*. Symbols: a, frontal lobe; b, olfactory fossa. Scale bar equals 2 cm. B, Least-squares fit between $\log_{10}$ olfactory fossa breadth and log cranial breadth for nine extant species of diurnal platyrrhines (open circles). The nocturnal platyrrhine *Aotus* (filled circle) has a proportionally larger olfactory fossa than diurnal platyrrhines. *Tremacebus* (Early Miocene) (cross) has a relatively large olfactory fossa, but within the bounds of diurnal species. *Homunculus* (open triangles) plots within the diurnal range and is virtually identical to (overlaps) *Callicebus moloch*. Slope of the least-squares fit is 0.72 +/- 0.18; $p > 0.004$. Data from Kay *et al.* (2004).

rotations, molding the six-canal system into a conformation that is advantageous for detecting key aspects of the species' head movement signature. In this way, canal morphology offers clues about locomotion in extinct taxa that supplement information provided by postcranial remains.

To improve methods for inferring locomotor behavior from fossilized inner ears Malinzak *et al.* (in press) measured *in vivo* angular head velocities in 11 primate species crossing several orders of magnitude in body size and exhibiting diverse locomotor behaviors. Cranial CT scans of the same taxa were used to predict vestibular sensitivity to rotations in all directions. Malinzak *et al.* report that animals exhibiting rapid rotation minimize anisotropy in vestibular sensitivity, such as can be achieved by orienting ipsilateral canals at near-orthogonal angles. Here, we reconstruct agility in *Homunculus* based on the semicircular structure of MPM-PV 3501 and 3502, as deduced from CT scans.

16.3.4 Analysis of the masticatory apparatus

The morphology of the cheek teeth provides insight into the physical properties of the foods eaten by a primate (Kay, 1975; Rosenberger and Kinzey, 1976; Kay and Covert, 1984; Strait, 1991; Anthony and Kay, 1993; Williams and Covert, 1994; Fleagle *et al.*, 1997; Meldrum and Kay, 1997; Boyer, 2008). In particular, the relative length of six shearing crests on lower molar crowns varies in a predictable way with dietary differences.

One size-adjusted measure of molar shearing in primates is the "shearing quotient" (SQ). The SQ concept is based on interspecific comparisons in a log–log plot of summed lower first molar (m1) shearing crest length against m1 length. The line was assigned a slope of 1.0 (slope of isometry) and passed through the mean ln ML and mean ln SH for extant taxa. Residuals are taken about the line and compared to generate dietary inferences (see caption of Table 16.4 for more details). The line of fit for the primate sample is statistically indistinguishable from isometry. Residuals from isometric lines in logarithmic space are considered members of the Mosimann family of shape ratios, and thus can be used to identify individuals of the same crest development but different molar size (Jungers *et al.*, 1995).

The relative size of the postcanine tooth roots is thought to be indicative of the loads generated in mastication (Spencer, 2003). Perry *et al.* (2010) quantified postcanine tooth root size in a sample of nine extant platyrrhine species and *Homunculus* (MPM-PV 3501). To estimate masticatory leverage in *Homunculus*, we used a scaled composite of three specimens: one cranium (MPM-PV 3502), and two fragmentary mandibles (MACN-A 5969 and MPM-PV 3504). Details of the analytical procedure are found in Perry *et al.* (2010). Molar enamel thickness and the degree of dental wear were qualitatively assessed.

16.3.5 Sexual dimorphism

Sexual dimorphism in platyrrhine primates is commonly observed in the size of the canines. Male canines are larger than those of females (Kay *et al.*, 1987a). Determination of sexual dimorphism presents methodological problems for a sample of an extinct species because the sex of individual specimens cannot be determined reliably for species showing relatively low degrees of dimorphism. We used five indirect methods: extrapolation from the coefficient of variation, called the "CV method" (Plavcan, 1994); division of the sample into two groups based on the mean, the "mean method" (Plavcan, 1994); the "method of moments" (Josephson *et al.*, 1996); calculating the ratio of the largest and smallest values, using the "assigned resampling method" (Lee, 2001); and using the "binomial dimorphism index" (Reno *et al.*, 2003). Each of these methods tends to overestimate dimorphism when dimorphism is low. For this reason, estimates of dimorphism in *Homunculus* canines were compared to estimates of dimorphism in extant anthropoids calculated using the same methods as used for the fossils.

16.3.6 Locomotion

Postcranial morphology

Most of what we can say about the locomotor capabilities of *Homunculus* is based on one specimen from early collections. In 1892, Carlos Ameghino recovered several postcranial elements, including a complete right femur (MACN-A 5758), a complete left radius (MACN-A 5760), fragments of an ulna (MACN-A 5759), and the distal end of the right humerus (MACN-A 5761), all evidently in close physical association with a mandible and presumed to belong to the same individual (MACN-A 5757). MACN-A 5757 has recently been designated the neotype of the species *Homunculus patagonicus* (Tejedor and Rosenberger, 2008). This is one of the most complete associations of dental and postcranial material of any fossil platyrrhine (Fleagle and Tejedor, 2002; Meldrum, 1993). The femur, ulna, and humerus were briefly described in notes published by Florentino Ameghino in 1893 and have been described and figured in a variety of subsequent publications (Bluntschli, 1931) (Fig. 16.5). Unfortunately the original distal humerus appears to be lost, but there is a good cast of the bone in the collections of the Anthropology Institute in Zurich, Switzerland. The skeletal elements mentioned above were found at the site of Corriguen Aike (Marshall, 1976). This site, also called Puesto Estancia La Costa, is the same as where a nearly complete right humerus was recovered (MPM-PV 3500). The new specimen was found *in situ* in Tauber's Fossiliferous Level (FL) 5.3 (Vizcaíno *et al.*, 2010) but not in direct association with any other primate material. This stratigraphic level and the superadjacent FL 6 have yielded three crania attributed to *Homunculus patagonius* (CORD-PZ 1130, MPM-PV 3501, 3503, and 3505), so this specimen is provisionally assigned to that taxon.

Table 16.1. *Body mass estimates (in grams) reconstructed from fossil crania of* Homunculus patagonicus

Body mass predictor	MPM-PV 3502	MPM-PV 3501	MPM-PV 3503	CORD-PZ 1130	MPM-PV 5000
Principal Component 1 ($R^2 = 0.97$)					
CI lower limit	1024	1000			
CI upper limit	2377	2293			
*Skull length ($R^2 = 0.97$)					
CI lower limit	1106	1186		1005	
CI upper limit	2411	2411		2043	
Number of cranial dimensions measured	17	13	3	4	7
Mean cranial estimate $\pm$ standard error	2301 $\pm$ 534	2347 $\pm$ 678	2338 $\pm$ 1020	1854 $\pm$ 265	2291 $\pm$ 1033

Note: CI is the 95% confidence interval for the body mass prediction. Cranial dimensions come from Hartwig (1993). Species mean body masses from Smith and Jungers (1997). *$\log_{10}$ (BM) = [$\log_{10}$ (skull length) $\times$ 3.30] – 2.95.

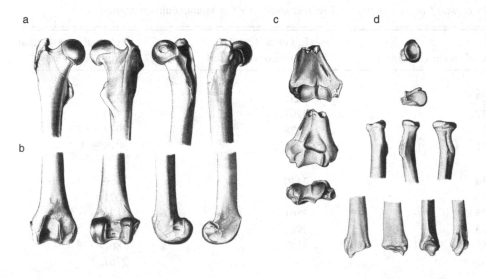

Fig. 16.5. Illustrations of postcrania of *Homunculus patagonicus* from Bluntschli (1931). a, b, Femur (MACN-A 5758). a, Anterior, posterior, medial, and lateral views. b, Anterior, posterior, medial, and lateral views. c, Humerus (MACN-A 5761): anterior, posterior and distal views. d, Radius (MACN-A 5760): various views of proximal and distal ends.

16.4 Results

16.4.1 Body size

Body mass from cranial dimensions Body mass was predicted based upon a principal components analysis of cranial measurements for two specimens (MPM-PV 3501, MPM-PV 3502). The factor loadings for these measurements have similar positive values, indicating that PC1 (which explains 87% of the total variance) is a good body mass predictor. Least-squares regressions of PC1 and $\log_{10}$ body mass (dependent variable) produced estimates of 1.5 kg for both specimens, consistent with the results for cranial length alone (Table 16.1). Total skull length can be measured in three of the *Homunculus* crania, yielding body mass estimates of 1.4 kg (CORD-PZ 1130), 1.5 kg (MPM-PV 3501), and 1.6 kg (MPM-PV 3502).

Estimated body masses for all available cranial dimensions were averaged together to form a mean estimate for each specimen. These estimates range from 1.85 kg $\pm$ 0.26 (CORD-PZ 1130) to 2.34 kg $\pm$ 0.68 (MPM-PV 3501).

Body mass from dental dimensions Body mass reconstructions from individual lower first molar specimens range between 1.87 kg (MLP 11–121, Meldrum and Kay's equation) and 2.65 kg (MACN SC 339, Conroy's anthropoid equation), with mean estimates of 2.4 kg from Conroy's equations and 2.1 kg based upon the Meldrum and Kay equation (Table 16.2). These estimates are slightly below the estimates of the body mass of *Homunculus* made by Conroy (1987), based on a more limited sample. The lower molar estimates calculated here exceed those derived

Table 16.2. *Body mass estimates (in grams) from lower first molar area of* Homunculus patagonicus

Specimen	M1 area (in mm)	Conroy (1987) monkey equation	Conroy (1987) anthropoid equation	Meldrum and Kay (1997) female platyrrhine equation
MACN-A 10403	17.16	2537	2577	2255
MACN-A 5757	16.45	2373	2410	2110
MACN-A 5969b	17.14	2531	2570	2250
MACN-SC 338	16.38	2357	2393	2095
MACN-SC 339	17.46	2606	2647	2317
MACN-SC 341	17.38	2587	2628	2300
MACN-SC 336	16.73	2438	2475	2167
MLP 11–121	15.25	2109	2475	1874
MLP 55-XII-13–156	15.88	2247	2280	1997
Mean ± SE	16.65 ± 0.23	2420 ± 52	2495 ± 38	2152 ± 47

Note: Conroy's (1987) monkey equation, $\ln BM = 1.561 * \ln(m1\ area) + 3.41$, correction factor (ratio estimator, RE) from Smith (1993) = 0.991; Conroy's (1987) anthropoid equation, $\ln BM = 1.570 * \ln(m1\ area) + 3.38$, RE = 1.011; Meldrum and Kay's (1997) female platyrrhine equation, $\ln (female\ BM) = 1.565 * \ln (m1\ area) + 3.272$.

Table 16.3. *Body mass estimates (in grams) from upper first and second molar area of* Homunculus patagonicus

Specimen	Area of molar (in mm)	Egi *et al.* (2004) anthropoid equation	Egi *et al.* (2004) total sample equation
M1			
MPM-PV 3503	28.23	3655	3320
MPM-PV 3501	26.47	3263	2974
MPM-PV-3502	26.31	3228	2943
Mean ± SE	27.01 ± 0.61	3382 ± 137	3079 ± 121
M2			
MACN-SC 342	20.49	2440	2169
MACN-SC 337	19.8	2333	2071
MACN-SC 334	21.25	2561	2279
MPM-PV 3503	23.73	2962	2646
MPM-PV 3501	25.15	3197	2862
MPM-PV-3502	24.53	3093	2767
Mean ± SE	22.49 ± 0.92	2764 ± 149	2466 ± 136

Note: Egi *et al.*'s (2004) anthropoid equation for M1, $\ln BM = 1.767 * \ln (M1\ area) - 4.555$, RE = 0.950; total sample equation for M1, $\ln BM = 1.713 * \ln (M1 area) - 4.535$, RE = 1.012; anthropoid equation for M2, $\ln BM = 1.319 * \ln (M2\ area) - 3.071$, RE = 1.037; total sample equation for M2, $\ln BM = 1.353 * \ln (M2\ area) - 3.348$, RE = 0.980.

from the PCA of the two most complete crania (~1.5 kg, see Table 16.1); however, they do not markedly differ from the mean estimates for each cranium, based upon available cranial dimensions (~1.9 kg to 2.4 kg).

Estimates calculated for the upper M1 and M2 areas use Egi *et al.*'s (2004) equations (Table 16.3). Upper molar dimensions provide estimates ranging from 2.5 to 3.4 kg, significantly higher than estimates from the lower molar, cranial, or (as will be shown below) postcranial dimensions

(Fig. 16.6). This result may indicate upper molar hypertrophy in *Homunculus*.

Body mass from postcranial dimensions A well-preserved right femur of *Homunculus* (MACN-A 5758) had previously been described as too large to belong to this genus (Rosenberger, 1979). However, comparison of second molar area to femur length indicates that the size of the femur is consistent with that of the associated mandible (MACN-A 5757)

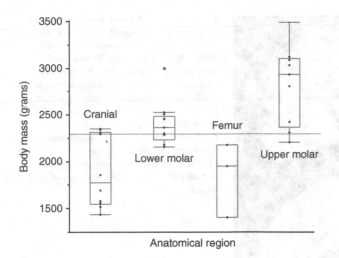

Fig. 16.6. Body mass estimates for *Homunculus patagonicus*, by anatomical region. Box and whisker plots where the center horizontal line for each box is the median, and the bonding box represents the limits of the first and third quartiles. The gray line crossing all anatomical regions is the median estimate using all data. The asterisk represents the estimated body mass of MPM-PV 3708, an unnamed new species. Refer to Tables 16.1–16.3 for further details.

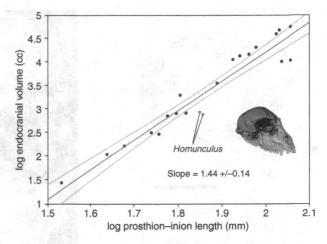

Fig. 16.7. Least-squares fit of $\log_{10}$ endocranial volume versus $\log_{10}$ skull length (prosthion–inion length) for 19 species of extant platyrrhines. Data from Isler *et al.* (2008). *Homunculus* (indicated by inset reconstruction, and labeled by lines to the large open circles) has a proportionally very small brain.

(Tejedor and Rosenberger, 2008). Ciochon and Corruccini (1975) analyzed the dimensions of the femur using PCA. Although body mass reconstruction was not a specific target of their study, this specimen is observed to lie between extant platyrrhines *Cebus* and *Chiropotes* on PC1, which in this analysis is an overall size axis. This placement would be consistent with a rough estimate of 2 to 2.5 kg, based on the body weights of the extant taxa.

The length of the *Homunculus* femur (110.8 mm) produces a body mass estimate of 1.4 kg using Runestad's equation for the total primate sample. This body mass is notably smaller than would be indicated by Ciochon and Corruccini's (1975) PCA, and is more consistent with the PCA based on cranial dimensions. The surface area and volume of the femoral head produce estimates of 2.2 kg and 1.9 kg, respectively. These values are more consistent with the dental dimensions and the means of individual cranial estimates, described above.

The combined skeletal elements known for *Homunculus* provide a range of body mass estimates comparable to that of extant pitheciin genera *Pithecia* or *Cacajao*. Cranial estimates for MPM-PV 5000, the type specimen of *H. blakei*, fall comfortably within the range of estimates for other *Homunculus* crania. Furthermore, the degree of size variation observed in the dental dimensions (with *H. blakei* included) does not exceed that expected for a sample of moderately sexually dimorphic species of platyrrhine. Given that the estimated body size of *Homunculus* exceeds the upper body size limit of extant insectivorous primates (Kay, 1975; Gingerich *et al.*, 1982; Kay and Covert, 1984), we can rule out a large insectivorous component for the species.

One mandibular specimen (MPM-PV 3708) from Puesto Estancia La Costa is a distinct and larger species of *Homunculus*, yet to be named. MPM-PV 3708 is indicated by the asterisk in Fig. 16.6.

16.4.2 Brain size and sensory capabilities

Endocranial volumes (ECV) of the two suitable specimens (MPM-PV 3501 and MPM-PV 3502) are very similar: 19.5 cc and 20.2 cc, respectively. A logistic bivariate plot with prosthion-inion length as the independent variable and endocranial volume as the dependent variable indicates that *Homunculus* falls below the lower limit for ECV in extant anthropoids (Fig. 16.7), but within the envelope of extant strepsirrhines (data from Isler *et al.* (2008) not shown in Figure).

Homunculus is the second Miocene stem platyrrhine reported to have a very small brain for its body size (*Chilecebus* being the other; Flynn *et al.*, 1995; Sears *et al.*, 2008). It therefore appears that living platyrrhines evolved larger brains relative to body size than stem platyrrhines. A parallel increase in encephalization also occurred in catarrhine primates, as the early Oligocene stem catarrhine *Aegyptopithecus* also had a relatively small brain (Radinsky, 1979; Simons *et al.*, 2007). The small size of the brain in *Homunculus* is partly an effect of the small size of the frontal lobe of the cortex, as illustrated in lateral profile (Fig. 16.8).

Several hypotheses have been offered to explain the general phenomenon of increased brain size in anthropoids, amongst which are dietary shifts (the brain is energetically expensive: Aiello and Wheeler, 1995), the evolution of more effective strategies for acquiring hidden or embedded resources (extractive foraging: Parker and Gibson, 1977), and evolving patterns of social behavior (e.g. Machiavellian intelligence:

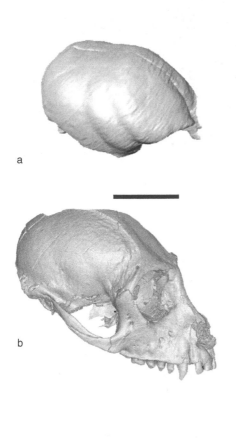

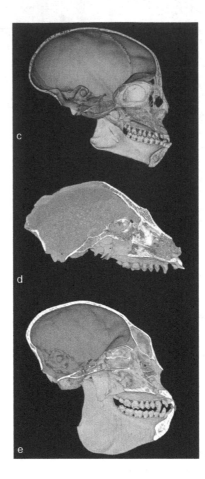

Fig. 16.8. a, b, Anterolateral views of an endocranial reconstruction and the cranium of *Homunculus patagonicus* (MPM-PV 3502). Scale bar equals 20 mm. c–e, Midsagittal sections through the braincases of *Saimiri*, *Homunculus patagonicus*, and *Callicebus*, respectively. Among extant platyrrhines, *Saimiri* has a relatively large brain with a steep frontal profile whereas *Callicebus* has a relatively small brain with a low frontal profile. Note the very low frontal profile of *Homunculus* suggesting that the frontal cortex was small. Panels c–e are not to scale.

Byrne, 1997). These hypotheses may not be mutually exclusive and tend to produce similar predictions in comparative tests across primates and other mammals (Deaner *et al.*, 2000). Therefore, a precise adaptive explanation for the evolution of increased brain size in extant platyrrhines remains obscure.

Vision In proportion to skull length or palate length, the orbits of *Homunculus* are comparable to diurnal platyrrhines, catarrhines, and strepsirrhines, and outside that circumscribed by extant nocturnal primates. We conclude therefore that *Homunculus* spp. were diurnally active, as illustrated in Fig. 16.3. The details of the postorbital wall are well preserved in MPM-PV 3501, 3502, 5000, and CORD-PZ 1130. Postorbital closure in *Homunculus* is nearly complete, with only a small inferior orbital fissure (Tauber, 1991).

A distinctive feature of New World monkeys and anthropoids generally is the presence of enhanced visual acuity, where acuity is broadly defined as the ability to visually discriminate spatial details (Walls, 1942). Living diurnal anthropoids perform better on visual acuity tests than do diurnal strepsirrhines (reviewed in Kirk, 2006). Postorbital closure helps to stabilize the periorbital fascia and protects the eye from movements of the muscles of mastication

within the infratemporal and temporal fossae (Menegaz and Kirk, 2009). This increased stability assists visual acuity and therefore is functionally linked with other adaptations for more acute vision in anthropoids, including the presence of a retinal fovea and the absence of a *tapetum lucidum* (Cartmill, 1980; Ross and Hylander, 1996).

Anthropoids have a pit in their central retina ("fovea") containing a dense concentration of cone receptors (Kirk and Kay, 2004). These cones relay information to the brain via a correspondingly large population of retinal ganglion cells. As a result, all living anthropoids except *Aotus* have a greater number of optic nerve fibers leaving each retina and a larger cross-sectional area of the optic nerve than do living strepsirrhines (Kirk and Kay, 2004).

We examined two well-preserved optic canals in *Homunculus* MPM-PV 3502 and 3505. The optic foramen quotient of *Homunculus* falls within the range of extant diurnal anthropoids, suggesting that the visual acuity of *Homunculus* was comparable to that of diurnal platyrrhines. This is not especially surprising because the stem anthropoid *Simonsius grangeri* Gingerich, 1978 (Oligocene, Egypt) also had an enlarged optic canal and presumably had very high visual acuity like extant anthropoids (Bush *et al.*, 2004).

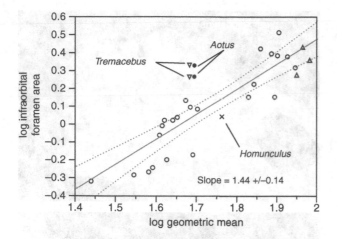

Fig. 16.9. Least-squares fit between $\log_{10}$ infraorbital foramen area and $\log_{10}$ geometric mean of cranial breadth and cranial length for 26 species of diurnal platyrrhines (open circles). Two species of the nocturnal platyrrhine *Aotus* plot above this line, having proportionately larger IOFs than diurnal platyrrhines. *Homunculus* plots with diurnal species. *Tremacebus* (represented by right and left IOFs) plots with *Aotus*. Folivorous species, indicated by open triangles do not differ from frugivorous species. Slope of the least-squares fit is 1.44 +/- 0.14. Data for extant platyrrhines from Muchlinski (2010).

Touch The infraorbital foramen transmits the infraorbital nerve returning sensory information from the vibrissae and skin of the snout, together with the vasculature that supplies the same region. The foramina for branches of this nerve, artery, and vein are small and occasionally multiple in several specimens of *Homunculus*. Muchlinski (2010) examined the ecological correlates of infraorbital foramen size (IOF) in primates and reported that more frugivorous species have relatively greater IOF size than their more folivorous or faunivorous close relatives. However, the relationship between the two variables is not sufficiently strong to make an inference for an extinct species. Muchlinski found no consistent relationship between relative IOF size and either arboreality versus terrestriality or nocturnality versus diurnality. Nevertheless, *Aotus*, the only nocturnal platyrrhine, has a relatively larger IOF than any other platyrrhine (Fig. 16.9). Several specimens of *Homunculus* preserve the infraorbital foramen, which in each case is small like that of diurnal platyrrhines. Interestingly, the IOF of Colhuehuapian *Tremacebus* is comparatively large and resembles nocturnal *Aotus*.

Olfaction The olfactory fossa of *Homunculus* appears to have been relatively small, like those of extant diurnal platyrrhines and catarrhines, whereas nocturnal *Aotus* has a proportionately larger fossa and olfactory bulbs (the latter is also true for the reconstruction of these features for early Miocene *Tremacebus*; Fig. 16.4) (Stephan *et al.*, 1981; Kay *et al.*, 2004). Nocturnal primates as well as nocturnal carnivorans and rodents likewise have relatively larger olfactory

bulbs than their diurnal close relatives (Barton, 2006). The small size of *Homunculus'* olfactory fossa is consistent with its small orbits in suggesting diurnal habits.

Hearing Analysis of the inner ear morphology of *Homunculus* suggests that this taxon had both low- and high-frequency auditory sensitivity comparable to that of extant platyrrhines. As reported by Coleman *et al.* (2010), the tympanic membrane area in *Homunculus* is smaller than expected for a monkey of its size, but the area of the oval window (where the footplate of the stapes sits) also is relatively small. The two measurements yield similar predictions of low-frequency sensitivity (thresholds of ~32–36 dB SPL (sound pressure level) at 250 Hz). The cochlear length in *Homunculus* predicts somewhat higher low-frequency threshold (~24 dB SPL) but the ranges of all these estimates overlap one another and fall within the range of known low-frequency sensitivity in living platyrrhines.

Finally, stapedial footplate area is correlated with high-frequency sensitivity and the estimated hearing threshold for *Homunculus* (~13 dB SPL at 32 kHz) is similar to the values for extant *Saimiri* and *Aotus* and also falls within the range of strepsirrhines. In comparison, hearing thresholds for Old World monkeys are higher (22–39 dB at 32 kHz), indicating that cercopithecoids have poorer high-frequency hearing than either living or fossil platyrrhines (Coleman *et al.*, 2010).

16.4.3 Organs of balance
The semicircular canals lie close to right angles to one another: that is, they approach orthogonality. The rotational head speed of 135 degrees per second predicted using the equation of Malinzak *et al.* (in press). This indicates that *Homunculus* was very agile, as this is similar to the rotational head speeds of the Madagascar bamboo lemur *Hapalemur griseus* (146 degrees per second), and the sifaka, *Propithecus verreauxi* (123 degrees per second). The two are among the more agile extant leaping lemurs of similar body size.

16.4.4 Masticatory apparatus

Tooth eruption sequence A juvenile cranium (MPM-PV 3505) (Fig. 16.2) allows us to draw some conclusions about the upper tooth eruption sequence of *Homunculus*. Based on the degree of wear on the erupted teeth present, and on CT images of the positions of encrypted adult teeth, we infer an eruption sequence of dP2, dP3, dP4 for the deciduous premolars and M1, (I1/I2), M2, P2, P4, P3, (M3/C) for the permanent teeth.

The M3 of MPM-PV 3505 is not yet erupted whereas the adult incisors are already fully erupted. This eruption pattern resembles that of *Apidium*, an Early Oligocene African stem anthropoid (Kay and Simons, 1983). These findings do not support the hypothesis that the primitive tooth eruption pattern in platyrrhines is one of early molar eruption (Henderson, 2007). Henderson's hypothesis assumes that

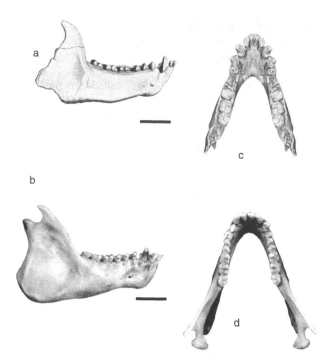

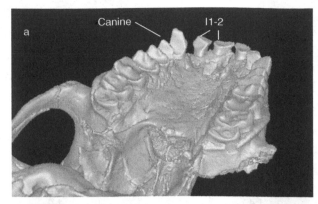

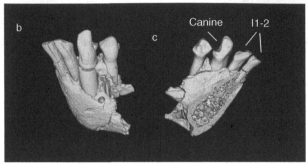

Fig. 16.10. a, Lateral view of *Homunculus patagonicus* mandible MACN-A 5757, neotype, Puesto Estancia La Costa, from Bluntschli (1931). b, Lateral view of *Aotus trivirgatus*. c, Occlusal view of *Homunculus patagonicus* MPM-PV 3504, Estancia La Costa, reconstructed by mirror imaging. d, Occlusal view of *Aotus trivirgatus*. Scale bars equal 10 mm.

Fig. 16.11. Upper and lower incisors and canines of *Homunculus patagonicus*. a, Palate of MPM-PV 3502. b, c, MPM-PV 3504, lower anterior teeth.

the eruption pattern in *Aotus* is the primitive one for platyrrhines. Indeed, to the degree that *Homunculus* was a folivore (see below), this specimen differs from the pattern seen in some folivorous strepsirrhines, which are characterized by early eruption of all the molars (Godfrey *et al.*, 2001, 2004).

Diet and feeding behavior The mandibular corpus of *Homunculus* is similar in depth to that of *Aotus* and houses stoutly rooted cheek teeth, suggesting repetitive loading during mastication of herbaceous vegetation (Fig. 16.10). The mandibular ramus is broad but not deepened as in many extant folivorous platyrrhines. The corpora are V-shaped in occlusal view. In MPM-PV 3504 the corpora are joined at a fully ossified symphysis that would have facilitated the recruitment of significant amounts of muscle forces from the non-chewing (contralateral) side of the jaw. The reconstructed locations and sizes of the attachments of the chewing muscles (Perry *et al.*, 2010) suggest that these muscles were not especially large in *Homunculus* compared with like-sized extant platyrrhines, nor was mechanical advantage particularly great.

The mandibular condyle of *Homunculus* is complete in only one specimen (MPM-PV 3708), which we regard as another species of larger body size (R. F. Kay and J. M. G. Perry, unpublished data). The condyle in this specimen has an extensive articular surface that likely served to distribute joint

reaction force so as to compensate for repetitive loading during incision or mastication. We cannot rule out the alternate possibility that a small number of heavy loading events (as in breaching failure-resistant seeds) selected for this morphology but this is unlikely, given our interpretation of the diet based on the molar structure.

Homunculus is known from a complete upper and lower incisor and canine series (MPM-PV 3502 and 3504; Fig. 16.11). The upper central incisor is spatulate and the cross-section of its root is labio-lingually broad, suggesting that it was optimized to resist powerful labio-lingual bending stresses engendered during ingestion. The upper lateral incisor is more pointed but functioned with the upper central in edge-to-edge incision. The lower incisors present a conjoined ribbon-like occlusal edge that occludes with the upper central and the mesial part of the upper lateral incisor, as in living platyrrhines like *Callicebus* or *Saimiri*.

Homunculus has moderately elongate lower incisors (from the cemento-enamel junction to the cusp tip), reminiscent of the condition seen in *Callicebus* or *Saguinus*, although not as extreme as in the callitrichine *Callithrix* or the pitheciins *Pithecia* and *Chiropotes*. The elongate incisors are used quite differently in callitrichins and pitheciins: in the former they are utilized in gouging and scraping bark to obtain tree gum (Byrne, 1997; Coimbra-Filho and

Mittermeier, 1976) whereas in the latter they are used to pry open hard-shelled fruits (Kinzey and Norconk, 1990; Norconk and Conklin-Brittain, 2004). Thus, the morphological and functional convergence on this incisal pattern is not associated with any particular dietary regime. Rather, it is more conservatively understood as an adaptation for producing a vertical load by means of an upward and forward jaw movement. *Homunculus* may have used its incisors for powerful incision, although whether for tree scoring, fruit husking, or some other purpose is not clear.

The canines of *Homunculus* are small relative to the molars, do not project far above the premolar tooth row, and are not separated from the incisors by a marked diastema. In these respects, the greatest structural similarity is with living *Callicebus* or *Aotus*, not with *Chiropotes*. In the latter, the canine is tusk-like and projecting, even in females.

Homunculus seems to have used its canines in conjunction with the incisors to separate a bite of food or substrate.

In summary, *Homunculus* had spatulate upper incisors and relatively narrow and procumbent lower incisors resembling those of extant *Callicebus* or *Saguinus*. These features suggest that bark stripping, tree gouging, or fruit husking were likely components of the feeding repertoire. The small canines would have been useful tools for food incision – separation of a bite – but not in the fashion seen in pitheciins or callitrichins in which that tooth is enlarged or incorporated in the gouging mechanism. Especially given their small roots (Perry *et al.*, 2010) relative to pitheciins, for example, these canines would not have been very useful tools for resisting heavy loads in food preparation.

Table 16.4 summarizes the species means for tooth length, shearing crest length, and the SQs for a broad

Table 16.4. *Shearing quotients and inferred diets of Early Miocene Patagonian platyrrhines*

Taxon	N	M1 length	Sum, M1 shear	Expected shear	Shear quotient	Major dietary feature
Callimico goeldii	3	2.60	5.48	4.72	16.14	Insects
Brachyteles arachnoides	9	7.22	15.19	3.10	15.93	Leaves
Alouatta palliata	10	6.92	13.91	12.56	10.76	Leaves
Alouatta caraya	6	6.72	13.09	12.20	7.33	Leaves
Alouatta fusca	6	6.70	12.94	12.16	6.42	Leaves
Aotus trivirgatus	10	3.06	6.16	5.55	10.92	Fruit/leaves
Saimiri sciureus	5	2.87	5.54	5.21	6.36	Insect/fruit
Lagothrix lagotricha	8	5.47	10.12	9.93	1.94	Fruit/leaves
Leontopithecus rosalia	5	3.09	5.62	5.61	0.22	Fruit/insects
Ateles geoffroyi	10	5.26	9.31	9.55	−2.47	Fruit
Callicebus moloch	10	3.18	5.50	5.77	−4.70	Fruit
Saguinas mystax	5	2.52	4.03	4.57	−11.88	Fruit/insects
Callithrix argentata	4	2.22	4.08	4.03	1.27	Fruit/gum
Cebuella pygmaea	4	1.78	3.26	3.23	0.92	Gum/fruit
Pithecia monachus	4	4.00	6.78	7.26	−6.60	Fruit
Cebus apella	5	4.79	7.71	8.69	−11.31	Fruit
Chiropotes satanas	5	3.64	5.50	6.61	−15.53	Seeds/fruit
Cacajao melanocephalus	2	3.97	5.90	7.20	−17.70	Seeds/fruit
Dolichocebus gaimanensis	1	3.91	6.58	7.10	−7.27	Fruit
Mazzonicebus almendrae	1	4.19	6.49	7.60	−14.61	Seeds/fruit
Soriacebus spp.	2	3.41	5.27	6.20	−15.24	Seeds/fruit
Carlocebus carmeni	2	4.77	7.76	8.66	−8.85	Fruit
Homunculus patagonicus	6	4.28	7.64	7.77	1.73	Fruit/leaves

Note: The estimate of shearing development is based on measurements of six lower molar crests (Kay, 1975). A line was fitted to a bivariate cluster of the natural log of M1 length (ln ML) versus the natural log of the sum of the measured shearing crests (ln SH). The line was assigned a slope of 1.0 (slope of isometry) and passed through the mean ln ML and mean ln SH for extant taxa. The equation expressing this line is: ln SH = 1.0(ln ML) + 0.596. For each taxon, the expected ln SH was calculated from this equation, and converted to a real number. The observed (measured) shearing for each species was compared with the expected in real space and expressed as a residual (shear quotient, or SQ): SQ = 100 * (observed − expected)/(expected). Extant and extinct taxa are listed separately according to dietary categories (Fleagle *et al.*, 1997). Diet is inferred for extinct taxa by comparison to a modern analog with a similar SQ. All measurements are in millimeters.

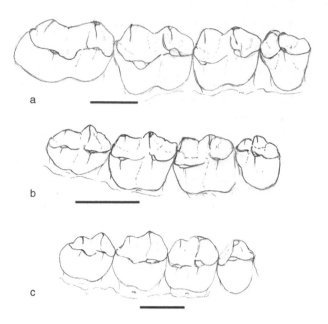

Fig. 16.12. Occlusolateral view of the right lower cheek teeth (p4–m3) of a, *Alouatta caraya* (a leaf-eating species); b, *Homunculus* spp. (MPM-PV 3708); and c, *Ateles geoffroyi* (a fruit-eating species), showing the range of development of the lingual shearing crests. Scale bars equal 5 mm. Drawing by R. F. Kay.

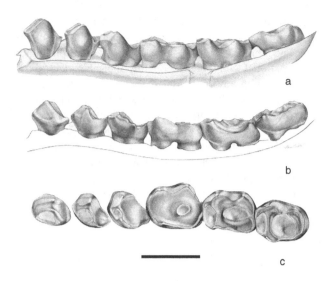

Fig. 16.13. The neotype of *Homunculus patagonicus* MACN-A 5757, illustrating the extreme wear typical of the species. a, Buccal, b, lingual, and c, occlusal views. From Tejedor and Rosenberger, 2008, by permission. Scale bar equals 5 mm.

selection of platyrrhine m1s. Extant platyrrhines that ingest large amounts of plant or animal structural carbohydrate (e.g. cellulose or chitin) have positive SQs and, therefore, relatively better-developed molar shearing crests than species that ingest less fibrous foods such as fruit. Further, SQs of seed-eating platyrrhines fall below or are at the low end of the range of the gum- and fruit-eating taxa (Fig. 16.12 illustrates these differences visually). The mean *Homunculus* SQ is 1.73 (Table 16.4). The living species closest in this respect to *Homunculus* have diets consisting of a mixture of leaves and fruit (for a summary of extant platyrrhine diets and shearing see Anthony and Kay, 1993). The limited data available for another slightly older Miocene Patagonian monkey *Carlocebus*, from the Pinturan fauna, suggest a diet with more fruit than that of *Homunculus* (SQ = –8.85). *Soriacebus*, another Pinturan monkey, has even less overall shearing and in combination with its anterior tooth specializations for fruit husking this indicates that it was probably a seed predator (Kay, 2010). Expanding our observations more broadly, Patagonian monkeys of the Colhuehuapian faunas of Patagonia are at least as diverse as those of the Pinturan. *Dolichocebus* molar shearing (SQ = –7.3) is more similar to *Carlocebus*, and *Mazzonicebus* (–14.6) resembles *Soriacebus* (–15.2).

The teeth of species that specialize in eating hard seeds or in splitting open tough, hard fruits have thicker enamel (Kay, 1981) or more specialized enamel structure (Martin *et al.*, 2003) than closely related frugivorous and folivorous species.

Precise measurement of enamel thickness was not undertaken in *Homunculus*, but tooth wear suggests that the enamel was quite thin in this species. Preliminary examination of broken enamel surfaces suggests that the enamel microstructure was not complex. There is no known evidence that *Homunculus* incorporated hard objects in its diet.

Homunculus specimens are notable for their extremely heavy tooth wear. A number of specimens have the enamel of the occlusal surface largely worn away (for example the neotype from Corriguen Aike figured by Tejedor and Rosenberger, 2008) (Fig. 16.13). Several specimens show a substantial wear gradient along the molar row, again suggesting that wear was rapid. Observation of many museum specimens of extant platyrrhines leaves the impression that tooth wear rates are greater in primate populations that inhabit dry forests or habitats with long seasonal dry intervals than in populations that inhabit humid forests. If so, the wear pattern of *Homunculus* could indicate that the environments where this genus occurred had deciduous forests with low rainfall and with considerable seasonal variation. It also remains a possibility that exogenous dust particles would have played a significant role in tooth wear, given the predominantly tuffaceous nature of the sediments where *Homunculus* is found.

Postcanine tooth root size and root surface area are relatively much greater in *Homunculus* than in any extant platyrrhine. The difference is especially striking for the molar tooth roots, which are extremely large. The great root surface area accommodated an extensive periodontal ligament, stabilizing the tooth against these heavy loads. Thus, *Homunculus* likely used heavy loads in mastication, perhaps to comminute tough foods or foods of very high yield strength.

Table 16.5. *"Method of Means" for estimating sexual dimorphism in the canines of* Homunculus patagonicus *compared with that of 24 species of extant platyrrhine monkeys*

METHOD	Maxillary			Mandibular		
Method of Means	Mesiodistal	Height	Buccolingual	Mesiodistal	Height	Buccolingual
Homunculus patagonicus	1.16	1.28	1.12	1.27	1.52	1.16
Extant platyrrhine max	1.39	2.74	1.42	1.35	1.74	1.50
Extant platyrrhine median	1.15	1.36	1.16	1.16	1.38	1.22
Extant platyrrhine min	1.08	1.13	1.09	1.05	1.09	1.07
Competition level 1 med	1.10	1.22	1.10	1.10	1.22	1.12
Competition level 1 max	1.23	1.38	1.23	1.21	1.34	1.18
Competition level 1 min	1.08	1.13	1.09	1.05	1.09	1.07
Competition level 2 med	1.15	1.61	1.18	1.16	1.44	1.22
Competition level 2 max	1.20	1.65	1.24	1.17	1.57	1.23
Competition level 2 min	1.14	1.29	1.15	1.14	1.24	1.14
Competition level 3 med	1.25	1.68	1.22	1.17	1.59	1.30
Competition level 3 max	1.34	1.76	1.28	1.23	1.74	1.41
Competition level 3 min	1.16	1.27	1.10	1.14	1.23	1.26
Competition level 4 med	1.31	1.67	1.35	1.28	1.49	1.37
Competition level 4 max	1.39	2.74	1.42	1.35	1.57	1.50
Competition level 4 min	1.12	1.33	1.30	1.25	1.44	1.29

Extant species are taken together or partitioned into four competition levels based on a literature survey (Plavcan and Van Schaik, 1995). Height is crown height. The method is explained in the text.

16.4.5 Sexual selection

Canine dimorphism varies considerably in extant platyrrhines, being highest in species of *Alouatta* and lowest in species of callitrichines, *Callicebus* and *Aotus*. In *Homunculus*, dimorphism is near the median value for extant platyrrhines.

Canine dimorphism in extant anthropoids has been consistently associated with variation in male–male competition (Kay *et al.*, 1987b; Plavcan and van Schaik, 1995) but is not sufficiently tightly correlated with competition levels to infer the level of competition precisely, except in extreme cases (i.e. when dimorphism is very high or very low). Estimates of canine dimorphism in *Homunculus* (Tables 16.5, 16.6) overlap the range of dimorphism seen in all competition levels, although the canine dimorphism is greater than that seen in most polyandrous and monogamous species and falls most comfortably among species that exhibit a low intensity of male–male competition, as in the spider monkey *Ateles*.

In short, *Homunculus* was characterized by having extremely small canine size coupled with modest canine dimorphism. This pattern suggests that *Homunculus* did not exhibit intense male–male competition for mates and sexual selection for large canines, nor did it exhibit equal degrees of selection for canine tooth size in males and females. Plavcan *et al.* (1995) demonstrated that relative male and female canine size in anthropoids is correlated with estimates of intrasexual competition. For males, such

competition is normally associated with access to mates, and hence sexual selection. Female competition is normally associated with access to resources. However, where competition occurs in the context of coalitions, selection for large canine size appears reduced in both sexes. The small canines of *Homunculus* certainly suggest a lack of selection for enhanced weaponry in either sex (assuming both sexes are represented, of course). However, the presence of modest canine dimorphism itself suggests that selection pressures are not equal between males and females. This pattern in *Homunculus* implies either minimal contest competition in both females and males, or a large role for coalitionary behavior in competition for resources or mates.

16.4.6 Locomotion

All previous assessments of the femur of *Homunculus* (MACN-A 5758; Fig. 16.14) suggest that this animal used leaping in its locomotor repertoire (Bluntschli, 1931; Ciochon and Corruccini, 1975; Ford, 1990; Meldrum, 1993). Proximally, the articular surface of the femoral head extends onto the posterior surface of the neck. The greater trochanter is broad and rugose and overhangs the femoral shaft anteriorly. This would have provided an extensive attachment for the vastus lateralis muscle, a powerful extensor of the knee (Meldrum, 1993). There is a prominent intertrochanteric line. All of these features are common among leaping primates (e.g. Fleagle, 1977;

Table 16.6. *Comparison of sexual dimorphism in the maxillary canines of Homunculus patagonicus to the range and medians of extant platyrrhine monkeys as a group, and divided into competition levels, using four different methods for estimating dimorphism.*

Binomial Dimorphism Index

	Method		
	Mesiodistal	Height	Buccolingual
Homunculus patagonicus	1.15	1.31	1.10
Extant Median	1.13	1.32	1.15
Extant Max	1.34	2.23	1.37
Extant Min	1.07	1.09	1.08
Competition level 1 Med	1.09	1.18	1.09
Competition level 1 Max	1.21	1.35	1.21
Competition level 1 Min	1.07	1.09	1.08
Competition level 2 Med	1.14	1.50	1.17
Competition level 2 Max	1.19	1.58	1.20
Competition level 2 Min	1.11	1.25	1.13
Competition level 3	1.23	1.55	1.20
Competition level 3 Max	1.29	1.64	1.24
Competition level 3 Min	1.14	1.23	1.09
Competition level 4 Med	1.28	1.60	1.31
Competition level 4 Max	1.34	2.23	1.37
Competition level 4 Min	1.11	1.28	1.26

Method of moments

	Method		
	Mesiodistal	Height	Buccolingual
Homunculus patagonicus	1.18	1.40	1.12
Extant Median	1.15	1.37	1.16
Extant Max	1.43	1.96	1.47
Extant Min	1.08	1.11	1.08
Competition level 1 Med	1.10	1.18	1.10
Competition level 1 Max	1.22	1.37	1.22
Competition level 1 Min	1.08	1.11	1.08
Competition level 2 Med	1.15	1.66	1.18
Competition level 2 Max	1.20	1.70	1.26
Competition level 2 Min	1.12	1.25	1.14
Competition level 3 Med	1.28	1.76	1.24
Competition level 3 Max	1.38	1.85	1.30
Competition level 3 Min	1.17	1.29	1.11
Competition level 4 Med	1.35	1.56	1.39
Competition level 4 Max	1.43	1.96	1.47
Competition level 4 Min	1.13	1.37	1.33

Assigned resampling method

	Mesiodistal	Height	Buccolingual
Homunculus patagonicus	1.12	1.22	1.08
Extant Median	1.12	1.32	1.12
Extant Max	1.36	2.11	1.37
Extant Min	1.05	1.07	1.07
Competition level 1 Med	1.08	1.12	1.07
Competition level 1 Max	1.17	1.32	2.28
Competition level 1 Min	1.05	1.07	1.07
Competition level 2 Med	1.12	1.53	1.14
Competition level 2 Max	1.16	1.55	1.19
Competition level 2 Min	1.10	1.22	1.11
Competition level 3 Med	1.21	1.64	1.19
Competition level 3 Max	1.33	1.69	1.24
Competition level 3 Min	1.13	1.23	1.08
Competition level 4 Med	1.27	1.58	1.30
Competition level 4 Max	1.36	2.11	1.37
Competition level 4 Min	1.10	1.31	1.29

CV method

	Mesiodistal	Height	Buccolingual
Homunculus patagonicus	1.16	1.41	1.08
Extant Median	1.13	1.40	1.16
Extant Max	1.42	4.01	1.46
Extant Min	1.05	1.08	1.06
Competition level 1 Med	1.08	1.20	1.09
Competition level 1 Max	1.24	1.43	1.24
Competition level 1 Min	1.05	1.08	1.06
Competition level 2 Med	1.14	1.69	1.18
Competition level 2 Max	1.21	1.77	1.24
Competition level 2 Min	1.12	1.30	1.15
Competition level 3 Med	1.26	1.72	1.23
Competition level 3 Max	1.37	1.87	1.31
Competition level 3 Min	1.14	1.27	1.08
Competition level 4 Med	1.33	1.77	1.38
Competition level 4 Max	1.42	4.01	1.46
Competition level 4 Min	1.11	1.33	1.30

Note: Ranges and medians are presented for extant species as a group ($N = 24$), and within four "competition levels" (Kay *et al.*, 1987a). Sample sizes for: Competition level 1: $n = 10$; level 2, $n = 4$; level 3, $n = 5$; level 4, $n = 5$. Abreviations: Med, median; Max, maximum; Min, minimum; Height is crown height. The various methods are explained in the text.

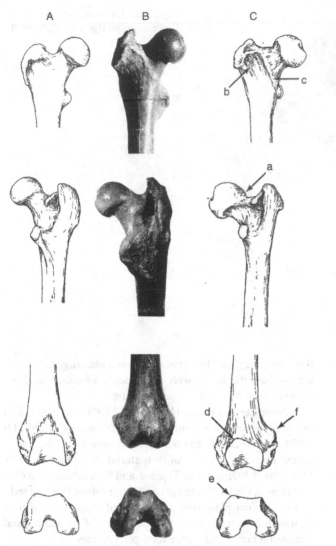

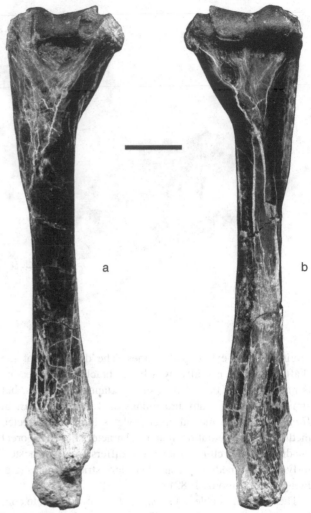

Fig. 16.14. Right femur of *Homunculus patagonicus* (MACN-A 5758) (B) compared with that of (C) *Pithecia pithecia* and (A) *Chiropotes satanas*, anterior, posterior, and distal views, after Fleagle and Meldrum (1988). *Homunculus* shares with *Pithecia* all the following characteristics related to leaping locomotion: a, femoral articular surface extends onto neck; b, greater trochanter overhangs anterior surface of femoral shaft; c, prominent intertrochanteric line; d, patellar groove extends far proximally onto femoral shaft; e, raised lateral lip of patella; f, pronounced patellar tubercle.

Fleagle and Meldrum, 1988). The distal epiphysis of the femur is relatively deep anteroposteriorly and narrow mediolaterally, a feature that improves the lever arm for the action of the knee extensors and is characteristic of many leapers (Fleagle, 1977; Fleagle and Simons, 1995). Distally, the patellar groove is deep and narrow. It extends more proximally than most platyrrhines and the lateral lip is raised. All of these features help prevent patellar dislocation during powerful hindlimb extension (Fleagle, 1977).

Fig. 16.15. Humerus of *Homunculus patagonicus* MPM-PV 3500. Posterior (a) and anterior (b) views. Scale bar equals 10 mm.

A right humerus (MPM-PV 3500) from Puesto Estancia La Costa (Fig. 16.15; Fleagle and Kay, 2006) lacks only the head and the most proximal part of the shaft. The distal part of the bone is virtually identical in both size and morphology to the distal fragment of a humerus (MACN-A 5761) found with the neotype specimen (Bluntschli, 1931; Tejedor and Rosenberger, 2008) (Fig. 16.5), so it most likely belongs to *Homunculus patagonicus*. Overall, the humeral shaft is relatively more robust than that of extant platyrrhines and more similar to that seen in some extant Malagasy lemurs such as *Varecia*, and in African Oligocene anthropoids *Aegyptopithecus* Simons, 1965 and *Apidium* Osborn, 1910.

A notable feature of the proximal shaft is the prominence of the deltopectoral and deltotriceps crests. On the posterior surface of the shaft the supinator crest is very long, carrying the origin for the brachioradialis muscle far proximally, as in *Apidium* and *Aegyptopithecus*, but unlike the more distal

Fig. 16.16. Life reconstruction of *Homunculus patagonicus* by Manuel Sosa.

origin in most extant platyrrhines. The distal end of the shaft is marked laterally by a broad brachialis flange that is much broader than that of most extant platyrrhines, but similar to that of many prosimians or *Aegyptopithecus*. In *Homunculus*, the medial epicondyle is large and directed medially, an orientation that is characteristic of arboreal quadrupeds and clinging taxa, but differs from the posteriorly directed medial epicondyle of terrestrial primates (e.g. Fleagle and Simons, 1982).

The humeral trochlea of *Homunculus* is cylindrical to conical with a weak medial lip. This shape is characteristic of arboreal quadrupeds or clingers and differs from the situation found in terrestrial quadrupeds or hominoids in which the medial trochlear lip is prominent. Neither humeral specimen shows clear evidence of a proximal extension of the anterior trochlear surface onto the proximal humeral shaft, a "clinging facet", characteristic of many primates (e.g. callitrichins, indriids) that regularly adopt clinging postures on vertical supports in which the coronoid process of the ulna is pressed against the supratrochlear region of the humerus (Szalay and Dagosto, 1980). The capitulum is rounded and separated from the trochlea by a wide *zona conoidea*. It is much broader anteriorly than posteriorly and there is a distinct capitular tail. These features suggest that the elbow of *Homunculus* was most commonly used in a flexed rather than extended posture. On the dorsal surface of the distal humerus, the olecranon fossa is relatively shallow, from which we infer that the ulna of *Homunculus* probably had a relatively long olecranon and moved on flexed elbows. There is no evidence of extensive buttressing of the capitulum as in many terrestrial quadrupeds and most cercopithecids (Bown *et al.*, 1982; Fleagle and

Simons, 1982). The humerus of *Homunculus* suggests arboreal quadrupedal habits with no evidence of either forelimb suspension or habitual clinging postures.

A complete left radius (MACN-A 5760) associated with the neotype of *Homunculus* was described by Bluntschli (1931; Fig. 16.5) but has not been discussed in most recent reviews of the *Homunculus* material (e.g. Ford, 1990; Meldrum, 1993; but see Tejedor and Rosenberger, 2008). The bone is relatively straight and more robust than the radii of most extant platyrrhines. The radial head is more oval than round, suggesting a limited range of pronation and supination compared with extant platyrrhines.

Limb proportions Meldrum (1993) remarked that the radius of *Homunculus* is much larger in proportion to the femur than in *Aotus* and closely resembles *Alouatta*. He concluded that *Homunculus* may have employed more forelimb-assisted climbing than *Aotus* and was similar to *Alouatta* and *Lagothrix*. Tejedor and Rosenberger (2008, Fig. 9) showed that relative to tooth size, the length of the radius was slightly above the regression line for a diverse group of platyrrhines, and the femur was below the regression line. This line of inference supports Meldrum's (1993) conclusions about the limb proportions.

Although the humerus described by Fleagle and Kay (2006) is clearly from a different individual than the radius and femur, it is virtually identical in size to the distal fragment of humerus associated with those specimens. If we use the reconstructed length for the new humerus to estimate the humerus length of the neotype specimen, we can calculate a humerofemoral index and a brachial index for *Homunculus*.

The estimated humero-femoral index of *Homunculus* of 83.6 is well above that of *Saguinus*, *Callithrix*, or *Callicebus*; slightly above that for *Saimiri*, *Chiropotes*, or *Cebus*; slightly below values for *Leontopithecus* and *Alouatta*; and well below the values for most atelines. The radial index of 103 is very high for a platyrrhine and is most similar to *Leontopithecus* and *Ateles*.

16.5 Conclusions

A flesh reconstruction of *Homunculus patagonicus* is given in Fig. 16.16. The combined cranial, dental, and skeletal elements of *Homunculus* provide a range of body mass estimates between 1.9 and 3.4 kg, comparable to those of extant pitheciin genera *Pithecia* or *Cacajao*. As indicated from the orbits, *Homunculus* was diurnal and had acute vision like most extant platyrrhines. A reduced infraorbital foramen suggests poorly developed vibrissae on the muzzle, possibly reinforcing the hypothesis that this taxon was diurnal. The olfactory fossae were small, indicating a modern anthropoid-like sense of smell. The diet consisted of a mixture of leaves and fruit. The extremely heavy tooth wear exhibited by many specimens suggest seasonal shifts towards a more abrasive diet, a hypothesis supported by the fluctuation in rainfall and temperature that these extratropical latitudes experience. Sexual dimorphism is observed in *Homunculus*, indicating that the species was social rather than solitary. The limb skeleton indicates that the species was wholly arboreal and utilized above-branch arboreal quadrupedalism with a crouched posture. Some leaping was likely a significant part in the locomotor repertoire, as indicated especially by the hindlimb and semi-circular canal architecture.

ACKNOWLEDGMENTS
We thank the Patagonia field crews, seasons 2003 to present, for their efforts in collecting Miocene mammals, including primates, from the Santa Cruz Formation of coastal Argentina. We thank the Battini family for their hospitality during the fieldwork. We thank the staff of the Museo Regional Provincial "Padre Manuel Jesús Molina" (Río Gallegos, Argentina), the staff of the National Museum of Natural History (particularly Linda Gordon) for allowing us to borrow specimens of platyrrhines for CT scanning and study. We thank the Museo Argentino de Ciencias Naturales "Bernardino Rivadavia" (Buenos Aires), and the Museo de La Plata for allowing us to study specimens in their care. Special thanks to Dr. Matthew Colbert and the staff of the University of Texas (Austin) MicroCT Facility for making the scans used in this study. This project was funded by PICT 26219 and 0143, CONICET PIP 1054, UNLP N 647 to Sergio Vizcaíno; NSF grants 0851272 and National Geographic grants to R. F. Kay and S. F. Vizcaíno; and Midwestern University funds to J. M. G. Perry. Manuel Sosa prepared the flesh reconstruction of *Homunculus* under the supervision of R. F. Kay.

Appendix 16.1   Specimens

Specimens collected by Museo de La Plata/Duke University expeditions

All specimens come from Santa Cruz Province, Argentina (see Vizcaíno *et al.*, Chapter 1: Figs. 1.1 and 1.2 and Appendix 1.1 for localities, and Matheos and Raigemborn, Chapter 4, Figs. 4.3 and 4.4 for stratigraphy). Stake levels are illustrated by Fleagle *et al.* (Chapter 3, Fig. 3.5) Fossiliferous level = FL.

Homunculus patagonicus

MPM-PV-3500, humerus. Locality and horizon: Puesto Estancia La Costa, FL 5.3, Estancia La Costa Member, Santa Cruz Formation. MPM-PV 3501, cranium. Locality and horizon: Puesto Estancia La Costa, FL 5.3, Estancia La Costa Member, Santa Cruz Formation. MPM-PV 3502, cranium. Locality and horizon: Killik Aike Norte, base of ash level KAN (Perkins *et al.*, Chapter 2), Santa Cruz Formation. MPM-PV 3503, cranium. Locality and horizon: Puesto Estancia La Costa, FL 5.3, Estancia La Costa Member, Santa Cruz Formation. MPM-PV 3504, mandible. Locality and horizon: Estancia La Costa, FL 3, Estancia La Costa Member, Santa Cruz Formation. MPM-PV 3505, juvenile cranium. Puesto Estancia La Costa, FL 5.3, Estancia La Costa Member, Santa Cruz Formation.

Homunculus sp. nov.?

MPM-PV 3708, mandible. Puesto Estancia La Costa, FL 7, Estancia La Costa Member, Santa Cruz Formation.

Specimens in previous collections

Homunculus patagonicus

Coastal Santa Cruz Formation, locality unknown: MACN-A 635, symphysis with roots of left and right i1–2 and right p2–3, alveoli for left and right canines, right p4 *in situ*. TYPE *Anthropops perfectus* Ameghino, 1891; MLP 66-V-2-2, mandibular fragment with roots for i2, c, p2–3, broken p4, broken m1; MLP 11–121, left dp4–m1 with an unerupted p4.

Santa Cruz Formation, Monte Observación (= Cerro Observatorio): MACN-A 5966, left mandibular fragment with m2, TYPE *Pitheculus australis* Ameghino, 1894; MACN-A 5969, left mandible with ascending ramus and partial condyle with m2 and alveolus for m3; MACN-A 8648, right dp4, initially named as a species of *Stilotherium* (*S. grande* Ameghino 1894) ascribed to *Homunculus grandis* (Hershkovitz, 1981); MACN-A 10403, left mandibular fragment with m1–m2; MACN SC 1149, right m2? (locality: Stake 147); MACN SC 274, M3 (no locality); MACN SC 275, left C (no locality); MACN SC 332, possible eroded lower incisor (locality: Stake 22); MACN SC 334, right M2 (locality: Stake 13); MACN SC 335, right C (locality: Stake 2); MACN SC 336, right mandibular fragment with m1–2 (locality: Stake 28); MACN SC 337, left M2 (locality: Stake 19); MACN SC 338, left m1 (locality: Stake 69); MACN SC 339 left mandibular fragment with m1–2 (locality: Stake 69); MACN SC 341, left m1 (locality: Stake 29); MACN SC 402, right c (locality: Stake 110); MACN SC 2916, left P3 (locality: Stake 206); MACN SC 2918, right m2 (locality: Stake 206); MACN SC 3026, right p4 (locality: Stake 7); MACN SC 3074, left m1 (locality: Stake 7); MACN SC 3089, right I1 (no locality); MACN SC 3090, right m2? (no locality); MACN SC 3112, left M2 (locality: Stake 19); MACN SC 3114, left i1 (locality: Stake 19); MACN SC 3200, right m2 (no locality); MACN SC 3116, left p2 (locality: Stake 25); MACN SC 342, right M2 (locality: Stake 54).

Monte León: MLP 55 XII 13–156, m1.

Corriguen Aike (= Puesto Estancia La Costa): MACN-A 5757, *Homunculus patagonicus* NEOTYPE, Mandible with very worn right i1–m3; left i1, p2–m3; MACN-A 5758, right femur; MACN-A 5759, ulna; MACN-A 5760, partial left radius; MACN-A 5761 (lost), distal right humerus; MACN-A 5968, frontal and facial fragment with left p2–m3; CORD-PZ 1130, partial skull.

Killik Aike Norte: MPM-PV 5000, face with forehead, orbits, snout, dental arcade, and roots of the pterygoid processes, right C, P2–3, root of P4, crowns of M1–3, left C, broken crowns of P2–3, roots of M1–3. TYPE *Killikaike blakei* Tejedor *et al.*, 2006. Level unknown; collected by J. Case (personal communication).

MPM-PV 1607, KAN ash level (Perkins *et al.*, Chapter 2) right M1–3.

Probably Killik Aike Norte: MACN-A 12498, right mandible with i2–m2, symphysis and roots of left i1–2. TYPE *Homunculus patagonicus* Ameghino, 1891 (now lost).

REFERENCES

Aiello, L. C. and Wheeler, P. (1995). The expensive-tissue hypothesis – the brain and the digestive-system in human and primate evolution. *Current Anthropology*, **36**, 199–221.

Anthony, M. R. L. and Kay, R. F. (1993). Tooth form and diet in ateline and alouattine primates: reflections on the comparative method. *American Journal of Science*, **283A**, 356–382.

Barton, R. A. (2006). Olfactory evolution and behavioral ecology in primates. *American Journal of Primatology*, **68**, 545–558.

Bluntschli, H. (1931). *Homunculus patagonicus* und die zugereihten Fossilfunde aus den Santa-Cruz-Schichten Patagoniens. *Gegenbaurs morphologisches Jahrbuch*, **67**, 811–892.

Bown, T. M., Kraus, M., Wing, S. *et al.* (1982). The Fayum primate forest revisited. *Journal of Human Evolution*, **11**, 624–648.

Boyer, D. M. (2008). Relief index of second mandibular molars is a correlate of diet among prosimian primates and other euarchontan mammals. *Journal of Human Evolution*, **55**, 1118–1137.

Bush, E. C., Simons, E. L. and Allman, J. (2004). High-resolution computed tomography study of the cranium of a fossil anthropoid primate, *Parapithecus grangeri*: new insights into the evolutionary history of primate sensory systems. *Anatomical Record*, **281A**, 1083–1087.

Byrne, R. W. (1997). Machiavellian intelligence. *Evolutionary Anthropology*, 172–180.

Cartmill, M. (1980). Morphology, function and evolution of the anthropoid postorbital septum. In *Evolutionary Biology of the New World Monkeys and Continental Drift*, ed. R. Ciochon and B. Chiarelli. New York: Plenum Press, pp. 243–274.

Ciochon, R. L. and Corruccini, R. S. (1975). Morphometric analysis of platyrrhine femora with taxonomic implications and notes on two fossil forms. *Journal of Human Evolution*, **4**, 193–217.

Coimbra-Filho, A. F. and Mittermeier, R. A. (1976). Exudate eating and tree gouging in marmosets. *Nature*, **262**, 630.

Coleman, M. N., Kay, R. F. and Colbert, M. (2010). Auditory morphology and hearing sensitivity in fossil New World monkeys. *Anatomical Record Part A: Discoveries in Molecular Cellular and Evolutionary Biology*, **293**, 1711–1721.

Conroy, G. C. (1987). Problems of body-weight estimation in fossil primates. *International Journal of Primatology*, **8**, 115–137.

Deaner, R. O., Nunn, C. L. and van Schaik, C. (2000). Comparative tests of primate cognition: different scaling methods produce different results. *Brain, Behavior and Evolution*, **55**, 44–52.

Egi, N., Takai, M., Shigehara, N. and Tsubamoto, T. (2004). Body mass estimates for Eocene eosimiid and amphipithecid primates using prosimian and anthropoid scaling models. *International Journal of Primatology*, **25**, 211–236.

Eldredge, N. (1985). *Unfinished Synthesis: Biological Hierarchies and Modern Evolutionary Thought*. New York: Oxford University Press.

Elton, C. S. (1927). *Animal Ecology*. New York: MacMillan.

Fleagle, J. G. (1977). Locomotor behavior and skeletal morphology of sympatric Malaysian leaf monkeys (*Presbytis obscura* and *Presbytis melalophos*). *Yearbook of Physical Anthropology*, **20**, 440–453.

Fleagle, J. G. (1990). New fossil platyrrhines from the Pinturas Formation, southern Argentina. *Journal of Human Evolution* **19**, 61–85.

Fleagle, J. G. and Kay, R. F. (2006). A new humerus of *Homunculus* from the Santa Cruz Formation (early–middle Miocene, Patagonia). *Journal of Vertebrate Paleontology*, **26**, 62A.

Fleagle, J. G. and Meldrum, D. J. (1988). Locomotor behavior and skeletal morphology of two sympatric pitheciine monkeys, *Pithecia pithecia* and *Chiropotes satanas*. *American Journal of Primatology*, **16**, 227–249.

Fleagle, J. G. and Simons, E. L. (1982). The humerus of *Aegyptopithecus zeuxis*: a primitive anthropoid. *American Journal of Physical Anthropology*, **59**, 175–193.

Fleagle, J. G. and Simons, E. L. (1995). Limb skeleton and locomotor adaptations of *Apidium phiomense*, an Oligocene anthropoid from Egypt. *American Journal of Physical Anthropology*, **97**, 235–289.

Fleagle, J. G. and Tejedor, M. F. (2002). Early platyrrhines of southern South America. In *The Primate Fossil Record*, ed. W. C. Hartwig. Cambridge: Cambridge University Press, pp. 161–173.

Fleagle, J. G., Kay, R. F. and Anthony, M. R. L. (1997). Fossil New World monkeys. In *Mammalian Evolution in the Neotropics*, ed. R. F. Kay, R. H. Madden, R. L. Cifelli and J. J. Flynn. Washington, D.C.: Smithsonian Institution Press, pp. 473–495.

Flynn, J. J., Wyss, A. R., Charrier, R. and Swisher, C. C. III. (1995). An early Miocene anthropoid skull from the Chilean Andes. *Nature*, **373**, 603–607.

Ford, S. M. (1990). Locomotor adaptations of fossil platyrrhines. *Journal of Human Evolution*, **19**, 141–173.

Gingerich, P. D., Smith, B. H. and Rosenberg, K. (1982). Allometric scaling in the dentition of primates and prediction of body weight from tooth size in fossils. *American Journal of Physical Anthropology*, **58**, 81–100.

Godfrey, L. R., Samonds, K. E., Jungers, W. L. and Sutherland, M. R. (2001). Teeth, brains, and primate life histories. *American Journal of Physical Anthropology*, **114**, 192–214.

Godfrey, L. R., Samonds, K. E., Jungers, W. L. and Sutherland, M. R. (2004). Ontogenetic correlates of diet in Malagasy primates. *American Journal of Physical Anthropology*, **123**, 250–276.

Hartwig, W. C. (1993). Comparative morphology, ontogeny and phylogenetic analysis of the platyrrhine cranium. Unpublished Ph.D. dissertation, University of California, Berkeley.

Henderson, E. (2007). Platyrrhine dental eruption sequences. *American Journal of Physical Anthropology*, **134**, 226–239.

Isler, K., Kirk, E. C., and Miller, J. M. A. *et al.* (2008). Endocranial volumes of primate species: scaling analyses using a comprehensive and reliable data set. *Journal of Human Evolution*, **55**, 967–978.

Josephson, S. C., Juell, K. E. and Rogers, A. R. (1996). Estimating sexual dimorphism by Method-of-Moments. *American Journal of Physical Anthropology*, **100**, 191–206.

Jungers, W. L., Falsetti, A. B. and Wall, C. E. (1995). Shape, relative size and size-adjustments in morphometrics. *Yearbook of Physical Anthropology*, **38**, 137–162.

Kay, R. F. (1975). The functional adaptations of primate molar teeth. *American Journal of Physical Anthropology*, **43**, 195–216.

Kay, R. F. (1981). The nut-crackers: a new theory of the adaptations of the Ramapithecinae. *American Journal of Physical Anthropology*, **55**, 141–152.

Kay, R. F. (2010). A new primate from the Early Miocene of Gran Barranca, Chubut Province, Argentina: Paleoecological Implications. In *The Paleontology of Gran Barranca: Evolution and Environmental Change through the Middle Cenozoic of Patagonia*, ed. R. H. Madden, G. Vucetich, A. A. Carlini and R. F. Kay. Cambridge: Cambridge University Press, pp. 220–239.

Kay, R. F. and Cartmill, M. (1974). Skull of *Palaechthon nacimienti*. *Nature*, **252**, 37–38.

Kay, R. F. and Cartmill, M. (1977). Cranial morphology and adaptations of *Palaechthon nacimienti* and other Paromomyidae (Plesiadapoidea ? Primates), with a description of a new genus and species. *Journal of Human Evolution*, **6**, 19–35.

Kay, R. F. and Covert, H. H. (1984). Anatomy and behaviour of extinct primates. In *Food Acquisition and Processing in Primates*, ed. D. J. Chivers, B. A. Wood and A. Bilsborough. New York: Plenum Press, pp. 467–508.

Kay, R. F. and Fleagle, J. G. (2010). Stem taxa, homoplasy, long lineages and the phylogenetic position of *Dolichocebus*. *Journal of Human Evolution*, **59**, 218–222.

Kay, R. F. and Kirk, E. C. (2000). Ostological evidence for the evolution of activity pattern and visual acuity in primates. *American Journal of Physical Anthropology*, **113**, 235–262.

Kay, R. F. and Simons, E. L. (1983). Dental formulae and dental eruption patterns in Parapithecidae (Primates, Anthropoidea). *American Journal of Physical Anthropology* **62**, 363–375.

Kay, R. F., Plavcan, J. M., Glander, K. E. and Wright, P. (1987a). Sexual selection and canine dimorphism in New World monkeys. *American Journal of Physical Anthropology*, **77**, 385–397.

Kay, R. F., Plavcan, J. M., Wright, P. C. and Glander, K. E. (1987b). Behavioral and size correlates of canine dimorphism in platyrrhine primates. *American Journal of Physical Anthropology*, **72**, 218.

Kay, R. F., Campbell, V. M., Rossie, J. B., Colbert, M. W. and Rowe, T. (2004). The olfactory fossa of *Tremacebus harringtoni* (Platyrrhini, early Miocene, Sacanana,

Argentina): implications for activity pattern. *Anatomical Record*, **281A**, 1157–1172.

Kay, R. F., Fleagle, J. G., Mitchell, T. R. T. *et al.* (2008). The anatomy of *Dolichocebus gaimanensis*, a primitive platyrrhine monkey from Argentina. *Journal of Human Evolution*, **54**, 323–382.

Kinzey, W. G. and Norconk, M. A. (1990). Hardness as a basis of fruit choice in two sympatric primates. *American Journal of Physical Anthropology*, **81**, 5–15.

Kirk, E. C. (2006). Eye morphology in cathemeral lemurids and other mammals. *Folia Primatologica*, **77**, 27–49.

Kirk, E. C. and Kay, R. F. (2004). The evolution of high visual acuity in the Anthropoidea. In *Anthropoid Origins: New Visions*, ed. C. F. Ross and R. F. Kay. New York: Kluwer/ Plenum Publishing, pp. 539–602.

Lee, S. H. (2001). Assigned resampling method: a new method to estimate size sexual dimorphism in samples of unknown sex. *Przegląd Antropologiczny – Anthropological Review*, **64**, 21–39.

Malinzak, M. D. (2010). Experimental analyses of the relationship between semicircular canal morphology and locomotor head rotations in primates. Unpublished Ph.D. dissertation, Duke University, Durham, NC.

Malinzak, M. D., Kay, R. F. and Hullar, T. E. (in press). Predicting locomotion from the primate semicircular canal system. *Proceedings of the National Academy of Sciences USA*.

Marshall, L. G. (1976). Fossil localities for Santacrucian (early Miocene) mammals, Santa Cruz Province, southern Patagonia, Argentina. *Journal of Paleontology*, **50**, 1129–1142.

Martin, L. B., Olejniczak, A. J. and Maas, M. C. (2003). Enamel thickness and microstructure in pitheciin primates, with comments on dietary adaptations of the middle Miocene hominoid *Kenyapithecus*. *Journal of Human Evolution*, **45**, 351–367.

Martin, R. D. (1990). *Primate Origins and Evolution: A Phylogenetic Reconstruction*. Princeton, New Jersey: Princeton University Press.

Meldrum, D. J. (1993). Postcranial adaptations and positional behavior in fossil platyrrhines. In *Postcranial Adaptations in Nonhuman Primates*, ed. D. L. Gebo. DeKalb, IL: Northern Illinois University Press, pp. 235–251.

Meldrum, D. J. and Kay, R. F. (1997). *Nuciruptor rubricae*, a new pitheciin seed predator from the Miocene of Colombia. *American Journal of Physical Anthropology*, **102**, 407–427.

Menegaz, R. A., and Kirk, E. C. (2009). Septa and processes: convergent evolution of the orbit in haplorhine primates and strigiform birds. *Journal of Human Evolution*, **57**, 672–687.

Muchlinski, M. N. (2010). Ecological correlates of infraorbital foramen area in primates. *American Journal of Physical Anthropology*, **141**, 131–141.

Norconk, M. A. and Conklin-Brittain, N. L. (2004). Variation on frugivory: the diet of Venezuelan white-faced sakis. *International Journal of Primatology*, **25**, 1–26.

Parker, S. T. and Gibson, K. R. (1977). Object manipulation, tool use and sensorimotor intelligence as feeding adaptations in cebus monkeys and great apes. *Journal of Human Evolution*, **6**, 623–641.

Perry, J. M. G., Kay, R. F., Vizcaíno, S. F. and Bargo, M. S. (2010). Tooth root size, chewing muscle leverage, and the biology of *Homunculus patagonicus* (Primates) from the late early Miocene of Patagonia. *Ameghiniana*, **47**, 355–371.

Plavcan, J. M. (1994). A comparison of four simple methods for estimating sexual dimorphism in fossils. *American Journal of Physical Anthropology*, **94**, 465–476.

Plavcan, J. M. and Van Schaik, C. P. (1995). Intrasexual competition and canine dimorphism in anthropoid primates. *American Journal of Physical Anthropology*, **87**, 461–477.

Radinsky, L. B. (1979). The fossil record of primate brain evolution. *James Arthur Lecture on the Evolution of the Human Brain. American Museum of Natural History*, **49**, 1–27.

Ré, G. H., Bellosi, E. S., Heizler, M. *et al.* (2010). A geochronology for the Sarmiento Formation at Gran Barranca. In *The Paleontology of Gran Barranca: Evolution and Environmental Change through the Middle Cenozoic of Patagonia*, ed. R. H. Madden, G. Vucetich, A. A. Carlini and R. F. Kay. Cambridge, Cambridge University Press, pp. 46–60.

Reno, P. L., Meindl, R. S., McCollum, M. A. and Lovejoy, C. O. (2003). Sexual dimorphism in *Australopithecus afarensis* was similar to that of modern humans. *Proceedings of the National Academy (USA)*, **100**, 9404–9409.

Rosenberger, A. L. (1979). Phylogeny, evolution and classification of New World monkeys (Platyrrhini, Primates). Unpublished Ph.D. thesis, City University of New York.

Rosenberger, A. L. and Kinzey, W. G. (1976). Functional patterns of molar occlusion in platyrrhine primates. *American Journal of Physical Anthropology*, **45**, 261–298.

Rosenberger, A. L., Cooke, S. B., Rímoli, R., Ni, X. and Cardoso, L. (2011). First skull of *Antillothrix bernensis*, an extinct relict monkey from the Dominican Republic. *Proceedings of the Royal Society, B, Biological Sciences*, **278**, 67–74.

Ross, C. (1996). An adaptive explanation for the origins of the Anthropoidea (Primates). *American Journal of Primatology*, **40**, 205–230.

Ross, C. and Hylander, W. L. (1996). *In vivo* and *in vitro* bone strain in owl monkey circumorbital region and the function of the postorbital septum. *American Journal of Physical Anthropology*, **101**, 183–215.

Ross, C. F., Hall, M. I. and Heesy, C. P. (2007). Were basal primates nocturnal? Evidence from eye and orbit shape. In *Primate Origins: Adaptations and Evolution*, ed. M. Ravosa and M. Dagosto. New York: Springer, pp. 233–256.

Runestad, J. A. (1994). Humeral and femoral diaphyseal cross-sectional geometry and articular dimensions in Prosimii and Platyrrhini (Primates) with applications for reconstruction of body mass and locomotor behavior in Adapidae (Primates; Eocene). Unpublished Ph.D. thesis, Johns Hopkins University.

Scott, W. B. (1928). Mammalia of the Santa Cruz beds. Primates. In *Reports of the Princeton University Expeditions to Patagonia, 1896–1899*. Vol. 6, *Paleontology III*, Part 5, Princeton: Princeton University Press, pp. 342–351.

Sears, K. E., Finarelli, J. A., Flynn, J. J. and Wyss, A. (2008). Estimating body mass in New World "monkeys" (Platyrrhini, Primates), with a consideration of the Miocene platyrrhine, *Chilecebus carrascoensis. American Museum Novitates*, 1–29.

Simons, E. L., Seiffert, E. R., Ryan, T. M. and Attia, Y. (2007). A remarkable female cranium of the early Oligocene anthropoid *Aegyptopithecus zeuxis* (Catarrhini, Propliopithecidae). *Proceedings of the National Academy of Sciences USA*, **104**, 8731–8736.

Smith, R. J. (1993). Bias in equations used to estimate fossil primate body mass. *Journal of Human Evolution*, **25**, 31–41.

Smith, R. J. and Jungers, W. L. (1997). Body mass in comparative primatology. *Journal of Human Evolution*, **32**, 523–559.

Snowdon, P. (1991). A ratio estimator for bias correction in logarithmic regressions. *Canadian Journal of Forestry Research*, **21**, 720–724.

Spencer, M. A. (2003). Tooth-root form and function in platyrrhine seed-eaters. *American Journal of Physical Anthropology*, **122**, 325–335.

Spoor, F., Garland, T., Krovitz, G. *et al.* (2007). The primate semicircular canal system and locomotion. *Proceedings of the National Academy of Sciences USA*, **104**, 10808–10812.

Stephan, H., Frahm, H. D. and Baron, G. (1981). New and revised data on volumes of brain structures in insectivores and primates. *Folia Primatologica*, **35**, 1–29.

Strait, S. G. (1991). Dietary reconstruction in small-bodied fossil primates. Unpublished Ph.D. dissertation, State University of New York, Stony Brook.

Szalay, F. S., and Dagosto, M. (1980). Locomotor adaptations as reflected on the humerus of Paleogene Primates. *Folia Primatologica*, **34**, 1–45.

Tauber, A. (1991). *Homunculus patagonicus* Ameghino, 1891 (Primates, Ceboidea), Mioceno Temprano, de la costa Atlantica Austral, prov. de Santa Cruz, República Argentina. *Academia Nacional de Ciencias de Córdoba, Argentina*, **82**, 1–32.

Tejedor, M. F. (2005). New specimens of *Soriacebus adrianae* Fleagle, 1990, with comments on pitheciin primates from the Miocene of Patagonia. *Ameghiniana*, **41**, 249–251.

Tejedor, M. F. and Rosenberger, A. L. (2008). A neotype for *Homunculus patagonicus* Ameghino, 1891, and a new interpretation of the taxon. *PaleoAnthropology*, 67–82.

Tejedor, M. F., Tauber, A. A., Rosenberger, A. L., Swisher, C. C. III and Palacios, M. E. (2006). New primate genus from the Miocene of Argentina. *Proceedings of the National Academy of Sciences, USA*, **103**, 5437–5441.

Vizcaíno, S. F., Bargo, M. S., Kay, R. F. *et al.* (2010). A baseline paleoecological study for the Santa Cruz Formation (late-early Miocene) at the Atlantic Coast of Patagonia,

Argentina. *Palaeogeography, Palaeoclimatology, Palaeoecology*, **292**, 507–519.

Walls, G. L. (1942). *The Vertebrate Eye and its Adaptive Radiation*. New York: Hafner Publishing Company.

Williams, B. A. and Covert, H. H. (1994). New early Eocene anaptomorphine primate (Omomyidae) from the Washakie Basin, Wyoming, with comments on the phylogeny and paleobiology of anaptomorphines. *American Journal of Physical Anthropology*, **93**, 323–340.

Zachos, J., Pagani, M., Sloan, L., Thomas, E. and Billups, K. (2001). Trends, rhythms, and aberrations in global climate 65 Ma to present. *Science*, **292**, 686–693.

17 A review of the paleoenvironment and paleoecology of the Miocene Santa Cruz Formation

Richard F. Kay, Sergio F. Vizcaíno, and M. Susana Bargo

Abstract

The paleoenviroment and paleoecology of the Santa Cruz Formation (SCF) is summarized, combining the data from the chapters of this book and new examination of the community structure of the vertebrate fauna using modern analogs. Emphasis is placed on the SCF outcrops along the coastal Atlantic between about 50.3° and 51.6° S and their faunas (~17.9 to 16.2 Ma; Santacrucian SALMA). New data on the sedimentology, the ichnology, and the flora and fauna of the SCF is particularly strong for the lower parts of the SCF south of the Río Coyle (FL 1–7). FL 1–7 (~17.4 to 17.5 Ma) is analogous to a single modern fauna of limited geographic and temporal scope. As paleolatitude during Santacrucian times was the same as that of today, FL 1–7 was extratropical and had highly seasonal day lengths. The Andes had not risen to a sufficient altitude to block westerly winds and moisture from reaching the Atlantic coast. New dates for FL 1–7 indicate that the mid-Miocene global climatic cooling had not yet begun. Several taxa recovered at FL 1–7 or in nearby penecontemporaneous levels (e.g. palm trees, the frog *Calyptocephalella*, the lizard *Tupinambis*, the anteater *Protamandua*, and the primate *Homunculus*) strongly indicate that the climate of FL 1–7 was much warmer and wetter than today. The overall mammalian species richness and niche composition, expressed as percentages of arboreal or scansorial, frugivorous, and grazing, suggest that overall rainfall was in the range of 1000 to 1500 mm per annum. Occurrence of trees and forest-dwelling birds and mammals (porcupines, spiny rats, sloths, scansorial marsupials, and monkeys) supports this conclusion. The occurrence of calcareous root casts in paleosols indicates high seasonality in rainfall with cool wet winters and dry warm summers. Grasses were also present, and a number of vertebrate taxa (giant terrestrial birds, many notoungulates, glyptodonts, and armadillos) appear to have been adapted to open environments. Consideration of sedimentologic, ichnologic, floral, and faunal elements taken together suggests a landscape for FL 1–7 consisting of a mosaic of open temperate humid and semi-arid forests, with ponds in some areas and seasonal flooding in others, no doubt promoting the formation of marshlands with a mixture of grasses and forbes.

Resumen

En este capítulo se resume el paleoambiente y paleoecología de la Formación Santa Cruz (SCF), combinando los datos de los capítulos de este libro y un nuevo análisis de la estructura de la comunidad de la fauna de vertebrados utilizando análogos modernos. Se hace hincapié en los afloramientos de la SCF a lo largo de la costa atlántica entre ~ 50,3 ° y 51,6 ° S y sus faunas (~17,9 a 16,2 Ma; SALMA Santacrucense). Los nuevos datos sobre la sedimentología, icnología y la flora y la fauna son particularmente fuertes para las partes bajas de la SCF al sur del Río Coyle (FL 1–7). FL 1–7 (~17,4 – 17,5 Ma) es análoga a una fauna moderna de limitada distribución geográfica y temporal. Puesto que la paleolatitud en tiempos Santacrucenses fue la misma que la de hoy, FL 1–7 era extratropical y la duración del día muy estacional. Los Andes no se habían elevado lo suficiente para bloquear los vientos del oeste y evitar que la humedad llegue a la costa atlántica. Nuevos fechados para FL 1–7 indican que el enfriamiento climático global del Mioceno Medio aún no había comenzado. Varios taxones recuperados en FL 1–7 o en niveles cercanos penecontemporaneos (e.g., palmeras, la rana *Calyptocephalella*, el lagarto *Tupinambis*, el oso hormiguero *Protamandua* y el primate *Homunculus*) indican claramente que el clima de FL 1–7 era mucho más cálido y húmedo que en la actualidad. La riqueza de especies de mamíferos y la composición de nicho total, expresados como porcentajes de arborícolas o trepadores, frugívoros y pastadores sugieren que la precipitación total estuvo en el rango de 1000 a 1500 mm por año. La presencia de árboles y de aves y mamíferos que habitan en bosques (puercoespines, ratas espinosas, perezosos, marsupiales trepadores y monos) apoya esta conclusión. La aparición de moldes calcáreos de raíces en paleosuelos indica alta estacionalidad de las precipitaciones, con inviernos fríos húmedos y veranos secos y cálidos. Se registran también gramíneas y una serie de taxones de vertebrados (aves terrestres gigantes, muchos notoungulados, gliptodontes y armadillos) que estarían adaptados a ambientes abiertos. El examen de los elementos sedimentológicos, icnológicos, de la flora y la fauna, en conjunto, sugiere para FL 1–7 un paisaje consistente en un mosaico de bosques templado húmedos abiertos y semiáridos, con lagunas en algunas zonas e inundaciones temporales

Early Miocene Paleobiology in Patagonia: High-Latitude Paleocommunities of the Santa Cruz Formation, ed. Sergio F. Vizcaíno, Richard F. Kay and M. Susana Bargo. Published by Cambridge University Press. © Cambridge University Press 2012.

331

en otras, sin duda promoviendo la formación de marismas con una mezcla de herbáceas y gramíneas.

17.1 Introduction

The chapters of this book have focused on the continental flora and fauna of the Santa Cruz Formation (SCF) and its stratigraphic context. Our efforts over the past 10 years have been to recover more and relatively complete specimens of this extraordinary fauna from rocks of SCF cropping out along the Atlantic Ocean between about 50.3° and 51.6° S. Further excellent exposures were collected along the north shore of the estuary of the Río Gallegos, and north from the Río Coyle to Monte León (Fig. 17.1, and Vizcaíno *et al.*, Chapter 1: Fig. 1.1).

The purpose of this chapter is to bring together the various strands of evidence on the ecology of southern Patagonian during Santacrucian times. Using the information from different sources, we evaluate the paleocommunity and infer the environmental conditions that then existed. In short, what was the environment in Patagonia in the late Early Miocene and how was the paleocommunity structured?

We are not the first to have tried to do this, but previous efforts have been hampered in several ways:

1. Older collections lacked stratigraphic resolution. This was true even for those specimens with precisely known provenance because a refined chronologic framework was unavailable. Earlier works on the fauna had to fall back on generalizations about taxa described from the entire

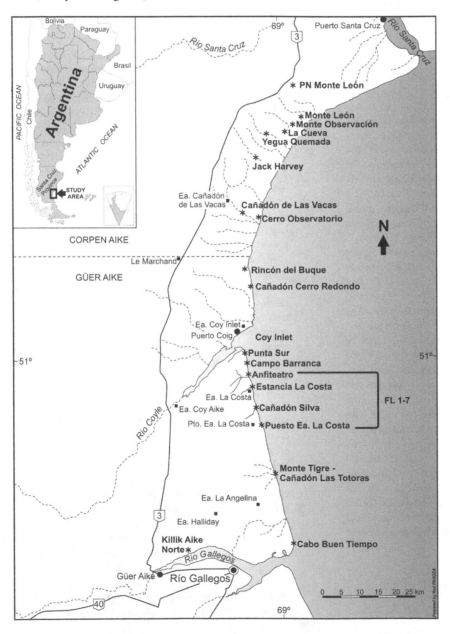

Fig. 17.1. Map of the study area in the coastal Santa Cruz Formation (Santa Cruz Province, Argentina) with the fossil localities (see Vizcaíno *et al.*, Chapter 1, Appendix 1.1 for further information).

Santa Cruz Formation. Coalescence of faunas from the formation as a whole into a single taxonomic list produces an over-count of the number of taxa living at any given time. On occasion, samples were combined that concatenate different times, environments, and climates, as well as possibly confusing ancestral-descendant taxa (i.e. anagenetic lineage) with alpha species richness (Croft, 2001). Some faunas at different localities were treated as temporally separate because the relative stratigraphic horizons were misunderstood (Tauber, 1997a, b). We now know that a number of stratigraphic levels south of the Río Coyle are actually laterally continuous with one another and penecontemporaneous (Perkins *et al.*, Chapter 2; Fleagle *et al.*, Chapter 3).

2. The precise geochronology was uncertain, so the tendency was to place the available data into broader global schemes or regional tectonic events. Much importance has been placed on the timing of the Mid-Miocene Climatic Optimum (MMCO) (Zachos *et al.*, 2001) and the possibility that faunal differences between the lower Estancia La Costa Member and upper Estancia La Angelina Member of the SCF either reflect the end of the MMCO in combination with a lowering of sea level between 16.5 and 15.5 Ma (Tauber, 1999), or relate to Andean uplift and the consequent formation of rain shadow aridity along the southern Patagonian coast (Pascual, 1984a, b).

3. Insufficient effort was made to use total evidence – some placed emphasis on certain elements of the fauna, others on peculiarities of the flora, with reliance on one or another "indicator" taxon to infer ancient climate or biotic community structure as a whole (see below).

4. While considerable effort has been expended to understand the paleobiology of some taxonomic groups (sloths, armadillos, ungulates), other groups have been understudied from an adaptive perspective (marsupials, rodents, primates).

5. Lacking precise information about the paleobiology of individual taxa, analysis of the overall niche structure of the mammalian community was likewise imprecise, a situation exacerbated by the limited comparative data on the niche structure of extant mammalian communities of South America.

In this volume, we are able to overcome several of the deficiencies mentioned above by bringing together new and revised information on the temporal interval of deposition, the temporal relationship among the various geographically separate faunal localities, and the relationship between similarly aged faunas and climatically significant regional and global events. The authors of the chapters have added considerably to our knowledge base, working to a great extent on the particular hypotheses listed in Vizcaíno *et al.* (2010).

The latter authors identified working hypotheses about the Santacrucian mammalian community to be tested using an ecomorphological approach, of the sort used by R. F. Kay and R. H. Madden on the La Venta fauna (Middle Miocene, Colombia; Kay and Madden, 1997a, b) based on detailed functional analysis in a well-defined phylogenetic framework. Vizcaíno *et al.* (2010) commented that the coexistence of so many kinds of herbivores implies fine partitioning of vegetation as a food source, as reflected in wide morphological variation in the masticatory apparatus. They inferred that there was strong competition for plant resources in environments ranging from arboreal to subterranean. They also reconstructed the niche space of the main lineages of Santacrucian herbivores and carnivores that are the subjects of other chapters in this volume. Finally, they compared fossil predator/prey diversity and the distribution of body mass values with local (i.e. a community restricted to one locality or a restricted area) recent faunas, to explore the predator/prey relationship and body mass pattern.

In this chapter we emphasize a distinction, often underappreciated, between two sorts of information. First, we endeavor to reconstruct paleoecology, as understood to mean the use of abiotic and biotic evidence to reconstruct environmental parameters of the SCF, such as mean annual rainfall and temperature. Second, we want to understand aspects of the paleoecology of individual species based upon their adaptive characteristics such as body size, diet, locomotion, and substrate preference (paleoautecology), and to build from this the overall niche characteristics of the SCF fauna as a whole (paleosynecology) (Gastaldo *et al.*, 1996). In this chapter, we summarize information from the previous chapters of this volume about the geologic and biotic evidence for Patagonian climate in the late Early Miocene. We also summarize the biotic data pertinent to reconstructing the mammalian community structure in one tightly constrained temporal interval in the SCF. To further accomplish these goals, we have gathered new composite data on the niche structure of extant mammalian faunas living in a variety of environments. The data is used to discover underlying similarities with, and differences between, the extant faunas and our late Early Miocene biotic community using information about individual species and synthetic techniques including cenogram analysis (Legendre, 1986; Croft, 2001) and niche structure analysis, e.g. total species richness, predator/prey ratios, percentages of arboreal species, etc. (Kay and Madden, 1997a, b; Croft, 2001).

17.2 Background

17.2.1 Physical evidence for paleoclimate and paleoenvironment

Geologic setting The Santa Cruz Formation (SCF) is a series of Early Miocene continental rocks deposited in the

Austral Geologic Basin of southern Patagonia. Its geographic limits extend from Río Sehuén and Río Chico in the north southward to the Straits of Magellan (Vizcaíno *et al.*, Chapter 1: Fig. 1.1). The formation stretches from the Atlantic coast westward to the foothills of the Andes (Malumián, 1999). Among the more complete sections of the formation are those in the east at Monte León (~50.25° S, 69° W) and in the west near Lago Posadas and Lago Pueyrredón (~48° S, 72° W).

Near the Atlantic coast at Monte León, the SCF rests on the shallow marine sediments of the Monte León Formation. An ^{40}Ar/^{39}Ar date from near the top of the Monte León Formation is 19.33 Ma (Fleagle *et al.*, 1995). At Monte León, the SCF is ~200 m thick and spans the temporal interval between 17.9 Ma and 16.2 Ma, as extrapolated from a series of ^{40}Ar/^{39}Ar dates, paleomagnetics, and sedimentation rates (see Fig. 2.2 in Perkins *et al.*, Chapter 2). Gravel deposits called "Rodados Patagonicos" unconformably overlie most coastal Atlantic exposures of the SCF.

In the western Patagonian steppe, south of Lago Posadas and Lago Pueyrredón, the SCF rests on nearshore marine rocks of the Centinela Formation. Here, the SCF sediment thickness reaches nearly 500 m. The base of the SCF at Lago Posadas may be older than about 22.4 Ma (Blisniuk *et al.*, 2005, 2006), an age that is further constrained with a date of 25.0 Ma (^{87}Sr/^{86}Sr) reported by Parras *et al.* (2008) from the underlying Centinela Formation south of El Calafate. Perkins *et al.* (Chapter 2) have a different interpretation about the age of the base of the Lago Posadas SCF, based on different assumptions about sedimentary rates. The SCF extends upward to about 14.2 Ma; a minimum age of 12.1 Ma is established by an overlying plateau basalt (Blisniuk *et al.*, 2005).

The Pinturas Formation (PF) is found northward in Santa Cruz Province along the piedmont of the Andes to about 46° S. The PF is perhaps 18.5 Ma at its base (Perkins *et al.*, Chapter 2) making it mainly older than coastal exposures of the SCF, but temporally equivalent to the lower part of SCF at Lago Posadas. Isolated exposures of rocks of similar age (18.7 to 19.7 Ma) and containing a similar fauna are found as far north as 42° S at Gran Barranca, Chubut Province, in the San Jorge Geologic Basin (Ré *et al.*, 2010).

Paleolatitude Despite considerable westward movement of the South American plate, the latitude of Patagonia has scarcely changed since the Miocene. On the other hand, substantial uplift of the southern Andes has occurred since the Oligocene. This was a consequence of the subduction of the Nazca and Antarctic plates beneath the South American Plate in the Peru–Chile Trench. Beginning at 28–26 Ma, there was an increase in the spreading rates and a change of the geometry of plate convergence. Rapid acceleration of

the uplift occurred when the Chile Rise (where spreading is occurring between the Nazca and Antarctic plates) came into contact with the Peru–Chile Trench around 14 Ma at 54–55° S, and migrated rapidly northward by 8° of latitude within a few million years (Blisniuk *et al.*, 2006). The deformation and surface uplift in the southern Patagonian Andes on the west side of the continent produced a thick clastic wedge to the east in the Austral Basin beginning in the Late Oligocene and continuing to the present. Representing the Late Oligocene and Early Miocene are nearshore marine rocks of the Centinela Formation and Monte León Formation (Guerstein *et al.*, 2004). These marine beds are paraconformably overlain by Early Miocene fluvial deposits of the PF and SCF.

Regional climate The events just described are thought to have had a profound effect on regional climate. Because of the latitudinal stability of Patagonia, the prevailing westerly winds that bring moisture onto the continent were similar in the Miocene to those of today. But the formation of a high mountain barrier had a strong effect on the east–west distribution of rainfall, producing a severe orographic rain shadow. In a simulation of the relative importance of factors that can affect rainfall in the southern Andes, Lenters and Cook (1995) observed that with sea surface temperatures held constant the preponderant factor influencing rainfall distribution in southern Patagonia is orographic.

Blisniuk *et al.* (2005) examined stable isotopes of oxygen and carbon from Miocene paleosols in a stratigraphic profile of the SCF south of Lago Posadas on the eastern flank of the Andes at ~47.4° S. Pedogenic δ^{18}O was used as a signal for the altitude of the Andes at the time of deposition. Topography has a differential effect on the isotopes of oxygen in precipitation. As a topographic barrier lifts an air mass, the air cools and expands. This causes more rainfall to occur on the windward side of a mountain than on its lee side, a rain-shadow effect. Furthermore, the condensed precipitation falling on the windward side has a depleted component of the heavier isotopes of oxygen in the water, so there is a progressive decrease in δ^{18}O in precipitation on the lee side with increasing local elevation. This record of depletion is reflected in the soil carbonates in paleosols formed at the time of deposition. The 500-m Lago Posadas section of the SCF spans the temporal interval from ~22 to 14 Ma. Blisniuk *et al.* (2005) report a decrease in δ^{18}O values from the lower to the upper part of the section, most of which occurs after about 14.5 Ma (Blisniuk *et al.*, 2005). They interpret the data as indicating a surface uplift of 1.2 ± 0.5 km between ~17 and 14 Ma (but most appreciably after 14.5 Ma) with increased rainfall on the western side of the Andes and decreased rainfall on the eastern side.

Blisniuk *et al.* (2005) also examined the data from carbon isotopes in pedogenic carbonate. The carbon isotope value of pedogenic carbonate is largely controlled by the average carbon isotope composition of vegetation that grew on it. At the time interval in question, C3 plants dominated plant communities. Values of δ^{13}C in soil carbonates with C3 plants range from –22‰ to –8‰ and are typically between –12‰ and –10‰. Values close to –22‰ are formed under closed canopy forests (Cerling and Quade, 1993) whereas those close to –8‰ are associated with water stress and perhaps aridity (Farquhar *et al.*, 1989). At Lago Posadas, mean levels of δ^{13}C are –9.8‰ in the lowest 100 m of the section, consistent with a pure C3 ecosystem but one not likely to have been associated with a closed canopy. Between 100 m and 200 m, δ^{13}C values drop to –11.8‰, suggesting increased precipitation in the eastern foreland during this time interval. Higher levels of δ^{13}C (–8.1‰) suggest greater water stress (i.e. more aridity) on C3 plants, presumably a rain-shadow effect by 14.5 Ma (Blisniuk *et al.*, 2005).

Global climatic overprint So far we have discussed data suggesting a strong regional tectonic effect on climate change in the SCF with evidence for a rain shadow produced by Andean uplift especially beginning around 14.5 Ma. The timing of Andean orographic aridity in southern Patagonia corresponds to other global phenomena that would have had local effects. According to Zachos *et al.* (2001), Antarctic continental ice-sheets first appeared in the earliest Oligocene and persisted until about 26 to 27 Ma, when a warming trend reduced the extent of Antarctic ice. From this point until the Middle Miocene (~15 Ma), global ice volume remained low, and bottom water temperatures trended slightly higher. The warm climatic phase peaked in the MMCO (17 to 15 Ma), and was followed by a gradual cooling and re-establishment of a major ice-sheet on Antarctica. Thus the timing of worldwide cooling corresponds roughly to the timing of the decreased δ^{18}O in Lago Posadas soil carbonates. Following a 2 Ma interval of equable climate, at about 14–15 Ma, southern Patagonian climate experienced global cooling and regional aridity effects.

17.2.2 Biotic evidence for paleoclimate and paleoenvironment

Bryan Patterson and Rosendo Pascual (1968, 1972) probably attempted the first systematic review of the evolution of the Cenozoic vertebrate faunas of South America, with attention to its environmental significance. In their commentary about the Santacrucian fauna they stated:

> The mammals of southern Patagonia, through the Miocene, suggest a climate sufficiently genial to permit such now mainly tropical animals as porcupines, echimyids, dasyproctids, anteaters and primates to

flourish there. The molluscs of the Patagonian Miocene are in accord on this point. The environment suggested by the mammalian faunas throughout much of this stretch of time is a woodland and savanna one that graded northward into the rain forest, woodland, and savanna of the tropical zone, then no doubt more extensive than at present. Patterson and Pascual (1972: 251).

In 1986, Pascual gathered the then flourishing new community of Cenozoic vertebrate paleontologists from La Plata and Buenos Aires in a symposium on the evolution of Cenozoic vertebrates of South America. In the concluding contribution to that symposium (Pascual, 1986) commented that during Ortiz-Jaureguizar's (1986) "Patagonian Faunistic Cycle" (including the Deseadan, Colhuehuapian, and Santacrucian Ages), the presence at the most extreme southern latitudes of a variety of browsers and/or omnivorous forms, and among them arboreal forms like monkeys or suspected arboreal ones like sloths, demonstrated the vast southern extension of favorable environmental conditions, typical of parkland-savanna, with an optimal balance between grasslands and forests.

In a landmark contribution, Pascual and Ortiz-Jaureguizar (1990) examined faunal change among South American Cenozoic mammals based on the percentages of herbivorous species with different cheek-tooth crown heights. As chronologic units they used South American Land Mammal Ages (SALMA); they included the Deseadan, Colhuehuapian, and Santacrucian mammal faunas collectively as the "Patagonian Faunistic Cycle" (as named earlier by Ortiz-Jaureguizar, 1986), and recognized two subcycles within it – Deseadan and Pansantacrucian. The latter encompasses the Colhuehuapian and the Santacrucian SALMAs. From low-crowned and rooted to high-crowned, rootless, and ever-growing cheek teeth, they recognized four categories: brachydont, mesodont, protohypsodont, and euhypsodont. According to their analysis, by the beginning of the Patagonian Faunistic Cycle (Late Oligocene) many families of mammals had evolved protohypsodont and euhypsodont cheek teeth, a phenomenon they attributed to the increase in the number of grazing species coevolving with the spread of grasslands at middle Patagonian latitudes. Further changes noted between the Deseadan and Pansantacrucian subcycles included a decrease in the percentage of brachydont genera, an increase in mesodont genera, and a slight decline in protohypsodont forms, but stasis in the euhypsodont taxa. They attributed these changes to a shifting balance of grassland and woodland habitats provided by a "park savanna."

More recently, Ortiz-Jaureguizar and Cladera (2006) summarized the paleoenvironmental changes of southern South America (i.e. south of 15° S) through the Cenozoic, emphasizing the relationships between biomes and the geological forces that, through different climatic-environmental factors, have driven its evolution. Their review of faunal evidence from the SCF suggests that prevailing climatic conditions in Patagonia in the mid-Miocene were warmer and more humid than those of today and that these conditions began to shift towards cooler and more arid conditions in the Middle Miocene.

Vizcaíno *et al.* (2010) observed that such analyses, while broadly useful, lack the kind of stratigraphic and chronologic precision so valuable for an understanding of ecological conditions at a single place and within a narrow range of time. In this latter sense, Adán Tauber's work is significant. In the late 1990s and based on his own stratigraphic work, fossil collections, and taxonomic identifications, Tauber (1997a, b, 1999) recognized two members (lower Estancia La Costa and upper Estancia La Angelina) in the coastal SCF between the Río Coyle and the Río Gallegos, and 22 fossiliferous levels (FL) (see Matheos and Raigemborn, Chapter 4, for an overview). In the biostratigraphic approach (Tauber, 1997a) provisionally proposed two biozones within lower Estancia La Costa Member, the *Protypotherium attenuatum* zone (lower, FL 3–7) and the *Protypotherium australe* zone (upper, FL 8–10), noting that for a formal definition of these biozones it would be necessary to confirm its regional applicability with a more complete paleontological record. Tauber and others (Krapovickas *et al.*, 2008) extended the *P. attenuatum* biozone to encompass FL 1 to 7 and the *P. australe* from FL 5.3 to 10, again with the caveat that there was a need to establish the geographic distribution of the *Protypotherium* spp. more certainly to certify their value as characteristic fossils of the biozones. Cassini *et al.* (Chapter 14) suggest that there is insufficient documented morphologic difference to justify the distinction between these two species. Because of these uncertainties, and because we are not concerned in this chapter with possible biotic shifts between the two members of the formation, we will not consider these biozones in this chapter.

Tauber (1997b, 1999) followed Pascual and Ortiz-Jaureguizar (1990) in the observation that a peculiarity of the Santa Cruz Formation is the contrasting presence of indicators of different climatic and environmental conditions. On the one hand, he interpreted elements of the fauna, such as the occurrence of primates, and echimyid and erethizontid rodents, as indicators of warm and humid conditions and forests; on the other hand, the presence of gypsum crystals, mud cracks, and other sedimentological features suggested to him the presence of open environments in relatively drier conditions. Based on the presence or absence of faunal elements in the different levels and inferring habits on the basis of phylogenetic affinities with living taxa, Tauber (1997b, 1999) concluded that the climate had become less humid and more "open" (that is, with fewer trees), from the lower to the upper levels of the Estancia La Costa Member, including also the lower levels of the upper Estancia La Angelina Member. Tauber (1999) correlated his

proposed climate shift with some regional or global "causes," such as the end of the MMCO in combination with a lowering of sea level between 16.5 and 15.5 Ma, although not with another possible climate modifier, the Andean uplift occurring just to the west in Patagonia.

Tauber also provided a discussion on the value of the armadillos as indicators of climatic conditions. Beyond the fact that Tauber (1999) questioned previous interpretations by Vizcaíno (1994) based on Tauber's erroneous review of the literature (Vizcaíno, 2001), he accepted that *Stegotherium tessellatum* Ameghino, 1887 would have tolerated well-defined seasonal changes. He also mentioned that *Proeutatus lagena* Ameghino, 1887 and *Proeutatus* cf. *deleo*, recorded in the lowest levels of the Santa Cruz Formation in the area of study, were slightly larger than *Proeutatus oenophorus* Ameghino, 1887 from the upper levels, suggesting less favorable conditions for these upper levels.

Taking a different approach, Croft (2001) used cenogram analysis (a plot of vertebrate body sizes within a community; Valverde, 1964) to interpret paleoenvironmental conditions for some of the best-known South American fossil mammal assemblages from the Eocene to the Pleistocene. These were then compared with more traditional interpretations (based on herbivore craniodental and postcranial adaptations) to judge congruence between the different methods of paleoenvironmental reconstruction. Croft (2001) included two faunas from the Early Miocene Santacrucian Age: Tauber's (1997a) *Protypotherium attenuatum* and *Protypotherium australe* biozones. He found them to be very similar, suggesting that no environmental characteristics distinguish them. The Santacrucian habitat is usually interpreted as "mixed" because of the presence of at least one arboreal primate (*Homunculus* Ameghino, 1891) in addition to many (supposed) savanna-adapted mammals, but Croft's cenogram statistics suggested that the region was wetter and less open than previously thought (e.g. by Pascual and Ortiz-Jaureguizar, 1990). The lack of differentiation between the two faunas is notable because it contrasts with the suggestion of Tauber (1997a, 1999) that the climate had become less humid and more "open" from the lower to the upper levels of the Estancia La Costa Member of the SCF. If anything, the analysis of Croft (2001) suggests slightly greater rainfall in the upper (*Protypotherium australe* Ameghino, 1887) zone, as indicated by the slope in his cenogram plot for the medium-sized mammals.

Vizcaíno *et al.* (2006) used armadillos as a proxy to estimate environmental conditions for the coastal SCF. Given the relatively limited geographic range of the Santacrucian specimens that provide the basis for their estimate of species richness, it would appear that the list of armadillos from the Santacrucian would more closely approximate alpha (local) rather than gamma (regional) richness.

For example, an analysis of the distribution of the living armadillos (Wetzel, 1985) reflects that the highest gamma richness today occurs in a more or less restricted region east of the Andes in central South America, between 12° and 32° S latitude. This area coincides with the Chaqueña Province of the biogeographic division of the Neotropical Region proposed by Cabrera and Willink (1980). Depending on the accuracy of the distribution maps, marginal areas of the Amazonia, Yungas, Cerrado, Paranense, Puneña, Monte, and Espinal Provinces are also involved. The first four belong to the Amazonic Domain of the proposed division, characterized by a dense vegetational cover, abundant and diverse flora and fauna that developed in a warm and humid climate, and limited seasonality in rainfall. The last three provinces, together with the Chaqueña Province, belong to the Chaqueño Domain, with predominant xerophytic vegetation in a continental climate with low to moderate rainfall, cool winters, and hot summers. In the Chaqueña Province, seven genera and 11 species of armadillos belonging to the five tribes recognized by Wetzel (1985) are recorded.

The taxa represent the complete range of size and digging capacities of living armadillos, including the ambulatory tolypeutines, the digging euphractines, dasypodines, and priodontines, and the subterranean chlamyphorines. If the metabolic requirements and other related biologic parameters such as population density were comparable to those of the living faunas, the diversity recorded is consistent with the environmental interpretation of Tauber (1997b, 1999) of open vegetation in relatively dry conditions with marked seasonality for the upper levels of the Estancia La Costa Member and lower levels of the Estancia La Angelina Member.

Knowledge of the evolution of the Early Miocene paleoflora of Patagonia (Barreda and Palazzesi, 2007), up to the time of this volume, is consistent with the coexistence of closed and open environments. Forests would have persisted as riparian or gallery forests with decreasing species richness across extra-Andean Patagonia until about the Middle Miocene; in the late Early Miocene drier conditions would have prevailed in lowland areas, and the contraction of humid elements coincides with the expansion of xerophytic taxa. In the time of deposition of the SCF, Patagonia must have represented two units (Barreda *et al.*, 2007): the *Nothofagidites* unit in the southwest was dominated by elements of the austral forests (e.g. Nothofagaceae, Podocarpaceae, Araucariaceae, Misodendraceae, Menianthaceae, Rosaceae, Cunoniaceae); and the Transitional, in central and south-eastern Argentina, defined by a mixed of Neotropical and Austral components.

17.2.3 Paleoecologic hypotheses

Croft's predator–prey hypothesis Croft (2006) proposed the predator–prey diversity hypothesis stating that predator

species richness causally limits prey richness, and that the effect is more pronounced in open habitat faunas than in closed ones. Support for the hypothesis consists of three lines of evidence. (1) The diversity of mammalian predators correlates significantly with the diversity of medium-sized mammals (500–8000 g, as originally defined by Legendre, 1986); (2) there are fewer prey species in open environments than in closed environments; (3) the relationship between the numbers of predators and prey is causal because predator diversity is not correlated with other habitat variables such as total annual rainfall or vegetation structure.

Croft extrapolated his findings to the extinct faunas of South America in general and to the late Early Miocene Santa Cruz faunas in particular. Before the Great American Biotic Interchange (GABI), South America's guild of mammalian secondary consumers did not include the Carnivora, the mammalian clade that dominates that guild in South America today. Croft noted that the number of mammalian predator species evidently was lower than expected in faunas that pre-date GABI. He proposed that pre-GABI faunas could have had greater-than-expected numbers of medium-sized mammals as a result of this apparently low diversity of mammalian predators, and that this effect would have been more pronounced in open habitats than in closed ones. Consequently, with respect to the SCF faunas, the large number of medium-sized primary consumers does not necessarily indicate that there was a high level of rainfall; rather it is a consequence of there being fewer predators than today. Cenogram analysis, he noted, may lead to the erroneous conclusion that the habitat was more "closed" and had more rainfall than it in fact had. To our knowledge, no one has determined whether the predator to prey ratios of extant faunas vary as a function of rainfall or some other environmental factor, so the question remains as to whether the predator to prey ratios of mammals in pre-GABI faunas do indeed agree with the expectations of the paleoenviroment. One objective of our analysis is to determine if this is so.

Fariña's paleocommunity model In a now-classic study, Fariña (1996) analyzed mammalian paleocommunity structure using a general ecological relationship between population density and body size (Damuth, 1981, 1987, 1992). Fariña assessed the abundance of herbivorous and carnivorous mammals larger than 10 kg listed for the South American Late Pleistocene fauna from Luján (Buenos Aires Province, Argentina) (Tonni *et al.*, 1985). Vizcaíno *et al.* (2010) used Fariña's method to examine the paleoecological characteristics at two important localities in the SCF in evaluating one aspect of the study of Croft (2001, 2006): his hypothesis that there is a depauperate carnivore paleoguild for some Santacrucian faunas. The Fariña model identifies an apparent imbalance at several well-sampled

SCF localities wherein the secondary productivity of the ecosystem far exceeds the energetic requirements of the carnivores present. Indeed, their analysis suggests that secondary productivity is at least three times that of the carnivore to herbivore ratios seen in modern faunas. In this chapter, we address some of the aspects of Fariña's model, as applied by Vizcaíno *et al.* (2010), but we do so incompletely, because we have not yet completed work on a database to examine its validity for extant continental faunas. Further synthetic work is being conducted by us and by Mariana Di Giacomo (personal communication, 2011).

Niche metrics and niche space Kay and Madden (1997a, b) examined how the mammal record of the Middle Miocene of La Venta, Colombia, compared with those of the living communities in terms of niche structure, and assessed what aspects of niche structure can be used to infer the paleoclimate and paleoecology. This approach seeks to identify non-taxonomic compositional similarities based on niche structure caused by underlying biotic adaptation. Among other features, Kay and Madden examined the relationship between total species richness and the macroniche composition of diet, locomotor, and body-size classes among non-volant mammalian faunas in tropical South America. They applied this model to reconstruct the paleoclimate of the Colombian assemblage. Here, we refine and extend these tools to examine the paleosynecology of the Santacrucian fauna based upon niche-structure comparisons including total species richness, the proportions of species that are primary versus secondary consumers, the size distributions of herbivorous species, and the percentages of mammalian species that are frugivorous, grazing, and arboreal.

17.3   Methods: a localized approach

The focus in this chapter is the especially rich faunas of the coastal Atlantic localities Anfiteatro, Estancia La Costa, Cañadón Silva, and Puesto Estancia La Costa covering a distance of ~15 km north to south (Fig. 17.1). In his geological survey of this part of the coast, Tauber (1994) established several fossil levels (FL) 1–7 in this region (see Matheos and Raigemborn, Chapter 4, Figs. 4.3 and 4.4). Based on a concept that the SCF beds dip gently to the southeast (Ameghino, 1906), Tauber concluded that these are successively younger levels from north to south. However, new evidence has emerged based on the tephrostratigraphic correlations reported by Perkins *et al.* (Chapter 2: Fig. 2.2), who have established that up to three of the same levels crop out at all these localities and are very close in age. On this basis the composite fauna of the above-mentioned localities is treated as a single paleofauna, which we call FL 1–7.

In this volume we have gathered a much broader array of data – encompassing studies of the ichnology, sedimentology, paleobotany, and the invertebrate and vertebrate record. These data have been placed within a more refined stratigraphic framework (Matheos and Raigemborn, Chapter 4, as modified by Perkins *et al.*, Chapter 2 and Fleagle *et al.*, Chapter 3). We seek to identify areas of agreement and conflict among the paleoenvironment (including climate) and paleobiologic inferences of various authors, and to identify areas where more research is needed. Also, our approach is to concentrate on a narrow stratigraphic interval rather than a longer temporal interval. The faunas comprising the Santacrucian Land Mammal Age span 1.7 Ma or more of Earth history during which time important regional and global climatic changes were occurring, so treating them as one paleocommunity for the purposes of a paleoecological study may include, as one fauna, species that lived in different places during a long temporal interval.

For heuristic purposes, we identify two separate analytical approaches to the reconstruction of paleoenvironment and paleoecology at FL 1–7. These could be called *specific* and *synthetic*.

With specific approaches, we examine certain abiotic and biotic indicators, be they ichnofossils or biotic taxa or groups of taxa with similar habits, comparing their abundance of distribution today with what it was in Santacrucian times. For example, what are the possible ecological "causes" for the presence of myrmecophagus mammals (be they anteaters or armadillos) in the Santacrucian when they do not occur in Patagonia today? Why are there no crocodiles in the Santacrucian? What is the implication of the presence of palm fossils so far south?

We also undertake several synthetic approaches. We reconstruct a niche structure for FL 1–7 by identifying the number of species present, the body size, locomotion, and diet of the species. We have assembled similar kinds of data for the extant fauna of South America, choosing extant communities that inhabit regions of the continent from the equator to the high latitudes, and sampling environments that vary in rainfall, temperature, and vegetation composition. At the same time we do not ignore important overall differences between Santacrucian and modern faunas that might be informative for understanding how biotic conditions compare between the two. Synthetic approaches that investigate paleoenvironment and/or paleosynecology described below include niche structure analysis, cenogram analysis, and Croft's predator–prey hypothesis.

For these three analyses, the overall structure of the non-volant mammalian fauna from four localities (Anfiteatro, Estancia La Costa, Cañadón Silva, and Puesto Estancia La Costa) that include FL 1–7 of the Santa Cruz Formation is presented in Table 17.1. We compare this fauna with that of 25 modern mammalian faunas from South America listed in

Table 17.1. *The fauna of Fossiliferous Levels 1–7 (FL 1–7), Santa Cruz Formation*

Taxon		ANF	ELC	CS	PLC	FL 1–7	Body mass (kg)	Mass category	Diet category	Locomotor or substrate preference
Incertae sedis	*Necrolestes patagonensis*				1	1	<0.10	I	I(F)	T(F)
Hathliacynidae	*Cladosictis patagonica*	1	1	1	1	1	6.6	III	V	A(T)
	Sipalocyon gracilis		1		1	1	2.11	III	V	A(T)
	Perathereutes pungens				1	1	1	II	V	A(T)
Borhyaenoidea	*Prothylacynus patagonicus*				1	1	31.79	IV	V	A(T)
	Lycopsis torresi		1		1	1	20.07	IV	V	A(T)
Borhyaenidae	*Arctodictis munizi*				1	1	50	IV	V	T(A)
	Borhyaena tuberata	1			1	1	36.4	IV	V	T(C)
	Acrocyon sectorius		1			1	11.49	IV	V	A(T)
Palaeothentiidae	*Palaeothentes minutus*				1	1	0.082	I	I(F)	T(A)
	Palaeothentes lemoinei				1	1	0.425	II	I(F)	T(A)
	Acdestis owenii	1			1	1	0.344	II	F(I)	?
Microbiotheriidae	*Microbiotherium tehuelchum*				1	1	0.061	I	I(F)	A
Peltephilidae	*Peltephilus pumilus*	1	1		1	1	11	IV	S(Tu)	T(F)
Dasypodidae	*Proeutatus oenophorus*	1	1		1	1	15	IV	S(L)	T(F)
	Stenotatus patagonicus				1	1	4	III	S(I)	T(F)
	Prozaedyus proximus	1		1	1	1	1	II	S(I)	T(F)
Propalaehoplophoridae	*Cochlops muricatus*	1			1	1	82.99	IV	L	T(A)
	Propalaehoplophorus australis				1	1	81.64	IV	L	T(A)
Myrmecophagidae	*Protamandua rothi*		1	1		1	4.00–5.00	III	MYR	A(T)
Megatherioidea	*Hapalops* sp. 1		1	1		1	46.29	IV	L	A(T)
	Hapalops sp. 2		1	1		1	27.71	IV	L	A(T)
	Pelecyodon cristatus		1	1		1	50	IV	L	A(T)
	Hyperleptus sp.				1	1	~40	IV	L	A(T)
Megalonychidae	*Eucholoeops fronto*		1		1	1	78	IV	L	A(T)
	E. ingens				1	1	76.88	IV	L	A(T)
Mylodontidae	*Nematherium* sp.		1		1	1	95.02	IV	G	A(T)
Astrapotheriidae	*Astrapotherium magnum*	1	?		1	1	921.32	VI	L/G	T(A)
Intheratheriidae	*Interatherium robustum*	1	1		1	1	2.38	III	G	T(C)
	Protypotherium australe		1		1	1	7.73	III	G	T(C)
Hegetotheriidae	*Hegetotherium mirabile*				1	1	7.71	III	G	T(C)
Homalodotheriidae	*Homalodotherium cunninghami*				1	1	405.08	V	L	T(A)
Toxodontidae	*Adinotherium ovinum*	1	1		1	1	100.29	V	G	T(A)
	Nesodon imbricatus		1		1	1	637.51	VI	G/L	T(A)
Proterotheriidae	*Diadiaphorus majusculus*				1	1	82.05	IV	L	T(C)
	Tetramerorhinus mixtum				1	1	35.06	IV	L	T(C)
	Thoatherium minusculum				1	1	24.2	IV	L	T(C)
Macraucheniidae	*Theosodon gracilis*				1	1	121.55	V	L	T(C)
	Theosodon lydekkeri				1	1	130.93	V	L	T(C)
Neoepiblemidae	*Perimys* spp.				1	1	0.32	I	?	T(C)

Table 17.1. (*cont.*)

Taxon		ANF	ELC	CS	PLC	FL 1–7	Body mass (kg)	Mass category	Diet category	Locomotor or substrate preference
Dasyproctidae	*Neoreomys australis*	1			1	1	7.12	III	F(L)	T(C)
Eocardiidae	*Eocardia montana*				1	1	1.4	III	G	T(C)
	Eocardia fissa				1	1	2.84	III	G	T(C)
Echimyidae	*Spaniomys modestus*				1	1	0.65	I	F(L)	A
Erethizontidae	*Steiromys duplicatus*				1	1	14.17	IV	L	A(T)
	Steiromys detentus				1	1	<10	III	L	A(T)
Acaremyidae	*Sciamys* sp.	1			1	1		I	F(I)	?
Platyrrhini	*Homunculus patagonicus*		1		1	1	2.7	II	F(L)	A
	Homunculus sp. nov.				1	1	4	II	F(L)	A
	Total species	11	18	6	43	49				

Notes: **Body mass categories**: **I**, 10 to 100 g; **II**, 100 g to 1 kg; **III**, 1 to 10 kg; **IV**, 10 to 100 kg; **V**, 100 to 500 kg; **VI**, > 500 kg. **Dietary categories**: **V**, vertebrate prey; **S(I)**, scavenging and insects; **S(Tu)**, scavenging and tubers; **S(L)**, scavenging and browse; **MYR**, termites and ants; **I(F)** insects and fruit or nectar; **F(I)**, fruit with invertebrates; **S**, small seeds of grasses and other plants; **F(L)**, fruit with leaves; **L**, leaves (= dicot leaves, buds shoots); **G**, grass stems and leaves (graze). **Locomotor or substrate preferences categories**: **A**, arboreal; **A(T)**, arboreal and terrestrial (scansorial); **T(A)**, terrestrial and ambulatory; **T(C)**, terrestrial and cursorial; **T(F)**, terrestrial and fossorial; **SAq**, semi-aquatic. ANF, Anfiteatro; ELC, Estancia La Costa; CS, Cañadón Silva; PLC, Puesto Estancia La Costa; FL, Fossiliferous Levels 1–7, taxon occurs in at least one of the localities.

Table 17.2.[1] For the taxonomic allocations of extant faunas, we follow Wilson and Reeder (2005). The sampling areas of these faunas represent a wide range of mean annual rainfall and the faunas range in latitude from the equator to ~45° S. At one extreme, in the eastern lowlands of Ecuador, rainfall is approximately 3500 mm per year with no appreciable dry season. At the other extreme are localities in Patagonian Argentina with rainfall <200 mm per annum, where it is dry the year around. Mean annual temperature (MAT) ranges from 7.5 to 28.2 °C. Because the Santa Cruz Formation's depositional environment indicates a lowland with little topographic relief (Matheos and Raigemborn, Chapter 4), and because faunal composition is affected by altitude (Ojeda and Mares, 1989), our sample does not include faunas living in areas above 1000 m in elevation.

For all statistical analyses we use JMP Pro 9.0.0 for the MacIntosh.

17.3.1 Niche characteristics of extant and Santa Cruz Formation faunas

Each extant species is assigned a body mass, locomotor behavior, and diet category from the literature on mammalian ecology. For body mass we recognize six categories: (I) 10 to 100 g, (II) 100 g to 1 kg, (III) 1 to 10 kg, (IV) 10 to

100 kg, (V) 100 to 500 kg, and (VI) > 500 kg. Six locomotor or substrate preferences are recognized, following Fleming (1973) and Andrews *et al.* (1979): A, arboreal; A (T), arboreal and terrestrial (scansorial); T(A), terrestrial and ambulatory; T(C), terrestrial and cursorial; T(F), terrestrial and fossorial; SAq, semi-aquatic. We use 11 diet categories: V, vertebrate prey; S(I), scavenging and insects; S (Tu), scavenging and tubers; S(L), scavenging and browse; MYR, termites and ants; I(F) insects and fruit or nectar; F (I), fruit with invertebrates; S, small seeds of grasses and other plants; F(L), fruit with leaves; L, leaves (= dicot leaves, buds, shoots not including grass, which has a very high silica content); G, grass stems and leaves (graze).

We use four indices devised by Kay and Madden (1997a, b) to express the number of species within a guild (that is, with a particular niche specialization or within a body size range) relative to total number of species.

1. **Frugivore index** expresses the proportion of frugivorous species to the total number of plant-eating species in a fauna:

$$100 * [(F(I) + S + F(L))/(F(I) + S + F(L) + L + G)]$$

2. **Browsing index** expresses the proportion of browsing or leaf-eating species to the total number of herbivorous plant-eating species in a fauna:

$$100 * [(L)/(L + G)]$$

[1] The full datasets upon which these analyses are based are available upon request to RFK.

Table 17.2. *Extant South American faunal communities*

Locality (fauna)	State / Province, Country	Latitude/ longitude	Average altitude or altitude range (m)	Annual rainfall (mm)	Vegetation (estimated length of dry season in months)	References
Guatopo	Miranda, Venezuela	10° N 66° W	250–1430	1500	Semideciduous, submontane to montane forest (6 months)	(Eisenberg *et al.*, 1979)
Masaguaral	Guarico, Venezuela	8° 34′ N 67° 35′ W	75	1250	Subtropical vegetational mosaic high savanna (6 months)	(Eisenberg *et al.*, 1979)
Puerto Páez	Apure, Venezuela	6° 23′ N 67° 29′ W	76	1500	Seasonally flooded high grass savanna with scattered patches of low forest and palms (6 months)	(Handley, 1976)
Puerto Ayacucho	Amazonas, Venezuela	5° 15′ N 67° 40′ W	99–195	2250	Savannas of the Rio Orinoco and evergreen forest/savanna mosaic	(Handley, 1976)
Esmeralda	Amazonas, Venezuela	3° 05′ N 65° 35′ W	130–1830	2000	Nearly continuous evergreen forest in valley up to low dense montane forest	(Handley, 1976)
Manaus	Amazonas, Brazil	2° 30′ S 60° W	10	2200	Primary upland terra firme forest; (3 months)	(Malcolm, 1990)
Belém	Pará, Brazil	1°2 7′ S 48° 29′ W	10	2600	Vicinity of Belém, now urban and suburban (2 months)	(Pine, 1973)
Caatingas	Exu, Pernambuco, Brazil	7° 31′ S 40° 00′ W	200	<500	Semi-arid caatinga (>7 months)	(Mares *et al.*, 1981, 1985; Streilein, 1982)
Federal District	Brasilia, Brazil	15° 57′ S 47° 54′ W	1100	1586	Seasonal xerophyllous savanna grasslands and gallery forests (5 months)	(Mares *et al.*, 1989a)
Acurizal	Mato Grosso, Brazil	17° 45′ S 57° 37′ W	100–900	1120	Pantanal; pastures, secondary forest, cerrado and deciduous forests (7 months)	(Schaller, 1983)
Chaco	Salta Province, Argentina	22° 24′ S 63° W	200–500	700	Subtropical, drought-resistant, thorn forest (9 months)	(Mares *et al.*, 1989b; Ojeda and Mares, 1989)
Transitional Forest	Salta Province, Argentina	22–24° S 64° W	350–500	700–900	Transitional deciduous forest with trees 20 to 30 m tall	(Mares *et al.*, 1989b; Ojeda and Mares, 1989)

Table 17.2. (*cont.*)

Locality (fauna)	State / Province, Country	Latitude/ longitude	Average altitude or altitude range (m)	Annual rainfall (mm)	Vegetation (estimated length of dry season in months)	References
Low Montane	Salta Province, Argentina	22–24° S 64° W	500–1500	800	Lower montane moist forest (2 months)	(Mares *et al.*, 1989b; Ojeda and Mares, 1989)
Cocha Cashu	Madre de Dios, Perú	12° S 70° W	400	2000	Lowland floodplain rainforest (3 months)	(Janson and Emmons, 1990)
Río Cenepa (Alto Marañón)	Amazonas, Perú	4° 47′ S 78° 17′ W	210	2880	Abandoned fields, secondary regrowth riparian forest, undisturbed humid forest	(Patton *et al.*, 1982)
Ecuador Tropical	Oriente, Ecuador	1° N–5° S 75°–78° W	800–1000	1795–4795	Amazonian lowland evergreen rainforests	(Albuja, 1991)
Río Teuco	Salta Province, Argentina	24° 12′ S, 62°40′ W	190–240	600–700	Chaco, xeric thorn scrub	(Mares *et al.*, 1989b)
SE Tucuman Chaco	Tucuman Province, Argentina	27° 52′ S 65° 35′ W	450–600	400–600	Thorn forest Chaco	(Barquez *et al.*, 1991)
Southern Tucuman	Tucuman Province, Argentina	27° 47′ S 65° 10′ W	320–370	600–800	Subtropical forest	(Barquez *et al.*, 1991)
Nahuel Huapi Humid forest	Río Negro Province, Argentina	41° S 71° 30′ W	~1200	1500–3500	Humid forest	(Grigera *et al.*, 1994)
Nahuel Huapi steppe	Río Negro Province, Argentina	41° S 71° 30′ W	~1000	600–1000	Steppe	(Grigera *et al.*, 1994)
NE Misiones Province	Misiones Province, Argentina	26° S 54° W	255–500	1600–2000	Rainforest	(Redford and Eisenberg, 1992)
Peninsula Valdés	Chubut Province	42°S 64°W	0–60	200–205	Monte: temperate tussock grass steppe/scrubland	(Nabte *et al.*, 2009)
Patagonian steppe	Chubut Province, Argentina	43° S 68° W	350–700	100–200	Monte: temperate tussock grass steppe/scrubland	(Redford and Eisenberg, 1992)
Pampas	Argentine pampas regional fauna	34° S 58° W	50	800–1000	Temperate grasslands	(Redford and Eisenberg, 1992)

Table 17.3. *Niche metrics of extant South American faunas and Santa Cruz Fossil Levels 1–7*

Locality	Rainfall	Rainfall group narrow	Number of primary consumers	Number of predators	Predator/ total prey ratio	Arboreality index	Browsing index	Frugivore index
Acurizal	1120	2	24	18	0.75	25.0	75.0	65.2
Belem	2600	3	35	27	0.77	36.7	81.8	69.4
Caatingas	400	0	9	14	1.56	21.7	0.0	60.0
Chaco	700	1	16	20	1.25	6.9	40.0	6.3
Chubut	150	0	19	16	0.84	5.7	40.0	21.1
Cocha Cashu	2000	3	50	21	0.42	48.6	80.0	80.0
Ecuador Tropical	3295	3	52	31	0.60	46.4	77.8	82.7
Esmeralda	2000	3	49	18	0.37	54.4	100.0	89.1
Federal District, Brazil	1586	2	38	28	0.74	19.9	78.6	58.8
Guatopo	1500	2	26	15	0.58	36.6	75.0	84.0
Manaus	2200	3	37	16	0.43	50.0	83.3	83.3
Masagural	1250	2	19	11	0.58	31.0	80.0	70.6
Misiones	2100	3	36	32	0.89	24.3	75.0	52.9
Nahuel Huapi humid forest	2500	3	10	12	1.20	6.8	55.6	10.0
Nahuel Huapi steppe	800	1	10	10	1.00	0.0	28.6	30.0
Pampas	900	1	16	16	1.00	7.8	57.1	17.6
Peninsula Valdés	202	0	10	12	1.20	9.1	25.0	20.0
Puerto Ayacucho	2250	3	32	13	0.41	45.6	100.0	82.8
Puerto Paiz	1500	2	17	6	0.35	29.2	83.3	57.1
Río Cenepa	2880	3	39	23	0.59	42.7	85.7	82.1
Río Teuco	650	1	15	15	1.00	8.3	41.7	20.0
Salta low montane	800	1	15	12	0.80	22.2	62.5	46.7
Salta transitional forest	800	1	20	26	1.30	17.4	50.0	40.0
Tucuman subtropical forest	650	1	9	16	1.78	14.0	50.0	55.6
Tucuman thorn forest	500	0	10	16	1.60	9.6	0.0	30.0
Santa Cruz FL 1–7			32	17	0.53	22.4	65.4	18.8

3. **Arboreality index** is used to express the proportion of arboreal species to the total number of non-volant species:

$$100 * [(A + 0.5A(T))/(A + A(T) + SAq + T)]$$

4. **Predator/prey ratio** expresses the proportion of secondary consumers:

[*Number of insectivores (including anteaters)* + *carnivores* + *scavengers*]/*number of herbivores*

Table 17.3 summarizes the data on species richness, predator and prey richness, and three indices and one ratio for the 25 extant faunas and the non-volant mammal species from FL 1–7 of the Santa Cruz Formation. FL 1–7 species are assigned diet, locomotor, and body size classes in Table 17.1 based on the results of studies in this volume (see Chapters 10 to 16 for details).

Each of the above descriptors of extant mammalian macroniche structure is examined using simple least-squares regression.

17.3.2 Cenogram analysis
Legendre (1986) devised an analytical procedure for reconstructing paleoenvironmental variables from body size

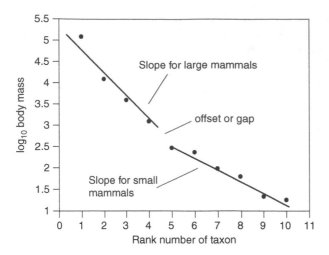

Fig. 17.2. Bivariate plot of $\log_{10}$ body mass versus species rank for primary consumers at Peninsula Valdes. The plot is called a cenogram. Three variables are thought to be correlated to environmental variables: slope for medium-sized mammals (>500 g; <250 kg), slope for small mammals (<500 g), and the size of the offset between the two distributions.

distributions among extant non-volant mammalian primary consumers. He plotted the logarithm of species body mass on the rank of the species, with the resultant plot being a cenogram, as illustrated in Fig. 17.2. These lines can be represented by regression equations of $\log_{10}$ body mass versus species rank. Legendre (1986) derived three empirical rules from the shape of the cenograms, subsequently refined and quantified (Gingerich, 1989; Gunnell, 1994; Croft, 2001).

1. The slope for the medium-sized mammals (those with mass between 500 g and 250 kg) is more pronounced in more arid environments.
2. The slope for the small mammals is related to temperature. The more pronounced the slope the colder the temperature.
3. The size of a gap or offset occurring between small- and medium-sized species (500 g cut-off) is related to how "open" the environment is: larger gaps are related to more open environments.

To carry out cenogram analyses of the extant and extinct faunas we calculated least-squares regression lines of $\log_{10}$ body mass versus species rank, from largest to smallest species. We determined the slope for the medium-sized mammals (those with mass between 500 g and 250 kg), the slope for the small mammals (<500 g), and the size of the gap or offset between small- and medium-sized species (500 g cut-off). Data summarizing the cenogram statistics for the extant faunas and FL 1–7 are presented in Table 17.4.

17.3.3 Predator versus prey

Our additional data on extant faunas mentioned above allow us to further evaluate the predator versus prey hypothesis. The newly revised faunal lists for the SCF further permit an examination of whether Santacrucian faunas conform to expectations of extant faunas or represent an outlier demanding explanation.

To evaluate Croft's (2006) hypothesis, we gathered data on the number of predators and the richness of medium-sized mammalian prey species (500–8000 g) for the extant faunas for which annual rainfall and vegetation structure are known. We further examine whether the Santa Cruz fauna departs significantly from that expected based on the extant communities.

17.3.4 Combined analyses

It is possible, even likely, that niche metrics, cenogram results, and predator/prey ratios may be highly intercorrelated and not independent estimators of rainfall. To examine this possibility, we used two multivariate techniques: principal components analysis (PCA) and discriminant function analysis (DFA). PCA does not assume the existence of groups *a priori* in the data, whereas DFA is a test of the ability of the data to discriminate among predetermined groups, in this case rainfall categories (Table 17.3). The PCA analysis also gives us a way of saying which, among all the extant faunas sampled, most closely resembles FL 1–7, and as such serves as a basis for assessing the overall synecology of FL 1–7.

17.4 Results

17.4.1 Specific approaches

Trace fossil evidence

Krapovickas (Chapter 6) reports a variety of ichnofossils from FL 1–7. Several notable features are of interest in our reconstruction of paleoenvironment and paleoecology. Krapovickas does not mention the presence of traces attributable to termite activity reported by Tauber (1994, 1996: 8) at Cañadón Silva. In any case, the absence of termites in the ichnofossil record does not exclude the presence of termites, since extant termites have a variety of nests. Some species have extensive underground nests and build aboveground mounds. Others build arboreal nests. The presence of a small myrmecophagus species (*Protamandua* Ameghino, 1904) suggests that arboreally nesting termites (at least) were present (see below). Interestingly, underground termite nests have been reported in the Pinturas Formation (Bown and Laza, 1990).

Although it does not fall within the range of a "biotic" indicator, we should mention in passing that Krapovickas (Chapter 6) reports the presence of pedogenic carbonate in

Table 17.4. *Variables for cenogram analysis*

Locality	Absolute latitude	Rainfall	Mean annual temperature	Primary consumers <500 g, semilog slope	Primary consumers 0.5 kg to 250 kg, semilog slope	log delta between +/- 500 g	Number of primary consumers <500 g	Number of primary consumers >500 g	Average log delta between all primary consumers	Vegetation classification
Acurizal	18	1120	22.5	-0.248	-0.129	0.245	7	17	-0.164	2
Belem	1	2600	28.2	-0.091	-0.082	0.385	15	20	-0.104	1
Caatingas	8	400	22.5	-0.222	-0.394	0.404	5	4	-0.340	4
Chaco	22	700	22.5	-0.239	-0.197	0.416	7	9	-0.230	4
Chubut	42	150	7.5	-0.092	-0.366	0.623	13	6	-0.213	6
Cocha Cashu	12	2000	22.5	-0.064	-0.063	0.065	20	30	-0.072	1
Ecuador	1	3295	25	-0.053	-0.067	0.111	24	28	-0.066	1
Esmeralda Tropical	3	2000	22.5	-0.0431	-0.065	0.342	24	25	-0.065	1
Federal District, Brazil	16	1586	20.5	-0.061	-0.142	0.26	24	14	-0.096	3
Guatopo	10	1500	22	-0.114	-0.107	0.276	12	14	-0.130	1
Manaus	3	2200	26.6	-0.095	-0.076	0.185	15	22	-0.093	1
Masagural	9	1250	26.2	-0.152	-0.164	0.356	7	12	-0.135	3
Misiones	26	2100	25	-0.112	-0.084	0.127	15	21	-0.109	2
Nahuel Huapi humid forest	41	2500	7.5	-0.208	-0.409	1.266	6	4	-0.391	1
Nahuel Huapi steppe	41	800	7.5	-0.115	-0.612	0.864	6	4	-0.396	5
Pampas	34	900	17.5	-0.187	-0.463	0.29	10	6	-0.270	5
Peninsula Valdés	42	202	7.5	-0.267	-0.65	0.623	6	4	-0.426	6
Puerto Ayacucho	5	2250	27	-0.075	-0.15	0.378	14	7	-0.104	3
Puerto Paiz	6	1500	20.8	-0.17	-0.288	0.451	9	8	-0.243	3
Río Cenepa	4.7	2880	26	-0.094	-0.083	0.172	14	25	-0.040	1
Río Teuco	24	650	22.5	-0.354	-0.184	0.416	5	10	-0.264	4

Table 17.4. (cont.)

Locality	Absolute latitude	Rainfall	Mean annual temperature	Primary consumers <500 g, semilog slope	Primary consumers 0.5 kg to 250 kg, semilog slope	log delta between +/– 500 g	Number of primary consumers <500 g	Number of primary consumers >500 g	Average log delta between all primary consumers	Vegetation classification
Salta low montane	23	800	22.5	−0.219	−0.205	0.416	6	9	−0.236	1
Salta transitional forest	23	800	22.5	−0.157	−0.114	0.416	8	12	−0.156	2
Tucuman subtropical forest	28	650	22.5	−0.203	−0.441	1.226	6	3	−0.420	2
Tucuman thorn forest	28	500	22.5	−0.274	−0.246	0.923	5	5	−0.339	4
Santa Cruz FL 1–7	51	–	–	−0.149	−0.083	0.25	9	23	−0.111	–

Vegetation classification: 1, Evergreen forest; 2, Cerrado and deciduous forest; 3, high grass savanna with forest patches; 4, semi-arid thorn forest; 5, temperate grassland, steppe; 6, temperate tussock grass steppe/scrubland.

the form of calcareous rhizoconcretions from all FL 1–7 localities, interpreted as indicating moderately well-drained soils developed under drier climatic conditions. Cerling and Quade (1993) observe that soil carbonates form in soils with a net water deficit, generally in soils where precipitation is less than 1000 mm per annum. This interpretation is at variance with our reconstruction of rainfall >1000 mm from the mammal data (see below). Indeed, soil carbonate formation is not always restricted to regions with less than 1000 mm per annum. Calcitic rhizoliths have been reported in soils of the Mississippi River floodplain, where annual rainfall is relatively high (1500 mm) (Farrell, 1987). Thus, the key to forming calcite in and around root channels appears to be episodic drying of the soil for a sufficiently long period (Kraus and Aslan, 1993). If precipitation were concentrated during the winter (colder) months, a summer dry season could stabilize paleosol carbonates, even if total precipitation was high. Interestingly, in Patagonia today, the rain comes in June and July when the temperature is the lowest. So, the low rainfall, dry austral summer, and high-rainfall wet austral winter could have lead to the formation of soil carbonate.

Plant fossils

Brea *et al.* (Chapter 7) reconstruct the vegetation of the Estancia La Costa Member of the Santa Cruz Formation. These authors describe the vegetation of the ELC Member as ranging from an open temperate semi-arid forest to a temperate, warm humid forest (temperate referring to markedly seasonal climates, with seasonality in temperature and/or water availability). Their integrated analysis of floristic proxy data suggests that southeastern Patagonia experienced seasonally low levels of precipitation that may have limited plant growth in some seasons and promoted the presence of grasses along with shrubs and trees. Proxies from wood anatomy at the Punta Sur locality (stratigraphically lower in the Estancia La Costa Member than our fossil localities) give an estimated MAT ranging from 9 to 19 °C with an annual temperature fluctuation of ~4 °C. Rainfall for Punta Sur was estimated at 869 mm (with an estimated range up to 1809 mm). Brea *et al.* (Chapter 7) note that the greater abundance of tropical panicoid grasses at Estancia La Costa suggest warmer, more humid conditions. The identification of seven tree families, also documented in the phytolyth record at Estancia La Costa, is the first record of a diverse late Early Miocene forest on the Atlantic coast supporting the idea that Patagonian forests occurred much farther east than today. They record palm phytoliths at Punta Sur. This may mean that warmer conditions prevailed, since this palm record is 15° of latitude farther south than the southernmost record of palms today.

Vertebrates

Amphibians The record of amphibians in SCF is scant. However, the presence of *Calyptocephalella* Strand, 1928 in the Estancia La Costa locality is an important climate indicator (Fernicola and Albino, Chapter 8). Its presence is the southernmost record. Today, *Calyptocephalella* occurs from Coquimbo (at approx. 29° S) to Puerto Montt (40° S) (MAT 10.65 °C; rainfall 1803 mm), in Chile (Veloso *et al.*, 2008). The presence of this animal seems to indicate the presence of permanent lowland lakes, ponds, and quiet streams, possibly developed in a forested area (Fernicola and Albino, Chapter 8).

Reptiles Amongst the scant remains of reptiles, the occurrence of *Tupinambis* Daudin, 1802 and "colubrids" represents the southernmost record in their respective evolutionary histories, suggesting that in the late Early Miocene warmer and probably more humid conditions than those prevailing at present existed in southern Patagonia (Fernicola and Albino, Chapter 8). The *Tupinambis* specimens collected so far are from Monte León at an unknown stratigraphic level. Today the southern limit of *Tupinambis* is ~40° S (Albino, 2011; Embert *et al.*, 2009). Extant tupinambines do not tolerate temperatures below 14 °C (Brizuela, 2010).

If we are to reconstruct the climate of the Santacrucian as warm and humid, then the absence of crocodiles, particularly Alligatoridae, is surprising. Several genera of alligatorids reach as far south today as 30° S (Medem, 1958; Plotkin *et al.*, 1983). But the last record of an alligator in southern Patagonia is Middle Eocene (Gasparini *et al.*, 1986). Speculatively, there may have been dispersal barriers in the dry mid-latitudes that excluded alligators from southward dispersal into suitable habitats in the Early Miocene of Patagonia.

Likewise, turtles have not been recovered from the Santacrucian. The Pelomedusoidea and Meiolaniidae were present in Patagonia in the early Tertiary but disappear after the Casamayoran (late Middle Eocene). Tortoises (Testudininae) progressively invaded what is the present territory of Argentina from the north and appear in several Colhuehuapian localities (Early Miocene), after which they disappear. The general explanation offered is that tortoises were extirpated from today's southern Argentina because of climatic cooling and drying linked with the rising Andes and northward retreat of the tropical belt (Auffenberg, 1974). Extant South American tortoises are for the most part subtropical to tropical in distribution and are common in subhumid to arid grasslands and savanna habitats, though there are a few species that prefer mesic tropical forests (Auffenberg, 1974). Their farthest southern distribution today occurs in Buenos Aires Province (Gasparini *et al.*, 1986).

Birds Amongst the known birds, a number of predatory ground dwellers are represented (phorusrhacids and a seriemas). Degrange *et al.* (Chapter 9) suggest that these taxa would have been best suited for life in open areas because of their cursorial abilities. In the absence of placental carnivores, phorusrhacids and seriemas, along with several groups of marsupials (see below), occupied the large-predator niche in the Santacrucian paleocommunity. A herpetothere falconiform, *Thegornis musculosus* Ameghino, 1895, is also recorded among birds (Noriega *et al.*, 2011). This species is closely related to the living Laughing Falcon *Herpetotheres*, common in tropical and subtropical zones of South America at forest edges, open forests, and mixed palm savanna and forest habitats. The close phylogenetic affinities suggest that *Thegornis* was a forest dweller. Degrange *et al.* (Chapter 9) suggest that Santacrucian habitats were Chaco-like, being characterized by seasonality in temperature and rainfall and the presence of alternating areas of herbaceous vegetation with shrubby or forested areas. Unfortunately, the bird record so far does not identify any species of frugivores. Birds make a significant contribution to the frugivore niche in southern temperate forests today (see below).

Marsupials Abello *et al.* (Chapter 10) summarize the SCF non-carnivorous marsupials, including four Microbiotheria (four species of *Microbiotherium* Ameghino, 1887) and ten Paucituberculata (five species of *Palaeothentes* Ameghino, 1887, *Stilotherium dissimile* Ameghino, 1887, two species of *Acdestis* Ameghino, 1887, and *Abderites meridionalis* Ameghino, 1887). In FL 1–7 area, only four species (*Microbiotherium tehuelchum* Ameghino, 1887, *Palaeothentes minutus* Ameghino, 1887, *Palaeothentes lemoinei* Ameghino, 1887, and *Acdestis owenii*) are recorded (Table 17.1). The only extant representative of Microbiotheriidae is *Dromiciops*. The taxon is nocturnal, insectivorous, and arboreal with a partially prehensile tail and lives in bamboo thickets of Valdivian temperate rainforests. It hibernates, allowing it to survive the cold winters as far south as Chiloe Island (Redford and Eisenberg, 1992). It is tempting to extrapolate these habits to the Santacrucian *Microbiotherium*, but such claims should be viewed with caution, as there is very little available other than dental, mandibular, and facial remains. Based on dental remains, Abello *et al.* (Chapter 10) suggest that *M. tehuelchum*, was a small insectivore.

Among the medium-sized paucituberculatans, *Palaeothentes* is known from postcranial remains, which suggest that it was an agile cursorial species with leaping abilities, and probably terrestrial (Abello *et al.*, Chapter 10). Dietary specializations range from insectivores (*P. minutus* and *P. lemoinei*) to insectivores-frugivores (*A. owenii*). The substrate preference of *A. owenii* can be inferred only from their dental specializations for eating more fruit, which

may be related to more arboreal or at least scansorial habits. As already noted, Abello *et al.* (Chapter 10) conclude that the range of ecologies of these taxa would best indicate forested semitropical habitats. The range of ecologies described appears most similar to those observed in forested environments in southeastern Australia (Kay and Hylander, 1978; Strait *et al.*, 1990; Dumont *et al.*, 2000).

Larger marsupials of the SCF reviewed by Prevosti *et al.* (Chapter 11) consist of 11 sparassodont (Hathliacynidae and Borhyaenoidea) species that appear to have been strict carnivores with a wide range of locomotor abilities (from scansorial to terrestrial), and body masses (1 kg to greater than 50 kg). A dietary separation was proposed, with most hathliacynids being less hypercanivorous than borhyaenoids, and the larger borhyaenids extreme hypercarnivorous, and probably scavengers.

In FL 1–7 eight species are recorded: *Cladosictis patagonica* Ameghino (1887), *Sipalocyon gracilis* Ameghino, 1887, *Perathereutes pungens* Ameghino, 1891 (hathlyacinids), *Prothylacynus patagonicus* Ameghino, 1891, *Lycopsis torresi* Cabrera, 1927 (borhyaenoids), *Borhyaena tuberata* Ameghino, 1887, *Arctodictis muñizi* Mercerat, 1891, and *Acrocyon sectorius* Ameghino, 1887 (borhyaenids) (Table 17.1). Prevosti *et al.* (Chapter 11) note that the diversity of taxa observed in FL 1–7 is similar to that observed in present and past placental hypercarnivore communities. Following Prevosti *et al.* (Chapter 11), the taxonomic richness of Santacrucian hypercarnivores is substantial and appears to run contrary to the hypothesis offered by Croft (2001) that the Santacrucian carnivore niche is "depauperate."

Armadillos and glyptodonts Vizcaíno *et al.* (Chapter 12) analyze the cingulates of the SCF indicating that at least five genera of armadillos and four genera of glyptodonts were potentially sympatric. For the FL 1–7 coastal area they mention the presence of three species of armadillos (*Proeutatus oenophorus* Ameghino, 1887, *Prozaedyus proximus* Ameghino, 1887, and *Stenotatus patagonicus* Ameghino, 1887) and two of glyptodonts (*Propalaehoplophorus australis* Ameghino, 1887 and *Cochlops muricatus* Ameghino, 1889) (Table 17.1). With respect to limb adaptations and substrate use, these Santacrucian armadillos were good diggers but not fossorial to the extent found in some specialized living taxa. Glyptodonts were moderately large-sized (~80–90 kg) and ambulatory forms.

In contrast to the locomotor uniformity, cingulates show more variation regarding dietary habits. Within armadillos, the masticatory apparatus denotes a broader range of scavenging and utilization of underground food resources, such as tubers, than in the living armadillos. Glyptodonts also exhibit variation in the feeding apparatus suggesting a broad niche breadth: *Propalaehoplophorus australis* would have been a highly selective feeder, while *Cochlops muricatus* was probably a less selective feeder, both in moderately open habitats.

Considering the taxonomic richness of Santacrucian armadillos compared with the distribution of living species and the ecomorphological diversity of glyptodonts, their environment is interpreted as a mixture of open and relatively closed vegetation in relatively dry conditions, perhaps bushlands or dry forests, physiologically (although not taxonomically) agreeing in general with the modern Chaqueña biogeographic province.

Anteaters A notable presence with climatic implications is the record of semi-arboreal anteaters in the SCF, including one unquestionable species, *Protamandia rothi* Ameghino, 1904, and a myrmecophagid indet. (Bargo *et al.*, Chapter 13). *Protamandua rothi* is recorded in FL 1–7 coastal area and is of the size of the living *Tamandua*, well adapted for climbing, i.e. probably semi-arboreal, and for scratch-digging the substrate in feeding on ants or termites. The specialized feeding habits of vermilinguans are indicative of subtropical and warm temperate environments, because they would have depended on a year-round availability of social insects.

Three extant specialist myrmecophagous mammals have similar southern range limits between 28° and 34° S: the giant armadillo *Priodontes*, and two anteaters *Myrmecophaga* and *Tamandua*. The southerly limits of the mammalian distributions are strikingly similar to those of the most common termites upon which they prey (Constantino, 2002; Redford, 1984): the most southerly distributions of these termites occur in southeast Brazil or Uruguay between 27° S and 34.5° S. Some of these termites build nests underground or in mounds; others build arboreal nests. The southern limits of today seem to be related to rainfall and mean annual temperature.

There are also latitudinal abundance gradients of termites. Termites are most diverse in the tropics. By some estimates termites and ants, social bees, and social wasps comprise three-quarters of the soil fauna in the central Amazonian rainforest (Fittkau and Klinge, 1973; Wilson, 1987). Whilst they decline in richness outside the tropics, generally accompanied also by a reduced biomass, they continue to be important in subtropical ecosystems and occur as far south as 40–45° S. Sanderson (1996) summarized overall termite biomass as an order of magnitude different between the central Amazon and Río Negro Province at 40° S. Soil-feeding species are especially significant contributors to richness in rainforests. These species show a much more rapid drop-off in richness with latitude than wood-feeding species (Bignell and Eggleton, 2000).

The same variation in biomass occurring between the latitudes and the tropics is also seen between more and less

humid regions of the tropics. Melo and Bandeira (2004) reported termite density in northeast Brazilian Caatinga. For example, in highland humid forest, surrounded by the Caatinga, the values of density and biomass were 22 and 18 times higher, respectively, than in drier Caatinga environments. They attributed these differences to the plant productivity of the ecosystem. Species feeding on forest-litter were little represented in Caatinga environments, because of the adverse dry conditions of the Caatinga during much of the year. Melo and Bandeira (2004) noted a closer resemblance of the Caatinga environment to that of the Cerrado than to dense forest ecosystems. Thus, species adapted to a Cerrado or Caatinga-type environment would not be unexpected for FL 1–7.

Sloths The SCF demonstrates an unparalleled richness of sloths consisting of 11 genera encompassing basal Megatherioidea, Megalonychidae, and Mylodontidae (Bargo *et al.*, Chapter 13). However, in the coastal FL 1–7, five genera are recorded: *Hapalops* Ameghino, 1887, *Pelecyodon* Ameghino, 1891, *Hyperleptus* Ameghino, 1891, (megatherioids), *Eucholoeops* Ameghino, 1887 (megalonychid), and *Nematherium* Ameghino, 1887 (mylodontid). In general these sloths were moderately large-sized forms (~70–80 kg), the largest reaching about 100 kg (e.g. *Nematherium*), with a locomotor pattern very different from that of living sloths (the latter exhibit extreme arboreality and suspensory adaptations). Their locomotor mode more nearly resembled that of extant vermilinguans or pangolins. The morphology implies well-developed digging capabilities, but a mixture of terrestrial and scansorial activities was also possible. Megatherioid sloths, especially *Eucholoeops* (Bargo *et al.*, 2009), were most likely leaf eaters, and the primary method of food reduction must have been by shearing or cutting (with a predominance of orthal movements). The mylodontid *Nematherium* appears to have fed on more three-dimensional foods such as plant underground storage organs. Semi-arboreal habits for at least some of the sloths (e.g. *Hapalops*) indicate they lived in forests or that forested areas were present nearby. This environment is consistent with the feeding habits proposed for megatherioids, i.e. leaf eating, although the existence of open environments cannot be ruled out based on the feeding and locomotor habits of mylodontids.

"Ungulates" A great variety of "ungulates" is recorded in the SCF, including seven clades: Hegetotheriidae, Interatheriidae, Toxodontidae, and Homalodotheriidae (Notoungulata), Proterotheriidae and Macraucheniidae (Litopterna), and Astrapotheriidae (Astrapotheria) (Cassini *et al.*, Chapter 14). They fall into a broad range of body sizes, and exhibit diverse locomotor and feeding strategies.

In the FL 1–7 coastal area, 12 species are recorded: one astrapothere, *Astrapotherium magnum* (Owen, 1853); three typotheres (Hegetotheriidae + Interatheriidae), *Hegetotherium mirabile* Ameghino, 1887, *Interatherium robustum* (Ameghino, 1891), and *Protypotherium australe* Ameghino, 1887; two toxodontids, *Nesodon imbricatus* Owen, 1846 and *Adinotherium ovinum* Owen, 1853; one homalodotheriid, *Homalodotherium cunninghami* Flower, 1873; three proterotheriids, *Diadiaphorus majusculus* Ameghino, 1887, *Tetramerorhinus mixtum* (Ameghino, 1894), and *Thoatherium minusculum* Ameghino, 1887; and two macraucheniids, *Theosodon lydekkeri* Ameghino, 1887 and *Theosodon gracilis* Ameghino, 1891 (Table 17.1). Typotheres vary from 2 to 10 kg, proterotheriids from 20 to100 kg, while toxodonts and macraucheniids surpass 100 kg, and *Astrapotherium* exceeded 1000 kg. Locomotor behavior was mostly cursorial. Typotheres might have engaged in occasional digging, but this group also includes the most agile and fastest forms, perhaps more so than in the proterotheriids. *Interatherium* shows evidence of swimming or climbing capabilities. *Homalodotherium* had a peculiar locomotion, with the ability to adopt a bipedal posture to use the forelimbs to forage on trees. Finally, *Astrapotherium* may also have had semi-aquatic habits. According to the ecomorphological analysis, Cassini *et al.* (Chapter 14) conclude that notoungulates inhabited and/or fed in open habitats and were mainly grazers. On the other hand, all litopterns and astrapotheres inhabited and/or foraged in closed habitats. Among litopterns, proterotheriids were likely browsers, while macraucheniids had intermediate diets in mixed environments. *Astrapotherium* also had an intermediate diet, but in more closed environments. Finally, the relationship between the area of occlusal tooth wear and body size suggests that the Santacrucian "ungulates" had a basal metabolic rate equivalent to extant perissodactyls and artiodactyls: typotheres could compensate for reduced intraoral food processing with better digestive capabilities, while litopterns would have had better intra-oral food preparation and not as good digestive capabilities as notoungulates.

Rodents Candela *et al.* (Chapter 15) review the caviomorphs rodents of the SCF. In the FL 1–7 coastal area six genera, *Perimys* Ameghino, 1887 (Neoepiblemidae), *Neoreomys* Ameghino, 1887 (Dasyproctidae), *Eocardia* Ameghino, 1887, and *Spaniomys* Ameghino, 1887 (Eocardiidae), *Steiromys* Ameghino, 1887 (Erethizontidae), and *Sciamys* Ameghino, 1887 (Acaremyidae), and at least eight species are recorded (see Table 17.1). Amongst the eocardiids was *Eocardia fissa* Ameghino, 1891, a cursorial agile rodent, similar to *Pediolagus salinicola*, the living Chacoan cavy. The latter prefers open, semi-arid woodland and scrubland. It is diurnal and does not dig burrows, sheltering

instead in hollow dead logs or in burrows dug by other animals. Forbs and woody plants are the most important dietary items. Grass consumption is mostly limited to the rainy season, when grass regrowth becomes available on rangelands (Rosati and Bucher, 1992). A scansorial porcupine, *Steiromys duplicatus* Ameghino, 1887, is also common. Candela *et al.* (Chapter 15) suggest that the dasyproctid *Neoreomys australis* Ameghino, 1887, is a cursorial species similar to the living *Agouti* (= *Cuniculus*) *paca*. This accords with the conclusions of Townsend and Croft (2008) who suggested that the enamel microwear patterns of *Neoreomys* resemble those of pacas; pacas occupy a wide range of forests in moist areas, including gallery forests, and have a frugivorous diet (Eisenberg, 1989). The limit of its southern distribution, like that of primates, is ~32° S in the Cerrado biome.

Primates The occurrence of *Homunculus* Ameghino, 1891, as far south as ~50° S is notable (Kay *et al.*, Chapter 16). The farthest southern distribution of the order today is between 26° and 30° S. There is no generalization about diet or body size that can be found in the most southerly taxa today. Large-bodied folivorous/frugivorous *Alouatta* reaches almost 30° S in Uruguay; the frugivore and seed-predator *Cebus* has its southern limit in Misiones Province, Argentina at a similar latitude; and the small frugivorous-insectivorous *Aotus* crosses the southwestern Paraguay border into Argentina at ~26° S.

17.4.2 Synthetic approaches

Niche characteristics of extant South American communities and the Santacrucian

Species richness Kay and Madden (1997a, b) reported strong correlations between rainfall and several metrics of the trophic structure of non-volant mammal communities in tropical South America. Our larger dataset enables us to re-examine and refine some of their conclusions.

Of our 25 South American local faunas described in Tables 17.2 and 17.3, 20 are situated within the limits of the tropics (north of the Tropic of Capricorn) and 5 are not. The two sets of faunas behave very differently with respect to species richness. In the tropics, there is a strong relationship between rainfall and total species richness, whereas in more southerly faunas we find no relationship between rainfall and total species richness (Table 17.3, Fig. 17.3a). Considering the details of this relationship more finely, species richness of primary consumers is even more highly correlated with rainfall in the tropics whereas the richness of secondary consumers versus rainfall barely rises to the level of significance (Table 17.3, Figs. 17.3b, c). In both cases, there is no significant correlation among extratropical faunas. At FL 1–7 we identify 49 mammalian species. This

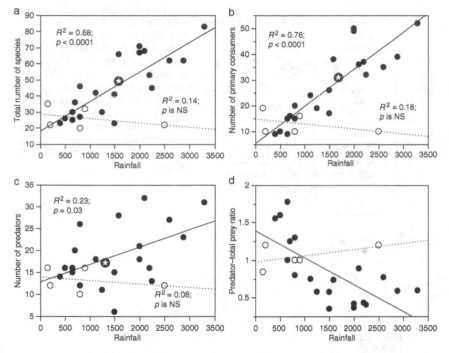

Fig. 17.3. Rainfall and species richness. a, Rainfall versus total species richness of non-volant mammals. P values are given; NS is "not significant." b, Rainfall versus species richness of non-volant mammalian primary consumers. c, Rainfall versus species richness of non-volant mammalian secondary consumers. d, Rainfall versus the ratio of predator to prey. Solid circles represent faunas within the tropical zone ($N = 20$). Open circles represent faunas outside the tropical zone ($N = 5$). Lines represent the least-squares fit for tropical species (solid lines) and extratropical species (dashed lines). Closed circle with star represents the position of the FL 1–7 fauna.

is far more richness than any of our extant Patagonian faunas irrespective of rainfall. If this fauna is gauged against the low-latitude faunas summarized in Fig. 17.3a, an inferred rainfall would be about 1500 mm/year; considering only primary or only secondary consumers yields rainfall estimates of ~1750 and 1250 mm, respectively.

Diet Among extant communities, the number and proportions of species within various dietary niches varies greatly relative to rainfall, again with the caveat that these findings work best in the tropics. The proportion of frugivores compared with other primary consumers is greater in wet than in dry climates (Fig. 17.4a). Likewise, the number of browsers relative to the number of grazers is greater in wet than in dry localities (Fig. 17.4b).

The browsing index at FL 1–7 would be typical of rainfall in the range of 1500 mm, which accords well with other estimates of rainfall using niche metrics (Fig. 17.4b). On the other hand, the frugivore index at FL 1–7 is very low (Fig. 17.4a). If FL 1–7 were a tropical fauna, the rainfall would be predicted to be less than 500 mm. This index is very similar to what we see for any high-latitude fauna irrespective of rainfall. We might speculate that the low number of "frugivores" is a consequence of the extreme seasonality of fruit production at these latitudes, and not an indicator of low rainfall as might be inferred by analogy with low-latitude faunas. Alternatively, the fruit may have been present but the frugivores were birds rather than mammals: it has been noted that more than 50% of the woody species of

the temperate forest produce fleshy fruits, but most of the seed dispersal is accounted for by just a few opportunistic species, mostly birds, many of which are migrant species (Aizen and Ezcurra, 1998). Such a situation, where birds were the dominant seed dispersers and many of them were migrant species, may also have occurred in the Miocene.

Substrate preference As indicated by the arboreality index, tropical localities with higher rainfall have more arboreal and scansorial species, but this relationship does not hold for extratropical localities (Fig. 17.4c). FL 1–7 again follows the pattern of a tropical fauna with ~1500 mm rainfall per annum.

Cenogram analysis

Table 17.4 presents the relevant data useful for testing the reliability of cenogram slopes and gaps for inferring rainfall, mean annual temperature, and environmental "openness."

Cenogram analysis makes three predictions:

1. When arranged by log size, from largest to smallest, the slope for the medium-sized mammals (those with mass between 500 g and 250 kg) is said to be more steeply negative in dry than in wet environments. For 20 tropical South American faunas we find this trend to be true. Localities with more rainfall have more shallow negative slopes, whereas those with steeper negative slopes have less rainfall (Fig. 17.5a). Outside the tropics, there is no significant relationship between rainfall and the slope for

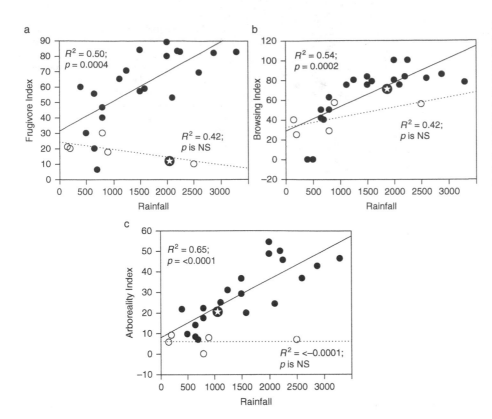

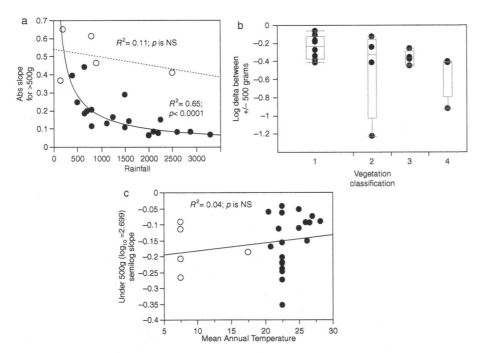

Fig. 17.4. Rainfall and niche indices. a, Rainfall versus frugivore index. b, Rainfall versus browser index. c, Rainfall versus arboreality index. Solid circles represent faunas within the tropical zone ($N = 20$). Open circles represent faunas outside the tropical zone ($N = 5$). Lines represent the least-squares fit for tropical species (solid lines) and extratropical species (dashed lines). Closed circle with star represents the position of the FL 1–7 fauna.

Fig. 17.5. Cenogram analyses for extant faunas of South America ($N = 25$). a, Absolute value (abs) of $\log_{10}$ slope of mammalian primary consumers >500 g; <250 kg. Solid circles represent faunas within the tropical zone ($N = 20$). Open circles represent faunas outside the tropical zone ($N = 5$). Lines for each group represent least-squares fit between rainfall and absolute slope in log space. b, Box and whisker plots of the $\log_{10}$ gap between species above and below 500 g by vegetation types 1–4 in the tropics. c, Absolute value of $\log_{10}$ slope of mammalian primary consumers less than 500 g versus mean annual temperature. A species with a body mass of 500 g has a $\log_{10}$ body mass of 2.699. For plots a and c, the values of explained variance (R^2) and significance (p, where NS means not significant) are presented.

medium-sized mammals (Fig. 17.5a). As can be seen in Fig. 17.5a, slopes become asymptotic with increased rainfall, so the slope values at FL 1–7 (−0.08, not shown) might suggest only that rainfall exceeded ~750 mm/year, but could have been far higher.

2. A cenogram predicts that the slope for the small mammals is related to MAT. The more pronounced (i.e. more negative) the slope, the colder the temperature. Our data show no significant correlation between the steepness of the small-mammal slope and MAT (Fig. 17.5b). Our data do demonstrate a weak but significant correlation between the small-sized mammal slope and rainfall (not predicted by the model), although, as with the slopes for the medium-sized mammals, while the two are correlated significantly, the variance explained is limited and the predictive power slight.

3. A third prediction is that the size of the gap or offset occurring between small- and medium-sized species (500 g cut-off) is related to the openness of the environment: larger gaps are related to more open environments. When we plot "gap" size against a range of tropical habitats from closed canopy rainforest to tropical savanna environments, we find no consistent relationship between the two (Fig. 17.5c). The size of the gap at FL 1–7 best fits with the distribution of closed environments (0.25, vegetation type 1) but overlaps the ranges of more open environments as well.

Using our dataset, it appears that all the cenogram "rules" derive from two simple generalizations. At least in the Neotropics, rainfall is inversely related to the numbers of species (richness) of mammalian primary consumers found in a habitat. Also, the size range of mammalian primary consumer species is similar irrespective of habitat, roughly 10 g to 1000 kg. If there are fewer species spanning the same size range in more arid habitats, this produces cenogram slopes plots that are more negative, as illustrated in Fig. 17.6.

Predator to prey ratios

Croft (2001) suggested that the number of mammalian predator species was lower than expected in the Santacrucian faunas by comparison with the number of prey species. As noted above, Croft's proposal has two components. First, he proposed that pre-GABI faunas could have had a greater-than-expected number of medium-sized herbivorous mammals as a result of this apparently low diversity of mammalian predators. Second, he suggested that this effect would have been more pronounced in open habitats than in closed ones. Our data suggest the reverse of Croft's prediction. Among extant mammal faunas of South America predator richness is proportionately higher, not lower, as rainfall decreases: drier localities have greater numbers of

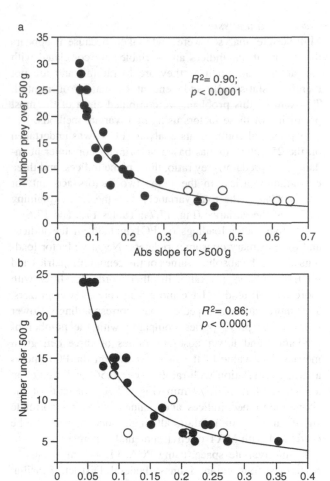

Fig. 17.6. Cenogram slopes and primary consumer richness for extant faunas of South America ($N = 25$). a, Absolute value (abs) of $\log_{10}$ slope of mammalian primary consumers >500 g; <250 kg versus the number of primary consumers greater than 500 g. Solid circles represent faunas within the tropical zone ($N = 20$). Open circles represent faunas outside the tropical zone ($N = 5$). b, Absolute value of $\log_{10}$ slope of mammalian primary consumers less than 500 g versus the number of primary consumers less than 500 g. Lines represent the least-squares fit in log space.

predator species for the number of prey species (Fig. 17.3d). The predator/prey ratio fit is much weaker than the others but yields a rainfall of >1000 mm. Given that species richness and other niche metrics all suggest similar levels of rainfall (i.e. >1000 mm), the predator to prey ratios of mammals in FL 1–7 are not apparently out of agreement with expectations. Of course, we cannot generalize these findings to other pre-GABI faunas. Prevosti *et al.* (Chapter 11) observe that species richness of predatory mammals was within the range seen in extant South American mammals. In short, our analysis agrees with that of Prevosti *et al.*: the mammalian carnivore niche was not clearly depauperate in the Santacrucian.

A combined analysis

Multivariate analyses were undertaken because it appears that many of our indices and variables are correlated with one another as much as they are to rainfall and are not therefore statistically independent estimators of rainfall. To examine this problem, we combined eight of the most promising of these factors using multivariate methods.

A principal components analysis (PCA) was undertaken on the 25 extant faunas based on primary consumer abundance, the predator/prey ratio, three niche indices, and three cenogram metrics. In the PCA, two factors account for nearly 82% of the total variance, with the first explaining 71.2% of the variance (Fig. 17.7a; Tables 17.5 and 17.6).

Positive factor loadings on PC1 include all the indices and total primary consumer richness. Negative factor loadings are for the absolute values of the cenogram metrics and for the predator/prey ratio. In other words, localities with more rainfall tend to have more frugivores, fewer grazers, and more arboreal species, but correspondingly lower numbers of prey species compared with the number of predators, and lower absolute values of three cenogram metrics. The value of the factor scores on the PC1 shows a strong correlation with rainfall ($r^2 = 0.56$) and predict an annual rainfall of 1579 mm (Fig. 17.7b). Several of the above-mentioned indices and richness measures correlate significantly with rainfall, although none appears to be correlated with MAT or environmental "openness."

In multivariate space, using PCA, FL 1–7 most closely resembles four extant faunas: Federal District, Brasilia; Puerto Páez, and Masaguaral, in the llanos of Venezuela, and Puerto Ayacucho, on the Río Orinoco, Venezuela. All four are subtropical sites with a vegetational mosaic of savannas and gallery forests with seasonal rainfall and a 5- to 6-month dry season. Another similar fauna with similar niche metrics is Acurizal, Brazil, presenting a mixture of cerrado and deciduous forests with a 7-month dry season. These localities share more or less the same vegetational mosaic of open intervals interspersed by gallery forests with palms adjacent to rivers, often with seasonal flooding but also with long dry intervals.

In all analyses, the high-latitude sites with low rainfall have niche metrics similar to those in the tropical zone, when matched for rainfall. However, the high-latitude and high-rainfall site of the Nahuel Huapi forest (northwest Patagonia, Argentina) is consistently an outlier. Despite high rainfall, this forest has very low species richness with few arboreal or scansorial species and very few frugivorous species. Examination of mammal distributions at other wet southerly sites shows similar richness values (Redford and Eisenberg, 1992), suggesting that Nahuel Huapi is not an unusual southern humid forest. Without the outlier Nahuel Huapi humid forest, r^2 for PC1 versus rainfall would be much higher at 0.77.

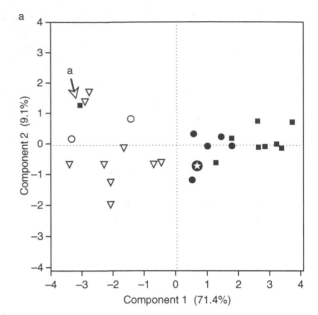

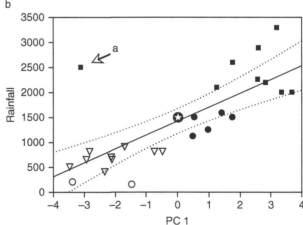

Fig. 17.7. Principal component analysis of eight variables and 25 extant mammal communities listed in Table 17.1. a, Bivariate plot of the first and second principal components. Least-squares regression to predict rainfall in mm from PC1 is: 322.7(PC1) + 1368. Predicted rainfall at FL 1–7 is 1512 mm. As indicated by the dashed lines, the 95% confidence interval for the estimate is large (~400 to 2100 mm). b, Plot of rainfall versus the first principal component (Prin1). Star is the fitted value for FL 1–7. Solid squares indicate rainfall exceeding 2000 mm/yr. Solid circles indicate rainfall between 1000 and 2000 mm/yr. Inverted open triangles have rainfall <1000 mm and >500 mm/yr. Open circles indicate less than 500 mm/yr. The symbol labeled "a" represents the outlier Nahuel Huapi humid forest fauna.

Using discriminant function analysis (DFA), we examined classification success of the same eight variables to assign extant localities to one of the following four rainfall categories: 0 = less than 500 mm per annum, 1 = between 500 and 1000 mm per annum, 2 = 1000 to 2000 mm per annum, and 3 = above 2000 mm per annum. DFAs for 25

Table 17.5. *Factor loadings (eigenvectors) for a principal components analysis on scaled correlations among eight variables*

Variable	Principal component 1	Principal component 2
Number of primary consumers	0.386	0.005
Arboreality index	0.388	0.031
Browsing index	0.363	0.186
Frugivore index	0.362	0.122
Predator/prey ratio	−0.355	−0.355
Absolute value of +/−500 g gap	−0.313	0.589
Absolute value of slope for >500 g	−0.334	0.489
Absolute value of slope for <500 g	−0.320	−0.320

Table 17.6. *Values for the first and second principal components in a principal components analysis of 25 extant faunas*

Locality	PC1	PC2
Acurizal	0.50	−1.15
Belem	1.77	0.20
Caatingas	−2.32	−0.64
Chaco	−2.11	−1.23
Chubut	−1.45	0.85
Cocha Cashu	3.35	−0.11
Ecuador Tropical	3.20	0.02
Esmeralda	3.70	0.76
Federal District, Brazil	1.43	0.26
Guatopo	1.77	−0.04
Manaus	2.84	−0.04
Masagural	0.99	−0.05
Misiones	1.26	−0.60
Nahuel Huapi humid forest	−3.10	1.30
Nahuel Huapi steppe	−2.80	1.73
Pampas	−1.67	−0.12
Peninsula Valdés	−3.37	0.20
Puerto Ayacucho	2.58	0.77
Puerto Paiz	0.55	0.36
Rio Cenepa	2.60	−0.08
Río Teuco	−2.10	−1.96
Salta low montane	−0.48	−0.60
Salta transitional forest	−0.73	−0.64
Tucuman subtropical forest	−2.93	1.41
Tucuman thorn forest	−3.45	−0.63
Santa Cruz FL 1–7	0.56	−0.78

Values for PC1 and PC2 are FL 1–7 extrapolated from extant factor loadings.

extant faunas show that the combined variables correctly assign localities to rainfall groups in 92% of cases (2 misclassifications of 25 are misclassified by one level) (Table 17.7). In three other cases, the assignment was correct but at less than the 95% level of confidence. Surprisingly, Santa Cruz fauna FL 1–7 was assigned to rainfall class 3 (>2000 mm rainfall per annum) at the 96% confidence level, which seems high based on all other findings.

17.5 Discussion and conclusions

In the previous sections we tried to compile and analyze the most relevant information from all the other contributions of this volume, as well as from many external sources, to generate the most complete reconstruction possible of the ecosystem that occurred in a defined area and a restricted time of the SCF. We intended to reconstruct a landscape (physiography and vegetation) and climate and to understand the overall niche characteristics of the fauna. We say we tried and we intended, because the quantity and diversity of evidence generated since we started the project and during the editing process of this volume is phenomenal, and we are still far from evaluating all the hypotheses that each and every chapter may produce in the near future.

However, we are confident that we are better positioned now than we were before this project to understand the paleoecology of at least one restricted area and time, thereby forming the framework for examining the changes in environments and biotas throughout the geographic and temporal range of the SCF (see below). The new data on the sedimentology, the ichnology, and the flora and fauna of the SCF summarized in this book are particularly strong for the lower parts of the SCF south of the Río Coyle. Through the efforts of geologists we now know that the localities collected most intensively – Anfiteatro, Estancia La Costa, Cañadón Silva, and Puesto Estancia La Costa – all sample a rather narrow interval of time between about 17.4 and 17.5 Ma (Perkins *et al.*, Chapter 2). For the purposes of this chapter, we concentrate on the levels originally designated as fossil levels 1–7 by Tauber (1994), which we regard as being a single suite that we call FL 1–7. It is worth remembering that continental drift of the South American plate over the past 20 million years has produced perhaps a northward movement of only a few degrees in the paleolatitude of the southernmost part of Patagonia, where the SCF crops out. Therefore the geographic location of the FL 1–7 localities was then essentially where it is today.

Table 17.7. *Discriminant function analysis using eight variables and indices to classify 25 extant faunas and FL 1–7 Santa Cruz Formation*

Locality	Rainfall category	Probability of correct assignment	Predicted rainfall	Probability (predicted)	Other possible assignments
Acurizal	2	0.9995	2	0.9995	
Belem	3	0.9932	3	0.9932	
Caatingas	0	1.0000	0	1.0000	
Chaco	1	0.9750	1	0.9750	
Chubut	**0**	**0.3553**	**1***	**0.6436**	
Cocha Cashu	3	1.0000	3	1.0000	
Ecuador Tropical	3	1.0000	3	1.0000	
Esmeralda	3	1.0000	3	1.0000	
Federal District, Brazil	2	0.9815	2	0.9815	
Guatopo	2	0.9958	2	0.9958	
Manaus	3	0.9993	3	0.9993	
Masagural	2	0.9976	2	0.9976	
Misiones	**3**	**0.4211**	**3**	**0.4211**	**1: 0.32 2: 0.26**
Nahuel Huapi humid forest	**3**	**0.5274**	**3**	**0.5274**	**1: 0.47**
Nahuel Huapi steppe	1	0.2101	0*	0.7895	
Pampas	1	0.9986	1	0.9986	
Peninsula Valdés	0	0.9913	0	0.9913	
Puerto Ayacucho	3	0.9610	3	0.9610	
Puerto Paiz	2	0.9902	2	0.9902	
Rio Cenepa	3	0.9622	3	0.9622	
Río Teuco	1	0.9479	1	0.9479	
Salta low montane	**1**	**0.5625**	**1**	**0.5625**	**2: 0.43**
Salta transitional forest	1	0.9951	1	0.9951	
Tucuman subtropical forest	1	0.9805	1	0.9805	
Tucuman thorn forest	0	0.9996	0	0.9996	
Santa Cruz FL 1–7	–		3	0.9558	

Faunas in bold are either incorrectly classified or classified with less than 95% likelihood.
Rainfall categories: 0: <500 mm rainfall/annum; 1: between 500 and 1000 mm; 2: between 1000 and 2000 mm; 3: > 2000 mm.
*indicates misclassified.

17.5.1 Describing the landscape and climate

The physiography, sedimentology, and distribution of trace fossils indicate that FL 1–7 represent paleosols formed in ash falling into a fluvial system dominated by sheet-flooding, and overbank-flooding from laterally stable channels (Matheos and Raigemborn, Chapter 4; Krapovickas, Chapter 6). In particular, the ichnofossils of FL 1–7 indicate water table conditions with moderately drained soils that were at least seasonally waterlogged. The presence of the frog *Calyptocephalella* seems to indicate permanent lowland lakes, ponds, and quiet streams, possibly developed in a forested area (Fernicola and Albino, Chapter 8). Although not yet certainly recorded at FL 1–7, remains of a number of

waterfowl indicate the presence of flooded areas or permanent water bodies in others parts of the SCF.

The flora described by Brea *et al.* (Chapter 7) consists of forest coexisting with open areas with grasses. The forests are similar to those found today in several regions of the Southern Hemisphere including central Chile, southeastern Australia, and New Zealand, with the novel record of palms 15° of latitude farther south than its southernmost record in South America today.

Different elements of the vertebrate fauna indicate the coexistence of open and forest habitats. For instance, among birds the falconid *Thegornis* suggests forested habitats. Likewise, the medium- to giant-sized terrestrial avian predators would have been better suited for life in open areas because of their cursorial abilities (Degrange *et al.*, Chapter 9). Among the marsupials, the range of ecologies of the small insectivorous and frugivorous forms would be best explained as indicating forested semitropical habitats (Abello *et al.*, Chapter 10) while the larger and hypercarnivorous sparassodonts include a range of scansorial and terrestrial species (Prevosti *et al.*, Chapter 11). Obviously the more scansorial taxa suggest the presence of some forested patches. Among xenarthrans, the remarkable taxonomic richness of Santacrucian armadillos is complemented by the wide ecomorphological diversity of glyptodonts that may well have inhabited a mixture of open and relatively closed vegetation in relatively dry conditions, perhaps bushlands or dry forests (Vizcaíno *et al.*, Chapter 12). At least some sloths and anteaters likewise may have been arboreal or scansorial (Bargo *et al.*, Chapter 13), again indicating nearby forested areas. Among ungulates, the comprehensive ecomorphological analysis provided by Cassini *et al.* (Chapter 14) classifies all litopterns and *Astrapotherium* as dwelling and/or foraging in closed habitats, and all notoungulates in open habitats. Among rodents, the scansorial porcupine *Steiromys duplicatus* is common, suggesting forested habits. The eocardiid rodent *Eocardia fissa* and the dasyproctid *Neoreomys australis* suggest open habitats. *Eocardia fissa* has been likened to the Chacoan cavy, a species that prefers open, semi-arid woodland and scrubland. *Neoreomys australis* was a cursorial species, more similar to the living paca *Agouti* (= *Cuniculus*) *paca* (Candela *et al.*, Chapter 15). Notably, pacas today occupy a wide range of forests in moist areas, including gallery forests, and their distributional limit today is ~32° S in the Cerrado biome. The primate *Homunculus* was clearly an arborealist committed to forested areas (Kay *et al.*, Chapter 16).

We have multiple lines of evidence to infer the climate that dominated this landscape. As mentioned above, the timing and pattern of the uplift of the Andes had a profound local effect on rainfall beginning in the mid-Miocene. The isotopic record of the sediments temporally equivalent to FL 1–7 shows little change in oxygen isotopic composition (pedogenic $\delta^{18}O$) through most of this time interval, indicating that insufficient Andean uplift had occurred to induce aridity by a rain-shadow effect until <15 Ma (Blisniuk *et al.*, 2005), so no part of the Estancia La Costa Member would have experienced discernable climatic change owing to regional tectonics. The record of carbon isotopes tells the same story. As indicated by pedogenic $\delta^{13}C$, C3 plants dominated the landscape and an apparent decline in $\delta^{13}C$, suggesting aridity-induced change, did not occur until <15 Ma (Blisniuk *et al.*, 2005).

The plant proxies for rainfall at the lower Punta Sur locality suggest 869 ± 940 mm per annum, and the greater abundance of tropical panicoid grasses suggests more humid conditions for FL 1–7 (Brea *et al.*, Chapter 7).

Global climatic effects also would have been in play. A warm phase in the Cenozoic record peaked in the MMCO between 17 and 15 Ma and was followed by a gradual cooling and reestablishment of major ice-sheets on Antarctica (Zachos *et al.*, 2001). Thus, at about 14–15 Ma, southern Patagonian climate experienced a combination of global cooling and regional aridity. But for the coastal Atlantic record of Santacrucian mammals, which samples a time before 15 Ma, the MMCO was still in full force.

Again, proxies for estimating MAT from the lower Punta Sur locality range from 9 to 19 °C with an annual temperature fluctuation of ~4 °C, but the greater abundance of tropical panicoid grasses suggests warmer (and more humid) conditions for FL 1–7 (Brea *et al.*, Chapter 7).

Among the fauna, the presence of the lizard *Tupinambis* suggests warm humid conditions. Extant *Tupinambis* does not tolerate temperatures below 14 °C. The presence of myrmecophagous specialists like *Protamandua* (Vermilingua) is indicative of subtropical and warm temperate environments on account of the environmental limitations of social insects such as termites. Although Krapovickas (Chapter 6) does not describe termite nests among the ichnofossils (contrary to previous reports by Tauber, 1996), the possibility exists that there were termites building nests above ground in trees, which, in turn reinforces the possibility that the vermilinguan *Protamandua* may have been arboreal or at least scansorial. We come finally to the primates, for which the far southern occurrence is almost 20° farther south than their current distribution. The presence of *Homunculus* certainly guarantees warm climatic conditions with enough rainfall to produce forested areas. No platyrrhine primate, living or fossil, is known to have lived in environments without trees.

Climate interpretation requires consideration of possible seasonality. Paleolatitude alone tells us something about it. We mentioned above that the area of the FL 1–7 localities was then essentially where it is today, geographically speaking. Therefore we have reason to believe that the area of the FL 1–7 localities was extratropical during Santacrucian times and would have had highly seasonal day lengths. Westerly winds at Patagonian latitudes have been the primary source of precipitation in the area throughout the time interval relevant to our study, and it is reasonable to suppose that rainfall would have been seasonal and Mediterranean – cool wet winters with dry warm summers. Although, as described by Brea *et al.* (Chapter 7), equivalent forests today develop in regions with pronounced maritime influence on their climates, especially the moderation of seasonal extremes, the reconstruction of the vegetation of the Estancia La Costa Member of the Santa Cruz Formation includes a range of open temperate semi-arid forests and

Fig. 17.8. Artist's reconstruction of life at FL 1–7, Santa Cruz Formation, late Early Miocene, Argentina.

Fig. 17.8. (*cont.*)

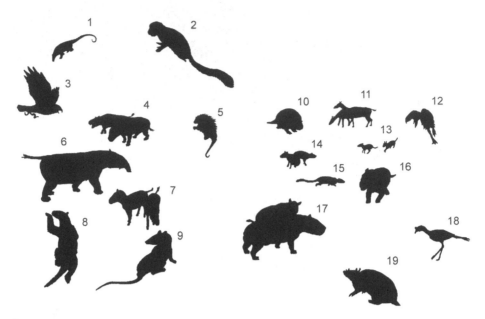

Fig. 17.8. key (*cont.*)
1. *Protamandua rothi* (Myrmecophagidae), 2. *Homunculus patagonicus* (Platyrrhini), 3. *Thegornis musculosus* (Falconidae), 4. *Nesodon imbricatus* (Toxodontidae), 5. *Steiromys duplicatus* (Erethizontidae), 6. *Astrapotherium magnum* (Astrapotheriidae), 7. *Diadiaphorus majuscules* (Proterotheriidae), 8. *Eucholoeops fronto* (Megalonychidae), 9. *Palaeothentes lemoinei* (Palaeothentidae), 10. *Propalaehoplophorus australis* (Propalaehoplophoridae), 11. *Theosodon gracilis* (Macraucheniidae), 12. *Phorusrhacos longissimus* (Phorusrhacidae), 13. *Protyphotherium australis* (Interatheriidae), 14. *Neoreomys australis* (Dasyproctidae), 15. *Interatherium robustum* (Interatheriidae), 16. *Bohryaena tuberata* (Borhyaenidae), 17. *Adinotherium ovinum* (Toxodontidae), 18. *Cariama santacrucensis* (Cariamidae), 19. *Peltephilus pumilus* (Peltephilidae).

temperate, warm humid forests (temperate referring to markedly seasonal climates, with seasonality in temperature and/or water availability). According to Brea *et al.* (Chapter 7), this part of Patagonia experienced seasonally low levels of precipitation that may have limited plant growth and promoted the presence of grasses along with shrubs and trees. This interpretation is consistent with the presence of soil carbonates in the form of rhyzoliths that imply seasonal variation in the level of the water table.

Our synthetic approaches yield similar kinds of paleoclimatic inferences to those derived from biotic elements considered separately. Most of the niche metrics derived from extant Neotropical communities suggest a wet environment – large alpha species richness of primary consumers, the frequent occurrence of arboreal and scansorial species, and the low percentage of grazing species among the primary consumers. One exception is the surprisingly low percentage of frugivores, which may be explained by the far southerly location of our localities constraining them to have been highly seasonal, at least in the light cycle. As well, the frugivore guild in extant temperate austral faunas is dominated by birds, especially migrant ones, for which we have no satisfactory record in the Miocene.

The cenogram analysis based on extant communities is less informative. We do find, as predicted from the cenogram model, that extant mammal communities with more

rainfall have more shallow negative slopes for the mammals in the size range from 500 g to 250 kg, whereas those with steeper negative slopes have less rainfall. However, given that only 47% of the variance in slope is explained by rainfall, the predictive value of this relationship is limited. Conclusions about mean annual temperature based on cenogram slopes for mammals less than 500 g are not supported by our data.

Unsurprisingly, our findings are not particularly different from those of Croft (2001) who used an overlapping dataset. Croft examined only the first and third predictions and found limited utility for these statistics. Rodríguez (1999) used a similar quantitative procedure to test the validity of these "rules" using a dataset of 92 mammalian communities, only two of which are South American. His results differ slightly from ours: he was unable to confirm that larger mammal slopes are more steeply negative in more arid environments, nor that the slope of smaller mammals was related to temperature. He did confirm that a gap in medium-sized species is related to the vegetation structure, but only in tropical communities, a finding not supported by our data.

The records mentioned so far all seem to point in a single direction – higher temperatures and more rainfall than prevail today. As yet, we do not have nearly as good a record in other fossil levels and have not tried to establish whether there was faunal change within the SCF.

In summary, the evidence so far analyzed indicates that the fauna of FL 1–7 localities developed in an area where forest similar to those existing today in the Andes and the piedmont with the range of latitudes of Patagonia coexisted with grasslands, and with permanent and temporary water bodies. The area was more humid and warmer than today, with rainfall rates of more than 1000 mm, wet winters and dry summers, mean annual temperatures >14 °C, and marked seasonality in day length.

17.5.2 What is the appropriate modern analog for the paleoecosystem?

The extant community metrics examined above (including the indices, species richness and predator/prey ratios) perform poorly when it comes to estimating paleoenvironmental parameters of extant southern temperate forest communities. Our representative fauna, the Nahuel Huapi humid forest, stands as an exception to all the generalizations offered by the niche metrics. It seems that mammalian communities in the far southern latitudes of South America today are not very productive. This finding serves as an added signpost allowing us to be that much more certain of our conclusions about SCF mammal environments, for how could the Santacrucian mammalian fauna have been so productive if the climate were not much warmer than today?

In some ways this is in accordance with the previous approach to the paleoecology of the SCF by Vizcaíno et al. (2010) mentioned in the Background section (17.2.3). These authors examined the paleoecological characteristics at two important localities in the SCF, Campo Barranca (CB) and Puesto Estancia La Costa (PLC). Campo Barranca is clearly stratigraphically lower than FL 1–7, and PLC includes the most productive of Tauber's fossil levels (FL 5.3, 6, and 7) included in FL 1–7. Using a general ecological relationship between population density and body size (Damuth, 1981, 1987, 1992; Fariña, 1996), Vizcaíno et al. (2010) identified an apparent imbalance at CB and PLC localities wherein the apparent secondary productivity of the ecosystem exceeds, by at least three times, the energetic requirements of its carnivores, when compared with ratios seen in modern faunas. That is, there is much more herbivore biomass than carnivores to consume it. As noted above, this is roughly in accordance with Croft's carnivore depauperate carnivore guild hypothesis.

However, Prevosti and Vizcaíno (2006) and Vizcaíno et al. (2010) discussed several caveats to the application of the model they used. Several factors (climate, prey density and availability, presence of competitors and predators, epidemics, and population genetic diversity) affect the densities of predators and prey on a local scale. The latter authors found a positive relationship between carnivore density and prey abundance for several carnivores in different habitats. Furthermore, Prevosti et al. (Chapter 11)

observe that known species richness of predatory mammals at FL 1–7 was within the range seen in extant South American mammals, and, as described above, our analysis agrees with that of Prevosti et al.: the mammalian carnivore niche was not clearly depauperate in the Santacrucian.

There is yet another aspect of the work of Vizcaíno et al. (2010) that must be considered. The primary productivity was calculated by analogy with extant environments of the Chaqueña Biogeographic Province. In this contribution we decided against applying the calculations to the southern humid temperate forest reconstructed here (Brea et al., Chapter 7) as an analog to estimate primary productivity, because several of the necessary methodological improvements are still in progress; for instance, we cannot include animals less than 10 kg (Mariana Di Giacomo, personal communication, 2011). At issue here is why the authors chose the Chaqueña Province as a modern analog. They based their interpretation on Vizcaíno et al. (2006) who found that the high taxonomic richness of armadillos of the SCF is much greater than in modern Patagonia and other regions, and only comparable to that of the Chaqueña Province. In other cases the Cerrado or Caatinga biomes were proposed as appropriate analogs. For instance, both are mentioned above in relation to the proposed abundance of termites, and Candela et al. (Chapter 15) observes that the taxonomic composition and paleobiological characteristics of Santacrucian rodents denote the presence of mixed woodland habitats with probable seasonality in rainfall patterns, similar to the Brazilian Cerrado biome.

These comparisons draw attention to the value of the vertebrates as proxies to reconstruct vegetation. In our case, if we were to choose a single element (e.g. microbiotheriids: Abello et al., Chapter 10) as the appropriate indicator, we might conclude that the vegetation was southern humid temperate forest, albeit developed in warmer conditions and spread far beyond its present range. However, individual elements of the fauna, particularly those with close living relatives, may lead us to a wrong conclusion. Thus, the point is that, except for cases where we can postulate extreme specificity (for example, the living three-toed sloth *Bradypus* feeds only on leaves of certain plants; Chiarello, 2008), vertebrates are indicating more the kind of substrate (e.g. ground, trees) and the major features of the environment (e.g. open, closed) than the taxonomic constitution of the flora.

17.5.3 Paleoecological hypothesis

In addition to the hypotheses indicated at the beginning of this chapter, other chapters of this volume have dealt with many paleoecological hypotheses, either in relation to the reconstruction of the environment or to the way the birds and mammals interacted with it or with one another. Many of these hypotheses, particularly those related with the paleoautoecology and paleosynecology, were proposed by Vizcaíno *et al.* (2010).

These hypotheses and others have been tested to different degrees. We have established that the landscape in which the Santacrucian fauna of FL 1–7 lived consisted of a mosaic of open temperate humid and semi-arid forests, with lakes in some areas and seasonal flooding in others, no doubt promoting the formation of marshlands with a mixture of grass and forbes. The chapters present robust evidence that allows us to place almost every bird or mammal into this environmental mosaic and to specify substrate preference and use (Fig. 17.8). We can also decide where they fit in the food web and who preyed upon whom (Prevosti *et al.*, in press). Although our understanding is restricted geographically and temporally, we have a much-improved understanding of the paleoecology of the SCF.

There are still many hypotheses that remain to be tested. For instance, among those enunciated by Vizcaíno *et al.* (2010), although probably much of the information needed is in the chapters, we have not discriminated how competition for plant resources was partitioned between ground dwellers – including some armadillos, glyptodonts, probably some sloths, ungulates, and some rodents – and tree dwellers – including most sloths, other rodents, and primates.

17.5.4 Next steps in the study of paleoecology of the Santacrucian

Several neglected areas remain to be developed and improved upon. For example, it will be essential to perform detailed taphonomic studies to evaluate the genesis of the assemblages. To this end, we will begin more focused efforts to recover small vertebrates through screening of sediment at key points in the stratigraphic section. Additional tools for paleoenvironmental studies will need to be refined. For example, it will be useful to extend the faunal analysis of on-crop biomass (Fariña's model) to all homeothermic vertebrates large and small and to evaluate how to get reliable estimates of basal metabolism rates for phorusrhacid birds. It is also crucial to evaluate how low metabolism and/or the phylogenetic signal affect density.

The results in this volume set the stage for future studies of the Santa Cruz Formation *sensu lato* that would bring a clearer understanding of biologic change that occurred in the Early and Middle Miocene of Patagonia. Some localities to the north and the west are ripe for this kind of approach, and many of them have not been revisited or systematically sampled for 65 to 100 years. Examples are Barrancas Blancas (on the south bank of the Río Santa Cruz) and Rincón del Buque (north to the Río Coyle), which have not been systematically collected since Ameghino's time (end of the nineteenth century) and since the 1940s, respectively. At Lago Posadas, the fossiliferous SCF extends upward into the Colloncuran age and downward into the Pinturan or even the Colhuehuapian, and over a much wider geographic range of localities in the SCF and its partially coeval Pinturas Formation.

A critical aspect of our studies is to expand and refine the biostratigraphic framework in which the successive Santacrucian faunas evolved, work begun so well by Fleagle, Perkins and their collaborators (Chapters 2 and 3). In this way it will be possible to separate the effects of local paleoenvironmental effects on community structure from those introduced by the sampling of different intervals of geologic time. The chronostratigraphic framework provided in Chapter 2 will be absolutely essential to accomplish this goal and needs to be refined and expanded.

ACKNOWLEDGMENTS

The work summarized here owes its debt to the many people and funding agencies already cited in the Preface and Introduction to this volume. We thank A. Abba, U. Pardiñas, and D. Udrizar, who helped to collect information on recent faunas from Patagonia. For careful reading and comments on this manuscript we thank especially G. De Iuliis, G. Cassini, and M. Di Giacomo.

REFERENCES

Aizen, M. A. and Ezcurra, C. (1998). High incidence of plant–animal mutualisms in the woody flora of the temperate forest of southern South America: biogeographical origin and present ecological significance. *Ecología Austral*, **8**, 217–236.

Albino, A. M. (2011). Evolution of the Squamata reptiles in Patagonia based on the fossil record. *Biological Journal of the Linnean Society*, **103**, 441–457.

Albuja, L. (1991). Lista de vertebrados del Ecuador. III. Mamíferos. *Politécnica*, **16**, 163–201.

Ameghino, F. (1906). Les formations sédimentaires du Crétacé supérieur et du Tertiaire de Patagonie avec un parallele entre leurs faunes mammalogiques et celles de l'ancien continent. *Anales del Museo Nacional de Buenos Aires, Série 3*, **15**, 1–568.

Andrews, P., Lord, J. M. and Nesbit-Evans, E. M. (1979). Patterns of ecological diversity in fossil and modern mammalian faunas. *Biological Journal of the Linnean Society*, **11**, 177–205.

Auffenberg, W. (1974). Checklist of fossil land tortoises (Testudinidae). *Bulletin, Florida State Museum Biological Sciences*, **18**, 121–251.

Bargo, M. S., Vizcaíno, S. F. and Kay, R. F. (2009). Predominance of orthal masticatory movements in the early Miocene *Eucholaeops* (Mammalia, Xenarthra, Tardigrada, Megalonychidae) and other megatherioid sloths. *Journal of Vertebrate Paleontology*, **29**, 870–880.

Barquez, R. M., Mares, M. A. and Ojeda, R. A. (1991). *Mamíferos de Tucumán*. Norman, OK: University of Oklahoma Press.

Barreda, V. and Palazzesi, L. (2007). Patagonian vegetation turnovers during the Paleogene–Early Neogene: origin of arid-adapted floras. *The Botanical Review*, **73**, 31–50.

Barreda, V., Anzótegui, M. L., Prieto, A. R. *et al.* (2007). Diversificación y cambios de las angiospermas durante el Neógeno en Argentina. *Publicación Especial Ameghiniana 50° Aniversario*, **11**, 173–191.

Bignell, D. E. and Eggleton, P. (2000). Termites in ecosystems. In *Termites: Evolution, Sociality, Symbioses, Ecology*, ed. T. Abe, D. E. Bignell and M. Higashi. Dordrecht: Kluwer Academic Publishers, pp. 363–387.

Blisniuk, P. M., Stern, L. A., Chamberlain, C. P., Idelman, B. and Zeitler, P. K. (2005). Climatic and ecologic changes during Miocene surface uplift in the Southern Patagonian Andes. *Earth and Planetary Science Letters*, **230**, 125–142.

Blisniuk, P. M., Stern, L. A., Chamberlain, C. P. *et al.* (2006). *Links between Mountain Uplift, Climate, and Surface Processes in the Southern Patagonian Andes*. Berlin/ Heidelberg: Springer, pp. 429–440.

Bown, T. M. and Laza, J. (1990). A fossil nest of a Miocene termite from southern Patagonia, Argentina, and the oldest record of the termites in South America. *Ichnos*, **1**, 73–79.

Brizuela, S. (2010). *Los lagartos continentales fósiles de la Argentina (excepto Iguania)*. Unpublished Doctoral thesis, Universidad Nacional de La Plata, Argentina.

Cabrera, A. L. and Willink, A. (1980). *Biogeografíaa de América Latina*. Organización de los Estados Americanos (OEA).

Cerling, T. E. and Quade, J. (1993). Stable carbon and oxygen isotopes in soil carbonates. In *Climate Change in Continental Isotopic Records*, ed. P. K. Swart, K. C. Lohmann, J. McKenzie and S. Savin. Washington: American Geophysical Union. Geophysical Monographs 78. pp. 217–231.

Chiarello, A. G. (2008). Sloth ecology: an overview of field studies. In *The Biology of the Xenarthra*, ed. S. F. Vizcaíno and W. L. Loughry. Gainesville, FL: University of Florida Press, pp. 269–280.

Constantino, R. (2002). The pest termites of South America: taxonomy, distribution and status. *Journal of Applied Entomology*, **126**, 355–365.

Croft, D. A. (2001). Cenozoic environmental change in South America as indicated by mammalian body size distributions (cenograms). *Diversity and Distributions*, **7**, 271–287.

Croft, D. A. (2006). Do marsupials make good predators? Insights from predator–prey diversity ratios. *Evolutionary Ecology Research*, **8**, 1193–1214.

Damuth, J. D. (1981). Population density and body size in mammals. *Nature*, **290**, 699–700.

Damuth, J. D. (1987). Interspecific allometry of population density in mammals and other mammals: the independence of body mass and population energy use. *Biological Journal of the Linnean Society*, **31**, 193–246.

Damuth, J. D. (1992). Taxon free characterization of animal communities. In *Terrestrial Ecosystems Through Time. Evolutionary Paleoecology of Terrestrial Plants and Animals*, ed. A. K. Behrensmeyer, J. D. Damuth, W. A. DiMichele *et al.* Chicago: Academic Press, pp. 183–200.

Dumont, E. R., Strait, S. G. and Friscia, A. R. (2000). Abderitid marsupials from the Miocene of Patagonia: an assessment of form, function, and evolution. *Journal of Paleontology*, **74**, 1161–1172.

Eisenberg, J. F. (1989). *Mammals of the Neotropics, I, Panama, Colombia, Venezuela, Guayana, Suriname, French Guiana*. Chicago: University of Chicago Press.

Eisenberg, J. F., O'Connell, M. A. and August, P. V. (1979). Density, productivity, and distribution of mammals in two Venezuelan habitats. In *Vertebrate Ecology in the Northern Neotropics*, ed. J. F. Eisenberg. Washington, D.C.: Smithsonian Institution Press, pp. 187–207.

Embert, D., Fitzgerald, L. and Waldez, F. (2009). *Tupinambis merianae*. IUCN Red List of Threatened Species. Version 2011.1.

Fariña, R. A. (1996). Trophic relationships among Lujanian mammals. *Evolutionary Theory*, **1**, 125–134.

Farquhar, G. D., Ehleringer, J. R. and Hubick, K. T. (1989). Carbon isotope discrimination and photosynthesis. *Plant Molecular Biology*, **40**, 503–537.

Farrell, K. M. (1987). Sedimentology and facies architecture of overbank deposits of the Mississippi River, False River region, Louisiana. In *Fluvial Sedimentology*, ed. F. G. Ethridge, R. M. Flores and M. D. Harvey. Society of Economic Paleontologists and Mineralogists Special Publication **39**, 111–120.

Fittkau, E. J. and Klinge, H. (1973). On biomass and trophic structure of the central Amazonian rain forest ecosystem. *Biotropica*, **5**, 2–14.

Fleagle, J. G., Bown, T. M., Swisher, C. C, III. and Buckley, G. A. (1995). Age of the Pinturas and Santa Cruz formations. *VI Congreso Argentino de Paleontologia y Bioestratigrafia, Actas*, 129–135.

Fleming, T. H. (1973). Numbers of mammalian species in north and central American forest communities. *Ecology*, **54**, 555–563.

Gasparini, Z., de la Fuente, M. and Donadío, O. (1986). Los reptiles Cenozoicos de la Argentina: implicancias paleoambientales y evolución biogeográfica. *IV Congreso Argentino Paleontologia y Bioestratigrafia*, pp. 119–130.

Gastaldo, R. A., Savrda, C. and Lewis, R. D. (1996). *Deciphering Earth History: A Laboratory Manual with Internet Exercises*. Raleigh, NC: Contemporary Publishing Company.

Gingerich, P. D. (1989). New earliest Wasatchian mammalian fauna from the Eocene of northwestern Wyoming: composition, and diversity in a rarely sampled high-floodplain assemblage. *University of Michigan Papers in Paleontology*, **28**, 1–97.

Grigera, D., Ubeda, C. A. and Cali, S. (1994). Caracterización ecológica de la asamblea de tetrápodos del Parque y Reserva Nacional Nahuel Huapi, Argentina. *Revista Chilena de Historia Natural*, **67**, 273–298.

Guerstein, G. R., Guler, M. V. and Casadío, S. (2004). Palynostratigraphy and palaeoenvironments across the Oligocene–Miocene boundary within the Centinela Formation, southwestern Argentina. In *The Palynology and*

Micropalaeontology of Boundaries, ed. A. B. Beaudoin and M. J. Head. *Geological Society of London Special Publications*, **230**, 325–343.

Gunnell, G. F. (1994). Paleocene mammals and faunal analysis of the Chappo type locality (Tiffanian), Green River basin, Wyoming. *Journal of Vertebrate Paleontology*, **14**, 81–104.

Handley, C. O. Jr. (1976). *Mammals of the Smithsonian Venezuelan Project*, **20**, 1–91.

Janson, C. H. and Emmons, L. H. (1990). Ecological structure of the nonflying mammal commuity at Cocha Cashu Biological Station, Manú National Park, Peru. In *Four Neotropical Rainforests*, ed. A. H. Gentry. New Haven: Yale University Press, pp. 314–338.

Kay, R. F. and Hylander, W. L. (1978). The dental structure of mammalian folivores with special reference to Primates and Phalangeroidea (Marsupialia). In *The Biology of Arboreal Folivores*, ed. G. G. Montgomery. Smithsonian Institution Press, Washington DC, pp. 173–192.

Kay, R. F. and Madden, R. H. (1997a). Mammals and rainfall: paleoecology of the middle Miocene at La Venta (Colombia, South America). *Journal of Human Evolution*, **32**, 161–199.

Kay, R. F. and Madden, R. H. (1997b). Paleogeography and paleoecology. In *Mammalian Evolution in the Neotropics*, ed. R. F. Kay, R. H. Madden, R. L. Cifelli and J. J. Flynn. Washington, D.C.: Smithsonian Institution Press, pp. 520–550.

Krapovickas, J., Tauber, A. A. and Rodríguez, P. E. (2008). Nuevo registro de *Protypotherium australe* Ameghino, 1887: implicancias bioestratigráficas en la Formación Santa Cruz. *Proceedings of the XVII Congreso Geológico Argentino*, 1020–1021.

Kraus, M. and Aslan, A. (1993). Eocene hydromorphic paleosols: significance for interpreting ancient floodplain processes. *Journal of Sedimentary Petrology*, **63**, 453–463.

Legendre, S. (1986). Analysis of mammalian communities from the late Eocene and Oligocene of southern France. *Palaeovertebrata*, **16**, 191–212.

Lenters, J. D. and Cook, K. H. (1995). Simulation and diagnosis of the regional summertime precipitation climatology of South America. *Journal of Climate*, **8**, 2988–3005.

Malcolm, J. R. (1990). Estimations of mammalian densities in continuous forest north of Manaus. In *Four Neotropical Rainforests*, ed. A. H. Gentry. New Haven, CT: Yale University Press, pp. 339–357.

Malumián, N. (1999). La sedimentación y el volcanismo Terciarios en La Patagonia extraandina. In *Geología Argentina*, ed. R. Caminos. Buenos Aires: Instituto de Geología y Recursos Minerales, pp. 557–612.

Mares, M. A., Willig, M. R., Streilein, K. E. and Lacher, T. E. J. (1981). The mammals of northeastern Brazil: A preliminary assessment. *Annals of the Carnegie Museum*, **50**, 81–137.

Mares, M. A., Willig, M. R. and Lacher, T. E. J. (1985). The Brazilian Caatinga in South American zoogeography: Tropical mammals in a dry region. *Journal of Biogeography*, **12**, 57–69.

Mares, M. A., Braun, J. K. and Gettinger, D. (1989a). Observations on the distribution and ecology of the mammals of the Cerrado grasslands of central Brazil. *Annals of the Carnegie Museum*, **58**, 1–60.

Mares, M. A., Ojeda, R. A. and Barquez, R. M. (1989b). *Guide to the Mammals of Salta Province, Argentina*. Norman, OK: University of Oklahoma Press.

Medem, F. (1958). The crocodilian genus *Paleosuchus*. *Fieldiana*, **39**, 227–247.

Melo, A. C. S. and Bandeira, A. G. (2004). A qualitative and quantitative survey of termites (Isoptera) in an open shrubby Caatinga in Northeast Brazil. *Sociobiology*, **44**, 707–716.

Nabte, M. J., Saba, S. L. and Monjeau, A. (2009). Mamíferos terrestres de la Península Valdés: lista sistemática comentada. *Mastozoología Neotropical*, **16**, 109–120.

Noriega, J. I., Areta, J. I., Vizcaíno, S. F. and Bargo, M. S. (2011). Phylogeny and taxonomy of the Patagonian Miocene Falcon *Thegornis musculosus* Ameghino, 1895 (Aves: Falconidae). *Journal of Paleontology*, **85**, 1089–1104.

Ojeda, F. A. and Mares, M. A. (1989). A biogeographic analysis of the mammals of Salta Province, Argentina. *Special Publications, Museum of Texas Tech University*, **27**, 1–66.

Ortiz-Jaureguizar, E. (1986). Evolución de las comunidades de mamíferos cenozicos sudamericanos. Un estudios basado en técnicas de análisis multivariados. *IV Congreso Argentino de Paleontología y Bioestratigrafía*, pp. 191–208.

Ortiz-Jaureguizar, E. and Cladera, G. A. (2006). Paleoenvironmental evolution of southern South America during the Cenozoic. *Journal of Arid Environments*, **66**, 498–532.

Parras, A., Griffin, M., Feldmann, R. *et al.* (2008). Correlation of marine beds based on Sr- and Ar-date determinations and faunal affinities across the Paleogene/Neogene boundary in southern Patagonia, Argentina. *Journal of South American Earth Sciences*, **26**, 204–216.

Pascual, R. (1984a). La sucesión de las edades-mamífero, de los climas y del diastrofismo sudamericanos durante el Cenozoico: fenómenos concurrentes. *Anales Academia Nacional de Ciencias Exactas*, **36**, 15–37.

Pascual, R. (1984b). Late Tertiary mammals of southern South America as indicators of climatic deterioration. *Quaternary of South America and Antarctic Peninsula*, **2**, 1–30.

Pascual, R. (1986). Evolución de los vertebrados Cenozoicos: sumario de los principales hitos. *IV Congreso Argentino de Paleontología y Bioestratigrafía, Mendoza*, pp. 209–218.

Pascual, R. and Ortiz-Jaureguizar, E. (1990). Evolving climates and mammal faunas in Cenozoic South America. *Journal of Human Evolution*, **19**, 23–60.

Patterson, B. and Pascual, R. (1968). Evolution of Mammals on Southern Continents. The Fossil Mammal Fauna of South America. *The Quaterly Review of Biology*, **43**, 409–451.

Patterson, B. and Pascual, R. (1972). The fossil mammal fauna of South America. In *Evolution, Mammals, and Southern Continents*, ed. A. Keast, F. C. Erk and B. Glass. Albany: State University of New York Press, pp. 247–309.

Patton, J. L., Berlin, B. and Berlin, E. A. (1982). Aboriginal perspectives of a mammal community in Amazonia Peru: Knowledge and utilization patterns among the Aguaruna Jivaro. In *Mammalian Biology in South America, Special*

Publication Series, *Pymatuning Laboratory of Ecology*. Vol. 6, ed. M. Mares and H. H. Genoways. Pittsburgh: University of Pittsburgh Press, pp. 111–128.

Pine, R. H. (1973). Mammals (exclusive of bats) of Belém, Pará, Brazil. *Acta Amazonica*, **3**, 47–79.

Plotkin, M., Medem, F., Mittermeirer, R. and Constable, I. (1983). Distributon and conservation of the black caiman (*Melanosuchus niger*). *Advances in Herpetology and Evolutionary Biology: Essays in Honor of Ernest E. Williams*. Vol. 1. Cambridge, MA: Museum of Comparative Zoology, pp. 695–703.

Prevosti, F. J. and Vizcaíno, S. F. (2006). The carnivore guild of the late Pleistocene of Argentina: paleoecology and carnivore richness. *Acta Paleontologica Polonica*, **51**, 407–422.

Prevosti, F. J., Forasiepi, A. M. and Zimicz, N. (in press). The evolution of the Cenozoic terrestrial mammalian predator guild in South America: competition or replacement? *Journal of Mammalian Evolution*, DOI 10.1007/s10914-011-9175-9.

Ré, G. H., Bellosi, E. S., Heizler, M. *et al.* (2010). A geochronology for the Sarmiento Formation at Gran Barranca. In *The Paleontology of Gran Barranca: Evolution and Environmental Change through the Middle Cenozoic of Patagonia*, ed. R. H. Madden, G. Vucetich, A. A. Carlini and R. F. Kay. Cambridge: Cambridge University Press, pp. 46–60.

Redford, K. H. (1984). The termitaria of *Cornitermes cumulans* (Isoptera, Termitidae) and their role in determining a potential keystone species. *Biotropica*, **16**, 112–119.

Redford, K. H. and Eisenberg, J. F. (1992). *Mammals of the Neotropics. II. The Southern Cone, Chile, Argentina, Uruguay, Paraguay*. Chicago: University of Chicago Press.

Rodríguez, J. (1999). Use of cenograms in mammalian palaeoecology. A critical review. *Lethaia*, **32**, 331–347.

Rosati, V. R. and Bucher, E. H. (1992). Seasonal diet of the Chacoan Cavy (*Pediolagus salinicola*) in the western Chaco, Argentina. *Mammalia*, **56**, 567–574.

Sanderson, M. G. (1996). Biomass of termites and their emissions of methane and carbon dioxide: A global database. *Global Biogeochemical Cycles*, **10**, 543–557.

Schaller, G. B. (1983). Mammals and their biomass on a Brazilian ranch. *Arquivos Zoologica Sao Paulo*, **31**, 1–36.

Strait, S. G., Fleagle, J. G., Bown, T. M. and Dumont, E. R. (1990). Diversity in body size and dietary habits of fossil caenolestid marsupials from the Miocene of Argentina. *Journal of Vertebrate Paleontology* **10**, 44A.

Streilein, K. E. (1982). Ecology of small mammals in the semi-arid Brazilian Caatinga. I. Climate and faunal composition. *Annals of the Carnegie Museum*, **51**, 79–107.

Tauber, A. A. (1994). Estratigrafía y vertebrados fósiles de la Formación Santa Cruz (Mioceno inferior) en la costa atlántica entre las rías del Coyle y Río Gallegos, Provincia de Santa Cruz, República Argentina. Unpublished Doctoral thesis, Universidad Nacional de Córdoba, República Argentina.

Tauber, A. A. (1996). Los representantes del género *Protypotherium* (Mammalia, Notoungulata, Interatheriidae) del Mioceno temprano del Sudeste de la Provincia de Santa Cruz, República Argentina. *Academia Nacional de Ciencias de Córdoba*, **95**, 1–29.

Tauber, A. A. (1997a). Bioestratigtrafía de la Formación Santa Cruz (Mioceno inferior) en el extremo sudeste de la Patagonia. *Ameghiniana*, **34**, 413–426.

Tauber, A. A. (1997b). Paleocología de la Formación Santa Cruz (Mioceno inferior) en el extremo sudeste de la Patagonia. *Ameghiniana*, **34**, 517–529.

Tauber, A. A. (1999). Los vertebrados de la Formación Santa Cruz (Mioceno Inferior-Medio) en el extremo sureste de la Patagonia y su significado paleoecológico. *Revista Española de Paleontología*, **14**, 173–182.

Tonni, E. P., Prado, J. L., Menegaz, A. N. and Salemme, M. C. (1985). La Unidad Mamífero (Fauna) Lujanense. Proyección de estratigrafía mamaliana al Cuaternario de la Región Pampeana. *Ameghiniana*, **22**, 255–261.

Townsend, K. E. B. and Croft, D. A. (2008). Enamel microwear in caviomorph rodents. *Journal of Mammalogy*, **89**, 730–743.

Valverde, J. A. (1964). Remarques sur la structure et l'évolution des communautés de vertébrés terrestres. I. Structure d'une commumauté. II. Rapports entre prédateurs et proies. *La Terre et la Vie*, **111**, 121–154.

Veloso, A., Formas, R. and Gerson, H. (2008). *Calyptocephalella gayi*. IUCN Red List of Threatened Species. Version 2011.1.

Vizcaíno, S. F. (1994). Mecánica masticatoria de *Stegotherium tessellatum* Ameghino (Mammalia, Xenarthra) del Mioceno de Santa Cruz (Argentina) Algunos aspectos paleoecológicos relacionados. *Ameghiniana*, **31**, 283–290.

Vizcaíno, S. F. (2001). Aclaraciones al trabajo de A. Tauber (1999) "Los vertebrados de la Formación Santa Cruz (Mioceno inferior-medio) en el extremo sureste de la Patagonia y su significado paleoecológico." *Revista Española de Paleontología*, **16**, 346.

Vizcaíno, S. F., Bargo, M. S., Kay, R. F. and Milne, N. (2006). The armadillos (Mammalia, Xenarthra, Dasypodidae) of the Santa Cruz formation (early-middle Miocene). An approach to their paleobiology. *Palaeogeography, Palaeoclimatology, Palaeoecology*, **237**, 255–269.

Vizcaíno, S., Bargo, M., Kay, R. *et al.* (2010). A baseline paleoecological study for the Santa Cruz Formation (Late-Early Miocene) at the Atlantic Coast of Patagonia, Argentina. *Palaeogeography, Palaeoclimatology, Palaeoecology*, **292**, 507–519.

Wetzel, R. M. (1985). Taxonomy and distribution of armadillos, Dasypodidae. In *The Evolution and Ecology of Armadillos, Sloths, and Vermilinguas*, ed. G. G. Montgomery. Washington D.C.: Smithsonian Institution Press, pp. 25–46.

Wilson, D. E. and Reeder, D. M. (2005). *Mammal Species of the World: A Taxonomic and Geographic Reference*. Baltimore: Johns Hopkins University Press.

Wilson, E. O. (1987). Causes of ecological success. *Journal of Animal Ecology*, **56**, 1–9.

Zachos, J., Pagani, M., Sloan, L., Thomas, E. and Billups, K. (2001). Trends, rhythms, and aberrations in global climate 65 Ma to present. *Science*, **292**, 686–693.

Index